PHYSICAL METHODS OF CHEMISTRY
Second Edition

Volume II

ELECTROCHEMICAL METHODS

PHYSICAL METHODS OF CHEMISTRY

Second Edition

Edited by: Bryant W. Rossiter
John F. Hamilton

PHYSICAL METHODS OF CHEMISTRY

Second Edition

Edited by

BRYANT W. ROSSITER
and
JOHN F. HAMILTON

Research Laboratories
Eastman Kodak Company
Rochester, New York

Volume II
ELECTROCHEMICAL METHODS

A WILEY-INTERSCIENCE PUBLICATION

JOHN WILEY & SONS

New York • Chichester • Brisbane • Toronto • Singapore

Library of Congress Cataloging in Publication Data:

Author's name entry.
Physical methods of chemistry.

"A Wiley-Interscience publication."
Includes index.
Contents: —v. 2. Electrochemical methods.
1. Chemistry—Manipulation—Collected works.
I. Rossiter, Bryant W., 1931–   . II. Hamilton,
John F.
QD61.P47   1985   542   85-6386
ISBN 0-471-08027-6 (v. 2)

Printed in the United States of America

10 9 8 7 6 5 4 3 2 1

CONTRIBUTORS

ERIC R. BROWN, Research Laboratories, Eastman Kodak Company, Rochester, New York

ZBIGNIEW GALUS, Department of Chemistry, The University of Warsaw, Warszawa, Poland

WILLIAM E. GEIGER, Department of Chemistry, The University of Vermont, Burlington, Vermont

M. DALE HAWLEY, Department of Chemistry, Kansas State University, Manhattan, Kansas

ROBERT KALVODA, J. Heyrovsky Institute of Physical Chemistry and Electrochemistry, Czechoslovak Academy of Sciences, Prague, Czechoslovakia

THOMAS R. KISSEL, Research Laboratories, Eastman Kodak Company, Rochester, New York

RICHARD L. MCCREERY, Department of Chemistry, Ohio State University, Columbus, Ohio

LOUIS MEITES, Chemistry Department, George Mason University, Fairfax, Virginia

ROYCE W. MURRAY, Department of Chemistry, The University of North Carolina at Chapel Hill, Chapel Hill, North Carolina

LUBOMÍR POSPÍŠIL, J. Heyrovsky Institute of Physical Chemistry and Electrochemistry, Czechoslovak Academy of Sciences, Prague, Czechoslovakia

JAMES R. SANDIFER, Research Laboratories, Eastman Kodak Company, Rochester, New York

MICHAEL SPIRO, Department of Chemistry, Imperial College of Science and Technology, London, United Kingdom

ANTONÍN A. VLČEK, J. Heyrovsky Institute of Physical Chemistry and Electrochemistry, Czechoslovak Academy of Sciences, Prague, Czechoslovakia

JIŘÍ VOLKE, J. Heyrovsky Institute of Physical Chemistry and Electrochemistry, Czechoslovak Academy of Sciences, Prague, Czechoslovakia

PREFACE TO PHYSICAL METHODS OF CHEMISTRY

This is a continuation of a series of books started by Dr. Arnold Weissberger in 1945 entitled *Physical Methods of Organic Chemistry*. These books were part of a broader series, *Techniques of Organic Chemistry*, and were designated as Volume I of that series. In 1970, *Techniques of Chemistry* became the successor to and the continuation of the *Techniques of Organic Chemistry* series and its companion, *Techniques of Inorganic Chemistry*, reflecting the fact that many of the methods are employed in all branches of chemical sciences and the division into organic and inorganic chemistry had become increasingly artificial. Accordingly, the fourth edition of the series entitled *Physical Methods of Organic Chemistry* became *Physical Methods of Chemistry*, Volume I in the new *Techniques* series. That last edition of *Physical Methods of Chemistry* has had wide acceptance and it is found in most major technical libraries throughout the world. This new edition of *Physical Methods of Chemistry* will consist of eight or more volumes and is being published as a self-standing series to reflect its growing importance to chemists worldwide. This series will be designated as the second edition (the first edition, Weissberger and Rossiter, 1970) and will no longer be subsumed within *Techniques of Chemistry*.

This edition heralds profound changes in both the perception and practice of chemistry. The discernible distinctions between chemistry and other related disciplines have continued to shift and blur. Thus, for example, we see changes in response to the needs for chemical understanding in the life sciences. On the other hand, there are areas in which a decade or so ago only a handful of physicists struggled to gain a modicum of understanding but which now are standard tools of chemical research. The advice of many respected colleagues has been invaluable in adjusting the contents of the series to accomodate such changes.

Another significant change is attributable to the explosive rise of computers, integrated electronics, and other "smart" instrumentation. The result is the widespread commercial automation of many chemical methods previously learned with care and practiced laboriously. Faced with this situation, the task of a scientist writing about an experimental method is not straightforward.

Those contributing to *Physical Methods of Chemistry* were urged to adopt as their principal audience intelligent scientists, technically trained but perhaps inexperienced in the topic to be discussed. Such readers would like an introduction to the field together with sufficient information to give a clear understanding of the basic theory and apparatus involved and the appreciation for the value, potential, and limitations of the respective technique.

Frequently, this information is best conveyed by examples of application, and many appear in the series. Except for the purpose of illustration, however, no attempt is made to offer comprehensive results. Authors have been encouraged to provide ample bibliographies for those who need a more extensive catalog of *applications*, as well as for those whose goal is to become more expert in a *method*. This philosophy has also governed the balance of subjects treated with emphasis on the *method*, not on the results.

Given the space limitations of a series such as this, these guidelines have inevitably resulted in some variance of the detail with which the individual techniques are treated. Indeed, it should be so, depending upon the maturity of a technique, its possible variants, the degree to which it has been automated, the complexity of the interpretation, and other such considerations. The contributors, themselves expert in their fields, have exercised their judgment in this regard.

Certain basic principles and techniques have obvious commonality to many specialties. To avoid undue repetition, these have been collected in Volume I. We hope they will be useful on their own and will serve as reference material for other chapters.

We are deeply sorrowed by the death of our friend and associate, Dr. Arnold Weissberger, whose enduring support and rich inspiration had motivated this worthy endeavor through four decades and several editions of publication.

BRYANT W. ROSSITER
JOHN F. HAMILTON

Research Laboratories
Eastman Kodak Company
Rochester, New York
March 1986

PREFACE

This volume begins with a general treatment of electrochemical theory common to all electroanalytical methods. Chapter 1, "Choosing and Performing an Electrochemical Experiment," is designed to assist those less familiar with the practice of electrochemistry in selecting particular electrodes or methods appropriate to a given need or problem. The more experienced electrochemist will find this chapter contains useful reference material, but will likely proceed directly to a method of choice. Chapter 7, "Spectroelectrochemistry," is new to the series and the technique involves a spectroscopic probe to monitor electrochemical events *in situ* with an intimate coupling of the electrochemical and spectrochemical process. This powerful new technique provides structural information about solution or absorbed species as well as redox information of the various chemical components. The remaining chapters contain the most recent information about the practice of various electrochemical techniques that over the years have proven to be of value in a wide variety of situations and applications.

We acknowledge our deep gratitude to the contributors who have spent long hours over manuscripts. We greet previous contributors, Dr. Eric R. Brown, Professor Louis Meites, Professor Royce W. Murray, and Dr. Michael Spiro, and welcome several new contributors to Volume II: Professor Zbigniew Galus, Professor William E. Geiger, Professor M. Dale Hawley, Professor Robert Kalvoda, Dr. Thomas R. Kissel, Professor Richard L. McCreery, Professor Lubomír Pospíšil, Dr. James R. Sandifer, Professor Antonín A. Vlček, and Professor Jiří Volke.

We are also extremely grateful to the many colleagues from whom we have sought counsel on the choice of subject matter and contributors. We express our gratitude to Mrs. Ann Nasella for her enthusiastic and skillful editorial assistance. In addition, we heartily thank the specialists whose critical readings of the manuscripts have frequently resulted in the improvements accrued from collective wisdom. For Volume II they are Dr. E. S. Brandt, Dr. E. R. Brown, Mrs. A. Kocher, Dr. J. R. Lenhard, Dr. G. L. McIntire, Dr. F. D. Saeva, and Dr. J. R. Sandifer.

BRYANT W. ROSSITER
JOHN F. HAMILTON

Rochester, New York
March 1986

CONTENTS

PHYSICAL METHODS OF CHEMISTRY

Second Edition

Volume II

ELECTROCHEMICAL METHODS

Chapter **1**

CHOOSING AND PERFORMING AN ELECTROCHEMICAL EXPERIMENT

William E. Geiger and M. Dale Hawley

1   INTRODUCTION

Electrochemical techniques have been developed to a high degree and have helped address a myriad of problems involving redox chemistry. With the versatility and benefits of electrochemistry increasingly recognized over the past two decades, more nonspecialists have used this method on chemical problems. This chapter is intended to provide an aid to those who are not greatly experienced in electrochemical methods, but who seriously want to begin using these techniques effectively. The organization of the chapter derives from practical considerations, revolving around the questions most frequently facing a new investigator: Which technique should be used for a particular application? What solvent and supporting electrolyte should be used? What are the advantages of different metals as working electrodes? Which reference electrodes are most suitable? How should the electrochemical cell be designed? How are solution resistance problems best dealt with?

Although we have striven for a systematic approach, space limitations necessitate a fairly cursory look at these questions. Voltammetric techniques are discussed in detail in other chapters of this volume, and many of the practical experimental aspects have been reviewed more extensively elsewhere. In an effort organized around experimental rather than theoretical considerations, there is bound to be a substantial personal component to the treatment, based on the experiences and prejudices of the authors. It is hoped, however, that the nature and level of the treatment will provide a suitable introduction to those beginning electrochemical work.

1.1   Reversibility of Redox Couples

A typical electrochemical reaction is given in (1). In this model, the oxidized and reduced forms of the redox couple, Ox and Red, respectively, are inter-related by a transfer of n electrons,

$$\mathrm{Ox} + ne^- \underset{}{\overset{k_s}{\rightleftharpoons}} \mathrm{Red} \underset{}{\overset{k_c}{\rightleftharpoons}} \mathrm{Z} \tag{1}$$

at the electrode surface. The primary electrode product, Red, may not be stable, and may undergo follow-up homogeneous chemical reactions in solution to give another product(s) Z. There may also be chemical reactions coupled to Ox, not shown in (1), which determine the availability of Ox to undergo the reduction. The rate constants of the heterogeneous reactions and homogeneous coupled reactions determine the reversibility of the redox couple. The standard heterogeneous electron-transfer rate constant, k_s or k^0, is the value of this quantity at the standard potential, E^0. In a perfectly reversible system, k_s is large enough that Ox and Red remain in equilibrium at the electrode surface, with the equilibrium value determined by the Nernst equation, (2):

$$E_{app} = E^0 - 2.3 \frac{RT}{nF} \log \frac{a_{Red}}{a_{Ox}}$$
(2)

where $2.3 \, RT/F = 0.059$ V at 298 K.

In practice, the observation of "Nernstian" behavior depends not only on the value of k_s but also on the time scale of the experiment that probes the concentrations of Ox and Red. Dc polarography is a much longer time-scale experiment than is, for example, ac polarography, since the time scale of the former is determined by drop time of the mercury electrode and of the latter by the inverse of the ac frequency. Consequently, even moderate charge-transfer kinetics can make a system appear Nernstian in dc polarography; such an *electrochemically reversible* system need have only a k_s greater than about 2×10^{-2} cm/s [1]. On the slow end of the charge-transfer phenomenon, systems with rate constants below about 3×10^{-5} cm/s are termed *electrochemically irreversible*, and rate constants between these limits yield *quasi-reversible* processes.

Quite distinct from the charge-transfer step are considerations of the stability of the primary electrolysis product, Red. If Red can be quantitatively reoxidized back to Ox, the couple is said to be *chemically* reversible. Again, this depends on the time scale of the experiment. An ac polarographic experiment might probe the stability of Red over the time frame of a millisecond, whereas bulk coulometry provides similar information over much longer periods, approximately 0.5 h. A system that is reversible in both the electrochemical and chemical sense is termed *totally* reversible.

1.2 Goals of the Experiment

Perhaps the earliest question to be asked by the experimentalist should concern the goals of the experiment, for they determine the techniques chosen. Some of the more common goals of electrochemical work are: qualitative or quantitative analysis of an inorganic ion or organic compound, at moderate to trace concentrations; determination of the *n*-value of a redox process and the reversibility (and E^0 value) of the couple; measurement of electron-transfer rates; mechanistic studies, especially probing the existence of radical ions or other transient intermediates; probing of the chemical and physical properties of electrogenerated compounds (coupling spectroscopy with electrochemistry, probing the reactions of electrogenerated species); synthesis of compounds by electrochemical methods.

Most electrochemical methods can yield at least some information about any or all of these questions. But each method, as well as each set of experimental conditions, is optimum for helping with one or another aspect of the problem.

2   ELECTROANALYTICAL TECHNIQUES

2.1   The General Problem

The fundamental quantities in electroanalytical chemistry are potential, current, concentration, and time. The relationship among these quantities will often be complex and dependent upon which of these parameters are being controlled, the mode of mass transfer, the kinetics of both the heterogeneous electron-transfer reaction and the homogeneous chemical reactions that accompany electron transfer, and the geometries of the working electrode and the cell. In this short overview of several of the more important electroanalytical techniques, numerous simplifying assumptions must be made. Unless it is stated otherwise, it will be assumed that: (1) the rate of electron transfer between the electroactive species and the electrode surface is rapid; (2) neither the oxidized nor the reduced form of the redox couple is involved in any solution reaction; (3) neither form of the redox couple is adsorbed on the electrode surface; (4) the working electrode is planar; and (5) the thickness of the solution layer immediately adjacent to the working electrode surface is small with respect to the perpendicular distance between the planar electrode surface and the cell wall. This last restriction rules out the important class of electrochemical methods involving thin-layer electrodes.

The three-electrode configuration is standard in the modern potentiostats and galvanostats that are used for the control of potential and current, respectively. The electrodes consist of the (1) working electrode at which the redox process of interest occurs, (2) the auxiliary electrode, which is the second current-carrying electrode in the cell, and (3) the reference electrode, which carries no cell current and to which the potential of the working electrode is referred. Unless coulometric electrolyses are being performed, it is usually unnecessary to isolate the working and auxiliary electrodes.

The electrochemical response is also dependent upon the mode of transport. To minimize the electrolysis times in coulometric electrolyses, fresh solution is continually brought to the large working electrode by forced convection. This may involve stirring the solution, either by mechanical means or by an inert gas; rotation or vibration of the working electrode; or a combination of solution stirring and movement of the electrode surface. The mode of mass transfer in the remaining electrochemical methods of interest to us is diffusion. Migration is specifically eliminated as an important mode of mass transfer by the addition of a large excess of inert supporting electrolyte, for example, Et_4NClO_4.

The reversible, one-electron reduction of species A to its stable anion radical, $A^{\mp}$ (3), causes A to be depleted in the layer of solution that is immediately adjacent to the electrode surface.

$$A + e^- \rightleftharpoons A^{\mp} \tag{3}$$

The electrolytic reduction of A creates a concentration gradient that causes A to diffuse from the bulk of solution to the electrode surface. If diffusion is the only mode of mass transfer by which A is transported to the electrode surface, then the partial differential equation that describes the concentration of A as a function of both time t and distance x from the electrode is

$$\frac{\partial C_A(x, t)}{\partial t} = D_A \frac{\partial^2 C_A(x, t)}{\partial x^2} \tag{4}$$

where D_A is the diffusion coefficient for species A and has the units of cm²/s [2]. To solve this partial differential equation, appropriate boundary and initial conditions that are unique for each electrochemical technique must be specified. If the solution is initially homogeneous with respect to species A, then the initial condition for all methods is simply $C_A(x, t=0) = C_A^b$, where C_A^b is the bulk concentration of species A. In addition, because thin-layer techniques are excluded by restriction number 5 above; that is, the thickness of the solution layer immediately adjacent to the working electrode surface is small with respect to the perpendicular distance between the planar electrode surface and the cell wall, the boundary condition for all t as x approaches infinity will be $C_A(x=\infty, t) = C_A^b$. This leaves only the boundary condition at $x=0$ to be specified for each electroanalytical method.

2.2   Chronoamperometry

In the chronoamperometric method, the potential is stepped from a value where no faradaic process occurs to a potential sufficiently negative that the concentration of species A at the electrode surface will be effectively zero (Figure 1.1a). Since this makes the applicable boundary condition at the electrode surface $C_A(x=0, t>0) = 0$, (4) can now be solved by appropriate methods to give the concentration of species A for all values of x and t (5) [1]:

$$C_A(x, t) = C_A^b \, \text{erf}(\lambda) \tag{5}$$

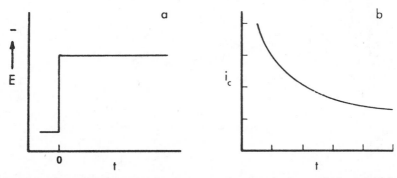

Figure 1.1   (a) The forcing function for chronoamperometry: At time $t=0$, the potential of the working electrode is stepped from a value where no faradaic process occurs to a value sufficiently negative that the concentration of species A at the electrode surface is rendered zero. (b) The response function: If the reduction of A to A^- is diffusion controlled, the current i is predicted by the Cottrell equation (7) to be inversely proportional to $t^{1/2}$. In aprotic solvent systems and with unshielded planar electrodes, reasonable agreement of experimental data with the model can often be obtained for t in the range $10^{-3}\,\text{s} \leqslant t \leqslant 10\,\text{s}$.

where $\lambda = x/2(D_A t)^{1/2}$ and erf(λ) is the error integral defined as

$$\text{erf}(\lambda) = \frac{2}{\pi^{1/2}} \int_0^\lambda e^{-z^2} \, dz \tag{6}$$

Equation (5) has been plotted for $D = 10^{-5}$ cm^2/s and for values of t equal to 10^{-1} s, 1 s, and 10 s (Figure 1.2). It should be noted from these two curves that the concentration gradient, $\partial C_A(x, t)/\partial x$, decreases with increasing t for all values of x. In addition, the thickness of the solution layer in which the concentration of species A is affected significantly by the electrode surface is given approximately by $2(D_A t)^{1/2}$ [when the argument of the error function in (5) is 1, that is, $x = 2(D_A t)^{1/2}$, the value of the error function is 0.85].

The instantaneous current that will be observed in this experiment is given by the expression

$$i = nFAD_A \left. \frac{\partial C_A(x, t)}{\partial x} \right|_{x=0} \tag{7}$$

where A is the electrode area, n is the number of electrons, and the other terms have their usual significance. Differentiation of (5) with respect to x, subsequent evaluation of the derivative at $x = 0$, and substitution into (7) affords the Cottrell equation (8):

$$i = \frac{nFAD_A^{1/2}C_A^b}{(\pi t)^{1/2}} \tag{8}$$

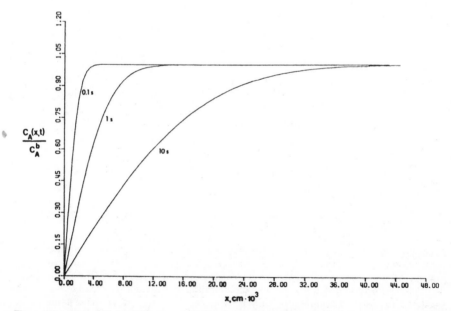

Figure 1.2   Plot of $C_A(x, t)/C_A^b$ versus distance x from a planar electrode surface for $t = 10^{-1}$ s and 1 s for the diffusion-controlled, chronoamperometric reduction of A.

Note that for a diffusion-controlled electrode process $it^{1/2}$ is predicted to be a constant. Experimentally, adherence to the Cottrell equation is good at unshielded, planar electrodes as long as $2(D_A t)^{1/2} < 0.05r$, where r is the radius of the planar electrode. For a planar electrode in which $A = 0.20\ \text{cm}^2$, $it^{1/2}$ can be expected to vary less than 10% in most solvent–electrolyte systems for $10^{-3}\ \text{s} < t < 3\ \text{s}$.

The chronoamperometric method can be used to measure any one of the variables in the Cottrell equation (8), provided all remaining quantities are known. One of the most useful applications of this method involves the determination of n in an electrode reaction. Frequently, homogeneous chemical reactions that are coupled to the heterogeneous electron-transfer process produce additional electroactive species. If the electrode area and the concentration are known and the diffusion coefficient can be estimated from results on another well-behaved, similar-sized molecule, then the variation of n can be studied as a function of time t. The results of such a study can be used to elucidate important steps in the overall electrode process as well as the kinetics of the coupled chemical reactions.

2.3   Chronopotentiometry

In the chronopotentiometric method, the cell current i (Figure 1.3a) is controlled by a galvanostat while the cell potential is monitored as a function of time t (Figure 1.3b). If the current is held constant during either a reduction or an oxidation, the corresponding boundary condition at the electrode surface

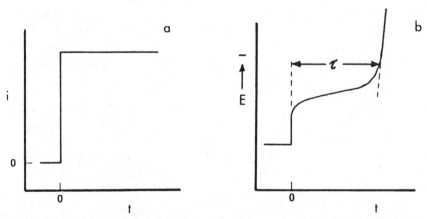

Figure 1.3   (a) The forcing function for chronopotentiometry: At time $t = 0$, a constant current is applied, which causes species A to be reduced to its anion radical. (b) Response function: The transition time τ is elapsed time required to reduce the concentration of species A to zero at the electrode surface. Since A is reduced reversibly to A^-, the Nernst equation is obeyed in the time range $0 \leqslant t \leqslant \tau$. When all A has been consumed at the electrode surface ($t > \tau$), the electrode will rapidly assume a potential that causes the next most readily reduced component in the solvent–electrolyte system to undergo reduction.

will be

$$i = \text{constant} = nFAD_A \left. \frac{\partial C_A(x, t)}{\partial x} \right|_{x=0} \qquad (9)$$

The quantity of interest, the transition time τ, is the elapsed time between the initiation of the constant-current experiment and the time at which the concentration of the electroactive species reaches zero at the electrode surface. Solution of partial differential (4) with (9) as the boundary condition at $x = 0$ leads to the Sand equation (10):

$$i\tau^{1/2} = \frac{nFA(D_A\pi)^{1/2}C^b}{2} \qquad (10)$$

Historically, chronopotentiometry and reverse-current chronopotentiometry were among the first electroanalytical methods to be used in quantitative kinetics studies of complex electrode processes [3]. Although the more recently developed cyclic voltammetric (discussed later) and chronoamperometric methods have led to decreased use of chronopotentiometric methods, sufficient mechanistic problems still arise where only chronopotentiometric methods are applicable that a knowledge and understanding of these methods remain important [4].

2.4   Single-Sweep Peak and Cyclic Voltammetry

Voltammetric techniques involve the measurement of cell current as a function of the electrode potential. In the single-sweep method, the potential between the working and reference electrodes is varied linearly in only one direction; an isosceles triangular waveform is used as the forcing function in the cyclic voltammetric experiment (Figure 1.4a). A typical response curve for the reversible reduction of A to its stable anion radical A^- is shown in Figure 1.4b. The potential scan is initiated in the negative-going direction from a potential that is too positive to cause the reduction of A (Point 1 in Figure 1.4). Since the original solution is assumed to contain A as the only electroactive species, the faradaic current at this point will be zero.

The working electrode will eventually attain a sufficiently negative potential such that A is reduced rapidly to A^- (Point 2). The maximum in the cathodic current (Point 3) is the result of an increasingly more negative working electrode potential that facilitates the reduction of A, being countered by the depletion of A in the vicinity of the electrode surface because of the redox process. When the potential of the working electrode is sufficiently negative that the concentration of A at the electrode surface is effectively zero, the current will be diffusion controlled and its magnitude will be dependent upon the rate at which A is replenished from the bulk of solution by diffusion (Point 4).

The direction of the potential scan is reversed at the cathodic switching limit, $E_{\lambda,c}$ (Point 5). As the linearly varying working electrode potential is made more positive, a point will be reached (Point 6) when the reoxidation of A^- to A occurs at a significant rate. The peak anodic current for the reoxidation

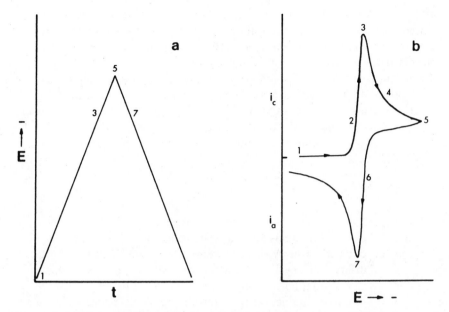

Figure 1.4 (a) The forcing function for the cyclic voltammetric experiment. (b) Response curve for the reversible reduction of A to $A^{\overline{\cdot}}$. The cyclic voltammetric scan is initiated from the potential at point 1 and proceeds first in the negative-going direction. The scan direction is reversed at point 5.

of $A^{\overline{\cdot}}$ (Point 7) is located at a potential $(E_{p,a})$ that is approximately 57 mV more positive than the peak cathodic potential $(E_{p,c})$. The occurrence of the anodic peak is the result of the same factors that afforded the cathodic peak; that is, the increasingly more positive potential that facilitates the reoxidation of $A^{\overline{\cdot}}$ is opposed by the depletion of $A^{\overline{\cdot}}$ in the diffusion layer.

The boundary condition at the electrode surface for the reversible, cyclic voltammetric reduction of A is the Nernst equation [(2) rewritten for the present reaction in (11)], where $E = E_{\text{initial}} - vt$, in which v is the scan rate in volts per second and t is the elapsed time.

$$E = E^0 - \frac{RT}{nF} \ln \frac{a_{A^{\overline{\cdot}}}}{a_A} \tag{11}$$

When the partial differential equation (4) is evaluated under the boundary and initial conditions for this experiment [5], the expression obtained for the peak cathodic current is given by (12).

$$i_p = 0.4463 nFAC_A^b \left(\frac{nF}{RT} \right)^{1/2} D_A^{1/2} v^{1/2} \tag{12}$$

Note that $i_p/v^{1/2}$ will be a constant for a diffusion-controlled reaction. Cyclic voltammetry is most useful as an analytical tool when the concentration of the species that is to be oxidized or reduced (the electroactive species) is in the range of 10^{-5}–10^{-2} M. The lower limit arises when the nonfaradaic current becomes the predominant fraction in the total cell current. The nonfaradaic current is caused by the charging of the electrode's electrical double layer as the potential of the electrode is varied linearly (discussed later).

From an analytical viewpoint, the location of the peak potential can be used in the qualitative identification of the electroactive species while the peak current can be used in its quantitative determination [(12)]. When the diffusion coefficients for the oxidized and reduced forms are also equal, then it can be shown that $E_{1/2}$, the half-wave potential in polarographic experiments, is related to $E^{0'}$ by

$$E_{1/2} = E^0 + \frac{RT}{nF} \ln \left(\frac{D_{A^-}}{D_A} \right)^{1/2} = E^{0'} \tag{13}$$

Since $E_{1/2} = E_{p,c} + 28.5$ mV for a reversible, one-electron reduction process, peak potentials for such systems also have thermodynamic significance.

Although cyclic voltammetry is used extensively for the purposes mentioned above, the real utility of the cyclic voltammetric method is realized when homogeneous chemical reactions accompany electron transfer. For example, when the product of the electrode reduction reaction undergoes an irreversible homogeneous chemical reaction to produce an electroinactive species, Z (Scheme I), the magnitude of the anodic peak for the reoxidation of A^- on the reverse, positive-going sweep will be decreased by the amount of A^- consumed by the chemical reaction.

$$A + e^- \rightleftharpoons A^-$$
$$A^- \xrightarrow{\ k\ } Z$$

Scheme I

From studies involving variations in the cyclic voltammetric scan rate v and the concentration of the initial electroactive species, the rate constant for the solution reaction can be determined [5]. The reader is referred to Chapter 4 for applications of this important method to specific chemical systems.

2.5   Polarography

Polarography is a voltammetric technique that involves a dropping mercury electrode as the working electrode. The lifetime t_{max} of each mercury drop is a function of both the mercury flow rate m and the electrode potential and normally varies from 2 to 8 s. The input waveform to the potential control device is a linearly varying voltage (Figure 1.5a). The output response (Figure 1.5b) for the reversible reduction of A to A^- is a saw-toothed, S-shaped curve. The sawtooth shape of the curve is the result of the periodic increase in the electrode area as the mercury drop grows and the subsequent falling of the drop when the mass of the drop, mt, exceeds the effect of the mercury surface tension. The overall S shape of the curve obeys the Nernst equation when the

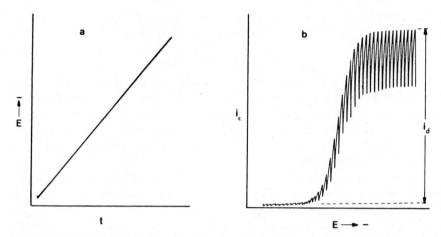

Figure 1.5 (a) The forcing function for dc polarography: The scan rate typically is 0.1 V/min. (b) Response curve: The current is recorded as a function of the applied voltage E. The potential at the point $i_d/2$ is $E_{1/2}$; i_d is the diffusion-limited current and occurs when the concentration of the electroactive species at the electrode surface is effectively zero. Each sawtooth on the voltammetric response curve is the result of the periodic growth and fall of the mercury drop.

electroactive species is reduced reversibly. When $E \ll E^0$, the concentration of the electroactive species at the electrode surface will be effectively zero. The limiting current will then become diffusion controlled and independent of the applied potential.

The relatively rapid growth of the mercury drop allows the electrode to be treated mathematically as if it were an expanding plane. This modification to the original problem involving semi-infinite linear diffusion of the electroactive species to a planar electrode surface (4) leads to the partial differential equation

$$\frac{\partial C_A(x, t)}{\partial t} = D_A \frac{\partial^2 C_A(x, t)}{\partial x^2} + \frac{2x}{3t} \frac{\partial C_A(x, t)}{\partial x} \tag{14}$$

With an appropriate change of variable and the assumption that the electrode area is sufficiently negative such that the concentration of A at the electrode surface is zero, it can be shown that the current is given by the expression

$$i = \frac{\sqrt{7/3} nFAD^{1/2}C_A^b}{(\pi t)^{1/2}} \tag{15}$$

Note that with the exception of the $(7/3)^{1/2}$ term, (15) is identical to the previously derived Cottrell equation (8). Since the area of the mercury drop is a function of time, an expression for the area in terms of t is desirable. If the mercury drop is treated as if it were a perfect sphere, then the mass of the mercury drop, mt, is given by $\frac{4}{3}\pi r^3 d$, where d is the density of mercury. Since $A = 4\pi r^2 = 4\pi(3mt/4\pi d)^{2/3} = 8.5 \times 10^{-3} m^{2/3}t^{2/3}$, the instantaneous current i at time t in the diffusion-controlled range will be

$$i = 706nm^{2/3}t^{1/6}C^bD^{1/2} \tag{16}$$

If the average current, i_{av}, is calculated throughout the lifetime of the drop t'_{max} (16), which is known widely as the Ilkovič equation, becomes

$$i_{av} = 607 n m^{2/3} t_{max}^{1/6} C^b D^{1/2} \tag{17}$$

The current is measured in amperes when the concentration is expressed in terms of moles per cubic centimeter and in microamperes when the concentration is expressed in terms of millimoles per liter. The test for diffusion control of the limiting current involves changing the height of the mercury column, h. Since $i \propto m^{2/3} t^{1/6} \propto h^{2/3} h^{-1/6} \propto h^{1/2}$, $i/h^{1/2}$ will be independent of h when the process is diffusion controlled.

The ready ease with which mercury is oxidized limits the polarographic method mainly to cathodic processes. The useful concentration range for the electroactive species extends approximately from 5×10^{-5} to $10^{-2} M$. The lower limit arises when the nonfaradaic current caused by the charging of the electrical double layer becomes large when compared to the faradaic current that results from the electroreduction of the electroactive species. Since any change in either the electrode's area or its potential necessitates the flow of nonfaradaic current, the polarographic, cyclic voltammetric, and chronopotentiometric methods all share this inherent limitation. If electrochemical methods are to be used for the study of more dilute solutions, procedures must be devised whereby the faradaic and nonfaradaic currents can be separated and measured. The next section focuses on several methods by which this separation of currents has been achieved.

2.6   Pulse Voltammetry Methods

One simple model of the electrochemical cell involves a capacitor C to represent the working electrode's double-layer capacitance and a series resistor R to represent the solution resistance (Figure 1.6). Since $E = iR + Q/C$, then for a potential step of magnitude E, the current at any time t will be given by

$$i = \frac{E_e^{-t/RC}}{R} \tag{18}$$

Figure 1.6   Simple model of the electrochemical cell.

For an electrode capacitance of 5 μF (a typical electrode with an area of 0.25 cm^2), a solution resistance of 5 Ω (an aqueous solvent–electrolyte system), and a potential step of 1 V, the nonfaradaic current at 10^{-5}, 10^{-4}, and 10^{-3} s will be 1.34×10^{-1}, 3.66×10^{-3}, and 8.50×10^{-19} A, respectively! In contrast, if an electroactive species were present, the faradaic current for a diffusion-controlled process (see the Cottrell equation) is proportional to $t^{-1/2}$ and at 10^{-3} s would still be 10% of that at 10^{-5} s. The much more rapid decay of the nonfaradaic current than the faradaic current in this time range suggests a plausible solution to the problem of distinguishing between the two types of currents: apply a step potential to the working electrode, wait for the nonfaradaic current caused by the charging of the electrical double layer to decay to an insignificantly small fraction of the total current, and then measure the more slowly decaying faradaic current.

2.6.1 Square-Wave Voltammetry

The forcing function consists of a staircase waveform upon which a square-wave signal has been superimposed [6] (Figure 1.7). The step size ΔE_s of the staircase waveform is usually a few tenths of a millivolt while the magnitude of the square wave ΔE typically is in the range of 5–50 mV. The frequency usually ranges from 10 to 200 Hz. The sweep rate may be altered by changing either the step size or the frequency.

The working electrode for square-wave voltammetry can be any inert stationary electrode of fixed area. Since the potential sweep is discontinuous, the charging current decays relatively rapidly to an insignificantly small value a few milliseconds after the application of the potential step. When the decay of the charging current is complete, the current may then be sampled (see dots on Figure 1.7) to arrive at the faradaic current that arises from the electrode reaction. The final current waveform is the difference between the currents on the negative-going and positive-going half-cycles. For a reduction, when $E \gg E_{1/2}$, the electrode potential at every point on the square wave will be too positive to cause reduction of species A. However, when $E \sim E_{1/2}$, a potential in the negative-going direction is predicted by the Nernst equation to cause A to be reduced to A$^-$, while a potential step in the positive-going direction will cause the electrolytically generated A$^-$ to be reoxidized to A. Since the magnitudes of the faradaic currents are proportional to the changes that occur in the concentrations of A and A$^-$ as a result of the potential steps, the differential current will be maximum when $E \sim E_{1/2}$. When $E \ll E_{1/2}$ and the diffusion coefficients of A and A$^-$ are equal, the concentrations of A and A$^-$ at the electrode surface will be effectively zero and C_A^b, respectively. Because the differential current will again be very small when $E \ll E_{1/2}$, the plot of the differential current as a function of E will be bell-shaped and symmetrical with a peak at approximately $E_{1/2}$.

The peak current is a linear function of the concentration and the square root of the square-wave frequency and a more-complex function of the square-wave amplitude, measurement time, and staircase amplitude. The detection

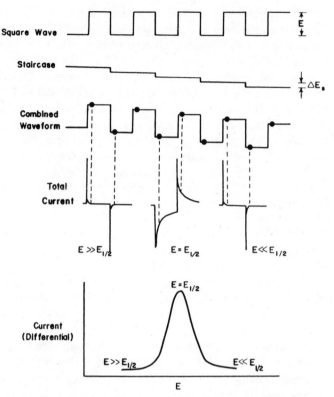

Figure 1.7   Square-wave voltammetry. A square wave, 5 mV $< \Delta E_s <$ 50 mV, is superimposed on a staircase waveform (0.05 mV $< \Delta E_s <$ 1 mV) of the same frequency (10 Hz $< v <$ 200 Hz). The scan rate is determined by the step size of the staircase waveform and its frequency. The current is sampled as soon as the charging current caused by the potential pulse has decayed to an insignificantly small fraction of the total current. The points at which the current is sampled are indicated by the black dots on the combined input waveform. The differential current is measured each cycle and is the difference in total current at corresponding points on the positive-going and negative-going pulses. When the pulse step on the combined waveform is small, the potential at which the maximum differential current is obtained is approximately $E_{1/2}$.

limits for Tl(I), Cd(II), and In(III) have been shown by Ramley and Krause [6] to be 1.1×10^{-7}, 4.7×10^{-8}, and 2.5×10^{-8} M, respectively. In addition to its excellent sensitivity, the square-wave voltammetric method is also well suited to the study of complex mixtures of electroactive species. Whereas the presence of a large amount of a readily reduced component might obscure the presence of a small amount of a more difficultly reduced species, the differential method of square-wave voltammetry will provide discernible peaks for each species with each peak height being proportional to the concentration of that particular electroactive species.

2.6.2   Pulse Polarography Methods

Whereas a stationary electrode with a fixed electrode area is used in square-wave voltammetry, pulse polarographic methods [6, 7], as the name implies, use the dropping mercury electrode as the working electrode. Because the area of the dropping mercury electrode increases during the life of the drop, a finite amount of charging current will be required at all times to charge the electrode's increasing capacitance. Since the *growth rate* of the mercury drop decreases throughout the life of the drop, the instantaneous charging current, i_c, can be obtained by differentiating $Q = C_i A(E_z - E)$ with respect to time, where C_i is the capacitance per cm^2 and E_z is the point of zero charge:

$$\frac{dQ}{dt} = i_c = 5.67 \times 10^{-3} m^{2/3} t^{-1/3} C_i (E_z - E) \tag{19}$$

Since the faradaic and charging currents will be proportional to $t^{1/6}$ and $t^{-1/3}$, respectively, the contribution of the nonfaradaic charging current to the total current will be minimum when $t = t_{max}$. Accordingly, the techniques illustrated here, normal pulse and differential pulse polarography (Figures 1.8 and 1.9,

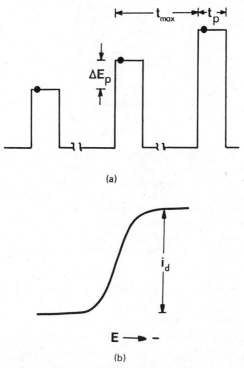

(a)

(b)

Figure 1.8   (a) The forcing function of normal pulse polarography: The sweep rate will be a function of both the increase in step size ΔE_p and the drop life of the mercury drop t_{max}. Typically, 0.5 mV $< \Delta E_p < 4$ mV, 1 s $< t_{max} < 4$ s, while the pulse length t is in the range 10 ms $< t_p < 100$ ms. The current is sampled as soon as the charging current caused by the potential pulse has decayed to an insignificantly small fraction of the total current. The points at which the current is sampled are indicated by the black dots on the input waveform. The mercury drop is dislodged mechanically by a solenoid-operated hammer when the potential pulse is terminated. (b) Response function: $E_{1/2}$ is the potential at which $i = i_d/2$. The record current i_d at any point is the difference in current at potential ΔE_p and $E_{initial}$.

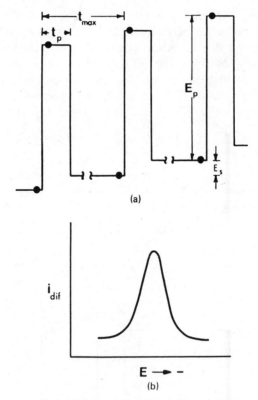

(a)

(b)

Figure 1.9 (a) The forcing function of differential pulse polarography: The sweep rate is a function of both step size E_s and the drop life of the mercury drop t_{max}. Typically, $0.5\,mV < E_s < 4\,mV$, $1\,s < t_{max} < 4\,s$, and $5\,mV < E_p < 100\,mV$. The current is sampled shortly before and after the application of the potential pulse; the points at which the current is sampled are indicated by the black dots. The duration of the potential pulse t_p normally is in the range $10\,ms < t_p < 100\,ms$. (b) Response function: The maximum differential current i_{dif} occurs when $E \sim E_{1/2}$. Since the ratio of the concentrations of the oxidized and reduced forms of the reversible redox couple is then determined by the Nernst equation, substantial differential currents are observed only in the vicinity of $E_{1/2}$.

respectively), feature the application of a potential pulse just before the mercury drop is dislodged by some mechanical means (typically $1\,s < t_{max} < 4\,s$, where t_{max} is the drop life).

Normal pulse and differential pulse polarography provide S-shaped (Figure 1.8b) and bell-shaped (Figure 1.9b) voltammograms, respectively. While normal pulse polarography can be expected to provide somewhat larger difference currents and inherently greater sensitivity, the latter method is better suited for studying complex samples that contain several electroactive components in widely different amounts. The lower detection limit for well-behaved systems (e.g., Cd) is approximately $10^{-7}\,M$.

When the polarographic methods are compared to square-wave voltammetry, the latter method is generally superior. Because of the greater sweep rate in square-wave voltammetry, the instrument time for a square-wave voltammetric analysis will be less than that of either of the polarographic techniques. In addition, square-wave voltammetry can be used with all stationary electrodes, which increases significantly the number of potential compounds that may be studied both oxidatively and reductively. The most important advantage of the polarographic methods occurs when the electrode process yields a product that fouls the electrode surface. With the polarographic methods, the electrode fouling problem will be minimized because of the frequent, periodic replacement of the working electrode surface.

2.7   Combining Electrochemistry With Spectroscopy

The development of cyclic voltammetry and other related large-amplitude techniques (e.g., reverse-current chronopotentiometry) was in response to the need for additional information concerning the fate of the product(s) of an electron-transfer reaction. While the success of these methods in the study of relatively complex electrode reactions was immediate, electrochemical detection of intermediates and products is limited to electroactive species. Furthermore, electrochemical methods often provide little detail concerning the structures of the electroactive species and their specific interactions with components of the solvent–electrolyte system. The need for additional chemical information pertaining to the electrode products quickly led to hybrid methods that exploit the differences in properties between the starting material and its reaction products.

Since free radicals and radical ions are often formed in organic-electrochemistry, one of the successful hybrid methods has combined electrochemistry and electron spin resonance [8]. The hyperfine splitting can be used to identify the radical intermediate or product, to study the distribution of the unpaired electron density within the radical, and to determine the radical species' structure. As an example, the effect of hydroxylic solvents upon the solvation of nitroaromatic anion radicals and the twisting of the intro group from the plane of the benzene ring were studied by ESR using the *intra muros* electrochemical generation technique, wherein the radical is electrochemically generated within the walls of the sample cavity [8]. The use of ESR in the identification of radical intermediates and its role in the elucidation of an overall electrode process are illustrated by the Kitagawa, Layloff, and Adams study in which halogenated nitrobenzene anion radicals were found to undergo loss of halide ion and to afford the nitrobenzene anion radical as a final product [8]. One caveat must be offered. Since most ESR–electrochemical studies make no attempt to determine the concentration of the radical that is being detected, a stable radical arising from a minor reaction channel might be interpreted inadvertently to lie in the principal reaction channel. All mechanistic interpretations must be made with care and caution and, if possible, should not be offered until confirming information is obtained by independent experiments.

The combination of molecular spectroscopy and electrochemistry, which is referred to as spectroelectrochemistry [9], has been particularly versatile and useful. The earliest studies involved the absorption of ultraviolet and visible radiation by intermediates and products that were electrogenerated at optically transparent electrodes (OTEs). Optically transparent ($20 < \% T < 85$) electrodes are often either a very thin (10–500 nm) conducting film of Pt, SnO_2, or Hg-coated Pt that is deposited on a transparent support, or a very fine micromesh metal that is both electrically conducting and inert (e.g., Au, with 100–2000 wires/in.) (see also Section 3.4.3).

Two major cell types are used with OTEs. In thin-layer spectroelectro-chemistry the thickness ($\sim 30\,\mu m$) of the thin layer of solution confined next to the OTE is approximately that of the diffusion layer [$x = 2(Dt)^{1/2}$] after several seconds of electrolysis. The optical path, as shown in Figure 1.10a, passes perpendicularly through the OTE and the thin layer of solution. Thin-layer OTEs have been used to obtain the absorption spectra of electrode products to determine, in combination with electron mediators, formal reduction potenti-als of species that undergo slow heterogeneous electron transfer, and to study the rates of relatively slow homogeneous reactions that involve an electro-generated reactant.

The second cell for spectroelectrochemistry is similar to a conventional electrochemical cell and has a solution thickness that always greatly exceeds the maximum diffusion layer thickness (Figure 1.10b). The OTE generally serves as one side of the cell; the optical path is perpendicular to the OTE and the diffusion layer. If the wavelength of the monochromatic radiation is such that the product of the electrode reaction absorbs, the increase in the absorbance will be a function of the rate at which the product is electrogenerated. In the

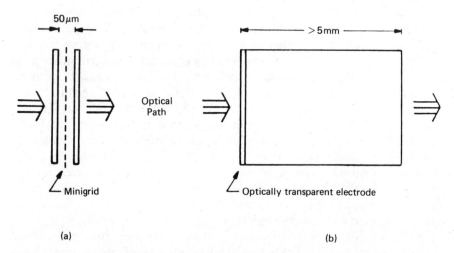

(a)   (b)

Figure 1.10   (a) Thin-layer spectroelectrochemical cell using an inert, optically transparent, mini-grid working electrode. (b) Conventional cell with an optically transparent electrode serving as one of the cell's sides. The optical path is perpendicular to the OTE surface.

case of a diffusion-controlled process, the absorbance A is given by

$$A = \frac{2\varepsilon C^b D^{1/2} t^{1/2}}{\pi^{1/2}} \tag{20}$$

where C^b and D refer to the starting material and ε is the molar absorptivity of the electrode product. If the electrode product is now consumed by a follow-up chemical reaction, the decrease in the absorbance will be a function of the rate of the homogeneous chemical reaction. The value of the rate constant may be obtained experimentally by fitting the experimental A versus t curve to a dimensionless working curve of A versus kt that is obtained by a digital simulation of the reaction model [10]. This method is capable of monitoring certain second-order reactions that are diffusion controlled. Because uncompensated iR loss is a much less serious problem in the conventional cell type than in thin-layer cells, the full-sized cell is required when kinetic studies of rapid homogeneous chemical reactions are performed.

The number of potential combinations of electrochemical and spectroscopic methods is large. Several particularly effective combinations include Raman and resonance Raman, fluorescence, and internal and specular reflection. The interested reader is referred to [9] and the references contained therein for applications of these hybrid spectroscopic–electrochemical methods to the study of specific chemical systems.

2.8 Coulometric Methods

With the exception of the thin-layer cell, which was used in conjunction with certain optically transparent electrodes, all of the above electrochemical methods had a small electrode-area to solution-volume ratio (A/V). In addition, the only mode of mass transfer permitted was diffusion. As a result of these restrictions, the fraction of the electroactive species consumed in any series of diffusion-controlled electrochemical experiments can be calculated to be insignificantly small. If the goal of the experiment is to effect the complete electrochemical transformation of reactant into product in a realistically short time, then the microelectrode–semi-infinite linear diffusion restrictions must be lifted. The exhaustive electrolysis methods, which will be discussed briefly, feature larger A/V ratios and rapid stirring of the solution to increase the rate of the electrochemical reaction. To prevent both the transport of an electrode product from one current-carrying electrode to the other and its subsequent electrolysis, an electrically conducting diaphragm is used to separate the products of the working and auxiliary electrodes.

If the potential of the working electrode is controlled, it can be shown that the relationship between the instantaneous current i and the other experimental parameters is given by

$$i = i_{init} e^{-pt} \tag{21}$$

where i_{init} is the current at $t = 0$ and p is the cell constant that includes A/V and the mass-transfer rate. Controlled-potential electrolysis is especially useful when the power requirement is relatively modest and control of the working electrode potential is required to minimize the occurrence of secondary electrode processes. A secondary electrode reaction might occur, for example, if the desired product of the initial electrode reaction was electroactive at a more negative potential and was inadvertently reduced because of inadequate potential control of the working electrode. The secondary reaction would not only reduce the yield of the desired product, but would also cause a corresponding decrease in the current efficiency. Because it is difficult to build potentiostats that have high-current, high-voltage capabilities and to control the potential of the working electrode accurately in high-current-density cells, controlled potential coulometry is used principally for electrolyses that involve relatively small amounts of substrate.

Practically all large-scale industrial electrosyntheses employ control of the cell current. The working electrode potential in these processes will remain nearly constant if the electroactive species is replaced at the rate at which it is consumed by the electrolysis. Controlled-current electrolyses are most successful when the electrode products are electroinactive and are readily removed from the electrolyte. Some important industrial processes include the electrolysis of brine to give chlorine and sodium hydroxide, the hydrodimerization of acrylonitrile to form adiponitrile, the electrosynthesis of tetraalkyl lead compounds, and electrochemical fluorination.

3   EXPERIMENTAL ASPECTS

3.1   Cell Design and Resistance Effects—Dealing With Resistance Problems

Design of an electrochemical cell is generally governed by the requirements (a) to eliminate unwanted contaminants; (b) to provide a symmetrical electric field and uniform current distribution of the working electrode; and (c) to minimize ohmic drop between the working electrode and the reference electrode.

Requirement (a) must address two problems: contamination of the electrolysis solution by external atmosphere (specifically, oxygen and water from ambient air) and contamination from electrode products produced at the auxiliary electrode. The latter is a problem only in coulometric electrolyses and can be eliminated by use of divided cells in which a diaphragm (e.g., glass frit or permeable membrane) separates the working and auxiliary compartments. The classical H-cell arrangement (Figure 1.11) was designed with this thought in mind. Elimination of moist air is necessary not only because oxygen is electroactive, but also because many electrode products (or reactants) react with oxygen or water. Traditionally, air is eliminated by bubbling an inert gas (nitrogen or argon is suitable) through the test solution prior to electrolysis and then maintaining an inert atmosphere blanket during the experiment. In recent years use of inert atmosphere glove boxes to house electrochemical cells has

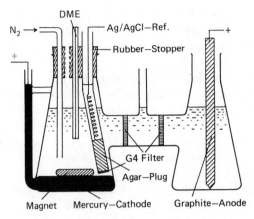

Figure 1.11 Typical H-cell for polarography or bulk electrolysis, allowing for physical separation of cathodic and anodic compartments. Reprinted from M. M. Baizer, Ed., *Organic Electrochemistry*, Dekker, New York, 1972, p. 177, with permission.

increased. This gives one the ability to sample or work up electrolysis solutions without exposing them to air. The most rigorous method to eliminate air contamination is through vacuum-line techniques, and several different cell designs have been reported, ranging from fairly simple [11] to rather complex [12].

Requirements (b) and (c) involve the relative placement of the working, auxiliary, and reference electrodes (only three-electrode cell arrangements are considered here—there is no longer any reason to perform two-electrode measurements). From the viewpoint of current and potential distribution, except for some microelectrodes, it is desirable to have one of two arrangements for the working and auxiliary electrodes: either concentric cylinders or parallel planes. Other arrangements are compromises that are more likely to lead to errors during large-electrode or high-current applications, because ohmic potential drop (*iR* loss), can lead to significant errors in setting or measuring the applied potential E_{app}. Consider the H-cell arrangement (Figure 1.11) in which a fixed potential is applied by the potentiostat to the cell. The actual working electrode potential E_{work} is related to E_{app} and the *iR* drop through the equation

$$E_{app} = E_{work} - E_{ref} + iR \tag{22}$$

In this equation E_{ref} is the reference electrode potential, i is the current flowing from the working electrode, and R is the resistance of the solution between the working and reference electrodes. With the arrangement of electrodes in Figure 1.11, not all parts of the working electrode are at the same potential; rather, the potential varies as the *iR* drop varies between the reference electrode and a particular point on the working electrode surface. In many cases these potential gradients can reach several hundred millivolts. In considering this problem, Harrar and Shain [13] have recommended that the reference electrode be placed on a line *between* the working and auxiliary electrodes. In this way the actual electrode potential will not *exceed* (e.g., be more negative for a reduction process) that of the control (or nominal) potential. If the reference probe is removed from this line, for example, on the far side of the working electrode away from the auxiliary compartment, the working electrode will have sections at which the actual potential is more negative for a reduction or more positive for an oxidation than the control potential. Then, if reduction involves two closely spaced waves (approximately 300 mV or less apart), electrolysis at the potential nominally on the plateau between the two waves could unwittingly give products from the second reduction process [13, 14]. Cells with more ideal electrode geometries minimize the problem of working electrode potential variations. Figure 1.12 shows such an arrangement for a mercury pool electrode; if a cylindrical platinum gauze working electrode is used, the auxiliary electrode frit and the reference probe are placed inside the cylinder.

Even more accurate control of electrode potential is needed in voltammetric experiments, in which potential shifts of as little as 20 mV may be used as criteria for diagnosis of electrode mechanisms. Consider a solution of 0.5 M Bu_4NClO_4 in THF in which the working and reference electrodes are 2 cm apart and a current of 50 μA is flowing. Using the specific resistance for that solution reported by House and co-workers [15] a resistance of about 10^3 Ω

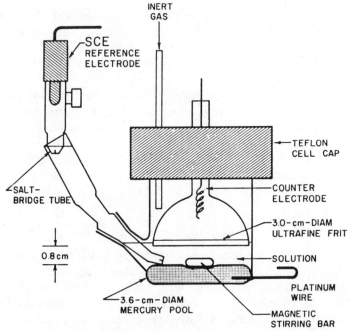

Figure 1.12 A well-designed cell for coulometric electrolysis at a mercury pool electrode. Reprinted from [16], p. 150, with permission.

is anticipated, leading to an ohmic loss of 50 mV, clearly an unacceptable error for most applications.

Placing the working and reference electrodes closer together reduces the ohmic drop, which is uncompensated by the potentiostat. The most effective way of accomplishing this is by using a Luggin probe [16, 17], a glass piece connecting on one side to the reference electrode and on the other side to a narrow opening of about 1 mm diameter, which is quite close to the working electrode (Figure 1.13). The Luggin probe is filled with the same electrolyte as used in the test solution, and the reference electrode is inserted into the probe, while being careful to isolate the reference electrode by a salt bridge, frit, and so on. Any residual ohmic loss will be determined by the solution resistance between the working electrode and the tip of the Luggin probe, even though the reference electrode proper may be considerably upstream. This device could lower the ohmic loss to a few millivolts in the THF example given above.

Optimum placement of the Luggin probe is not a trivial problem. If it is too close to the working electrode, changes in spatial distribution of the electric field and current inhomogeneities may result. If it is too far away, it will not be effective in minimizing iR loss. This problem has been treated [17] for a spherical microelectrode (approximating a hanging mercury drop or dropping mercury electrode). When the working electrode/Luggin probe distance was greater

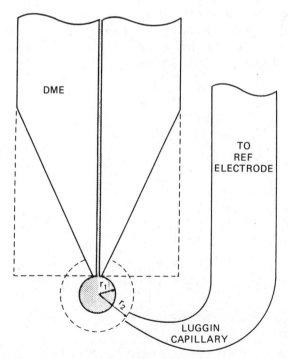

Figure 1.13 A Luggin capillary reference probe in association with a dropping mercury electrode. Reprinted from [16], p. 119, with permission.

than about 5 mm, the uncompensated iR loss was approximately constant. Since moving the Luggin probe closer than the diameter of the probe tip is not recommended, the optimum separation would seem to be 1–2 mm. This analysis assumes a working electrode having a radius of about 0.5 mm and an optimum placement of the auxiliary electrode, and the reader is referred elsewhere for more details [17]. Based on the foregoing considerations, a good cell design for voltammetry studies might have the electrode arrangement shown in Figure 1.14; in this arrangement the auxiliary electrode is wound around the working electrode probe, and the Luggin probe minimizes the resistance effects.

There are many cell designs developed for special uses, such as vacuum-line work, electrochemistry–ESR spectroscopy, and optical spectroelectrochemistry; reviewing them is beyond the scope of this chapter. The book by Sawyer and Roberts [16] offers leading references [18].

Ohmic potential drop may or may not significantly affect experimental results. It is unlikely to affect coulometric electrolyses involving a substance giving only a single wave. Likewise, considerable uncompensated resistance can be tolerated in many chronoamperometry studies, in which the current is usually monitored at high overpotentials (considerably past the E^0 value). On the other hand, iR loss can lead to deceptive results in experiments involving rapidly changing potentials near the E^0 value, such as cyclic voltammetry or ac polarography.

The effects of ohmic drop on quantitative cyclic voltammetry studies have been particularly well documented [19–22]. One important example involves peak separation measurements in cyclic voltammetry (CV), which are used qualitatively to diagnose the reversibility of a charge-transfer step and quanti-

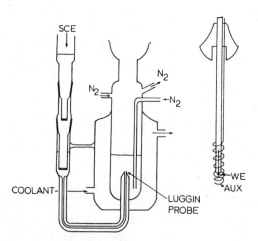

Figure 1.14 A cell designed with good relative placement of the working, auxiliary, and reference electrodes. The working electrode probe is inserted into the top of the cell, within a few millimeters of the Luggin probe. This cell was designed for low-temperature use and provides for thermal isolation of the reference electrode. Reprinted from [76], with permission.

tatively to measure heterogeneous charge-transfer rates. Non-Nernstian redox couples display a peak separation greater than $60/n$ mV and Nicholson [23] has calculated the relationship between ΔE_{pk} and a dimensionless parameter ψ, which is related to the k_s value (k_s is the standard heterogeneous electron-transfer rate) by

$$\psi = k_s/\sqrt{a\pi D} \tag{23}$$

where $a = nFv/RT$. The calculated working curve is shown in Figure 1.15. At higher scan rates (lower ψ), ΔE_{pk} increases. However, this is also true for a perfectly Nernstian system in which there is error because of ohmic loss: peaks spread apart because of uncompensated iR drop as the scan rate increases. Figure 1.15 shows that effects of slow charge transfer and iR loss on the scan rate dependence of ΔE_{pk} are similar and one would be hard-pressed to distinguish between the two effects. Undoubtedly many reported k_s values measured by CV are below the true value because of improper attention to ohmic loss.

How therefore can resistance effects be recognized and either eliminated or at least properly treated? As a good beginning point, measurements should be performed on model systems for which the charge-transfer mechanisms and electron-transfer rates are known, and any deviations from that behavior should be evaluated in light of possible resistance effects. Evans and co-workers [22] have provided a good model system for linear scan and cyclic voltammetry data, p-nitrotoluene in $CH_3CN/0.083\ M\ Bu_4NClO_4$. This compound undergoes a very rapid one-electron charge transfer, and peak separations of about 60 mV were observed at a hanging mercury drop electrode, even at scan rates exceeding 100 V/s.

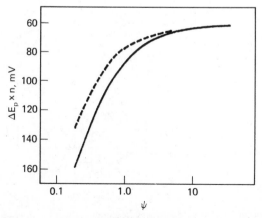

Figure 1.15 Working curve showing relationship between cyclic voltammetry peak separations and the dimensionless parameter ψ, which is defined either in (23) (solid line) from the effects of slow electron transfer, or from $\psi = 1/(nF/RT)nFA \times (\pi aD_0)^{1/2}CR_u$, in which R_u is the uncompensated resistance. Reprinted from [23], with permission.

If resistance effects are spotted, they can be minimized by a several-pronged approach. First, since iR loss is proportional to current flow, it will also be proportional to the concentration of the electroactive compound and the area of the working electrode. Therefore, low concentrations should be used and large electrodes should be avoided. Of course, if concentrations are too low, the charging current may become quite appreciable compared to the faradaic current. We seldom use concentrations below about 2×10^{-4} M. Second, the Luggin probe reference electrode tip should be properly employed, as discussed previously. Third, positive feedback iR compensation can be used, since most commercially available instruments now have this feature. In this procedure a voltage signal is fed back into the instrument's control amplifier to compensate for the ohmic loss in solution. The signal is obtained by passing the current from the current amplifier through a variable resistor (Figure 1.16); the positive feedback is increased by changing the value of the resistor. In principle, the

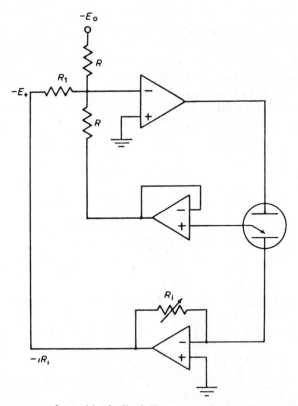

Figure 1.16 Arrangement for positive-feedback iR compensation in a three-electrode potentiostat arrangement. Compensation voltage is fed back from the current-follower amplifier into the summing point of the control amplifier, and is proportional to the value of R_i. E_0 may be an external voltage source such as an offset voltage or a triangular waveform. Reprinted from [24], with permission.

potentiostat will oscillate if too much iR compensation is applied, and this "built-in" protection usually works. However, there are circumstances, especially involving high values of damping capacitors in instruments, under which overcompensation can be achieved without viewing potentiostat oscillation. Evans and co-workers reported that they were able to achieve spurious 30 mV peak separations rather than the 60 mV expected for the p-nitrotoluene system quoted above if damping capacitors were added before employing positive feedback [22]. Thus this approach must be used with caution. A thorough discussion of the positive feedback question, as well as other aspects of ohmic loss, has been presented by Britz [24]. Whatever the approach employed, the experimentalist must always check out procedures by looking at model systems.

3.2 Solvent/Supporting Electrolyte Considerations

3.2.1 General Comments

The electrolyte medium, which consists of the solvent and the supporting electrolyte, exerts a major influence on the nature of the electrochemical process. The chemical properties of the electrolyte medium affect the electrochemical reaction mechanism in the same way solvents affect normal reaction chemistry. An additional constraint, however, is added by the requirement that the physical properties of the electrolyte must also suit the needs of the electrochemical measurement.

An inclusive list of desirable properties for an electrochemical solvent is impractical because changes in experimental goals sometimes require changes in solvent properties. However, for most applications, the desired solvent should:

1. Dissolve the test compound but be otherwise inert to it (important exceptions include studies in pH-buffered aqueous media, in which reactions of the electroactive compound with hydrogen ions are integral to the redox process).

2. Have highly conducting salt solutions to minimize measurement errors caused by resistance effects.

3. Be pure or conveniently purified, that is, free of electroactive impurities or other impurities that could react with the electrolysis product(s).

4. Have a favorable potential "window," that is, an adequate range of potentials over which it does not give rise to appreciable cathodic or anodic "background" currents.

5. Have a liquid range appropriate for the temperature range to be investigated.

6. Have relatively nontoxic vapors.

Additionally, there are a variety of other factors of lower general importance, including viscosity, volatility, acidity, tendency to poison catalysts in controlled-atmosphere boxes, compatibility with interfacing spectroscopic measurements, and so on.

Table 1.1 Some Important Properties of Common Nonaqueous Electrochemical Solvents

Solvent	Liquid Range (°C)	Dielectric Constant	Viscosity (mPa·s)	Typical Electrolytes
Acetone	−95 to +56	21	0.32 (25°)	Et_4NClO_4, Bu_4NPF_6, $NaClO_4$
Acetonitrile	−45 to +82	36	0.33 (25°)	Bu_4NPF_6, $LiClO_4$
Dichloromethane	−97 to +40	8.9	0.39 (30°)	Bu_4NPF_6, Bu_4NClO_4, Bu_4N (halide)
Dimethylformamide	−61 to 153	37	0.80 (25°)	Bu_4NPF_6, $LiCl$, $NaClO_4$
Dimethylsulfoxide	+18 to 189	47	1.99 (25°)	Bu_4NPF_6 (or BF_4)
Tetrahydrofuran	−108 to +66	7.6	0.55 (20°)	$LiClO_4$, Bu_4NPF_6 (or BF_4), $NaClO_4$

Table 1.2 Potential Windows[a] for Selected Electrolytes and Electrodes

Solvent	Electrolyte	Window	Electrode
CH_2Cl_2	Bu_4NPF_6	+2.0 to −2.0	Pt
		+0.9 to −2.0	Hg
THF	Bu_4NPF_6	+1.3 to −2.4	Pt
		+0.8 to −2.9	Hg
MeCN	Bu_4NPF_6	+0.6 to −2.8	Hg
		+2.5 to −2.0	Pt
Acetone	Bu_4NPF_6	+0.6 to −2.5	Hg
	Bu_4NPF_6	+1.4 to −2.1	Pt
DMF	Bu_4NPF_6	+0.5 to −2.8	Hg
		+1.5 to −2.8	Pt
	$NaClO_4$	+0.5 to −2.0	Hg
		+1.6 to −1.6	Pt
DMSO	Bu_4NPF_6	+0.3 to −2.8	Hg
	$KClO_4$	+0.2 to −1.9	Hg

[a]Potentials given versus aqueous SCE.

A summary of some important properties of widely used electrochemical solvents is given in Tables 1.1 and 1.2. The potential windows are taken either from Sawyer and Roberts [25], Mann [26], or from personal experience. The numbers should be considered as qualitative guides, since quantitative values would require reporting current densities. The two solvents in the tables with the lowest dielectric constants, tetrahydrofuran and dichloromethane, give electrolyte solutions that are quite poorly conducting and subject to rather large ohmic losses. Nevertheless, they are relatively nonreactive, and this makes them attractive for some studies. When using these solvents, particular attention should be paid to resistance effects (Section 3.1).

3.2.2 Influence on Electrochemical Measurements and Mechanisms

The electrolyte medium can affect the electrochemical experiment in three principal ways: namely, in the mass-transport process (how the electroactive compound is transported to the working electrode), in the electron-transfer step, and in reactions of electrode products with either the solvent itself or its impurities (such as, H_2O).

The mass-transport effect is based on the viscosity of the solvent and on the diffusion coefficient of the test compound. For coulometric electrolyses in which rapid transport of the electroactive compound to the electrode is desired to reduce electrolysis time, a low-viscosity solvent is preferred. For voltammetry experiments in nonstirred solutions (e.g., chronoamperometry and cyclic voltammetry) a high-viscosity solvent is preferable to maximize the time over which a static diffusion layer exists at the electrode. In practical terms, however, choice of an electrolyte system is seldom made on the basis of viscosity effects.

3.2.3   Electrolyte Effects on the Electron-Transfer Step

The electrolyte medium affects both the thermodynamics and kinetics of the electron-transfer step. A sophisticated understanding of these effects requires an in-depth knowledge of the electrode–solution interface [27]. The electrical "double-layer" is a relatively simple representation of this interface (Figure 1.17).

In this model, supporting electrolyte ions are attracted by coulombic forces to a mean distance from the electrode called the outer Helmholtz plane (OHP), with further ordering of the counterion at greater distances from the electrode. Solvent molecules occupy the space between the OHP and the electrode surface. A molecule that is subject to strong attractive forces from the metal electrode may penetrate the OHP into a plane closer to the electrode called the inner Helmholtz plane (IHP) and is said to be specifically adsorbed. Such molecules or ions exhibit redox properties different from their "soluble" counterparts because they are subject to increased electrical fields at these shorter distances from the electrode. The occurrence of adsorbed species can complicate electrochemical work; thus, adsorption is often seen as an unwanted phenomenon. However, some redox mechanisms, such as the reduction of hydrogen ions, depend critically on the presence of adsorbed reactants or intermediates and must be considered seriously. Fortunately, as discussed in other chapters in this volume, there are simple experimental tests for the presence of adsorbed species, and the reader is referred to those treatments for recommended procedures.

The electrical double-layer structure influences the *kinetics* of electron-transfer reactions and even, in some cases, the *mechanism* of the charge transfer. The former arises from the fact that the concentration distribution of electroactive species and the electric field itself are a function of the double-layer structure. The latter effect, on the redox mechanism, involves the mutual interactions of electroactive compound, solvent, electrode, and supporting electrolyte ions, and is necessarily quite complex. Of particular note is a situation in which the electrolyte ion participates in the charge-transfer step through an inner-sphere mechanism. *Homogeneous* inner-sphere electron-transfer processes have been widely studied and involve a mechanistic path in which the two reactant molecules share a common ligand [28]. A heterogeneous analogue of this is one in which a ligand bridge forms between the electrode and the electroactive species during the charge-transfer step. Usually the bridging ligand is an ion that happens to be specifically adsorbed on the electrode [29]. The reduction of In^{3+} at a mercury electrode provides a classic example of this effect [30]. In perchlorate or nitrate media, the In^{3+}/In couple is highly irreversible, as evidenced in Figure 1.18 by the drawn-out shape of the rising portion of the

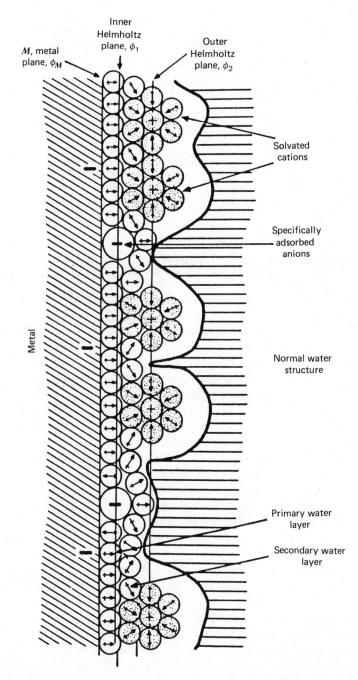

Figure 1.17 A model of the electrical double layer. Reprinted from D. M. Mohilner, in A. J. Bard, Ed., *Electroanalytical Chemistry*, Dekker, New York, Vol. 1, 1966, p. 246, with permission.

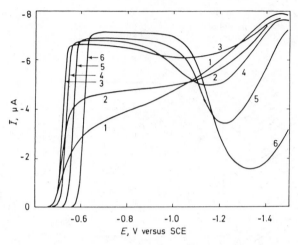

Figure 1.18   Traces of current maxima of dc polarograms of $1.0 \times 10^{-3}\,M$ In $(NO_3)_3$ with $1.0\,M$ $NaNO_3$ as supporting electrolyte (Curve 1). Subsequent traces taken in presence of added KSCN at concentrations of $1.0 \times 10^4\,M$ (2), $2.0 \times 10^{-3}\,M$, (3), $4.0 \times 10^{-2}\,M$ (4), $2.0 \times 10^{-1}\,M$ (5), and $1.00\,M$ (6). Reprinted from [30], with permission.

polarogram. Upon addition of a small amount of KSCN, the wave becomes reversible, presumably because of enhancement of charge transfer through the bridging interaction between In^{3+} and adsorbed SCN^-. Further addition of SCN^- does not affect the reversibility of the couple, but does affect the E^0 value (Curves 4–6 in Figure 1.18) because of In^{3+}/SCN^- equilibria in bulk solution.

Inner-sphere processes have not been found as commonly in organic or organometallic electrode processes. Instead, outer-sphere mechanisms (those in which there is no ligand or atom transfer to or from the electrode) predominate. In outer-sphere processes, the influence of the electrolyte medium on charge-transfer kinetics arises from the way in which the electrical field is affected by the ionic structure of the double layer [30], and from the influence of ion-pairing effects of an electrolyte ion with an electrode reactant or product. Ion-pairing effects on electron-transfer rates of organics have not been widely studied, but there are indications that even counterions usually treated as weakly interacting (such as R_4N^+) may influence charge-transfer rates [31].

More thoroughly investigated are the effects of solvent and supporting electrolyte on the thermodynamics of electrode processes, that is, their E^0 values. Again, ion pairing may influence the E^0 values, but the major effect is usually caused by solvation changes. It is the change in free energy of solvation between reactant and product that determines the contribution of solvent to the thermodynamic redox potential. For example, in the one-electron reduction of aromatic hydrocarbons, HC,

$$HC + e^- \rightleftharpoons HC^-$$

it has been shown [32] that the observed half-wave potentials are related to solvation changes by (24):

$$E_{1/2} = EA + \Delta G_{solv} + const \qquad (24)$$

in which EA is the electron affinity of the hydrocarbon in the gas phase and ΔG_{solv} is the difference in solvation energy between the neutral hydrocarbon and its anion radical. Solvent-induced E^0 shifts can normally be rationalized on the basis of a bulk property of the solvent. Take the reduction of HC as an example. One expects that ΔG_{solv} will be dominated by the energy of solvation of the anion and, indeed, as the dielectric constant of the solvent increases, a positive shift of E^0 values is observed [32]. Conversely, the reduction of solvated *cations* becomes more difficult (negative E^0 shift) with increasing solvent dielectric constant [33]. However, dielectric constant is not a universally successful guide to solvation strengths, and other schemes have been proposed to account more quantitatively for solvation effects, especially for solvents acting as Lewis bases. These schemes include the Gutmann donor number [34] and the Z-parameter of Kosower [35]. Some aspects of their application have been reviewed [36, 37]. Each of these attempts at quantifying solvation effects is successful in explaining trends in certain cases, especially those involving a homologous series of compounds in which the redox process appears to affect the bulk ordering of the solvent around the molecule, rather than specific solvation of the product or reactant. However, because specific solvation is so widespread, it is often hard to predict solvent effects. For example [33], although it is easier to reduce most simple metal ions in CH_3CN compared with H_2O, Ag^+ or Cu^+ are *harder* to reduce in CH_3CN by over 600 mV, reflecting the increased stability of acetonitrile complexes of silver and copper, compared with their aquated analogues.

The most common error in interpreting solvent effects on E^0 potentials is improper attention to liquid junction potentials. Junction potentials arise because the reference electrode is separated from the test solution by a device such as a salt bridge or cracked glass. The differential mobilities of supporting electrolyte ions across the barrier lead to the junction potential, and these mobilities change as the solvent is changed [38]. Although changes in liquid junction potential can be minimized by using a supporting electrolyte in which both ions are large and nonspecifically solvated, such as $[R_4P]^+[(C_6H_5)_4B]^-$ [39], with more common supporting electrolytes, changes of 100 mV or more are possible when changing solvents. One proposed approach to obviate this effect involves referring potentials to a reference redox couple of a compound that presumably has an E^0 independent of solvent. Large, symmetrical ions have been favored for this purpose, and several model systems have been proposed, including Rb^+/Rb [40], $(\eta^5 - C_5H_5)_2Fe^+/(\eta^5 - C_5H_5)_2Fe$ [41], and $(\eta^6 - C_6H_6)_2Cr^+/(\eta^6 - C_6H_6)_2Cr$ [42]. However, because the solvation inde-pendence of the reference couple has been assumed and not demonstrated, this approach yields only an *estimate* of correction for liquid junction potential.

Hence, it is questionable to attach thermodynamic significance to changes in E^0 values of less than 100 mV in different solvents.

3.2.4   Post-Electrolysis Reactions With Solvent

One of the most important and, usually, troublesome properties of a solvent is the possibility that it may react with the primary electrode products. Rapid expansion of interest in nonaqueous electrochemistry coincided with efforts to find media in which organic ion radicals were stable enough to allow study of their reaction pathways. Since organic radical anions and cations are particularly prone to addition or loss, respectively, of protons, it is important to control the acidity of the electrolyte medium. Modified arguments may be made for organometallic or inorganic compounds. In aqueous electrochemical studies this is accomplished by buffering the solution to a given pH. Careful study of pH effects has proven to be a powerful tool for deciphering organic redox mechanisms [43]. However, it has often been desirable to eliminate altogether, or as much as possible, the availability of protons in the electrolyte, for example, to stabilize radical anions. Since no practical electrochemical solvent is truly aprotic, protonation reactions are governed by the acidity of solvent protons or by the presence of residual amounts of water or other acidic impurities. It is often quite difficult to determine the chemical species responsible for protonation of electrogenerated radical anions. Coupled with the fact that the purity of solvents can vary so much from one laboratory to another, we are restricted to giving some general guidelines for choosing particular nonaqueous solvents. Many aspects of the purification and preferred uses of nonaqueous solvents have been treated in detail elsewhere [26, 44, 45].

Carefully purified CH_3CN is a good solvent for both oxidations and reductions. Water is difficult to eliminate as an impurity in acetonitrile, and water levels of several millimolar are common even in freshly purified solvent. Scavenging of trace water by using activated alumina or trifluoroacetic anhydride *in situ* has been shown effective for stabilizing organic radical ions [46] in this solvent. Although DMF and DMSO are not necessarily drier when used under electrochemical conditions, they tend to be better at stabilizing anion radicals—in at least one study, because the activity of water in DMF or DMSO is lower than in acetonitrile [47]. This is presumably caused by the better hydrogen-bonding abilities of the former two solvents, which leads to an effective tying up of the water. Other solvents more recently investigated for stabilization of organic radical ions are liquid ammonia for anions [11, 48] and liquid sulfur dioxide for cations [49]. Both are very promising.

Whereas solvent acidity is most important in organic electrode processes, it appears to be a solvent's *coordinating* ability that is most important in choosing solvents for inorganic or organometallic studies. CH_3CN is still a popular solvent because of its wide potential window and fairly highly conducting electrolyte solutions. However, acetonitrile and other solvents with Lewis base properties often react with metal compounds to either displace a ligand or to occupy an open coordination site. Since metal coordination preferences often change with changes in oxidation state, not surprisingly solvent loss or gain often accompanies electrochemical redox processes [50]. Three solvents popular for minimizing the coordination effect are dichloromethane, THF, and acetone. Although they have the benefit of a lower coordination tendency than CH_3CN, they have the problem of low dielectric constants (Table 1.1) and highly resistive electrolyte solutions. When working with these solvents, careful attention must be paid to problems of ohmic loss. Despite the measurement difficulties with these solvents, they have been widely employed, for often one or another will be key to stabilizing a particular electrode product. THF appears to be a good solvent for stabilizing reduced metal complexes. Dichloromethane, the least coordinating of the six solvents discussed in this section, has been popular since the early 1960s for metal complex redox studies [51]. Acetone is the most suitable of the poorly coordinating solvents for low-temperature studies [52].

3.3 Reference Electrodes

3.3.1 General Considerations

Most investigators who perform cyclic voltammetric, chronoamperometric, and the other electrochemical experiments described above will be interested only in the chemical reaction that occurs at the working or indicator electrode. Since the ease of an electron's removal from or addition to an electroactive species is a function of the energy of the electrons within the working electrode, all thermodynamic and many kinetic measurements require that the potential of the working electrode be known and controlled with respect to the fixed potential of a reference electrode. While the internationally accepted primary reference is the standard hydrogen electrode (SHE),

$$2H^+ + 2e^- \rightleftharpoons H_{2(g)}$$

where all components are present at unit activity, the difficulty of constructing, using, and maintaining this electrode makes its everyday use unattractive. Unfortunately, the number of other redox couples that may be used satisfactorily in the reference half-cell is limited. Although certain experimental conditions may dictate additional requirements for the reference electrode, a suitable couple must be electrochemically and chemically reversible, stable and unreactive with respect to other species in the solvent–electrolyte system, and return to the equilibrium potential after polarization in either direction. These requirements imply that the Nernst equation be obeyed with respect to some component of the solvent–electrolyte system. This section will describe briefly the redox couples that have been used satisfactorily for the reference half-cell in a variety of aqueous and nonaqueous systems.

3.3.2 Aqueous Solution

The most commonly used reference electrode in aqueous solutions is the calomel electrode, which consists of a mercury paste, mercurous chloride (calomel), and potassium chloride in contact with a saturated aqueous solution of potassium chloride. Electrical contact with the paste is made by an amalgamated platinum wire, while electrolytic contact with the external solution is generally made with porous vycor glass, a porous asbestos fiber, cellulose pulp, or a ceramic frit.

The mercury–mercurous chloride electrode is an example of an electrode of the second kind (25), that is, an electrode in which the response of the metal ion couple (Hg/Hg_2^{2+}) is influenced by the activity of the anion (Cl^-) that forms a slightly soluble salt (26)–(28):

$$Hg_2Cl_{2(s)} + 2e^- \rightleftharpoons 2Hg_{(l)} + 2Cl^- \tag{25}$$

$$Hg_2Cl_{2(s)} \rightleftharpoons Hg_2^{2+} + 2Cl^-$$
$$K_{sp,Hg_2Cl_2} = [Hg_2^{2+}][Cl^-]^2 = 10^{-17.88} \tag{26}$$

$$E = E^0_{Hg_2^{2+}/Hg} - \frac{0.059}{2} V \log \frac{1}{[Hg_2^{2+}]}$$
$$= E^0_{Hg_2^{2+}/Hg} - \frac{0.059}{2} V \log \frac{[Cl^-]^2}{K_{sp,Hg_2Cl_2}} \tag{27}$$

$$= E^0_{Hg_2Cl_2/Hg} - 0.059V \log [Cl^-] \tag{28}$$

The passage of current through this half-cell could cause a change in the chloride activity and, thus, a corresponding change in the half-cell potential (28). To minimize this potentially deleterious effect, a relatively concentrated aqueous solution of potassium chloride should be used as the electrolyte. While most workers prefer a saturated KCl solution, this could lead to clogging of the junction with the external solution if KCl were to precipitate for any reason. Accordingly, 3.5 M KCl is recommended for the electrolyte fill solution by some workers, or NaCl may be employed in place of KCl.

The calomel electrode is stable at room temperature, but exhibits a significant temperature hysteresis effect when the temperature exceeds 50°C [53]. Another useful reference electrode of the second kind, which is more stable than the calomel electrode at higher temperature, is the silver–silver chloride electrode. As with the calomel electrode, the potential of this half-cell is a function of the standard electrode potential of the slightly soluble AgCl and the activity of silver ion and, through the K_{sp} relationship, the activity of the chloride ion (29)–(31).

$$AgCl_{(s)} + e^- \rightleftharpoons Ag_{(s)} + Cl^- \tag{29}$$

$$AgCl_{(s)} \rightleftharpoons Ag^+ + Cl^-$$
$$K_{sp,AgCl} = [Ag^+][Cl^-] = 10^{-9.75} \tag{30}$$

$$E = E^0_{Ag^+/Ag} - 0.059V \log - \frac{1}{[Ag^+]}$$
$$= E^0_{AgCl/Ag} - 0.059V \log [Cl^-] \tag{31}$$

3.3.3 Nonaqueous Solvent Systems—Electrodes of the Second Kind

Relatively few reference electrodes of the second kind that were well-behaved in aqueous media function equally satisfactorily in nonaqueous solvents. In silver halides, formation constants for the reaction $AgX_{(s)} + X^- \rightleftharpoons AgX_2^-$ are in excess of one in such frequently used solvents as acetonitrile, dimethylformamide (DMF), and dimethylsulfoxide (DMSO) [54]. This leads to extensive formation of AgX_2^- when X^- is present, thereby creating a liquid junction potential because of the unequal mobilities of AgX_2^- and X^-. In Hg_2X_2, decomposition of the mercurous halide in the presence of X^- (32) has reportedly occurred in numerous solvent systems, including acetonitrile and DMF.

$$Hg_2X_{2(s)} + X^- \rightarrow HgX_3^- + Hg \tag{32}$$

An electrode of the second kind that is reversible in DMF is based on the half-cell $Cd(Hg)|CdCl_{2(s)}$, $CdCl_2 \cdot H_2O_{(s)}$, $NaCl_{(s)}$ in DMF [55]. The solubility of $CdCl_2$ in DMF is reported to be 0.06–0.10 M and is only slightly dependent on chloride ion [56]. The reproducibility of this half-cell is excellent, especially when a definite hydrate equilibrium for the $CdCl_2$ is maintained. Unfortunately, the use of $Cd(Hg)|CdCl_2$ as an electrode of the second kind is less satisfactory in other frequently used solvents. The solubility of $CdCl_2$ is high in DMSO and, although $CdCl_2$ has low solubility in acetronitrile, it becomes quite soluble when Cl^- is present [54].

The most widely used nonaqueous reference electrode is based on the Ag/Ag^+ couple. Silver ion is normally furnished by soluble salts, such as $AgClO_4$ and $AgNO_3$, and is generally maintained at relatively high concentration, for example, 0.01–0.1 M. As long as the components of the solvent–electrolyte system are not oxidized by Ag^+, stable, reversible behavior for the Ag/Ag^+ couple is observed in most solvent systems. Some of the solvents in which the Ag/Ag^+ couple has been used include acetone, acetonitrile, diethyl ether, dimethoxyethane, DMF [57], DMSO, hexamethylphosphorotriamide (HMPA), propylene carbonate, pyridine, and tetrahydrofuran [54].

The most frequently used reference electrode in nonaqueous solvent systems is the saturated calomel electrode (SCE). This electrode is normally connected to the nonaqueous solution under study by means of a salt bridge that contains a nonaqueous solution of the supporting electrolyte, for example, $(n\text{-Bu})_4NClO_4$ in DMF. Although the nature of the salt bridge can have a significant influence on the reference electrode potential, the liquid junction potential is usually quite stable and reproducible. Precipitation of KCl at the nonaqueous–aqueous junction is a potential problem with certain nonaqueous solvents and is especially acute when a low dielectric constant solvent that is immiscible with water is used, such as, dichloromethane. Frequent checks of the reference electrode stability, its resistance, and its capacitance are required when the SCE is used under these conditions [58]. Although prolonged operation of the SCE in a nonaqueous solvent can lead to a small shift in the electrode potential and a slight increase in the resistance, the equilibrium values prior to operation in the nonaqueous solvent are usually reestablished after storage of the electrode in saturated aqueous KCl for several hours.

3.3.4   Dual Reference Electrodes

Low resistance in the reference electrode is particularly important when electrode reactions are studied by rapid scan cyclic voltammetry in nonaqueous solvents at low temperature. To compensate electronically for the ohmic potential loss between the reference and working electrodes, the time constant $\tau_r = R_r C_s$ that results from the combination of a stray capacitance C_s and of the resistance R_r of the reference electrode must be kept small. Excessively large reference electrode resistance is characterized by loss of potential control on the rising portion of the double-layer charging i–t curve and incomplete compensation [59]. One manifestation of incomplete compensation is an anodic–cathodic cyclic voltammetric peak separation that exceeds 60 mV for a redox couple when heterogeneous electron transfer is known to be rapid. The need for accurate potential control is especially important when extraction of kinetics information from cyclic voltammetric peak shapes is attempted.

The resistance problem attendant with most reference electrodes of the first and second kinds may be circumvented by using a dual reference electrode of the type shown in Figure 1.19 [59, 60]. The plantinum wire and the series capacitor shunt the standard reference electrode for the high-frequency components of the applied signal while a well-poised dc potential is maintained by the standard reference electrode. To illustrate the effectiveness of the dual reference electrode system, the cyclic voltammetric reduction of 1 mM fluorenone at 100 V/s in HMPA–0.1 M (n-Bu)$_4$NBF$_4$ afforded a separation of

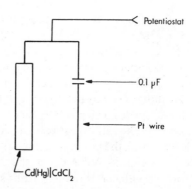

Figure 1.19   Double reference electrode for electrochemical studies in DMF.

only 60 mV between the anodic and cathodic peaks. In contrast, a separation of 310 mV was obtained when a conventional Ag/Ag^+ (0.01 M in HMPA) reference electrode was used under identical conditions [59].

3.3.5 Comparison of Reference Electrodes

To compare emf data that have been measured with respect to different reference electrode systems, the conversion of data from one reference electrode scale to a second reference electrode scale is required. If the solvent–electrolyte system is not permitted to change, then the conversion is given by (33),

$$E_{\text{couple versus Ref 2}} = E_{\text{couple versus Ref 1}} - E_{\text{Ref 2 versus Ref 1}} \tag{33}$$

where $E_{\text{couple versus Ref 2}}$ and $E_{\text{couple versus Ref 1}}$ are the potentials of the redox couple of interest with respect to reference electrodes 2 and 1, respectively. For example, if one wishes to restate the cathodic cyclic voltammetric peak potential of the fluoreneone/fluorenone anion radical couple, measured with respect to the reference electrode $Cd(Hg)|CdCl_{2(s)}$, $NaCl_{(s)}$ in DMF ($E_{\text{fluorenone versus Cd(Hg)}}$ = -0.56 V), in terms of the SCE ($E_{\text{SCE versus Cd(Hg)}} = 0.75$ V), then

$$E_{\text{fluorenone versus SCE}} = E_{\text{fluorenone versus Cd(Hg)}} - E_{\text{SCE versus Cd(Hg)}} \tag{34}$$

$$= -0.56 \text{ V} - 0.75 \text{ V} = -1.31 \text{ V}$$

The process is also illustrated in Figure 1.20.

A shift in the potential of the reference electrode may also result if the solvent and the composition of the salt bridge are changed. This shift is caused by changes in the degree of ion-pair formation and complexation, solvation, activity coefficients, and the liquid junction potential. As an example, the potential varies from 0.253 V for the cell

Aqueous SCE‖0.1 M $NaClO_4$, 0.01 M $AgNO_3$ in CH_3CN|Ag

to 0.300 V for the cell [61]

Aqueous SCE‖0.01 M $AgNO_3$ in CH_3CN|Ag

Obviously, details of the reference electrode construction and the composition of the salt bridge must be reported if reproducible results are to be obtained from one laboratory to the next.

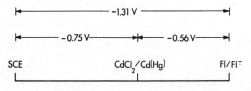

Figure 1.20 Depiction of the effect of the reference electrode on the measured cell potential.

3.3.6   Use of Internal Standards as Reference

Two situations arise frequently in nonaqueous electrochemical studies that require the use of an internal standard for the accurate reporting of reduction potentials:

1. When a pseudo-reference electrode is used, as with the silver wire pseudo-reference electrode in liquid ammonia.
2. When the reference electrode is unstable on a long-term basis and/or the liquid junction potential (Section 3.2.3) is not reproducible.

These difficulties may be circumvented by using an internal reference redox system [62]. The provisional recommendations of the Commission on Electro-

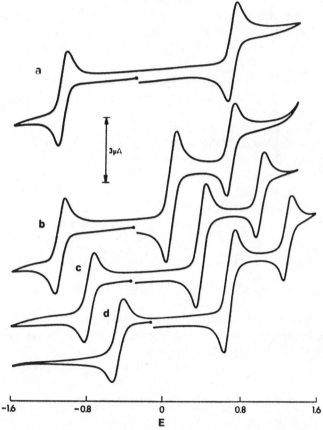

Figure 1.21   The cyclic voltammetric oxidation and reduction of 5 mM Ru(acac)$_3$ in CH$_3$CN–0.1 M (n-Bu)$_4$NClO$_4$ at a planar platinum electrode with (Curves b–d) and without (Curve a) ferrocene as an internal reference redox standard: (a) and (b) versus Ag/AgNO$_3$ (10^{-2} M); (c) versus SCE; (d) versus Cu wire. From [11]; reprinted, with permission, from R. R. Gagne, C. A. Koval, and G. C. Lisensky, *Inorg. Chem.*, **19**, 2855 (1980); copyright 1980 American Chemical Society.

chemistry allow the use of either the ferrocene/ferrocenium (Fc/Fc$^+$) couple or the bis(biphenyl)chromium(0)/bis(phenyl)chromium(I) ion (BCr/BCr$^+$) couple as the *reference redox system* [62]. The difference between these reference redox systems is nearly constant and independent of the solvent (1.124 ± 0.012 V for 22 solvents). Two couples are recommended rather than just one, because of the possibility for overlap between the electroactive species of interest and the reference redox system. It is sufficient in polarographic and voltammetric studies to add one form (Fc or BCr$^+$PPh$_4^-$) of the reference redox system; the other form of the redox couple will be formed at the working electrode during the experiment.

The internal reference redox system is illustrated in Figure 1.21 [62]. After the cyclic voltammogram for the electroactive species of interest has been recorded [e.g., tris(acetylacetonato)ruthenium(III), Ru(acac)$_3$ in CH$_3$CN], ferrocene is added and the cyclic voltammogram is again obtained in the potential range in which both ferrocene and the test compound are electroactive (Figure 1.21b). This method has the important advantage of allowing the use of any convenient reference electrode. Note in Figure 1.21 that, whereas the cyclic voltammetric peaks attributed to Ru(acac)$_3$ shift along the potential axis when the reference electrode is changed, the potential differences between the Ru(acac)$_3$ peaks and the Fc/Fc$^+$ reference redox system remain constant.

3.4   Working Electrodes

3.4.1   Inert Working Electrodes/Materials

The electrochemical cells that are discussed in Section 3.1 assume a three-electrode configuration consisting of an inert working electrode, an auxiliary current-carrying electrode, and a reference electrode. The electrochemical reaction of interest occurs at the working electrode at a potential that is either controlled or measured with respect to the reference electrode (Section 3.3). For most studies, the working electrode material must be electroinactive in the potential range of interest. This restriction, along with additional constraints that may be imposed by requisite electrode geometry and the conductivities of both the solution and the working electrode, often dictate the choice of electrode material and its design.

The two most frequently encountered working electrode geometries in voltammetric work are planar and spherical. From a practical mathematical standpoint, the treatment of an electrochemical problem involving electron transfer, diffusion to and from the working electrode surface, and the kinetics of coupled homogeneous chemical reactions is simpler for the planar configuration. Since the thickness of the diffusion layer is given approximately by $2\sqrt{Dt}$, and if it is assumed that $D = 1 \times 10^{-5}$ cm^2/s and that the radius of an unshielded planar electrode must exceed $2\sqrt{Dt}$ by at least an order of magnitude to obtain a satisfactory fit of experiment with theory, then the radius of the electrode must be

$$r \geqslant (6.3 \times 10^{-2} \text{ cm/s}^{1/2})t^{1/2}$$

For a typical planar electrode with an area of 0.25 cm^2, satisfactory fit of data to a one-dimensional diffusion model would be expected for $t \leqslant 5$ s.

The material used for the working electrode may also dictate the geometry. With mercury, a spherical geometry in the form of a dropping mercury electrode for polarography and a hanging mercury drop for voltammetric studies is usually adopted. The size of the hanging mercury drop can be controlled within the limits imposed by the mass of the drop and the surface tension of mercury, while the drop size in the polarographic experiment can be controlled by a solenoid-operated hammer that periodically dislodges the growing drop. Solid microspherical electrodes, such as, platinum, are used extensively in qualitative cyclic voltammetry studies and when electrode areas smaller than those furnished by the commercial manufacturers of planar electrodes are desired. Details for the preparation of microelectrodes have been described [63].

MERCURY

If the working electrode is to be used for reductions in aqueous solution or in nonaqueous solvents that contain strong oxygen- and nitrogen-centered acids, the electrode material of choice probably would be mercury. The inherently slow rate of hydrogen evolution on mercury permits a cathodic limit of approximately -1.0 V versus SCE in 1 M $HClO_4$ and -2.5 V in 0.1 M NaOH. In the absence of strong proton donors in dipolar, aprotic solvents such as N,N-dimethylformamide and acetonitrile, a cathodic limit of approximately -3.0 V is attainable when $(n\text{-Bu})_4NClO_4$ is used as the supporting electrolyte (Section 3.2.4). The ready ease of mercury oxidation, especially when anions are present that precipitate or complex Hg(I), such as Cl^-, limits the use of mercury in anodic studies.

Mercury's high hydrogen overpotential makes the reduction of many other difficult-to-reduce species possible. Unfortunately, the absence of response to weak proton donors that may be present in nonaqueous solvent systems may also give misleading results. As an example, the reduction of many organic materials affords anion radicals that react readily with adventitious proton donors such as water (Section 3.2.4). Thus, if the goal of the experiment is to characterize the reactivities of the anion radicals, the presence of electroinactive proton donors might mask other important reaction channels. The second principal problem with mercury as an electrode material is its reactivity. In addition to the relative ease of mercury oxidation in the presence of certain anions, mercury is a good trap for certain radicals. The ready capture of electro-generated benzyl radicals to give dibenzyl mercury in yields that exceed 40% [64] demonstrates this potential limitation for the use of mercury as an electrode material. Mercury may also react with organometallic electrode reactants and products [65].

PLATINUM AND GOLD

The heterogeneous rate constants for hydrogen evolution at platinum and gold cathodes greatly exceed that for mercury. Accordingly, the use of platinum and gold as cathode materials is limited mainly to aprotic solvent systems. When the aprotic solvent-supporting electrolyte system has been carefully purified and the concentrations of water and other adventitious proton donors are low, a cathodic limit $[-3.0$ V versus SCE in DMF -0.1 M $(n$-Bu$)_4$NClO$_4]$ equal to that for mercury can be attained. Because both plantinum and gold are also less reactive toward electrogenerated intermediates and products than mercury, these materials are preferred by many electrochemists for cathodic studies in aprotic solvents. Although gold has several inherent advantages over platinum, for example, greater hydrogen overpotential and less adsorption of hydrogen, the ease of fabricating platinum into planar and microspherical electrodes accounts for platinum's greater popularity among electrochemists.

On the anodic side, all noble metals will form an oxide film or an oxygen layer in aqueous solution at relatively positive potentials. In addition, these electrode materials are susceptible to oxidation when anions such as the halides and cyanide are present. In the absence of these complexing anions in most aprotic solvents, both plantinum and gold perform well as anode surfaces.

CARBON

Its inertness, large potential window, low electrical resistance, ease of fabrication into electrodes, low cost, and reproducible electrode surface have made carbon [66] one of the most popular and useful electrode materials. While numerous forms of carbon electrodes have been reported, two specific types, the carbon paste and the glassy carbon, demonstrate the utility and versatility of this material.

The carbon paste electrode [67] is normally shaped into a circular disk and consists of a Teflon piece into which has been machined or drilled a well that holds the carbon paste and the electrical connector (such as, copper wire) that extends from the working electrode lead to the carbon paste. A suitable paste for aqueous work can be prepared by thoroughly mixing 15 g of finely ground carbon with 9 mL of mineral oil (Nujol). The resultant paste will have the consistency of peanut butter. Since the anodic potential limit and low residual current of the carbon paste electrode are superior to those for platinum and gold, the carbon paste electrode finds extensive use in anodic studies. If fouling of the electrode surface should occur, the surface can be easily renewed by replacing the thin layer of carbon paste that contacts the solution. If care is taken in preparing the electrode surface, the electrode area can be readily reproduced ($\pm 1\%$). On the cathodic side, the limit is set by the reduction of the residual oxygen dissolved in the paste.

Although a nonaqueous carbon paste electrode has been described, the vitreous or glassy carbon electrode offers superior performance in nonaqueous media. Glassy carbon is a proprietary material [68] that is highly conductive, anisotropic, impermeable to gases, and highly resistant to chemical attack. Relative to platinum, glassy carbon has a large overpotential for hydrogen evolution. This is an important advantage, since this means that glassy carbon can be used in nonaqueous solvent systems in the presence of many oxygen- and nitrogen-centered acids (e.g., 1,1,1,3,3,3-hexafluoro-2-propanol) that are electroactive on platinum. In addition, the reductions of many halogenated organic compounds are less complicated on glassy carbon than on platinum. This is believed to be because of reduced halide ion adsorption and/or reduced halide-ion-bridged electron transfer on the glassy carbon surface.

3.4.2   Chemically Modified Electrodes

When a working electrode is inserted into a conducting solution that contains an electroactive species, adsorption of a component of the solvent–electrolyte system, the electroactive species, and/or the product of the electron transfer onto the electrode surface often results. Since adsorption causes modification of the working electrode surface [69, 70], this phenomenon may have a significant effect on the redox behavior of the electroactive component. For example, the adsorption of iodide ion, which may be formed *in situ* as a product of the electrode process, causes the electroreduction of allyl iodide at a platinum surface to occur in either one or two steps, depending on the concentration of iodide ion [71]. The observation that adsorbed species may influence redox behavior has prompted extensive, systematic studies in attempts to prepare electrode surfaces that will function in novel ways.

Procedures for immobilizing reagents on an electrode surface include covalent bonding between the reagent and the electrode, deposition of the polymer or the preparation of the polymer on the electrode surface, and chemisorption. Chemisorption tends to produce the least stable surface with respect to repeated use and exposure to solution, since only about one monolayer of material can be adsorbed onto the surface. The polymer-coated surface, in which the equivalent of 100–1000 monolayers have been affixed to the surface by adsorption and insolubility of the polymer, is more reproducible than the monolayer surface, and, in addition, generally provides a larger electrochemical response because of its larger number of redox sites. Examples of redox polymers include poly(vinylferrocene), poly(vinylpyridine), and poly(nitrostyrene). Covalent attachment involves the formation of a chemical bond between the electrode surface and an appropriate functional group. In the case of metal oxide surfaces or an oxidized carbon surface, the linkage is usually made by an oxygen bond. A typical reaction sequence first involves reaction with an organic silane, followed by treatment with an appropriate reagent to give an amide, sulfonamide, or other functional group:

$$Pt \equiv\!\!-OH + (CH_3O)_3Si(CH_2)_3NH(CH_2)_2NH_2$$

$$\rightarrow Pt \equiv\!\!-OSi(CH_2)_3NH(CH_2)_2NH_2 \xrightarrow{\overset{\displaystyle O}{\overset{\|}{\text{(HOCR)}}}}$$

$$Pt \equiv\!\!-O\!-\!Si(CH_2)_2NH\overset{\displaystyle O}{\overset{\|}{C}}R$$

The functional group attached in the last step of this reaction sequence is often electroactive. For example, when R is a certain Ru(II) complex, the immobilized Ru(II) moiety can be reversibly oxidized to the Ru(III) complex [69]. The formal potential of the immobilized redox couple is nearly the same as that of the unattached couple. This means that electron transfer between the electrode and the immobilized redox couple occurs readily and that both the identity of the electrode material and the chemical linkage between the electrode and the functional group are relatively unimportant.

One of the interesting ways to exploit the properties of the chemically modi-fied electrodes is electrocatalysis. In this application, the immobilized redox couple acts as a fast electron-transfer mediator for a species in solution that would be oxidized or reduced slowly, if at all, at the unmodified electrode surface at the applied potential:

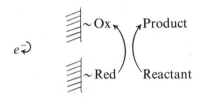

For example, Collman and co-workers [72] have reported the four-electron, mass-transfer-controlled reduction of O_2 to H_2O without significant H_2O_2 production. The surface of a graphite electrode was modified by the irreversible adsorption of a cofacial dicobalt porphyrin dimer that allows the reduction of oxygen to proceed in a manner that cannot be duplicated at an unmodified electrode surface.

3.4.3   Optically Transparent Electrodes

The need to monitor intermediates and products of electrode reactions by absorption spectroscopy has led to the development of two general classes of optically transparent electrodes [73] (see also Section 2.7). In the first class, a thin film of electrically conducting material (such as, Pt, Au, SnO_2, or mercury-coated Pt) is deposited on a transparent substrate (such as glass or quartz). Because the film thickness is quite thin (10–500 nm), transparencies in the range from 20 to 85% can be achieved. The second class of optically transparent electrodes is based on a fine micromesh (100–2000 wires/in.) of electrically conducting material (e.g., Au). Electrodes of the former type can be used in single potential step chronoamperometric studies to study short-lived intermediates that react by a first-order pathway with a rate constant as large as 10^6 s^{-1}. Both classes of electrodes are well suited to thin-layer spectrochemical cells (Section 2.7) and have been used in experiments to record absorption spectra of inter-mediates and products, determine n values, and study certain electron-transfer-mediated reactions of chemical and biochemical interest. Because of the thinness of both the film deposits and the minigrid wires, a substantive cell current can cause an appreciable iR drop across the electrode surface. This can have a deleterious effect if the electroactive species undergo multiple, close-spaced, electron-transfer reactions (Section 3.1).

3.5   Low-Temperature Electrochemistry

Briefly mentioned in Section 3.2.4 was the use of liquid ammonia as an electrochemical solvent. Occasional reports on the use of this solvent have appeared for years [74], but recent work on redox mechanisms of organic compounds has indicated its great utility as a medium for stabilization of anion radicals. For example, Smith and Bard [48] have shown that not only is the radical anion of nitrobenzene stable in lNH_3, but even its *dianion* can be isolated after cathodic reduction. The low acidity of ammonia accounts partially for the lack of protonation of $PhNO_2^{2-}$, but the subambient operating temperature ($-40°C$) is not to be overlooked, since it favors kinetic stabilization of the anions. In fact, a study by Saveant and Thiebault [75] has differentiated between the stabilizing effects of low temperature and low acidity for lNH_3. In a study of the reduction of halobenzophenones:

The anion radicals were no more stable in lNH_3 than in DMF at the same temperature, but dianions were considerably more stable in lNH_3. These observations were ascribed to the fact that the decomposition of the dianion occurs through rapid protonation reactions, whereas the rate-determining step in decomposition of the monoanions is unimolecular cleavage of the C—X bond.

These examples show that much information can be gained by working at reduced temperatures. Very frequently, processes that are chemically irreversible because of the follow-up reactions often accompanying the formation of radical anions or cations can be made to be reversible by slowing down post-electrode reactions. The benefits of studying the radicals themselves are obvious, but there are benefits also for mechanistic studies of reaction routes of the radicals, since reacting species can be deliberately added to the solution (e.g., electrophiles to solutions of anion radicals) and the response of the anion radical to that reagent can be measured. In short, the experimentalist may retain control of the kinetics of the reaction through temperature control.

A systematic theoretical and experimental study of low-temperature electro-chemistry was reported by Van Duyne and Reilley [76]. These authors evaluated several solvent systems, including alcohols, DMF, acetonitrile, and higher nitriles, for their suitability for low-temperature work. Table 1.3 combines some of their data with our own observations in giving low-temperature limits for certain electrolytes.

Irrespective of changes in redox mechanism that may arise from lowering the temperature, a number of other changes are expected involving the potentials of the redox process and the observed currents:

1. E^0 values are a function of temperature. The direction and magnitude of the shift are determined by the value of ΔS^0 for the redox couple. For a process in which ΔS^0 is governed by solvation changes, ΔS^0 will usually be negative for a reduction, giving a positive shift in E^0 as the temperature is lowered. Conversely, negative shifts in E^0 will normally be observed for oxidations at low temperatures. The amount of change has been found [76] to be about 0.3–0.4 mV/deg for several types of organic or organometallic compounds undergoing either reduction or oxidation.

2.   Apparent changes in E^0 value may arise from shifts in the E^0 of the reference electrode or through changes in the liquid junction potential. The latter has been reported to be small ($\sim 10^{-2}$ mV/deg) [77] but the former can be appreciable. To eliminate this source of error, it is recommended that the reference electrode be thermostated at room temperature, connection to the cell proper being maintained by a salt bridge or other suitable device.

3.   Diagnostics based on the value of $2.3RT/nF = 59/n$ mV (at 298 K) must be modified. Thus, at $-30°C$ (243 K), the theoretical value for the slope of a Nernstian polarographic wave {plot of $-E$ versus $\log[i/(i_d - i)]$} is $48/n$ mV and the separation between anodic and cathodic peaks in cyclic voltammetry is $49/n$ mV at this temperature. Similar adjustments must be made in mechanistic diagnostic criteria for peak potential–scan rate relationships in linear scan voltammetry. Techniques involving imposition of large overpotentials tend to

Table 1.3   Approximate Low-Temperature Limits for Common Electrochemical Solvent Systems

Solvent	Supporting Electrolyte	Temperature (°C)
Acetone	Bu_4NPF_6	−78
Acetonitrile	Bu_4NPF_6 or Et_4NClO_4	−45
Dichloromethane	Bu_4NPF_6	−60
Dimethylformamide	Bu_4NClO_4 or Bu_4NPF_6	−78
Ethanol	$LiClO_4$	−100
Propionitrile	Bu_4NClO_4	−100
Tetrahydrofuran	Bu_4NPF_6	−25

have parameters with less complex temperature dependence, so that methods such as double potential step chronoamperometry have been recommended for quantitative low-temperature work [76].

4.   Solution resistance increases at lower temperatures. An increase in ion pairing of the supporting electrolyte and the increased viscosity of the solvent contribute to the decrease in solution conductivity. The data of Van Duyne and Reilley [76] show, for example, that in lowering a DMF/0.1 M Bu$_4$NClO$_4$ solution from 300 K to 195 K ($-78°$C), the resistance increases by about a factor of 16. It is obvious that particular attention must be paid to minimization of ohmic loss in low-temperature work.

5.   Diffusion currents decrease as the temperature is lowered, because diffusion is an activated process. In typical systems, diffusion currents decrease by about 1.0% per degree [78].

6.   Low-temperature experiments are analogous to high-scan-rate experiments at room temperature. With follow-up reactions having an activation energy of 25–40 kJ/mole, for example, lowering the temperature by 30–40° lowers the rate of the follow-up reaction by about a factor of 10, and is equivalent to increasing the scan rate by the same amount.

When proper attention is given to cell design and other experimental factors, even relatively resistive solvents like dichloromethane can yield semiquantitative data. CH$_2$Cl$_2$ was not seen as a promising low-temperature solvent by Van Duyne and Reilley [76] because Bu$_4$NClO$_4$ supporting electrolyte precipitated as the temperature was lowered. However, tetrabutylammonium salts with other counterions, such as PF$_6^-$ or BF$_4^-$, retain reasonable solubility at reduced temperatures and voltammetric measurements have been successfully obtained. Lucigenin, a biacridinium ion,

undergoes two stepwise one-electron reductions: first to a cation radical and then to a neutral molecule [79]. As the reductions occur, the compound undergoes a conformational change to a twisted form. Cyclic voltammetry scans at $-50°$C in dichloromethane/Bu$_4$NBF$_4$ were necessary to provide definitive evidence about the stepwise reduction and isomerization process, since the two closely spaced waves were insufficiently resolved in DMF.

The reduction of the iron cluster Fe$_3$(CO)$_{12}$ was studied in CH$_2$Cl$_2$ [80] and shown to reduce in two one-electron steps. Cyclic voltammetry studies and controlled-potential electrolyses (coupled with ESR spectroscopy) at $-93°$C were critical to identification of the initial one-electron reduction product $Fe_3(CO)_{12}^-$. The supporting electrolyte was Bu_4NPF_6.

Representative studies making effective use of low-temperature voltammetry have been reported both for organic [81] and for metal-containing [82] systems.

3.6   Identification of Coulometric Products

One of the most important steps in all mechanistic studies of electrode processes is the identification of products and the determination of their distribution. Each product, including those formed in small amounts, supplies important information concerning possible intermediates and their modes of reaction. It is only after the actual intermediates are identified by electrochemical and/or spectroscopic methods and the kinetic data fit the proposed model that the characterization of the electrode process can be considered reasonably complete.

If the goal of the electrochemical experiment is the elucidation of the electrode process, recovery of each product in pure form is usually unnecessary. Separation and identification of the electrolyzed products are achieved most readily and most frequently by gas–liquid (GLC) [83] and high-performance liquid chromatography (HPLC) [84]. If a tetraalkylammonium salt was used as the supporting electrolyte, its removal in a GLC determination can be effected by the insertion of a small glass capillary, which is packed with glass wool, into the injection port. Identification of each product can often be made by comparison of its chromatographic behavior with that of the authentic compound on two or more stationary liquid phases. If the compound cannot be identified in this manner, the product may be isolated by preparative gas–liquid chromatography and subsequently identified by conventional spectroscopic methods [85, 86].

Many electrolyses afford products that are either too unstable thermally or have too low vapor pressures to allow separation by GLC. Accordingly, if one or more of the products contain chromophores that absorb in the ultraviolet or visible region, a method involving separation by HPLC and quantitative determination by molecular absorption spectroscopy can be a particularly effective and desirable alternative. Although the detection limit will be a function of the molar absorption coefficient of the absorbing species at the wavelength of choice, materials present at 10^{-5} M levels and greater can usually be determined accurately and reproducibly. Removal of the supporting electrolyte prior to analysis is usually unnecessary when reverse-phase methods are used.

When recovery of the product or products is required, separation of the product(s) from the supporting electrolyte and the solvent possibly can be achieved by: distillation [86]; pouring of the electrolyzed solution over ice and collection of the precipitated products [86]; evaporation of the solvent and precipitation of the products [64]; distillation of the solvent; and/or extraction with an immiscible solvent [86–89]. Separation of the several products by conventional large-scale chromatographic methods usually follows.

References

1. P. Delahay, *New Instrumental Methods in Electrochemistry*, Interscience, New York, 1954, Chapter 4.
2. P. Delahay, *New Instrumental Methods in Electrochemistry*, Interscience, New York, 1954, Chapter 3.
3. A. C. Testa and W. H. Reinmuth, *Anal. Chem.*, **32**, 1512, 1518 (1960); W. H. Reinmuth, *Anal. Chem.*, **32**, 1514 (1960).
4. D. E. Bartak, T. M. Shields, and M. D. Hawley, *J. Electroanal. Chem.*, **30**, 289 (1971).
5. R. S. Nicholson and I. Shain, *Anal. Chem.*, **36**, 706 (1964); J. E. B. Randles, *Trans. Faraday Soc.*, **44**, 327 (1948); A. Sevcik, *Collect. Czech. Chem. Commun.*, **13**, 349 (1948).
6. L. Ramley and M. S. Krause, Jr., *Anal. Chem.*, **41**, 1362, 1365 (1969); J. Osteryoung, *J. Chem. Educ.*, **60**, 296 (1983).
7. A. J. Bard and L. R. Faulkner, *Electrochemical Methods*, Wiley, New York, 1980, Chapter 5; J. B. Flato, *Anal. Chem.*, **44**, 75A (No. 11) (1972); E. P. Parry and R. A. Osteryoung, *Anal. Chem.*, **37**, 1634 (1964).
8. D. H. Geske and A. H. Maki, *J. Am. Chem. Soc.*, **82**, 2671 (1960); R. N. Adams. *J. Electroanal. Chem.*, **8**, 151 (1964); J. Q. Chambers, T. P. Layloff, and R. N. Adams, *J. Phys. Chem.*, **68**, 661 (1964); T. Kitagawa, T. P. Layloff, and R. N. Adams, *Anal. Chem.*, **35**, 1086 (1983).
9. W. R. Heineman, *J. Chem. Educ.*, **60**, 305 (1983); T. Kuwana and N. Winograd, "Spectroelectrochemistry at Optically Transparent Electrodes," in A. J. Bard, Ed., *Electroanalytical Chemistry*, Vol. 7, Dekker, New York, 1974, Chapter 1.
10. S. W. Feldberg, "Digital Simulation: A General Method for Solving Electrochemical Diffusion–Kinetic Problems," in A. J. Bard, Ed., *Electroanalytical Chemistry*, Vol. 3, Dekker, New York, 1969, Chapter 4.
11. A. Demortier and A. J. Bard, *J. Am. Chem. Soc.*, **95**, 3495 (1973).
12. J. D. L. Holloway, F. C. Senftleber, and W. E. Geiger, Jr., *Anal. Chem.*, **50**, 1013 (1978); J. L. Mills, R. Nelson, S. G. Shore, and L. B. Anderson, *Anal. Chem.*, **43**, 157 (1971).
13. J. E. Harrar and I. Shain, *Anal. Chem.*, **38**, 1148 (1966).
14. M. M. Baizer, J. P. Petrovich, and J. Tyssee, *J. Electrochem. Soc.*, **117**, 173 (1970).
15. H. O. House, E. Feng, and N. P. Peet, *J. Org. Chem.*, **36**, 2371 (1971).
16. D. Sawyer and J. Roberts, *Experimental Electrochemistry for Chemists*, Wiley, New York, 1974, pp. 118–123.
17. W. Schaap and P. McKinney, in G. J. Hills, Ed., *Polarography 1964*, Interscience, New York, 1966, pp. 197–214.
18. W. Schaap and P. McKinney, *Experimental Electrochemistry for Chemists*, Wiley, New York, 1974, pp. 160–162.
19. C. P. Andrieux, L. Nadjo, and J. M. Saveant, *J. Electroanal. Chem.*, **26**, 147 (1970).
20. J. C. Imbeaux and J. M. Saveant, *J. Electroanal. Chem.*, **28**, 325 (1970).
21. J. C. Imbeaux and J. M. Saveant, *J. Electroanal. Chem.*, **31**, 183 (1971).
22. P. E. Whitson, H. W. Vanden Born, and D. H. Evans, *Anal. Chem.*, **45**, 1298 (1973).
23. R. S. Nicholson, *Anal. Chem.*, **37**, 1351 (1965).
24. D. Britz, *J. Electroanal. Chem.*, **88**, 309 (1978).
25. W. Schaap and P. McKinney, *Experimental Electrochemistry for Chemists*, Wiley, New York, 1974, Chapter 4.
26. C. K. Mann, "Nonaqueous Solvents for Electrochemical Use," in A. J. Bard, Ed., *Electroanalytical Chemistry*, Vol. 3, Dekker, New York, 1968, pp. 57–134.
27. P. Delahay, *Double-Layer and Electrode Kinetics*, Wiley-Interscience, New York, 1965; E. Gileadi, *Electrosorption*, Plenum, New York, 1967.

28. R. D. Cannon, *Electron Transfer Reactions*, Butterworths, London, 1980, Chapter 5.
29. M. J. Weaver and F. C. Anson, *J. Am. Chem. Soc.*, **97**, 4403 (1975); V. I. Kravtsov, *J. Electroanal. Chem.*, **69**, 125 (1976); M. J. Weaver, *Inorg. Chem.*, **18**, 402 (1979).
30. N. Tanaka, *Pure Appl. Chem.*, **44**, 627 (1975).
31. A. J. Fry, C. S. Hutchins, and L. L. Chung, *J. Am. Chem. Soc.*, **97**, 591 (1975).
32. M. E. Peover, "Electrochemistry of Aromatic Hydrocarbons," in A. J. Bard, Ed., *Electroanalytical Chemistry*, Dekker, New York, 1967, Chapter 1.
33. J. F. Coetzee, D. K. McGuire, and J. L. Hendrick, *J. Phys. Chem.*, **67**, 1814 (1963).
34. V. Gutmann and E. Wychera, *Inorg. Nucl. Chem. Lett.*, **2**, 257 (1966).
35. M. Mohammed and E. M. Kosower, *J. Phys. Chem.*, **74**, 1153 (1970).
36. B. Kratochvil, *CRC Crit. Rev. Anal. Chem.*, **1**, 415 (1971).
37. D. Sawyer and J. Roberts, *Experimental Electrochemistry for Chemists*, Wiley, New York, 1974, p. 172.
38. J. J. Lingane, *Electroanalytical Chemistry*, 2nd ed., Interscience, New York, 1958, p. 60.
39. E. Grunwald, G. Baughman, and G. Kohnstam, *J. Am. Chem. Soc.*, **82**, 5801 (1960).
40. J. F. Coetzee and W-S. Siao, *Inorg. Chem.*, **2**, 14 (1963).
41. H. M. Koepp, H. Wendt, and H. Strehlow, *Z. Elektrochem.*, **64**, 483 (1960).
42. V. Gutmann, G. Gritzner, and K. Dankogmuller, *Monatsh. Chem.*, **104**, 990 (1973).
43. P. Zuman and C. L. Perrin, *Organic Polarography*, Interscience, New York, 1969; P. Zuman, *Electrochim. Acta*, **21**, 687 (1976).
44. D. Sawyer and J. Roberts, *Experimental Electrochemistry for Chemists*, Wiley, New York, 1974, pp. 203–210.
45. R. N. Adams, *Electrochemistry at Solid Electrodes*, Dekker, New York, 1969, pp. 29–36.
46. O. Hammerich and V. D. Parker, *Electrochim. Acta*, **18**, 537 (1973).
47. J. R. Jezorek and H. B. Mark, Jr., *J. Phys. Chem.*, **74**, 1627 (1970).
48. O. R. Brown, R. J. Butterfield, and J. P. Millington, *Electrochim. Acta*, **27**, 1655 (1982), and references therein; A. J. Bard, *J. Am. Chem. Soc.*, **97**, 5203 (1975); W. H. Smith and A. J. Bard, *J. Electroanal. Chem.*, **76**, 19 (1977).
49. L. A. Tinker and A. J. Bard, *J. Am. Chem. Soc.*, **101**, 2316 (1979).
50. K. M. Kadish and M. M. Morrison, *J. Am. Chem. Soc.*, **98**, 3326 (1976).
51. F. Rohrscheid, A. L. Balch, and R. H. Holm, *Inorg. Chem.*, **5**, 1542 (1966).
52. A. M. Bond, R. Colton, and J. J. Jackowski, *Inorg. Chem.*, **14**, 274 (1975).
53. A. K. Covington, in R. A. Durst, Ed., *Ion-Selective Electrodes*, NBS Spec. Publ. No. 314, Washington, D.C., 1969, Chapter 4.
54. J. N. Butler, "Reference Electrodes in Aprotic Organic Solvents," in P. Delahay and C. W. Tobias, Eds., *Advances in Electrochemistry and Electrochemical Engineering*, Vol. 7, Interscience, New York, 1970, pp. 77–175.
55. L. W. Marple, *Anal. Chem.*, **39**, 844 (1967).
56. J. N. Butler, "Reference Electrodes in Aprotic Organic Solvents," in P. Delahay and C. W. Tobias, Eds., *Advances in Electrochemistry and Electrochemical Engineering*, Vol. 7, Interscience, New York, 1970, p. 129.
57. C. K. Mann and K. K. Barnes, *Electrochemical Reactions in Nonaqueous Systems*, Dekker, New York, 1970, p. 181.
58. K. M. Kadish, S-M. Cai, T. Malinski, J-Q. Ding, and X-Q. Lin, *Anal. Chem.*, **55**, 163 (1983).
59. D. Garreau, J. M. Saveant, and S. K. Binh, *J. Electroanal. Chem.*, **89**, 427 (1978).
60. C. C. Herrmann, G. G. Perrault, and A. A. Pilla, *Anal. Chem.*, **40**, 1173 (1968).
61. R. C. Larson, R. T. Iwamoto, and R. N. Adams, *Anal. Chim. Acta*, **25**, 371 (1961).
62. G. Gritzner and J. Kuta, *Pure Appl. Chem.*, **54**, 1527 (1982); R. R. Gagne, C. A. Koval, and G. C. Kisensky, *Inorg. Chem.*, **19**, 2854 (1980).

63. D. Sawyer and J. Roberts, *Experimental Electrochemistry for Chemists*, Wiley, New York, 1974, pp. 60–100.
64. J. Grimshaw and J. S. Ramsay, *J. Chem. Soc. B*, 60 (1968); O. R. Brown, H. R. Thirsk, and B. Thornton, *Electrochim. Acta*, **16**, 495 (1971).
65. R. E. Dessy, F. E. Stary, R. B. King, and M. Waldrop, *J. Am. Chem. Soc.*, **88**, 471 (1966); D. N. Hendrickson, Y. S. Sohn, W. H. Morrison, Jr., and H. B. Gray, *Inorg. Chem.*, **11**, 808 (1972); A. M. Bond, R. Cotton, and J. J. Jackowski, *Inorg. Chem.*, **18**, 1977 (1979); S. W. Blanch, A. M. Bond, and R. Cotton, *Inorg. Chem.*, **20**, 755 (1981).
66. G. Dryhurst, and D. L. McAllister, "Carbon Electrodes," in P. T. Kissinger and W. R. Heineman, Eds., *Laboratory Methods in Electroanalytical Chemistry*, Dekker, New York, 1984, Chapter 10.
67. R. N. Adams, *Anal. Chem.*, **30**, 1576 (1958).
68. C. E. Plock, *J. Electroanal. Chem.*, **18**, 289 (1968).
69. R. W. Murray, *Acc. Chem. Res.*, **13**, 135 (1980).
70. A. J. Bard, *J. Chem. Educ.*, **60**, 302 (1983).
71. A. J. Bard and A. Merz, *J. Am. Chem. Soc.*, **101**, 2959 (1979).
72. J. P. Collman, M. Marrocco, P. Denisevich, C. Koval, and F. C. Anson, *J. Electroanal. Chem.*, **101**, 117 (1979).
73. W. R. Heineman, *J. Chem. Educ.*, **60**, 305 (1983).
74. O. R. Brown, Ed., *Electrochem. Spec. Period. Report*, **4**, 55 (1974).
75. J. M. Saveant and A. Thiebault, *J. Electroanal. Chem.*, **89**, 335 (1978).
76. R. P. Van Duyne and C. N. Reilley, *Anal. Chem.*, **44**, 142, 158 (1972).
77. A. J. de Bethune, T. S. Lichts, and N. Swenderman, *J. Electrochem. Soc.*, **106**, 616 (1959).
78. L. Meites, *Polarographic Techniques*, 2nd ed., Interscience, New York, 1965, p. 139.
79. E. Ahlberg, O. Hammerich, and V. D. Parker, *J. Am. Chem. Soc.*, **103**, 844 (1981).
80. D. Miholova, J. Klima, and A. A. Vlcek, *Inorg. Chim. Acta*, **27**, L67 (1978).
81. S. F. Nelsen, E. L. Clennan, and D. H. Evans, *J. Am. Chem. Soc.*, **100**, 4012 (1978); V. D. Parker, *Acta Chem. Scand. B*, **35**, 123 (1981); A. Demortier and A. J. Bard, *J. Am. Chem. Soc.*, **95**, 3495 (1973).
82. R. L. Deming, A. L. Allred, A. R. Dahl, A. W. Herlinger, and M. O. Kestner, *J. Am. Chem. Soc.*, **98**, 4132 (1976); A. M. Bodn, B. S. Grabaric, and J. J. Jackowski, *Inorg. Chem.*, **17**, 2153 (1978).
83. J. G. Lawless, D. E. Bartak, and M. D. Hawley, *J. Am. Chem. Soc.*, **91**, 7121 (1969).
84. R. N. McDonald, F. M. Triebe, J. R. January, K. J. Borhani, and M. D. Hawley, *J. Am. Chem. Soc.*, **102**, 7867 (1980).
85. W. M. Moore, A. Salajegheh, and D. G. Peters, *J. Am. Chem. Soc.*, **97**, 4954 (1975).
86. M. M. Baizer and J. L. Churma, *J. Org. Chem.*, **37**, 1951 (1972).
87. S. Wawzonek, R. C. Duty, and J. H. Wagenknecht, *J. Electrochem. Soc.*, **111**, 74 (1964).
88. H. Lund and N. J. Jensen, *Acta Chem. Scand. B*, **28**, 263 (1974).
89. W. M. Moore and D. G. Peters, *Tetrahedron Lett.*, 453 (1972).

Chapter **2**

POTENTIOMETRY: OXIDATION—REDUCTION, pH MEASUREMENTS, AND ION-SELECTIVE ELECTRODES

Thomas R. Kissel

1   INTRODUCTION

1.1   Definition and Historical Perspective

Potentiometry is that branch of electroanalytical chemistry in which equilibrium interfacial distributions of ions or electrons are measured by means of zero-current, open-circuit electrode potential differences. These measured potential differences can be used either to extract thermodynamic information about the chemical systems employed or to quantify individual species in the systems.

Potentiometry is a classic analytical technique with roots before the turn of this century but having three principal embodiments until the middle 1960s. These were oxidation–reduction measurements with metal electrodes, pH measurements with the hydrogen ion-sensitive glass electrode, and halide ion quantification using the familiar metal/metal halide indicator electrodes. Cell potentials were measured with classic null-type, slide-wire potentiometers or analog-output, vacuum-tube pH meters. Several developments in the late 1960s led to a virtual rebirth of interest in potentiometry. Practical membrane-based electrodes highly selective toward individual ions such as fluoride [1], calcium [2], and potassium [3] ushered in the ion-selective electrode era. Simple and inexpensive electrode formats including solid matrix [4, 5], plastic [6], or coated wire [7] not only allowed widespread commercialization of selective indicator electrodes, but also enabled workers to evaluate possible ion-selective chemistries easily in their own laboratories. By coupling chemical and biochemical reactions that produce or consume ions with ion-selective electrodes, potentiometric devices responsive to gaseous or neutral species became feasible. Finally, the advent of the transistor amplifier, integrated circuits, and microprocessor techniques led to sophisticated potential-measuring systems like the modern digital, microprocessor-equipped pH–pIon meters and fully automated potentiometric analyzers.

1.2   The Literature of Potentiometry

The intent of this chapter is to present an overview of the state of the art in potentiometry. As such, it can approach neither the detail nor the depth that is available in many treatments of specific parts of this field. A list of recommended additional treatments, both classic and recent, is given in the general section of the reference list. Because of the exponential growth of the potentiometric literature engendered by ion-selective electrodes, it is difficult by manual methods to maintain complete awareness, particularly if applications as well as theoretical considerations are included. However, with the advent of computerized searches, one finds it as much a problem to cull out unwanted keyword hits from the myriad applications papers as it is to avoid misses on key concept papers. A guide through this maze is provided by a series of biennial reviews [8] on ion-selective electrodes (including pH measurements). Buck and colleagues [9] have collated their contributions from five of these reviews into a comprehensive list covering the decade 1968–1978. Because oxidation–reduction potentiometry has lost favor to both voltammetric and ion-selective electrode techniques, no specific, comprehensive modern literature source exists. However, some applications are to be found in several recurring reviews [10, 11]. Finally, several journals relevant to the potentiometric field should be acknowledged [12–14].

2   THEORETICAL AND GENERAL CONSIDERATIONS

2.1   Interfacial Potential Differences and the Electrochemical Potential

The movement of charged species across conductive interfaces creates interfacial potential differences that, under suitable conditions, can be combined to yield meaningful cell voltages in potentiometry. An interface is defined here as a contact between two bulk phases. It may take many forms, including metal with metal, metal with aqueous or molten electrolyte, and crystalline solid or organic liquid with aqueous electrolyte. Interfaces may be conductive, meaning ions and electrons can migrate across, or they may be blocked. Truly blocked interfaces allow no charged species movement. Although potential differences can be generated at these interfaces by processes like ion adsorption or capacitive effects, they are not generally as useful in potentiometry. At conductive interfaces, charged species may move across the contact to relax any bulk-phase energy differences and thereby achieve equilibrium. At equilibrium there is no further net movement of charge, although microscopic reversibility allows for equal and opposite charge movement. The equilibrium distribution of charged species across all the potential differences in a cell creates a useful potential, which can be measured under zero-current conditions, that is, if the process of measurement does not itself alter the distribution equilibrium.

The parameter that permits quantitative description of this interfacial equilibrium distribution process is termed the *electrochemical potential*. It is defined for an individual charged species as the partial molar free energy of a system according to

$$\tilde{\mu}_i = \left(\frac{\partial G}{\partial n_i}\right)_{T,P,n_j \neq 1} \tag{1}$$

where G is the total free energy of the system, n_i is the number of moles of the ith component, and T,P are the indicators for an isothermal, isobaric process. The electrochemical potential of a species can be fully expressed as the addition of the chemical potential and the electrostatic energy component due to the electrical interaction within the phase as

$$\tilde{\mu}_i = \mu_i + z_i F\phi = \mu_i^0 + RT \ln a_i + z_i F\phi \tag{2}$$

where μ_i is the chemical potential consisting of a standard chemical potential μ_i^0, and a term logarithmic in ion activity. (The relation of ion activity, an effective ion concentration, to stoichiometric ion concentration by way of an activity coefficient will be described in a later section.) Note that the activity term multipliers are the gas constant and the temperature in degrees Kelvin. The electrostatic contribution is the product of the species charge z_i, the Faraday constant F, and the inner or Galvani potential of the phase ϕ.

The $\tilde{\mu}_i$ for an individual charged species is a mathematical concept that cannot be measured exactly, since, for example, one cannot add only cations or anions of a salt to a phase in order to monitor free energy changes. However, it is a very useful concept for mechanistic description of charge exchange at interfaces. Some properties of the electrochemical potential can be listed:

1. The sum of the electrochemical potentials of the cation and anion of a salt equals the salt chemical potential, or

$$\tilde{\mu}_+ + \tilde{\mu}_- = \mu(\text{salt}) = \mu^0(\text{salt}) + RT \ln a_+ a_- \tag{3}$$

This is equivalent to stating that within a conducting phase the influence of the phase potential ϕ is canceled.

2. The activity of electrons in a metal is considered invariant so that the electron electrochemical potential in a metal is

$$\tilde{\mu}_{e^-} = \mu_{e^-}^0 - F\phi \tag{4}$$

3. At equilibrium between two phases electrochemical potentials of all trans-

portable (reversible) charged species are equal:

$$\tilde{\mu}_i(1) = \tilde{\mu}_i(2) = \tilde{\mu}_i(n) \tag{5}$$

where the parentheses indicate the distinct phases.

Using (5) one can set up expressions for interfacial potential differences at a variety of interfaces. Buck [15] has done this rigorously for all relevant potentiometric interfaces. For example, a wire of metal M [or a metal amalgam M(Hg) for metals unstable in air] dipping into a dilute aqueous solution of a salt containing its univalent cation M^+ defines an interface

$$M/M^+X^-(aq) \tag{I}$$

where the slash indicates a boundary between the two phases. The Roman numeral will be used to denote a cell construction. The metal can be described as a neutral phase consisting of M^+ cations surrounded by an equivalent number of electrons. Since the electron does not normally exist in the free hydrated state in aqueous solution, only the M^+ species are considered to cross the interface. Upon first contact, the higher chemical potential of M^+ in the metal drives movement of M^+ into solution. Since bulk electroneutrality cannot be violated, this movement is soon opposed by an electric field generated at the interface by the charge separation. When the chemical and electrical forces balance out, equilibrium is achieved and for species M^+

$$\tilde{\mu}_{M^+}(\text{metal}) = \tilde{\mu}_{M^+}(\text{soln}) \tag{6}$$

Expanding further from (2) and allowing barred quantities to indicate the metal phase,

$$\bar{\mu}^0_{M^+} + RT \ln \bar{a}_{M^+} + F\bar{\phi} = \mu^0_{M^+} + RT \ln a_{M^+} + F\phi \tag{7}$$

which can be rearranged to yield the potential difference between metal and solution

$$\bar{\phi} - \phi = \frac{\mu^0_{M^+} - \bar{\mu}^0_{M^+}}{F} + \frac{RT}{F} \ln \frac{a_{M^+}}{\bar{a}_{M^+}} \tag{8}$$

One can recognize the embryonic form of the Nernst equation describing the potential difference at this interface. From (1) the metal is in equilibrium with its components

$$\bar{\mu}_M = \bar{\mu}^0_M + RT \ln \bar{a}_M = \bar{\mu}_{M^+} + \bar{\mu}_{e^-} \tag{9}$$

which, upon expanding terms, solving for $\bar{\mu}^0_{M^+}$, and substituting into (8), yields

$$\bar{\phi} - \phi = \frac{\mu^0_{M^+} + \bar{\mu}^0_{e^-} - \bar{\mu}^0_M}{F} + \frac{RT}{F} \ln \frac{a_{M^+}}{\bar{a}_{M^+}} \tag{10}$$

Allowing the activity of the pure metal to be unity and the standard chemical potentials to be constant results in

$$\bar{\phi} - \phi = k + \frac{RT}{F} \ln a_{M^+} \tag{11}$$

Thus this interfacial potential difference follows the M^+ activity in solution at equilibrium. One cannot measure $\Delta\phi$ in an absolute sense, because the standard chemical potentials are not absolutely defined. Finally, it should be restated that the bulk of the metal and aqueous phases are still neutral. The potential difference between phases occurs in the so-called space–charge region at the interface. This region can be quite limited in thickness (a few Debye lengths), and as a consequence both the quantity of M^+ transferred and the time required to reach equilibrium are vanishingly small.

Figure 2.1 presents a similar treatment based on the $\tilde{\mu}_i$ concept for a number of different interface types. For example, consider the case in Figure 2.1c where the calcium ion can be exchanged into a membrane phase containing some quantity of retained negative exchanger sites X^- by

$$Ca^{2+} + 2\bar{X}^- \rightleftharpoons \overline{CaX_2} \tag{12}$$

where bars indicate membrane species. The presence of the negative fixed sites prevents membrane uptake of anions, at least when the fixed-site concentration is much greater than the solution concentration. The interface is then termed permselective, that is, exchangeable only for counterions to the fixed-site charges (in this instance cations). It may also be deemed a selective interface if, for reasons generated by its membrane chemistry, the Ca^{2+} exchange process is much favored over that for other cations. For this interface one can define an ion extraction coefficient if one allows the number of X^- sites in the membrane to be fixed

$$K_{ext} = \frac{\bar{a}_{Ca^{2+}}}{a_{Ca^{2+}}} = K_{ex}(a_X)^2 \tag{13}$$

Here K_{ex} is the normal equilibrium constant for (12). By equating the $\tilde{\mu}_i$ for Ca^{2+} at equilibrium, as was done for the case of M^+ before, one obtains

$$\bar{\phi} - \phi = \frac{\mu^0_{Ca^{2+}} - \bar{\mu}^0_{Ca^{2+}}}{2F} + \frac{RT}{2F} \ln \frac{a_{Ca^{2+}}}{\bar{a}_{Ca^{2+}}} \tag{14}$$

If the unmeasurable membrane activity $\bar{a}_{Ca^{2+}}$ can be replaced, as was done for $\bar{a}_{M^+}$ in (10) and (11), then this interface again is seen to be dependent on solution Ca^{2+}. By substituting (13) into (14) it is seen that the single-ion extraction coefficient involving net charge transfer is potential dependent

$$K_{ext} = \exp\left[\frac{\Delta\mu^0}{RT} - \frac{2F\Delta\phi}{RT}\right] \tag{15}$$

If another cation Z^{2+} were present in the aqueous solution along with Ca^{2+}, then the equilibrium governing the ion-exchange process is

$$Z^{2+}(aq) + \overline{Ca^{2+}} \rightleftharpoons \bar{Z}^{2+} + Ca^{2+}(aq) \tag{16}$$

with the attendant equilibrium constant

$$K_{Ca^{2+}/Z^{2+}} = \frac{(\bar{a}_{Z^{2+}})(a_{Ca^{2+}})}{(\bar{a}_{Ca^{2+}})(a_{Z^{2+}})} \tag{17}$$

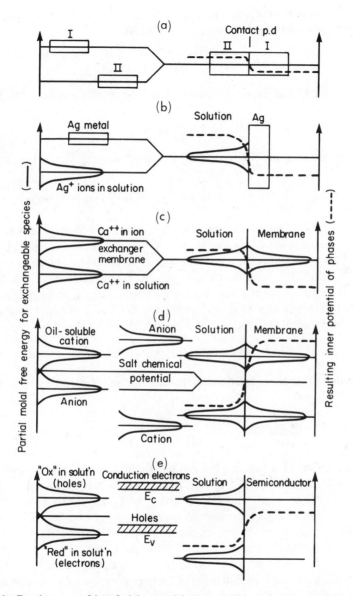

Figure 2.1 Development of interfacial potentials by reversible exchange equilibria. *Left side:* energies of exchangeable species prior to contact; *right side:* energies and developed potentials after contact. (a) Two metals that exchange electrons; (b) Ag metal and Ag$^+$-containing electrolyte. Gaussian curves indicate thermal broadening of ionic energy levels in an electrolyte; (c) a cation in an electrolyte (Ca^{2+}) and a liquid or solid ion exchanger containing the same counterion (Ca^{2+}); (d) an immiscible extraction interface, electrolyte/organic phase across which both cations and anions equilibrate; (e) an intrinsic semiconductor and an electrolyte containing a redox couple. Reproduced from [15], with permission.

Since (16) does not involve net charge transfer, the equilibrium constant is not potential dependent (the $F\phi$ terms cancel) so that

$$K_{Ca^{2+}/Z^{2+}} = \exp\frac{(\mu_{Z^{2+}}^0 - \mu_{Ca^{2+}}^0)}{2F} - \frac{(\bar{\mu}_{Z^{2+}}^0 - \bar{\mu}_{Ca^{2+}}^0)}{2F} \tag{18}$$

If the $\tilde{\mu}_{Ca^{2+}}$ interfacial equality is invoked and the membrane interior is neutralized by

$$\bar{a}_X = \bar{a}_{Ca^{2+}} + \bar{a}_{Z^{2+}} \tag{19}$$

the final expression for the interfacial potential is

$$\bar{\phi} - \phi = \frac{\mu_{Ca^{2+}}^0 - \bar{\mu}_{Ca^{2+}}^0}{2F} + \frac{RT}{2F} \ln\left[\frac{1}{\bar{a}_X}(a_{Ca^{2+}} + K_{Ca^{2+}/Z^{2+}} a_{Z^{2+}})\right] \tag{20}$$

Now the interfacial potential is responsive to solution activities of both species, and the potential is ideally "selective" to Ca^{2+} only if the ion-exchange constant is zero. This is an expression equivalent to the Nicolsky equation used with ion-selective electrodes, in that the $K_{Ca^{2+}/Z^{2+}}$ takes the form of a "selectivity coefficient," which indicates the relative influence of the interfering ion Z^{2+}.

Two other kinds of interfacial system are extremely important in potentiometry, the classical electrodes of the "zeroth" kind

$$\text{Electrode material (Pt, Au, C)|Ox, Red(aq)} \tag{II}$$

used in redox potentiometry, and the metal/metal salt electrodes of the "second" kind

$$\text{Metal M}\left|\begin{array}{c}\text{Insoluble Metal Salt}\\ MX_s\end{array}\right| M^+Y^-\text{(aq)} \tag{III}$$

the most familiar of which are the Ag/AgCl wire and the calomel Hg/Hg_2Cl_2 system. (Note that the M/M^+ system treated previously is considered an electrode of the "first" kind, and that in this traditional electrode classification scheme the ordinal descriptor refers to the number of interfaces in the system at which reversible ion transport occurs.) Both (II) and (III) can easily be treated by means of the electrochemical potential concept.

Although the electrode materials in (II) are not really inert in all chemical environs, it is assumed here that none of the common aqueous ions can cross the interface. Instead, the electron is the reversible charged species occurring in the electrode or as a "virtual" species in solution through the Ox and Red, which can accept or donate an electron according to the familiar half-reaction:

$$Ox^{z+} + ne^- \rightleftharpoons Red^{(z-n)+} \tag{21}$$

Equating electrochemical potentials on both sides of (21),

$$\tilde{\mu}_{Ox} + \tilde{\mu}_e(\text{in electrode material}) = \tilde{\mu}_{Red} \tag{22}$$

Expansion and collection of terms yields

$$zF\phi - (z-n)F\phi - nF\bar{\phi} = (\mu^0_{\text{Red}} + \mu^0_{\text{Ox}} - \bar{\mu}^0_e) + RT \ln \frac{a_{\text{Red}}}{a_{\text{Ox}}} \tag{23}$$

which takes the form of the Nernst equation for a redox couple

$$\bar{\phi} - \phi = \Delta\phi = E^0_{\text{Ox/Red}} - \frac{RT}{nF} \ln \frac{a_{\text{Red}}}{a_{\text{Ox}}} \tag{24}$$

Note that in normal usage the $\bar{\mu}^0_e$ term drops out because of a similar term arising in the contact potential of the connecting wire material (usually Cu). The traditional method of deriving the Nernst equation for redox systems involves reversible cell discharge of standard half-reactions and will be presented in Section 3. However, it is not as useful in describing ion-selective membrane potentials, where half-cell reactions may be quite unrealistic. The $\tilde{\mu}_i$ concept is thus more general.

The workable double-interface electrodes of type (III) possess metal-salt phases that are conductive by virtue of mobile interstitial ions or ionic impurities native or doped into the phase. The salt may be in the form of a single crystal, a pellet of pressed microcrystals, or an amorphous coating on the metal that might be generated by chemical or electrolytic means. At equilibrium the electrochemical potential of the mobile (or exchangeable) ion is equal across all interfaces:

$$\tilde{\mu}_{\text{M}^+,\text{metal}} = \tilde{\mu}_{\text{M}^+,\text{salt}} \tag{25}$$

$$\tilde{\mu}_{\text{M}^+,\text{salt}} = \tilde{\mu}_{\text{M}^+,\text{soln}} \tag{26}$$

Addition of (25) and (26) effectively removes the interposing salt phase from the overall potential difference so that

$$\phi_{\text{metal}} - \phi_{\text{soln}} = \frac{\mu^0_{\text{M}^+} + \mu^0_{e^-} - \mu^0_{\text{metal}}}{F} + \frac{RT}{F} \ln (a_{\text{M}^+,\text{soln}}) \tag{27}$$

This is the same result obtained for the single interface (I). By substitution of the MX solubility equilibrium

$$K_{\text{sp}} = (a_{\text{M}^+})(a_{\text{X}^-}) \tag{28}$$

into (27) the system is seen to be responsive to anion in addition to M^+ by

$$\Delta\phi = k' - \frac{RT}{F} \ln (a_{\text{X}^-}) \tag{29}$$

Here the constant term k' differs from that in (11) by an $RT \ln K_{\text{sp}}/F$ term. Electrode systems of this type can be used as ion-selective systems for either the cation or the anion species, and in addition as reference electrodes in potentiometric cells. Addition of a third interface by way of a second insoluble salt

$$M | MX_{(s)} | NX_{(s)} | N^+Y^- \text{ or } R^+X^-(\text{aq}) \tag{IV}$$

to follow aqueous N^+ or X^- is possible (electrode of the third kind) but these systems tend to exhibit slow response and dependence on solution oxygen presence.

2.2   Diffusion and Junction Potentials

In addition to the generation of potentials by interfacial equilibrium processes, one encounters steady-state, nonequilibrium diffusion potentials in potentiometry. These diffusion potentials occur because of differences in ionic mobilities as salt components respond to concentration gradients within bulk phases or mixture zones of two similar phases. Local electric fields (and so potentials) are generated if cation charge movement occurs at a different rate than anion charge movement; these fields serve to equalize the charge mobilities so that the total "salt" can diffuse under electroneutral conditions at zero applied current.

The expression for the diffusion potential can be obtained from thermodynamics by rearranging (1) to apply to charged species movement according to

$$dG = \sum_i \frac{t_i}{z_i} e\tilde{\mu}_i \tag{30}$$

where t_i represents the transference number of an ion defined as

$$t_i = \frac{|z_i| u_i C_i}{\sum_j |z_j| u_j C_j} \tag{31}$$

Here u_i is an ion mobility (in $cm^2 \, V^{-1} \, s^{-1}$) related to the diffusion coefficient D_i and the equivalent ion conductance λ_i by

$$u_i = \frac{\lambda_i}{F} = \frac{D_i F |z_i|}{RT} \tag{32}$$

Thus in (30) the t_i/z_i term represents the number of moles of charge moved by the "ith" species, and under electroneutral conditions

$$\sum_i t_i = 1 \tag{33}$$

within a phase or across a phase junction. Consider the junction region between two aqueous solutions depicted in Figure 2.2. Here a potential arises to oppose the hypothetical faster transference of cationic species. To fully describe this potential one divides the junction region into a number of infinitesimal volume elements in which (30) is to apply. The total "junction potential" then is the integral over all such elements

$$\int_1^2 dG = 0 = \sum_i \int_1^2 \frac{t_i}{z_i} d\tilde{\mu} \tag{34}$$

where the total ΔG must be zero because no net charge is transferred. The

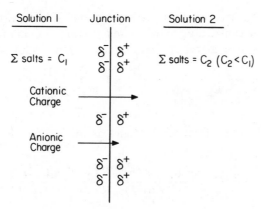

Figure 2.2 A hypothetical liquid junction in which cationic charge transference is greater than anionic charge transference. A junction potential arises at the interface to equalize the transference difference, allowing electroneutral salt flow.

μ_i^0 terms cancel since the phases are the same so that

$$E_j = \phi(2) - \phi(1) = \frac{-RT}{F} \sum_i \int_1^2 \frac{t_i}{z_i} \, d\ln a_i \tag{35}$$

This is the general equation for the junction potential E_j. An equivalent form can be obtained by combination of the force–flux system of Nernst and Planck,

$$J_i = -u_i C_i \frac{d\tilde{\mu}_i}{dx} \tag{36}$$

with the electroneutrality condition

$$I = F \sum_i z_i J_i = 0 \tag{37}$$

In these equations J_i is the ion flux and I the current.

Calculations of the junction potential according to (35) can only be approximate because of uncertain variations in ion mobility and activity with concentration, and because for the general multi-ion case the method of forming the junction may influence the resultant ion transference. By assuming linear concentration profiles in the junction region, constancy of mobilities and activity coefficients, and a well-defined junction (no convective mixing), Henderson [16] was able to integrate (35) to obtain

$$E_j = -\frac{\sum_i |z_i| u_i [C_i(2) - C_i(1)]}{\sum_i z_i^2 u_i [C_i(2) - C_i(1)]} \frac{RT}{F} \ln \frac{\sum_i z_i^2 u_i C_i(1)}{\sum_i z_i^2 u_i C_i(2)} \tag{38}$$

between aqueous solutions 1 and 2. Planck [17] first showed that for a constrained junction (defined by a permeable barrier such as a glass frit) containing only univalent species, exact linear concentration profiles exist, and modification of his solution can be used to calculate E_j for any system of ions having one class of cations (of like charge z_n) and one class of anions (of like charge z_x) by

$$E_j = \frac{\bar{u}_m - \bar{u}_x}{|z_n|u_m + |z_x|u_x} \frac{RT}{F} \ln \frac{\sum_i C_i(1)}{\sum_i C_i(2)} \tag{39}$$

where mean mobilities of cations ($\bar{u}_m$) and anions ($\bar{u}_x$) are E_j dependent and defined by

$$\bar{u}_i = \frac{\sum u_i C_i(2) \exp (FE_j/RT) - \sum u_i C_i(1)}{\sum C_i(2) \exp (FE_j/RT) - \sum C_i(1)} \tag{40}$$

This solution is thus transcendental, but is amenable to numerical evaluation by computer or hand calculators. Morf [18] has compared E_j calculations by means of the Henderson and Planck equations for simple salt junctions of the kind

$$MX(C_2)\|KCl, KNO_3(C_1, Cl^-/NO_3^- = 4) \tag{V}$$

with the double verticals here (and in subsequent cells) indicating a liquid junction. The two estimations differ significantly only when the C_2/C_1 ratio is large or when one ion of the salt MX is very mobile (H^+, OH^-).

The usual procedure in estimating E_j by (38) is to employ molar concentrations for the boundary solution ions; and for the ion mobilities in the junction region, values derived from equivalent ion conductivity data at infinite dilution are employed. Alternative calculation methods using ion activities and conductance data at actual boundary solution concentration have sometimes been invoked, but since the Henderson model enabling integration of (35) already assumes concentration-independent transference and activity coefficient, it is unclear that these alternatives are preferable. A comparison of calculated and experimental values for flowing junctions of single salts in cell (VI),

$$Ag|AgCl|MCl(C_1)\|NCl(C_2)|AgCl|Ag \tag{VI}$$

is given in Table 2.1. Note that for the experimental potential to represent only E_j, the a_{Cl} must be assumed identical on both sides of the cell so that electrode potentials [recall (29)] cancel.

Junction potentials are important in potentiometry for the following reasons:

1. Many reference electrodes like the saturated calomel (SCE) or the saturated Ag/AgCl (SSCl) systems contain a liquid junction in their construction. The overall potentials then contain the E_j component, which, even if operationally well behaved, is neither exactly calculable nor unambiguously measurable.

2. Operational artifacts at such reference electrode junctions, including excessive salt leakage, junction clogging, and spurious values in organic

Table 2.1 Comparison of Calculated and Measured Values of E_j at $25°C^a$

Junction	$(C_1)=(C_2)$	E_j (mV), Measured	E_j (mV), Calcd (λ_i^0)	E_j (mV), Calcd (λ_i at C)
HCl:KCl	0.1 M	26.78	26.85	28.52
HCl:NH$_4$Cl	0.1 M	28.40	26.88	28.57
HCl:NaCl	0.1 M	33.09	31.22	33.38
HCl:LiCl	0.1 M	34.86	33.65	36.14
KCl:NaCl	0.1 M	6.42	4.36	4.86
LiCl:NH$_4$Cl	0.1 M	-6.93	-6.77	-7.57
HCl:KCl	0.01	25.73	26.85	27.48
HCl:NaCl	0.01	31.16	31.22	32.02
KCl:NaCl	0.01	5.65	4.36	4.54

^{a}Data compiled from [19], pp. 236 and 339. Calculations are according to (38) with $u_i = \lambda_i/F$ and λ_i^0 the conductivity values at infinite dilution.

solvent or colloidal systems, are the most common error source in potentiometry.

3. Ion-selective membranes can be considered as junctions between aqueous bathing solutions. The internal junction potentials that may result (commonly called internal diffusion potentials) can contribute to the overall response characteristics of the membrane.

Consideration of point 1 requires that variations in E_j be minimized across different test media, since absolute values are unknown. Two means of accomplishing this are common. The first is to employ so-called salt bridges containing high concentrations of equitransferent salts. These salts, examples of which are KCl, KNO$_3$, NH$_4$Cl, NH$_4$NO$_3$, and LiCl$_3$COO, have nearly equal cation and anion mobilities. Inspection of (38) reveals that if C_2 is a concentrated equitransferent salt and sample C_1 a moderately dilute test solution ($C_2 \gg C_1$), the influence of the prelogarithmic test ion transference is small; thus for a univalent C_2

$$E_j \approx \left[\frac{u_2^+ - u_2^-}{u_2^+ + u_2^-}\right] \frac{RT}{F} \ln \frac{\sum z_i^2 u_i C_i(1)}{\sum u_i C_i(2)} \tag{41}$$

Since u_2^+ approaches u_2^- the magnitude of E_j is reduced, and its variation remains low as long as $\sum t_1 \ll \sum t_2$. The salt bridge may take the form of a true interposed junction as in (VII)

$$\text{Text } C_1 \left\|\begin{matrix}\text{Equitransferent}\\ \text{MX}\end{matrix}\right\| \text{Text } C_3 \tag{VII}$$

or the more familiar reference electrode form (VIII)

$$\text{Reference Electrode} \left| \begin{array}{c} \text{Equitransferent} \\ \text{MX} \end{array} \right| \right| \text{Test } C_1 \qquad \text{(VIII)}$$

The salt may be dissolved in solution, or placed in a variety of viscous matrices such as gels and porous polymers. In the former case a wide range of materials and shapes can be used to define the junction. Examples of some common modes are shown in Figure 2.3. Since junction reproducibility and stability are desirable, junctions that are constantly renewed are optimum for the most accurate quantitative work. These include sleeve and open-flowing modes. Unfortunately, these exhibit the highest leakage rate of salt into the test solution; placement downstream in a flow system or use of double junctions like (VII) with MX an innocuous salt are alternatives. Additionally, although no time dependence of E_j appears in (38), drift in E_j can occur as water and/or salt flow change boundary concentrations. Convective effects due to concentrated bridge salt solution flowing into dilute sample solutions can be minimized by placing the junction at the bottom of the test solution with the opening facing up as shown for several configurations in Figure 2.3.

The second "constant" E_j method involves defining the bulk ionic matrix with indifferent salts so that variations in measured test ions have little influence on total junction transference. This swamping can be achieved by addition of the indifferent salt to all samples, or may be possible without sample alteration if the natural matrix shows minimal ionic variation. Examples sometimes sug-

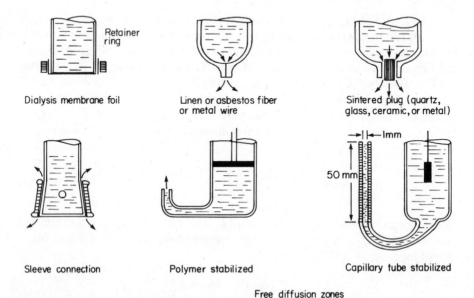

Figure 2.3 Possible salt-bridge/sample-solution contacts. From [20], with permission.

Table 2.2 Effect of Salinity on the Liquid Junction Potential Between Seawater and Saturated KCl or Seawater and 35°/$_{oo}$ Salinity Seawater[a]

Salinity, °/$_{oo}$	Liquid Junction Potential E_j (mV)	
	Saturated KCl	Seawater, °/$_{oo}$
10	+1.1	+7.0
15	+0.6	+4.8
20	+0.3	+3.2
25	0.0	+1.9
30	−0.2	+0.9
35	−0.5	0.0
40	−0.7	−0.7
45	−0.9	−1.4

30, 35, 40 } Open water range

[a]Taken from [21], with permission. Values are calculated at 25°C according to the Henderson equation. The salinity unit °/$_{oo}$ indicates solute concentration in parts per thousand by weight.

gested as fitting the latter case include seawater (Table 2.2) and some biological fluids such as serum and cerebrospinal fluid.

2.3   Membrane Potentials, Cells, and Cell Potentials

The two potential-generating processes just described for charged species, interfacial exchange equilibrium and steady-state diffusion, can be combined to provide a description of ion-selective membrane potentials. Consider the membrane as a water-insoluble bulk phase separating two aqueous solutions. The membrane material may be organic liquid, crystalline solid, amorphous solid such as glass, or a similar nature. If the two membrane–solution interfaces are reversible to ions in solution (i.e., if they are able to achieve ion-exchange equilibrium at the interfaces), then one can describe two-phase boundary potentials with the $\tilde{\mu}_i$ equality discussed previously. Ideally, these interfaces are also permselective (reversibility for ions of only one sign) and perfectly selective (interfacial equilibrium determined by only one of all possible ions of like sign). In addition to the two interfacial potentials there may exist an internal diffusion potential within the membrane, if ion transference differences arise within the phase. This total membrane system is depicted in Figure 2.4. Conceptually, the membrane potential E_M is then the summation of the two boundary potentials and the internal diffusion potential. Consider, for example, that the membrane in Figure 2.4 is the Ca^{2+} ion-exchanger system described in Figure 2.1c and (16)–(20). Thus the aqueous solutions contain both Ca^{2+} and Z^{2+} and permselectivity is provided by the membrane-confined anionic charge sites. The two phase-boundary potentials can be described by allowing $\tilde{\mu}_{Ca^{2+}}$

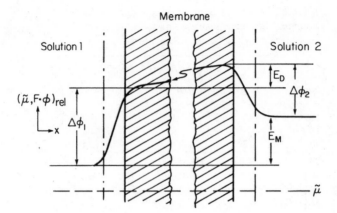

Figure 2.4 Schematic representation of equilibrium at an ion-conducting membrane: $\Delta\phi_{1,2}$ are interfacial potentials, E_D is the possible diffusion potential within the membrane, and E_M the total membrane potential. Reproduced from [20], with permission.

equality at equilibrium according to

$$\Delta\phi_1 = k' + \frac{RT}{2F} \ln \frac{a_{Ca^{2+}}(1)}{\bar{a}_{Ca^{2+}}(1)} \tag{42}$$

and

$$\Delta\phi_2 = k'' + \frac{RT}{2F} \ln \frac{\bar{a}_{Ca^{2+}}(2)}{a_{Ca^{2+}}(2)} \tag{43}$$

where the k signifies collection of μ^0 terms and overbars signify membrane activities at the respective solution–membrane boundaries. If the membrane anionic sites are fixed, the only possible internal charge diffusion arises because of differing $\bar{u}_i$ for Ca^{2+} and Z^{2+}. This diffusion can be described by (38) as a potential-generating process

$$\Delta\phi_d = \frac{RT}{2F} \frac{\bar{u}_{Ca^{2+}}\bar{a}_{Ca^{2+}}(1) + \bar{u}_{Z^{2+}}\bar{a}_{Z^{2+}}(1)}{\bar{u}_{Ca^{2+}}\bar{a}_{Ca^{2+}}(2) + \bar{u}_{Z^{2+}}\bar{a}_{Z^{2+}}(2)} \tag{44}$$

where the prelogarithmic term is unity because of the fixed nature of the sites [if the sites are mobile they must be included in the diffusion equation and (44) then becomes more involved]. Combining (42)–(44) with the overall membrane ion-exchange constant $K_{Ca^{2+}/Z^{2+}}$ defined in (17) leads to cancellation of the unmeasurable $\bar{a}_i$ terms and an expression for the membrane potential given by

$$E_M = \Delta\phi_1 + \Delta\phi_d + \Delta\phi_2 = \frac{RT}{2F} \ln \left[\frac{a_{Ca^{2+}}(1) + (\bar{u}_{Z^{2+}}/\bar{u}_{Ca^{2+}})K_{Ca^{2+}/Z^{2+}}a_{Z^{2+}}(1)}{a_{Ca^{2+}}(2) + (\bar{u}_{Z^{2+}}/\bar{u}_{Ca^{2+}})K_{Ca^{2+}/Z^{2+}}a_{Z^{2+}}(2)} \right] \tag{45}$$

If solution 2 is allowed to be of constant composition, an internal reference solution that may contain only Ca^{2+}, then (45) becomes

$$E_M = k + \frac{RT}{2F} \ln \left[a_{Ca^{2+}}(1) + k^{pot}_{Ca^{2+}/Z^{2+}}a_{Z^{2+}}(1) \right] \tag{46}$$

where $k_{Ca^{2+}/Z^{2+}}^{pot}$ is the *potentiometric selectivity coefficient*, defined for this system by the product of the membrane mobility ratio and the overall ion-exchange constant. In general an equation similar to (46) can be derived for all membrane types; the k^{pot} will be made up of mobility and equilibrium terms peculiar to each system (see Section 5 for a complete list). In addition, more than one ion may pose an interference so that a general form of (46) for a "primary" measured ion a_i and any number of interfering ions a_j should be expressed as

$$E_M = k + \frac{RT}{z_i F} \ln \left[a_i + \sum k_{i/j}^{pot} (a_j)^{z_i/z_j} \right] \tag{47}$$

By enclosing the ion-selective membrane, the internal reference solution, and some internal reference electrode within a suitable structure like a glass or plastic cylinder one obtains an ion-selective electrode (ISE) as shown in Figure 2.5. The internal reference electrode potential is designed to be invariant in response to the constant internal solution; in some instances direct reference electrode–membrane contact (Figure 2.5a) is feasible.

Complete potentiometric cells consisting of ion or redox-responsive indicator

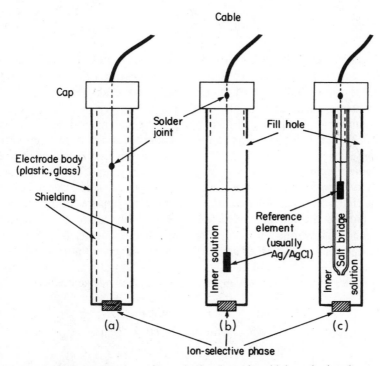

Figure 2.5 Possible constructions of ion-selective electrodes with ion-selective phrases: (a) direct contact of the metallic conductor with the active phase; (b) contact through an inner reference solution and an internal reference element; (c) contact through an inner solution and an internal reference element with salt bridge. Fill hole often omitted in (b). Reproduced from [20], with permission.

electrodes and reference electrodes can now be described. Potentiometric cells are classified broadly into two types depending on the presence or absence of a liquid junction. Junction cells (or cells with transference) can be depicted generally as

$$\text{Cu Wire} \bigg| \text{Reference Electrode} \bigg|\bigg| \begin{matrix} \text{Test} \\ \text{Solution} \end{matrix} \bigg| \text{Indicator Electrode} \bigg| \text{Cu Wire} \qquad \text{(IX)}$$

so that the measured cell potential according to IUPAC convention becomes

$$E_{\text{cell}} = E_{\text{right}} - E_{\text{left}} = E_{\text{IND}} + E_j - E_{\text{REF}} \qquad (48)$$

assuming the Cu lead contact potentials cancel out. If care is taken with the cell/test solution design, then E_j and E_{REF} are constant, and using (11), (24), (29), (46), or the like for E_{IND}, one obtains either

$$E_{\text{cell}} = k + \frac{RT}{nF} \ln \frac{a_{\text{Ox}}}{a_{\text{Red}}} \qquad (49)$$

for redox cells or

$$E_{\text{cell}} = k' + \frac{RT}{z_i F} \ln a_i \qquad (50)$$

for ion-selective electrode systems. Even though (50) implies that individual ion activities are measurable from cell potentials, the presence of the E_j term in the cell constant k' precludes an exact calculation of the absolute a_i, and only relative values are available. Some of the common reference electrodes employed in junction cells are listed in Table 2.3. The classic monograph by Ives and Janz [22] on this subject remains the best source for details on preparation and responses of reference electrodes. Nonpolarizability (high exchange current density); noncontamination of test solution; stability; and low temperature, light, and oxygen sensitivity are all desirable characteristics. As mentioned before, the characteristics of the liquid junction are often more variable than the electrode part of the reference electrode. Johnson and associates [26] have studied junction variability in several convenient reference electrode designs.

The second cell type is termed a junctionless cell or cell without transference. Here the salt bridge is omitted and the reference potential is set by the activity of an ion in the test solution, which the reference electrode senses:

$$\text{Cu Wire} \bigg| \begin{matrix} \text{Reference} \\ \text{Electrode} \\ \text{for } j \end{matrix} \bigg| \begin{matrix} \text{Test} \\ \text{Solution} \\ (\sum a_i) + a_j \end{matrix} \bigg| \begin{matrix} \text{Indicator} \\ \text{Electrode} \\ \text{for some } i \end{matrix} \bigg| \text{Cu Wire} \qquad \text{(X)}$$

By suitable design of test solution one may keep a_j effectively constant so that (49) or (50) is again obtained. The same precaution about relation of E_{cell} to a_i prevails, however, since the constant k term now also contains the single-ion activity term $(RT/z_j F) \ln a_j$. Alternatively, j may be made the salt counterion

Table 2.3 Reference Electrodes in Junction Cells[a]

Electrode	Salt-Bridge Solution	Reversible Ion	General Characteristics	Reference
Ag/AgCl	KCl; 3.5 $M \rightarrow$ sat. + AgCl, sat.	Cl^- (Ag^+)	Good temperature stability; Ag^+, Cl^- contamination and AgCl clogging of junctions possible; compact	[23]
Tl(Hg)|TlCl (Thalamid)	KCl; 3.5 $M \rightarrow$ sat. + TlCl, sat.	Cl^- (Tl^+)	Better than calomel at high temperature; large negative standard potential	[20]
(Hg)|Hg_2Cl_2 +(Hg) paste (calomel)	KCl; 1 $M \rightarrow$ sat. + Hg_2Cl_2, sat.	Cl^- (Hg_2^{2+})	Widely used; large temperature hysteresis; Hg_2^{2+} contamination	[24]
(Hg)|Hg_2SO_4 +(Hg) paste	Sat.; K_2SO_4	SO_4^{2-} (Hg_2^{2+})	No Cl^- contamination; larger and less stable E_j; Hg_2^{2+} contamination	—
Pt|I^-, I_3^- || redox	KCl; 3.5 $M \rightarrow$ sat. (double junction)	Redox couple	No AgCl junction clogging; temperature coefficient can be designed to match dE_{IND}/dT so $dE_{cell}/dT \approx 0$	[25]

[a]For preparation and response details in full, see [22]. Other references are recent discussions of response characteristics.

71

of the varying test ion i so that for a salt MX with a_m the test ion one finds

$$E_{cell} = E_{IND} - E_{REF} = k_m + \frac{RT}{z_m F} \ln a_m - \left(k_x - \frac{RT}{z_x F} \ln a_x \right)$$

$$= k' + \frac{RT}{F} \ln (a_m)^{1/z_m}(a_x)^{1/z_x} \tag{51}$$

This version of the junctionless cell is then responsive to the salt or mean activity defined by

$$a_{\pm}^z(mx) = (a_m)^{z_x}(a_x)^{z_m} \tag{52}$$

where

$$z = z_x + z_m \tag{53}$$

Thus for a univalent salt (51) can be written

$$E_{cell} = k' + \frac{2RT}{F} \ln a_{\pm}(mx) \tag{54}$$

and the sensitivity ideally becomes 118.32 mV per decade change in salt activity at 25°C. Most any reversible electrode, either ion-selective or of the "nth" kind, can be employed as the reference. Two precautions must normally be followed: alterations in free a_j by processes such as ion association must be accounted for and care must be taken with measurement instrumentation if both indicator and reference electrodes have high impedance (low exchange current density) like the pH and pNa glass electrodes. Several pH–pIon meters now commercially available can be used successfully with two high impedance electrode sources.

2.4   Activity Coefficients and Activity Standards

As outlined in the previous section individual electrodes respond to single-ion activities, but measurable cell potentials reflect only mean (salt) activities or inexact combinations of single-ion activities and E_j parameters. This situation has several consequences in potentiometric analysis. The first is that single-ion activity standards to be employed for cell activity calibration are not thermodynamically available and so must be defined by convention. The second is that any potentiometric analysis seeking results in a concentration domain must provide for proper transformation from the activity domain, either by estimation of activity coefficients or by adjustment of the ionic test medium to ensure their constancy.

The relation of stoichiometric concentration to ionic activity can be expressed similarly for any of the three common concentration scales:

$$(molar)\quad a_{i,c} = y_i C_i \tag{55}$$

$$(molal)\quad a_{i,m} = \gamma_i m_i \tag{56}$$

$$\text{(mole fraction or rational) } a_{i,x} = f_i \chi_i \tag{57}$$

where the unitless activity coefficient defines the deviation from ideal solute behavior for the ion (or salt) at finite concentrations. Because of their charged nature, electrolytes exhibit variations in the activity coefficients that are dependent on the total ionic strength of the media defined on the molal scale by

$$I_m = \tfrac{1}{2} \sum z_i^2 m_i \tag{58}$$

Since molality and molarity are virtually equivalent in dilute aqueous solution, the molar-based I_c is often employed; but for proteinaceous, mixed solvent or high concentration aqueous cases (58) is more suitable. On any of the scales the dilute solution limit is established as

$$\lim_{I \to 0} (y_i, \gamma_i, f_i) = 1 \tag{59}$$

It is to be appreciated that the activity coefficients in (55)–(57) can be expressed as single-ion or mean-salt quantities such as

$$\gamma_{\pm}^z = (\gamma_+)^{z^-} (\gamma_-)^{z^+} \tag{60}$$

The mean-salt parameters for pure single-salt solutions are experimentally available from isopiestic or potentiometric techniques; elaborate compilations are now available from workers at the National Bureau of Standards (NBS) [27]. Starting with the work of Debye and Hückel efforts have been made to account for the observed single-salt data by empirical or theoretical means over wide concentration ranges, to account for salt behavior in mixed electrolyte solution, and to apportion the mean activity coefficients into single-ion values by various conventions. The whole subject has been treated exhaustively [28]. Table 2.4 lists several of the more important methods with comments about their utility. Note the general form descriptive of all the treatments:

$$\log \gamma_{\pm} = \frac{|z^+ z^-| a \sqrt{I}}{1 + b \sqrt{I}} + c \sqrt{I} \tag{61}$$

where parameter a is solvent dependent, parameter b both solvent and ion-size dependent, and parameter c (which may itself contain terms in I) descriptive of the total media. At low ionic strength (< 0.01) one finds $\gamma_{\pm}$ virtually independent of ion identity, but at higher concentrations the media-specific b and c terms must be incorporated.

Electrolyte mixtures have traditionally been treated empirically by the Harned rule

$$\gamma_{\pm,i}(\text{mixt}) = \gamma_{\pm,i}(\text{pure}) + \sum_{j \neq i} \alpha_{ij} \chi_j \tag{62}$$

where the α's are interaction coefficients and $\gamma_{\pm,i}(\text{pure})$ is taken at the same ionic strength as that of the total mixture. An alternative procedure is to employ the fitting parameters from the Pitzer or Bromley treatments in Table 2.4 as weighting coefficients in a mixture calculation. Pytkowicz [30] and other

Table 2.4 Common Methods for Calculation of Aqueous Activity Coefficients

Method	Form of Equation	Adjustable Parameters	Comments	Reference
Debye–Hückel	$$-\log \gamma_{\pm} = \frac{A|z^+ z^-|\sqrt{I_m}}{1 + Ba^0 \sqrt{I_m}}$$	A, B solvent dependent; a^0 the distance of closest approach of ion pair or crystallographic ion radius of single ions .	$Ba^0 = 0$, limiting law, good to $I \leqslant 0.01$. $Ba^0 = 1.5$, Bates–Guggenheim, used for γ_{Cl^-} in pH convention, good to $I \leqslant 0.1$. Assigned a^0 values for each ion to calculate γ_i (Kielland). $Ba^0 = 1$, simple extended DH, good to $I \leqslant 0.1$.	[29]
Simple DH extensions	$$-\log \gamma_{\pm} = \mathrm{DH} + \beta I$$	β a function of medium and ion identity, empirical constant often employed	$Ba^0 = 1$, $\beta = -0.3$, Davies equation, good to 2% up to $I = 0.1$. Ba^0 and β adjustable parameters fitted for each salt to fit NBS $\gamma_{\pm}$ data, good to $I \leqslant 1.0$.	[30] [31]
Stokes–Robinson hydration theory	$$-\log \gamma_{\pm} = \mathrm{DH} + \frac{h}{2}\log a_w + \log\left[1 + (2-h)\frac{m}{55.5}\right]$$	h is hydration number of solute; a_w is water activity of solution	Can be used for $\gamma_{\pm}$ up to $\sim I = 5$, suggested convention for γ_i sets $h_{Cl^-} = 0$.	[32]
Pitzer–Kim	$$-\log \gamma_{\pm} = \mathrm{DH} + [\beta^0 + \beta' \exp(-\alpha\sqrt{I})]m$$	β^0, β' differ for each salt, α a constant	Good to high I for over 200 salts; for salt mixtures use same β^0, β' weighted by salt fraction.	[33]
Bromley	$$-\log \gamma_{\pm} = \mathrm{DH} + \frac{(0.06 + 0.6B)|z^+ z^-|I}{\left(1 + \frac{1.5}{|z^+ z^-|}I\right)^2} + BI$$	B characteristic of each salt	Empirical fit good to 5% at $I \leqslant 6$. Ion association causes errors; can be applied to mixtures.	[34]

74

marine chemists who must deal with concentrated aqueous mixtures at $I = 0.7$ have recently shown that Harned's rule can be derived from an ion-association model. In this model the activity coefficient of a salt in a mixture is identical with that of the pure salt at the effective ionic strength of the mixture I_e

$$I_{m,e} = \tfrac{1}{2} \left[\sum_i |z_i|^2 m_{i,F} + \sum_p |z_p|^2 m_p \right] \tag{63}$$

where F denotes free dissociated salt and p charged ion-paired salt. Thus the total amount of ionic strength can be decreased by ion association of all possible salts according to the stoichiometric equilibrium association constant K^*, related to the thermodynamic constant K^0 by

$$K_{mx}^* = \frac{m_{mx}}{m_m m_x} = K_{mx}^0 \frac{\gamma_m \gamma_x}{\gamma_{mx}} \tag{64}$$

These workers find that even NaCl shows evidence of ion pairing in aqueous systems when this treatment is employed.

Several conventions for splitting $\gamma_\pm$ into its single-ion components have been proposed. The pH convention of Bates and Guggenheim assumes that for ionic strength of less than 0.1 the chloride single-ion γ in simple buffer solution can be given by

$$-\log \gamma_{Cl^-} = \frac{A\sqrt{I}}{1 + 1.5\sqrt{I}} \tag{65}$$

This convention is then used to define an operational pH scale whereby $-\log a_{H^+}$ is "measurable" (see Section 4 for details), and pa_H standards can be prepared. Attempts to apply similar treatments to other salts and higher concentrations, thereby obtaining pM^+ or pX^- standards, are still under scrutiny. Table 2.5 gives values obtained for some simple salts by the Stokes–Robinson hydration theory approach, the one most often employed. Solutions with similarly "defined" single-ion activities are available from the NBS.

For the analyst interested in practical concentration-obtaining procedures the best method is to level the ionic strength by means of a high concentration of inert salt (often given the acronym TISAB, or total ionic strength adjusting buffer). Often for crude work the activity is assumed equal to concentration and any corrections are ignored. This is particularly true in pH and redox measurements. For ISE work under nonleveled I_m conditions, transformation from aqueous activity to concentration is accomplished by selecting a method in Table 2.4, especially under dilute $(I < 0.1)$ aqueous conditions. Mixed solvent cases require special attention and will be treated in each of the subsequent sections. Finally, since potentiometric measurements yield free activity, in thermodynamic studies caution must be exercised in apportioning changes in activity to either the activity coefficient or the equilibrium process under study. Thus, a change in cell potential even in junctionless mode cannot distinguish a $\Delta\gamma$ from an ion-binding phenomenon. Careful attention must be paid to system

Table 2.5 Single-Ion Activity Coefficients Based on the Stokes–Robinson Hydration Theory[a]

Concentration	Salt	γ_+	γ_-
0.1 molal	KF	0.775	0.775
	NaCl	0.783	0.773
	KCl	0.773	0.768
	$CaCl_2$	0.269	0.719
	NaBr	0.788	0.776
	KBr	0.769	0.775
	$CaBr_2$	0.283	0.729
1.0 molal	KF	0.645	0.645
	NaCl	0.697	0.620
	KCl	0.623	0.586
	$CaCl_2$	0.263	0.690
	NaBr	0.739	0.639
	KBr	0.639	0.596
	$CaBr_2$	0.378	0.748

[a]Taken from [32], with permission.

definition; some workers prefer to make "effective" thermodynamic measurements at some defined standard state (e.g., in 1 M $NaNO_3$ or 1 M $NaClO_4$ or with $I_{m,T}$ at some defined value).

2.5   General Potentiometric Techniques, Instrumentation, and Applications

A generalized breakdown of modern potentiometric procedures is given in Table 2.6. Specific examples of each procedure can be found in several compilations [35, 36]. These general classifications are, of course, arbitrary but serve to outline the field as a whole. The term *direct potentiometry* is a source of some confusion. It is meant here to indicate an analysis method whereby each sample yields only one potentiometric signal; this signal is related to analyte concentration by separate calibration. The sample itself can be neat, diluted, or altered by chemical reaction before potentiometric measurement. Others include the ISE single- or multiple-point standard addition technique under the heading of direct potentiometry. In the clinical field the term *direct potentiometry* has come to indicate that the sample is undiluted; analysis of a diluted sample unfortunately being termed *indirect potentiometry*.

Direct potentiometric techniques (in the usage defined here) are generally favored because of their simplicity. However, indirect titrimetric procedures can often yield better precision and in many instances neither calibration nor ideal electrode sensitivity behavior are necessary. Few techniques other than potentiometry can successfully indicate free ion activity. This indication can be viewed

Table 2.6 General Classification of Potentiometric Techniques

Purpose	Classification	Methods and Comments	Examples
Analysis	Direct	Calibration with standards to bracket the unknown. May be batch, flow, or flow injection mode. Sample may be diluted or undiluted. Sample may be presented after suitable reaction, or as effluent from a separation technique.	pH measurement Redox potential measurement pIon measurement
	Indirect	Titration with single or multiple additions of reagents that alter the potentiometric signal. May be batch, flow, or flow injection mode. Actual analyte may only be related by equilibrium to potentiometrically sensed species. Includes null-point potentiometry where standard is added into second cell to match potential of unknown cell.	Acid–base, redox, or ISE potentiometric titrations ISE standard addition or subtraction method Precipitation, compleximetric titrations with ISE end point indicator
Fundamental information	Equilibrium	Thermodynamic information obtained, usually from direct methods but sometimes from titration experiments.	Measurements of E^0, pK_a, $K_{association}$, K_{sp}, $\gamma_{\pm}$. Also ΔG^0, ΔH^0, ΔS^0 possible
	Rate	Reaction kinetic studies, often in stopped-flow or rapid-mixing experiments.	Rate constants up to about 10^8 L/mol·s measurable

as a curse when one desires total stoichiometric ion concentration, or as a blessing when one desires fundamental information about a chemical system. The many thermodynamic measurements derived from potentiometry (especially from pH and ISE measurements) are testimony to the latter.

As with most analytical methods the potentiometric field has been strongly influenced by the advent of the microprocessor and computer-based techniques. It is possible to develop a hierarchy of this influence for any given potentiometric analysis. This must again be arbitrary but is attempted in Table 2.7. The Table 2.7 listing is purposely left general; specific (sometimes commercially available) examples of each class could be given but would easily become outdated. The hierarchy extends from the all-manual potentiometric analysis by means of a pH/mV meter to the totally automatic, dedicated analysis system.

The modern (field effect transistor-) FET-amplifier pH/mV meters with digital display can be purchased in a variety of sizes and configurations ranging from the familiar boxlike unit to small, inexpensive, hand-held, battery-operated, barred-shaped devices. All possess adequate single-ended (and in some cases, double-ended) high impedance inputs, and the "research" grade units yield outputs good to 0.1 mV or 0.001 pH unit. Burton [37] has discussed circuit design and parameter adjustment (i.e., slope, offset, temperature control) for the FET-amplifier pH/mV meter. Programs for potentiometric data workup by off-line computer have been advanced to a sophisticated stage. Many lists exist for exact, statistically robust estimations of complex formation equilibria on mainframe systems with high level languages [38] or on desktop microcomputers in BASIC [39]. Potentiometric titration data can be analyzed for simultaneous derivation of end point, electrode calibration parameters, and pK_a/E^0 information [40]. On-line data storage and calculation are also feasible, since many of the FET-amplifier meters have digital output options. By adding commercially available electrode multiplexers, a small "computerized" potentiometric system can be built around a conventional commercial pH meter.

Real-time data analysis with the possibility of computer control or feedback attains the next level of sophistication. The simplest example is the microprocessor-based pH–pIon/mV meter, available now from most of the common pH meter manufacturers. The onboard microprocessor software can store and calculate data required for many standard procedures: direct potentiometric calibration and measurement in pH or pIon units; simple, indirect methods like standard addition, standard subtraction, and analate addition; temperature correction and system diagnostic messages; and blank correction for low concentration analysis. A limited degree of experimental control is also possible with these meters, including microprocessor determination of equilibrium potentials from drifting electrode signals and prediction of the successive volumes of standard required to obtain equally spaced potentials in multiple addition techniques. Input/output functions range from alphanumeric digital displays with a few push buttons to complete keyboards with CRT and printer outputs. Moody and Thomas [41] have analyzed the working capabilities of one commercial microprocessor that is typical of the group.

Table 2.7 Levels of Automation in Potentiometric Analysis

Classification	Sample Preparation (Dilution etc.)	Sample Delivery to Cell	Cell Potential Measurement	Data Output	Data Analysis Method	Comment
0	Manual	Manual	Traditional pH/mV meter (FET amplifier)	Manual or analog	Manual	Traditional pH, redox analysis
1	Manual	Manual	Traditional pH/mV meter (FET amplifier)	Manual or analog	Computer	Off-line calculation of direct or indirect potentiometric results
2	Manual	Manual	Traditional pH/mV meter (FET amplifier)	Digitized	Computer	On- or off-line data workup
3	Manual	Manual	Dedicated electrometer; may be multiplexed	Digitized	Computer	Typical of noncommercial research system; may be computer control
4	Manual	Manual	Microprocessor-based pH/mV meter	Digital display	Computer onboard the pH/mV meter	ISE standard addition titration, pIon measurement, pH with stored calculated parameters
5	Automatic	Automatic	Traditional pH/mV meter	Analog	Manual	Traditional flow system or titrator with motorized burette
6	Automatic	Automatic	Dedicated electrometer	Digitized	Computer	Off-line calculation of flow, flow injection, or titration results
7	Automatic	Automatic	Dedicated electrometer	Digital display	Computer onboard instrument	Automatic titrator with feedback and real-time output; dedicated potentiometric analyzers as in clinical field; computer control of entire experiment

A final level of complexity is generated with complete automation and computer control of all phases of the potentiometric measurement. Three examples are prevalent: dedicated home-built research systems, commercial analyzers such as are found in the clinical or industrial field, and automatic titrators. Most researchers in the ISE field have constructed their own automatic systems. Martin and Freiser [42] describe such a system, designed for both potentiometric titrations and calibration/response evaluation of new ISE chemistries. A schematic of this system is given in Figure 2.6, the basic components of which are a digital burette, a commercial FET-amplifier meter with electrode multiplexer, and a 24K microcomputer with associated floppy-disk drive. Software modifications would enable redox titrations, metal–ligand complex studies, or automatic ISE selectivity coefficient determinations to be performed. Hardware modifications like multiple sampling by carousel or parallel flow cells are also easily envisioned to replace the discrete batch mode shown in the schematic. Commercial manufacturers have applied these ideas to the design of instruments dedicated to defined potentiometric analyses. While often sacrificing flexibility the instruments are valuable where multiple samples must be analyzed with high throughput and minimal labor. About 30 different clinical analyzers now exist for potentiometric measurement of either

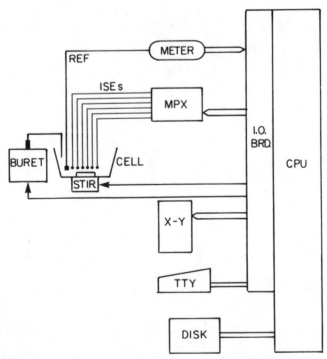

Figure 2.6   Schematic diagram of a discrete automatic potentiometric analysis system. From [42], with permission.

blood-gas quantities (pH, pCO_2) or inorganic electrolytes in body fluids (Na^+, K^+, Cl^-, CO_2, Ca^{2+}); Covington has listed characteristics of a select few [43]. Also, several manufacturers (such as Technicon, NOVA Biomedical, and Ionics) offer potentiometric flow systems that can be tailored to automatic analysis of specific ions in industrial or water-quality situations.

Titrators are a specialized form of a completely automated potentiometric instrument on which development has proceeded for many years. Svehla [44] has reviewed the general subject in a book, and Smit and co-workers [45, 46] have specifically discussed computer-controlled titrations. Hardware advances include the precision digital burette with interchangeable snap-in reagent holders, automatic burette filling, and automatic coupling to sample carousels. Microprocessor control of titrant addition mode, determination of optimum electrode reponse time, and exact statistical end point calculation are common. Titrant addition can be continuous, stepwise by volume, or stepwise by potential. The titration may be stopped at a predefined pH/mV end point, at a predefined titrant volume, or when a preset number of end points have been found. The end point calculation mode can be chosen by inflection point, by the inter-polation method, or by exact equivalence point. User input options vary from keyboard to mark-sensed control cards. Bender and Kujawa have described the capabilities of one modern commercial system in detail [47].

Applications of potentiometry are too numerous to even catalog here. Potentiometric techniques play an important role in biology and clinical chemistry, water-quality analysis, soil and seawater research, industrial toxi-cology, the pharmaceutical industry, and in fundamental studies of complexa-tion equilibrium. Recurring reviews [8, 11] or recent compilations [9, 36, 48] should be consulted for examples in specific areas.

3   OXIDATION—REDUCTION POTENTIOMETRY

3.1   Thermodynamics, Cell Potentials, and Redox Terminology

By using the electrochemical potential concept the Nernst equation was derived for redox systems according to (21)–(24). Alternatively, one may choose to proceed by applying thermodynamics to reversible galvanic cell potentials. Both formalisms produce the same result; the latter is also presented here since it introduces terminology more consistent with the total field of electrochemistry.

Consider the reaction between zinc metal and silver chloride,

$$Zn + 2AgCl \rightleftharpoons 2Ag + Zn^{2+} + 2Cl^- \tag{66}$$

At constant temperature and pressure the maximum net work available from the reaction is given by the free energy ΔG:

$$\Delta G = \Delta G^0 + RT \ln \frac{(a_{Ag})^2(a_{Zn^{2+}})(a_{Cl^-})^2}{(a_{Zn})(a_{AgCl})^2} \tag{67}$$

where ΔG^0 is the standard free energy obtained when all reactants and products are in their standard states (i.e., unit activity, a hypothetical state usually defined as a pure gas at 1 atmospheric pressure, pure liquids or solids at all temperatures, and solutes at infinite dilution). Since the free energy ΔG_{eq} available from a cell at thermodynamic equilibrium is zero, ΔG^0 is thereby defined for the reaction as

$$\Delta G^0 = -RT \ln K_{eq} \tag{68}$$

The ΔG for (66) also indicates the tendency for the reaction to proceed as written. By thermodynamic convention a negative value indicates (66) will proceed spontaneously, a positive value suggests the opposite.

One way to achieve reaction (66) is to dip zinc metal and a AgCl-coated silver wire into an aqueous solution of ZnCl salt of some concentration C, represented by the following cell diagram:

$$Cu|Zn|Zn^{2+}(C), Cl^-(2C)|AgCl|Ag|Cu \tag{XI}$$

The electromotive force, emf, or cell potential E of (XI) could be measured by shorting the Cu leads through an infinite resistance (an open circuit), or by adjusting an external battery source to a value E' exactly opposing E so that no current flow is observed. If reaction (66) changes direction when E is altered by an infinitesimally small amount E', then cell (XI) would be thermodynamically reversible. Also, the current discharge of cell (XI) through the infinite resistance would represent the maximum amount of electrical energy available from the cell reaction. This electrical energy is then related to the reaction free energy by

$$\Delta G = -nFE = -2FE \tag{69}$$

where n is the number of electrons passed per mole of reactants and the Faraday F is the charge per mole of electrons. By introduction of the sign in (69), one can relate the direction-sensitive ΔG to the electrostatic term E. Consistency with the thermodynamic convention then dictates that a positive cell potential indicates a spontaneous reaction (a negative ΔG). From (67)–(69) one obtains both the Nernst equation for this cell,

$$E = E^0 - \frac{RT}{2F} \ln \frac{(a_{Ag})^2(a_{Zn^{2+}})(a_{Cl^-})^2}{(a_{Zn})(a_{AgCl})^2} \tag{70}$$

where E^0 is the cell potential at standard state or the *standard potential*, and the relation of E^0 to the cell equilibrium constant

$$E^0 = \frac{RT}{nF} \ln K_{eq} = \frac{RT}{2F} \ln K_{eq} \tag{71}$$

Overall cell reactions like (66) can be considered as the sum of two redox half-reactions, and cell potentials the sum of two half-cell potentials. Current IUPAC usage dictates that half-reactions be portrayed as reductions,

$$Ox + ne^- \rightleftharpoons Red \tag{72}$$

To be consistent with sign conventions, the cell schematic for a reaction such as (66) is written so that the half-reaction appearing as a reduction in the overall equation becomes the right (positive, cathodic) electrode. For instance, for (66)

$$2AgCl + 2e^- \rightleftharpoons 2Ag + 2Cl^-$$ (73)

is schematically placed on the right

$$Cl^-(2C)|AgCl|Ag$$ (XII)

with a half-cell potential given by

$$E_{AgCl/Cl^-} = E^0_{AgCl/Cl^-} - \frac{RT}{2F} \ln \frac{(a_{Ag})(a_{Cl^-})^2}{(a_{AgCl})^2}$$ (74)

The half-reaction appearing as an oxidation is placed on the left (negative, anodic) side of the schematic

$$Zn|Zn^{2+}(C)$$ (XIII)

but its half-cell potential is still written as a reduction

$$Zn^{2+} + 2e \rightleftharpoons Zn$$ (75)

described by

$$E_{Zn^{2+}/Zn} = E^0_{Zn^{2+}/Zn} - \frac{RT}{2F} \ln \frac{(a_{Zn})}{(a_{Zn^{2+}})}$$ (76)

Combination of (XII) and (XIII) yields the schematic (XI) corresponding to reaction (66), with the overall cell emf given by the convention

$$E_{cell} = E_{right} - E_{left}$$ (77)

whereby (70) is obtained. Conversely, one may easily write the correct chemical reaction for any cell schematic, taking care that the stoichiometry follows from a common n value.

Half-cell potentials pertain to a single electrode and so are not measurable. However, by choosing one half-cell reaction to have some conventionally defined half-cell potential, a scale or ordering of all possible half-cell reactions is possible. The standard hydrogen electrode (SHE), sometimes called the normal hydrogen electrode (NHE), is the chosen half-cell. It consists of the hypothetical half-cell

$$Pt|H_2(a=1)|H^+(a=1)$$ (XIV)

with a standard half-cell potential $E^0_{H^+,H_2}$ defined to be 0 volts at all temperatures. The half-reaction for this cell is then

$$2H^+ + 2e \rightleftharpoons H_2$$ (78)

Combination of (XIV) written on the left with any other solution–electrode redox couple written on the right

$$Pt|H_2(a=1)|H^+(a=1), \text{ solution electrode}$$ (XV)

allows one to measure the *electrode potential* of the redox couple in question or the emf of the half-cell reaction

$$Ox + ne \rightleftharpoons Red \tag{79}$$

pertinent to the solution–electrode couple. The electrode potential may also be termed the *reduction potential* for this couple and is given by

$$E_{Red} = E_{Ox/Red} = E_{Ox/Red}^0 - \frac{RT}{nF} \ln \frac{a_{Red}}{a_{Ox}} \tag{80}$$

If species are in their standard states, then *standard electrode potentials, standard reduction potentials,* or *standard cell emf's* are defined. For instance the standard electrode potential of the Cd^{2+}/Cd couple would be the emf of the cell

$$Pt|H_2(a=1)|H^+(a=1), Cd^{2+}(a=1)|Cd \tag{XVI}$$

Tables of standard potentials have been compiled; the most recent is that of Antelman and Harris [49], which covers the literature up to 1981. An abbreviated tabulation of electrode potentials and standard electrode potentials for the common reference electrodes is presented in Table 2.8.

Direct measurement of standard potentials usually proceeds by comparison of the new half-cell reaction to that of a previously measured couple. Thus the SHE is not used, and in fact is not a realizable electrode at all. To illustrate, the E^0 of the $Cl^-/AgCl/Ag$ couple can be obtained from the cell

$$Pt|H_2(g, 1\ atm)|H^+, Cl^-(m)|AgCl|Ag \tag{XVII}$$

whose cell reaction is given by

$$\tfrac{1}{2}H_2(g, 1\ atm) + AgCl \rightleftharpoons Ag + H^+(m) + Cl^-(m) \tag{81}$$

and whose cell potential is

$$\begin{aligned} E &= E_{AgCl/Ag}^0 - E_{H^+/H_2}^0 - \frac{RT}{F} \ln \frac{(a_{Cl^-})(a_{H^+})(a_{Ag})}{(a_{AgCl})(a_{H_2})^{1/2}} \\ &= E_{AgCl/Ag}^0 - \frac{RT}{F} \ln (a_{Cl^-})(a_{H^+}) \end{aligned} \tag{82}$$

when species at unit activity are considered. Recalling the definition of $\gamma_{\pm}$,

$$E_{AgCl/Ag}^0 = E + \frac{2RT}{F} \ln m_{HCl} + \frac{2RT}{F} \ln \gamma_{\pm} \tag{83}$$

and by estimating $\gamma_{\pm}$ from methods given in Table 2.4 one can extrapolate dilute solution data to find the E at $m_{HCl} = 0$, which is E^0. The AgCl/Ag couple can then be employed with any subsequent new couple. It is to be appreciated that any E_j appearing in E^0 measurements of this type introduces a further uncertainty. Standard potentials may also be obtained by independent estimation of K_{eq} coupled with (71), by calculation/measurement of ΔH^0 and ΔS^0 to yield ΔG^0 and thus E^0 by way of (69) or by voltammetric procedures. Often

Table 2.8 Electrode Potentials and Standard Electrode Potentials of Some Useful Reference Electrodes at 25°C in Aqueous Solution

Electrode	Half-Reaction	Other Conditions	E°(V)	E(V)
$Ag^+|Ag$	1. $Ag^+ + e \rightleftharpoons Ag$	Standard states	0.7991	—
$Cl^-|AgCl|Ag$	2. $AgCl + e \rightleftharpoons Ag + Cl^-$	$\longrightarrow$	0.2224	—
$Cl^-|Hg_2Cl_2|Hg|Pt$	3. $Hg_2Cl_2 + 2e \rightleftharpoons 2Hg + 2Cl^-$		0.2682	—
$SO_4^{2-}|Hg_2SO_4|Hg|Pt$	4. $Hg_2SO_4 + 2e \rightleftharpoons 2Hg + SO_4^{2-}$		0.6158	—
$H^+|H_2(g)|Pt$	5. $2H^+ + 2e \rightleftharpoons H_2(g)$	pH 7, $a_{H_2} = 1$	—	−0.414
$H^+|H_2(g)|Pt$	6. $2H_2O + 2e \rightleftharpoons 2OH^- + H_2(g)$	$a_{OH^-},\ a_{H_2} = 1$	—	−0.828
$KCl(sat.)|AgCl|Ag$	7. Same as 2		—	0.197
$KCl(sat.)|Hg_2Cl_2|Hg|Pt$	8. Same as 3		—	0.241
$KCl(1\ M)|Hg_2Cl_2|Hg|Pt$	9. Same as 3		—	0.280
$KCl(0.1\ M)|Hg_2Cl_2|Hg|Pt$	10. Same as 3		—	0.334

redox couples may appear in E^0 compilations strictly from theoretical considerations rather than by means of laboratory measurement.

The value of the E^0 table lies in the amount of useful thermodynamic and electrochemical information therein. For instance, a relative scale of oxidizing and reducing power is indicated; couples with large positive E^0 are strong oxidizing agents (a strong affinity for electrons), while large negative values indicate strong reducing power. In postulating the thermodynamic feasibility of complete cell reactions, the E^0 separation between two couples yields information about the K_{eq} for the unknown reaction (for two one-electron couples, a K_{eq} of 10^6 is obtained with a ΔE^0 of 0.36 V). Of course, the actual reactant and product concentrations can also be factored into this calculation by the use of (80) for each couple.

One difficulty in using standard potential tables to predict cell reaction feasibility is that under laboratory conditions of pH, O_2 pressure, ionic strength, complexing agents, and so on the actual active redox couple species is uncertain and thus may have a half-cell potential decidedly different from E^0. The concept of a half-cell *formal potential* $E^{0\prime}$ was introduced to address this problem. The formal potential of a couple (versus the SHE) is that potential determined experimentally with unit formula weights of both reduced and oxidized species, and with all other solutes at defined concentrations. A comparison of E^0 and $E^{0\prime}$ for a few couples is given in Table 2.9. Note, for example, the decreased tendency for the Pb^{2+}/Pb couple to undergo reduction in the 2 M NaAc solution ($E^{0\prime} < E^0$) because of the combined effects of acetate complexation and lowered activity coefficient. Although strictly valid only for the described experimental conditions, the $E^{0\prime}$ compilation adds additional practical information to a standard potential list.

Measurement of effects like complexation, pH, and solubility equilibria may be possible as an alternative to the formal potential concept. For metals and metal-salt precipitates, addition of the half-reaction reduction equilibrium

$$m M^{n+} + mne \rightleftharpoons mM \tag{84}$$

with the solubility equilibrium

$$M_m X_p \rightleftharpoons m M^{n+} + p X^{x-} \tag{85}$$

yields the overall metal-salt expression

$$M_m X_p + mne \rightleftharpoons mM + p X^{x-} \tag{86}$$

The standard potentials of (86) and (84) are then related by

$$E^0_{X^-/MX} = E^0_{M^+/M} + \frac{RT}{mnF} \ln K_{sp} \tag{87}$$

where K_{sp} is the solubility product of (85). Likewise for a complexation equilibrium that competes with the electron for the free metal ion in (84), the E^0 is altered so that

$$E^0_{M(L)/M} = E^0_{M^+/M} - \frac{RT}{nF} \ln K_A \tag{88}$$

Table 2.9 Comparison of Standard Potentials and Formal Potentials[a]

Half-Reaction	Standard Potential, E^0_{Red} (V)	Formal Potential (V)			
		1 M HCl	1 M HClO$_4$	1 M H$_2$SO$_4$	Other Solutions
$Ce^{4+} + e = Ce^{3+}$	1.61	1.28	1.7	1.44	1.61(1 M HNO$_3$)
$Co^{3+} + e = Co^{2+}$	1.84		1.95	1.80	1.80(1 M HNO$_3$)
$Cr^{3+} + e = Cr^{2+}$	−0.41	−0.38			
$Cu^{2+} + e = Cu^{+}$	0.153	0.45			
$Fe^{3+} + e = Fe^{2+}$	0.771	0.700	0.75	0.68	0.61(0.5 M H$_3$PO$_4$) / 0.56(0.1 M HCl)
$Fe(CN)_6^{3-} + e = Fe(CN)_6^{4-}$	0.36	0.71	0.72	0.72	0.48(0.01 M HCl) / 0.46(0.01 M NaOH)
$2H^{+} + 2e = H_2(g)$	0.000	−0.005	−0.005		−0.005(1 M HNO$_3$)
$Hg_2^{2+} + 2e = 2Hg(l)$	0.789	0.274	0.776	0.674	0.281(1 M KCl) / 0.241(sat. KCl)
$Hg_2Cl_2 + 2e = 2Hg + 2Cl^{-}$	0.268				0.281(1 M KCl) / 0.334(0.1 M KCl)
$MnO_4^{-} + 4H^{+} + 3e = MnO_2(\beta) + 2H_2O$	1.68		1.60		{1.60(1 M HNO$_3$) / 1.65(0.5 M H$_2$SO$_4$)
$Pb^{2+} + 2e = Pb(s)$	−0.126		−0.14	−0.29	−0.32(2 M NaAc)
$PbO_2 + 4H^{+} + 2e = Pb^{2+} + 2H_2O$	1.455		1.47	1.628	
$Sn^{4+} + 2e = Sn^{2+}$	0.15	0.14			
$Tl^{+} + e = Tl(s)$	−0.3363	−0.551	−0.33	−0.33	
$Tl^{3+} + 2e = Tl^{+}$	1.25	0.77	1.26	1.22	1.23(1 M HNO$_3$)
$V^{3+} + e = V^{2+}$	−0.255	−0.28	−0.27	−0.27	

[a]Taken from [50], with permission.

where K_A is the association constant for the overall complexation of M^{n+} with ligand L. Of course, the pH of the solution will also alter the potential of any half-cell reaction in which H^+ is involved. Thus, some metal–metal oxide systems like the antimony and palladium electrodes have been suggested as pH sensors since they roughly follow the half-cell reaction

$$M_aO_b + 2bH^+ + 2b(e) \rightleftharpoons aM + bH_2O \tag{89}$$

where the oxide stoichiometry varies with each metal. However, instability of the oxide layer, susceptibility to other redox agents, and sensitivity to solution pO_2 are problems. The classic quinhydrone electrode (see [22] for complete details) is

$$a_{H^+}, \text{hydroquinone} + \text{benzoquinone (sat. quinhydrone)}|\text{Pt} \tag{XVIII}$$

and yields a potential given by

$$E_Q = E_Q^0 - \frac{RT}{2F} \ln \frac{\gamma_{QH_2}}{\gamma_Q} - 2.303 \frac{RT}{F} \text{pH} \tag{90}$$

which is analytic for pH assuming no hydroquinone dissociation, equality of the γ terms for the neutral species, and no E_j influence. The pH utility fails above pH 7–8, where the hydroquinone starts to dissociate according to

$$H_2Q \rightleftharpoons H^+ + HQ^- \rightleftharpoons H^+ + Q^{2-} \tag{91}$$

For such an organic system the potential can be generally expressed as

$$E = E_{Q/H_2Q}^0 + \frac{RT}{2F} \ln \frac{\gamma_{QH_2}}{\gamma_Q} + \frac{RT}{2F} \ln \left[a_H^2 + \frac{K_1 a_H}{\gamma_{HQ^-}} + \frac{K_1 K_2}{\gamma_{Q^{2-}}} \right] \tag{92}$$

and estimates for the acid dissociation constants K_1 and K_2 of (91) can be obtained from plots of E versus pH. Finally, pH sensitivity may occur in a redox couple whose nominal half-cell reactions do not include H^+. Usually metal ion hydrolysis is implicated; the dependence of the $Fe(CN)_6^{3-}/Fe(CN)_6^{4-}$ formal potential on HCl concentration in Table 2.9 is an example.

Simultaneous consideration of all possible factors that may influence electrode potential is complicated and done practically only by computer. For ease in visualizing the output, two- or sometimes three-dimensional plots of E versus pH, I, and various ligand species are often constructed. Högfeldt [51] has summarized the approach. As an example, Figure 2.7 shows a predominance area diagram for the Fe-H_2O system at 50°C. Here the electron "activity" pE has been defined from the electrode potential as

$$pE = -\log(a_{e^-}) = \frac{EF}{RT \ln 10} \tag{93}$$

which also defines a corresponding standard electron activity pE^0 by

$$pE^0 = \frac{\log K_{eq}}{n} = \frac{E^0 F}{RT \ln 10} \tag{94}$$

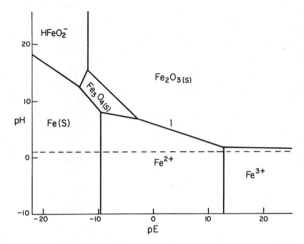

Figure 2.7   Predominance area diagram for the system Fe–H$_2$O–e^- at 50°C. From [51], with permission.

As one varies pH or pE in this system, various half-cell reactions and chemical species become important. For instance, along the dashed line at a constant pH of 1 only the Fe, Fe^{2+}, and Fe^{3+} species are observed. At neutral pH the couple

$$\tfrac{1}{2}Fe_2O_3(s) + 3H^+ + e^- \rightleftharpoons Fe^{2+} + \tfrac{3}{2}H_2O \tag{95}$$

is dominant from a pE of 0 to 10; it is indicated by line 1. Slices in pH or pE show dominant redox couples as species boundary lines are crossed. A third dimension in species concentration (to gauge γ effects) or ligand concentration can be added.

3.2   Redox Electrodes, Apparatus, and Techniques

Generally, indicator electrodes in redox measurements are made from noble metals like platinum or gold, or other "inert" materials like carbon. The criterion for utility is that electron exchange rather than ion exchange or electrode corrosion determines the potential. Many designs are available commercially; varying tip shapes or electrode holder materials (inert glass, epoxy, or plastic) are possible. Common configurations for commercial ORP (oxidation–reduction potential) electrodes include single or double indicator electrodes, combination electrodes with SCE or SSCl references, "rH" combinations of metal and pH glass, or triple configurations with metal, pH glass, and SCE or SSCl reference in one cylindrical body.

Attempts have been made to design redox-sensitive membranes analogous to ion-selective membranes, although none have been commercialized. The general scheme has been to incorporate into the membrane phase oil-soluble redox couples that might respond (i.e., exchange electrons with aqueous couples) like the noble-metal systems. The ideal would be to mimic the reversible electron transport apparently realized in biological membranes, but on thick (0.1–1 mm) phases that are physically robust. Thus Shinbo and associates [52] added dibutylferrocene to dioctyl phthalate in a poly(vinyl chloride) (PVC) matrix (similar to ISE–PVC membrane construction) and were able to follow the aqueous ferricyanide couple at 52 mV/decade. Liteanu and Hopirtean [53] pressed a pellet from a powder of the stable free radical, octachlorophenothiazine. When the pellet was placed in an electrode body it followed the Fe^{3+}/Fe^{2+} couple with about half the Nernstian sensitivity, and could be used successfully to follow several redox titrations. Some redox sensitivity has been observed for glasses doped with Fe_2O_3 or for certain tungsten bronzes [15]; one such glass has been commercialized by Russian workers as a substitute for Pt in solutions where the metal is poisoned by sulfide or subject to varying pO_2. Finally, redox reactions can be performed at semiconductor electrodes, but few instances of potentiometric sensing systems have been demonstrated.

Measurement apparatus for redox potentiometry is, for the most part, common to pH and ISE work. Thus pH/pIon meters usually can be used in the millivolt mode and can easily handle the low impedance redox electrode sources. About the only concern is the large potential range sometimes encountered with some couples. In addition, these meters normally have polarizing outputs available for performing titration with impressed currents of small magnitude. Similarly, commercial automatic titrators, discussed in Section 1, are compatible with redox titrations. Special precautions are needed if the titration must be done in the absence of oxygen. Several anaerobic cells for manual titrations have been described in the previous edition of this chapter [54]; in addition, most of the commercial titrators provide vessels that allow for continuous degassing.

A summary of potentiometric redox techniques is outlined in Table 2.10. The direct methods involve single cell potential measurements or a series of discrete measurements under different conditions. The attempts to use E (sometimes termed E_h to indicate the SHE as reference) as a monitor for electron activity, somewhat analogous to the use of pH for acidity, are common in soil and water-quality fields. If only thermodynamic considerations were important, all couples in a medium would attain this potential. However, kinetic and dissolution phenomena are often more important, so that oxidized and reduced species of a couple may not be present in the ratio dictated by E. In addition it may not even be possible to identify the major redox couple that causes changes in E.

Prediction of E^0 or $E^{0'}$ by direct measurement is still practiced on new couples. Equilibrium constants can also be derived, although titration methods are less tedious. A large amount of data has been collected in the biochemical field for metalloenzymes, cells, and metabolites; compilations are available [55]. The use of mediators is common. These are electrode-reversible couples that will attain the equilibrium potential generated by the electrode-insensitive biological couple. They often are colored dyes, so that both potential and spectra can be followed. Ideally the mediator has an E_{mx} (a biological system midpoint $E^{0'}$ where oxidized and reduced species are equal concentrations at some

Table 2.10 Techniques in Redox Potentiometry

Classification	Method	Purpose	Examples	Comments
Direct	Measurement of E	Fundamental	E of given medium	Information on oxidizing, reducing power of media like waters, soils
			E^0 or $E^{0'}$ of a given couple	Extrapolation of discrete E measurements for biological systems; an indirect redox mediator often employed
Indirect	Titration; measurement of E vs. titrant volume	Analytical or fundamental	Classic redox titrimetry with indicator and reference electrodes	Analysis with oxidizing or reducing agents or thermodynamic information like $E^{0'}$, n, K_A, K_{sp}, or ΔG^0
Indirect	Titration; measurement of E vs. titrant volume	Analytical	Zero-current bipotentiometric, one- or two-electrode polarized methods	Analysis for irreversible or slow-reacting couples
Indirect	Potentiometric stripping analysis, E vs. time	Analytical	Preconcentrate materials into Hg amalgams with reductive current, oxidize chemically to strip at zero current	Rapid analysis of trace metals in various media, alternative to anodic stripping voltammetry

defined pH $= x$) similar to that of the unknown couple, and thus functions like a pH indicator. Table 2.11 lists some of the more common mediators. Dutton [56] has reviewed potentiometric redox measurements on biological systems.

Potentiometric titrations define the major portion of redox measurements. Additions of reducing or oxidizing agents to an analyte redox couple alter the measurable cell potential according to (80). With accurately known titrants the change in potential at the equivalence point can be used to quantify either reduced or oxidized initial species. Alternatively, by following the shape of the curve, equilibrium information may be gained about the system. The familiar conventional potentiometric titration (treated in detail in the next section) employs one indicator electrode and a cell reference electrode under zero-current conditions. It is also possible to use two electrodes of identical material that have been treated differently, and monitor the potential differences at zero current. The advantages are cell simplicity and derivative-like end point changes. Kekedy and Popescu have discussed several embodiments [57]. When a slow approach to redox equilibrium is encountered, polarization of one or both indicator electrodes in a titration cell is suggested. Either dc or ac can be employed; the former with two identical working electrodes is often termed the bipotentiometric technique. Bishop and Cofre have compared methods of polarization for titrations over a range of kinetic behavior [58]. The recurring reviews of Stock [59] offer up-to-date discussion and applications of all the nonconventional redox titration methods.

A final indirect technique of interest is potentiometric stripping analysis. This is a hybrid voltammetric–potentiometric method wherein metal analytes

Table 2.11 Some Common Redox Mediators[a]

Component	Approximate E_{m7} (mV)	n Value
Potassium ferro/ferricyanide	430	1
2,3,5,6-Tetramethylphenylenediamine ("diaminodurol" or DAD)	260	2
N,N,N',N'-Tetramethylphenylenediamine (TMPD)	260	1
N-Methylphenazonium methosulfate (PMS)	80	2
N-Ethylphenazonium ethosulfate (PES)	55	2
N-Methyl-1-hydroxyphenazonium methosulfate (pyocyanine)	-34	2
2-Hydroxyl-1,4-naphthoquinone	-145	2
Anthraquinone-26-disulfonate	-185	2
Anthraquinone-2-sulfonate	-225	2
N,N-Dibenzyl-4,4-bipyridinium dichloride (benzyl viologen)	-311	1
N,N'-Dimethyl-4,4-bipyridinium dichloride (methyl viologen)	-430	1

[a]From [56], with permission.

of interest are reductively deposited into a Hg amalgam, and then stripped back into solution upon chemical oxidation by solution oxidizing components. Jagner has given a concise review of the technique [60]. The time necessary for complete reoxidation of any analyte M_i^{n+} from the amalgam $M_i(Hg)$ is approximated by

$$t_{i,\text{strip}} = kC_{M_i^n} + t_{\text{dep}}d\left(\sum_i D_{Ox}C_{Ox}\right) \qquad (96)$$

where t_{dep} is the time of reductive deposition, C_i are bulk species concentrations, D_{Ox} is the solution diffusion coefficient of oxidizing species, d the product of diffusion layer thicknesses during deposition and stripping, and k a constant. During the stripping the potential of the working electrode will be governed by each corresponding couple according to

$$E = E_{M_i/M_i(Hg)}^0 + \frac{RT}{nF}\ln\left[\frac{C_{M_i^2+}(\text{electrode surface})}{C_{M_i(Hg)}}\right] \qquad (97)$$

as the solution oxidizing action takes place. The experimental output of E versus time like that in Figure 2.8 then provides both qualitative (E) and quantitative (time length) information. Since the working electrode (normally a mercury-coated glassy carbon) is usually rotated and concentrations of solution oxidants like O_2 are high, (96) predicts very short stripping times. Indeed the output often must be computer-recorded since stripping times of milliseconds are possible for trace metal concentrations. Advantages of this technique over the corresponding anodic stripping voltammetric (ASV) method are that time is a precisely measurable parameter, electrode area is not critical, and there is

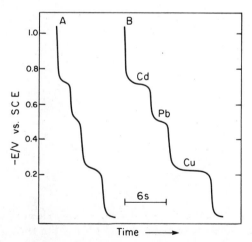

Figure 2.8   Potentiometric stripping curves registered after (A) 2 min and (B) 4 min of potentiostatic deposition at -1.10 V vs. SCE for a sample containing 1 mg/L of cadmium(II), lead(II), and copper(II), and 80 mg/L of mercury(II) in 1 M HCl. From [60], with permission.

little interference from bulk electroactive components that are not oxidized under the zero-current conditions of the stripping phase.

3.3 Redox Titrations

Conventional zero-current redox titrations employing an indicating electrode and a nonidentical reference electrode involve additions of a redox titrant to a sample couple according to

$$n_2 Ox_1 + n_1 Red_2 \rightleftharpoons n_2 Red_1 + n_1 Ox_2 \tag{98}$$

where subscript 2 indicates titrant and 1 the sample analyte, and where it is understood that the titration can be performed in either direction. The reaction (98) can be derived by summing the two half-reactions of titrant and sample:

$$Ox_1 + n_1 e \rightleftharpoons Red_1 \tag{99}$$

$$Ox_2 + n_2 e \rightleftharpoons Red_2 \tag{100}$$

From purely thermodynamic considerations a successful titration of Ox_1 requires that the equilibrium constant for (98) be at least 10^6 (or 10^{-6} if Red_1 is to be titrated), or that the $\Delta E^{0\prime}$ of the couples be greater than $0.18 (n_1 + n_2)/(n_1 n_2)$ V at 25°C.

Under ideal conditions one can segment the variation of measured potential into three regions during titrant volume addition. Before the equivalence point the sample couple will "poise" the indicator electrode potential so that the cell potential is given by

$$E = E_1^{0\prime} + \frac{RT}{n_1 F} \ln \frac{C_{Ox_1}}{C_{Red_1}} + E_j \tag{101}$$

where cell E_j is assumed constant and all other media-related factors like γ and pH are taken into $E^{0\prime}$. After complete titration of Ox, the only couple in significant concentration is that of the titrant species so the cell potential becomes

$$E = E_2^{0\prime} + \frac{RT}{n_2 F} \ln \frac{C_{Ox_2}}{C_{Red_2}} + E_j \tag{102}$$

At the equivalence point the potential is the arithmetic mean of the two formal potentials according to

$$E_{eq} = \frac{n_1 E_1^{0\prime} + n_2 E_2^{0\prime}}{(n_1 + n_2)} \tag{103}$$

Combination of the three cell potential expressions yields the familiar sigmoidal redox titration curve, as shown in Figure 2.9 for the titration of Fe(II) with Ce(IV) in 0.5 M sulfuric acid. Note that the Ce(IV) titrant abscissa is given as a stoichiometric percentage of the original Fe(II) sample concentration.

The titration mode offers increased precision over direct redox potentiometry because of the large change in potential at the equivalence point, and

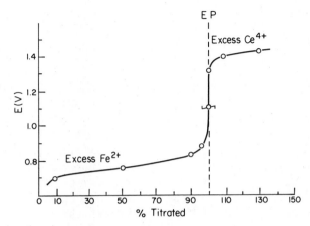

Figure 2.9 Titration of ferrous ion with ceric sulfate in 0.5 M H_2SO_4 with SHE reference electrode; Ce(IV) addition plotted as percent of initial Fe(II). From [61], with permission.

because titrant volume additions can be controlled rather precisely with automatic burets. In addition the shape of the titration curve given by (101)–(103) is independent of reactant concentrations for simple cases. In the two half-reaction regions defined by sample and titrant couples there is seen to be little variation in cell potential with titrant addition. In effect these are redox-buffered regions analogous to pH-buffered portions of acid–base titration curves. It can be shown that the maximum redox "poise" or buffer capacity of a redox couple is found at the formal potential of the couple.

In addition to adequate $\Delta E^{0\prime}$ several other factors must be addressed if a redox titration is to be performed and interpreted successfully:

1. For analytical titrations the analyte must be initially totally oxidized or reduced, and a prior oxidation/reduction step must be included if this is not true.

2. Ideally the kinetics of both the redox reaction and the electrode response are faster than the titrant mixing time. Slow redox kinetics may sometimes be overcome by electrode polarization, or by adding excess of titrant and back-titrating with a second reagent.

3. Sample, titrant, and products should be stable toward air oxidation; if not, titration-vessel purging of O_2 is necessary.

4. The titration medium should be controlled so that γ, pH, and any ligand complexation are kept constant. Alternatively, in fundamental studies these factors may be systematically varied to extract thermodynamic information. Ligands can also be intentionally added to the medium to mask redox interferences.

5. A means of equivalence point detection must be provided if this is not done potentiometrically.

6. Other redox-active couples, either analyte-independent interferences or consecutive electron-transfer steps of the analyte itself, must have adequate $E^{0'}$ separation from the primary analyte transition. If this is true then these species may also be quantitated during the titration.

For details of reagents and procedures specific to given assays several compilations can be consulted [62, 63]. Common prior oxidants include silver(II)oxide, ozone, and the halogens; examples of prior reductants are metal amalgams, SO_2, and stannous chloride. A more extensive list with original usage references can be found in [62]. Columns packed with redox polymers [64] are also useful for this purpose since they provide a more convenient separation of treated analyte and excess reagent than the boiling or filtration steps often required with the chemical agents. The number of possible titrants is very large and includes permanganate, I_2, Br_2, cerium(IV), chromium(VI), lead(IV) acetate, and the oxyhalogens as oxidants, and chromium(II), iron(II), titanium(III), vanadium(IV), arsenic(III), and tin(II) as common reductants. The monograph by Berka and associates [65] can be consulted for characteristics of newer redox titrants like chloramine T, hydrazine sulfate, cobalt(II), manganese(III), ascorbic acid, and N-bromosuccinimide. Stability of the redox titrant is of crucial concern, and coulometric generation of titrant species is to be recommended whenever possible. Standard reference materials for determining reagent titer are available from the NBS. These include potassium dichromate, arsenic trioxide, and sodium oxalate; Yoshimori [66] has outlined details of drying and weighing for accurate preparation.

With manual titration it is common to employ redox indicators to monitor the equivalence point. The substances undergo a one- or two-color change in the region of E_{eq}, and so are similar in principle to the use of indicators in acid–base titrations. The most general class of indicators are themselves redox couples that in trace concentration follow the cell potential dictated by the analyte–titrant system according to

$$E = E_{ind}^{0'} + \frac{RT}{n_{ind}F} \ln \frac{C_{ind}(Ox)}{C_{ind}(Red)} \quad . \tag{104}$$

where C_{Ox} and C_{Red} have different colors. A second class reacts specifically with principal reactant or product, like the classic I_2–starch system. In the general case described by (104), color discrimination is practically obtained at a 10/1 intensity ratio of the two forms, so that the useful indicator potential range is $59/n_{ind}$ mV about $E_{ind}^{0'}$. Sriramam [67] has treated the subject in a more rigorous fashion, accounting for varying molar absorptivities and intensity ratios. A brief list of redox indicators is shown in Table 2.12; the text edited by Bishop [68] can be consulted for extensive examples. Since indicator errors such as medium effects, interactions with analyte or titrant species, and simple blank equivalents can occur, potentiometric equivalence point estimates are normally preferred for automated titration.

The accuracy of the titration is usually governed by electrode and reaction

Table 2.12   Selected List of Redox Indicators[a]

Indicator	Color Change		$E^{0\prime}$ (V) at pH = 0
	Ox	Red	
Vanadium(II)-1,10-phenanthroline	Colorless	Blue-violet	0.14
Indigo monosulfonate	Blue	Colorless	0.26
Phenosafranine	Red	Colorless	0.28
Indigo tetrasulfonate	Blue	Colorless	0.36
Methylene blue	Green-blue	Colorless	0.36
1-Naphthol-2-sulfonic acid indophenol	Red	Colorless	0.54
Variamine blue	Colorless	Blue	0.59(pH 2)
2-Hydroxyvariamine blue	Colorless	Blue	0.56(pH 2)
			0.06(pH 12)
Diphenylamine (diphenylbenzidine)	Violet	Colorless	0.76±0.1
Diphenylaminesulfonic acid	Red-violet	Colorless	0.80±0.1
Erioglaucine	Red	Green	1.0
Fe 5,6-Dimethylphenanthroline	Pale blue	Red	0.975
p-Ethoxychrysoidine	Pale yellow	Red	1.0
[Fe(bipyr)$_3$]SO$_4$	Red	Pale blue	1.03
p-Nitrodiphenylamine	Violet	Colorless	1.06
[Fe(II)(*o*-phenanthroline)$_3$](ClO$_4$)$_2$ (ferroin)	Pale blue	Red	1.11
[Ru(bipyr)$_3$]SO$_4$	Green	Orange	1.24
[Fe(II)(nitro-*o*-phenanthroline)$_3$] (ClO$_4$)$_3$ (nitrophenanthroline)	Pale blue	Violet-red	1.25

[a]From [61], with permission.

system performance around the equivalence point and by the method used to estimate the equivalence point. The true theoretical equivalence point, defined in simple cases by expressions like (103), can differ from both the inflection point (point of maximum slope of the titration curve) and the end point. The end point is defined by some measured parameter like a potential or a redox indicator color change, and thus may be subject to determinate error. Ebel and Seuring [69] have reviewed methods of equivalence point estimation especially as they are embodied in automatic titrators. End point errors can often be accounted for by first titrating a known standard in the reaction matrix so as to obtain empirically the difference between the chosen end point and the true value. Various methods can be employed for inflection point determination. The most common employ several potential readings around the equivalence point to derive the maximum slope by first or second derivative or other inter-polation procedures. Microprocessor-controlled titrators can adjust titrant volume deliveries to yield equally spaced potential values [47]; this improves the accuracy of the interpolation methods.

Since indicator electrode drift or reaction system irreversibility can severely affect cell potential around the equivalence point (EP), prediction methods using other portions of the titration curve have been suggested. Linearizations of the redox titration curve according to the Gran function is possible. Equations (101) and (102) can be transformed into the forms

$$V \exp\left[\frac{n_1(E - E_1^{0\prime})}{k}\right] = F_1 \propto (V_{eq} - V) \text{ before EP} \tag{105}$$

$$\exp\left[\frac{-n_2(E - E_2^{0\prime})}{k}\right] = F_2 \propto (V - V_{eq}) \text{ after EP} \tag{106}$$

where V is titrant volume, V_{eq} is titrant volume at the equivalence point, k is F/RT, and F_i indicates the Gran function. Ideally, plots of F_1 or F_2 against titrant volume intersect the volume axis at the equivalence point, and so potentials around the equivalence point are not required. An example of Gran plots calculated for titration of Fe(II) with dichromate at pH 0, 1, and 2 is shown in Figure 2.10. Here F_1 is defined like (105) with the Fe(II)/Fe(III) couple, F_3 is like F_1 except that the multiplier is $(V_0 + V)$, and F_2 refers to the post-equivalence chromate couple

$$HCrO_4^- + 7H^+ + 3e \rightleftharpoons Cr^{3+} + 4H_2O \tag{107}$$

so that

$$F_2 = V \exp\left[\frac{-3(E - E_{Cr}^{0\prime})}{k}\right] \propto (V - V_{eq}) \tag{108}$$

The calculation shows that at pH 2 the equivalence point transition of the traditional titration curve is poorly defined; in this case solid $Fe(OH)_3$ has begun to form. Yet either F_2 or F_3 [where the Fe(II)/Fe(III) couple is corrected for $Fe(OH)_3$ formation] is suitable for equivalence point detection. Gran plot curvature usually indicates that unaccounted factors like $\Delta\gamma$, electrode sensitivity

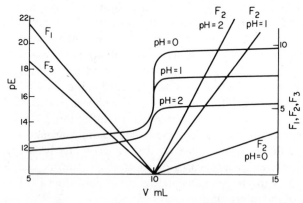

Figure 2.10 The pE and Gran functions plotted against titrant volume for the calculated redox titration of 100 mL of 0.06 M Fe(II) with V mL of 0.1 M dichromate at pH 0, 1, and 2. F_1 refers to the Fe(II) couple at pH 0 or 1; F_3 the same couple at pH 2, and F_2 the dichromate couple. The pE is defined as in (93). Taken from [38], with permission.

change, or additional equilibrium steps are operative. Modification to derive a more rigorous Gran expression is then suggested.

Finally, exhaustive curve fitting using all measured titration potentials is possible with a computer. Either weighted nonlinear regression by adjustment of parameters like E_{eq}, E^0, γ, and C_{Ox} is performed, or the inflection point is determined by curve fitting of all data with polynomials without regard to theoretical exactness. Although application of these methods to redox titrations is not widespread, in principle they are suitable. The Ebel and Seuring review [69] lists specific references for each method.

3.4   Redox Measurements in Nonaqueous Media

Potentiometric redox measurements are often performed in nonaqueous or mixed solvent media. This can entail extra experimental difficulties like solvent purification and handling, or even working at the high temperatures of molten-salt preparations. However, there may be important advantages to be gained. Solutes may be more soluble in such media, oxidizing or reducing power of a redox couple may be altered by interaction with solvent or by increased ion pairing, or there may simply be no alternative for certain chemical systems like the molten salts.

To interpret measured electrode potentials in nonaqueous media, the previous description of aqueous cell potentials and redox terminology must be expanded. For a solute like a neutral salt one can write

$$_s\mu_i = _s\mu_i^0 + RT \ln a_i^* \tag{109}$$

and

$$a_i^* = m_i \,_s\gamma_i \tag{110}$$

where asterisk or subscripts indicate the nonaqueous media. As the solute concentration approaches infinite dilution in the nonaqueous medium $_s\gamma_i$ becomes unity, but this value may not equal unity on the aqueous $_w\gamma_i$ scale. In addition, because of differences in solvation the standard chemical potentials may differ. One quantifies both these differences by defining the transfer activity coefficient $_m\gamma_i$, also known as the medium effect, by

$$\Delta G_t^0 \equiv _s\mu_i^0 - _w\mu_i^0 = RT \ln _m\gamma_i \tag{111}$$

so that

$$a_i = a_i^* \,_m\gamma_i \tag{112}$$

The $_m\gamma_i$ is then a measure of the standard free energy of transfer of the solute from water to the nonaqueous medium; and operationally it defines the relation between aqueous and nonaqueous activity scales. Recalling relation (69) one can also define

$$\ln _m\gamma_i \equiv nF \frac{(_wE_i^0 - _sE_i^0)}{RT} \tag{113}$$

Values of $_m\gamma_i$ (or the more commonly listed $\log {_m\gamma_i}$) for electrolytes can be measured by methods employed for $\gamma_\pm$ determination: solubility, vapor pressure, or even potentiometric procedures. Large negative $\log {_m\gamma_i}$ indicates that a salt prefers the nonaqueous medium; for AgI in the silver-solvating medium acetonitrile the $\log {_{AN}\gamma_{AgI}}$ is -1.7 as opposed to 2.9 in ethanol.

The problem of how to deal with the Nernst equation for a redox system in nonaqueous solvents (114) remains

$$_sE = {_sE^0_{Ox/Red}} + \frac{RT}{nF} \ln \frac{a^*_{Ox}}{a^*_{Red}} \tag{114}$$

For aqueous systems the E^0 scale was defined against the arbitrary zero potential value assigned to the SHE. One can similarly assign an arbitrary potential to a defined reference electrode system in the medium of interest, allowing a redox scale to be built. However, as many such scales would be obtained as there were solvent systems, and there would still be no way to compare potentials of identical couples in different media. Instead, some couples have been suggested as solvent-independent reference systems to be assigned the value of zero volts. These include the Cs/Cs^+, the Rb/Rb^+, and the ferrocene/ferrocenium$^+$ couples [70]; all involve ions of large size and low polarizability to reduce anomalous solvation effects. (The SHE is a poor choice for such a couple precisely because of strongly solvent-dependent solvation of the hydrogen ion.)

Alternatively one could define by convention some value of a single ion $_m\gamma_i$ that would enable a scale of $_m\gamma_j$ to be built since

$$\log {_m\gamma_{MX}} = (\log {_m\gamma_M} + \log {_m\gamma_X}) \tag{115}$$

where MX represents any electrolyte, and $_m\gamma_{MX}$ is measurable. In particular, values of $_m\gamma_{H^+}$ are sought because definition of a pa^*_H scale according to

$$pa_H = pa^*_H {_m\gamma_{H^+}} \tag{116}$$

would relate to the water scale, and because it might be possible to use the aqueous SHE potential as the absolute reference for all solvents by

$$_wE^0(i, s) = {_sE^0(i, S)} + {_wE^0(H^+, S)} = {_sE^0(i, S)} + \frac{RT}{F} \ln {_m\gamma_{H^+}} + {_wE^0(H^+, H_2O)} \tag{117}$$

Here the standard potential of the SHE in solvents on the aqueous scale would be $0.0591 \log {_m\gamma_{H^+}}$ at $25°C$, since $_wE^0(H^+, H_2O)$ is defined as zero volts. The currently favored approach to $_m\gamma_{H^+}$ assumes that "reference electrolyte" like tetraphenylarsonium tetraphenylborate possesses equal cation and anion solvation in any solvent so that

$$\log {_m\gamma_{Ph_4As}} = \log {_m\gamma_{BPh_4}} \tag{118}$$

According to (115) the individual $_m\gamma_i$ are then assigned by measurements on the salt, and calculation routes to all other individual $_m\gamma_i$ (including $_m\gamma_{H^+}$) are available. Table 2.13 lists single-ion $\log {_m\gamma_i}$ for several ions of particular interest

Table 2.13 Transfer Activity Coefficients of Single Ions ($\log\ _m\gamma_i$) Based on the Tetraphenylarsonium Tetraphenylborate Reference Electrolyte (25°C, Molal Scale, Reference Solvent Water)[a]

Ion	CH_3OH	C_2H_5OH	CH_3CN	DMF	DMSO	NMePy	PC	$HCONH_2$
Ph_4As^+ or BPh_4^-	−4.2	−3.7	−5.8	−6.7	−6.5	−7.0	−6.1	−4.1
H^+	1.8	1.7	8.0	−2.5	−3.3	—	—	—
Ag^+	1.2	0.8	−3.9	−3.0	−5.9	−5.3	2.9	−2.7
Cl^-	2.1	3.4	7.3	8.1	6.8	9.7	6.7	2.5
Rb^+	1.7	2.7	1.1	−1.8	−1.9	−1.8	−0.4	−0.9

[a]Data taken from [70], with permission. NMePy = N-methyl-2-pyrrolidone, PC = propylene carbonate.

obtained by this approach; Marcus [71] has compiled an exhaustive data base in which other approaches are included.

Since no convention has been universally adopted, one cannot yet define $_sE_i^0$ with respect to any other solvent than that in which it was measured. However, measurements of $_sE_i^0$ within a solvent according to cells like

$$Pt|H_2(g)|HX(m), \text{ solvent } S|AgX|Ag \qquad (XIX)$$

with extrapolation to $m=0$ are common, and compilations can be found [49]. Precaution must be taken with these measurements if appreciable ion association occurs because of low solvent dielectric constant. Once $_sE_i^0$ values are established, then predictions about redox reaction feasibility are possible, and values of $\log {}_m\gamma_i$ for salts are calculable if corresponding aqueous data exist.

Practical experimentation for nonaqueous redox potentiometry is driven chiefly by characteristics of the medium. Nonaqueous solvents with appreciable Lewis acid or base character, or those possessing dielectric constants above 25, are suitable choices for acceptable solute and electrolyte solubilities and reasonable electron-transfer rates. In systems like the hydrocarbons, conductivities and electron-transfer rates are very slow; and because reactions take place chiefly between associated ion pairs the usual formalism of half-cell reduction reactions may be inadequate. Rumeau [72] has discussed redox reactions in such solvents in some detail. Some solvents, like liquid ammonia or the amines, are able to solvate the electron, and so reduction potentials like (119) must be considered:

$$M^+ + Se^- \rightleftharpoons M + S \qquad (119)$$

where Se^- indicates the solvated electron. Many common aqueous oxidizing and reducing agents are insoluble in organic solvents, and alternate forms must be chosen (for instance, use of triphenylmethylarsonium permanganate instead of the potassium salt in solvents like chloroform and nitrobenzene). As mentioned previously, favorable solvation of oxidized or reduced forms of a couple may be used to advantage in certain solvents. Kratochvil [73] has summarized experimental redox parameters in acetic acid, acetonitrile, DMF, DMSO, and several other solvents.

Working indicator electrodes are normally chosen from the same group defined for aqueous systems: platinum, gold, or carbon. It is often advisable to establish the time response required to obtain reversible behavior according to the Nernst equation at these electrodes, sluggish behavior becoming more evident as solvent dielectric falls. A variety of reference electrodes has been employed including the Ag/Ag^+ couple with $AgNO_3$ or $AgClO_4$; the metal amalgams $M(Hg)/M^+$, where M is an alkali metal; the $Pt/I_3^-, I^-$ couple; and the aqueous SCE (see [22] for more detail). Often they are separated from the test solution by an additional bridge or frit containing a halide salt in the solvent media, but large junction potentials are still common and caution must be used in comparative work between different solvents.

Although not strictly classifiable as redox measurements, investigations of analytical cell potentials or standard potentials in molten salts or with solid electrolytes can be mentioned here. Valuable thermodynamic information on molten salts can be obtained from E^0 determinations on cells such as

$$\text{Reference electrode} \left| \begin{array}{c} \text{Reference} \\ \text{ion,} \\ \text{molten} \\ \text{solvent} \end{array} \right| \begin{array}{c} \text{Conductive} \\ \text{junction} \\ \text{material} \end{array} \left| \begin{array}{c} \text{Indicator} \\ \text{ion,} \\ \text{molten} \\ \text{solvent} \end{array} \right| \text{Indicator electrode} \qquad \text{(XX)}$$

where the "solvent" is a defined molten-salt mixture such as a LiCl-KCl eutectic, and the indicated couples vary from M/M^+ to Pt/Ox,Red. Ives and Janz [22] should be cited for basic operating techniques in fused salts; a more recent review of applications is given in [74]. Solid electrolyte materials have taken on great importance in recent studies of fuel cells and battery systems. Potentials reversible to ions can be obtained from various conductive solid electrolytes like $M_2O \cdot Al_2O_3$ (β-alumina), MAg_4I_5 (M = Rb, K, NH_4), and $MBiF_4$ (M = K, Rb, Tl), so that voltage measurements can provide useful thermodynamic data. In addition, oxygen-sensitive oxide electrolytes like CaO-doped ZrO_2 yield potentiometric gas sensors according to

$$\text{Pt} \left| \begin{array}{c} \text{Gas,} \\ \text{known} \\ pO_2 \end{array} \right| \text{Oxide electrolyte} \left| \begin{array}{c} \text{Gas,} \\ \text{unknown} \\ pO_2 \end{array} \right| \text{Pt} \qquad \text{(XXI)}$$

An excellent summary of the field is presented by Hagenmuller and VanGool [75].

3.5   Applications

Analytical redox titrations still find favor in the electroplating and pharmaceutical industries, and to a limited extent, in water-quality analysis. Ion-selective electrode and voltammetric techniques have replaced many of the classical, wet redox assays. Standard potential measurements, however, have seen increased application, particularly in the nonaqueous systems. For example, Das and co-workers [76] determined ($_sE^0 - _wE^0$), or ΔG_t^0, directly by use of the symmetric junctionless cell

$$\text{Ag} | \text{AgX} | \text{MX(m), S} | \text{M(Hg)} | \text{MX(m), } H_2O | \text{AgX} | \text{Ag} \qquad \text{(XXII)}$$

for alkali halides MX in aqueous mixtures S of 20–80% acetonitrile by extrapolation to $m = 0$. By measurements of $_m\gamma$ for Ph_4AsBPh_4 in the solvent system they also derived a set of single-ion $_m\gamma_i$, which corroborated the poor solvating ability of CH_3CN for the alkali cations and halide anions. In contrast, the favorable solvation of Cu(I) by this solvent was employed by Verma and Sood [77] in a typical example of an analytical nonaqueous redox titration. They titrated several mercaptopyrimidines with $Cu(II)ClO_4$ in acetonitrile using a Pt indicator electrode, a methanolic SCE as reference, and diphenylamine as a redox indicator. The increased oxidizing power of the Cu(II)/Cu(I) couple [stronger than Ce(IV)/Ce(III) in this solvent] yielded a sharp inflection point of 0.2–0.4 V, with a precision of 0.2–0.6%. In a modern aqueous redox application Karlberg and Thelander [78] have described an analysis of reducing substances like ascorbate and Fe(II) based on reaction with a stream of Ce(IV) solution in a flow injection module. Downstream changes in potential at a Pt/SCE electrode pair followed millimolar concentrations in reducing agent at a throughput of 45–60 samples per hour with a precision of 1.2%. Additional examples of redox system applications can be found in several of the cited references [10, 11, 36, 61, 63].

4   pH AND ACID—BASE MEASUREMENTS

4.1   Concepts and Operational Definition of pH

It has been demonstrated that no potentiometric cell can yield unequivocal values of single-ion activities. Therefore, the quantity pH, rigorously defined as $(-\log a_{H^+})$, is not measurable. Yet a potentiometrically determined scale of acidity would be extremely valuable in characterizing the chemistry of solution media. A way around this apparent dilemma has evolved by way of an operational definition of pH [79], which provides a stable and reproducible measurement scale in combination with the modern H^+-sensitive glass electrode. The operational definition is coupled to a pH measuring cell of the sort

$$\begin{array}{c|c|c} \text{Reference} & \text{Standard S} & \text{Indicator electrode} \\ \text{electrode} & \text{or} & \text{for } H^+ : \text{glass} \\ & \text{Solution X} & H_2, \text{ and so on} \end{array} \qquad \text{(XXIII)}$$

whereby one obtains

$$\text{pH}(X) = \text{pH}(S) - \frac{E_x - E_s}{RT \ln 10/F} \qquad (120)$$

so that the relative difference in potential then indicates a relative difference in pH. The pH(X) is not $(-\log a_{H,x})$ because of the presence of the uncertain E_j term assumed constant in (120); even if $[E_j(X) - E_j(S)]$ were zero, the $a_{H,x}$ derived would still be tied by convention to the scale decided upon for $a_{H,s}$. However, in dilute aqueous solutions with an equitransferent salt bridge as part of the reference electrode, pH(X) may be regarded as "approaching" the $(-\log a_{H,x})$ as best defined by modern solution theory.

The procedure for establishing the value of pH(S) for any one standard was developed by workers at the NBS and is now widely accepted. Measurements in the junctionless Harned cell

$$\begin{array}{c|c|c|c|c} \text{Pt} & H_2(\text{g, 1 atm}) & \begin{array}{c} \text{Buffer} \\ \text{standard} \\ (m_S), m_{Cl} \end{array} & \text{AgCl} & \text{Ag} \end{array} \qquad \text{(XXIV)}$$

where m_{Cl} is provided by small amounts of alkali salts, can be described by

$$-\log a_H - \log \gamma_{Cl} = \text{p}(a_H \gamma_{Cl}) = \frac{E - E^0}{RT \ln 10/F} + \log m_{Cl} \qquad (121)$$

The E^0 is known from measurements with HCl alone in cell (XXIV), so that calculations of $p(a_H\gamma_{Cl})$ can be obtained at each m_{Cl} tested. Extrapolation to $m_{Cl}=0$ provides $p(a_H\gamma_{Cl})^0$, where the effect of chloride is removed. Then pH(S) is defined in the standard by

$$pH(S) = p(a_H\gamma_{Cl})^0 + \log \gamma_{Cl} \tag{122}$$

and the single-ion γ_{Cl} is obtained from the Bates–Guggenheim form of the Debye–Hückel equation applied at the standard I_m:

$$-\log \gamma_{Cl} = \frac{A\sqrt{I_m}}{1+1.5\sqrt{I_m}} \tag{123}$$

This form is assumed to represent the standard-independent chloride activity accurately at the $I_m < 0.1$ mol/kg characteristic of all the chosen standards. The total series of measurements is then repeated at several temperatures to derive a suitable $pH(S) = f(T)$ relationship.

Standards with pH(S) spanning the aqueous pH range are of course desirable. The original NBS recommendation, since adopted by many nations, was to designate seven buffer mixtures as pH(S) standards according to (121)–(123). The pH(S) values of these materials, which are available as certified SRMs from the NBS or as buffer solutions traceable to the NBS standards from various manufacturers, are given in Table 2.14. British and Japanese workers, however, accredit only one primary standard according to the procedure of (120)–(122). The pH(S) for this material, potassium hydrogen phthalate, is then used to assign secondary pH standards in the cell

$$\text{Pt}\left|\text{H}_2(\text{g, 1 atm})\right|\text{S}=\text{KHP}\left\|\begin{array}{c}\text{KCl,}\\ \geqslant 3.5\ M\end{array}\right\|\text{X}\left|\text{H}_2(\text{g, 1 atm})\right|\text{Pt} \tag{XXV}$$

Table 2.14 Primary Buffer Standards of the NBS pH Scale at Several Temperatures[a]

	pH(S) at		
Buffer Solution	15°C	25°C	37°C
KH tartrate (sat. at 25°C)	—	3.557	3.548
KH$_2$ citrate ($m=0.05$)	3.802	3.776	3.756
KH phthalate ($m=0.05$)	3.999	4.008	4.028
KH$_2$PO$_4$ ($m=0.025$), Na$_2$HPO$_4$ ($m=0.025$)	6.900	6.865	6.841
KH$_2$PO$_4$ ($m=0.008695$), Na$_2$HPO$_4$ ($m=0.03043$)	7.448	7.413	7.385
Na$_2$B$_4$O$_7$ ($m=0.01$)	9.276	9.180	9.088
NaHCO$_3$ ($m=0.025$), Na$_2$CO$_3$ ($m=0.025$)	10.118	10.012	9.910

[a]Some trailing zeros not shown, but all molalities known to four significant figures. From [79], with permission.

Table 2.15 Summary of the Two Approaches to pH Scales and Measurement[a]

	Multistandard	Single Standard
Number of primary standards	5–7	1
Number of secondary standards	Some at high and low pH	Infinity in principle, including high ionic strength buffers
Determination of $pa_H\gamma_{Cl}$	Harned cell for all primary standards	Harned cell for single primary standard
Bates–Guggenheim convention for γ_{Cl}	Applied to all primary standards at all T	Applied to single primary standard at all T
Determination of secondary standards (S_2)	Harned cell (pa_H) or operational cell	Operational cell with liquid junction formed with cylindrical symmetry
E_H versus pH(S)	Points may scatter and best slope may not be theoretical depending on number and which standards used	Single point, slope defined as theoretical
Determination of pH(X) with glass–calomel cell with liquid junction	Operational cell with one or two primary standards	Operational cell with one secondary standard, or two secondary (primary standard can be one of these)
Inconsistency test	Inconsistency arises from residual liquid junction potential or ion size incorrectly given	Cannot be inconsistent but pH(X) incorporates any residual liquid junction potential between S and X
Advantages	Two or more primary standards can be used to check glass electrodes; primary standards available as SRMs	Defined by single substance; high purity specification required on one substance only; two or more secondary standards can be used to check glass electrodes; secondary standards consistent, available as SRMs or user-determined; easy to change to different primary standard (additive correction)

Table 2.15   *(continued)*

	Multistandard	*Single Standard*
		and to special pH scales; requires no changes to take advantage of improvements in measurement
Disadvantages	Primary standards are inconsistent when measured in operational cell; disagreement about number of primary standards; uncertainty for accuracy class 0.02 (0.003 pH) depending on which primary standard used; residual liquid junction potential incorporated in pH(X)	Residual liquid junction potential incorporated in pH(X) and pH(S)

[a]Taken from [80], with permission.

Use of a thin junction capillary with cylindrical symmetry for the KCl bridge is assumed to reduce differences in $[E_j(X) - E_j(S)]$ across varying X. Differences between the two approaches have been reviewed by Covington [80]; these are highlighted in Table 2.15. The resulting ΔpH(S) offsets for the buffer standards are given in Table 2.16. For common laboratory usage with the glass/calomel electrode pair, where the pH accuracy attainable is usually 0.01–0.02 unit,

Table 2.16   Comparison of pH Standards of the British Single-Standard (BSI) Scale with Those of the NBS Multistandard Scale at 25°C[a]

Solution (Molality)	*BSI Value*	ΔpH *(NBS–BSI)*
KH tartrate (saturated at 25°C)	3.556	0.001
KH phthalate (0.05)	4.005	0.003
KH_2PO_4 (0.025), Na_2HPO_4 (0.025)	6.857	0.008
KH_2PO_4 (0.008695), Na_2HPO_4 (0.03043)	7.406	0.007
$Na_2B_4O_7$ (0.01)	9.182	-0.002
$NaHCO_3$ (0.025), Na_2CO_3 (0.025)	9.995	0.017

[a]The value for the common primary standard, KH phthalate, is different because of a slight error in the original NBS Harned cell measurement for this material. Taken from [79], with permission.

differences of these scales are of a pedantic nature. Only for careful thermo-dynamic work with junctionless cells, where interpretation of the pH as pa_H is desired, would the difference be important. The multistandard approach attempts to ascribe an interpretative meaning to pH(S), while the single standard assignment provides better consistency within the operational cell use. A provisional recommendation from IUPAC [81] essentially allows either procedure as long as proper documentation is given in any publication of pH(X) data.

The standard pH(S) scales are set up so that reproducible measurements can be obtained using cell (XXIII) for a variety of cases: high and low ionic strength aqueous, soil, protein, gel and water-solvent mixture pH(X) determinations. Although relative acidity is thereby indicated, the meaning of the pH(X) on the dilute aqueous scale is often uncertain. For interpretable pH(X) $\simeq -\log a_{H,X}$ to be properly obtained, new standards would have to be defined in the medium of choice. In certain cases other procedures can be followed. Often in constant-ionic-strength media, where solutes of interest play a minor role, conditional or apparent equilibrium constants can be derived using the operational pH. Alternatively, the empirical relation between added concentrations of strong acid or strong base and the operational pH can be used to derive pC_H from pH. In several media like seawater and physiological fluids, attempts have been made to define standards. If the constant ionic strength were high enough in such fluids, then cell (XXIII) could yield $pmH(X)$ from

$$pmH(X) = pmH(S) - \frac{(E_X - E_S)}{RT \ln 10/F} \tag{124}$$

where it is again assumed that the media diminish ΔE_j. Finally, in nonaqueous media, the $_m\gamma_{H^+}$ can provide the link to the aqueous scale. This procedure is discussed fully in the subsequent section on nonaqueous measurements.

4.2   Electrodes and Measurement Apparatus

The two nonglass pH sensors of traditional importance are the hydrogen electrode and the quinhydrone electrode. The latter is rarely used today for any purpose and will not be discussed further; [22] provides a complete description. The hydrogen electrode still serves a useful purpose in fundamental studies involving junctionless cells, of which E^0 measurements are an example. Basically, it consists of a Pt wire or foil fused into an inner glass cylinder, with provision for H_2 saturation often made by a sidearm in a second concentric outer glass cylinder as depicted in Figure 2.11. Several designs of the double-tube apparatus are commercially available. The wire is coated with Pt or Pd black by electrolysis in chloroplatinous or chloropalladous acid. The Pd black provides somewhat better resistance to reducible substances like the primary pH standard phthalate solution, and several designs for internal electrolytic H_2 regeneration exist for it [82]. More commonly, H_2 gas is provided from an external source having both a humidifier and a hot, catalytic metal O_2 scrubber.

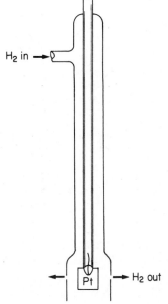

H₂ in →

← Pt → H₂ out

Figure 2.11 Hydrogen electrode of the immersible Hilde-brand type. Internal connection to the Pt may be made by Hg or by a solid conductive solder contact.

Ideally, the electrode response will be

$$E = \frac{RT}{F} \ln \frac{a_{H^+}}{\sqrt{p_{H_2}}} = \frac{RT}{F} \ln \frac{1}{\sqrt{p_{H_2}}} - \frac{RT}{F}(pH) \qquad (125)$$

where corrections of p_{H_2} to 1 atm under various conditions are given in tables like that found in Bates [24]. Disadvantages of this electrode include its cumbersome nature and slow response, the fact that the H_2 saturation can remove pH-determining gases like NH_3, CO_2, and SO_2 from the test solution, and the susceptibility to CN^-, S^-, O_2, and various other organic or inorganic oxidizing substances. Nevertheless, when used carefully, it can yield theoretical Nernstian pH response spanning the entire aqueous 0–14 range with a precision of ± 0.001 pH unit.

The pH glass electrode maintains its position as the most widely used pH sensor to date. It provides a convenient, fast, highly selective, redox-insensitive, and nearly universal means of pH measurement. Modern pH glasses generally consist of mixed compositions of Li_2O, BaO, La_2O_3, and SiO_2, whose proportions are varied to gain different characteristics (see the Nicolskii and Belyustin review [83] for more detail). The pH response of these glasses is generally thought to come from a combination of hydrogen ion-exchange equilibria at the fixed SiO^- sites and ionic interdiffusion potentials. Hydration of a 5–100 nm thick surface layer creates an active silicate gel layer in which the H^+ potential-determining processes occur; the major part of the 50–500 μm thick glass remains dry and impervious to H^+ transport (Figure 2.12). A compromise in

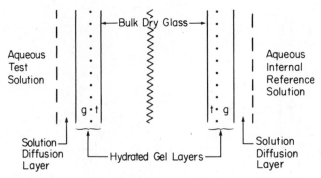

Figure 2.12 Cross-sectional schematic of pH glass regions (not to scale). Hydrated gel layers (5–100 nm) are thought to consist of a low-resistance gel region g and a higher-resistance transition region t. Virtually no H^+ conduction occurs across the 50–500 μm thick bulk glass region.

glass hygroscopicity is struck for properly functioning pH glass; neither un-hydratable glasses like Pyrex nor very soft soda glasses perform satisfactorily. Yet even Nernstian pH glass membranes gradually dissolve as the wetted gel layer moves inward at a steady state on both interfaces, and lifetimes of 1–2 years are often seen as a limit. In alkaline media, where the rate of silicate dissolution at the gel/solution boundary is enhanced, useful lifetimes may be less.

The exact electrode mechanism is not yet completely defined after more than 40 years of study [83, 84]. However, much of the equilibrium response data can be fitted to an equation of the form

$$E_g = E' + \frac{nRT}{F} \ln[(a_{H^+})^{1/n} + (K'a_j)^{1/n}] \tag{126}$$

where n is an empirical constant dependent on glass composition, and E' an internal potential containing the internal reference electrode response as well as a glass asymmetry potential. The K' term is usually described by the product of K_{ex} for the M^+-H^+ heterogeneous ion-exchange reaction on the silicate sites and a gel-layer mobility ratio $(\bar{u}_m/\bar{u}_{H^+})$ raised to the nth power. For modern commercial glasses in the pH range 1–12, $n = 1$ and $(K'a_j) \ll a_{H^+}$ so that ideal response is noted. Above this range the $(K'a_j)$ or selectivity coefficient for alkali metals is a factor, introducing an "alkaline" error in addition to the possibility of rapid silicate dissolution. Manufacturers often provide a nomogram for correction of this error; Bates [24] lists some values for common commercial electrodes. In the 0–1 pH range electrodes can also exhibit a negative "acid" error, thought to be due to Donnan exclusion breakdown by the strong acid anions.

Several other properties of glass pH electrodes are of importance in describing their function. No time dependence is provided in (126), and under ideal be-havior the equilibrium potential will be reached in 10–100 ms. Yet sluggish response of up to several minutes is noted for old electrodes, probably because of slow diffusion through a relatively thick leached gel layer. Even for new electrodes transients may be observed in the alkaline or acid regions depending upon solution ionic strength, buffer capacity, and alkali metal content. Glass resistivities are quite high, necessitating both thin electrode bulbs and high impedance measuring devices able to handle the $10-1000\ M\Omega$ source electrodes. Electrode resistance also increases in a nearly log-linear fashion with decreasing temperature, so that a low resistance glass might be chosen for low temperature studies. The asymmetry potential included in E' in (126) is defined by a nonzero cell potential when inner and outer electrode solutions are identical. It can vary between electrodes or drift with time for a given electrode, and arises because of glass structure differences after forming or upon solution chemical attack. This potential is usually accounted for by standard calibration or by extrapolation procedures.

Traditional internal reference electrodes like the SCE or a Ag/AgCl wire are combined with HCl or dilute buffered KCl internal solutions to set the electrode internal potential. The pH-measuring cell is then completed by an external reference like the SCE or a $4\ M$ KCl–Ag/AgCl system; if the external reference is integral to the body of the pH electrode, a "combination" pH electrode is obtained. The total cell expression for any unknown X is then

$$E_X = E_i + E^{0\prime} + E_j(X) + \frac{RT}{F}\,pH(X) \tag{127}$$

where E_i is defined to include the glass asymmetry potential and $E^{0\prime}$ is the standard potential of the external reference. Note that all terms in (127) can exhibit temperature dependence. The usual procedure is to choose an internal reference solution so that an isopotential point, a pH area where the total standard potential is temperature independent, is near mid-range pH 7. Furthermore, an internal solution for which dE_i/dT exactly opposes $d(E^{0\prime} + E_j)dT$ is ideal. Standardization at one temperature and pH(X) measurement at another temperature then becomes possible after manual or automatic compensation adjustment of the RT/F factor. In one commercial meter the glass electrode resistance itself is used to monitor temperature and so make an "internal" correction.

Nonequilibrium temperature-dependent effects are often observed in external electrode junctions containing solid AgCl, or in any reference element that is not completely immersed. Temperature hysteresis is also well documented for calomel systems above 75°C. In an attempt to address these problems a Pt-redox reference couple has been incorporated in the so-called Ross pH and reference electrodes [25]. Here an immobilized internal bulb solution having virtually no temperature sensitivity is fashioned by varying an I_3^-/I^- redox ratio so that its dE/dT bucks that of the internal pH buffer and asymmetry potential. The external reference is a double-junction arrangement consisting of another $Pt/I_3^-,I^-$ half-cell that is positioned in a concentric outer tube filled with $3\ M$ KCl. The advantages claimed include fast temperature-cycling response independent of hysteresis or precipitate solubility effects, freedom from AgCl clogging of the reference-sample junction, and reduced reference electrode drift since changes in salt-bridge KCl concentration cause only slight E_j variation in the absence of an internal Cl^- potential.

Glass electrodes can be bought or made in an almost limitless variety of shapes and sizes. Some of the more important categories are listed in Table 2.17. Capillary-flow types can be used with small samples and find important usage in commercial blood-gas analyzers. Of the many electrode shapes possible the flat and spear variations are two of the most useful. Many combination designs offer concentric or adjacent reference junctions, which usually are the source of any measurement artifacts. Process control electrodes represent efforts to design rugged systems that are free of maintenance. Extremes of pressure, temperature, and sample inhomogeneity present difficult problems for both glass indicator and reference electrodes. Several designs offer complete electrode body submersibility, self-cleaning by vibration or abrasion, or differential measurement to eliminate junction fouling. A whole technology has built up around glass microelectrodes for intracellular pH and other measurements; and an offshoot has been the commercial availability of 1–2 mm electrodes for laboratory analysis with small samples. Finally, efforts to eliminate the internal filling solution have continued for many years. A host of electronic, ionic, or mixed conductive materials has been employed but none have developed past the research or patent stage.

Many commercial glass electrodes are prehydrated and shipped in a package containing a wet sponge or gel. These electrodes should be ready to use, but dry electrodes must be hydrated in the laboratory. The hydration procedure varies with the glass type [88], but a minimum overnight exposure to distilled water, dilute HCl, or mildly acidic buffer solutions is recommended. Storage of such "conditioned" electrodes should be in aqueous solutions, again in distilled water or mildly acidic buffer. Satisfactory performance after a conditioned electrode dries out may sometimes be attained after long aqueous exposure, but hysteresis effects are more common. Equilibrium response to buffer standards should be obtained in less than 15 s for macroelectrodes; longer response time indicates either surface fouling or a thick hydrated layer of insensitive silicate. Regeneration by cycling between 0.1 M HCl and NaOH, or removal of films by solvent or pepsin-HCl (for protein) treatment is suggested. As a last resort, brief exposure to the etching action of NH_4HF_2 followed by a rinse in 5 M HCl can be implemented. Long exposure to alkaline solutions or nonaqueous solvents (see discussion in a following section) is to be avoided. Other practical usage suggestions are covered in several monographs [24, 89].

The weak point of the pH measuring cell is normally the reference electrode/ junction component. Anomalous stirring effects, potential drift, or non-Nernstian behavior can often be traced there. Illingworth [90] has studied the performance of the common porous-ceramic-plug junction in combination electrodes and found notable variation in pH prediction traceable almost solely to the poor junction performance, as Table 2.18 indicates. Such effects, which may not be evidenced by standardization with buffers of equal ionic strength, can be common in reference electrodes that allow minimal electrolyte

Table 2.17 Several Important Categories of Glass pH Electrode

Type	Applications	Commercial Availability	Comments	Reference
Capillary flow	Flow systems, small samples, blood pH	Yes	Reference electrode downstream, 10–100 μL samples possible, caution about cleaning, response time	[85]
Flat surface or spear-tip macroelectrodes	Gels, foods, films, small samples	Yes	May be pH only or combination with variety of reference designs	Electrode manufacturers' bulletins
Process control	Industrial stream, pH measure, and control	Yes	Rugged, completely enclosed designs; often can be abrasion cleaned	[86]
Microelectrodes (0.1 μm to 2 mm tip diameter)	Biology, small samples	Yes, from 1 mm and above; equipment for preparation of smaller electrodes is also available	Often high resistance and slow response times; can be single or double barrel with reference	[87]
Solid internal contact	Where heat, pressure, or inverted pH measurement required	No	Ag, AgX, or metal oxide conductive contacts; not as reliable as solution contacts	Citations in [8]

Table 2.18 pH Measurements with Combination Electrodes Having a Porous Ceramic Plug[a]

		pH Reading				
Buffer No......	1	4	5	6	9	
Composition...	50 mM Phthalate	500 mM Phosphate	50 mM Phosphate	5 mM Phosphate	10 mM Borate	
Electrode Sample						
"Used" electrodes in normal service ($n=30$)	4.045 ± 0.083	6.711 ± 0.113	6.865	6.860 ± 0.0133	9.049 ± 0.113	
With substitute reference half-cell ($n=18$)	4.013 ± 0.028	6.518 ± 0.014	6.865	7.061 ± 0.013	9.175 ± 0.051	
New electrodes with integral reference half-cell ($n=11$)	4.062 ± 0.063	6.594 ± 0.088	6.865	7.001 ± 0.057	9.079 ± 0.139	
"Best" pH estimate	4.008	6.520	6.865	7.070	9.180	

[a]Substitute half-cell is a free flowing Ag/AgCl junction with 3 M KCl. Best pH estimate derived from H_2 electrode/SCE cell at 25°C. From [90], with permission.

outflow. Thus the gelled or nonrefillable salt-bridge electrolytes are not recommended for the most careful work. Liquid salt-bridge junctions can be checked for resistance or nominal electrolyte flow rate against a new reference of the same type [89]. Anomalous values indicate that cleaning by pressurization or warming is required. The suspension effect, a marked change in measured pH upon addition of charged colloids or gels to aqueous solutions, has classically been ascribed to reference junction anomalies. A study by Brezinski [91] refutes this claim, although use of a properly functioning reference bridge is still deemed necessary.

The common pH glass electrode possesses the disadvantages of high impedance, breakability, low resistance to etching by alkali or HF media, the necessity of preconditioning and aqueous storage, and the inability to miniaturize in a simple way. Various H^+-sensitive electrodes employing materials other than glass have been proposed in recent years. Table 2.19 lists a few of the more common chemistries; several have been commercialized. The antimony and other metal oxides are reversible to H^+ by redox reactions involving OH^- or H^+. As such these electrodes are often sensitive to solution pO_2 and strong redox couples. Complexation with buffer anions is possible, particularly for the well-studied Sb electrode. However, electrode durability, lower impedance, and ease of miniaturization are attractive. Both the ZrO_2 and the Pd hydride systems have been suggested for high temperature ($>100°C$) measurements on geothermal brines or pressurized nuclear coolants. The former seems to be free of redox or pO_2 sensitivity but possesses high impedance at lower temperatures. Attempts to put neutral amines or charged liquid ion exchangers within plastic "ISE" membranes have met with some research success for pH measurements within about three units of neutrality. Possible advantages include ruggedness, small size, and freedom from the redox sensitivity shown by the metal or carbon-based electrodes. The H^+ ion-selective field effect transistor (ISFET) has a bare gate material, which after hydration exhibits a pH response. The mechanism is thought to be formation of SiO^- or $Si_3N_4O^-$ fixed sites, analogous to a thin glass gel-layer case [99]. (The ISFET device characteristics are described in the ISE section.) In general, electrodes with electron-transfer couples (like the M_xO_y type) or with surface sites (like the C fiber) exhibit low impedance and require no internal reference element. The "membrane" electrodes show higher impedances, often have to be preconditioned, and must be internally referenced, although solid internal contacts have been tried in place of solutions in almost all cases. No completely satisfactory replacement for the pH glass electrode has been found, but these electrodes do provide attractive alternatives in certain analytical situations.

The modern pH/mV meter has evolved to the point that the voltage measurement is of little concern in the pH cell, as was outlined in Section 2. Tests for proper functioning of the manual meters have been described [37, 89]. Most of the microprocessor-based meters have built-in checks for meter and even electrode response. The low cost of solid-state devices has allowed home-built electrometer systems to be easily constructed for special applications [101].

Table 2.19 Some Nonglass pH-Sensitive Electrodes

Electrode Type	Active Material	Impedance	Internal Reference Required	Commercial Availability	Comments	Reference
Antimony	Sb_2O_3	Low	No	Yes	Subject to redox agents and complexers	[92]
Metal oxide	M_xO_y	Low to high	No	No	M = Bi, Pd, Sn, Ir, Te, etc.	[93]
Ceramic oxide	Y-doped ZrO_2	High	Yes	No	High temperature pH sensor	[94]
Metal hydride	$Pd/Pd(H_2)$	Low	No	Yes	In essence a stable H_2 electrode	[95]
Ion-exchange membrane	Cation exchange resin	Low	Yes	Yes	For HF solution or other strong acids, pH ⩽ 4	[96]

Carbon fiber	COOH or quinone sites on carbon	Low	No	No	Shows redox sensitivity and initial pO_2 function	[97]
PVC membrane	Liquid exchanger	Low	Yes	No	$3 < pH < 10$; ISE-type fabrication	[98]
	Neutral carrier	High	Yes	No		
ISFET	SiO_2 or SiO_2/Si_3N_4	Low	No	No	Impedance matching by FET channel; can be ultramicro-device	[99]
Polymer	H^+ exchanger in polycarbonate	High	Yes	No	$4 < pH < 10$ for best results; carrier is phosphorylation uncoupler	[100]

4.3   Acid—Base Relationships in Aqueous Solution

4.3.1   The Acid Dissociation Constant

In aqueous solutions of acids or bases the Bronsted–Lowrey equilibrium concept is usually applied according to

$$H_xA^{z+1}+H_2O \overset{K_a}{\rightleftharpoons} H_{x-1}A^z+H_3O^+ \tag{128}$$

$$B^z+H_2O \overset{K_b}{\rightleftharpoons} BH^{z+1}+OH^- \tag{129}$$

where proton exchange occurs between the acid solutes HA or the basic solutes B and the amphiprotic solvent H_2O. The thermodynamic equilibrium constants K_a for acid and K_b for base can then be defined by

$$K_a = \frac{a_{H^+}a_A}{a_{HA}a_{H_2O}} \quad \text{and} \quad K_b = \frac{a_{OH^-}a_{BH}}{a_B a_{H_2O}} \tag{130}$$

where solute charges are implicit. In addition, the solvent water is understood to be governed by its own thermodynamic autoprotolysis or ionization constant K_w described by

$$K_w = \frac{a_{H^+}a_{OH^-}}{a_{H_2O}} \tag{131}$$

for the equilibrium reaction

$$2H_2O \overset{K_w}{\rightleftharpoons} H_3O^+ + OH^- \tag{132}$$

In dilute aqueous solutions a_{H_2O} is usually considered constant and so is often not expressed in (130) and (131), but it must be accounted for in concentrated salt or nonelectrolyte solutions. Modern practice now describes solute dissociation almost exclusively in terms of K_a, but a base dissociation K_b may be obtained by using the familiar K_w/K_a relation.

The K_a, K_b constants define acid or base strength of the solute with respect to that of the solvent H_2O. For dilute solute concentrations the range of acid/base strength is defined by the strongest acid (H_3O^+) or base (OH^-) that can exist in the solvent, which for $pK_w \approx 14$ becomes a pa_H of 0–14 at 25°C. Acids with $pK_a < 3$ or bases with $pK_a > 10$ ($pK_b < 4$) are usually considered as "strong" or completely dissociated according to (128)–(129). However, this designation is somewhat arbitrary since solute concentration may affect the extent of dissociation. Solutes may have multiple dissociable H^+ sites, which are designated by K_1, K_2, K_n, whose ordering is by acid strength with K_1 describing the strongest acid site. Values of K_a are temperature dependent, and this dependence varies with solute charge variations in (130). Some representative data for $pK_a = -\log K_a$ are given in Table 2.20. Efforts to obtain a general first-principal model for K_a dependence on temperature by combining electrostatic and chemical effects have been moderately successful [103], but empirical fitting to specific solute charge types is more common.

It is to be appreciated that pK_a values can be expressed on molal, molar, or mole fraction scales. The first two are most common and are virtually equivalent

Table 2.20 Temperature Dependence of pK_a for Some Common Species[a]

Species	Charge of Acid Form	pK_a at								ΔpK_a (40–10°C)
		0°C	10°C	20°C	25°C	30°C	40°C	50°C	60°C	
Acetic acid	0	4.781	4.762	4.756	4.756	4.757	4.769	4.787	4.812	+0.007
Ammonia	+1	10.081	9.730	9.401	9.246	9.093	8.805	8.540	8.288	−0.925
Boric acid	0	9.508	9.379	9.278	9.234	9.195	9.128	9.077	9.031	−0.251
Glycine	$+1 \ pK_1$	2.443	2.398	2.364	2.351	2.340	2.325	2.319	2.324	−0.073
Glycine	$+ -pK_2$	10.497	10.193	9.910	9.780	9.652	9.412	9.189	8.983	−0.781
Water	0	14.944	14.535	14.167	13.996	13.833	13.535	13.262	13.107	−1.00

[a]pK_a on molal scale. Data condensed from [102], with permission.

for dilute solutions; literature compilations generally are presented in molal units. In addition to the thermodynamic expression for K_a several other forms of approximate constants are found:

$$K^* = \frac{(m_{H^+})(m_A)}{(m_{HA})} = K_a \frac{\gamma_{HA}}{\gamma_H \gamma_A} \qquad (133)$$

$$K_a' = a_H \frac{(m_A)}{(m_{HA})} = K_a \frac{\gamma_{HA}}{\gamma_A} \qquad (134)$$

The K_a' is often employed on the molar scale in the Henderson–Hasselbach form

$$pK_a' = pH - \log \frac{C_A}{C_{HA}} \qquad (135)$$

where it is assumed that the operational $pH = -\log a_H$, and that the activity-coefficient term is implicit in the approximate constant.

Exact potentiometric K_a determinations can be made in the Harned cell arrangement

$$Pt|H_2(gas)|m_{H^+}, m_A, m_{HA}, m_{Cl}|AgCl|Ag \qquad (XXVI)$$

by extrapolation of the term $\log(\gamma_{Cl}\gamma_{HA}/\gamma_A)$ to zero ionic strength. Similar procedures with the pH glass electrode as indicator in the pH range 2–12 have been employed. Although the evaluation can be repeated at a number of different temperatures to obtain a separate empirical relation for $K_a = f(T)$, Covington and co-workers [104] have described a multiple linear regression procedure for extracting $K_a = f(I,T)$ in one step. With care pK accuracies of 0.001 are obtainable by the Harned cell method. Operational cells with a liquid junction reference may also be used to estimate pK_a. Most commonly this is done within the confines of a potentiometric titration. Here the operational pH is measured as, for instance, a strong base titrant is added to a weak acid solute. Then application of an equation like (135) allows multiple estimates of pK_a' to be made. These estimates are then corrected for γ terms using methods outlined in Table 2.4; and the titrations are often performed in constant-ionic-strength media to limit $\Delta\gamma$ and ΔE_j. The values of pK_a derived from the operational titration method can be accurate to within about 0.02 pK_a unit. May and co-workers [105] have outlined a computer program for the simultaneous derivation of pK_a's and cell parameters from such titration procedures. Other nonpotentiometric methods for estimation of pK_a values are common; a discussion is given in [102].

Individual compilations of aqueous pK_a values for organic acids [106], organic bases [107], and inorganic species [108] have been collected under the auspices of IUPAC. A listing of pK_a's for 400 compounds of pharmaceutical interest is also available [109]. Smaller compilations can be found in chemical or biochemical handbooks, and in the stability-constant listings where the pK_a's of the various organic and inorganic ligands are given with H^+ as the "metal."

There have been many attempts to relate solute structure and K_a value, often for the purpose of predicting solute acid strength without resorting to many laboratory measurements. The best success has been obtained with groups of structurally similar compounds, such as HA_1 and HA_2 among others, where HA_1 is considered the "reference acid." Since the K_a is a measure of the difference in standard free energy of solvation of acid and conjugate base forms according to

$$\Delta G^0_{HA/A} = G^0_{A,solv} - G^0_{HA,solv} = -RT \ln K_a \tag{136}$$

differences in K_{a_2} from the reference acid K_{a_1} reflect solvation differences in both acid and base forms for the test acid. The change in K_a values is then described by

$$\Delta\Delta G^0 = \Delta G^0_{HA_2/A_2} - \Delta G^0_{HA_1/A_1} = -RT \ln \frac{K_{a_2}}{K_{a_1}} \tag{137}$$

The most consistent linear–free energy relationship of this type is the Hammett equation

$$\Delta pK^R_x = \log \frac{K^R_x}{K^R_H} = \rho_R \sigma_x \tag{138}$$

for compounds in the series XC_6H_4R. The σ_x is a substituent effect in the benzoic acid series (R = COOH) relating changes in pK_a of the m- or p- substituted compound to that of the free acid according to

$$\sigma_x = \Delta pK^{COOH}_x = \log \frac{K^{COOH}_x}{K^{COOH}_H} \tag{139}$$

The ρ_R or reaction constant defines the proportionality in (138) for each given R but is independent of substituent x. Values of σ_x are usually independent of the nature of the ionizable group R, so that if ρ_R is known for a solute like a benzoate or a phenylacetic acid, a whole series of pK^R_x can be estimated from (138). Treatment of the same kind has been applied to solutes where the ionizable group R is attached directly to a substituent Z, as in the aliphatic carboxylate or amine series. The Taft relation is then used to quantitate the pK_a differences between solutes according to

$$\Delta pK^R_z = \rho'_R \sigma'_z \tag{140}$$

where σ'_z is termed the polar substituent effect and ρ'_R the polar reaction constant. The Hammett, Taft, and other related formalisms have proved their worth for a large number of cases, but no completely general relationship of pK_a to simultaneous changes in solute shape, ionizable group, and substituent is available. Further detail on this topic is provided in most physical organic textbooks, and Perrin and co-workers have devoted a book to the subject [110].

4.3.2   Buffers

Inspection of the expression given for pK'_a in (135) reveals that additions of strong acid or base x to a solution containing HA and A change the concentration of both components according to $\log[(C_A \pm x)/C_{HA} \mp x]$. Thus the pH of the solution changes by this ratio rather than directly, and the HA/A pair provides a pH buffering action in the solution. Both solvent water and all dissolved HA/A pairs can act as buffers in aqueous solution. A quantitative measure of the effectiveness of any buffer is given by the buffer capacity β where

$$\beta \equiv \frac{dC_{OH}}{dpH} \tag{141}$$

This parameter, also known as the buffer index or buffer value, expresses the differential change in pH observed as molar increments of strong base are added to a solution. For a monobasic weak acid–conjugate base pair at a concentration $C_T = (C_A + C_{HA})$ one can derive from simple equilibrium principles

$$\beta = 2.303 \left[\frac{K_a' C_T C_{H+}}{(K_a' + C_{H+})^2} + (C_{H+}) + \frac{K_w}{C_{H+}} \right] \tag{142}$$

The second two terms in the brackets represent buffering by H^+ and OH^- of the solvent water, and can be neglected in the range $3 < pH < 11$. Then it is found that the maximum value of β occurs at $C_H = K_a'$ (or $pK_a' = pH$ if activity effects are ignored), leading to

$$\beta_{max} = 0.576 C_T \tag{143}$$

Thus the maximum β is independent of the K_a' value to a first approximation. Table 2.21 shows the variation of β as the pH varies around the buffer pK_a'; a crude rule of thumb defines adequate buffering action to occur at a pH within ± 1 unit from the pK_a'. In general the buffer capacity is an additive function of all the buffers present in a solution. It can be calculated from equilibrium principles, or measured as the inverse of the differential slope at any point in a titration curve. Tsuji [112] has recently described a differentiator–integrator circuit that extracts and displays the β values during a normal acid–base potentiometric titration.

Three additional descriptive properties of buffers are common: the $\Delta pH_{1/2}$ or dilution value, the salt effect, and the temperature coefficient dpH/dT. The first two originate principally from the variation of activity coefficients of the buffer components as the total solute concentration changes. The $\Delta pH_{1/2}$ is defined by

$$\Delta pH_{1/2} \equiv (pH)_{Ci/2} - (pH)_{Ci} \tag{144}$$

and can be quantitatively expressed by differentiation with respect to volume of the equation relating the pH equilibrium to the $d \log \gamma/dI$ obtained from the Debye–Hückel formulation. Bates [24] has derived the resultant dpH/dV expressions for several HA/A charge types. Similarly, addition of neutral salt or simple buffer concentration increases affect the pH because of $d \log \gamma/dI$ changes that may be different for each HA/A charge type. The combined influence of this salt effect and the dilution value $\Delta pH_{1/2}$ are shown for several buffers in Table 2.22. The third buffer property dpH/dT stems from both the dpK_a/dT relation described previously and from any $d \log \gamma/dT$ displayed by the solute. Table 2.23 lists the temperature coefficients for many of the common buffer systems.

Table 2.21 Dependence of Buffer Capacity, β, of HA/A Buffer on pH[a]

pH = pK_a Minus	%β_{max}	pH = pK_a Plus	%β_{max}
2.0	3.9	0.1	98.6
1.9	4.9	0.2	94.8
1.8	6.1	0.3	88.9
1.7	7.7	0.4	81.4
1.6	9.6	0.5	73.0
1.5	11.9	0.6	64.1
1.4	14.7	0.7	55.4
1.3	18.2	0.8	47.2
1.2	22.3	0.9	39.7
1.1	27.2	1.0	33.0
1.0	33.0	1.1	27.2
0.9	39.7	1.2	22.3
0.8	47.2	1.3	18.2
0.7	55.4	1.4	14.7
0.6	64.1	1.5	11.9
0.5	73.0	1.6	9.6
0.4	81.4	1.7	7.7
0.3	88.9	1.8	6.1
0.2	94.8	1.9	4.9
0.1	98.6	2.0	3.9
0.0	100.0		

[a]From [111], with permission.

Table 2.22 Comparison of Salt Effects and Dilution Values[a]

Buffer Solution	Initial I	Final I	Salt Effect, Δpa_H	$\Delta pH_{1/2}$ Observed	$\Delta pH_{1/2}$ Calculated
KH phthalate, 0.05 M	0.053	0.106	−0.044	0.057	0.061
KH$_2$PO$_4$, 0.025 M; Na$_2$HPO$_4$, 0.025 M	0.1	0.2	−0.088	0.088	0.096
KH$_2$PO$_4$, 0.01 M; Na$_2$HPO$_4$, 0.01 M	0.04	0.08	−0.074	0.073	0.079
KH$_2$PO$_4$, 0.005 M; Na$_2$HPO$_4$, 0.005 M	0.02	0.04	−0.057	0.059	0.065
Na$_2$B$_4$O$_7$, 0.01 M	0.02	0.04	−0.014	0.004	0.018

[a]From [24], with permission.

Table 2.23   Change of pa_H with Temperature at 25°C for Several Buffers[a]

Solution	$\partial pa_H/\partial T$ pa_H units K^{-1}
Hydrochloric acid, 0.1 M	+0.0003
Potassium tetraoxalate, 0.05 M	+0.001
Potassium hydrogen tartrate, 0.03 M	−0.0014
Potassium hydrogen phthalate, 0.05 M	+0.0012
Acetic acid, 0.1 M; sodium acetate, 0.1 M	+0.0001
Acetic acid, 0.01 M; sodium acetate, 0.01 M	+0.0002
Potassium dihydrogen phosphate, 0.025 M; disodium hydrogen phosphate, 0.025 M	−0.0028
Diethylbarbituric acid, 0.01 M; sodium diethylbarbiturate, 0.01 M	−0.0144
Ammonia, 0.1 M; ammonium chloride, 0.1 M	−0.0303
Borax, 0.01 M	−0.0082
Sodium bicarbonate, 0.025 M; sodium carbonate, 0.025 M	−0.0090
Trisodium phosphate, 0.01 M	−0.026
Sodium hydroxide, 0.1 M	−0.0332

[a]From [24], with permission.

The choice of buffer systems can be dictated by HA/A reactivity with the chemical system studied, by complexation tendencies with metals in the media, or even by stability parameters, in addition to the pK_a and other properties already described. Formulation and other pertinent details are listed for a large number of common buffers in [24] and [111]; shorter compilations are found in chemical handbooks. Specialized buffer systems include those of constant ionic strength, those with volatile components that may be easily removed to alter pH after desired reactions are run, and universal buffers, which consist of multiple HA/A pairs that provide a nearly constant β across a wide pH range. Good and co-workers [113] have identified a number of buffers covering the physiological pH range 6–8.5 that are useful in biochemical studies. Many of these buffers are now commercially available.

4.3.3   Acid–Base Titrations

Potentiometric titrations retain an importance in modern acid–base techniques. For analytical purposes, standardized solutions of acidic or basic titrants are added to unknown solute solutions, and the change in potential or in some solution property at the equivalence point is used for quantitation. Alternatively, the shape of the titration curve can be used to extract information about the solutes such as K_a or metal–ligand complexation values.

The curves obtained upon titration of monoprotic weak acids with a strong base possess the familiar sigmoidal shape, with a loss in discernible pH change at the equivalence point being noted as the acid becomes weaker. Multiprotic or polyprotic species and acid mixtures will show multiple equivalence point breaks if the adjacent pK_a separation is greater than about four units. However, automatic and microprocessor-controlled titrators can mathematically derive equivalence point information even where no discernible "breaks" occur in the titration plot. For example, with microprocessor adjustment of volume to yield constant increments of electrode potential a titroprocessor was able to ascertain the three equivalence points of citric acid ($pK_1 = 3.08$, $pK_2 = 4.74$, $pK_3 = 5.40$) to better than a percent [47]. Increased accuracy would still be achieved, of course, if pK_a separation were greater. For minimal error in monoprotic species titrations the K_a/C_b (for weak bases) or $1/(C_a K_a)$ (for weak acids) terms should be minimized. Reference [103] gives a complete discussion of conditions necessary to reduce titration errors.

For equivalence point estimation the potentiometric glass electrode potential variation is to be preferred wherever rapid, reliable electrode performance is possible. For manual data collection and calculation this was often prohibitive, but with computerized titrators it is routine. Even hand calculators have enough power to calculate reasonably accurate equivalence point potential data; Clarke [114] outlines several potentiometric titration programs in Reverse Polish calculator notation. Although many nonpotentiometric equivalence point estimation techniques are possible, the indicator methods are still most prominent. An indicator should be chosen with a high molar absorptivity and a $pK_{ind} \pm 1$ unit from the intended equivalence point, attention being paid to possible salt effects on γ_{ind} or interactions with solute components like surfactants and polyelectrolytes. Bates's book [24] should be consulted for a thorough treatment of indicator equilibria.

Potentiometric titrations are preferably performed with concentrated titrants and with titrand media having added concentrations of neutral salts. In this way dilution errors and variations in E_j or γ are minimized. In fact, large additions of salt or nonelectrolyte can often be employed to improve equivalence point detection by either reducing a_{H_2O} or raising γ terms for titrand species. Christian [115] has given a review of such procedures. Absorption of atmospheric CO_2 can cause determinate errors in acid–base titrations, and removal by N_2 purging or correction by blank titration or Gran treatment of the data is suggested. Wozniak and Nowogrocki [116] have made a study of the effect of varying purging times on CO_2 removal and have suggested correction factors to be added to the exact equivalence point expressions.

The study of methods of equivalence point estimation is very well developed for acid–base potentiometric titrations; the Ebel and Seuring review [69] outlines the principal variants. Exact Gran functions can be derived from equilibrium and mass-balance considerations. Midgely and Torrance [36] have shown that for a weak acid HA/A pair titrated with a strong base the function is

$$F = \frac{h(V_0 + V)(10^{-pH} - K_w 10^{pH})/\gamma_H + C_b V_b R}{h - R} \tag{145}$$

where V_b is volume of titrant C_b added, h is the number of dissociable protons of the acid, and R is a function of the stepwise pK_a values for the acid given by

$$R = \frac{\sum_{i}^{n} (i)10^{(pK_i - ipH)}\gamma_A/\gamma_i}{1 + \sum_{1}^{n} 10^{(pK_i - ipH)}\gamma_A/\gamma_i} \tag{146}$$

The function given by (145) is evidently nonlinear but can be evaluated by computer. Often approximations of the total function are employed in linear regions before or after the equivalence point according to

$$F = \frac{V_b}{K_a} 10^{-pH} \text{ before} \tag{147}$$

$$F = \frac{(V_0 + V_b)}{C_b} K_w 10^{-pH} \text{ after} \tag{148}$$

Procedures for approximating potentiometric titration curves by multiparametric fitting have been given [39]; in the most general case the electrode response slope, the pK_a's, and the titrand concentration are all derived simultaneously from the data. With the advent of microcomputers it has become feasible to treat titration curves and all other acid–base calculations by a single unified expression in a_{H+}; the traditional student approximation methods ignoring γ_{H+} and segmenting regions of the titration curve are no longer necessary. Scarano and co-workers [117] have outlined a BASIC program for evaluating such an a_{H+} function, and show its successful application to titration curves, to distribution calculations relating species to pH, and to buffer calculations relating pH to solute content.

Several instrumental variations in acid–base titrations have been suggested. The first is the use of linear-response reagents for direct analytical determination of acid–base solutes [118]. Instead of a strong-acid or strong-base titrant, the delivered reagent is a buffer mixture adjusted to possess constant β over a given pH range. Then the pH in the titrand will change linearly with reagent addition by an amount proportional to initial titrand solute concentration. The advantage of the technique is the simplicity engendered by the linear response of the pH glass indicator. A second instrumental variation is the incorporation of potentiometric titrations into flow systems [35]. Here the advantage sought is increased sample throughput. Titrations may be performed by varying the reagent volume flow with respect to a constant sample volume flow rate. In the "triangle-programmed" case the reagent flow rate is first increased past the equivalence point and then decreased back past the equivalence point, the time between equivalence points used to extract solute concentration. An even higher throughput can be achieved by allowing the sample stream to set a constant flow rate and injecting small-volume plugs of concentrated titrants, defining the flow-injection analysis (FIA) technique. Precautions about sample or titrant dispersion, tubing geometry, and pH glass indicator response time have been outlined [119]. Often an indicator-based end point detection is warranted at throughputs too high for optimum glass electrode response.

4.4 pH in Nonaqueous Media

4.4.1 Solvents and the pH Measurement

Definition of a universal pH convention relating potentiometric measurements to hydrogen ion activity in mixed or purely nonaqueous solvents has not yet been attempted. Part of the difficulty lies in the large array of possible solvent types, and the wide variety of chemical properties found within this array. Even classification of solvent type is somewhat arbitrary. Table 2.24 lists some useful solvents grouped by one suggested classification. Amphiprotic solvents possess both acidic and basic properties and are characterized by an autoprotolysis constant K_s defined like that for H_2O in equation (132). Within this class, solvents more acidic than water are termed protogenic, and those more basic assessed as protophilic. The dipolar aprotic solvents have effectively no tendency toward H^+ donation or acidity, but large dipole moments and reasonably high dielectric constants. Here the protophilic solvents are better hydrogen bond acceptors than the protophobic type. The inert solvents are both poor acids or bases and suppressors of solute ionization tendencies in light of their low dielectric constants, and acid–base transfer is thought to occur only between solutes.

Acid–base reactions can be described in the amphiprotic solvents in the same way as for water, that is, by the dissociation equilibria

$$HA + HS \xrightleftharpoons{K_a} H_2S + A \qquad (149)$$

$$B + HS \xrightleftharpoons{K_a} S + HB \qquad (150)$$

here left unsigned for generality. The solvated proton H_2S becomes the measured "acid" species and the S or lyate the basic component. The acid–base strength of solutes then depends on the relative acid–base strength of the solvent. Organic acid solutes too weak to dissociate in water may do so in a more strongly basic SH like ethylenediamine. Conversely, the mineral acids, which are completely dissociated or "leveled" in water, may be differentiated in an acidic solvent like glacial acetic acid. The K_a again becomes characteristic of solute acid strength; some values are given in Table 2.25 for various solvents.

In many of the aprotic solvents and in virtually all the inert solvents the meaning of acidity becomes less clear. The acid dissociation may be complicated by significant ion-pair formation according to

$$HA + SH \xrightleftharpoons{K_1} \underset{\text{ion pair}}{SH_2^+ A^-} \xrightleftharpoons{K_2} SH_2^+ + A^- \qquad (151)$$

where K_1 represents ionization and K_2 ion-pair dissociation so that the product $K_1 K_2$ forms the usual K_a. In addition, if the solvent possesses poor solvating ability for the anionic or cationic species formed by the dissociation, solvation by undissociated acid solute species may occur. Solvation of the anion by the neutral conjugate acid is termed homoconjugation according to

$$A^- + nHA \xrightleftharpoons{K_c} A(HA)_n^- \qquad (152)$$

while solvation by another species HR is termed heteroconjugation and is

Table 2.24 pK_s and Other Properties of Solvents at 25°C[a]

Solvent	pK_s	D[b]	μ[c]	η[d]
Amphiprotic				
Neutral				
Water	14.2	78	1.83	0.890
Ethylene glycol	15.8	38	2.28	16.9
Methanol	16.5	33	1.71	0.544
Ethanol	18.7	24	1.70	1.08
n-Propanol	19.2	20.3	1.68	2.00
Isopropanol	20.6	19.4	1.66	2.08
n-Butanol	21.6	17.5	1.66	2.56
tert-Butanol	(~22)	12	1.66	
Protogenic				
Formic acid	6.2	58	1.82	1.97
Acetic acid	14.5	6		1.23
Protophilic				
Formamide		109	3.73	3.30
Dimethyl sulfoxide	~33	46	3.96	1.96
Hexamethylphosphorous triamide		29.6	5.39	3.25
Ammonia	33	17		
Ethylenediamine	>33	13	1.99	1.54
Dipolar Aprotic				
Protophilic				
Dimethyl sulfoxide	~33	46	3.96	1.96
N,N-Dimethylformamide		37	3.86	0.796
Pyridine		12	2.23	0.83
Tetrahydrofuran		6.5	1.71	0.45
Dioxane		2.2	0.45	1.20
Protophobic				
Propylene carbonate		64	4.98	2.53
Sulfolane (30°C)		44	4.81	10.3
Acetonitrile	>33	36	3.92	0.345
Nitromethane		36	3.56	0.610
			(30°C)	
Acetone		21	2.88	0.304
Methyl isobutyl ketone		12	2.79	0.55
Inert				
1,2-Dichloroethane		10	1.4	0.73
Chloroform		5	1.15	0.59
p-Dichlorobenzene (53°C)		2.5	0	0.839
n-Octane		2	0	0.515
Benzene		2	0	0.60
Carbon tetrachloride		2	0	0.88

[a]From [120], with permission.
[b]D = Dielectric constant.
[c]μ = Dipole moment in Debye units.
[d]η = Viscosity, in centipoise.

Table 2.25 pK_a for Some Acids in Several Amphiprotic or Diprotic Apolar Solvents[a]

Acid	Water	EtOH	MeOH	DMSO	CH_3CN	DMF
HCl	−3.7	1.9	1.1	2.0	10.4	3.2
HNO_3	−1.8	4.0	3.2	1.4	10.5	—
H_2SO_4 pK_1	—	—	—	—	7.8	3.0
pK_2	2.0	—	—	14.5	25.9	17.2
Succinic pK_1	4.2	—	9.1	9.5	17.6	10.0
pK_2	5.6	—	11.5	16.7	29.0	17.2
Acetic	4.75	10.4	9.7	12.6	22.3	13.5
Benzoic	4.2	10.0	9.3	11.0	20.7	12.3
Salicylic	3.0	8.5	7.8	6.7	16.7	8.2
Picric	0.3	4.1	3.8	−1.0	11.0	—
Phenol	9.9	—	14.3	16.4	27.2	1.6

[a]DMSO = Dimethyl sulfoxide, DMF = N,N-dimethylformamide. Data from [120], with permission.

governed by a similar equilibrium. These latter equilibria are particularly troublesome in nonaqueous titrations because end points are smeared out by the competition between the basic titrant and the conjugates for the HA. In inert solvents these conjugations can proceed to the oligomer or even polymeric stages, and the existence of dissociated species of any kind is questionable. A summary of acid–base properties peculiar to each type of solvent class can be found in [120]; [121] outlines the application of the various acid–base theories (Bronsted, Lewis, and so on) in both aqueous and nonaqueous media.

There are three ways to proceed with the interpretation of pH measurements in nonaqueous solvents. In the following descriptions of the methods, it is assumed that stable reversible indicator and reference-electrode behavior are observed. The first method proceeds by deriving an independent pH* scale for each independent solvent system. Suitable pH*(S) values are ascribed to standard buffers dissolved in the solvent by using the usual Harned cell approach outlined in (121)–(123). Then pH*(X) measurements obtained in operational cells like (XXIII) define acidity in the solvent scale established by pH*(S) if the usual assumption about constancy of E_j is made. Buffer standards have been derived for quite a few solvent systems; several are presented in Table 2.26. While this method can yield operationally consistent acidity information, the relation of pH* to either the aqueous pH scale or to a pH* scale in any other solvent is undefined.

Another common procedure is simply to employ the NBS-based aqueous pH(S) standards as cell calibrators for the pH*(X) measurements. The meaning of the resulting cell potential is totally undefined, especially in light of the possible large E_j variance between aqueous and solvent media. At best the

Table 2.26 pH*(S) Values in Several Nonaqueous Media at 25°C[a]

Solvent	Buffer Composition (Molal Concentration)	pH*(S)
50% MeOH	0.05 Acetic acid, 0.05 Na acetate, 0.05 NaCl	5.493
	0.05 NaH succinate, 0.05 NaCl	5.666
	0.02 KH_2PO_4, 0.02 Na_2HPO_4, 0.02 NaCl	7.884
MeOH	0.01 Salicylic acid, 0.01 Na salicylate	7.48
	0.01 Succinic acid, 0.01 LiH succinate	8.67
50% EtOH	0.01 Citric acid, 0.02 KH_2 citrate	4.07
	0.0053 Na tetraborate	10.60
DMSO	0.005 2,6-Dihydroxybenzoic acid, 0.005 Et_4N salt	3.10
	0.001 Acetic acid, 0.001 Cs acetate	11.56
DMF	0.0088 2,6-Dinitrophenol, 0.0082 Et_4N salt	4.53
	0.0002 m-Nitrophenol, 0.0002 K salt	15.41
Acetone	0.0001 o-Nitroaniline, 0.001 ClO_4 salt	4.82
	0.001 Picric acid, 0.001 Et_4N salt	9.20
Acetonitrile	0.005 Picric acid, 0.005 Bu_4N picrate	10.89
Pyridine	0.01 $HClO_4$	2.66

[a]Additional data are listed in [24] and [122].

measurements are reproducible; they may also give relative information about "acidity" for varying unknown solutes at constant solvent composition. Often the cell millivolt reading is employed rather than the pH output, since the "pH" can easily exceed the common 0–14 range. Figure 2.13 shows typical "pH" and voltage ranges observed in various solvents when the pH cell–meter combination has an aqueous isopotential point at pH 7 = 0 V.

The relation of pH measurements in different media could be standardized if a defined single reference state or scale were agreed upon. If this were to be the aqueous scale, the pH could be related to the pH* as follows

$$pH \simeq pa_{H^+} = -\log m_H({}_w\gamma_H) = pa_H^* - \log {}_s\gamma_H \simeq pH^* - \log {}_s\gamma_H \qquad (153)$$

where the ${}_s\gamma_H$ is the proton-transfer activity coefficient defined in (111). Although values for ${}_s\gamma_H$ like those in Table 2.13 can be derived from the Ph_4As/BPh_4 reference salt assumption, this convention has not been universally agreed upon. It is to be appreciated that the operational pH* approaches pa_H^* in solvents containing dilute solutes, just as was defined for water. In actual practical usage the exact relation between aqueous and nonaqueous operational scales must include the E_j component

$$pH_w(X) - pH^*(X) = -\log {}_s\gamma_H + \frac{\Delta E_j F}{RT} \ln 10 \qquad (154)$$

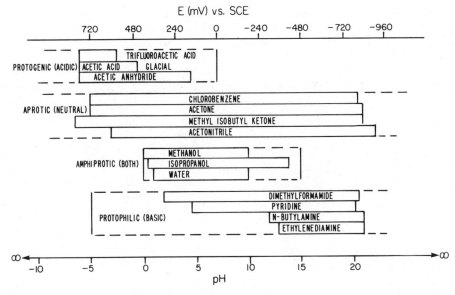

Figure 2.13   pH and voltage ranges observed for various solvents. From [89], with permission.

If derived values of $_s\gamma_H$ are employed, then estimations of E_j in various media are possible. Table 2.27 lists values estimated by this procedure for the E_j between buffer solutions and saturated KCl for water–ethanol mixtures. An effective change in ΔpH of better than a full unit is calculated from the E_j variation between 0 and 100% ethanol.

4.4.2   Experimental Techniques

The glass electrode is the most commonly employed indicator electrode in nonaqueous solvents, although the H_2 electrode is still used for E^0* measurements. In addition, the metal-oxide and carbon electrodes described in Table 2.19 are occasionally employed in titration work. Two references [24, 89] outline practical details for glass electrode handling in various solvents. Obviously, there are differences in glass hydration possible between solvent and water-based media. Drifting potentials may occur as solvent dehydrates the aqueous gel layer; periodic changes in aqueous media are recommended. Etching in bifluoride is sometimes employed to reduce the thickness of the gel layer and thus the hydration–dehydration time in the solvent. Even with a stabilized hydration state the electrode response may be slow because of low effective dielectric constant in the solvent-saturated gel layer. A lower resistance, narrow-range pH electrode is therefore often chosen. To reduce the magnitude of the large asymmetry potential brought about by internal aqueous and external solvent solutions a nonaqueous internal is sometimes desired. Glass electrodes having various solvent-based internal solutions are commercially available. Finally, alkaline attack by cations like Na, K, and Li can be enhanced

Table 2.27 Average ΔE_j Between Buffer Standards and Saturated Aqueous KCl in Water–Ethanol Solvent at 25°C[a]

X_{EtOH}	ΔE_j (mV)
0	0
0.07	5
0.163	-2
0.298	-27
0.519	-37
0.696	-33
1.0	-73

[a]$X =$ Mole fraction, ΔE_j calculated using reference salt assumption for $_s\gamma_H$. From [79], with permission.

in nonaqueous media. For titrations with strong nonaqueous bases, quaternary ammonium base cations are recommended instead.

Reference electrodes for use in nonaqueous solvents were discussed in the redox section. Use of a methanolic KCl solution instead of a saturated aqueous KCl bridge (for SCE and Ag/AgCl) is common; an alternative is to employ a double-junction arrangement with an R_4NClO_4 or other salt dissolved in the sample solvent as the outside solution making contact with the sample. Complete removal of water from the reference electrode filling solution(s) is particularly desirable when using 100% solvent; even traces of water in dipolar aprotic media can alter measured potentials markedly, since H_2O can act as a facile heteroconjugating species. Other examples of nonaqueous reference electrodes are found in [22].

Sample conductivity can be very low in some of the aprotic or neutral solvents if only acid–base solutes are employed. To avoid this source of potential noise, additions of inert soluble electrolyte salts like the tetraalkylammonium halides or perchlorates are common. Solvent purification to remove acidic or basic impurities or traces of water is often essential in titration studies. Purification recipes have been suggested for each type of solvent [120, 123].

4.4.3 Nonaqueous Titrations

Analytical determinations by acid–base titration in nonaqueous solvents have lost favor in recent years with the increasing application of liquid chromatographic techniques. Since the same solvents can often be employed, the traditional solubility advantage enjoyed by titrimetric methods is no longer unique. Interest in solute–solvent interactions in nonaqueous media is still very active, however, and titrimetry can often provide a simple approach to phenomena like ion pairing and homoconjugation. One continuing problem in researching an analytical method is the large array of solvent–titrant combinations that are possible. Some order has been brought to the task by the classification scheme according to solvent type (Table 2.24), but specific solvent systems within each class present particular advantages and disadvantages that often make comparisons difficult. The practical text by Fritz [124] still provides one of the best initial overviews of the field. Many of the individual solvents are treated exhaustively in separate chapters in [120] and [125]. For the most current studies the recurring reviews by Kratochvil should be consulted [10].

For simple quantification of a single acid–base solute, or additive determination of total acid or base, a leveling solvent may be chosen. Thus acetic acid serves to increase basicity of organic solutes too weakly basic to be titrated in water, and by using a strong-acid titrant the large end point ΔE characteristic of strong-acid–strong-base titrations is observed. However, for differentiating solute mixtures or separating out multiple dissociations of a single solute an aprotic solvent with a large potential range (small K_s) is more suitable. For instance, in acetic acid a mixture of tributylamine and N-ethylaniline yields only one end point upon titration with perchloric acid. In acetonitrile, a weakly basic solvent with a large potential range, the two separate breaks can be observed [124].

In general, $HClO_4$ dissolved in acetic acid or dioxane is a first choice for titration of bases. Solvent media for these titrations include acetic acid (often with acetic anhydride to remove traces of H_2O), nitromethane, or acetonitrile, CH_3COOH being the most common because its solvent behavior has been very well characterized [120]. For quantification of acids a basic titrant system like tetraethylammonium hydroxide in 2-propanol or benzene-methanol is usually chosen. Titration solvents vary from the amphiprotic 2-propanol to the dipolar aprotic DMF, acetone, and acetonitrile. The glass electrode functions well in a surprising number of these solvent systems, so potentiometric detection methods are possible. In the inert solvents, however, the electrodes may not be well behaved, and indicator end point methods must be employed. A wide variety of indicators is available [120]; crystal violet (for acids), thymol blue (for bases), or the nitrophenol–nitroaniline group are common. Simple working recipes for both acidimetric and basic titrations are given in [124].

Potentiometric curve shapes may be quite complicated if association and conjugation effects are in force. For many of the amphiprotic solvents, however, the familiar sigmoid-shaped curves are obtained. Some effort has been devoted to relating the half-neutralization potential (HNP) of solutes in nonaqueous media to the aqueous pK_a for the same solute [124]. Success in these efforts could allow end point separation prediction in solvent media to be based on the large amount of aqueous pK_a data [106, 107]. However, in aprotic media the homoconjugation effects can cause the potential to change widely in the HNP region in contrast to the normal "buffer" plateau, as Figure 2.14 illustrates. Of

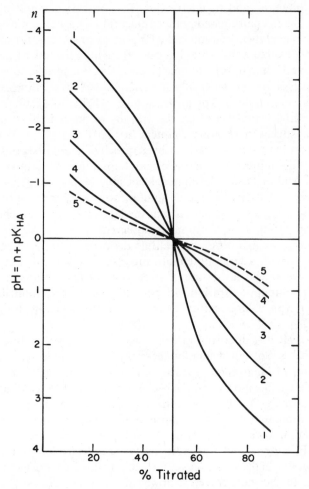

Figure 2.14 Calculated potentiometric titration curve of HA with Et_4NOH for various values of the homoconjugation term $C_{HA}K_C = k$. (1) $k = 1000$, (2) $k = 100$, (3) $k = 10$, (4) $k = 1$, (5) K_C and $k = 0$, the simple buffer case. From [120], with permission.

course, if all important equilibria can be determined, the shape of the titration curve can be calculated exactly by using the master equation in a_{H^+} as outlined in the preceding discussion of aqueous titrations. Velinov and Budevsky [126] have used such an approach to obtain the solvent K_s and the solute K_a by computerized regression of a weak-solute–strong-titrant system in which only solute and titrant initial concentrations are known. Kolthoff and Chantooni [120] have described the a_{H^+} master equations for many of the commonly observed nonaqueous acid–base systems, from which curve shapes can be simulated by computer.

Mixed amphiprotic solvent systems are popular because of handling ease and facile glass electrode response. Often addition of acetone, MeOH, or EtOH to a weak protolyte in water will sharpen the end point enough so that the titration can be performed. Alternatively, a two-phase titration can be employed. In this technique a water-immiscible solvent like chloroform is added to the aqueous system so that neutral protolyte species are extracted from the aqueous phase. In this way the separation of charged protolytes from neutral solutes of similar pK_a can be effected. Cantwell and colleagues [127] have outlined theoretical principles and working conditions for several titrations of this kind.

4.5   Applications in Specialized Media

In several specialized media a relatively constant solution matrix has led to definition or investigation of special pH measurement scales. For instance, in D_2O a pD scale analogous to the pH scale has been set by the Harned cell (XXIV) procedure with the substitution of a deuterium gas electrode for the H_2 electrode. Two buffer standards have been defined [128]: the equimolal 0.025 m KD_2PO_4/Na_2DPO_4 buffer with pD(S) = 7.428 at 25°C and the equimolal 0.025 m $NaDCO_3/Na_2CO_3$ buffer with pD(S) = 10.736 at the same temperature (see [128] for data at other temperatures). The same protonated SRM salts listed in Table 2.14 are employed; the small error in isotopic exchange is included in the defined pD(S) uncertainty limit of 0.01 unit. Relation of pH and pD measurements is not strictly possible, since the arbitrary E^0 for both D_2 and H_2 electrodes is assigned to be zero volts. Comparison of the pD(S) with the pH(S) of the phosphate and carbonate buffers shows the former to be about 0.6 unit higher. Bates [24] has shown that the pH glass electrode responds accurately in D_2O, and that as an empirical procedure use of aqueous pH standards and the glass/aqueous SCE pair in D_2O will yield an operational pH about 0.45 unit less than the true pD.

Seawater can be classified on average as a nearly constant ionic strength medium ($I_m \sim 0.7$), and junction potential variations are small as shown in Table 2.2. The use of the NBS-based pH(S) standards (at $I \sim 0.1$) to standardize the operational glass/SCE cell for this medium has been questioned [21], since both the large $\Delta E_j(X-S)$ and the calculation of γ_{Cl} according to the pH convention at this high ionic strength prevent a meaningful "a_H" from being obtained. Two other scales have been proposed. The first is a total hydrogen ion concentration scale defined by

$$pH_t(SWS) = -\log(C_{H^+} + C_{HSO_4^-}) \qquad (155)$$

where $C_{HSO_4^-}$ is the amount of complex formed at the sulfate level of seawater (normally 30 mM). An equimolar tris buffer in a synthetic solution of average seawater salt composition is titrated with HCl to provide the "standard" for the scale. The advantage is that a concentration is obtainable from the usual operational pH cell, and the fact that $\Delta E_j(X-S)$ is minimized. The HSO_4^- equilibrium is obviously dependent on concentration, temperature, and pressure and so complicates the interpretation of the measured results. A second procedure is to treat seawater as a medium (like D_2O or any defined nonaqueous solution) in which E^0 and buffer standards can be assigned using the Harned cell procedure, a difference being that a pm_H scale replaces the usual pH scale in light of the constant γ_{H^+}. An equimolar tris buffer in a synthetic seawater again is used as a standard. All three measurement methods have their proponents [21], and the pitfalls of assigning a conventional pH scale in even a well-defined medium are highlighted by the seawater case. Nevertheless, as Bates has recently summarized [129], measurements on the three scales are suitable for reproducible pK and acidity determinations, and the results can be interconverted with simple constants. Culberson [21] has discussed the instrumental aspects of the seawater pH measurement, highlighting the design and response of pressurized glass/reference cells.

The very narrow pH range of human blood (about 0.06 unit at 37°C) dictates that extreme care be taken with the measurement in this medium. The modern generation of automatic blood-gas analyzers, which use the pH and pCO_2 measurements to calculate a variety of acid–base parameters, has reduced the imprecision in pH to 0.005 unit or better [130]. In general, a rigorously thermostated pH glass-capillary electrode is combined with a Ag/AgCl/KCl reference electrode of varying junction design, and sample sizes of $100\,\mu L$ or less are common. Calibration is performed with two pH(S) solutions from the NBS primary standard scale: the 0.025 m equimolal KH_2PO_4/Na_2HPO_4 buffer (pH = 6.838 at 37°C) and the 1/3.5 phosphate mixture (pH = 7.382 at 37°C). Both buffers possess an I of about 0.1 m, whereas the normal blood sample has $I \simeq 0.16\,m$. This difference yields a displacement of the measured pH from the "true" pa_{H^+} of about 0.01–0.02 unit, due chiefly to the $\Delta E_j(X-S)$ since the Bates–Guggenheim convention for γ_{Cl^-} is not seriously compromised at this I. Thus the clinical measurement is not totally accurate, but is a conventional one where high reproducibility is stressed. Bates and co-workers [131] have studied both tris-based buffers and additions of NaCl to the phosphate buffers as ionic-strength-matching alternatives to the current calibrators. The former can cause junction-potential anomalies at linen-fiber and dialysis-membrane junctions, and has been given only secondary standard status at the NBS [128]. The NaCl-phosphate buffers do provide a pH measurement more consistent with the aim of obtaining pa_H in blood, but there does not seem to be a movement to switch by the clinical community [132]. There are often shifts observed in relative predicted pH by instruments of different manufacturers, however. These shifts (0.01–0.02 pH unit) are thought to result from the different range of KCl concentration used in the salt bridges. Since saturated KCl may produce protein precipitation at the junction interface, leading to apparent suspension effects in the measurement, a range of KCl concentration from 0.6 to 3.5 M is now employed. In light of equitransference principles at least a 2 M solution is preferred. Large errors in the blood pH measurement can nearly always be traced to the reference electrode, since Covington and associates [133] have shown that the pH glass performs to near-theoretical expectations (± 0.003 unit) when tested in junctionless cells in physiological solutions. Gradual coating of the glass by protein can reduce electrode response time, but suitable cleaning protocol can prevent serious errors. Complete review of this topic is afforded by [132] and [134].

5   ION-SELECTIVE ELECTRODES

5.1   Terminology and Classification Scheme

The functional response of ion-selective electrodes was shown in Section 2 to arise from potential-generating processes at ion-selective membranes [Figures 2.4 and 2.5, and (42)–(47)]. In specific ideal cases a response equation of the form of (47) can be derived for each membrane type, although operationally the electrode response may not always exactly conform to this behavior. Additional terminology employed to describe ion-selective electrode response and usage is outlined in Table 2.28 and Figure 2.15. Some standardization has been brought to this terminology by the IUPAC commission on analytical nomenclature [135], but the literature is not yet completely consistent in this regard. The list in Table 2.28 is by no means exhaustive, and other, more specific terminology will be introduced as required.

An ordered classification of all the ISE types is a formidable task because electrode distinctions tend to blur across membrane, format, and response characteristics. The suggested IUPAC grouping is followed in Table 2.29 with some modification; examples of each type are given. In the crystalline solid-membrane electrodes an insoluble inorganic (or rarely, organic) compound defines the active material, which can act as its own matrix or be dispersed in an inert binder. These electrodes can be used in any of the ISE formats depicted in Figure 2.5 and are often called "all-solid-state" electrodes if they possess a metallic internal contact. However, the term solid state could justifiably apply

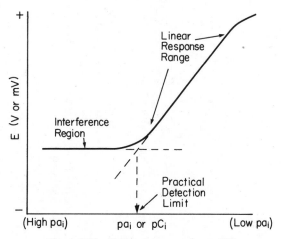

Figure 2.15   Calibration curve for an ISE.

Table 2.28 Some Common ISE Terminology

Descriptor	Definition	Symbol	Comments or Examples
Carrier (ionophore)	The active component in a mobile site membrane	$\bar{S}, \bar{I}, \bar{X}$	Valinomycin, a neutral K^+ carrier; R_4N^+, a charged anion carrier
Drift	Slow nonrandom change in ISE potential after practical response time	—	At fixed solution composition and temperature
Dynamic response time (transient)	Initial rapid change in ISE potential before practical response time	t_N ($N =$ seconds)	Depends on solution and membrane characteristics
Interference	Species that affects ISE response to primary analyte ion a_i	a_B, a_j	Electrode type alters ISE response; method type alters solution a_i
Linear range	Calibration response region of constant sensitivity	See Fig. 2.15	Usually bounded by detection limit and high a_i roll off
Matrix	Inert support for active components in ISE membrane	—	PVC, silicone rubber, porous glass frit, and so forth

Term	Definition	Symbol	Notes
Permselectivity	Ability of ISE membrane to reject ions of sign opposite to primary a_i	—	Co-ion $t_i=0$; if violated, then sensitivity $<$ Nernstian
Practical detecton limit	a_i At intersection of response and interference regions of calibration	See Fig. 2.15	Also, a_i when Nernstian response line is $18/z_i$ mV from calculated curve
Practical response time	When ISE potential is within 1 mV from final potential; t_{90} also used	—	No definitive parameters yet decided upon
Potentiometric selectivity coefficient	Operationally defined ISE selectivity parameter— see (47)	k_{ij}^{pot}	Smaller k_{ij}^{pot} means ISE more selective for a_i over a_j; not a true constant, may be $f(a_i)$ or $f(a_i/a_j)$
Selectivity	Ability of ISE to sense a_i over all other like-signed a_j	—	Not to be confused with permselectivity
Sensitivity (slope)	$E/\Delta pa_i$ of linear-response region in mV/(decade a_i)	S	Nernstian value at 25°C is $(59.16/z_i)$ mV/decade; may be sub-Nernstian or super-Nernstian for various reasons

Table 2.29 . Classification of Ion-Selective Electrodes

Electrode Class	Active Membrane Material	Examples
Primary Electrodes		
Solid-membrane electrodes	Crystalline (single, mixed, or poly)	
	Homogeneous	LaF_3, $AgCl/Ag_2S$, (no inert matrix)
	Heterogeneous	AgI in silicone rubber (has inert matrix)
	Noncrystalline	
	Glass	Na, K glass electrodes
	SiO_2, Si_3N_4	Gateless ISFET
	Organic	Ion-exchange resin electrode
Liquid-membrane electrodes	(May be unsupported or supported with an inert solid matrix)	
	Neutral carrier	Valinomycin/PVC K^+ ISE
	Charged carrier	Ph_4B^-, R_4N^+ in cellulose filters
Sensitized Electrodes		
Gas-sensing electrodes	Gas-permeable membrane (with electrolyte, ISE)	pCO_2, pNH_3 ISE
	Air gap (with electrolyte, ISE)	Same
Biosensitive electrodes	Enzyme(s) in matrix on ISE	Enzyme electrodes for urea, glucose
	Whole cells in matrix on ISE	Cysteine electrode
	Tissues slices on ISE	Glutamate electrode

to almost any of the primary electrode types fashioned without an internal solution. Also, by virtue of a solid matrix support, the liquid-membrane electrodes can become mechanically robust and are then not "liquid" in the usual sense. The sensitized electrodes form a general class in which a chemical reaction or gas equilibration of the analyte produces a change in concentration of the sensed primary ion. This indirect measurement allows quantification of many nonionic substituents using simple ISE base electrodes like the pH glass electrode. The ISFET devices (to be described later) could really be placed in almost all the electrode categories. The bare or gateless ISFETs have SiO_2 or Si_3N_4 insulating layers, selective for H^+ and Na^+, and can be considered homogeneous solid-membrane electrodes. However, most of the common ion-selective membranes of the other classes can be deposited on top of the gate layers to produce ISFETs that really represent all the electrode classes. In addition, a FET with a Pd metal gate (an IGFET) can be made sensitive to gases such as H_2 and H_2S.

This produces a gas-sensitive device that is a primary electrode rather than a reaction-sensitized indirect electrode. As a result of these overlaps the ISFETs are discussed separately in the format section.

5.2   Characteristics of Primary Ion-Selective Electrodes

5.2.1   Solid Membrane—Crystalline

The electrodes of this class are based on insoluble salts of the general type MX in which the potential-generating process is ion-exchange equilibrium between solution and the mobile ions in the salt (be they M^+ or X^-). The similarity between this class of ISE and the simple M/MX electrodes of the second kind discussed earlier should be evident. Both electrode variations normally possess primary M^+ cation sensitivity, and each can be used to quantitate the anion X^- by using the K_{sp} relation [see (25)–(29)]. The difference between electrode variations lies more in salt morphology than in principle, electrodes of the second kind often being more porous.

The solid-membrane crystalline electrodes can be composed of single-salt crystals, polycrystalline pressed pellets, or mixed crystals with a second precipitated salt. An inert matrix like silicone rubber, polyethylene, or PVC may be used to bind the active insoluble salt. Since all these compositions rely on surface ion-exchange equilibrium by active salt MX, the electrode form is dictated more by convenience and mechanical stability. Single crystals are difficult to prepare, and the LaF_3-based fluoride electrode is the only commercial electrode fashioned in this way. Most of the other membranes consist of a pressed or sintered pellet of precipitated polycrystals. Desirable membrane properties include insolubility, mechanical stability, nonporosity, conductivity, and fast ion-exchange rate. The number of salts that possess these properties is limited: the rare-earth fluorides; the halides of Ag, Pb, Hg, and Tl; the sulfides (and possibly the selenides or tellurides) of Ag, Pb, Hg, Zn, Cu, and Cd; and a few silver salts like cyanide and thiocyanate [15]. When a mechanically unstable or poorly conducting MX is chosen, it may be admixed with Ag_2S to improve membrane properties. Since Ag_2S is so insoluble ($pK_{sp} \sim 51$) it can often be considered an inert matrix for the more soluble MX. Surveys of the MX types and the various membrane formulations are available in several reviews [9, 136].

The common commercial electrodes of this class are listed in Table 2.30. Note that all the silver-salt electrodes are designated as anion electrodes even though Ag^+ is the primary salt-mobile ion that undergoes ion exchange. Evidence of response to pure solutions of either cation or anion is shown in "foldover" plots like that of Figure 2.16. The usual format for these electrodes is the barrel-shaped, plastic-body design with either internal solution or solid internal contact as shown in Figure 2.5. In the Ruzicka-Selectrode version of this format [138], a universal solid conductive internal contact of hydrophobic graphite is placed in the usual plastic barrel. Then a powder containing a particular MX is rubbed onto the graphite base, creating an X-selective membrane that can be replaced in favor of another M'X' system by simple abrasive

Table 2.30   Common Crystalline Solid-Membrane Electrodes[a]

Primary Sensed Ion	Active Material	Membrane Forms[b]	Working pa_i Range	Chief Interferences
F^-	LaF_3	S	0–6	OH^-
$Cl^-(Ag^+)$	AgCl	MP, H	0–5	Br^-, I^-, S^{2-}, CN^-
$Br^-(Ag^+)$	AgBr	MP, H	0–6	I^-, S^{2-}, CN^-
$I^-(Ag^+)$	AgI	MP, H	0–7	$S^{2-}, S_2O_3^{2-}, CN^-$
$S^{2-}(Ag^+)$	Ag_2S	P	0–7	Hg^{2+}
$CN^-(Ag^+)$	AgCN	MP	2–6	I^-, S^{2-}
$SCN^-(Ag^+)$	AgSCN	MP	0–5	Br^-, CN^-, S^{2-}
Cu^{2+}	CuS	MP	0–7	Ag^+, Hg^{2+}, high Cl^-
Cd^{2+}	CdS	MP	0–7	Ag^+, Hg^{2+}, Cu^{2+}
Pb^{2+}	PbS	MP	0–7	Ag^+, Hg^{2+}, Cu^{2+}

[a]All are commercially available with either internal solutions or solid internal contacts.
[b]S = single crystal, MP = mixed precipitate with Ag_2S in matrix-free membrane, H = precipitate in inert binder like silicone rubber, P = polycrystalline.

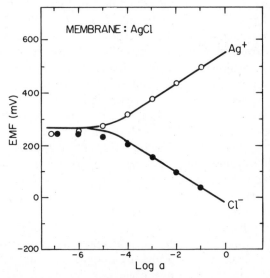

Figure 2.16   Cell potential for a silver chloride membrane electrode in halide (●) or silver nitrate (○) solutions. Reproduced from [137], with permission.

removal after an analysis. A variety of other shapes and sizes of solid-membrane crystalline electrodes is possible [9], but the barrel type is the principal commercial form.

In general the equilibrium K_{sp} of the MX salt is an important parameter that governs several electrode response characteristics. For instance, in the ideal case the electrode detection limit for M^+ or X^- is set by the solubility of the membrane material itself. Thus the theoretical detection limit for a sensed ion is the nth root of the K_{sp}, where n represents the number of the ions available from a molecule of the salt. An example is the response of the Ag_2S electrode to pAg^+ as depicted in Figure 2.17, where a pAg^+ of about 23 is sensed under carefully controlled conditions. However, in many other real cases the detection limit is raised above that predicted from the K_{sp}. Phenomena such as ion adsorption, ion occlusion, soluble crystal coprecipitation during electrode formation, or even solution artifacts like container adsorption and anion oxidation can contribute. Since varying precipitation conditions can produce mixed or undefined MX/NY crystalline stoichiometries, the thermodynamic K_{sp} for pure MX may also be inappropriate for electrode detection limit estimation.

A second K_{sp}-related parameter is electrode selectivity as represented by the potentiometric selectivity coefficient in (47). Generally electrode interferences consist of ions that form more insoluble salts than MX, complexing agents that can dissolve the membrane, or very strong redox agents. For the first case the interference proceeds through ion-exchange conversion of the MX surface to the more insoluble NX or MY. In the ideal equilibrium case no internal diffusion potentials arise and the k_{ij}^{pot} in (47) is given by the K_{sp} ratio (here shown for the more common interference by anion Y):

$$k_{X,Y}^{pot} = \frac{K_{sp}(MY)}{K_{sp}(MX)} \tag{156}$$

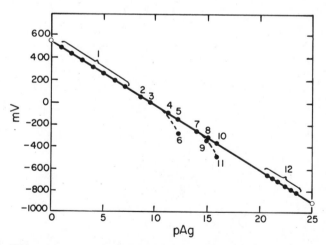

Figure 2.17 Wide-range response of a silver sulfide membrane electrode: (1) solutions of $AgNO_3$ ($\mu=0.1$), 10^{-1} to 10^{-7} M; (2) 0.1 M NaCl; (3) 1 M NaCl, saturated with AgCl; (4) 0.1 M NaBr; (5) 1 M NaBr saturated with AgBr; (6) 1 M NaBr, unsaturated; (7) 0.01 M NaI; (8) 0.1 M NaI, saturated with AgI; (9) 0.1 M NaI, unsaturated; (10) 1 M NaI, saturated with AgI; (11) 1 M NaI, unsaturated; (12) 0.01 M Na_2S ($\mu=0.11$), pH 7.0 to pH 12.0. From J. Vesely, O. J. Jensen, and B. Nicolaisen, *Anal. Chim. Acta*, **62**, 1 (1972), with permission.

Under carefully controlled conditions, using freshly polished nonporous electrodes in well-stirred solutions of large volume, the silver halide electrodes follow (156) with reasonable accuracy, as Figure 2.18 depicts. However, deviations from this theoretical behavior are often observed, and the apparent $k_{X,Y}^{pot}$ variation under different experimental conditions remains an active area of investigation for the solid-membrane crystalline electrodes. The deviations may vary among individual electrodes but mixed-crystal formations, diffusional effects in porous electrode surface layers, or simple interfering-ion depletion from low volume solutions have been suggested to render (156) invalid. More detail is given in the discussion following each particular electrode.

In the presence of a soluble, strong complexing agent for one of the ions MX in the membrane, the electrode potential may be altered by the reduction in the surface ion activity. Usually the reaction with a ligand L

$$\overline{MX} + nL^- \rightleftharpoons ML_n^{(n-1)-} + X^- \tag{157}$$

will produce a surface activity of X^- (or M^+) that is dependent on the mass transport of ML_n, L^-, and X^- between the bulk solution and the electrode surface. The electrode potential will then take on the form [15]

$$E = k + \frac{RT}{F} \ln \gamma_x \left[C_x + \frac{D_L d_x \gamma_x C_L}{n D_x d_L} \right] \tag{158}$$

where C_i are bulk concentrations, D_i are solution diffusion coefficients, and d_i the Nernst film diffusion thicknesses, which can in turn be a function of the hydrodynamics around the electrode. If $C_x = 0$ and the hydrodynamics are controlled, the electrode can be made responsive to L; and assays for CN^-,

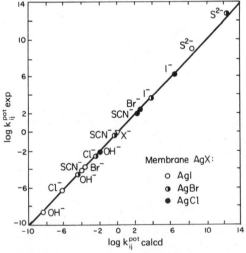

Figure 2.18 Comparison of the experimental and calculated anion selectivity coefficients of different silver halide membrane electrodes. From [137], with permission.

$S_2O_3^{2-}$, and thiols have been suggested with silver halide electrodes [9]. More complicated complexation behavior, like irreversible film formation by surfactants or proteins, has been noted and can lead to complete loss of electrode response. Restoration to virgin MX surface by abrasive polishing is possible for many of the electrodes in this class.

Response time is a parameter that has been actively studied in recent years for all the ISE classes. For the ideal crystalline solid-membrane electrodes in solutions of primary ions in the Nernstian activity range the process of solution film diffusion governs the response [136] so that

$$E(t) - E(\infty) = k + \frac{RT}{F} \ln \left\{ 1 - \left[1 - \frac{a_i(0)}{a_i(\infty)} \right] \exp \left(\frac{-t}{\tau} \right) \right\} \qquad (159)$$

where $a_i(0)$ and $a_i(\infty)$ are starting and ending step activities and the time constant τ is given by $d_i^2/2D$. Both stirring-dependent and activity-step direction-dependent response time (slower going from high to low activities) are predicted by (159); studies by Pungor and Toth [136] on the AgI/Ag_2S electrode shown in Figure 2.19 highlight such behavior. Deviations from this behavior are noted whenever membrane processes like slow surface reactions or membrane film diffusion occur on the same time scale as solution stagnant-layer diffusion. Exact quantitation of response time under these conditions is made difficult by the many possible contributing phenomena and by the lack of knowledge about transport parameters within the membrane phase.

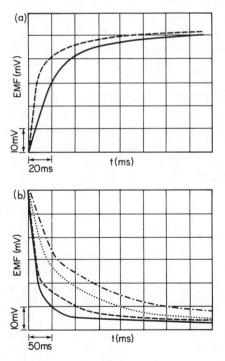

Figure 2.19 Response-time study of an AgI/Ag_2S electrode. Solution jet impinging on a 12 mm diameter membrane: (a) $a_1^0 = 1$ mM KI, $a_1^f = 10$ mM KI, (- - -) jet flow rate $V = 117$ mL/min, (——) $V = 54$ mL/min; (b) $a_1^0 = 10$ mM KI, $a_1^f = 1$ mM KI; (---) $V = 30$ mL/min, (···) $V = 54$ mL/min, (- - -) $V = 117$ mL/min, (——) $V = 138$ mL/min. From [136], with permission.

LaF$_3$ ELECTRODE

The fluoride electrode [1] is one of the most selective and useful devices in the ISE group. The LaF$_3$ single crystal, which may be doped with other rare-earth ions like europium to increase conductivity, exhibits an apparent pK_{sp} of about 24.5, yielding a detection limit of $pF^- \sim 8$. The fluoride ion is the mobile species in the crystal; response to La^{3+} is not observed because of the probable slow exchange rate for this ion. The excellent selectivity arises from the low K_{sp} and the inability of most anions to duplicate the F^- lattice mobility. Hydroxide ion is the major anionic interference so that the optimum pH range for this electrode is 5–8 (the low pH limit stems from a solution interference brought upon by HF and HF_2^- formation). The nature of the OH^- interference is a complicated combination of membrane and solution mobility ratios (u_{OH^-}/u_{F^-}) and metathetical ion-exchange reactions that can produce species of the type $La(OH)_N^{(3-N)+}$. It is found to vary with both stirring and F^- concentration, and is still a subject of active research interest [139]. Complex-forming anions like citrate, acetate, and other carboxylates, while not behaving as strict electrode interferences, are often observed to increase electrode response times. Apparently, a "hydrated" gel layer can form in the presence of these anions or at low concentrations of pure MF solutions, analogous to the pH glass electrode. Although response times in the linear concentration range are very rapid (0.5–30 s) under well-stirred conditions, a much longer approach to E_{eq} is noted at low concentrations. Research investigations of electrode response time [140] and impedance characteristics [141] can be consulted for further detail.

Solution internals of NaF-NaCl bathing a Ag/AgCl wire are traditional for the F^- electrode, but all-solid contacts have been successfully demonstrated [142]. The barrel-shaped electrode format is employed almost exclusively; but micro or flow-through designs have been fabricated over the years by various research groups [9]. The analytical applications of this electrode span many areas, and several compilations are available [8, 9, 48]. Direct potentiometry is normally performed with the aid of a total ionic strength adjusting buffer (TISAB) based on citrate or an EDTA variant like CDTA. This buffer sets ionic strength and pH, and serves to free any F^- complexed by solution Al^{3+} or Fe^{3+}. Indirect titrations for La, Bi, Al, Th, and other metals are also possible. Finally, a host of metal-fluoride complexation or solubility equilibria are amenable to study with this electrode.

AgX AND Ag$_2$S ELECTRODES

These electrodes have been more thoroughly studied than any of the other solid-membrane devices. All possess Ag^+ crystal mobility caused by Frenkel-type lattice defects. Although second-kind electrodes work by the same equilibrium principles, the move to crystalline or pressed-pellet AgX membranes was an attempt to obtain a more nonporous and mechanically stable form for the salts. The mixed Ag$_2$S/AgX crystals or the Pungor and associates [4] electrodes with AgX embedded in a silicone rubber matrix were developed because of the inability to obtain a satisfactory pure polycrystalline membrane for AgI. The common wisdom about second-kind AgX electrodes was that their porosity exposed bare silver to solution oxidizing agents, causing enhanced redox sensitivity. However, as the crystalline membrane electrodes (supposedly free from redox sensitivity) evolved from internal solution contacts to all-solid-state Ag contacts, the "cure" seems to have disappeared. Ironically, Harzdorf [143] has recently shown that second-kind electrolytic silver halide electrodes prepared under controlled current-density conditions show less redox sensitivity than the all-solid-state AgX/Ag_2S devices.

As discussed before, the K_{sp} value for the particular AgX provides a good first approximation for electrode detection limit, and the K_{sp} ratio also an estimate of the interference effect from any anion Y^-. With the mixed-crystal AgX/Ag_2S electrodes the response may vary widely with the conditions of the simultaneous precipitation. Mixed compounds like Ag_3SBr and Ag_3SI may form during the membrane pressing [15] so that the true stoichiometry can be indeterminate. Selectivity coefficients differing from the K_{sp} ratio can be observed; several references define the current theories behind this discrepancy [136, 139, 144]. Response time for new or freshly polished electrodes should be rapid and follow the form of (159), sluggish behavior usually indicating porous surface film formation. Studies of dynamic response times near the detection limit or in the mixed interference region have been made in Pungor's laboratory [145]. The AgI electrode is interesting because it often exhibits super-Nernstian (at low I^-) or sub-Nernstian (at very high I^-) behavior because of adsorption or AgI_2^- complex formation, respectively. Thus electrode-conditioning history or solution volume may become important when interpreting the response at low concentrations [146]. The Ag_2S electrode is by far the most selective of this group; only Hg(II) cation or very high concentrations of strong complexing agents interfere. Usually solution measurement artifacts like air oxidation, pH, or adsorption cause more of a problem than electrode response parameters. Applications lists and format variation descriptions for the AgX-Ag_2S membrane electrodes are available [8, 9, 48], and a review chapter has been prepared [147].

Cu, Pb, AND Cd ELECTRODES BASED ON MS/Ag$_2$S

These systems follow the divalent M^{2+} activity through the ion-exchange equilibrium with the Ag^+ ion:

$$M^{2+} + Ag_2S \rightleftharpoons MS + 2Ag^+ \qquad (160)$$

Therefore the a_{Ag^+} appearing in the Nernst relation can be replaced by

$$E = k + \frac{RT}{2F} \ln\left[(a_{M^{2+}}) \frac{K_{sp}(Ag_2S)}{K_{sp}(MS)} \right] = k' + \frac{RT}{2F} \ln(a_{M^{2+}}) \qquad (161)$$

While these electrodes are quite useful and reasonably selective against other metal cations (Zn^{2+}, Ni^{2+}, Mn^{2+}, Co^{2+}, etc.), nonideal response is prevalent because of the tendency to form nonequilibrium mixed-crystal stoichiometry. The Cu(II) electrode has been studied the most actively; VanDerLinden [148] has summarized the work in his and other laboratories. A main component in the CuS/Ag_2S pellet seems to be $Ag_{1.5}Cu_{0.5}S$, and the relations

$$4Ag_{1.5}Cu_{0.5}S \rightleftharpoons 3Ag_2S + CuS + Cu^{2+} + 2e^- \qquad (162)$$

$$Cu_2S \rightleftharpoons CuS + Cu^{2+} + 2e^- \qquad (163)$$

seem to be important in light of electrode redox sensitivity and the variable complexing ligand sensitivity that depends upon electrode history. Even electrodes prepared from Cu(I) salts show both Cu(I) and Cu(II) sensitivity [149]. Response to high levels of Cl^- or EDTA-type ligands has also been observed, possibly because of metal complex formation or ligand-corrosion mechanisms like that described by (158). Yet in suitable metal-ion buffer solutions the Cu(II) electrode demonstrates Nernstian response in the range pCu 3–19 [150]. The Pb and Cd ISEs show similar phenomena relatable to the surface chemistry. Some ESCA work and other fundamental studies relate surface heterogeneity to solution response for both the Pb [151] and Cd [152] systems. Both electrodes possess a workable pH range of 3–8, the upper limit being dictated by hydroxy complexes. Their use in indirect titrations is more common than direct potentiometric procedures; the Pb electrode is quite useful as an end point indicator for indirect sulfate and phosphate determinations with Pb^{2+} titrants. Both Se and Te have been employed in research work as the salt anion in electrodes for these three metals, but advantages over the corresponding sulfide salts have not been consistently demonstrated. Applications lists can be found in [8, 9, 48].

OTHER ELECTRODE SYSTEMS

A host of other insoluble salts has been tested over the years in an attempt to gain electrodes for various ions. A new MX salt may be used to prepare a homogeneous membrane, a mixed-crystal membrane employing Ag_2S or other salts, or a heterogeneous electrode in which a silicone rubber or other matrix is included. Examples include $BaSO_4$ (for Ba^{2+} or SO_4^{2-}), $Ag_2S/PbS/PbHPO_4$ (for HPO_4^{2-}), $PbMoO_4$ (for molybdate), LaF_3/CaF_2 (for Ca^{2+}), $BiPO_4$ (for Bi^{3+} or PO_4^{3-}), and $Co_3(PO_4)_2$ (for Co^{2+}). Compilations of other proposed electrode systems of the sort are found in [8, 9, 136]. Over 100 compounds of the MS/Ag_2S type have been screened [148]. As evidenced by the limited commercial availability of electrodes other than LaF_3, AgX, Ag_2S, and the (Pb, Cd, Cu)S/Ag_2S group, additional practical chemistries have not been found. Chief drawbacks are poor reproducibility and competing selectivities for many common ions. For instance, no completely acceptable phosphate or sulfate electrodes exist today because of the inability to prevent large chloride interferences, although a PO_4^{3-}-sensitive electrode has been marketed [148]. For any proposed MX electrode, the fundamental selectivity performance with respect to NY is usually limited by the K_{sp} ratio.

5.2.2   Solid Membrane—Noncrystalline

GLASS ELECTRODES

The early realization that the alkaline errors of pH glass electrodes described by (126) might lead to cation-selective glasses caused an extensive study of glass formulation and properties [153]. Out of this work came two principal variants: the pNa and pCation glass electrodes. Table 2.31 lists some characteristics of these electrodes. The sodium electrode differs from the pH formulation by addition of Al_2O_3 sites. It possesses good selectivity over most uni- and divalent cations, H^+ and Ag^+ being notable exceptions. The H^+ response sets the lower pH limit for the pNa electrode, and the Ag^+ response has been used to some analytical advantage. Aside from the Na-selective glass, no other formulation has been found to yield a highly selective pM electrode. The K^+, Li^+, or NH_4^+ formulations all exhibit $k_{M/N}^{pot}$ of around 0.1 with respect to Na^+ or each other. This does not negate their utility in controlled analytical situations; the Beckman 39137 microcation glass electrode has proved to be a ubiquitous analytical tool for a number of years. Efforts to develop alkaline-earth or other multivalent-selective glass electrodes have been unsuccessful, chiefly because of poor selectivity over the alkali metals. In fact the development of new glass formulations slowed during the 1970s but may again become active with the increased investigation of semiconductor or chalcogenide glasses (doped Se, Sb, Sn oxides, or mixed metal sulfides), which exhibit combined electronic–ionic conductivities [15, 83].

Response characteristics and laboratory usage of the pM glass electrodes are similar to the pH glass electrode. The hydrated gel layer, the high resistance and asymmetry potential, the necessity of preconditioning in aqueous solution, and the sluggish response behavior of old or film-coated electrodes are common to all the glass electrodes. The k^{pot} is dictated by a combination of ion-exchange and gel-layer mobility terms. In addition, dynamic response transients that relax after about a minute can often be observed in the mixed interference region. The K^+- and NH_4^+-generated transients in the pNa electrode response are of this type and have been characterized qualitatively [154]. Solid internal contacts of Ag, AgCl, or other metals, and bulb shapes other than the traditional sphere are possible for the pM glasses [9]. Applications of these glasses are widespread [8, 9, 154] but have been most firmly established in the clinical and biological areas. Here the pNa electrode either forms the basis for the Na measurement in serum, blood, and urine on many commercial analyzers or provides intracellular pNa information when used in a microelectrode form.

ION-EXCHANGE MEMBRANE ELECTRODES

The membranes in these electrodes are prepared from traditional organic, polymer-based, ion-exchange resins, either in homogeneous form or dispersed within inert matrices. The resin membranes are examples of classic fixed-site, water-hydrated matrices that remain permselective until bathing-solution

Table 2.31 Some Cation-Selective Glass Electrodes[a]

Electrode Type	Active Phase	Potential-Determining Ions	Working Range (M)	Selectivity Coefficient[b] k_{M-1}^{pot}	Recommended pH Range	Temperature Range (°C)	Electrical Resistance at 25°C (MΩ)	Recommended Reference Electrode	Manufacturer
pNa	NAS_{11-18}, $LAS_{26.2-12.4}$, etc.	$H^+ > Ag^+ > Na^+ > K^+$	$1-10^{-8}$	$Ag^+ \sim 500$; $H^+ \sim 10^3$; $K^+ \sim 10^{-3}$; $Li^+ \sim 10^{-3}$; $Cs^+ \sim 10^{-3}$; $Tl^+ \sim 2 \times 10^{-3}$; $Rb^+ \sim 3 \times 10^{-5}$; $NH_4^+ \sim 3 \times 10^{-5}$	7–10 ~ 4 pH units above pNa value	0–100	$\sim >100$	Ag/AgCl with double salt bridge (1 M NH_4NO_3)	Beckman, Corning, EIL, Ingold, Metrohm, Orion, Philips, Polymetron, Radelkis Radiometer, Schott & Gen Tacussel
pCation	NAS_{27-4}, $KABS_{20-5-9}$, etc.	$H^+ > Ag^+ > K^+$; $NH_4^+ > Na^+ >$ Li^+, Rb^+, Cs^+, Tl^+	1 to 5×10^{-6}	$Na^+ \sim 0.1$; $NH_4^+ \sim 0.3$; $Rb^+ \sim 0.5$; $Li^+ \sim 0.05$; $Cs^+ \sim 0.03$ (normalized to $K^+ = 1$)	7–13 (for pK) 4–10 (for pAg) ~ 2 pH above pK value	0–100	$\sim >100$	Ag/AgCl with double salt bridge (1 M Li-tri-chloroacetate)	Beckman, Corning, EIL, Ingold, Philips, Tacussel

[a]From [20], with permission.
[b]According to manufacturer.

concentrations exceed molar levels. These electrodes have not enjoyed widespread use because little selectivity between ions is observed in the water-swollen fixed sites, and because reproducibility is often difficult to obtain. Their chief advantages are resistance to extremes of acid or alkali and a high conductivity. One commercial variant mentioned in Table 2.19 is used for pH measurements in HF and other corrosively acidic solutions. Pandey and Tripathi [155] have evaluated collodion, Dow-50, and Amberlite 1RC-50 cation-resin electrodes for Na^+, K^+, and NH_4^+ selectivity and other properties. As expected, minimal interion selectivity was observed although Nernstian response to each individual ion was obtained. A list of additional resin-membrane studies is given by Buck and associates [9].

5.2.3 Liquid-Membrane Electrodes

ELECTRODES BASED ON CHARGED CARRIERS

In contrast to the fixed-site solid-membrane electrodes the liquid-membrane, charged-carrier electrodes possess oleophilic, mobile sites, which can extract solution counterions into the membrane phase. Permselectivity is still provided by the charged nature of the membrane carriers, but selectivity arises from the competitive degree of extractability of various counterions. Figure 2.20 depicts the various equilibria that must be accounted for in the description of these electrode systems (in this case a cation-sensitive membrane). All membrane species, $\overline{S^-}, \overline{M^+}, \overline{N^+}, \overline{MS}, \overline{NS}$, are mobile and are included in the description of the total membrane potential. For proper electrode functioning the sites $\overline{S}$ should be water insoluble, and unless the $\overline{S}$ are themselves liquid an additional oleophilic solvent of low vapor pressure is necessary. Totally liquid systems can be employed but are impractical. Instead, a porous support or an inert polymer

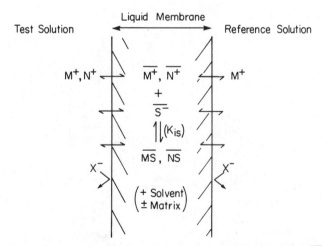

Figure 2.20 Schematic of a cation-sensitive, liquid ion-exchange membrane. $\overline{S^-}$ is the mobile charged carrier, K_{is} the membrane association constant for the ith cation.

support like PVC are commonly used as shown in Figure 2.21. The PVC variant (Figure 2.21a) eliminates the cumbersome manipulation of the ion-exchange liquid, and most commercial electrodes are now designed in this way. In addition the PVC membrane format is easily obtained in the laboratory by simple casting from a polymer solution of PVC and the carrier/solvent chemistry in an evaporable coating solvent like tetrahydrofuran or cyclohexanone. Small disks cut from the master membrane after drying are then solvent-bonded to PVC capillary tubes to form the electrode shell. Moody and Thomas [156] have discussed various simple techniques for preparing these electrodes.

The total equilibrium bi-ionic exchange process in the membrane

$$M^+(aq) + \overline{NS^-} \xrightarrow{K_{ex}} \overline{MS} + N^+(aq) \tag{164}$$

can also be described by a combination of single-ion extraction coefficients $k_i = \bar{a}_i/a_i$ and the membrane association constant K_{is} shown in Figure 2.20. Because of the many charged, mobile species involved in membrane equilibria an exact, general expression like (47) for the electrode potential of multivalent ions is not derivable. In the uni-univalent case these equilibria and the ion mobilities describe the membrane behavior exactly [15] if limiting cases are defined by the K_{is} parameter. When the membrane-extracted ions are completely dissociated (K_{is} small), then an expression like (47) is obtained with the k_{ij}^{pot} given by

$$k_{ij}^{pot} = \frac{u_j k_j}{u_i k_i} \quad \text{(dissociated case)} \tag{165}$$

Under conditions of complete association (large K_{is}) the response becomes

$$E = k' + \frac{RT}{F}(1-\tau) \ln\left[\sum_i (u_i + u_s)k_i a_i\right] + \frac{RT}{F}\tau \ln\left[\sum_i u_{is} K_{is} k_i a_i\right] \tag{166}$$

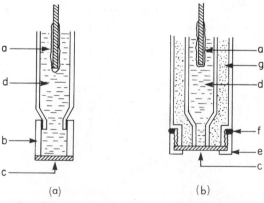

(a)   (b)

Figure 2.21 Construction of ion-selective, liquid-membrane electrodes. (a) Membrane chemistry contained within an inert PVC or silicone rubber matrix. (b) Membrane liquid contacts inner and outer aqueous solutions through a porous support. a—Ag/AgCl internal reference; b—plastic tube; c—matrix or porous support; d—inner filling solution; e—plastic cap; f—O-ring; g—liquid exchanger chemistry. From [148], with permission.

with the parameter τ being

$$\tau = \frac{u_s(u_{js}K_{js} - u_{is}K_{is})}{(u_i + u_s)u_{js}K_{js} - (u_j + u_s)u_{is}K_{is}} \tag{167}$$

Few cases have been studied in which the value of τ is such that both terms in (166) are important. When site mobility is small $\tau \to 0$ and (166) takes the usual Nicolsky form (47) with the k_{ij}^{pot} derived as

$$k_{ij}^{pot} = \frac{(u_j + u_s)k_j}{(u_i + u_s)k_i} \quad \text{(associated case, } u_s \to 0) \tag{168}$$

Conversely, if free site mobility is very large, $\tau \to 1$ and k_{ij}^{pot} in (47) will take the form

$$k_{ij}^{pot} = \frac{u_{js}K_{js}k_j}{u_{is}K_{is}k_i} \quad \text{(associated case, } u_s \text{ large)} \tag{169}$$

Simplistically these relations indicate dependence of k_{ij}^{pot} on purely solvent-related parameters for dissociated or low site mobility associated cases, whereas both solvent and site complexation strength affect k_{ij}^{pot} in (169). Laboratory testing of these relations has been limited by the large amount of membrane-specific terms required, but some examples have been published [157].

As might be expected from the complicated response indicated in (165)–(169), reported selectivity coefficients for liquid ion-exchange chemistries can vary with solvent and with interference/analyte concentration levels. The solvent (often termed solvent mediator) influence stems both from electrostatic (dielectric constant) and solvating considerations. Fujinaga and co-workers [158] describe a study on typical solvent effects. Efforts to correlate k_{ij}^{pot} with the measurable liquid extraction coefficient for various exchanger/solvent combinations have met with some success whenever the mobility terms in (165), (168), and (169) tend to cancel. In general, the liquid ion-exchange chemistries are not extremely selective, in part because the steric restrictions of an enforced "lattice" cavity are not available as they were in the solid-membrane electrodes.

For practical usage all membrane components should be water-insoluble, possess low vapor pressure, and preferably have high viscosities. In the PVC membrane format, the solvent mediator should readily plasticize the polymer. Sometimes a second "plasticizer" solvent is added to accomplish this end. There should be no need to condition a freshly prepared membrane if the site counterion is in the desired analyte form. However, artifacts and response changes are sometimes noted, possibly relatable to H_2O diffusion into the membrane phase. Buck and colleagues [159] studied the influence of H_2O permeation on a typical anion-sensitive exchanger system in PVC, and observed membrane structural changes with time. Many workers now precondition as a matter of course, particularly when using solid-contact internals. Electrode response time in the limit is governed only by solution diffusion, but in practice phenomena such as internal diffusion potential adjustment and chemistry-depleted stagnant membrane surfaces can push these times to minutes. Electrode lifetime is usually dictated by loss of the chemistry components to aqueous solution after prolonged usage. Qualitatively, the electrode detection limit is

Table 2.32 A Selection of Commercially Available Liquid Ion-Exchanger Electrodes^a

Electrode Type	Active Phase	Potential-Determining Ions	Working Range (M)	Selectivity Coefficient^b k_{M-1}^{pot}	Recommended pH Range	Temperature Range (°C)	MΩ at 25°C	Recommended Reference Electrode	Manufacturer
pCa, liquid	Ca salt of dialkyl phosphoric acid in dioctylphenyl-phosphonate	Zn^{2+}, Ca^{2+}, Fe^{2+}, Pb^{2+}	$1-10^{-5}$	$Zn^{2+} \sim 3.2$; $Ca^{2+} \sim 1.0$; $Fe^{2+} \sim 0.8$; $Pb^{2+} \sim 0.63$; $Cu^{2+} \sim 0.27$; $Ni^{2+} \sim 0.080$ $Sr^{2+} \sim 0.017$; $Mg^{2+} \sim 0.014$ $Ba^{2+} \sim 0.010$; $Na^+ \sim 10^{-3}$ $K^+ \sim 10^{-3}$	5–9	0–50 10–60	<25 <500	Normal Ag/AgCl	Orion Corning
pCa PVC matrix	Ca dioctyl-phenyl-phosphate Unknown	Same as above	$1-10^{-5}$	$Zn^{2+} \sim 1-5$; $Ca^{2+} \sim 1.0$; $Al^{3+} \sim 0.90$; $Mn^{2+} \sim 0.38$; $Cu^{2+} \sim 0.070$; $Fe^{2+} \sim 0.045$; $Co^{2+} \sim 0.042$; $Mg^{2+} \sim 0.032$; $Ba^{2+} \sim 0.020$; $Na^+ \sim 10^{-5}$; $K^+ \sim 10^{-6}$; $Li^+ \sim 10^{-4}$	5–11	0–60	~2		Radiometer HNU
pMe^{2+}, liquid (water hardness)	Ca salt of dialkyl phosphoric acid in decanol	Zn^{2+}, Fe^{2+}, Cu^{2+}, Ni^{2+}, Ca^{2+}, Mg^{2+}, Ba^{2+}, Sr^{2+}	$1-10^{-5}$	$Zn^{2+} \sim 3.5$; $Fe^{2+} \sim 3.5$; $Cu^{2+} \sim 3.1$; $Ni^{2+} \sim 1.35$; $Ca^{2+} \sim 1.0$; $Mg^{2+} \sim 1.0$; $Ba^{2+} \sim 0.94$; $Sr^{2+} \sim 0.54$; Na^+, $K^+ \sim 0.01$	Same as above	0–50 10–60	<25 <500	Same as above	Orion Corning
pClO$_4$, liquid	Fe(o-phen)$_3^{3+}$ in p-nitrocymene	ClO_4^-, OH^-	$0.1-10^{-5}$	$OH^- \sim 1.0$; $I^- \sim 1.2 \times 10^{-2}$; $NO_3^- \sim 1.5 \times 10^{-3}$; $Br^- \sim 5.6 \times 10^{-4}$.	3–10	0–50	~25	Same as above	Orion

			Detection limit	Interferences					Manufacturer[b]
pNO$_3$, liquid	Ni(o-phen)$_3^{2+}$ in p-nitrocymene	ClO$_4^-$, I$^-$, ClO$_3^-$, NO$_3^-$	1–10^{-5}	ClO$_4^-$ ~10^3; I$^-$ ~20; ClO$_3^-$ ~2; Br$^-$ ~0.9; S^{2-} ~0.57; OAc$^-$ ~5.1×10^{-4}; HCO$_3^-$ ~3.5×10^{-4}; F$^-$ ~2.5×10^{-4}; Cl$^-$ ~2.2×10^{-4}; SO$_4^{2-}$ ~1.6×10^{-4}	3–10	0–50	~25	Same as above	Orion
	Tridodecyl-hexadecyl-ammonium nitrate in n-octyl 2-nitrophenyl ether			NO$_2^-$ ~6×10^{-2}; CN$^-$ ~2×10^{-2}; HCO$_3^-$ ~2×10^{-2}	1–9	10–60	<500		Corning
pNO$_3$ PVC matrix	Tetradodecyl-ammonium nitrate	ClO$_4^-$, I$^-$, ClO$_3^-$, NO$_3^-$	1–10^{-5}	Cl$^-$ ~6×10^{-3}; OAc$^-$, CO$_3^{2-}$, S$_2$O$_3^{2-}$, SO$_3^{2-}$ ~6×10^{-3}; F$^-$ ~9×10^{-4}; SO$_4^{2-}$ ~6×10^{-4} H$_2$PO$_4^-$, PO$_4^{3-}$ ~3×10^{-4} HPO$_4^{2-}$ ~8×10^{-5}	3–11	5–40	~25		EIL
	Unknown				1–9	0–50	~25		Philips HNU
pBF$_4$ liquid	Ni(o-phen)$_3^{2+}$ in p-nitrocymene	I$^-$, BF$_4^-$	0.1–10^{-5}	I$^-$ ~20; NO$_3^-$ ~0.1; Br$^-$ ~4×10^{-2}; OAc$^-$; HCO$_3^-$ ~4×10^{-3}; F$^-$, Cl$^-$, SO$_4^{2-}$ ~10^{-3}	2–12	0–50	~25	Same as above	Orion

[a] From [20], with permission.
[b] According to the manufacturer.

determined by ion-exchanger aqueous solubility, but other parameters like ion mobilities have been suggested to influence the limit [160].

A list of several commercial liquid-exchange systems is presented in Table 2.32. Most are now available in the PVC plastic membrane form. The pCa systems are important in the water-quality, clinical, and biomedical fields. Note that a solvent mediator switch from the Ca^{2+}-solvating dioctylphenylphosphonate to the weakly coordinating decanol eliminates selectivity over Mg^{2+}; the water-hardness or divalent cation sensor is the result. Moody and Thomas [161] have reviewed pCa design considerations, and the principles employed in the development of this electrode are useful in guiding work on other chemistries. The pNO_3 electrodes, while not highly selective, provided an important tool for easy quantitation of this common ion. The pNO_3 chemistry variants based on the quaternary ammonium anion exchangers highlight the general utility that the R_4N^+ compounds provide. A host of similar anion-sensitive electrodes has been built on the commercial compound Aliquat 336 (trioctylmethylammonium chloride) exchanged into the analyte anion form either before or after PVC membrane casting. The anion selectivity generally follows the series $F^- < Cl^- < Br^- < NO_3^- < I^- < SCN^- < ClO_4^- < PF_6^-$, the larger and more polar anions being more favored. Anion electrodes based on Aliquat 336 (which can be replaced by less readily available quaternary ammonium, phosphonium, or arsonium exchangers) have been reported for salicylate, $ZnCl_4^-$, $AuCl_4^-$, ClO_4^{2-}, CO_3^{2-}, MnO_4^-, nicotine, atropine, and many other anions. A high degree of selectivity among these very oleophilic species is rarely obtained, so an analytical situation with the extractable anion in a Cl^-, F^-, or SO_4^{2-} matrix is most desirable.

A similarly useful cation-exchanger system analogous to the quaternary salts has not been identified, although the tetraphenylborate exchanger is fairly general. Actually, any oleophilic cation or anion can serve as the membrane exchanger species. Dyes like Brilliant Green or methyl violet, charged surfactants like dodecyl sulfate or cetyltrimethylammonium chloride, or more exotic species like $MoO(SCN)_5^-$ and the ferrocenium cation have been tried. Antibiotic compounds that possess carboxylic acid groups like monensin and grisorixin can also be considered in this class, since the carboxylate proton dissociates upon cation coordination, thus forming a neutral complex. A more complete list of electrodes can be found in the applications reviews [8, 9, 48]. Solid internal contacts like the Selectrode hydrophobized graphite or even bare Ag, Cu, and Pt wires are common for this electrode class. The latter are termed coated-wire electrodes and are discussed separately in the format-related section.

ELECTRODES BASED ON NEUTRAL CARRIERS

This electrode class is one of the most interesting of all the ISE categories. The neutral carrier ISE chemistry is composed of an uncharged, water-insoluble ionophore dissolved in an organic solvent of low vapor pressure and water solubility. This chemistry is today almost exclusively contained within a polymeric support like PVC, so that the overall electrode configuration is that given in Figure 2.21a. A large variety of natural and synthetic carriers have been employed, principally in fashioning cation-sensitive electrodes. Several of the more common compounds are shown in Figure 2.22 (note that monensin, previously mentioned as a liquid ion exchanger, is included here). It is evident that all the compounds in Figure 2.22 are large ring or long open-chain structures having a preponderance of ether, hydroxy, or carbonyl oxygen atoms. These atoms can solvate unhydrated cations by ion-dipole coordination, allowing entry of the cations into low dielectric organic solvents. Poonia and Bajaj [162] have recently reviewed the coordinating forces at work between alkali and alkaline earth cations and ionophores of this kind. Simon and co-workers pioneered the incorporation of these neutral carriers into functional electrodes; his review with Morf [163] should be consulted for a full description of these systems. In addition to low water solubility, successful ionophores possess good selectivity and fast ion-exchange kinetics in the extraction equilibrium

$$M^+ + n\overline{S} \underset{}{\overset{K_{ex}}{\rightleftharpoons}} \overline{(MS_n)^+} \qquad (170)$$

The latter two characteristics can be contradictory: a compound with a tight, well-defined cavity usually possesses good steric interion selectivity but may have very slow complexation/decomplexation rates that are evidenced in electrode drift. Conversely, an acyclic ionophore may not be practically selective even though approach to ion-exchange equilibrium is fast.

The fundamental response of the neutral carrier cation electrodes has been an area of active interest for many years. Much attention has been focused on the identity of the immobile anionic species that must be present in the bulk-membrane interior to maintain electroneutrality. It cannot be an anion from solution, since permselectivity would then be violated and sensitivity would be sub-Nernstian. Anionic impurity sites arising from the membrane solvent or from the "inert" matrix have been suggested but can be ruled out in Nernstian systems of high purity. Simon's group [164] maintains that an interaction with bathing-solution water is operative and for the K^+-valinomycin chemistry has proposed the reaction

$$K^+(aq) + \overline{Val} + \overline{H_2O} \rightleftharpoons \overline{KVal^+} + \overline{OH^-} + H^+(aq) \qquad (171)$$

in which relatively immobile hydroxide ions act like membrane fixed sites. Regardless, if permselectivity is maintained, the electrode response in the bi-ionic univalent case takes the form [15, 163]

$$E = k' + \frac{RT}{F} \ln[K_{ex,i} a_i + K_{ex,j} a_j] \qquad (172)$$

where the overall extraction coefficients for each ion describe (170). The $K_{ex,i}$ can also be expressed as

$$K_{ex,i} = \beta_{is}^w k_{is} \left(\frac{c_s}{k_s}\right)^n \qquad (173)$$

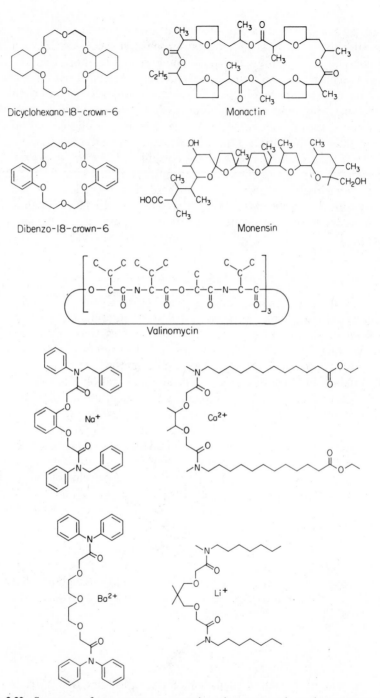

Dicyclohexano-18-crown-6

Monactin

Dibenzo-18-crown-6

Monensin

Valinomycin

Na⁺

Ca²⁺

Ba²⁺

Li⁺

Figure 2.22 Structures of some common neutral carrier compounds used to prepare cation-selective electrodes. From [20], with permission.

in which β_{is}^w is the aqueous complexation constant, k_{is} the partition coefficient of the charged complex, and c_s, k_s the ionophore concentration and partition coefficient raised to the nth [from (170)] power. The familiar k_{ij}^{pot} of the Nicolsky relation (47) may then be given by

$$k_{ij}^{pot} = \frac{\beta_{js}^w k_{js}(c_s/k_s)^{n'}}{\beta_{is}^w k_{is}(c_s/k_s)^n} \tag{174}$$

The only solvent-related influence in (174) is the complex partition coefficient k_{xs}. For ionophores with structured enclosed cavities like valinomycin, where n and k_{xs} are likely to be similar for like-charged cations, the choice of membrane solvent may be immaterial. For acyclic carriers, or in the cases of different ion coordination or valences, solvent dependence may be expected. It has been shown [163] that divalent/univalent cation selectivity can be improved by using a membrane solvent of higher dielectric constant.

Another response phenomenon related to permselectivity is the often-observed electrode sensitivity to high concentrations of polar solution anions. Here the permselectivity of carrier and solvent (from whatever source) is not enough to prevent anion extraction, and a diffusion-potential response is operative. The electrode potential then takes the form

$$E = k' + \frac{u_{is} - u_x}{zu_{is} + u_x} \frac{RT}{F} \ln a_i + \frac{nu_{is}}{zu_{is} + u_x} \ln \alpha \tag{175}$$

for cations of charge z^+ and anions of univalent negative charge. The α is a complicated function of membrane site concentration and IX salt extraction that varies with the degree of association of $\overline{IS^+}$ and $\overline{X^-}$ (a complete case-by-case description is found in [163]). Under the influence of the anion effect the cation sensitivity given by (175) may be sub-Nernstian or even anionic at high salt concentrations with polar membrane solvents. Remedies for this effect include the use of nonpolar solvents and the purposeful addition of lipophilic anionic sites like Ph_4B^- to the membrane. In the latter case permselectivity is increased although selectivity may possibly be affected. Figure 2.23 gives an example of the thiocyanate anion effect on the Na-selective chemistry of Figure 2.22.

Several other practical usage considerations for neutral carrier electrodes are somewhat different from those presented for liquid ion-exchange systems. Without added membrane anions, electrode impedances may approach the high levels found with glass electrode systems, and adequate shielding is then important. Dynamic response times may be lengthened over the ion-exchanger case because of the appearance of membrane properties in the rate constant τ; fast stirring and nonpolar membrane solvents are recommended [165]. Ligand concentration may be important in selectivity considerations if ions of different valence or coordination are to be discriminated, and in general a 0.1–1 mM membrane loading is chosen if solubility permits. From the previous discussions it is evident that solvent polarity is an important parameter; nonpolar solvents are preferred on all counts except in designing divalent-selective electrodes.

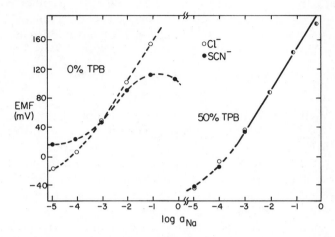

Figure 2.23 The SCN⁻ anion effect on a Na⁺ neutral carrier ISE based on the Na⁺ ligand of Figure 2.22. Membrane solvent: *o*-nitrophenyl octyl ether: 0, 50% TPB: mol% of the ligand concentration for tetraphenylborate addition to the membrane. From [163], with permission.

Electrode detection limits and lifetimes are governed by solvent and carrier partitioning; Oesch and Simon have presented a detailed summary of these factors for various solvent/carrier systems as they relate to macro-, micro-, and flow-stream electrode designs [166].

A list of the more common neutral carrier electrodes is presented in Table 2.33. Note that all these electrodes are cation-sensitive; no practical anion-responsive system has yet been devised although several possibilities have been treated by Eisenman and colleagues [168]. The valinomycin-based K⁺ ISE is the most widely employed electrode and is noteworthy for its remarkable selectivity over Na⁺. A principal application driving the design of several of the other systems was the need for highly selective sensors in the clinical field, where the H⁺, Ca²⁺, Na⁺, Li⁺, and Mg²⁺ measurements are desired. A review of the importance of the neutral carrier electrodes in the biomedical field is available [167]; several of the most selective systems have been incorporated into commercial blood analyzers. Other chemistries not described in Table 2.32 include those suggested for H⁺ (see Table 2.19), Mg²⁺, Ba²⁺, UO₂²⁺, Sr²⁺, Cd²⁺, and for enantiomers of chiral quaternary ammonium salts. The chronological review [8, 9] should be consulted for a more complete list and for original reference papers describing carrier systems not listed in Table 2.33. A few of the synthetic neutral carriers have been incorporated into commercial electrodes; more commonly, new compounds like the crown ethers, the cryptands, the spherands, and the acyclics are evaluated at the research level using the simple PVC-matrix electrode design [156].

Table 2.33 A Selection of Neutral Carrier Electrodes[a]

Electrode Type	Carrier	Solvent Type	pa_i Range	Selectivity Coefficient k_{M-1}^{pot}	pH Range	Manufacturer
pK	Valinomycin	Nonpolar	0–6	$Cs^+ \sim 0.4$; $NH_4^+ \sim 0.01$; $Rb^+ \sim 2$; $Sr^{2+} \sim 0.001$; Li^+, $Ba^{2+} \sim 0.0001$; H^+, Na^+, $Ca^{2+} \sim 0.00001$	2–11	Philips, Orion Radelkis, others
pNH_4	Nonactin Monactin	Nonpolar	0–6	$K^+ \sim 0.1$; $H^+ \sim 0.01$; Rb^+, Cs^+, $Li^+ \sim 0.004$ $Na^+ \sim 0.002$ All $M^{2+} < 0.0001$	4–10	Philips
pCa	See Fig. 2.22	Polar $\pm Ph_4B$	0–6	$Sr^{2+} \sim 0.01$; $H^+ \sim 10$ Li^+, $Cs^+ \sim 0.001$ Na^+, K^+, Mg^{2+}, Ba^{2+}, $Zn^{2+} \sim 0.0001$	3–10	Philips
pNa	See Fig. 2.22	Nonpolar	0–5	H^+, $K^+ \sim 0.5$; Ca^{2+}, $Mg^{2+} \sim 0.001$; Li^+, Cs^+, $NH_4^+ \sim 0.1$	4–12	—
pLi	See Fig. 2.22	Nonpolar	1–5	$H^+ \sim 1$; Na^+, $NH_4^+ \sim 0.05$ K^+, $Rb^2 \sim 0.005$ Cs^+, Ca^{2+}, Mg^{2+}, Ba^{2+}, $Sr^{2+} \sim 0.0001$	3–11	Philips

[a]All chemistries in PVC matrix with liquid solution internals. Data compiled from [20], [163], and [167].

5.3   Characteristics of Sensitized Ion-Selective Electrodes

5.3.1   Gas-Sensing Electrodes

The indirect gas-sensing electrodes rely on diffusion of a gaseous sample component into a dilute inner electrolyte layer, where a change in pH or ion content is measured with a sensing electrode. For example, the carbon dioxide electrode responds by the reaction

$$CO_2 + H_2O \rightleftharpoons HCO_3^- + H^+ \tag{176}$$

If the bicarbonate level in the inner solution, usually about 10 mM, is not altered significantly by the absolute amount of CO_2 diffusing in as gas, then

$$a_{H^+} = K_A \frac{a_{CO_2}}{a_{HCO_3^-}} = k' a_{CO_2} \tag{177}$$

where K_A is the acid dissociation constant of the carbonic acid. Then response of the CO_2 electrode will follow an internal pH glass electrode response given as

$$E = k'' + \frac{RT}{F} \ln a_{CO_2} \tag{178}$$

where k'' includes the glass electrode parameters like E^0 and asymmetry potential, the internal reference potential, and the k' from (177). Relation of the potential to a partial pressure of the analyte gas, for example, pCO_2, is accomplished by use of the Henry's law relation suitable for each gas.

Typical electrode constructions are depicted in Figure 2.24. The membrane type uses a thin gas-permeable polymer membrane to separate sample solution from inner electrolyte. Common membrane materials include microporous Teflon film, or homogeneous polymers such as silicone rubber and polyethylene

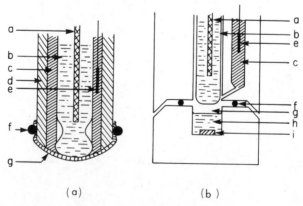

(a)   (b)

Figure 2.24 Gas-sensing electrodes: (a) gas-permeable membrane electrode; (b) air-gap electrode. a—inner reference electrode; b—filling solution of ISE; c—bulk internal electrolyte; d—shaft; e—reference electrode; f—O-ring; g—gas-permeable membrane/air gap; h—sample solution; i—stirring bar. From [148], with permission.

in which gases actually dissolve before permeation. All successful membrane materials are virtually impermeable to liquid water and ionic species, thus providing good selectivity over acid–base solutes that otherwise might alter inner solution pH. A second construction is the air-gap electrode of Figure 2.24b, in which an integral sample container positions an airspace between the sample and the inner electrolyte. The latter is usually wicked onto the glass electrode surface with the aid of a wetting agent or is supported by a thin porous sponge. The air gap serves as the "membrane" in this configuration, which has the advantages of fast response and freedom from membrane fouling but is more cumbersome to employ. Neither construction requires an external reference electrode, since an internal combination pH cell forms the measuring system. Most commonly the inner electrolytes contain an alkali chloride salt in addition to the ionic counterpart to the entering gaseous species, and an Ag/AgCl wire is then chosen as the inner reference electrode.

Table 2.34 lists commercially available gas sensors and some of their operating characteristics; both membrane and air-gap constructions can usually be found for each chemistry. The sensors for HF, H_2S, and HCN do not employ a pH glass electrode but rather the ISEs sensitive to the ionic counterpart of the diffusing gaseous species according to

$$HF + H_2O \rightleftharpoons H_3O^+ + F^- \quad (F^- \text{ ISE}) \tag{179}$$

$$H_2S + H_2O \rightleftharpoons H_3O^+ + HS^- \quad (Ag_2S \text{ ISE}) \tag{180}$$

$$HCN + H_2O \rightleftharpoons H_3O^+ + CN^- \quad (Ag_2S \text{ or AgCl ISE}) \tag{181}$$

Interferences for gas-sensing electrodes include sample volatiles with acid–base character or with redox power sufficient to alter internal electrolyte composition. Since compounds of this sort are limited in number, the gas electrodes possess a high degree of selectivity.

The operating range of the gas-sensing electrodes is determined by a number of factors, but inner electrolyte concentration is the predominant one. If too dilute, the inner ionic component (i.e., NH_4^+, HCO_3^-, HSO_3^-, etc.) concentration will be altered by incoming sample gas and the k' in (177) will no longer be constant. If too concentrated, then a significant partial pressure of diffusing gas will already exist because of simple dissociation equilibria and the inner electrolyte will act as a buffer with less than full Nernstian pH response. In addition, the electrode dynamic response time may be important in detecting low sample gas concentrations. Such factors as gaseous membrane permeabilities, slow electrolyte film reactions, sluggish glass electrode response, or equilibration between thin-film and bulk inner electrolyte areas may affect response time. Several authors [169, 170] have recently attempted to quantify these effects. Slow bulk/thin-film electrolyte equilibria can lead to prominent hysteresis effects in addition to lengthy response times. If sample and inner electrolyte differ greatly in osmotic pressure, then diffusion of water vapor can cause electrode drift by dilution or concentration of the thin-film layer. Dilution of concentrated samples is often recommended to remove this error source.

Table 2.34 Gas-Sensing Electrodes[a]

Indicated Gas	Indicator Electrode	Membrane	Reaction Solution $(M)^{b,c}$	Detection Limit (M)	Optimum pH	Interferences	Manufacturer
CO_2	pH glass electrode	Micropore filter[d] (1.5 μm)	10^{-2} NaHCO$_3$ 10^{-2} NaCl	$\sim 10^{-5}$	<pH 4		Ingold, Radiometer
NH_3	Same as above	0.1 mm Micropore Teflon film	10^{-2} NH$_4$Cl 0.1 KNO$_3$	$\sim 10^{-6}$	>pH 12	Volatile amines	Beckman, EIL, Orion, Radelkis, HNU
SO_2	Same as above	0.025 mm Silicone rubber	10^{-3} NaHSO$_3$ pH 5	$\sim 5 \cdot 10^{-6}$	<pH 0.7	Destroy Cl$_2$, NO$_2$ with N$_2$H$_4$; HCl, HF, acetate	EIL
NO/NO_2	Same as above	0.025 mm Micropore polypropylene	0.02 NaNO$_2$ 0.1 KNO$_3$	$\sim 10^{-6}$	<pH 0.7	Remove SO$_2$ with CrO$_4^{2-}$; CO$_2$	EIL, Orion, HNU
HF	Fluoride electrode	Micropore filter[d]	1 H$^+$	$\sim 10^{-3}$	<pH 2		Orion
H_2S	Ag$_2$S membrane	Same as above	Citrate buffer pH 5	$\sim 10^{-8}$	<pH 5	Reduce O$_2$ with ascorbic acid	Orion
HCN	Same as above	Same as above	10^{-2} KAg(CN)$_2$	$\sim 10^{-7}$	<pH 7	Remove H$_2$S with Pb^{2+}	Orion

[a]From [20], with permission.
[b]With air-gap electrodes an additional 0.1% of a nonionic wetting agent is added such as Victarwet (Stauffer Chemical Corp.).
[c]Membrane-covered sensors may also contain an agent to increase the viscosity.
[d]Such as Fluoropore with 1 μm pore size and a polyethylene backing (Millipore).

164

Attention must also be paid to the proper functioning of the inner sensor electrode. The pH glass inner sensor electrode should not be allowed to become dehydrated, and highly alkaline internal solutions are to be avoided. Slow response time at mid-range sample gas concentrations usually signals either membrane fouling or pH glass electrode problems.

Variations from the usual barrel-shaped format have included microelectrode and flow-through designs. The former are particularly useful in biological applications; a CO_2 electrode with a sensing-tip diameter of a few millimeters is commercially available. Inner ISE sensors other than those discussed have been tried on a research basis. For instance, Meyerhoff and Fraticelli [171] developed an NH_3 sensor with a nonactin-based neutral carrier membrane as the inner ISE and successfully automated its use in a flow system. Specific applications of gas-sensing electrodes can be found in the applications references [8, 9, 48]. The NH_3 and CO_2 electrodes are probably the most widely employed. In addition to primary analytical applications these two electrodes are often the base sensors for the biosensitive electrode systems discussed next. Riley [172] has presented a thorough review of the gas-sensing electrode technology.

5.3.2   Biosensitive Electrodes

This class of indirect ISE sensors combines a biocatalytic reaction that produces or consumes a measurable species with a base electrode taken from the primary ISE grouping. The general scheme is represented by

$$\text{Analyte} + \text{Biocatalyst} \rightarrow \text{Products} \pm \text{Ion} \tag{182}$$

where the ionic species is finally detected by the ISE. Most of the analytes studied to date have been biochemicals like enzyme substrates, amino acids, hormones, nucleotides, and drugs, although inorganic species can certainly be incorporated in such schemes. The first examples of this sensor were the enzyme electrodes, in which a single enzyme served as the biocatalyst. However, many different sorts of biocatalysts have now been studied, ranging from single enzymes to multiple enzymes and on up in complexity to tissue slices, bacteria, or whole cells. Table 2.35 lists just a few examples from the many such electrodes that have been proposed. An attempt is usually made to fashion an integral electrode by positioning the biocatalyst near the ISE surface, but often more practical analytical results are obtained if the biocatalytic reaction and sensing ISE are separate and coupled by a flow stream. Many analytical biosensor systems have been suggested with the Clark O_2 electrode or the Pt/H_2O_2, H_2O system as base sensors. Since these are amperometric devices they will not be discussed here, but the reaction principle given in (182) still follows.

The general principles for biosensitive electrode construction and use are illustrated by the enzyme electrodes. Figure 2.25 pictures three common device configurations. All contain a reactive layer in which an enzyme is contained by mechanical, physical, or chemical means. Diffusion of substrate into the reactive layer initiates enzymatic catalysis, and the resulting product ion buildup or

Table 2.35 Examples of Biosensitive Electrodes

Electrode Type	Analyte	Biocatalyst	Sensed Species	Sensor	Working Range (mM)	Reference
Enzyme	Urea	Urease	NH_3	NH_3 gas	0.1–10	[173]
	Penicillin	Penicillinase	H^+	pH glass	0.1–10	[173]
	Glucose	Glucose oxidase + peroxidase	I^-	Ag/AgI	1–100	[173]
	Tyrosine	Tyrosine-decarboxylase	CO_2	CO_2 gas	0.25–10	[173]
Dual enzyme	D-Gluconate	Gluconate kinase + 6-phosphogluconate dehydrogenase	CO_2	CO_2 gas	0.1–2.5	[173]
Tissue slice	Adenosine	Mouse intestine	NH_3	NH_3 gas	0.1–10	[174]
Bacterial	L-histidine	Pseudomonas	NH_3	NH_3 gas	0.03–3	[175]
Whole cell	Thyroxine	Liver microsomes	I^-	Ag/AgI	0.0001–0.01	[176]
Immuno	Lipid antibodies, complement	Lipid antigen liposomes containing $(C_5H_{11})_4N^+$	$(C_5H_{11})_4N^+$	$(C_5H_{11})_4N^+$ liquid membrane	—	[177]
Immuno	Estradiol	Peroxidase-labeled anti-estradiol	I^-	Ag/AgI	0.04–40 (nM)	[178]

166

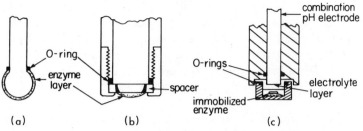

Figure 2.25   Enzyme electrode configurations: (a) glass-membrane electrode type; (b) gas-sensing electrode type; (c) air-gap type. Reprinted from [173], with permission.

depletion is then sensed by the electrode. The pH and gas-sensing ISEs are the most common internal sensors since many enzymatic reactions yield NH_3 and CO_2 as primary or secondary products. Large enzyme loadings are desirable to overcome activity losses during preparation and use, and to ensure that the steady-state reaction is not limited by enzymatic turnover. Since the combined Nernst film and enzyme membrane diffusion of substrate usually limit device response time, high stirring rates and thin enzyme membranes are desirable. Even so, response times in minutes are often observed, and a product washout step must often be provided between assays. Working substrate ranges generally lie between 0.1 and 50 mM, with the lower limit set by long response time and base ISE detection limits. Attention must also be paid to the often contradictory optimum conditions for both ISE sensor and enzymatic reaction. For instance, many NH_3-producing enzymatic reactions have neutral pH optima while the NH_3 gas sensor works best at alkaline pH. A successful analysis may then require a compromise of a slightly alkaline pH, separation of reaction and sensor with the addition of base to reaction products, or employment of a different sensor ISE like the nonactin NH_4^+-sensing neutral carrier membrane. Another fact about the enzyme (and other biosensitive) electrode systems is noteworthy: addition of the biocatalytic layer rarely improves the innate ISE selectivity. As a result, these electrodes can only be as selective as the catalytic layer and the base electrode allow. Finally, enzyme stability is often a limiting factor in electrode performance. Continual loss of reactivity may require frequent recalibration or membrane layer replacement. It may be because of this fact that stand-alone enzyme electrodes are presently being marketed by only one firm. A more thorough review of the enzyme electrode field can be found in [173].

Efforts to improve the biocatalytic-layer stability and selectivity led to the tissue, bacterial, and whole-cell electrodes. Rechnitz pioneered much of this work and has summarized the area in a general article [179]. Using more complicated reactive layers can provide pathways for determination of substrates for which no stable single enzymes are available. In addition, all necessary enzymatic cofactors are normally present in the natural reaction matrix provided by these layers, and stability is often enhanced. It may even be possible to regenerate the electrode catalytic layer *in situ*, as has been demonstrated for several of the bacterial electrodes sensitive to amino acids. Drawbacks to these electrodes include cumbersome electrode geometries, limited availability of suitable biocatalytic-layer material, and competing metabolic pathways in the natural media that often reduce selectivity. As Arnold and Rechnitz [174] have shown for the adenosine-sensitive tissue electrode, it may be possible to turn off these competing pathways selectively by addition of inhibitors to the reaction media.

Efforts have been made to quantitate very low concentration biomolecules like hormones and drugs with these biosensitive electrodes. Usually an amplification step is necessary given the micromolar or greater detection limit of the base ISE sensors. One possibility is to load liposomes with a large amount of an impermeable, easily sensed ion like a quaternary ammonium species. If the analyte can be made to selectively lyse the vesicles, enhanced sensitivity can be gained; but, unfortunately, few molecules possess this lysing ability. Attempts to measure a potential change due to the antigen–antibody reaction of a variety of proteins and drug haptens have been made. The electrodes can either be the indirect biosensitive type or in the primary class where a surface potential change caused by membrane loading of antigen/antibody component is measured. To date only limited success has been achieved because of sensitivity, selectivity, and reproducibility considerations.

The biosensitive electrodes have seen far more research application than use as practical commercial devices. Often a new device is tested successfully in aqueous buffer systems but has serious limitations for use in biological media or for *in vivo* measurements. Full commercial development of these electrodes must await advances in practical use and design. Vadgama [180] has assessed the limited impact of the enzyme electrodes, the first of the biosensitive systems, in the biomedical field. Other than the amperometric sensors for glucose, uric acid, and lactate, only the potentiometric urea electrode has seen widespread commercial utilization. Yet the analytical promise of simple potentiometric electrodes for such a wide variety of biomolecules will undoubtedly continue to fuel research on these devices.

5.4   Other ISE Formats and Devices

5.4.1   Nontraditional Formats

Modifications of the traditional barrel-shaped ISE format are possible. As was discussed for the glass pH electrode case, the desired goals are miniaturization and elimination of the liquid internal-reference solution. Although the patent and research literature are filled with possible ISE format variations, very few commercial examples have appeared as distinct electrode units. Most of the crystalline solid-membrane electrodes like AgCl and Ag_2S are available with solid internal contacts made from any Ag epoxy or solder at which mixed electronic–ionic conduction appears to occur. In principle the same format should work for the noncrystalline solid electrodes like the pH and pNa glasses, but few commercial examples exist despite the many patents on metal, metal halide salt, or carbon internal layers. A graphite-impregnated Teflon reference layer forms the universal base for the Ruzicka Selectrode materials. Ion sensitivity is provided by rubbing the electroactive material (silver halides, Ag_2S, DuS, CdS) onto the base or by adding a PVC-ionophore membrane disk (often with the interposition of a calomel paste). All the liquid-membrane electrodes (ion-exchange and neutral carrier) can be miniaturized by incorporation into micropipet tips; but an internal reference solution is usually provided. Numerous examples are given in a review by Walker [181].

Many of the format alterations appear in the ISE sensors employed in the various clinical analyzers now being marketed. Examples include cylindrical flow-through PVC-ionophore membranes with and without internal solutions; dip-type sensors consisting of PVC-ionophore membranes coated over silver wires; and flat, disposable ISEs having dry internals consisting of gelatin–metal halide salt mixtures. All these formats are designed to reduce sample volume requirements to the 10–500 μL range desirable in clinical situations. Usually the electrode designs are such that the specialized apparatus found in the various analyzers is required to obtain satisfactory potentiometric results.

5.4.2   Coated-Wire Electrodes

This class of alternative-format ISEs pioneered by Freiser and co-workers [182] has enjoyed enough application to be identified as a separate ISE category. The coated-wire electrodes (CWEs) consist of ion-selective chemistries in a polymeric matrix coated directly over a metallic wire conductor. No attempt is made to define a thermodynamic, ion-reversible internal reference by the usual solution or solid salt means. Many metals have been used as internal conductors, but Ag, Cu, and especially Pt are favored. Virtually all the liquid-membrane ISE chemistries can be incorporated in this format; the liquid ion-exchanger Aliquat 336 being particularly popular. Polymers of choice are polyvinyl chloride (PVC), polymethyl methacrylate (PMMA), or other hydrophobic solvent-soluble matrices. Unless the ion-selective chemistry possesses plasticizing ability (like the liquid exchanger, Aliquat) a plasticizer solvent is also included in the formulation. The coating "recipe" of chemistry, plasticizer, and polymer matrix dissolved in an evaporable coating solvent like tetrahydrofuran or cyclohexanone is virtually identical with that used to fabricate PVC liquid membranes for traditional barrel formats.

Electrode preparation proceeds by first cleaning the wire with mechanical polishing and solution washes. The clean, exposed end of the wire is then dipped into or sprayed or painted with the coating melt. Air drying followed by repeated coatings then follows, thicknesses of 0.1–1 mm usually being satisfactory. Uncoated-wire portions are then masked with potting epoxy, Teflon tape, Parafilm, or like nonconductive, water-impermeable material. Normally a preconditioning step of several hours or more in a dilute solution of the intended analyte must be provided. Storage is either in conditioning solution or in air if a conditioning step is performed after each long dry period. For detailed coating recipes one may consult the published literature, but as a starting point a weight percent ratio of 40:40:10 for polymer:plasticizer:ion-selective component is often usable. For the liquid exchangers like Aliquat or the tetraphenylborates, the counterion of choice may be introduced in a precoating ion-exchange equilibration or by conditioning the coated electrode in a concentrated salt solution of the analyte.

In general, CWE response characteristics like detection limit, selectivity coefficients, and lifetimes are similar to the PVC-membrane barrel-format electrodes with internal solution. Rather noticeable potential drifts occur during the CWE conditioning phase, probably related to the internal potential generation. Although purely capacitative coupling of the membrane to the metal conductor is theoretically possible, it appears instead that the oxygen/water half-cell is the internal potential-generating mechanism. Proof of this supposition accumulates from electrode potential changes with either test solution pO_2 or with wire-surface pretreatment by voltammetric cycling procedures [183]. Conditioning time to relatively stable potentials also increases with coating thickness, consistent with the premise that H_2O and O_2 must diffuse to the metal contact for internal potential generation.

The advantages of the CWE technology are low cost and simplicity. Any test ion-selective, liquid-membrane chemistry can be quickly evaluated in this format. Applications of the technology have been numerous but mainly based at a research level. Various anionic drug-sensitive electrodes have been prepared from Aliquat CWEs [184], although selectivity among different drug species is relatively poor. The same strategy has produced CWEs sensitive to anionic surfactants, amino acids, and multivalent cations present as SCN^- or Cl^- anionic complexes [182]. All the neutral-carrier chemistries can presumably also be incorporated in this format, but only the valinomycin-based K^+ electrode has been well studied to date. Miniaturization with very fine wire or with small, thin internal metal sheets should also be possible with the CWE technology.

5.4.3   ISFETs

The ion-selective field effect transistor (ISFET) is a device that has drawn increased attention in the past few years. This device is basically a combination of an ion-selective membrane and an ungated field effect transistor; a schematic is shown in Figure 2.26. The electric field at the insulator region next to the membrane modulates the current flow I_D between source and drain n-type semiconductor regions. This modulation occurs because of changes in the effective gate voltage formed by the combination of external poising voltage V_G, reference electrode potential E_R, and membrane surface potential E_m. Although a complete derivation is beyond the scope of this chapter, the ISFET response can be approximated by

$$I_D = kV_D \left(V_G - k' \pm \frac{RT}{z_i F} \ln a_i - E_R - \frac{V_D}{2} \right) \qquad (183)$$

where a_i is the solution activity of the ion to which the ISFET membrane

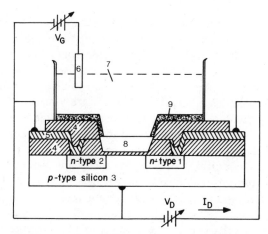

Figure 2.26   Schematic diagram of an ISFET: (1) drain, (2) source, (3) substrate, (4) insulator, (5) metal lead, (6) reference electrode, (7) solution, (8) membrane, (9) encapsulant. Parts (6), (7), and (8) replace the metal gate of a common MOSFET. Reproduced from [185], with permission.

responds and V_D is the drain voltage. Thus at constant V_D, V_G, and E_R measurements of the drain current will follow solution activity just as the E_{cell} for a normal ISE does. Janata and Huber [185] have described the conditions under which (183) is valid in their general review of ISFET principles. The ISFET dispenses with the internal reference as does the coated-wire electrode format, but here a truly blocked interface appears to be operative in that the insulator layer allows no charge (electrons or ions) to penetrate it.

The initial ISFET fabrications had no ion-selective membrane layers; surface hydration of insulator materials like SiO_2, Si_3N_4, and aluminosilicate apparently can produce fixed SiO^- sites, which provide H^+ or Na^+ sensitivity. These "bare-gate" ISFETs thus behave analogously to glass electrodes, and as such are susceptible to changes in response with hydration time. The most stable of these devices is the Si_3N_4 ISFET, which apparently hydrates to a relatively constant $Si_3N_4(O)_n$ composition. To improve stability and provide a wider range of ion sensitivity, most ISFETs are now prepared with an added ion-selective layer. Almost all chemistries have been tried, the only restriction being that the addition of the selective layer should provide a means of adhesion (like solvent casting or thermal sputtering) to the FET surface. Liquid-membrane chemistries are normally solvent cast from a PVC melt, while solid crystalline materials are often dispensed in microcrystals in a heterogeneous membrane based on silicone rubber or other solvent-castable elastomers. The fabrication engineering of the ISFETs is a formidable challenge because of the small size, the requirement of pinhole-free membranes of reproducible thickness, and the necessity of preventing solution leakage at the nonactive portions of the device. Employment of light-curable photoresist technology has been demonstrated for a K^+ sensor construction [186], and may point to a mass-production pathway. The placement of a polyimide mesh on the ISFET surface has also been shown to virtually eliminate solution leakage, presumably by providing anchor points for the PVC membrane layer [187]. It is probably the difficulty of large-scale production that has kept this kind of sensor from commercial application to date.

Equilibrium response characteristics like sensitivity, operating range, and selectivity for ion-selective chemistries in ISFETs are similar to what is observed for conventional formats. Either the drain current I_D is measured as a function of solution a_i, or the devices are operated in constant-current mode whereby variable a_i levels change the V_G by means of a feedback mode. Reference [185] provides measuring circuit descriptions for each case. Device stability is well over a month, the limit usually being set by either chemistry component leach-out from the membrane or loss of encapsulant integrity. An initial solution hydration period seems to be required to limit drift, as was true for CWEs. In the case of ion-sensitive chemistries with very high resistance like the neutral carrier membranes, response times can be shortened over conventional format ISEs because the RC time constant of the conventional ISE shielding cable is not operative. However, most ISFETs show similar response times to conventional ISEs because solution layer or membrane layer kinetics usually govern in both formats. Temperature sensitivity can be more severe in the ISFET format because of the FET operating characteristics, but this can often be compensated in a differential mode by using a second FET on the same microchip. Noise compensation by this differential procedure is also possible, and this has demonstrated advantages for intracellular ion measurements [188].

ISFET applications have been limited to the research level to date, but this position will undoubtedly change as manufacturing difficulties are overcome. The ultimate use of the device might be an inexpensive, single-use, multiple-ion sensor requiring only microliter solution volumes. Several hand-made reusable multiple-ion sensors have already been described, and suitable on-chip reference ports are possible as shown in Figure 2.27. Coupling of an ISFET with a bio-catalytic layer is feasible; so-called enzyme field effect transistors, ENFETs, have been demonstrated for urea and penicillin based on the enzyme electrode principle. Use of ISFETs in flow analysis schemes like flow-injection analysis

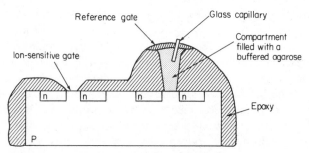

Figure 2.27 Schematic of a combination ISFET having sensor and reference gates on the same microchip. Reproduced from [185], with permission.

(FIA) may be attractive because the small ISFET size can help ease sample-dispersion and dead-volume problems [189]. Finally, attempts have been made to convert the charge-sensitive FET construction into an immunosensor capable of measuring the antigen–antibody reaction at the monolayer level [190], although success has been limited.

5.5   Important ISE Operating Characteristics

5.5.1   Selectivity Coefficient

Ion-selective electrode interferences by like-signed ions are most often characterized by means of the $k_{i/j}^{pot}$ parameter introduced in (46) and (47). The selectivity coefficient is not usually a constant, and so should not be used to quantitatively subtract a "blank" term from an ISE measurement. Instead, it can give an indication of the analytical utility of a proposed assay using an ISE as detector. A point of confusion to those unfamiliar with ISE technology is that the k^{pot} *decreases* as the ISE becomes more selective; most of the analytical literature finally being consistent on this matter. Electrode manufacturers usually list the $k_{i/j}^{pot}$ for each of several j interferences on any electrode; to be totally informative the method of measurement and interfering concentrations should also be listed. Pungor and co-workers [191] have compiled perhaps the most extensive listing of k^{pot} values for many different ion-selective chemistries. Measurement of $k_{i/j}^{pot}$ can be done by using separate-solution or mixed-solution techniques. In the separate solution case illustrated in Figure 2.28, response to individual analyte and interferent solutions is monitored. The k^{pot} can be extracted from any two potentials according to

$$\log k^{pot} = \frac{(E_2 - E_1)zF}{2.3\,RT} = \frac{\Delta E}{S} \tag{184}$$

or from any equipotential point (such as E_2 in Figure 2.28) where

$$k_{E_2}^{pot} = \frac{a_m}{a_I} \tag{185}$$

As can be noted in Figure 2.28, differences in electrode slope for analyte and interferent lead to differing k^{pot} estimates depending upon the potentials chosen for calculation. The mixed-solution method is more strongly recommended for k^{pot} estimation. Electrode response is followed for a series of analyte solutions containing a constant activity of interference, as shown in Figure 2.29. The k^{pot} is calculated as in (185), with the measured ion activity a_m taken from the point of intersection of interferent and Nernstian response lines. Alternatively, an a_m is taken where the potential of the experimental curve lies $18/z$ mV above the extrapolated Nernstian line. A family of k^{pot} values may be obtained depending upon the activity of interfering ion chosen. Yet this method possesses operational utility if k^{pot} measurement conditions are matched to the expected analysis

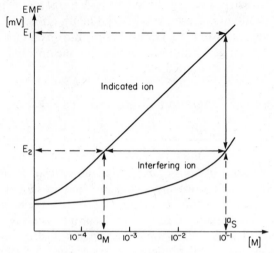

Figure 2.28 Determination of the selectivity coefficient by the separate-solution method. The k^{pot} is derived from (184) or (185). From [20], with permission.

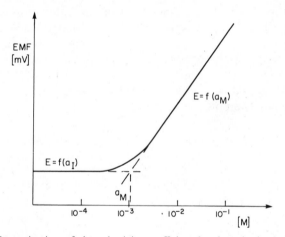

Figure 2.29 Determination of the selectivity coefficient by the mixed-solution method. The solutions contain a fixed amount of the interferent a_I at varying analyte a_M; (185) is used to extract the k^{pot}. From [20], with permission.

conditions. Another variation of the mixed-solution method allows the interfering-ion activity to vary at fixed analyte activity. It is especially useful for "at-a-glance" plots showing the pH range of an ISE.

5.5.2   Precision and Accuracy

From rearrangement and differentiation of the Nernst equation the da_i/dE for an ISE is found to be approximately $4z\%$ per millivolt at $25\,^\circ$C, independent of activity range. Thus cell potentials must be known to 0.25 mV if an assay precision of 1% is desired for a univalent ion (0.125 mV for a divalent ion). This precision level is about the limit attainable in direct ISE potentiometry without taking tedious operating precautions; it corresponds to ± 0.005 pH unit on the pH measurement scale.

Durst [192] has discussed limitations on ISE accuracy. It is important to recall that the electrodes respond to free ion activity. If analyte concentrations are desired, attention must be paid to sample ionic strength, the presence of inert volume-occupying agents like proteins, and possible analyte ion speciation under the influence of pH or complexing ligands. Apart from the effects of like-signed ions electrodes may suffer interference from species that adsorb at surfaces like surfactants or biomolecules, or from ions of opposite sign that violate permselectivity, as in the case of hydrophobic anion effects with neutral carrier membrane electrodes. Of course, attention must also be paid to the total ISE cell, since reference electrode difficulties outlined in the pH measurement section apply here as well. When one draws up a long list of possible ISE error sources, the simplicity and utility of the technology can sometimes appear diminished. The summary point simply is that one must pay attention to the analytical situation even with ISEs. Too often in the author's experience those unfamiliar with ISE technology assume a simple "dip-and-read" of a sensor into any unknown matrix will produce satisfactory results.

5.5.3   Time-Related Response Parameters

Three regions of time-related response are of interest with ISEs: the dynamic response-time region from activity change to equilibrium response, which includes all transient behavior (milliseconds to several minutes); the slow drift region in solutions of constant composition (minutes to hours); and the very long-term region representing usable ISE lifetime (months to years). Investigation of transient ISE response still represents an active research area in which newer techniques like impedance measurements [193] and flow-injection methods [194] are being applied. While many processes can affect this behavior, solution and membrane-diffusion layer phenomena are usually governing. Reduction of the influence of the Nernst solution film by vigorous stirring or jet-flow techniques can reveal the differences in membrane-diffusion processes peculiar to each type of ISE chemistry. However, relation of observed time constants to specific molecular processes is not yet completely possible. As a result no intrinsic "response time" can be stated for a given electrode, even though some manufacturers are beginning to list a range of response times expected for a properly functioning ISE. The slow drift in ISE response usually indicates changes in membrane or internal solution composition like hydration, dissolution, and so forth. While annoying, it may not be avoidable and is usually dealt with by frequent recalibration. Completion of these slowly proceeding compositional changes can lead to cessation of workable ISE response. Oesch and Simon [166] have related neutral carrier electrode lifetime to diffusion and solubility of plasticizer and ionophore, for example. As a crude rule of thumb the minimum expected solution lifetimes of ISEs are a year for static systems and six months for flow-stream sensors.

5.5.4   Other Operating Considerations

Complete operating procedures for all electrodes in each measurement mode (direct, titration, and so on) cannot be discussed here. Most of the general ISE references contain usage sections; the texts of Cammann and of Bailey are especially practical in this regard. One important parameter of general application is ISE cell temperature control. Temperature sensitivities of ISE sensor, reference electrode, liquid junction (if present), and even sample speciation equilibria must be considered. For careful work calibrators and samples must be at the same temperature, and the total cell temperature controlled to within $\pm 0.1°C$. For routine laboratory investigation control to $\pm 0.5-1°C$ is usually sufficient.

5.6   Analytical Techniques and Applications

5.6.1   Direct Potentiometry

In this most common ISE technique the sample activity or concentration is derived from a calibration curve like that shown in Figure 2.15. Usually two calibrators suffice to define the electrode response in the linear range; bracketing of the sample is of course desirable. Samples may be neat, or diluted with inert salts, buffers, or demasking agents; the "direct" simply refers to the method of calibration followed by separate sample measurement. If solution ionic strength and junction potential do not change significantly over the intended analyte assay range, then response will be linear in analyte concentration. Often this is not the case and response rolloff is noticed at high analyte concentrations. In order to linearize the calibration in this situation, a suitable activity-coefficient function is used to calculate analyte activity (often the $\gamma_\pm$ calculated in Table 2.4 will suffice). Although individual ion-activity standards have been suggested for pNa, pK, pF, and pCl [32], no defined calibration standards analogous to pH standard buffers have yet been universally adopted.

The direct potentiometric technique requires more knowledge about the sample matrix than other ISE methods, since for accurate assays calibrators and samples should be similar in composition. This technique does provide the fastest and most simple ISE analysis method, however. With modern microprocessor-based millivolt meters set up for readout in pIon units, stored calibration parameters may be used to obtain results directly [41]. Although assays in the linear response range are most desirable, extension of the direct ISE techniques to trace levels represented in the nonlinear response regions are possible with care. Midgely [195] has given several examples of successful ISE assays in this region.

Uncountable applications of the direct ISE analysis techniques are found in a wide variety of industrial, biological, and water-monitoring situations. The cited applications reviews can be consulted for extensive listings [8, 9, 11, 36, 48]. A noteworthy commercial application of this technique lies in the clinical measurement of electrolytes in human fluid samples. About 15 different manu-facturers market potentiometric analyzers for various ions from the group of biological interest (Na^+, K^+, Cl^-, CO_2, H^+, Ca^{2+}), featuring small sample sizes of less than 500 μL and a high degree of microprocessor-controlled auto-mation. Some of the systems dilute all fluids with an ionic-strength-adjusting buffer, but many perform an undiluted measurement on whole blood or serum. At present the latter systems report sample concentration by assuming E_j and γ_i remain constant over the relatively narrow clinical analyte ranges. However, the potential exists for actually reporting and using a sample activity, if agree-ment on suitable activity standards were to be reached and E_j variation is indeed small. This would be the first instance other than the pH measurement where the truly unique ability of the ISE as an activity-sensing device would be commercially exploited.

5.6.2   Indirect Techniques

In these methods the change in potential of a starting solution is measured upon addition of a second solution. The methods range from single or multiple additions involving a primary sensed ion to traditional potentiometric titrations in which the object may even be to quantify an ion not directly sensed by the ISE. These methods are chosen when sample matrix cannot be easily matched by synthetic calibrators, in order to extend the analytical utility of ISEs to species for which no direct measurement is possible, or simply to gain the increased accuracy and precision obtainable with multiple potential readings or large end point potential changes. The whole subject has been reviewed by Mascini [196].

In the standard addition variations a change in potential is related to a change in concentration of a primary sensed ion by

$$\Delta E = \pm S \log\left[\left(\frac{C_x + \Delta C}{C_x}\right)\frac{\gamma_x' k_x'}{\gamma_x k_x}\right] + \Delta E_j \qquad (186)$$

where C_x is the unknown concentration, γ_x the activity coefficient, k_x the fraction of free measured ion in complex-containing solutions, and primed quantities indicate the solution after addition. In the original standard addition experiment a known volume V_s of standard C_s is added to a volume V_x of the unknown. If the amount of C_s added is small, then the ΔE_j and $\gamma_x k_x$ terms fall out, and one may derive the unknown C_x according to

$$C_x = C_s\left(\frac{V_s}{V_x + V_s}\right)\left[10^{\Delta E/S} - \frac{V_x}{V_x + V_s}\right]^{-1} \qquad (187)$$

Variations on this method include standard subtraction, in which a known amount of complexing or precipitating agent is added, and analate addition or subtraction, where sample solution is added to standard. Most of the modern microprocessor-based millivolt meters enable simple performance of these incremental techniques by keyboard addition of C_s, S, and the volumes with subsequent automatic readout of C_x. The single-point incremental methods require knowledge of the electrode slope S. By using multiple additions it becomes possible to extract electrode E^0, S, and C_x by computer-calculated regression of the data. Linear regions of multiple addition techniques can also be treated by a suitable Gran function

$$F = (V_x + V_T)10^{\Delta E/S} \tag{188}$$

from whose slopes and intercepts C_x can be derived. Special Gran antilog paper or Gran slide rulers are available that make calculations unnecessary. The multiple addition methods are obviously more time consuming than single-point increments, but one gains precision and possible diagnostic information about system response. For each incremental method a standard or sample addition must be chosen so as to obtain large enough millivolt changes without drastically altering the measured solution matrix (such as E_j, γ, pH).

Ion-selective electrodes have long been employed as sensors in potentiometric titrations, the two most common variants being precipitation and compleximetric titrations. The large potential excursions available at the equivalence point yield precision and accuracy advantages over other ISE methods; the electrodes do not even have to respond in a Nernstian fashion. A large range of ions for which there are no practical direct ISE sensors, including La^{3+}, Al^{3+}, Co^{2+}, V^{2+}, and Zr^{4+}, can be analyzed by these titration methods. Titration applications and procedures are discussed in many of the references already cited [20, 35, 36, 44, 69, 192]. Horvai and Pungor [197] have addressed the precision available in ISE titration methods. As with pH and redox titrations the behavior of electrode and chemistry around the equivalence point and the mathematical attempts to locate the equivalence point are important considerations. Equivalence point artifacts, like occlusion or supersaturation in precipitation titrations and low sensed-ion levels in compleximetric titrations, can sometimes be avoided by the Gran procedure. Gran functions like (188) can be derived for each titration variant, and application of the functions to several points before and after the equivalence point allows linear extrapolation for V_e. Midgely and Torrance [36] have presented both simple and rigorous forms of the Gran function for the potentiometric titration variants, and have precautioned against possible curvature sources such as γ and K_{sp} changes. Computer-controlled titration systems that can be interrogated to find the optimum region for application of the Gran procedure have also been suggested [198].

Two final indirect ISE methods that are sometimes useful are the null-point and the differential-titration methods. In the null-point method two identical ISEs are placed in sample and reference compartments that are separated by a frit or capillary. Some typical cell arrangements are shown in Figure 2.30. The basic idea is to add a known solution to the reference ISE side until zero millivolt is obtained between the two sides. Multiple additions allow a statistically sound extrapolation to the null point. The ideal situation occurs when the analyte is a minor component of the sample matrix so that E_j and γ considerations do not apply. Advantages of the technique include a reduced reliance on Nernstian

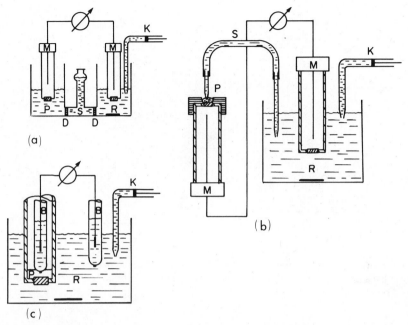

Figure 2.30 Cell constructions for null-point potentiometry with ion-selective electrodes. (a) Macroconstruction with two identical ISEs (M); the sample solution (P) is separated from the reference solution (R) by a salt-bridge (S) and two frits (D) with standard added by burette (K). (b) Microconstruction for small samples (P). (c) Semimicroconstruction using a dismountable ISE barrel and two identical reference electrodes (B). From [20], with permission.

electrode sensitivity, the ability to make the assay without altering the actual sample fluid, and increased accuracy for very dilute solutions. In the differential-titration procedure, responses of two identical electrodes are separated in time by suitable capillary spacing. The difference in potential between the two electrodes is monitored as the sample part of the cell is titrated, with the end point indicated as a peak in the ΔE versus titrant volume plot. The advantage here is that electrode E^0 or slope drift changes are compensated during the titration. A microprocessor-controlled, automatic differential titrator for pH or ion-selective electrodes has been described [199].

5.6.3   ISEs in Flowing Systems

The employment of ISEs in flowing analytical systems is attractive because of the higher sample throughput available and the possibility for downstream positioning of the reference electrode. Detector cells with limited dead space are relatively easy to design; several are shown in Figure 2.31. Electrodes can be used in continuous air-segmented flow systems [201] or in nonsegmental flow-injection systems [202]. Important phenomena to consider include potential variations due to pump oscillations, reduced linear ranges that occur because

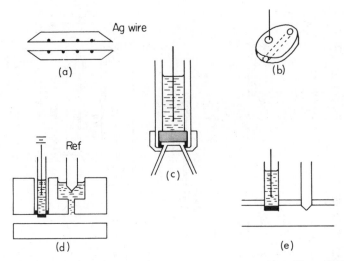

Figure 2.31 Flow-through potentiometric cells and electrodes: (a) capillary glass electrode; (b) capillary solid-membrane electrode; (c) flow-through cap electrode; (d) and (e) flow-through cells. From [200], with permission.

of sample dispersion in the flow stream, and electrode response time. In general electrode performance is fairly good even at high flow rates since the reagent stream provides a constant fresh bathing matrix and rapid flow can be used to reduce Nernst diffusion film thickness. Sample throughputs of up to 400 per hour have been achieved for both continuous and injection variants, although the 50–150 per hour range is more common. Solid-membrane electrodes like glass or crystalline types generally show fewer flow artifacts than PVC-based liquid-membrane electrodes; a description of a Cl^- ISE analysis in flow-injection mode outlines the system behavior [203]. Gozzi and Ferri [204] have described variations in selectivity, detection limit, and precision for a liquid ion-exchange electrode in a flow injection setup. Some variation can be traced to simple PVC membrane area changes with flow oscillations; solid contact internal reference layers or very low flow rates are two solutions to this problem currently employed in ISE-based clinical analyzers. All the ISE analytical techniques like direct potentiometry, incremental additions, or titrations can be successfully implemented in flowing systems. For titrations the mass flow rate of titrant or sample can be varied as an analogue of the volume addition in batch static procedures. Pungor and co-workers [200] discuss the various ways to perform titrations in flowing streams and list numerous applications of the technique.

5.6.4   Nonaqueous and Other Specialized Media

The application of ISEs in nonaqueous solvents has been limited to date but undoubtedly will increase as work on aqueous media continues to mature.

Pungor and co-workers have given several reviews on the subject including [205, 206]. One of the chief problems is that discussed for the pH electrode case: limited ability to fundamentally define what the electrodes are actually measuring. Effects like activity-coefficient variation, ion pairing, high resistance due to low solvent dielectric media, uncertainty in E_j at a bisolvent junction, and limited understanding of electrode response mechanisms act as roadblocks to the analyst. Yet the majority of the solid-membrane electrodes can be made to respond remarkably well in a wide variety of mixed or nonaqueous media. General nonaqueous response has been studied for the F^- electrode [207], the Pb^{2+} electrode [208], the silver halide electrodes [209], the Cu(II) electrode [210], and the cation glass electrodes [211]. On the whole, usable response is noted, particularly if 10–20% water is added to the pure nonaqueous solvent. Artifacts like super-Nernstian response or electrode–solvent interaction [as found for acetonitrile and the Cu(II) electrode] can sometimes be observed, however. Of course limited application of the liquid-membrane electrodes is found, since electrode system components like ionophore, solvent, and plastic matrix tend to dissolve in organic media.

The lack of defined activity standards puts a decided limitation on direct potentiometric use of ISEs in nonaqueous media. Efforts have been made to measure transfer activity coefficients (Table 2.13) with ISEs; and a defined set of $_m\gamma_i$ would then allow correspondence with the aqueous activity scale. More

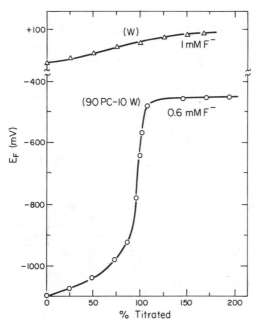

Figure 2.32 Comparison of the potentiometric titration of fluoride ion with lanthanum ion in water and in propylene carbonate containing 10 mol% (2 vol%) water. From [207], with permission.

often indirect titration procedures are employed. In some cases, like precipitation titrations, it may even be advantageous to work with organic solvent media, since the lowered K_{sp} values can both sharpen equivalence points and also reduce solubility-limited electrode detection limits. Figure 2.32 shows this effect for the titration of F^- with La^{3+} using the F^- electrode as end point detector. The same precautions about reference electrode choice and cell designs that were cataloged in the nonaqueous pH and redox sections apply equally well to the ISE case. Junctionless cells with Ag or Ag/AgCl wires, or junction-type reference electrodes with solvent-containing salt bridges are common.

Several aqueous media with particular matrix compositions like constant ionic strength have seen application of specialized ISE methodology. Although the principles and electrodes are the same as discussed here, operational procedures are tailored to suit the particular matrix. The following cited reviews can be consulted for ISE application in seawater [21], clinical medicine [212], pharmaceutical analysis [213], fermentation media [214], and environmental monitoring [215].

6   THE FUTURE OF POTENTIOMETRY

With the rapid expansion of ISE technology in the past decade now slowing, the total field of potentiometry seems again to be reaching a mature plateau. Potentiometric redox measurements continue to decline in favor; no major review or text has appeared in this area for nearly five years. pH and ISE techniques remain an important part of the analyst's repertoire.

While predictions are completely arbitrary, it appears that certain areas of this field will receive increased attention in the immediate future. Advances in *theory* are required in microscopic explanation of electrode response, in definition of pH and pX standard scales in both aqueous and nonaqueous media, and in the general knowledge about ion–ion and ion–solvent interaction. Progress in *materials research* leading to new electrodes is likely in the semiconductor–chalcogenide area, in neutral carrier chemistry to fill out missing cation- and possibly anion-sensitive electrodes, and in biosensor–immunoelectrode studies. Continued trends in *device engineering* like ISFET manufacturing and multiplexing, design of automatic, intelligent flow titration systems with microsensors, and complete abolition of internal reference solutions could be forthcoming. Perhaps the goal of disposable, factory-calibrated, multi-ion microsensors having all measuring and software data reduction within the devices will be realizable (the dip-and-read device?). It appears now that chemical and analytical principles rather than engineering considerations would provide the limits on such devices.

References

1.   M. S. Frant and J. W. Ross, *Science,* **154**, 1553 (1966).
2.   J. W. Ross, *Science,* **156**, 1378 (1967).
3.   L. A. Pioda, V. Stankova, and W. Simon, *Anal. Lett.,* **2**, 665 (1969).

4. E. Pungor, J. Havas, and K. Toth, *Z. Chem.*, **5**, 9 (1965).
5. J. Ruzicka and J. C. Tjell, *Anal. Chim. Acta*, **49**, 346 (1970).
6. G. J. Moody, R. B. Oke, and J. D. Thomas, *Analyst*, **95**, 910 (1970).
7. R. W. Cattrall and H. Freiser, *Anal. Chem.*, **43**, 1905 (1971).
8. M. E. Meyerhoff and Y. M. Fraticelli, *Anal. Chem.*, **54**, 27R (1982) (biennial reviews in April of even-numbered years).
9. R. P. Buck, J. Thompson, and O. Melroy, in H. Freiser, Ed., *Ion-Selective Electrodes in Analytical Chemistry*, Vol. 2, Plenum, New York, 1980, Chapter 4.
10. B. Kratochvil, *Anal. Chem.*, **54**, 105R (1982) (biennial reviews in April of even-numbered years).
11. "Applications Reviews," *Anal. Chem.*, **53**, 5 (1981) (reviews in April of odd-numbered years).
12. G. J. Moody and J. D. Thomas, Eds., *Ion-Selective Electrode Reviews*, Pergamon, Oxford–New York, 1979ff.
13. S. Middlehoek, Ed., *Sensors and Actuators*, Elsevier, Amsterdam, 1980ff.
14. H. K. Lonsdale, Ed., *Journal of Membrane Science*, Elsevier, Amsterdam, 1976ff.
15. R. P. Buck, "Theory and Principles of Membrane Electrodes," in H. Freiser, Ed., *Ion-Selective Electrodes in Analytical Chemistry*, Vol. 1, Plenum, New York, 1978.
16. P. Henderson, *Z. Phys. Chem.*, **59**, 108 (1907).
17. M. Planck, *Ann. Physik*, **39**, 161 (1890).
18. W. E. Morf, *Anal. Chem.*, **49**, 810 (1977).
19. D. A. MacInnes, *The Principles of Electrochemistry*, Dover, New York, 1961, pp. 236 and 339.
20. K. Cammann, *Working with Ion-Selective Electrodes*, Springer-Verlag, Berlin–New York, 1979, p. 40.
21. C. Culberson, "Direct Potentiometry," in M. Whitfield and D. Jagner, Eds., *Marine Electrochemistry*, Wiley-Interscience, New York, 1981.
22. D. Ives and G. Janz, *Reference Electrodes*, Academic, New York, 1961.
23. D. Brezinski, *Anal. Chim. Acta*, **134**, 247 (1982).
24. R. Bates, *Determination of pH: Theory and Practice*, 2nd ed., Interscience, New York, 1973.
25. J. Ross, *Chem. Eng. News*, **59**(20), 54 (1981).
26. K. Johnson, R. Voll, C. Curtis, and R. Pytkowicz, *Deep-Sea Res.*, **24**, 915 (1977).
27. W. Hamer and Y. Wu, *J. Phys. Chem. Ref. Data*, **1**, 1047 (1972) and following papers in *J. Phys. Chem. Ref. Data* by various authors; **6**, 385 (1977); **7**, 263 (1978); **8**, 923 and 1005 (1979); **9**, 513 (1980); and **10**, 1, 671, and 765 (1981).
28. R. Pytkowicz, Ed., *Activity Coefficients in Electrolyte Solutions*, Vols. 1 and 2, CRC, Boca Raton, FL, 1979.
29. H. Harned and B. Owen, *The Physical Chemistry of Electrolyte Solutions*, 3rd ed., Reinhold, New York, 1958, Chapters 12 and 14.
30. R. Pytkowicz, *Activity Coefficients in Electrolyte Solutions*, Vol. 2, CRC, Boca Raton, FL, 1979, Chapter 1.
31. P. Meier, *Anal. Chim. Acta*, **136**, 363 (1982).
32. A. Covington, "pX Standards," in A. Covington, Ed., *Ion-Selective Electrode Methodology*, Vol. 1, CRC, Boca Raton, FL, 1979.
33. K. Pitzer and J. Kim, *J. Am. Chem. Soc.*, **96**, 5701 (1974).
34. L. A. Bromley, *AIChE J.*, **19**, 313 (1973).
35. K. Toth, G. Nagy, and E. Pungor, "Analytical Methods Involving Ion-Selective Electrodes," in A. Covington, Ed., *Ion-Selective Electrode Methodology*, Vol. 2, CRC, Boca Raton, FL, 1979.

36. D. Midgely and K. Torrance, *Potentiometric Water Analysis*, Wiley, New York, 1978.
37. P. Burton, "Instrumentation for Ion-Selective Electrodes," in A. Covington, Ed., *Ion-Selective Electrode Methodology*, Vol. 1, CRC, Boca Raton, FL, 1979.
38. D. Dyrsson, D. Jagner, and F. Wengelin, *Computer Calculation of Ionic Equilibria and Titration Procedures*, Wiley, New York, 1968.
39. A. Zuberbuhler and T. Kaden, *Talanta*, **29**, 201 (1982).
40. P. Linder and R. Torrington, *Talanta*, **29**, 249 (1982).
41. G. Moody and J. Thomas, *Lab. Pract.*, **28**, 125 (1979).
42. C. Martin and H. Freiser, *Anal. Chem.*, **51**, 803 (1979).
43. A. Covington, *Lab. Pract.*, **31**, 240 (1982).
44. G. Svehla, *Automatic Potentiometric Titrations*, Pergamon, Oxford, 1978.
45. J. Smit and H. Smit, *Anal. Chim. Acta*, **143**, 45 (1982).
46. J. Smit, H. Smit, H. Steigstra, and U. Hannema, *Anal. Chim. Acta*, **143**, 79 (1982).
47. J. Bender and E. Kujawa, *Can. Res.*, **14**, 29 (1981).
48. G. Moody and J. Thomas, "Applications of Ion-Selective Electrodes," in H. Freiser, Ed., *Ion-Selective Electrodes in Analytical Chemistry*, Vol. 1, Plenum, New York, 1980.
49. M. Antelman and F. Harris, *The Encyclopedia of Chemical Electrode Potentials*, Plenum, New York, 1982.
50. R. Bates, "Electrode Potentials," in I. Kolthoff and P. Elving, Eds., *Treatise on Analytical Chemistry*, 2nd ed., Part 1, Vol. 1, Interscience, New York, 1978.
51. E. Högfeldt, "Graphic Presentation of Equilibrium Data," in H. Freiser, Ed., *Ion-Selective Electrodes in Analytical Chemistry*, Vol. 2, Part 1, Plenum, New York, 1980.
52. T. Shinbo, M. Sugura, and N. Kamo, *Anal. Chem.*, **51**, 100 (1979).
53. C. Liteanu and E. Hopirtean, *Talanta*, **24**, 589 (1977).
54. S. Wawzonek, "Potentiometry: Oxidation–Reduction Potentials," in A. Weissberger and B. Rossiter, Eds., *Physical Methods of Chemistry*, Vol. 1, Part 2, Wiley-Interscience, New York, 1971.
55. G. Fasman, Ed., *Handbook of Biochemistry and Molecular Biology*, 3rd ed., Vol. 1, CRC, Cleveland, OH, 1976, pp. 121–150.
56. P. L. Dutton, *Methods Enzymol.*, **54**, 411 (1978).
57. L. Kekedy and A. Popescu, *Talanta*, **22**, 135 (1975).
58. E. Bishop and P. Cofre, *Analyst*, **103**, 162 (1978).
59. J. Stock, *Anal. Chem.*, **54**, 1R (1982).
60. D. Jagner, *Analyst*, **107**, 593 (1982).
61. J. Goldman, "Titrimetry: Oxidation–Reduction Titration," in I. Kolthoff and P. Elving, Eds., *Treatise on Analytical Chemistry*, Vol. 2, Part 1, Interscience, New York, 1979.
62. W. Wagner and C. Hull, *Inorganic Titrimetric Analysis*, Dekker, New York, 1971.
63. F. Welcher, Ed., *Standard Methods of Chemical Analysis*, Krieger, Huntington, NY, 1975.
64. H. Cassidy, *J. Polym. Sci., Part D*, **6**, 1 (1972).
65. A. Berka, J. Vulterin, and J. Zyka, *Newer Redox Titrants*, Pergamon, Oxford, 1965.
66. T. Yoshimori, *Talanta*, **22**, 827 (1975).
67. K. Sriramam, *Talanta*, **23**, 864 (1976).
68. E. Bishop, Ed., *Indicators*, Pergamon, Oxford, 1972.
69. S. Ebel and A. Seuring, *Angew. Chem. Int. Ed. Engl.*, **16**, 157 (1977).
70. O. Popovych, "Transfer Activity Coefficients (Medium Effects)," in I. Kolthoff and P. Elving, Eds., *Treatise on Analytical Chemistry*, Vol. 1, Part 1, Interscience, New York, 1978.

71. Y. Marcus, *Pure Appl. Chem.*, **55**, 977 (1983).
72. M. Rumeau, "Redox Systems in Nonaqueous Solvents," in J. Lagowski, Ed., *The Chemistry of Nonaqueous Solvents*, Vol. 4, Academic, New York, 1976.
73. B. Kratochvil, *CRC Crit. Rev. Anal. Chem.*, **1**, 415 (1971).
74. P. Pemsler, J. Braunsteen, and K. Nobe, Eds., *Proceedings of the International Symposium on Molten Salts*, Electrochemical Society Press, Princeton, NJ, 1976.
75. P. Hagenmuller and W. VanGool, Eds., *Solid Electrolytes*, Academic, New York, 1978.
76. K. Das, A. Das, and K. Kundu, *Elec. Chim. Acta*, **26**, 471 (1981).
77. B. Verma and K. Sood, *Talanta*, **26**, 906 (1979).
78. B. Karlberg and S. Thelander, *Analyst*, **103**, 1154 (1978).
79. R. Bates, *CRC Crit. Rev. Anal. Chem.*, **10**, 247 (1981).
80. A. Covington, *Anal. Chim. Acta*, **127**, 1 (1981).
81. A. Covington, R. Bates, and R. Durst, *Pure Appl. Chem.*, **55**, 1467 (1983).
82. J. Dobson, *Platinum Met. Rev.*, **25**, 72 (1981).
83. B. Nikolskii, A. Belyustin, *Zh. Anal. Khim.*, **35**, 2206 (1980); *Trans. Ed. Russ. J. Anal. Chem.*, **35**, 1435 (1981).
84. G. Johansson, B. Karlberg, and A. Wikby, *Talanta*, **22**, 953 (1975).
85. "Radiometer Electrode Bulletin," Radiometer Co., Copenhagen, Denmark, 1981.
86. F. Shinskey, *pH and pIon Control in Process and Waste Streams*, Wiley, New York, 1973.
87. R. Thomas, *Ion-Sensitive Intracellular Microelectrodes*, Academic, London, 1978.
88. B. Karlberg, *Talanta*, **22**, 1023 (1975).
89. C. Westcott, *pH Measurements*, Academic, New York, 1978.
90. J. Illingworth, *Biochem. J.*, **195**, 259 (1981).
91. D. Brezinski, *Talanta*, **30**, 347 (1983).
92. S. Glab, G. Edwall, P. Jongren, and F. Ingman, *Talanta*, **28**, 301 (1981).
93. W. Grubb and L. King, *Anal. Chem.*, **52**, 270 (1980).
94. L. Niedrach, *J. Electrochem. Soc.*, **127**, 2122 (1980).
95. D. Macdonald, P. Wentrcek, and A. Scott, *J. Electrochem. Soc.*, **127**, 1745 (1980).
96. T. Eriksson and G. Johanssen, *Anal. Chim. Acta*, **63**, 445 (1973).
97. V. Jennings and P. Pearson, *Anal. Chim. Acta*, **82**, 223 (1976).
98. D. Ammann, F. Lanter, R. Steiner, P. Schulthess, Y. Shijo, and W. Simon, *Anal. Chem.*, **53**, 2267 (1981).
99. J. Janata and R. Huber, "Chemically Sensitive Field Effect Transistors," in H. Freiser, Ed., *Ion-Selective Electrodes in Analytical Chemistry*, Vol. 2, Plenum, New York, 1980.
100. O. LeBlanc, J. Brown, J. Klebe, L. Niedrach, G. Slusarczuk, and W. Stoddard, *J. Appl. Physiol.*, **40**, 644 (1976).
101. R. Edstrom, *J. Chem. Educ.*, **56**, A169 (1979).
102. D. Rosenthal and P. Zuman, "Acid–Base Equilibria, Buffers, and Titrations in Water," in I. Kolthoff and P. Elving, Eds., *Treatise on Analytical Chemistry*, Vol. 2, Part 1, Interscience, New York, 1979.
103. M. Blandamer, J. Burgess, P. Duce, R. Robertson, and J. Scott, *Can. J. Chem.*, **59**, 2845 (1981).
104. H. Butikofer, A. Covington, and D. Evans, *Electrochem. Acta*, **24**, 1071 (1979).
105. P. May, D. Williams, P. Linder, and R. Torrington, *Talanta*, **29**, 249 (1982).
106. G. Kortum, W. Vogel, and K. Andrussow, *Dissociation Constants of Organic Acids in Aqueous Solution*, Butterworths, London, 1961.
107. D. Perrin, *Dissociation Constants of Organic Bases in Aqueous Solution*, Butterworths, London, 1972.

108. D. Perrin, *Dissociated Constants of Inorganic Acids and Bases in Aqueous Solution*, Butterworths, London, 1969.

109. D. Newton and R. Kluza, *Drug Intell. Clin. Pharm.*, **12**, 546 (1978).

110. D. Perrin, B. Dempsey, and E. Serjeant, *pK_a Prediction for Organic Acids and Bases*, Chapman and Hall, London, 1981.

111. D. Perrin and B. Dempsey, *Buffers for pH and Metal Ion Control*, Chapman and Hall, London, 1974.

112. K. Tsuji, *Agric. Biol. Chem.*, **46**, 677 (1982).

113. W. Ferguson, K. Braunschweiger, W. Braunschweiger, J. Smith, J. McCormick, C. Wasmann, N. Jarvis, D. Bell, and N. Good, *Anal. Biochem.*, **104**, 300 (1980).

114. F. Clarke, *Calculator Programming for Chemistry and the Life Sciences*, Academic, New York, 1981, Chapter 3.

115. G. Christian, *CRC Crit. Rev. Anal. Chem.*, **5**, 119 (1975).

116. M. Wozniak and G. Nowogrocki, *Talanta*, **28**, 575 (1981).

117. E. Scarano, L. Campanella, P. Naggar, and R. Belli, *Ann. Chim.*, **72**, 157 (1982).

118. T. Damakos and J. Havas, *Talanta*, **24**, 335 (1977).

119. A. Ramsing, J. Ruzicka, and E. Hansen, *Anal. Chim. Acta*, **129**, 1 (1981).

120. I. Kolthoff and M. Chantooni, "General Introduction to Acid–Base Equilibria in Nonaqueous Organic Solvents," in I. Kolthoff and P. Elving, Eds., *Treatise on Analytical Chemistry*, Vol. 2, Part 1, 2nd ed., Interscience, New York, 1979.

121. H. Finston and A. Rychtman, *A New View of Current Acid–Base Theories*, Wiley, New York, 1982.

122. Z. Stransky and E. Kozakova, *Chem. Listy*, **73**, 337 (1979).

123. J. Coetzee, *Recommended Methods for Purification of Solvents and Tests for Impurities*, Pergamon, New York, 1982.

124. J. Fritz, *Acid–Base Titrations in Nonaqueous Solvents*, Allyn and Bacon, Boston, MA, 1973.

125. J. Lagowski, Ed., *The Chemistry of Nonaqueous Solvents*, Academic, New York, 1978, Vols. 1–5.

126. G. Velinov and O. Budevsky, *J. Electroanal. Chem.*, **95**, 73 (1979).

127. F. Cantwell, R. Hux, and S. Puon, *Anal. Chem.*, **52**, 2388 (1980).

128. R. Durst and J. Cali, *Pure Appl. Chem.*, **50**, 1485 (1978).

129. R. Bates, *Pure Appl. Chem.*, **1**, 229 (1982).

130. J. Kofstad, *Scand. J. Clin. Lab. Invest.*, **41**, 409 (1981).

131. R. Bates, C. Vega, and D. White, *Anal. Chem.*, **50**, 1295 (1978).

132. O. Siggaard-Anderson, Ed., "Blood pH, Carbon Dioxide, Oxygen, and Calcium Ion," Proc. 5th Meeting IFCC Expert Panel on pH and Blood Gases, Priv. Press, Copenhagen, Denmark, 1981.

133. R. Sprokholt, A. Maas, M. Rebelo, and A. Covington, *Anal. Chim. Acta*, **139**, 53 (1982).

134. O. Siggaard-Anderson, *The Acid–Base Status of the Blood*, 4th ed., Williams and Wilkins, Baltimore, MD, 1974.

135. G. Guilbault, *Pure Appl. Chem.*, **53**, 1907 (1981).

136. E. Pungor and K. Toth, "Precipitate-Based Ion-Selective Electrodes," in H. Freiser, Ed., *Ion-Selective Electrodes in Analytical Chemistry*, Vol. 1, Plenum, New York, 1978.

137. W. Morf, G. Kahr, and W. Simon, *Anal. Chem.*, **46**, 1538 (1974).

138. J. Ruzicka and C. Lamm, *Anal. Chim. Acta*, **54**, 1 (1971).

139. A. Hulanicki and A. Lewenstam, *Anal. Chem.*, **53**, 1401 (1981).

140. R. Hawkings, L. Corriveau, S. Kushneriuk, and P. Wong, *Anal. Chim. Acta*, **102**, 61 (1978).

141. J. Mertens, P. VanDenWinkel, and J. Vereeken, *J. Electroanal. Chem.*, **85**, 277 (1977).

142. T. Fjeldly and K. Nagy, *J. Electrochem. Soc.*, **127**, 1299 (1980).

143. C. Harzdorf, *Anal. Chim. Acta*, **136**, 61 (1982).

144. J. Sandifer, *Anal. Chem.*, **53**, 312 (1981).

145. E. Lindner, K. Toth, and E. Pungor, *Anal. Chem.*, **54**, 72, 202 (1982).

146. E. Harsani, K. Toth, L. Polos, and E. Pungor, *Anal. Chem.*, **54**, 1094 (1982).

147. R. Buck, "Crystalline and Pressed Powder, Solid-Membrane Electrodes," in A. Covington, Ed., *Ion-Selective Electrode Methodology*, Vol. 1, CRC, Boca Raton, FL, 1979.

148. W. VanDerLinden, "Ion-Selective Electrodes," in G. Svehla, Ed., *Comprehensive Analytical Chemistry*, Vol. 11, Elsevier, Amsterdam, 1981, pp. 330–334.

149. H. Hulanicki and A. Lewenstam, *Talanta*, **23**, 661 (1976).

150. A. Avdeef, J. Zabronsky, and H. Stuting, *Anal. Chem.*, **55**, 298 (1983).

151. E. Pungor, K. Toth, G. Nagy, L. Polos, M. Ebel, and I. Wernisch, *Anal. Chim. Acta*, **147**, 23 (1983).

152. W. VanDerLinden and R. Oostervink, *Anal. Chim. Acta*, **108**, 169 (1979).

153. G. Eisenman, Ed., *Glass Electrodes for Hydrogen and Other Cations*, Dekker, New York, 1967.

154. A. Covington, "Glass Electrodes," in A. Covington, Ed., *Ion-Selective Electrode Methodology*, Vol. 1, CRC, Boca Raton, FL, 1979.

155. S. Pandey and S. Tripathi, *J. Electroanal. Chem.*, **135**, 25 (1982).

156. G. Moody and J. Thomas, "Polyvinyl Chloride Matrix Membrane Ion-Selective Electrodes," in A. Covington, Ed., *Ion-Selective Electrode Methodology*, Vol. 1, CRC, Boca Raton, FL, 1979.

157. C. Fabiani, P. Danesi, G. Scibona, and B. Scuppa, *J. Phys. Chem.*, **78**, 2370 (1974).

158. H. Hara, S. Okazaki, and T. Fujinaga, *Anal. Chim. Acta*, **121**, 119 (1980).

159. R. Buck, D. Mathis, and F. Stover, *J. Memb. Sci.*, **4**, 395 (1979).

160. D. Mathis, R. Freeman, S. Clark, and R. Buck, *J. Memb. Sci.*, **5**, 103 (1979).

161. G. Moody and J. Thomas, *Ion-Sel. Electrode Rev.* **1**, 3 (1979).

162. N. Poonia and A. Bajaj, *Chem. Rev.*, **79**, 389 (1979).

163. W. Morf and W. Simon, "Ion-Selective Electrodes Based on Neutral Carriers," in H. Freiser, Ed., *Ion-Selective Electrodes in Analytical Chemistry*, Vol. 1, Plenum, New York, 1978.

164. A. Thoma, A. Viviani-Nauer, S. Arvanitis, W. Morf, and W. Simon, *Anal. Chem.*, **49**, 1567 (1977).

165. E. Lindner, K. Toth, E. Pungor, W. Morf, and W. Simon, *Anal. Chem.*, **50**, 1627 (1978).

166. U. Oesch and W. Simon, *Anal. Chem.*, **52**, 692 (1980).

167. P. Meier, D. Amman , W. Morf, and W. Simon, "Liquid-Membrane Ion-Selective Electrodes and Their Biomedical Applications," in J. Koryta, Ed., *Medical and Biological Applications of Electrochemical Devices*, Wiley, New York, 1980.

168. G. Eisenman, R. Margalit, and K. Kuo, in D. Lubbers, H. Acker, R. Buck, G. Eisenman, M. Kessler, and W. Simon, Eds., *Progress in Enzyme and Ion-Selective Electrodes*, Springer-Verlag, Berlin, 1981, p. 1.

169. T. Donaldson and H. Palmer, *AIChE J.*, **25**, 143 (1979).

170. M. Jensen and G. Rechnitz, *Anal. Chem.*, **51**, 1972 (1979).

171. M. Meyerhoff and Y. Fraticelli, *Anal. Chem.*, **53**, 992 (1981).

172. M. Riley, "Gas-Sensing Probes," in A. Covington, Ed., *Ion-Selective Electrode Methodology*, Vol. 2, CRC, Boca Raton, FL, 1979.

173. R. Kobos, "Potentiometric Enzyme Methods," in H. Freiser, Ed., *Ion-Selective Electrodes in Analytical Chemistry*, Vol. 2, Plenum, New York, 1980.
174. M. Arnold and G. Rechnitz, *Anal. Chem.*, **53**, 515 (1981).
175. R. Walters, B. Moriarty, and R. Buck, *Anal. Chem.*, **52**, 1680 (1980).
176. M. Meyerhoff and G. Rechnitz, *Anal. Lett.*, **12**, 1339 (1979).
177. K. Shiba, Y. Umezawa, T. Watanabe, S. Ogawa, and S. Fujiwara, *Anal. Chem.*, **52**, 1610 (1980).
178. J. Boitieux, C. Lemay, G. Desmet, and D. Thomas, *Clin. Chim. Acta*, **113**, 175 (1981).
179. G. Rechnitz, *Science*, **214**, 287 (1981).
180. P. Vadgama, *J. Med. Eng. Tech.*, **5**, 293 (1981).
181. J. Walker, "Single Cell Measurement with Ion-Selective Electrodes," in J. Koryta, Ed., *Medical and Biological Applications of Electrochemical Devices*, Wiley, New York, 1980.
182. H. Freiser, "Coated-Wire Ion-Selective Electrodes," in H. Freiser, Ed., *Ion-Selective Electrodes in Analytical Chemistry*, Vol. 2, Plenum, New York, 1980.
183. M. Majzurawska and A. Hulanicki, *Anal. Chim. Acta*, **136**, 395 (1982).
184. L. Cunningham and H. Freiser, *Anal. Chim. Acta*, **139**, 97 (1982).
185. J. Janata and R. Huber, "Chemically Sensitive Field Effect Transistors," in H. Freiser, Ed., *Ion-Selective Electrodes in Analytical Chemistry*, Vol. 2, Plenum, New York, 1980.
186. C. Wen, I. Lauks, and J. Zemel, *Thin Solid Films*, **70**, 333 (1980).
187. G. Blackburn and J. Janata, *J. Electrochem. Soc.*, **129**, 2580 (1982).
188. A. Haemmerli, J. Janata, and H. Brown, *Anal. Chem.*, **52**, 1179 (1980).
189. A. Ramsing, J. Janata, J. Ruzicka, and M. Levy, *Anal. Chim. Acta*, **118**, 45 (1980).
190. S. Collins and J. Janata, *Anal. Chim. Acta*, **136**, 93 (1982).
191. E. Pungor, K. Toth, and A. Hrabeczy-Pall, *Pure Appl. Chem.*, **51**, 1913 (1979).
192. R. Durst, "Sources of Error in Ion-Selective Electrode Potentiometry," in H. Freiser, Ed., *Ion-Selective Electrodes in Analytical Chemistry*, Vol. 1, Plenum, New York, 1978.
193. R. Buck, *Ion-Sel. Electrode Rev.*, **4**, 3 (1982).
194. A. Haemmerli, H. Brown, and J. Janata, *Anal. Chim. Acta*, **144**, 115 (1982).
195. D. Midgely, *Analyst*, **105**, 417 (1980).
196. M. Mascini, *Ion-Sel. Electrode Rev.*, **2**, 17 (1980).
197. G. Horvai and E. Pungor, *Anal. Chim. Acta*, **116**, 87 (1980).
198. J. Frazer, W. Selig, and L. Rigdon, *Anal. Chem.*, **49**, 1250 (1977).
199. N. Busch, P. Freyer, and H. Szameit, *Anal. Chem.*, **50**, 2166 (1978).
200. E. Pungor, K. Toth, and G. Nagy, "Analytical Methods Involving Ion-Selective Electrodes," in A. Covington, Ed., *Ion-Selective Electrode Methodology*, Vol. 2, CRC, Boca Raton, FL, 1979.
201. P. Alexander and P. Seegopaul, *Anal. Chem.*, **52**, 2403 (1980).
202. M. Trojanowicz and W. Matuszewski, *Anal. Chim. Acta*, **138**, 71 (1982).
203. M. Trojancwicz and W. Matuszewski, *Anal. Chim. Acta*, **151**, 77 (1983).
204. D. Gozzi and T. Ferri, *J. Electroanal. Chem.*, **109**, 213 (1980).
205. E. Pungor and K. Toth, "Ion-Selective Electrodes in Nonaqueous Solvents," in J. Lagowski, Ed., *The Chemistry of Nonaqueous Solvents*, Vol. 5A, Academic, New York, 1978.
206. E. Pungor, K. Toth, and P. Gabor-Klatsmanyi, *Hung. Sci. Instrum.*, **49**, 1 (1980).
207. J. Coetzee and M. Martin, *Anal. Chem.*, **52**, 2412 (1980).
208. S. Chaudhari and K. Cheng, *Mikrochim. Acta*, **2**, 411 (1979).
209. L. Bykova, N. Kazayran, E. Pungor, and N. Chernova, in E. Pungor and I. Buzas, Eds., *Ion-Selective Electrodes*, (Conference at Budapest, 1977), Elsevier, Amsterdam, 1978, pp. 281–289.

210. J. Coetzee and W. Istone, *Anal. Chem.*, **52**, 53 (1980).
211. L. Mukherjee, *Electrochim. Acta*, **22**, 1255 (1977).
212. D. Band and T. Treasure, "Ion-Selective Electrodes in Medicine and Medical Research," in A. Covington, Ed., *Ion-Selective Electrode Methodology*, Vol. 2, CRC, Boca Raton, FL, 1979.
213. V. Cosofret, *Ion-Sel. Electrode Rev.*, **2**, 159 (1980).
214. D. Kell, *Process Biochem.*, **15**, 18 (1980).
215. H. Herman, *Environ. Sci. Res.*, **13**, 103 (1978).

Bibliography

General Potentiometry

A. Bard and L. Faulkner, *Electrochemical Methods: Fundamentals and Applications*, Wiley, New York, 1980, Chapters 1 and 2.

H. Harned and B. Owen, *The Physical Chemistry of Electrolytic Solutions*, 3rd ed., Reinhold, New York, 1958, Chapters 10–15.

D. Ives and G. Janz, *Reference Electrodes*, Academic, New York, 1961.

D. MacInnes, *The Principles of Electrochemistry*, Dover, New York, 1961, Chapters 8–17.

D. Midgely and K. Torrance, *Potentiometric Water Analysis*, Wiley, New York, 1978.

R. Robinson and R. Stokes, *Electrolyte Solutions*, 2nd ed. rev., Butterworths, London, 1970.

Oxidation–Reduction

R. Bates, "Electrode Potentials," in I. M. Koltholf and P. Elving, Eds., *Treatise on Analytical Chemistry*, Part 1, Vol. 1, 2nd ed., Interscience, New York, 1978.

J. A. Goldman, "Oxidation–Reduction Equilibria and Titration Curves," in I. M. Koltholf and P. Elving, Eds., *Treatise on Analytical Chemistry*, Part 1, Vol. 3, Interscience, New York, 1983.

M. Rumeau, "Redox Systems in Nonaqueous Solvents," in J. Lagowski, Ed., *The Chemistry of Nonaqueous Solvents*, Vol. 4, Academic, New York, 1976.

pH and Acid–Base

R. Bates, *Determination of pH: Theory and Practice*, 2nd ed., Interscience, New York, 1973.

R. Bates, *CRC Critical Reviews in Analytical Chemistry*, **10**, 247 (1981).

J. Fritz, *Acid–Base Titrations in Nonaqueous Solvents*, Allyn and Bacon, Boston, MA, 1973.

I. Koltholf and M. Chantooni, "General Introduction to Acid–Base Equilibria in Non-aqueous Organic Solvents," in I. M. Koltholf and P. Elving, Eds., *Treatise on Analytical Chemistry*, Part 1, Vol. 2, 2nd ed., Interscience, New York, 1979.

D. Rosenthal and P. Zuman, "Acid–Base Equilibria, Buffers, and Titrations in Water," in I. M. Koltholf and P. Elving, Eds., *Treatise on Analytical Chemistry*, Part 1, Vol. 2, 2nd ed., Interscience, New York, 1979.

C. Westcott, *pH Measurements*, Academic, New York, 1978.

Ion-Selective Electrodes

P. Bailey, *Analysis with Ion-Selective Electrodes*, 2nd ed., Heyden, London, 1980.

K. Cammann, *Working with Ion-Selective Electrodes*, Springer-Verlag, Berlin–New York, 1979.

A. Covington, Ed., *Ion-Selective Methodology*, Vols. 1 and 2, CRC, Boca Raton, FL, 1979.

H. Freiser, Ed., *Ion-Selective Electrodes in Analytical Chemistry*, Vols. 1 and 2, Plenum, New York, 1980.

W. Morf, *Studies in Analytical Chemistry: The Principles of Ion-Selective Electrodes and of Membrane Transport*, Elsevier, Amsterdam, 1981.

Chapter **3**

VOLTAMMETRY WITH STATIONARY AND ROTATED ELECTRODES

Zbigniew Galus

1   INTRODUCTION

In the past decade there have been significant developments in electro-analytical methods. New methods have been invented and older ones improved. Significant progress has been made also in voltammetry with stationary and rotated electrodes, with rotated electrode voltammetry developing more rapidly, especially those methods using ring-disk electrodes and moving solutions.

This progress, "hydrodynamic voltammetry," covers both rotated electrodes and stationary electrodes with moving solutions.

The development of these voltammetry methods through the late 1960s is comprehensively presented by Piekarski and Adams [1] in the former edition of this Series and by Adams in his well-known monograph [2]. The basis of both methods is discussed in [3].

The early 1970s saw discussion of the theoretical problems of the rotating-disk technique, its application by Pleskov and Filinovskii [4], and application of the ring-disk technique by Albery and Hitchman [5].

The publication by Bard and Faulkner [6] presents voltammetry with rotated and stationary electrodes, in addition to other electrochemical techniques, with an extensive bibliography on these techniques.

2   VOLTAMMETRY WITH STATIONARY ELECTRODES

2.1   Conditions of Stationary Electrode Voltammetry (SEV) Experiment

A three-electrode system is normally used in voltammetry. The working electrode reaction is either electroreduction or electrooxidation. Assuming that neither the substrate nor the product of this reaction is associated with a chemical reaction, electroreduction is represented for the one-step, first-order process by

$$Ox + ne \rightleftharpoons Red \tag{1}$$

As in many other analytical electrochemical methods the reactant and product are transported to and from the electrode surface by diffusion. When either reactant or product is occasionally involved in a chemical reaction, this transport may be limited by the rate of the chemical step. Sometimes Ox and/or Red may be adsorbed on the electrode.

Other transport types, such as migration and convection, should be excluded. In the experimental solution, the influence of migration is minimized by using an electrolyte concentration on the order of 10^{-2} mol/dm^3 or greater. The addition of such a salt also increases solution conductivity.

In SEV the working electrode potential changes linearly in time with respect to a constant-potential reference electrode.

Initiating the electrochemical process (1) under the influence of an applied potential produces a concentration gradient. Since the Ox reactant concentration at the electrode surface drops (Figure 3.1a), Ox is transported to the electrode from the bulk of the solution.

The region adjacent to the electrode surface concentration is dependent on the electrode potential and also on distance from the electrode surface, as depicted in Figure 3.1a. If a point in this region is considered, the concentration at this point decreases with time, since the concentration gradient decreases as electrolysis develops.

The dependence of the current on potential in SEV exhibits a peak. The occurrence of this peak current is readily understood if the change of the depolarizer concentration gradient at the electrode surface in time (corresponding to different potentials) in the voltammetric experiment is considered. Such gradients are shown in Figure 3.1a. If considered at the electrode surface, they are linearly related to current i flowing across the circuit [3]:

$$i = nFAD_{Ox} \left(\frac{\partial C_{Ox}}{\partial x} \right)_{x=0} \tag{2}$$

In (2) $(\partial C_{Ox}/\partial x)_{x=0}$ is the concentration gradient of a reactant Ox, D_{Ox} is its diffusion coefficient, A is the electrode surface, F is the Faraday constant, and n is the number of electrons transferred in the elementary reaction (1).

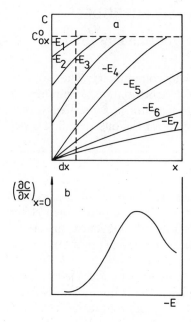

Figure 3.1 (a) The dependence of Ox concentration on the distance from the electrode, x, for different potentials. Negative values of the potentials increase from E_1 to E_7. More-negative potentials correspond to longer times of electrolysis. The change of concentration gradient with E in the region dx is easily observed. (b) Concentration gradients at the electrode surface plotted versus potential of the electrode. This figure is based on part (a).

If gradients such as those in Figure 3.1a are plotted as a function of potential, the curve shown in Figure 3.1b results.

When recording the current–potential curve, the solution must be quiet. This condition is more important when the working electrode potential is changed slowly. However at a very low scan rate, below 10^{-3} V/s, the influence of external vibrations and of natural convection, which may change the shape of i–E dependency from a peak to a wavelike shape, may be observed. Since the following equations were derived under the assumption of diffusional transport, this very low scan rate should not be used because obtained results are not quantitatively elaborated. Higher scan rates are preferred; however, very high scan rates result in a high capacitive-to-faradaic current ratio and the faradaic component measurement of the current is less precise. These problems will be examined when linear diffusion is discussed.

2.1.1   Electrodes

The working electrodes used in SEV have various shapes and are constructed from various materials having conducting or semiconducting properties. Such an electrode should have a rather small surface to ensure low currents in the circuit. Frequently the surface area of the working electrode varies between 10^{-2} and 10^{-1} cm^2.

When using solid materials the working electrode may be a rod with a plane exposed to the solution and with insulated edges and a mantle, which eliminates peripheral contributions (Figure 3.2a) in such a way that diffusion to the electrode is perpendicular to the surface of a plane.

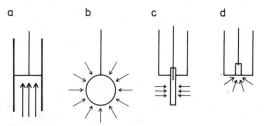

Figure 3.2 Schematic representation of diffusion to (a) plane, (b) spherical, (c) cylindrical, and (d) microdisk electrodes.

Frequently, when mercury is used as an electrode material, it is a hanging mercury drop electrode with spherical diffusion symmetry (Figure 3.2b).

For practical reasons small wires are also sometimes used with cylindrical diffusion (neglecting the contribution originating from the tip of the electrode, Figure 3.2c).

Although in practice the electrode shapes mentioned are probably most frequently used, electrodes of other shapes are also possible (Figure 3.2d). However, since the theory of electrochemical reactions occurring under stationary electrode voltammetry conditions is mostly limited to conditions of either linear or spherical diffusion, these electrode shapes are preferred.

Results obtained with other electrode shapes, especially in complicated electrode reactions, may be difficult to interpret; consequently, the application of such electrodes should be limited to quantitative analysis when only the dependence of the current on the reactant concentration is being studied.

However, even when using such electrodes, the theory derived for linear diffusion may be applied if the change of the potential during recording of the current–potential curve is rapid. This is illustrated by an insulator-surrounded disk electrode, which is shown in Figure 3.3 together with diffusion layers that develop at two different times of electrolysis (or different scan rates in SEV).

A central part of the diffusion field with a radius equal to that of the electrode r_0 corresponds to the diffusion layer observed in linear diffusion. With short electrolysis times (high scan rates) diffusion from a linear region predominates. This model illustrates that even at the same scan rate (identical thickness of the diffusion layer) the nonlinear contribution (from Field b) is more significant if the radius of the electrode is smaller than that assumed in Figure 3.3.

For longer electrolysis times (lower scan rates), the diffusion layer is larger (Curve 2 in Figure 3.3) and the nonlinearity contribution to total diffusion is much higher. In general, when the linear diffusion field is approximately 100 times larger than the nonlinear field (Field b in Figure 3.3), the edge effects can be neglected and the theory derived for linear diffusion conditions applied.

This theory may also be applied to the results obtained from spherical electrodes. Quantitative formulation of conditions justifying such a procedure is

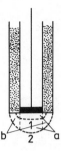

Figure 3.3 The schematic view of the disk electrode with developing diffusion layers. Curves 1 and 2 schematically show the limits of the diffusion layer at short and long times of electrolysis, respectively. (a) Points on the field of linear diffusion and (b) points on the field of edge diffusion.

presented in the section on spherical diffusion. The mercury electrodes used in SEV usually have a spherical shape, since they are constructed as hanging mercury drops. Such electrodes were described in the late 1950s [7] and subsequently introduced as an analytical tool to many laboratories. Hanging mercury drop electrodes are manufactured by Metrohm (Switzerland) and Radiometer (Denmark). In recent years Princeton Applied Research (PAR) Corporation (USA) introduced the static mercury drop electrode (Model 303).

Advantages of this electrode were presented by Peterson [8]. It may be used in polarography as a dropping mercury electrode with a controlled drop time, but it may also serve as a hanging mercury drop electrode with a constant controlled surface area when the mercury flow from the capillary is stopped by a special magnetic device. The electrode produced by Tesla Laboratorni Přistroje (Czechoslovakia) has similar features.

Various solid electrodes, such as graphite, paraffin-impregnated graphite, glassy carbon, and platinum-gold, are offered by PAR.

Metrohm also offers carbon paste electrodes (Adams-type [9]), especially useful for stripping voltammetry analysis.

Glassy carbon electrodes (GCE) useful for SEV and also for stripping analysis are produced by Bruker Analytik GMBH (West Germany).

The surface of a hanging mercury drop electrode is easily renewed—a very important feature when impurities from the solution are adsorbed onto the electrode or when the product of the electrode reaction covers the electrode surface and inhibits further processing. Under such circumstances, solid electrodes must be carefully cleaned after each experiment.

Since the inhibiting compounds present on the electrode surface are frequantly organic in nature, heating the electrode between individual measurements [10] by passing current from an external current source decomposes the adsorbed compound. Carbon paste electrode surfaces are easily renewed.

No strict limitations are imposed on selecting a reference electrode. However, the ions of the reference electrode should not interact with species of the studied solution. Usually electrodes such as calomel, mercurous sulfate, or silver–silver chloride are used. In this respect the reader may consult specialized books [11].

When working with nonaqueous solvents or with melts, other reference electrodes typically used for studied media are applied. Discussion of this technique is beyond the scope of this chapter.

2.1.2   Apparatus

SEV experimental equipment is relatively simple and readily available on the market. It is composed of several principal parts, the first of which is an electrolytic cell with three electrodes. The two-electrode system is only allowed at very low currents. A function generator changes the working electrode potential linearly with time. The scan rate is usually increased to thousands of volts per second. A potentiostat maintains the working electrode potential at the programmed value. The use of fast potentiostats in SEV is especially important (a) in measurements carried out in nonaqueous solvents, where the background electrolyte solubility may be quite limited, thereby decreasing the conductance of a solution, and (b) when high scan rates are used.

In the two-electrode system having a high circuit resistance the working electrode potential will not assume a programmed value. Because the potential of the well-constructed reference electrode is constant during the experiment, the working electrode potential will be different from the programmed value by an iR value.

This difference is larger when the iR is larger; thus, experiments with relatively high current flowing through the circuit require a good potentiostat.

The iR drop yields broader SEV current peaks and in the case of a cathodic reaction the peak potential is shifted to a more cathodic value (more anodic in the case of an oxidation reaction).

Exact control of the potential is especially important in measurements of the electrode kinetics, because then an uncompensated iR drop may be considered as an overpotential of the electrochemical reaction. An error on the order of several millivolts in an uncontrolled potential may significantly change, in particular cases, the heterogeneous rate constant.

Better control of potential is ensured by applying potentiostats with positive feedback compensation [6, 12–14]. In a three-electrode system the working electrode potential is kept at the programmed value against the reference electrode; this is achieved by passing the current through the working and auxiliary electrodes.

The scheme of such a circuit used in SEV is shown in Figure 3.4. Amplifier 1, which acts as a voltage follower, controls the potentials of the electrode. Since the working electrode is at virtual ground, the difference between the working electrode potential and the reference electrode potential is equal to the input voltage.

The current that flows across the working electrode is supplied by the auxiliary electrode; current does not flow through the reference electrode because of the very high input impedance of amplifier 1.

An iR drop sometimes may be very significant and may be comparable with the output voltage of the amplifier. In the two-electrode system the distance between both electrodes should be small to minimize an iR drop. In the three-

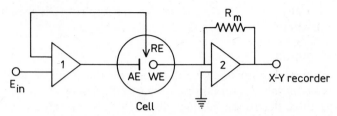

Figure 3.4 A schematic representation of a circuit used for SEV; AE, RE, and WE denote the auxiliary, reference, and working electrodes, respectively. Amplifier 1 is the control unit and 2 is a current-to-voltage converter. R_m determines the current sensitivity.

electrode system the positioning of the working and reference electrodes is not so critical; however, it is very important in solutions of low conductivity, because then a potential drop may be very large. In such solutions, where the gradient of the electrical potential is large, the difference between working electrode potential and reference electrode potential should be kept at a minimum. With the working electrode in the center, the auxiliary electrode should be on the opposite side with the reference electrode in between.

The potentiostat cannot compensate an iR drop of the working electrode and that resulting from the resistance between the working electrode and the reference electrode. As previously mentioned better compensation may be achieved using a circuit with a positive feedback. The application of this circuit frequently causes oscillations. However, these oscillations may be eliminated if a well-constructed, high quality circuit is used [15]. The use of an iR compensation with a positive feedback is especially useful when working with nonaqueous solvents.

When an iR drop is significant and the working electrode has a large surface area, the current density at different parts of the electrode varies; consequently, the potential will vary over the electrode. Because of this, mutual positioning of electrodes in the cell becomes important [16], especially in solutions of low conductivity.

In the working electrode arrangement careful attention should be given to the assumptions of the theory eventually used when calculating different parameters from experimental results. For example, the electrode cannot be placed close to the cell wall or to another nonconducting body.

The last component of SEV equipment is an X-Y recorder, or in the case of higher scan rates, an oscilloscope.

The block diagram of the apparatus used in SEV experiments is shown in Figure 3.4. This apparatus can be constructed in the laboratory from individual components or it may be purchased fully constructed. Bruker Analytik produces an E 310 Modular Research Polarograph, which also performs SEV experiments. Information is stored in a digital memory of 1024×8 bits and either simultaneously displayed on an oscilloscope or later on an X-Y recorder. Metrohm (Switzerland) produces a three-electrode polarograph, the E506 Polarecord.

Princeton Applied Research (USA) manufactures a multipurpose Model 370

Electrochemistry system, as well as a Polarographic Analyzer Model 174A that may be used for SEV experiments. The latter is a low-cost, standard industrial instrument.

Polarographs convenient for recording SEV curves have been produced by Tacussel (France) and by Radelkis (Hungary).

Stationary electrode voltammetry apparatus are described by many authors [6, 13, 17].

2.2   Theory of Stationary Electrode Voltammetry

The theoretical description of the current recorded in SEV as a function of the potential is obtained by solving Fick's Second Law of diffusion.

In linear diffusion

$$\frac{\partial C_{Ox}(x, t)}{\partial t} = D_{Ox} \frac{\partial^2 C_{Ox}(x, t)}{\partial x^2} \tag{3}$$

$$\frac{\partial C_{Red}(x, t)}{\partial t} = D_{Red} \frac{\partial^2 C_{Ox}(x, t)}{\partial x^2} \tag{4}$$

where $C_{Ox}(x, t)$ and $C_{Red}(x, t)$ are Ox and Red concentrations, respectively, and are dependent on the distance from the electrode x and time t. D_{Ox} and D_{Red} are diffusion coefficients of the Ox and Red form, respectively.

Spherical diffusion places an additional term on the right-hand side of (3)

$$\frac{\partial C_{Ox}(x, t)}{\partial r} = D_{Ox} \left[\frac{\partial^2 C_{Ox}(r, t)}{\partial r^2} + \frac{2}{r} \frac{\partial C_{Ox}(r, t)}{\partial r} \right] \tag{5}$$

When cylindrical rather than a spherical diffusion symmetry is considered, $1/r$ in place of $2/r$ should appear in the second term on the right-hand side of (5).

Problems related to (3)–(5) are discussed by Crank [18] and others.

2.2.1   Simple Reversible Electrode Processes

LINEAR DIFFUSION

If a simple electrode reaction, such as that represented by (1) with the rate limited by diffusion of Ox from a semi-infinite field with a linear symmetry, is considered, then (3) is solved by assuming proper initial and boundary conditions.

One of the boundary conditions at $x=0$ is the Nernst equation, since it was assumed above that the process is reversible (with rate limited by diffusion)

$$\frac{C_{Ox}}{C_{Red}} = \exp \left[\frac{nF(E - E^0)}{RT} \right] \tag{6}$$

where E^0 is the standard potential of the system studied and E is the time-dependent potential of the electrode.

In principle, though impractical, this potential could change with time according to different functions; however, in practice, the change is linear,

$$E = E_i - vt \tag{7}$$

where E_i is the initial potential of the electrode (at $t=0$), t is the time elapsing from the beginning of the electrolysis, and v is the potential scan rate expressed in volts per unit of time (usually seconds). The experimenter adjusts and controls v, which may be changed within wide limits and which will be further discussed in a later section.

Using boundary condition (6) with (7) and other boundary and initial conditions, the solution of the problem is obtained and usually expressed in the form of a time-dependent current.

This solution has no analytical form and may be expressed

$$i = nFAC_{Ox}^0 \sqrt{\pi D_{Ox} a} \chi(at) \tag{8}$$

where C_{Ox}^0 is the Ox concentration in the bulk, $a = nFv/RT$, A is the electrode surface, and $\chi(at)$ is the current function which was calculated by Ševčik [19] and Randles [20] as early as 1948, and later by Reinmuth [21], Matsuda [22], Gokhstein [23], and de Vries and van Dalen [24]. Nicholson and Shain [25] applied numerical methods to reach a solution.

Figure 3.5 shows the dependence of the function $\chi(at)$ on potential, similar in shape to the curve given in Figure 3.1b.

Since in (8) only $\chi(at)$ is potential dependent, while other parameters are constant (if v is adjusted), the shape of the dependence shown in Figure 3.5 determines the shape of the recorded current–potential curves. The shape of this curve was discussed earlier in conjunction with Figure 3.1b.

Since the maximum value of the $\sqrt{\pi}\chi(at)$ function equals 0.4463, (8) written for a maximum current (peak current) will be

$$i_p = 0.446 \frac{n^{3/2} F^{3/2}}{R^{1/2} T^{1/2}} A D_{Ox}^{1/2} C_{Ox}^0 v^{1/2} \tag{9}$$

At temperature of 25°C this equation will be

$$i_p = 2.69 \times 10^5 n^{3/2} A D_{Ox}^{1/2} C_{Ox}^0 v^{1/2} \tag{9a}$$

where the peak current is expressed in amperes, A in cm^2, C_{Ox}^0 in mol/dm^3, the reactant diffusion coefficient D_{Ox} in cm^2/s, and the scan rate v in V/s.

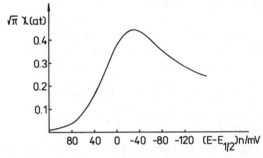

Figure 3.5 The dependence of the function $\sqrt{\pi}\chi(at)$ on potential expressed versus the reversible half-wave potential $E_{1/2}$.

Equation (9) is often called the Randles–Ševčik equation and points to the analytically important conclusion that the peak current is linearly dependent on the reactant concentration.

This equation is obeyed under the condition of linear diffusion from a semi-infinite field. However, as already mentioned, it may be used to interpret results obtained with electrodes of other shapes, if the electrode surface is not very small and the scan rate is sufficiently high.

The lowest limit of v used in practice is about 10^{-3} V/s [26]. Below this limit, natural convection and external interferences may play an important role and create significant deviation from the behavior predicted by (9) or (9a).

However, there is theoretically no limit imposed on very high scan rates; in practice, two problems should be considered: (a) increase of the capacitive current with v and (b) the possibility of distortion of recorded current–potential curves by ohmic drop.

Since the dependence of the capacitive current i_c on v may be described by the following equation [27, 28]

$$i_c = C_d v A \tag{10}$$

where C_d is the differential capacity of the electrical double layer, an increase of v results in a larger increase of the capacitive rather than the faradaic current according to

$$\frac{i_c}{i_p} = \text{const} \times v^{1/2} \tag{11}$$

Because high scan rates are sometimes required, for example, in kinetic studies, to overcome these problems and to keep the i_c/i_p ratio as low as possible, it is advisable to use a rather high concentration of the reactant in such studies. This procedure requires a good potentiostat to avoid distortion of the working electrode potential by the ohmic drop.

In practice, scan rates as high as 5×10^4 V/s were used [29, 30]; however, only scan rates of about 2×10^3 V/s may be employed with confidence.

Band-pass limitations and maximum usable scan rates in the study of faradaic processes were considered by Garreau and Savéant [31], who concluded that curve distortion should fall in the range of experimental error up to about 10^3 V/s.

SPHERICAL DIFFUSION

With spherical or cylindrical diffusion of the reactant to the electrode the dependence of the current on potential is more complicated. A solution to this problem was presented by Frankenthal and Shain as early as 1956 [32]. Later, Nicholson and Shain [25] solved this problem in a different way, proposing the following equation, valid at 25°C for the peak current,

$$i_p = 2.69 \times 10^5 n^{3/2} A D_{Ox}^{1/2} C_{Ox}^0 v^{1/2} + \frac{0.72 \times 10^5 n A D_{Ox} C_{Ox}^0}{r_0} \tag{12}$$

where r_0 is the radius of the electrode.

From this equation it follows that the first term represents the current observed in linear diffusion, whereas the second is an additional increase of the current caused by the nonlinearity of the diffusion.

Since the second term in (12) is independent of the scan rate, it may be neglected at higher v, if the following condition is fulfilled:

$$3.71 r_0 \left(\frac{nv}{D_{Ox}}\right)^{1/2} \gg 1 \tag{13}$$

Then (8) and other equations derived with the assumption of linear diffusion may be used to interpret results obtained with spherical electrodes. Considering (13), the peak current may be much higher in spherical diffusion compared with linear diffusion. This is especially true when both r_0 and v are low.

A similar pattern is observed with cylindrical diffusion symmetry [33]. However, the increase of the peak current caused by diffusion nonlinearity is not exceptional.

The dependence of the peak current on $(1/r_0)(D_{Ox}/nv)^{1/2}$ is shown in Figure 3.6 in the case of both cylindrical and spherical diffusion.

VOLTAMMETRY WITH MICROELECTRODES

In recent years *in vitro* and *in vivo* electrochemical experiments with microelectrodes, whose development was stimulated by their use in bioelectrochemical work, have frequently been conducted. Initiated and developed by Adams and co-workers [34], they recorded, for example, current–potential curves of a drug injected into the rat caudate nucleus with the use of a caudate-implanted syringe-electrode assembly [35]. Figure 3.7 shows such cyclic voltammetric curves.

The voltammetric curves recorded using a stationary, well-shielded microelectrode with strictly linear diffusion have the same shapes as those curves

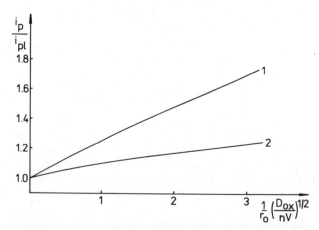

Figure 3.6 The dependence of the normalized peak current (i_{pl} is the peak current in the case of a linear diffusion) on $(1/r_0)(D_{Ox}/nV)^{1/2}$ parameter in the case of spherical (Curve 1) and cylindrical (Curve 2) diffusion.

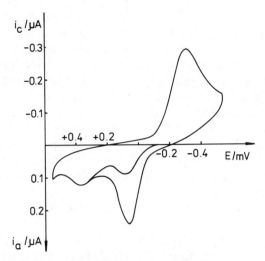

Figure 3.7 Cyclic voltammogram from syringe–electrode assembly implanted in the rat caudate nucleus after injection of the paraquinone of 6-hydroxydopamine (30 μg in 1 μL).

recorded with large surface electrodes. However, if a microelectrode has the form shown in Figure 3.1d, as is frequently the case, and if its surface is very small, the dependence of the current on potential may not exhibit a peak, especially at low scan rate. Such behavior was experimentally observed [36–38], and a typical curve is given in Figure 3.8. Currently, theoretical descriptions are unavailable for current–potential dependencies recorded using such electrodes; however, because of similarities among equations describing chronoampero-metric curves recorded with spherical electrodes [39] and microdisk electrodes [40–43], this similarity may also be assumed in voltammetry with stationary electrodes.

Consequently, the equation describing the peak current, or rather, the plateau of the limiting current, should be

$$i_p = 2.69 \times 10^5 n^{3/2} A D_{Ox}^{1/2} C_{Ox}^0 v^{1/2} + \frac{\text{const} \times n A D_{Ox} C_{Ox}^0}{r_0} \qquad (14)$$

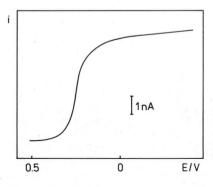

Figure 3.8 The current–potential curve of 4.0 mM K$_3$[Fe(CN)$_6$] in 1.0 M KCl recorded with the use of the microelectrode. Radius of the carbon fiber electrode equals 5.1×10^{-4} cm. Scan rate 0.1 V/s (according to [36]).

where r_0 is the radius of a microdisk electrode in centimeters, while the constant (const) should be near 10^5 when the units are concordant with those used in (9a).

When considering the formation of the current–potential dependency, similar to (14), one may express the total current as a sum of two terms, the first of which (a) is linearly dependent on $v^{1/2}$ and the second of which (b) is dependent on r_0 at a fixed concentration of the reactant.

These current components are schematically shown in Figure 3.9 as a function of a potential. In this figure $i/A = a + b(1)$ corresponds to the situation met when using typical HMDE. The other situation $i/A = a + b(2)$ illustrates the case of microelectrodes. With very small r_0 the $b(2)$ term is very large in comparison with a, and the current–potential curves should exhibit a limiting current plateau rather than a peak.

However, a wavelike current–potential dependence may be observed only if the radius of the microelectrode is well below 10^{-2} cm. Only then does the steady state resulting from the domination of the second term on the right-hand side of (14) exist. As it follows from (14) it should be observed more easily at a low scan rate.

Also, current–potential curves recorded by using spherical microelectrodes should have a similar form. To observe such curves without a peak current the inequality $3.7r_0(nv/D)^{1/2} \ll 1$ should be applied (r_0 is the radius of the spherical electrode).

Obviously not only the peak current, but also the total current–potential curve shape, are changed.

THIN-FILM AND THIN-LAYER STATIONARY ELECTRODE VOLTAMMETRY

In the process just discussed a reactant diffused to the electrode surface from a semi-infinite diffusion field. However, a reactant may be placed in a cell with a thickness on the order of 10^{-3} cm. One wall of such a cell constitutes the working electrode. In this case a reactant may quickly reach the electrode surface.

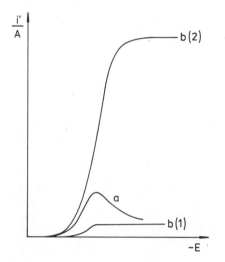

Figure 3.9 SEV current density potential curves: a is a linear diffusion term and the b terms reflect nonlinear diffusion—$b(1)$ when the radius of the electrode is relatively large and $b(2)$ for very small electrodes.

Since the diffusion layer thickness δ is [44]

$$\delta = \sqrt{\pi D t} \tag{15}$$

then, assuming $\delta = 10^{-3}$ cm equal to the thickness of a thin-layer cell and $D = 10^{-5}$ cm^2/s, one obtains $t = 3.2 \times 10^{-2}$ s.

These simple considerations show that under assumed conditions practically all substance dissolved in the solution may react at the electrode in less than 1 s. Such thin-layer cells are currently used in different electrochemical experiments [45, 46]. They are also used in SEV. Under selected conditions the equations that describe the current as a function of potential may be relatively simple. Assuming that the electrode reaction proceeds reversibly according to (1) with both reactants Ox and Red soluble in the solution phase and that the scan rate is sufficiently low and the thin layer is sufficiently thin, the current–potential curve is described by

$$i = \frac{n^2 F^2 C^0 A l v}{RT} \frac{\exp\left[\dfrac{nF}{RT}(E - E^0)\right]}{\left\{1 + \exp\left[\dfrac{nF}{RT}(E - E^0)\right]\right\}^2} \tag{16}$$

The analysis of (16) shows that the current reaches a maximum value at the formal potential E^0 and is given by

$$i_p = \frac{n^2 F^2 C^0 A l v}{4RT} \tag{17}$$

where l in these equations is the thickness of a thin-layer cell.

Current drops symmetrically at potentials less and more cathodic than E^0. Its shape is shown schematically in Figure 3.10a. Such peaks are quite narrow as

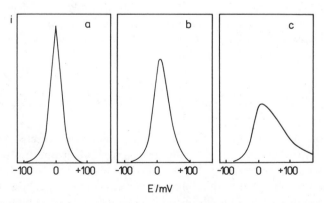

Figure 3.10 Theoretical current–potential curves obtained by using a planar, thin mercury-film electrode. Scan rates: (a) 0.125 V/min, (b) 0.5 V/min, and (c) 2 V/min. Thickness of the film, 100 μm (according to [51]).

the result of a finite diffusion field, and the reaction is nearly completed within about $\pm 2RT/nF$.

In a simple reversible reaction the peak of the anodic oxidation of the product formed in a cathodic run also occurs at $E = E^0$.

The condition of applicability of (17) will be discussed further in this section.

The theory and applications of linear sweep voltammetry (LSV) of thin-layer solutions were reviewed by Hubbard and Anson [45] and by Hubbard [46]. Thin-layer cells and experimental conditions are also described by these authors.

Considering the more recent development of mercury film electrodes and their broad application in analytical practice [47–50], the theory of reversible oxidation processes of metals from a thin mercury film is briefly discussed. This theory was presented by de Vries and van Dalen [51–53]; their treatment is strict but quite complex.

The shapes of voltammetric curves change when either the thickness of the thin layer, l, or the scan rate is changed. Figure 3.10 shows several theoretical voltammetric curves observed at different scan rates.

Under the limiting conditions de Vries and van Dalen derived the following equation for a peak current valid at 25°C:

$$i_p = 1.116 \times 10^6 n^2 AlC^0 v \tag{18}$$

This equation shows a linear dependence of the peak current on scan rate v rather than on $v^{1/2}$ as observed during diffusion from a semi-infinite field. Equation (18) is valid if the H parameter,

$$H = \frac{l^2 nFv}{DRT} \tag{19}$$

does not exceed 1.6×10^{-3}.

When the H parameter is within the limits $1.6 \times 10^{-3} < H < 0.144$, (18) is obeyed within the limits of 0.2%.

As a consequence of a rather high diffusion rate under thin-layer conditions, a diffusion coefficient (D) of reactant does not appear in (18).

Since the product AlC^0 represents the total amount of electroactive substance present in a thin layer, (18) may be written more concisely:

$$i_p = 11.56nQ \tag{18a}$$

where Q is the total charge that flows from a totally oxidized or reduced thin-layer substance.

Equation (18a) may be used successfully in organic electrochemistry to determine n, the number of electrons transferred when one molecule or ion interacts with the electrode.

The changes observed in the shapes of current–potential curves when going from a semi-infinite to a finite diffusion field may be expressed by the difference between the peak and $E_{1/2}$ potential, which for a reversible reaction is

$$n(E_p - E_{1/2}) = -1.43 + 29.58 \log H \quad \text{(in mV)} \tag{20}$$

This equation is valid if $H < 0.4$ and it predicts dependence of the peak potential on the thickness of a thin layer. The change of either l or v by one order of magnitude shifts E_p by $59/n$ and $59/2n$ mV, respectively.

In this case the recorded peaks are also much sharper, and their width measured at one-half the peak current $(b_{1/2})$ for $H < 1.6 \times 10^{-3}$ is

$$nb_{1/2} = 75.53 \text{ mV} \tag{21}$$

whereas in the case of hanging drop electrodes with $r_0 = 5 \times 10^{-2}$ cm, $nb_{1/2} \cong 200$ mV.

These sharp, narrow peaks enable easier separation of individual processes when a mixture of several reactants with similar electrode reaction potentials is studied. Stromberg and co-workers [54] discussed the problem of the "separability" of voltammetric peaks in terms of their different potentials and widths. Batley and Florence [55] performed an experimental comparison of the separation of anodic peak currents of oxidation of amalgams of Tl and Pb as well as Cd and In.

Stojek and Kublik [56] discussed the possibility of using thin-film mercury electrodes to improve resolution of voltammetric peaks. Variations of film thickness and scan rate may improve the separation of the neighboring peaks, but only if an unequal number of electrons are involved in both electrode reactions. They [56] also used silver-based mercury-film electrodes for the anodic stripping analysis of lead and copper, and compared their experimental results with theoretical predictions given for such a case by de Vries and van Dalen [51–53].

Behavior similar to the above also occurs if the electroactive substance and the product of the electrode reaction are immobilized on the electrode surface. Such immobilization may be accomplished, for instance, by chemisorption [57]. Because in such a case reactants are kept at the electrode surface, one observes a peak similar to that observed in thin-layer SEV with very thin-layer solutions. These problems are related to so-called chemically modified electrodes [58] and will be briefly discussed in Section 2.2.4.

CONSECUTIVE ELECRON TRANSFERS

For a two-step process one has

$$\text{Ox} + n_1 e \rightleftharpoons \text{Red}_1 \tag{22}$$

$$\text{Red}_1 + n_2 e \rightleftharpoons \text{Red} \tag{23}$$

This problem was generally discussed for stationary electrode voltammetry by Gokhstein and Gokhstein [59] and later given detailed theoretical consideration by Polcyn and Shain [60]. In a reversible reaction two independent cathodic peaks are observed, provided the reduction potentials differ by at least $118/n$ mV. If the difference is smaller, the two peaks coalesce to form one broad peak.

To determine the second current peak when two peaks are observed, the current of the first process at the potentials of the second peak must be shown, and may be calculated based on the function $\chi(at)$. The change of this current with potential, as a base for the measurement of the second current peak, is depicted in Figure 3.11.

Obviously, if Red_1 is more readily reduced than Ox, one peak corresponding to transfer $n_1 + n_2$ electrons is observed.

Electrode reactions of several depolarizers occur in separate peaks when the difference between electrode reaction potentials of the individual depolarizers exceeds $120/n$ mV.

Additionally, determination of the second and further peaks is rather difficult. When this difference is smaller than $120/n$ mV, the system can be analyzed by computer. Gutknecht and Perone [61] made an accurate determination of the currents of In(III) and Cd(II) electroreduction in 1 M HCl, even though the difference between the peak potentials was only 40 mV.

2.2.2   Irreversible Electron Transfers

The border between reversible and irreversible reactions cannot be drawn very precisely and depends on the standard rate constant of the system studied and also on the scan rate.

To observe irreversible behavior the following inequality should be fulfilled [62]:

$$k_s \left(\frac{1}{D_{Ox}^{1/2}}\right)^\beta \left(\frac{1}{D_{Red}}\right)^\alpha \leqslant a \times 10^{-2(1+\alpha)} \tag{24}$$

where k_s is the standard rate constant of the electrode reaction and α and β are transfer coefficients of the cathodic and anodic reactions, respectively. Explained differently, the reversible or irreversible behavior of a given system depends on the relative rates of the electrode process and on the mass transport rate [3], as depicted in Figure 3.12.

The reversible behavior is observed at such a low mass transport rate (scan rate) that this process controls the rate of the overall reaction (Figure 3.12, left part). The irreversible behavior occurs when $k \ll \bar{v}$ (Figure 3.12, right part). The two regions are separated by the region of quasi-reversibility in which $\bar{v}$ is similar to k.

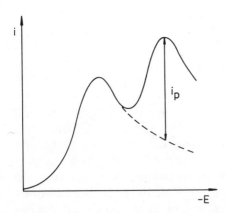

Figure 3.11 The schematic way of determining the height of the second voltammetric current peak. The dashed line in the region of the second peak may be theoretically calculated.

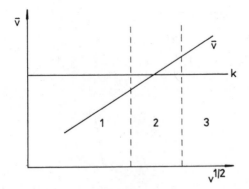

Figure 3.12 The dependence of the mass-transport rate ($\bar{v}$) on the square root of the scan rate ($v^{1/2}$) at some constant potential of the working electrode. The rate of the charge-transfer step is constant (k), because the potential of the electrode is not changed. (1) Region of reversibility; (2) region of quasi-reversibility; and (3) region of irreversibility.

Under such conditions the rate of the electron-transfer step controls the rate of the overall process

$$\text{Ox} + ne \xrightarrow{k_{fh}} \text{Red} \tag{25}$$

where k_{fh} is the rate constant of the cathodic reaction at some potential E. Also, the dependence of the current on potential exhibits a peak; however, the curve for peak current is usually more drawn-out than that for diffusion-controlled processes.

The theory of such processes occurring at planar electrodes is discussed in [62–66]. The current–potential curve shapes for totally irreversible reactions are dependent on the function $\chi(bt)$ [25] in the following way:

$$i = nFAC^0_{Ox}\sqrt{\pi D_{Ox}b}\chi(bt) \tag{26}$$

Figure 3.13 shows the function $\chi(bt)$ and $b = \alpha n_\alpha Fv/RT$. The expression for the peak current is

$$i_p = 0.496n\frac{F^{3/2}}{(RT)^{1/2}}A(\alpha n_\alpha)^{1/2}D^{1/2}_{Ox}v^{1/2}C^0_{Ox} \tag{27}$$

As with reversible reactions, there is also a linear dependence of i_p on the bulk reactant concentration C^0_{Ox} and on the scan rate to half-power.

The main difference between this and the reversible reaction is the dependence of the peak current on the $(\alpha n_\alpha)^{1/2}$ term instead of on the $n^{3/2}$ observed in reversible reactions [n_α is the number of electrons transferred in the rate-determining step, while n is the total number of electrons transferred in reaction (25)]. If αn_α is low, the peak current may be considerably lower than the peak of a reversible reaction under similar conditions.

Unlike the reversible process, the peak potential is now dependent on the

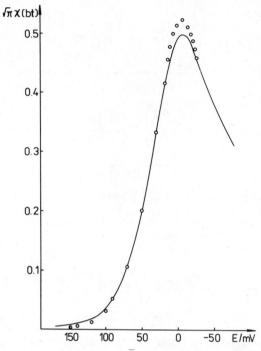

Figure 3.13 The dependence of the function $\sqrt{\pi}\chi(bt)$ on potential (solid line). Potential of the electrode is given in a scale $(E - E^0)\alpha n_\alpha + (RT/F) \ln [(\pi D_{Ox}b)^{1/2}/k_s]$. Circles (○) show the dependence according to [36].

scan rate

$$E_p = E^0 - \frac{RT}{\alpha n_\alpha F} \left[0.78 - \ln k_s + \ln \sqrt{\frac{D_{Ox}\alpha n_\alpha F v}{RT}} \right] \qquad (28)$$

where k_s is the standard rate constant related to k_{fh}, which appears in (25), by the equation

$$k_{fh} = k_s \exp \left[\frac{-\alpha n_\alpha F(E - E^0)}{RT} \right] \qquad (29)$$

where E^0 is the formal potential of the system studied.

The current–potential curve shapes for totally irreversible reactions are independent of the scan rate but depend on the αn_α parameter

$$E_p - E_{p/2} = -1.857 \frac{RT}{\alpha n_\alpha F} \qquad (30)$$

This equation provides a simple way of determining αn_α, an important parameter for elucidating electrode reaction mechanisms.

Irreversible processes occurring at spherical electrodes were also considered

[21, 67]. As in reversible reactions, an increase of the peak current compared with that seen for the linear-diffusion case under similar conditions is observed. To some extent the current–potential curve shapes may also be changed.

Equation (28) may be used to determine the standard rate constant of the electrode reaction. Before applying this equation one must be sure that the electrode reaction is totally irreversible, which may be conveniently checked by applying cyclic voltammetry. The separation of the anodic and cathodic peak potentials should exceed $250/n$ mV for totally irreversible behavior.

Under such conditions the rate constant may also be determined by analysis of the current at the foot of the voltammetric curve.

The current is then given by a simple equation [62],

$$i_c = nFAC^0_{Ox}k_{fh} \tag{31}$$

Since k_{fh} is exponentially dependent on overpotential, combining (31) and (29) yields

$$\ln i_c = \ln[nFAC^0_{Ox}k_s] - \frac{\alpha n_\alpha F}{RT}(E - E^0) \tag{32}$$

If the standard potential of the system studied is known, k_s is determined by extrapolation of the current to $(E - E^0) = 0$.

A theory of quasi-reversible electrode reactions developed by Matsuda and Ayabe [62] is rather complicated and not easily applied in practice.

Such processes are better handled by cyclic voltammetry, especially when determining the electrode kinetics of such reactions [68].

Although (32) may be used to study electrode kinetics, only the rising part of the current–potential curve may be used in analysis. Also, an approach based on (28) has some limitations. Since the peak potential is usually not as precisely determined as the half-peak potential $(E_{p/2})$, especially when voltammetric curves are broad, k_s may be determined using $E_{p/2}$ based on

$$E_{p/2} = E^0 + \frac{0.04236}{\alpha n_\alpha} - \frac{RT}{\alpha n_\alpha F}\ln\left(\frac{D_{Ox}\alpha n_\alpha Fv}{RT}\right)^{1/2} - \frac{RT}{\alpha n_\alpha F}\ln k_s \tag{33}$$

Wrona [69] suggested another approach to k_s calculations.

The current function $\chi(bt)$, as described by Nicholson and Shain [25], is dependent on the G parameter,

$$G = (E - E^0)\alpha n_\alpha + \frac{RT}{F}\ln\left[\frac{(\pi D_{Ox}b)^{1/2}}{k_s}\right] \tag{34}$$

Because the current is related to the function $\chi(bt)$ according to (26), dividing (26) by (27) gives

$$\chi(bt)\sqrt{\pi} = 0.496i/i_p \tag{35}$$

Except for the foot (from 0 to 4%) and the region of the peak (from 96 to 100%), the function $\chi(bt)\sqrt{\pi}$ is described with satisfactory accuracy by

$$\chi(bt) = A\exp[-B(G + 0.005)^C] \tag{36}$$

where A, B, and C have the following values: $A = 0.5208$, $B = 116.0$, and $C = 1.6582$. By using experimental data and (36) the values of the G function may be calculated.

Combining (29) and (34) gives

$$k_{fh} = \frac{1}{(\pi D_{Ox} b)^{1/2}} \exp\left(-\frac{FG}{RT}\right) \tag{37}$$

To obtain k_{fh} values αn_α values should be known, which may be accurately obtained from the dependence of the G function on potential (34). By using this approximate function, it is possible to evaluate the potential dependence of the rate constant of the studied reaction; this may also be useful for computer analysis of empirical voltammetric curves. This method was examined by studying the kinetics of several systems. Rate constants determined showed satisfactory agreement with those found by earlier procedures [69].

Also, in irreversible reactions, both the height of a peak and to some extent its shape may be changed if the diffusion symmetry is not strictly linear.

Spherical diffusion is especially important. A theoretical description of the irreversible reactions occurring under conditions of spherical diffusion was given by Reinmuth [64] and de Mars and Shain [67], but the practical application of their calculations was not very convenient. For this reason Nicholson and Shain [25] expressed the current by an equation similar in form to that given for reversible reactions

$$i = nFAC_{Ox}^0 D_{Ox} b^{1/2} \pi^{1/2} \chi(bt) + \frac{nFAD_{Ox}C_{Ox}^0 \phi(bt)}{r_0} \tag{38}$$

Both functions $\chi(bt)$ and $\phi(bt)$ are tabulated. While the function $\chi(bt)$ has a peaklike shape, $\phi(bt)$ changes with potential in a manner similar to a polarographic wave, reaching a plateau at potentials several tens of millivolts more negative than the peak potential.

The expression for the peak current is

$$i_p = 0.496nFA(\alpha n_\alpha)^{1/2}\left(\frac{F}{RT}\right)^{1/2} v^{1/2} D_{Ox}^{1/2} C_{Ox}^0 + \frac{0.694nFAD_{Ox}C_{Ox}^0}{r_0} \tag{39}$$

As in the reversible case, here too the influence of the second spherical term increases as the scan rate and radius of the electrode decrease.

2.2.3   Semi-Integral and Semi-Differential Stationary Electrode Voltammetry and Other Varients of SEV

An interesting improvement of voltammetry with stationary electrodes, which computes convolution integrals of the type

$$I(t) = \frac{1}{\pi^{1/2}} \int_0^t \frac{i(v)}{(t-v)^{1/2}} \, dv \tag{40}$$

directly from the experimental current–time curves, was proposed by Savéant and co-workers [70]. Similar ideas were developed independently by Oldham

[71]. $I(t)$ may be considered a semi-integral of $i(t)$

$$\frac{d^{-1/2}}{dt^{-1/2}} i(t) = I(t) \tag{41}$$

This integral is calculated by converting the analog output signal into a digital form and then performing a computer calculation of the convolution integral I. The main advantages of this method (Imbeaux and Savéant [72]), which may be called "convolution stationary electrode voltammetry," are listed below:

1. Improvement in accuracy through use of all the information contained in a single polarization curve rather than only that involved in the peak values.
2. Elimination of a severe limitation of SEV, in a study of potential-dependent phenomena.
3. Simplification of mechanistic analysis when secondary chemical reactions are involved.
4. Simplification of analysis of the reactions controlled by charge-transfer kinetics.
5. Significant simplification of the correction for ohmic drop.

The kinetics of the electrode reactions may be readily derived by the following equation [72]:

$$\ln k = \ln D^{1/2} - \ln \left\{ \frac{I_l - I\left[1 + \exp\left(\frac{F(E - E_{1/2})}{RT}\right)\right]}{i} \right\} \tag{42}$$

where i is the current observed in SEV, the convoluted current I is as in (40), and I_l is the limiting value of the convoluted current. Equation (42) corresponds to the quasi-reversible case. For a totally irreversible reaction (42) simplifies to

$$\ln k = \ln D^{1/2} - \ln\left(\frac{I_l - I}{i}\right) \tag{43}$$

This method of analysis was applied [73] in the study of the charge-transfer kinetics of the electroreduction of *tert*-nitrobutane in acetonitrile (AN) and dimethylformamide (DMF). The apparent standard rate constant k_s^{app} in DMF was within the limits $(4.2-4.8) \times 10^{-3}$ cm/s, a value approximately two times lower than the one reported earlier [74] which was determined differently. In AN, k_s^{app} varied [73] from 6.2×10^{-4} to 8.7×10^{-4} cm/s. These values are considerably lower than those later reported for this system by Corrigan and Evans [75].

Also the approach based on the calculation of the semi-differential of the current, e, was proposed [76–78]:

$$e = \frac{d^{1/2}i}{dE^{1/2}} \tag{44}$$

Bond [79] has compared these techniques with other electroanalytical methods.

Also other modifications of SEV were described. The staircase variant proposed by Barker [80] should be mentioned. In this technique the potential is applied in steps ΔE instead of linearly. Each step lasts for a period Δt. The current, which may be measured at the end of each time interval and plotted as a function of the potential, exhibits a curve similar to that recorded using normal SEV. However, in this case the double-layer charging current is largely eliminated. Consequently, the method is quite sensitive in analysis and offers the possibility of analyzing substances with concentrations on the order of 10^{-7} M.

This method was developed by Mann [81] and Nigmatullin and Vyaselev [82]. The theory of this method for the reversible electrode reactions was elaborated by Christie and Lingane [83]. Staircase voltammetry with varied current sampling times was also developed [84, 85].

Also derivative [86–88] and differential [89, 90] SEV were used mostly to improve analytical performance.

2.2.4   Electrode Processes of Adsorbed Reactants

Since adsorption of organic compounds on electrode surfaces occurs frequently [91], the change of the voltammetric characteristic induced by adsorption is briefly discussed.

In a reversible electrode reaction with reversible adsorption of an electroactive substance, several cases may be distinguished by (a) the extent of adsorption and (b) the substance, Ox or Red, undergoing adsorption. A theoretical consideration of such situations was presented by Wopschall and Shain [92].

If Ox is strongly adsorbed onto the electrode, two peaks are observed. The first corresponds to electroreduction of Ox diffusing from the bulk to the electrode and occurs at a reversible potential of the Ox/Red couple; at more negative potentials, electroreduction of the adsorbed species is observed (Figure 3.14).

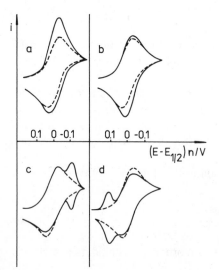

Figure 3.14 Typical cyclic voltammetric curves in an electrode reaction complicated by adsorption. Weak adsorption of substrate (Curve a) and of product (Curve b). Strong adsorption of substrate (Curve c) and of product (Curve d). Dashed lines indicate curves of the diffusion-controlled processes.

When a product of electroreduction adsorbs at potentials more positive than the reversible potential of the Ox/Red couple (unaffected by adsorption), a peak corresponding to the reaction

$$Ox + ne \rightleftharpoons Red_{ads} \tag{45}$$

is observed. At a more cathodic potential, the product Red, which is not adsorbed on the surface because all available sites are already occupied, undergoes electroreduction.

From this information it follows that the adsorption current peak should be controlled by the maximum number of molecules that may be placed on the electrode surface, whereas the second peak should be proportional to the bulk concentration of the reactant. With strong adsorption and low reactant concentration, only one current peak, controlled by adsorption, should be observed.

This current peak may be described by a simple equation [93], valid when only Ox is electroactive

$$i_p = \frac{n^2 F^2}{4RT} Av\Gamma_{Ox} \tag{46}$$

where Γ_{Ox} is the amount of Ox adsorbed on the electrode surface.

The difference between the potentials of both current peaks depends on the free energy of adsorption divided by RT.

If the reactant or product is only weakly adsorbed, one peak should be observed; however, its height should be increased, because in addition to the substance reaching the electrode by diffusion, the charge resulting from the process of adsorbed substance will flow in the circuit.

Feldberg [94] considered in detail the effect of adsorption on the shape of SEV curves. Unlike Wopschall and Shain [92], he did not limit his considerations to adsorption governed by the Langmuir isotherm, but exemplified systems that are described by the Frumkin isotherm. Processes with a slow exchange of charge (irreversible) associated with reactant adsorption were also discussed by Feldberg [94].

With a strong adsorption of Ox and an irreversible electrode reaction, the maximum current of the adsorption peak is [93]

$$i_p = \frac{\alpha n_\alpha n F^2}{2.718RT} Av\Gamma_{Ox} \tag{47}$$

The difference between this peak potential and the standard potential of the Ox/Red couple is dependent on the rate constant of the electrode reaction [93],

$$E_p = E^0 + \frac{RT}{\alpha n_\alpha F} \ln \frac{RT}{\alpha n_\alpha F} + \frac{RT}{\alpha n_\alpha F} \ln \frac{k_s}{v} \tag{48}$$

Hulbert and Shain [95] presented a theoretical study of SEV processes in which reactant adsorption is not at equilibrium.

SEV is not recommended as a method of quantitatively studying adsorption reactants, because during measurement the potential of the electrode changes linearly with time. A method such as chronocoulometry [96–99], in which the substance adsorbed at a constant initial potential reacts very quickly after sudden application of a new potential, is recommended.

However, SEV may be successfully used for qualitatively or semiquantitatively detecting adsorption of reactants, either by observance of additional voltammetric peaks or, when only one peak is observed, by noting its unusual shape.

In recent years redox-modified electrodes have been prepared [58, 93]. Laviron presented [100] a multilayer model for the semiquantitative study of space-distributed, redox-modified electrodes. These considerations are valid for a redox polymer electrode and also for adsorption of an electroactive substance in several layers. This model was applied [101] to the study of SEV curves in a simple redox system.

The theory for a monolayer is directly applicable to the rapid rate of electron exchange inside the coating and is applied to the study of the multilayer adsorption of benzo[c]cinnoline. These problems were further developed by Laviron [102] and interactions between the molecules in the first layer were considered.

2.2.5   Application of SEV to the Study of Organic Systems

SEV is frequently used in elucidating the mechanisms of complex electrode reactions; however, cyclic electrode polarization is usually applied. Cyclic voltammetry problems are covered in [103]. This section presents a brief outline of the application of noncyclic, linear-scan SEV.

Piekarski and Adams [1] have primarily discussed anodic reactions; therefore, results of electroreduction studies of organic substances are preferentially discussed herein. SEV determines the reversibility of reactions studied, where the equation

$$E_{p/2} - E_p = \pm \frac{0.057}{n} \quad (\text{V}) \tag{49}$$

is used. The plus and minus signs apply to the cathodic and anodic reactions, respectively. This equation results from the description of the shape of the reversible voltammetric peak.

However, cyclic voltammetry with

$$E_{pa} - E_{pc} = \frac{0.058}{n} \quad (\text{V}) \tag{50}$$

is more frequently used to test the reversibility of the process studied, E_{pa} and E_{pc} represent anodic and cathodic peak potentials, respectively. For such reversible charge-transfer reactions, SEV may be used to exactly determine peak or half-peak potential ($E_{p/2}$ can be more precisely measured, as was recognized long ago [104]) to calculate the equilibrium constants of the ion-pair-formation reaction between free radicals produced electrolytically and cations of the background electrolyte.

Although in such studies polarography is used frequently [105–108], SEV may also be successfully applied. With SEV it was possible to show that the ion-pair formation may dramatically change the mechanism of electroreduction of azoxybenzene [109], nitrosobenzene [110], nitrobenzene, and 3-nitropyridine [111] in DMF.

The influence of ion pairing on the voltammetric characteristics of other organic substances was also studied. Avaca and Bewick [112] considered the influence of several cations on the electroreduction of anthracene in HMPA. The redox behavior of anthracene and 9,10-disubstituted anthracenes in acid and aprotic media was also studied [113].

Ryan and Evans [114] studied the electroreduction of benzil in DMF, DMSO, and AN in the presence of group IA and IIA metal cations. The same authors used SEV to study [115] the effect of sodium ions on the electrochemical reduction of diethyl formate in DMSO and AN. Evans and Griffith studied the effect of metal ions [116] and pH [117] on the electrochemical reduction of several heterocyclic quinones. The formation of insoluble triple ions of the dianion of p-nitrobenzene cathodically produced with Na^+ and K^+ was reported by Ahlberg and co-workers [118].

Cyclic voltammetry and, to some extent, SEV were used by Bard and co-workers to study the behavior of certain organic substances in liquid ammonia [119–121]. The electroreduction of benzophenone [119], nitrobenzene and nitrosobenzene [120], and diethyl fumarate and cinnamonitrite [121] proceeds in this medium in two one-electron steps with the formation of a free radical and a dianion.

The electroreduction of cyclooctatetraene in liquid ammonia at -38 and $-65°C$ was interpreted [122] as a two-step, two-electron transfer reaction forming strong ion pairing of the dianion.

Diethyl fumarate was studied in DMF at 200 K by Grypa and Maloy [123]. Three interrelated reduction processes to anion, dianion, and a dimeric product of the initial reduction were observed.

The formation of free radicals was also observed in the electrode reaction of other substances. Ouziel and Yarnitzky [124] used SEV in the electroreduction of camphorquinone in DMF on a mercury electrode. In the absence of proton donors two one-electron waves were observed. A proton donor-to-substrate ratio of 2 is required to suppress the two initial waves completely.

Hawley and co-workers studied the electroreduction of nitrosobenzene in DMF and AN [125] and of fluorenone and p-cyanoaniline in DMF at the Pt electrode [126]. In these cases corresponding radical anions are formed. In the absence of proton donors nitrosobenzene free radicals dimerize to a dianion intermediate, whereas radicals studied in the second work [126] decompose by carbon–hydrogen bond cleavage to give two conjugate bases of the starting materials. Using SEV [127] in an aqueous alkaline solution of low ionic strength electroreduction of aliphatic primary and secondary nitro compounds at the Hg electrode proceeds in two steps with the formation of anion free radicals in the first step. This method was also used to study the oxidation of phenoxy radicals in aprotic media [128].

The reaction of electrochemically generated thiantrene radical cation and dication with anisole and water in liquid SO_2 was studied [129].

The electrooxidation of several oxocarbon salts of croconic acid and its dicyanomethylene derivatives proceeds in DMF in two consecutive reversible one-electron transfers with the formation of stable radical anions and neutral croconates [130].

Robinson and Osteryoung [131] studied the oxidation of several aromatic hydrocarbons at room temperature in the molten salt system, aluminum chloride-n-butylpyridinium chloride. In all cases, cation radicals were formed at potentials linearly dependent on the value of the first ionization potential of hydrocarbons. Radicals had a significantly greater stability in this molten-salt medium than in acetonitrile.

Passivating films result [132] from the electrochemical oxidation of benzene and biphenyl in liquid SO_2 in the presence of quaternary ammonium perchlorate. When SEV was used to study the oxidation of a series of pyrrole—bipyrrole, terpyrrole, polypyrrole [133]—and some substituted pyrroles [134], the anodic peak potential within this oligomeric series was found to change linearly with the lowest absorption energy.

The oxidation of several polynuclear aromatic hydrocarbons in propylene carbonate was studied by Madec and Courtot-Coupez [135]; Jensen and Parker [136] reduced several aromatic hydrocarbons in several common solvents. Anion-free radicals and the dianions stable during the experiment were observed with a low scan rate.

Savéant and co-workers [137], using SEV, studied the reduction of benzaldehyde, benzophenone, and p-phenylbenzophenone in a benzoic acid buffer. They proposed a mechanism of the reactions studied based on the dependencies of E_p on various parameters and on the results of additional studies.

Yasukouchi and co-workers [138] observed two waves during the anodic oxidation of acridine in AN, which they explained were caused by the oxidation of nonprotonated and protonated acridine. The oxidation pathway for the formation of tetramer by an ECEC mechanism was proposed.

Several other electrochemical processes were studied recently: electroreduction of bianthrones in DMF [139], redox chemistry of reduced pterin species [140], typical aromatic amines at graphite electrodes [141], vitamin B_{12a}/vitamin B_{12r} couple on mercury and Pt electrodes [142], oxidation of 9-methyl uric acid over a wide pH range [143], and electrode behavior of ellipticine derivatives in DMF [144].

The mechanism of the electrode behavior of d^8 iron and cadmium nitrosyl compounds with the phosphorus ligand $Ph_2PCH_2CH_2PPh_2$ [145] was studied by SEV. The reduction was found to proceed in two reversible, one-electron steps leading through an intermediate to d^{10} anionic complexes. This method was also applied to study the thick deposits on the electrode of phenazine, azobenzene, and chloranil [146] and polymeric anthraquinone layers on a mercury electrode [147]. In the latter case the electrode surface was modified by a deposited layer.

Savéant and co-workers used convolution SEV to study various problems, including the intramolecular pinacolization of 1,3-dibenzoylpropane [148].

The role of the scan rate in kinetic analysis by convolution SEV was discussed [149] and evaluated over 3.5 orders of magnitude of v using the fluorenone–fluorenone anion couple in AN. Later this method was experimentally evaluated up to a scan rate of 2278 V/s using the same fast redox system as in DMF [150]. The effects of ohmic drop, double-layer charging, and band-pass limitations of the instrument were discussed.

Procedures for precise measurements of potentials used when working with resistive media were discussed by Ahlberg and Parker [151].

Alstad and Parker [152] proposed a new way to analyze voltammetric data, which helps in elucidating the mechanism of electrode reactions.

Also of interest is a review by Brainina and Vydrevich [153], on using SEV in the stripping analysis of solids, and the monograph by Vydra and associates [154].

2.2.6   Application of SEV to Bioelectrochemical Studies

Significant developments of *in vivo* electroanalytical studies have recently occurred. Such studies generally utilize solid electrodes primarily consisting of various carbons. SEV was applied in such experiments although chrono-amperometric analysis is now used more frequently.

This work was initiated over 10 years ago by Adams, who studied the anodic oxidation of some neurotransmitter substances. In connection with these and other studies, microelectrodes were developed. Some of these electrodes were placed with a miniature calomel electrode in the brain tissue of rats. They may be used with graphite paste or solidified graphite-epoxy resin for the anodic oxidation of electroactive substances present in the brain. Mainly catecholamines and their metabolites and serotonin-type compounds should be detectable. However, at the oxidation potentials of, for example, dopamine, ascorbic acid, which is present in the mammalian brain in about millimolar concentration, is also oxidized. Works related to these problems have been discussed and reviewed by Adams [155].

Another application of electroanalytical methods in problems related to bioelectrochemistry is the study of ion transfer across the interface between two immiscible electrolyte solutions. In such studies chronopotentiometry and polarography have been used, but in recent years SEV and cyclic voltammetry also play an important role.

SEV yields curves similar to those observed when electroreduction or electro-oxidation occurs. The equations given in Section 2.2.1 are also applicable to the study of such ion transfers.

Ion transfer across two interfaces was investigated for the water–nitrobenzene and water–1,2-dichloroethane systems.

SEV was used to study the transfer of picrates [156], tetrabutylammonium and tetraethylammonium ions [157], and alkali metal ions [158, 159] across a water–nitrobenzene interface.

Koczorowski and co-workers used SEV to study the transfer of cesium ion [160] and tetrabutylammonium and tetramethylammonium ions [161] across the water–1,2-dichloroethane interface, and Samec [162] applied convolution SEV and cyclic voltammetry in such studies. In addition, voltammetric studies of the reduction of DNA [163] and other biologically important species [164] were conducted.

2.3   Stationary Electrode Voltammetry in the Study of Electron Transfers Coupled to Chemical Reactions

Theoretical elaborations are available of electrode reactions complicated by chemical reactions that may precede or follow the charge-transfer step proper. Other types of reactions may also occur, such as ECE or catalytic regeneration of the electrode reaction substrate.

These investigations will be discussed below, arranged according to reaction. Because the importance of the theory of EC processes in electroanalytical practice is probably greater than that of other reactions, this presentation begins with them.

2.3.1   Charge Transfer Followed by Chemical Reactions
FIRST-ORDER CHEMICAL REACTIONS

The influence of the following chemical reaction on the electrode process results in the decrease of the concentration of a primary product. Consequently, in a reversible reaction, when the electrode is sensitive to a fast change of electroactive species concentration, the SEV curves change to less negative values for a cathodic reaction (less positive in the anodic reaction).

A theoretical elaboration of the charge-transfer step associated with the following first-order chemical reaction according to the scheme

$$Ox + ne \rightleftharpoons Red \underset{k_2}{\overset{k_1}{\rightleftharpoons}} A \qquad (51)$$

was presented by Nicholson and Shain [25]; k_1 and k_2 are rate constants of the chemical reaction in both directions, and A is a final product, which remains inactive at the potential of the electrode reduction.

Only reversible charge-transfer reactions were considered, because in irreversible electron transfers the overall process rate is limited by the charge-transfer rate and the following chemical reaction has no influence on the noncyclic SEV curves recorded.

This theory shows that the following chemical reaction may change the slopes of the recorded curves slightly and somewhat increase (up to approximately 10%) the peak heights compared with those unaffected by the reaction.

The most important effect of such reactions is the current–potential curve shift to less cathodic potentials in electroreduction and to less anodic potentials in electrooxidation.

This peak potential shift in the cathodic reaction is given by the following equation:

$$E_p^k = E_{1/2} - \frac{RT}{nF}\left[0.78 + \ln K \left(\frac{nFv}{(k_1+k_2)RT}\right)^{1/2} - \ln(1+K)\right] \quad (52)$$

It follows from this equation that a tenfold decrease in scan rate leads to a shift of the peak potential in the positive direction by $2.3RT/2nF$ volts for values of the $[K^2 nFv/(k_1+k_2)RT]^{1/2}$ parameter larger than 1.

For irreversible chemical reactions (52) simplifies to

$$E_p^k = E_{1/2} - \frac{RT}{nF}\,0.78 + \frac{RT}{2nF}\ln\frac{RT}{nF} + \frac{RT}{2nF}\ln\frac{k_1}{v} \quad (53)$$

In (52) and (53), $E_{1/2}$ is the reversible half-wave potential of the reaction unaffected by the chemical reaction.

A shift of E_p^k to less negative values with a decrease of v should also be observed in this case.

The behavior of such systems with slow charge-transfer processes (quasi-reversible) was discussed by Evans [165] and Nadjo and Savéant [166]. The latter authors also discussed diagnostic criteria in the extended treatment.

Recently, the orthogonal colloquation technique [167] was applied to solve this problem independently [168]. This technique, which offers a shorter simulation time, was used earlier [169] to simulate SEV and cyclic voltammetric curves.

Laviron [170] presented a theoretical study of a reversible electrochemical reaction followed by a chemical reaction for thin-layer SEV. Laviron and co-workers [171, 172] theoretically considered the reversible surface electrochemical step followed by a first-order surface chemical reaction.

Savéant and Tessier [173] considered, in the framework of convolution SEV, charge-transfer processes coupled with a following irreversible chemical reaction. For the first-order EC reaction the following equation was found:

$$E = E^0 + \frac{RT}{2nF}\ln k_1 + \frac{RT}{nF}\ln\frac{I_l - I}{i} \quad (54)$$

where I_l is the limiting value of the convoluted current I.

DIMERIZATION REACTIONS

In the 1960s Savéant and Vianello [174, 175] and Nicholson [176] elaborated upon the theory of SEV for a reversible electrode reaction followed by dimerization of a primary product. It was assumed that the electrode reaction should be reversible, that is,

$$k_s > 10(D_{Ox}C_{Ox}^0 k_1)^{1/2} \quad (55)$$

should be fulfilled [165]. In (55) D_{Ox} is the diffusion coefficient of a substrate, C_{Ox}^0 is its bulk concentration, and k_1 is the rate constant of the dimerization reaction.

An even more extensive theoretical investigation of such processes was carried out by Andrieux and associates [177].

Three different forms of the dimerization reaction were considered using free radicals as the primary products undergoing the dimerization reaction.

1. Simple dimerization of free radicals—DIM 1:

$$\text{Ox} + e \rightleftharpoons \text{Red} \qquad E^0_{\text{Ox/Red}} \tag{56}$$
$$2\,\text{Red} \xrightarrow{k_1} \text{A}$$

2. Dimerization of free radicals with a substrate Ox—DIM 2:

$$\text{Ox} + e \rightleftharpoons \text{Red}$$
$$\text{Red} + \text{Ox} \xrightarrow{k_1} \text{B} \qquad E^0_{\text{Ox/Red}} \tag{57}$$
$$\text{B} + e \rightleftharpoons \text{A} \qquad E^0_{\text{B/A}}$$

It is assumed that $E^0_{\text{Ox/Red}} < E^0_{\text{B/A}}$.

3. Dimerization of a two-electron reduction product with a substrate Ox— DIM 3:

$$\text{Ox} + 2e \rightleftharpoons \text{Red}$$
$$\text{Red} + \text{Ox} \xrightarrow{k_1} \text{A} \tag{58}$$

It was assumed that A is electrochemically inactive in the potential range considered.

The equations that describe the peak potential in such processes, as well as the criteria that distinguish among these three mechanisms, are given in Table 3.1. Numerical values in Table 3.1 refer to the one-electron process at 25°C.

In all cases the conditions of (55) should be obeyed. Savéant and co-workers [173, 178] considered DIM 1 in the framework of convolution SEV. The following equation of the voltammetric curve was presented

$$E = E^0 + \frac{RT}{3F} \ln\left(\frac{2k_1}{3FAD_{\text{Ox}}^{1/2}}\right) + \frac{RT}{F} \ln\left(\frac{I_l - I}{i^{2/3}}\right) \tag{59}$$

where I_l is the limiting value of the convoluted current I. The effect of scan rate and initial concentration, which are formally absent in this equation, exists through i, which is obviously dependent on $v^{1/2}$.

Different variants of dimerization reactions occurring under thin-layer conditions were considered by Laviron [170], who also considered [179, 180] dimerization reactions of adsorbed substances.

Many electrodimerization reactions proceed according to the reaction

$$2\,\text{Ox} + 2\text{H}^+ + 2e \rightarrow \text{DH}_2 \tag{60}$$

Savéant and co-workers [181, 182] theoretically elaborated on such processes, giving corresponding equations and useful diagnostic criteria.

Table 3.1 Properties of Voltammetric Peaks in the Case of Reversible Electrode Reactions with Fast and Irreversible Dimerization

Mechanism	Basic Equation	$\dfrac{\partial E_p}{\partial \log v}$ (mV)	$\dfrac{\partial E_p}{\partial \log C_{Ox}^0}$ (mV)	$E_{p/2} - E_p$ (mV)
DIM 1	$E_p = E^0 - 1.038\dfrac{RT}{F} + \dfrac{RT}{3F}\ln\left(\dfrac{RT}{F}\dfrac{k_1 C_{Ox}^0}{v}\right)$ $E_p = E^0 - 0.058 + 0.0197\log\left(\dfrac{k_1 C_{Ox}^0}{v}\right)$	-19.7	19.7	38.8
DIM 2	$E_p = E^0 - 0.456\dfrac{RT}{F} + \dfrac{RT}{2F}\ln\left(\dfrac{RT}{F}\dfrac{k_1 C_{Ox}^0}{v}\right)$ $E_p = E^0 - 0.059 + 0.0296\log\left(\dfrac{k_1 C_{Ox}^0}{v}\right)$	-29.6	29.6	58.3
DIM 3	$E_p = E^0 - 0.401\dfrac{RT}{F} + \dfrac{RT}{4F}\ln\left(\dfrac{RT}{F}\dfrac{k_1 C_{Ox}^0}{v}\right)$ $E_p = E^0 - 0.0338 + 0.0148\log\left(\dfrac{k_1 C_{Ox}^0}{v}\right)$	-14.8	14.8	29.3

APPLICATION TO ORGANIC SYSTEMS

Many researchers have used considerations presented in the sections dealing with first-order chemical and dimerization reactions in electroanalytical practice.

Andrieux and Savéant [183] studied electroreduction of several imines, mainly anils in AN and DMF. Depending on the structure of imine and solvent, either a single two-electron wave or two successive waves were observed. In the latter case the dimerization reaction was associated, at least partly, with a product of the first wave.

In the course of the electroreduction of activated olefins in solvents of low acidity [184, 185], the coupling of two anion radicals is the most probable reaction path in cases where the protonation reactions do not influence the overall kinetics. Using SEV, intramolecular pinacolization of 1,3-dibenzylopropane in AN was studied [186].

Laviron and Mugnier [187] studied the electrochemical reduction of *cis*-azobenzene in DMF. *cis*-Azobenzene radical anions obtained in the first one-electron step isomerize rapidly to *trans*-anions.

The mechanism of electroreduction of substituted acetophenone in AN and DMF was investigated by Andrieux and Savéant [182], while the electroreduction of carbonyl compounds was extensively studied by Savéant and associates [188, 189].

Lasia [190] studied the kinetics of the dimerization reaction of ion pairs of phthalic aldehyde and phthalic anhydride radical anions with alkali metal cations in DMF solution. SEV was used in addition to cyclic voltammetry.

Using thin-layer SEV, Vallat and Laviron [191] investigated the mechanism of oxidation of substituted triphenylamines. Tetraphenylbenzidine, which is formed during the oxidation, arises from the dimerization of two cation radicals. SEV and cyclic voltammetry were used in studies of the oxidation of iminobibenzyl and several related compounds in acetonitryle [192]. It is suggested that the mechanism of this oxidation is an ECE path with coupling of an initial one-electron product in a very fast reaction to form a dimeric compound, which is further oxidized.

Fawcett and Lasia [193], in their study of the mechanism of the electroreduction of aromatic aldehydes in DMF and AN in the presence of $N(C_2H_5)_4^+$, K^+, and Na^+ perchlorates, concluded that free radicals disappear in the dimerization reaction.

Thin-layer SEV was used to study the course of the reversible reduction of nitrosobenzene to phenylhydroxylamine [194]. At pH 12, the rate constant of the coupling reaction of nitrosobenzene with phenylhydroxylamine to azoxybenzene was 10^2 L/mol·s.

Procedures were proposed [195] for studying the electrochemical kinetics of systems for which the charge transfer is associated with follow-up chemical reactions by convolution SEV. The same technique was used [196] in the study of the electrohydrodimerization of activated olefin p-methylbenzidine-malononitrile. This method was shown to be more powerful than conventional SEV for discriminating among the numerous mechanistic possibilities. The application of convolution SEV to mechanistic analysis and rate determination has been discussed [197]. The practical use of this method was tested on the reductive pinacolization of acetophenone in AN using C^0, v, and the water content as operational parameters.

Parker [198] suggested rules that may be useful in the mechanistic analysis of complicated redox reactions.

Jensen and Parker [199] studied the kinetics of protonation of perylene and anthracene dianions by MeOH in THF.

Earlier elaborated theory [171] for surface chemical reaction following a charge-transfer step was used by Laviron [200] to study the reduction of azobenzene to hydrazobenzene followed in an acid medium by the benzidine rearrangement. A satisfactory agreement of theory and experiment was observed.

A simple method for determining the rate constant of chemical reactions following a reversible electrode reaction was proposed by Nadjo and Savéant [201]. By plotting E_p as a function of log v, one may observe a change of the slope of such dependence at some value of v. At a high scan rate, E_p should be independent of v.

The limiting scan rate v^* at which the slope of the E_p versus log v dependence changes is related to k_1:

$$k_1 = 0.52v^* \frac{nF}{RT} \tag{61}$$

This equation was applied to the study of a deactivation reaction of p-chlorobenzophenone free radicals in DMF with 0.1 M $(C_2H_5)_4NClO_4$ as a background electrolyte; k_1 was found to be 10 s^{-1}.

2.3.2 Charge Transfer With Cyclic Regeneration of the Reactant

In the first process of two principal cases discussed briefly in this section, the reaction is first order with respect to the primary product (so-called catalytic processes); in the second process, the primary product undergoes a disproportionation reaction.

REGENERATION OF THE REACTANT

The general scheme of such processes is

$$
\begin{array}{c}
\text{Ox} + ne \rightleftharpoons \text{Red} \\
\text{Red} + A \xrightarrow{k_1} \text{Ox}
\end{array}
\tag{62}
$$

In principle, both the charge-transfer step and the chemical step may be either reversible or irreversible. The oxidant A, in the chemical reaction, regenerates Ox. To simplify the mathematical treatment, it is usually assumed that the concentration of A is so large compared to that of Ox that in practice it does not change at the electrode surface in the course of the electrochemical experiment. Theoretical treatment of such processes, limited to irreversible chemical transformations, was given by Savéant and Vianello [202].

Later, Nicholson and Shain [25] presented a more extended treatment of that problem. With a fast, irreversible chemical reaction and reversible charge-transfer process [25]

$$
_k i = \frac{nFA(Dk_1)^{1/2}C_{Ox}^0(C_A^0)^{1/2}}{1 + \exp\left[\dfrac{nF(E - E_{1/2})}{RT}\right]}
\tag{63}
$$

Analysis of this equation shows that the shape of the current–potential curve is changed and is similar to a polarographic wave or to curves recorded with the use of the rotating-disk electrode.

The limiting value of the catalytic current $_k i_l$ corresponding to a plateau

$$
_k i_l = nFA(Dk_1)^{1/2}C_{Ox}^0(C_A^0)^{1/2}
\tag{64}
$$

is independent of the scan rate. This equation may be easily obtained by writing the general equation for the limiting current,

$$
i_l = \frac{nFAC_{Ox}^0 D}{\delta}
\tag{65}
$$

where δ is the thickness of the diffusion layer. Substituting δ with μ, the thickness of the kinetic layer,

$$
\mu = \left(\frac{D}{k_1 C_A^0}\right)^{1/2}
\tag{66}
$$

(64) is obtained.

The value $_ki_l$ may be much higher than the diffusion-limited peak current (9). The ratio of both currents is

$$\frac{_ki_l}{i_p} = 2.24 \left(\frac{k_1 C_A^0 RT}{nFv}\right)^{1/2} \tag{67}$$

The same problem (62), but with a reversible chemical reaction, was theoretically considered by Rampazzo [203]; and in this case, $(k_1 C_A^0)^{1/2}$ in (64) should be replaced by $(k_1 C_A^0 + k_2)^{1/2}$, where k_2 is the rate constant of the reverse reaction $Ox \xrightarrow{k_2} Red + A$.

The irreversible charge-transfer process coupled with the catalytic regeneration of a substrate was also considered by Nicholson and Shain [25]: as expected, when the regeneration of Ox is very effective, a current plateau is observed instead of a peak current of the value described by (64). However, the complete description of the current–potential curve is more complex than (63) and valid for reversible charge-transfer processes. A detailed description is given in [25].

CHARGE TRANSFER FOLLOWED BY PRIMARY PRODUCT DISPROPORTIONATION

This reaction may be represented by a set of equations

$$Ox + ne \rightleftharpoons Red$$
$$2\,Red \xrightarrow{k_1} Ox + A \tag{68}$$

and is theoretically described by Mastragostino and associates [204] for reversible charge-transfer reactions.

At 20°C E_p is dependent on various parameters in the following way:

$$E_p = E^0 - \frac{0.071}{n} - \frac{0.0197}{n} \log n + \frac{0.0197}{n} \log\left(\frac{k_1 C_{Ox}^0}{v}\right) \tag{69}$$

The values of parameters $\partial E_p / \partial \log C_{Ox}^0$ and $\partial E_p / \partial \log v$ are identical to those observed in the case of the disproportionation reaction—DIM 1. When studying disproportionation rates compared to a simple diffusion-controlled process, the increase of the peak current is more useful than the change of E_p potential.

The dependence of the function ψ_p, defined as

$$\psi_p = 0.446 _k i_l / i_p \tag{70}$$

on the parameter $\lambda = RTk_1 C_{Ox}^0 / nFv$ was theoretically calculated by Savéant and co-workers [204, 205]. Olmstead and Nicholson [206] extended such calculations to spherical electrodes. The dependence of the ψ_p function on λ is tabulated and may be used to determine k_1.

Laviron and co-workers [207] have considered a second-order regeneration reaction occurring in a thin layer of the solution.

APPLICATIONS

Diproportionation reactions were primarily studied. Najdo and Savéant [188] investigated the electroreduction of several aromatic carbonyl compounds in ethanolic alkaline media using SEV extensively. Kinetic studies show the presence of disproportionation in several cases as a rate-determining step.

Kudirka and Nicholson [208] applied cyclic voltammetry to the study of disproportionation of free radicals formed by the electroreduction of a series of sulfonephthalein acid–base indicators in aqueous solutions.

The theory described by Savéant and co-workers [204] was used by Mastragostino and Savéant [209] to analyze the disproportionation reaction of uranium(V) in 3.3 mol/dm^3 HClO$_4$ and 2.7 mol/dm^3 NaClO$_4$; k_1 was 3.8×10^4 L/mol·s.

Sipos and associates [210] studied the electroreduction of uranium(VI) in acidic perchlorate media. The formation of unstable uranium(V) during such a reaction and its further transformations were investigated by using SEV, also with a thin-layer cell. A similar study of uranium, in the presence of acetylacetone, was also carried out [211].

The catalytic reaction theory was applied [212] to determine the reaction rate between Ti(III) generated in the electroreduction process and hydroxylamine; k_1 was 41.2 ± 2.3 L/mol·s.

2.3.3   Chemical Reactions Coupled Between Two Electron Transfers (ECE Reactions)

Theoretically various chemical reactions may occur between two chargetransfer steps. Our considerations are limited to a rather simple set of equations; nonetheless, the main features of such processes are also presented in the discussion of this simple case.

ECE PROCESS WITH FIRST-ORDER CHEMICAL REACTIONS

The ECE reaction is very simply represented by

$$Ox_1 + n_1 e \rightleftharpoons Red_1 \quad E_1^0$$

$$Red_1 \xrightarrow{k_1} Ox_2 \tag{71}$$

$$Ox_2 + n_2 e \rightleftharpoons Red_2 \quad E_2^0$$

During the chemical process the product of the first electrochemical reaction Red$_1$ is transformed into a substance Ox$_2$, which can be further reduced to Red$_2$. E_1^0 and E_2^0 are the standard potentials of the two electrochemical reactions.

Such processes (71) frequently occur in electrochemical reactions (1) and their theoretical elaborations under SEV conditions were presented by Nicholson and Shain [213] and later by Mastragostino and associates [204], who considered a first- or pseudo-first-order irreversible chemical reaction.

Different variants of the electrode reactions were considered [213] assuming either reversibility or irreversibility of both charge-transfer steps. Considering both steps as fast, two main cases may be distinguished when one or two voltammetric peaks are observed, depending on the relative values of formal potentials of both steps. If E_1^0 is less cathodic than E_2^0, two separate peaks are observed, with the second peak height dependent on both the chemical reaction and the scan rates. When k_1/v is high, the height of the second peak is near or equal to that predicted by the Randles–Ševčik equation.

From the ratio of experimental peak height to that controlled by diffusion (either determined experimentally at low v or calculated using the Randles–Ševčik equation) the rate constant of the chemical step using the tabulated data of Nicholson and Shain [213] may be calculated. In a fast chemical step the properties of the first peak are concordant with those of a reversible charge-transfer process followed by a chemical reaction.

When E_2^0 is equal to or less negative than E_1^0 and k_1/v is high, only one peak is recorded and the number of transferred electrons equals $n_1 + n_2$. When k_1/v is very low, the peak current corresponds only to the first step of process (71).

The irreversibility of the charge-transfer steps changes the recorded peak shapes. The extended treatment of these processes, with consideration of both reversible and irreversible charge-transfer processes, was also presented by Najdo and Savéant [166]. The results of this work are difficult to present concisely; therefore, the literature should be consulted. In addition [166], the diagnostic criteria permitting proper mechanistic analysis of the reaction studied are given. Such analysis should be based on experiments conducted at different scan rates.

Nicholson and Shain [214] presented their theoretical results for an ECE reaction with $n_1 = n_2$ by the following empirical equation:

$$\frac{{}_k i_l}{i_p} = \frac{0.4 + k_1/a}{0.396 + 0.469 k_1/a} \tag{72}$$

in which ${}_k i_l$ and i_p denote peak currents measured at the same scan rate, for kinetically controlled and kinetically uncomplicated processes, respectively. The term i_p is either calculated or experimentally discovered by working at high scan rates.

Amatore and Savéant [215] considered ECE reactions with a more complicated chemical step.

APPLICATION TO ORGANIC SYSTEMS

SEV is not often used to study the kinetics of ECE reactions because it is more suited to general elucidation of electrode reactions. To determine the rate constant of kinetic processes, techniques such as chronoamperometry are preferred. Nonetheless, SEV was used by Nicholson and Shain [214] to study the electroreduction of p-nitrosophenol, yielding p-hydroxylaminophenol as a primary product, which is rapidly dehydrated to p-benzoquinonoimine, which in turn reacts under diffusion control to give p-aminophenol.

The rate constant found [216] agrees with that determined by other techniques. Leedy and Adams [217] studied the electroreduction of N,N-dimethyl-p-nitrosoaniline, which yields N,N-dimethyl-p-phenylenediamine along a reduction path similar to that described for p-nitrosophenol.

The possible participation of ECE mechanisms in the electroreduction of aromatic carbonyl compounds in alkaline media was discussed by Nadjo and Savéant [188].

The electroreduction of p-dinitrobenzene in acidic solutions at low temperature proceeds [218] through the intermediate N-p-nitrophenyl-N,N-dihydroxyamine. The dehydration of this amine leads to p-nitrosonitrobenzene, which is reduced to N-p-nitrophenylhydroxylamine. The chemical step in this ECE reaction involves the acid–base catalysis.

2.3.4   Chemical Reactions Preceding the Charge-Transfer Step

The electrode step proper may be preceded by different chemical reactions of the generation of reactant. Most extensively studied was a first-order chemical reaction preceding the charge-transfer step and consequently this process is discussed below.

Early theoretical elaboration of the reaction formulated by the equation

$$A \underset{k_2}{\overset{k_1}{\rightleftharpoons}} Ox + ne \rightarrow Red \tag{73}$$

was presented by Savéant and Vianello [219]. In (73) A is a substance inactive in the potential range in which Ox may react.

Nicholson and Shain [25] considered both a reversible and an irreversible charge-transfer step coupled with a reversible chemical reaction. The following equation is proposed for the reversible electrode reaction:

$$\frac{_k i_l}{i_p} = \frac{1}{1.02 + 0.471\sqrt{a}/K\sqrt{l}} \tag{74}$$

where $K = k_1/k_2$ and $l = k_1 + k_2$; i_p is the peak current that would be observed if the electrode reaction were not influenced by the chemical step. Equation (74) approximates a numerical elaboration with an error of about 1%.

When the kinetic effect is strong, the shape of the recorded current–potential curve is usually changed from a peaklike to a wavelike form; $_k i_l$ is then the current of a plateau.

In the case of the irreversible charge-transfer step the equation relating the kinetic current $_k i_l$ to i_p is similar to (74):

$$\frac{_k i_l}{i_p} = \frac{1}{1.02 + 0.531\sqrt{b}/K\sqrt{l}} \tag{75}$$

where b is $\alpha n_a Fv/RT$.

In this case, at some values of the parameter $K\sqrt{l}$, especially at higher scan rates, current–potential dependencies exhibit a wavelike shape. There is no exact theoretical treatment of quasi-reversible electrode reactions preceded by a chemical reaction.

In addition, the process with a monomerization chemical reaction of the type $A \xrightarrow{k_1} 2Ox$ preceding the charge-transfer step was also theoretically elaborated by Savéant and Vianello [175].

Although the theory of electrode processes with preceding chemical reactions was developed for SEV, this method was accidentally applied in electroanalytical practice [220].

3   VOLTAMMETRY WITH ROTATED ELECTRODES

Variously shaped stationary electrodes placed in solutions with forced convection [221–223], as well as vibrating [224] and rotated electrodes, were electroanalytically used. They were applied earlier in the form of wires by Nernst [225, 226]. Analytical application of such electrodes is possible because they yield reproducible limiting currents when used [227]. However, the use of such electrodes in physicochemical studies is possible only if the recorded currents are properly described theoretically.

In a general form, the equation for a current may be written

$$i = \frac{nFAD_{Ox}[C_{Ox}^0 - C_{Ox}(0)]}{\delta_{Ox}} \tag{76}$$

where δ_{Ox} is the thickness of the diffusion layer introduced by Nernst [225] and C_{Ox}^0 and $C_{Ox}(0)$ are the bulk and surface concentrations of Ox, respectively.

For the limiting current, (76) simplifies to

$$i_l = \frac{nFAD_{Ox}C_{Ox}^0}{\delta_{Ox}} \tag{77}$$

Rotated electrode applications in analysis are based on (77), which predicts a linear dependence of the limiting current on the concentration of the reactant. To use these electrodes nonanalytically it was necessary to develop more precise equations in which δ_{Ox} would be strictly described. Such equations were derived for electrodes in the form of a rotating disk.

Consideration of the general form of the convective diffusion equation shows [228] that one of a few electrode systems having uniform accessibility and steady-state currents is the rotating disk (the others are wall-tube and membrane electrodes). A uniform accessibility to the electrode means that the flux of reactant on the electrode surface is equal at each point on the surface. Then the current density is also uniform along the surface and may be easily calculated by dividing the total current by the area of the electrode. The interpretation of results obtained with such electrodes is much simpler.

The theory and practice of rotated disk voltammetry were covered in monographs [2–4, 6, 221] and review papers [229, 230].

3.1   Working Conditions of the Rotated Disk Electrode Voltammetry

A schematic view of the rotated disk electrode (RDE), together with a convective flow of the solution when the RDE is rotated about an axis perpendicular to the electrode surface, is shown in Figure 3.15a and 3.15b.

The edge of a disk should be well isolated so only the bottom surface of a conductor is exposed to the solution.

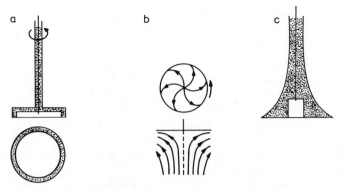

Figure 3.15 Schematic view of (a) the rotating disk electrode, (b) the flow of the solution to its surface, and (c) the RDE with a bell-like shape.

The shape of the mantle made of an insulator, surrounding the electrode, also influences performance, since its shape influences the hydrodynamic flow of the solution. This influence was studied by Riddiford and co-workers [231, 232] and Prater and Adams [233].

From these studies it was concluded that the insulating mantle should have a bell-like form (Figure 3.15c).

In the electrode reaction, an active part of the disk electrode is frequently made from platinum and gold; however, other conducting materials such as carbon paste or various kinds of carbon are also used [2, 230].

An insulating part may be made of a Teflon coating, glass, epoxy resins, or other materials. When the electrode is made of platinum wire (with a diameter of approximately 1 mm), it is useful to seal this wire into a soft glass tubing. However, to strengthen this construction it should be placed in a hole made in a Teflon-coated mantle [234]. Frequently, a disk made of conducting material is placed directly into a Teflon-coated hole. Such procedures are described for the preparation of both metal disk electrodes [235, 236] and carbon paste electrodes [237].

The surfaces of these materials at the bottom of the RDE should be well polished. The surface smoothness of the electrode is very important not only to decrease the capacitive current, but also to create better conditions for the laminar flow of the solution at the electrode surface.

To have such a smooth surface the disk must be ground and polished to a mirrorlike finish. Different procedures for the proper preparation of the RDE surface were developed in various laboratories.

The irregularities of the electrode surface should be much lower than the diffusion-layer thickness δ.

If the RDE radius is too small, diffusion from its edges occurs throughout the disk; therefore, the RDE radius should not be small. Theoretically (see Section 3.2.2 discussing the limitations in the use of RDE), it should be infinitely large. Practically, electrodes with a radius of 1 mm have been successfully applied [234].

The electrode should rotate with a constant well-controlled speed. The deviation from the vertical axis should also be as low as possible. The fulfillment of these conditions is very important to ensure proper functioning of the electrode.

The RDE should also be placed in a proper cell. Since the theory requires that the solution volume be infinitely large, studies were conducted to determine practical small volumes that were able to yield results concordant with theoretical predictions.

Prater and Adams [233] obtained results equally good when either a 100-mL beaker or a 9-L vessel were used as a cell.

Gregory and Riddiford [238] obtained good results when the distance between the bottom of a cell and the electrode surface exceeds 5 mm.

Immersion depth of the RDE in the solution was also studied [233]. This is not a critical parameter, and it is advisable that the immersion be small, slightly below the liquid surface.

To a large extent the good performance of the RDE depends on the rotation system. The rotation rate should be constant, with neither radial nor axial vibrations, when measurements are taken. It should be possible to easily change the rotation rate from one value to another.

The RDE was frequently mounted on the rotating shaft, with radial vibrations best minimized by self-locking cones. The rate of rotation may be determined by using a stroboscope.

Such an electrode may be connected to a circuit used in stationary electrode voltammetry, replacing the SEV working electrode. In this case the applied scan rate should be low (of the order of 0.1–0.2 V/min) and along with a constant well-known RDE rate of rotation the current–potential curves are recorded.

An alternative procedure applies different potentials to the RDE, and after a constant time lapse the current is measured. In this case, before reading the value of current after potential application, it is necessary to have a steady-state current. The changes of current occurring after a potential application result from the time needed to form a new concentration profile. The duration of such a transient is determined by the ratio δ^2/D and it is typically of the order of several seconds.

The other source of the current change is the charging of the double layer (usually very short) and some possible changes of the surface of the electrode.

RDE voltammetry apparatus may be constructed in the laboratory or purchased commercially. A comprehensive line of rotators with speeds ranging up to 10^4 rpm and accurate speed control are manufactured by Pine Instrument Company (USA), which also supplies interchangeable disk electrodes, Models DD 15 and DTJ 36. The Bruker E-R-5/E-RS-1 rotating disk electrode system comprises a motor of hollow rotor design coupled to a tachometer-generator. The electrode is driven by coupling sleeves and its electronic speed control uses the tacho-generator feedback principle, yielding a very wide dynamic range and accuracy. Metrohm (Switzerland) and Tacussel (France) also produce equipment for RDE voltammetry.

3.2 Theory of Rotating Disk Electrode Voltammetry

3.2.1 The Limiting Current

The exact expression for δ_{Ox} may be derived for some electrode geometries when the diffusion equation can be properly formulated. The most comprehensive theoretical studies were conducted for a rotating disk electrode (RDE). Figure 3.15 depicts such an electrode, together with the convective flow of solution when the RDE is rotated about an axis perpendicular to the electrode surface. The rotation of the disk causes the solution to move from the bottom of the cell to the electrode surface, where a fluid boundary layer is formed. This layer of the solution moves together with the rotating electrode surface.

This movement extends further away from the electrode, but it decays with the distance. As a result of the rotation of this layer and centrifugal forces, in a unit of time part of the solution leaves this layer while equal volume of the solution moves to this layer from the region situated at some distance from the electrode.

The thickness of the fluid boundary layer, δ_0, is

$$\delta_0 = 3.6 \left(\frac{\nu}{\omega}\right)^{1/2} \tag{78}$$

where ν is the kinematic viscosity of the solution, which for aqueous solutions is near to 10^{-2} cm^2/s and ω is the angular velocity of RDE; $\omega = 2\pi N$, N being the number of revolutions per second.

For the rotating disk electrode the equation for convective diffusion should be obeyed,

$$V_y \frac{dC_{Ox}}{dy} = D_{Ox} \frac{d^2 C_{Ox}}{dy^2} \tag{79}$$

where V_y is the convection flow rate perpendicular to the electrode surface.

Levich [239] first solved this equation with a properly formulated V_y and the necessary initial and boundary conditions and arrived at an equation corresponding to (76); however, δ, now described in detail, is equal to

$$\delta_{Ox} = 1.61 D_{Ox}^{1/3}\ \nu^{1/6} \omega^{-1/2} \tag{80}$$

Comparison of both thicknesses yields

$$\frac{\delta_{Ox}}{\delta_0} = 0.447 \left(\frac{D}{\nu}\right)^{1/3} \tag{81}$$

and shows that this ratio is independent of the rotation rate. As it follows from (81) δ_{Ox} is considerably smaller than δ_0, since D/ν for aqueous solution is of the order of 10^{-3}, being equal to about 5% of δ_0.

Combining (80) with (77) yields the following limiting-current expression:

$$i_l = 0.62nFAD_{Ox}^{2/3}v^{-1/6}\omega^{1/2}C_{Ox}^0 \tag{82}$$

Later Gregory and Riddiford [238], Newman [240], and Kassner [241] derived more precise and complicated expressions for δ_{Ox}. When $D/v = 10^{-3}$, Levich's δ_{Ox} is about 3% higher than that predicted by the latter authors. At higher D/v values, this error increases. For aqueous solutions v is about 10^{-2} cm^2/s and D is usually slightly below 10^{-5} cm^2/s; consequently, the error resulting from Levich's equation should be in the limit of routine experimental error. More sophisticated treatment is advised for more precise results.

3.2.2   RDE Limitations in the Analysis of Experimental Results

The RDE theory is developed for a rotating disk (plane) with a very large radius, which is placed in an infinitely large volume of solution [4].

Such assumptions are usually not true in practice. The application of the RDE with a small radius may lead to a so-called "edge effect," which results from the radial diffusion of a reactant at the edges of a disk in addition to the transport normal to the electrode surface. This effect develops at the disk edge in a ring with a thickness equal to δ, and is, therefore, especially large when the rotation rate is low. The higher ω, the smaller is the electrode radius that may be applied. Details of this problem are discussed by Smyrl and Newman [242].

The volume of the solution being studied is also important. Miller and Bruckenstein [243] have shown that when using electrodes with a radius of about 2–3 mm and a solution volume of about 0.5 cm^3, the RDE theory is still obeyed.

The maximal RDE eccentricity at which the theory of this electrode becomes invalid was theoretically and experimentally considered [244–246]. Furthermore, if the eccentricity, expressed as the ratio of the electrode radial deviation during rotation to the conducting part of the disk radius, is lower than 0.5–0.6, the limiting current should be unaffected.

The influence of nonideal electrode smoothness on i_l was studied using model electrodes on which some surface roughness was artificially introduced [247, 248]. These studies show that although surface roughness has a rather small influence on the limiting current, it enhances turbulent flow.

The turbulent flow of a fluid may also occur to RDE when the electrode is large and its rotation rate is high. The transition from laminar to turbulent flow occurs when the dimensionless Reynolds number Re, determined as

$$\text{Re} = \frac{\omega r_0^2}{v} \tag{83}$$

is about 10^5. Here r_0 is the radius of the RDE together with the insulating part. For instance, Daguenet and Robert [249] found that the turbulent flow develops at Re $= 2.7 \times 10^5$. A slightly lower value was determined by Gregory and associates [250].

With axial or radical RDE vibrations turbulence may appear at a lower Re value. Condition (83) imposes a limitation on the application of a very high rotation rate. In practice, rotation rates up to 100 rps are used, although an electrode with higher rotation rates was also constructed and applied [251]. Application of very low rates of rotation is also restricted. Obviously at very low ω, when the potential applied to the RDE is linearly scanned in time, the current–potential dependence could be recorded exhibiting a current peak, as in SEV. Natural convection at very low ω may also influence the results.

The lowest ω's that may be practically used are approximately described by the following condition:

$$3.6 \left(\frac{v}{\omega}\right)^{1/2} \approx r_0 \tag{84}$$

which means that when using a RDE with a radius of approximately 1 mm, the rate of rotation should be not lower than 3 rps.

Current distribution on the RDE surface was studied by several workers, particularly Newman [252]. Fur currents considerably lower than the limiting current, the local current density differs at various RDE regions.

Newman's theoretical ideas were experimentally verified [253]. Prater and co-workers [254], using the digital simulation method, calculated the current density along the disk surface.

3.2.3 Reversible Electrode Processes

Equation (82), which describes the limiting current, is valid independent of the rate of the electrode reactions studied. However, the shape of the current–potential curve and its position with respect to the formal potential of the reaction studied is a function of the reversibility of this process.

When both forms Ox and Red are initially present and are soluble in the solution, the following equation is valid [2, 3]:

$$E = E_{1/2} + \frac{RT}{nF} \ln \left(\frac{_c i_l - i}{i - _a i_l}\right) \tag{85}$$

In this equation

$$E_{1/2} = E_f^0 + \frac{RT}{nF} \ln \left(\frac{D_{Red}^{2/3} f_{Ox}}{D_{Ox}^{2/3} f_{Red}}\right) \tag{86}$$

where f_{Ox} and f_{Red} are activity coefficients of Ox and Red forms, respectively; $_c i_l$ and $_a i_l$ are limiting currents of the cathodic and anodic reactions, respectively.

When only one form, that is, Ox, is present in the solution before the electrochemical experiment, (85) simplifies to

$$E = E_{1/2} + \frac{RT}{nF} \ln \left(\frac{_c i_l - i}{i}\right) \tag{87}$$

Equation (87) has a form identical to that of the equation of a polarographic wave, only with a slightly different meaning of $E_{1/2}$. Schematic current–potential dependence, calculated by (85), is shown in Figure 3.16.

The shapes of both anodic and cathodic current–potential curves resemble

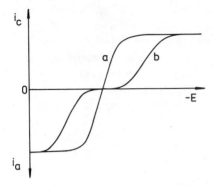

Figure 3.16   A schematic of the anodic–cathodic current–potential curves of (a) reversible and (b) irreversible systems recorded with the use of the RDE.

polarographic waves and are different from the shapes of curves recorded in SEV. Such a wavelike current–potential dependence may be readily explained. Both in rotating disk and polarographic methods a steady state exists—in polarography because of a periodic renewal of a mercury drop and in RDE as a result of a uniform rotation of the electrode.

In consequence, the diffusion-layer thickness in RDE voltammetry is independent of time in the experiment (at fixed ω) and so is the limiting current.

Equations (85) and (87) are frequently used to check the reversibility of the processes studied. For reversible reactions,

$$\frac{\partial E}{\partial \log\left(\dfrac{c i_l - i}{i - a i_l}\right)} \quad \text{and} \quad \frac{\partial E}{\partial \log\left(\dfrac{c i_l - i}{i}\right)}$$

should equal $2.3RT/nF$.

3.2.4   Irreversible Electron Transfers

The general equation valid for quasi-reversible and irreversible reactions was given by Delahay [255]

$$i = nFAk_s\left\{ \frac{C_{\mathrm{Ox}}^0 \exp\left[-\dfrac{\alpha nF}{RT}(E - E_f^0)\right] - C_{\mathrm{Red}}^0 \exp\left[\dfrac{(1-\alpha)nF}{RT}(E - E_f^0)\right]}{1 + \dfrac{k_s}{D}\left\{\exp\left[-\dfrac{\alpha nF}{RT}(E - E_f^0)\right] + \exp\left[\dfrac{(1-\alpha)nF}{RT}(E - E_f^0)\right]\right\}} \right\} \tag{88}$$

The following assumptions were made in the derivation of this equation: $\delta_{\mathrm{Ox}} = \delta_{\mathrm{Red}} = \delta$ and $D_{\mathrm{Ox}} = D_{\mathrm{Red}} = D$.

In a fully irreversible cathodic reaction, (88) simplifies to

$$E = E_f^0 + \frac{RT}{\alpha nF}\ln\left(\frac{D_{\mathrm{Ox}}}{k_s \delta_{\mathrm{Ox}}}\right) + \frac{RT}{\alpha nF}\ln\left(\frac{c i_l - i}{i}\right) \tag{89}$$

As for reversible reactions in this case the current–potential curve resembles a polarographic wave, but the curve is shifted to more negative (positive) potentials for cathodic (anodic) processes from the formal potential. Also, the increase of current with potential is usually not as steep as with fast electrode reactions (Figure 3.16, Curve b).

It follows from (89) that in a totally irreversible reaction $E_{1/2}^{irr}$ is

$$E_{1/2}^{irr} = E_f^0 + \frac{RT}{\alpha n F} \ln\left(\frac{D_{Ox}}{k_s \delta_{Ox}}\right) = E_f^0 + \frac{RT}{\alpha n F} \ln\left(\frac{D_{Ox}^{2/3} \omega^{1/2}}{1.61 k_s \nu^{1/6}}\right) \qquad (90)$$

Unlike the $E_{1/2}$ of reversible reactions, $E_{1/2}^{irr}$ is now a function of ω. The higher ω, the more negative (positive) is $E_{1/2}^{irr}$ of electroreduction (electrooxidation).

Equation (90) may be used to determine k_s and α either from the plot of $\log\left[(_c i_l - i)/i\right]$ versus E or from the dependence of $E_{1/2}^{irr}$ on $\log \omega$.

Other methods of calculating kinetic data were developed for quasi-reversible reactions. The following equation is valid for the cathodic process:

$$i_c = \frac{n F A D_{Ox} C_{Ox}^0}{\delta_{Ox} + D_{Ox}/k_{fh}} \qquad (91)$$

which, for high k_{fh} (when $\delta_{Ox} \gg D_{Ox}/k_{fh}$), simplifies to (77) for current controlled only by the mass-transfer rate, whereas in the case of very low $k_{fh}(\delta_{Ox} \ll D_{Ox}/k_{fh})$, (95) simplifies to (31).

Frequently, in the study of the kinetics of quasi-reversible reactions, the method proposed by Jahn and Vielstich [256] is used, and is based on

$$\frac{1}{i} = \frac{1}{n F A(k_{fh} C_{Ox}^0 - k_{bh} C_{Red}^0)} \left\{ 1 + \left(\frac{k_{fh} \delta_{Ox} \omega^{1/2}}{D_{Ox}} + \frac{k_{bh} \delta_{Red} \omega^{1/2}}{D_{Red}}\right) \frac{1}{\omega^{1/2}} \right\} \qquad (92)$$

A graph of $1/i$ versus $1/\omega^{1/2}$ is plotted for constant overvoltage values. From the slope of the straight line and from the value of the point of its intersection with the ordinate for $\omega = 0$, values of k_{fh} and k_{bh} are obtained.

If the reverse (oxidation) reaction can be neglected, (92) simplifies to

$$\frac{1}{i_c} = \frac{1}{n F A k_{fh} C_{Ox}^0} + \frac{1.61 \nu^{1/6}}{n F A D_{Ox}^{2/3} C_{Ox}^0} \cdot \frac{1}{\omega^{1/2}} \qquad (93)$$

and the plot of $1/i_c$ versus $1/\omega^{1/2}$ determines k_{fh}. Equation (93) may be written in a concise form,

$$\frac{1}{i_c} = \frac{1}{i_k} + \frac{1}{i_t} \qquad (94)$$

where i_k is the current component limited by the charge transfer rate, and i_t is limited by the transport rate. Randles [257] proposed the following equation:

$$-\log(n F k_{fh} C_{Ox}^0) = \log\left(\frac{1}{i} - \frac{1}{_c i_l}\right) - \left(\frac{1}{i} + \frac{1}{_a i_l}\right) \exp\left(\frac{n F n}{RT}\right) \qquad (95)$$

where $_c i_l$ and $_a i_l$ denote limiting cathodic and limiting anodic current, respectively. Using currents measured at different overvoltages η, as well as limiting currents of oxidized and reduced forms, one may determine k_{fh} at different potentials. Küta and Yeager [258] proposed the calculation method of standard rate constants of electrode reactions analogous to that elaborated by Koryta [259] for polarography.

Still another method of k_s determination was proposed by Małyszko [260]. It is based on

$$\frac{i_l}{nFAk_{fh}C_{Ox}^0} = \frac{i_l - i}{i} - \frac{i_l - i_r}{i_r} \tag{96}$$

where i_r is the current observed at the same potential at which the current i is measured in the reversible process.

Equation (96) is a more useful form of the equation originally derived by Jordan and Javick [261]. For the totally irreversible reaction (96) simplifies to

$$k_{fh} = \frac{i_l}{nFAC_{Ox}^0} \left(\frac{i}{i_i - i} \right) \tag{97}$$

A theoretical treatment of current–potential curves recorded with the use of RDE for simple redox reactions with both oxidized and reduced forms adsorbed on the electrode was given by Laviron [262]. It was assumed that the Langmuir isotherm is obeyed and that the adsorption rate is not the limiting factor. The effect of the electrode material on the heterogeneous rate constant of such adsorbed reactants was discussed.

A theory for RDE voltammetry with electron transfer mediated by redox polymer reactions was also presented [263], as well as electrode reaction mechanisms occurring at rotated Pt disk electrodes coated with thin films of the redox polymer [264].

3.2.5   Applications to Organic Systems

The RDE has been frequently used to determine the diffusion coefficient of various ions and substances, a use to which it is especially well suited. Well-controlled laminar transport insensitive to natural convection or accidental vibrations and the absence of maxima found in other voltammetries are advantages of this method. Several examples may be mentioned, such as the determination of the diffusion coefficients of hydroquinone and quinone in $2\,M$ KCl at 21°C: 0.91×10^{-5} and 1.1×10^{-5} cm^2/s [265]. Adams and coworkers reported the diffusion coefficient of o-dianisidine in $1\,M$ H$_2$SO$_4$ as 3.7×10^{-6} cm^2/s [237] and later that of N,N-dimethylaniline [266].

Lohmann and Mehl [267] determined diffusion coefficients for positive and negative ions of anthracene.

The diffusion coefficient of benzoic acid in aqueous solutions of sucrose and glycerine was determined by using a rotating disk made from this acid [268].

The RDE was also applied to determine diffusion coefficients in melts [269, 270].

In the study of the mechanism of organic electrode reactions the RDE was used to investigate the electroreduction of formaldehyde [271]. It was deduced that this reaction consists of reversible radical–anion formation followed by a rate-determining protonation. The kinetics of the anodic oxidation of hydroquinone and ethyl and butyl alcohols, as well as the cathodic reduction of quinone, were also studied [265].

Nesterov and Korovin [272] studied the oxidation of hydrazine in alkaline solutions using a nickel RDE. They showed that the reaction proceeds from the adsorbed state. Later the same substance was studied [273] using a platinum RDE. It was concluded that the oxidation is also a surface process involving a fast, three-electron transfer preceding a one-electron, rate-determining step. Harrison and Khan [274] also studied the electrooxidation of hydrazine in acid solutions.

Fleischmann and associates [275] studied the oxidation of hydrazine at an oxide-covered nickel anode. Oxidation of methylhydrazine on gold electrodes in acid solution was also investigated [276].

Březina and co-workers [277], who studied the oxidation of ascorbic acid at the Pt RDE, concluded that the reaction proceeds as a two-electron process with two reaction paths. For the anodic oxidation of aniline at the Pt RDE in acetonitrile, the path suggested [278] was formation of free cation radicals followed by a rapid deprotonation step with further oxidation of the radical $C_6H_5NH\cdot$.

Also the electroreduction of m-dinitrobenzene was studied in a NH_4NO_3-NH_3 solution at $0°C$ [279].

3.3   RDE Voltammetry in the Study of Electron Transfers Coupled to Chemical Reactions

3.3.1   Charge Transfer Followed by Chemical Reaction

FIRST-ORDER CHEMICAL REACTIONS

The electrode reaction represented by (51) was considered for the first time by Galus and Adams [280], who used an approximate method based on the reaction-layer concept. For an irreversible, following chemical reaction, the half-wave potential is described by

$$E^k_{1/2} = E_{1/2} + \frac{RT}{nF} \ln (1.61 v^{1/6} D_{Red}^{-1/6}) + \frac{RT}{2nF} \ln \left(\frac{k_1}{\omega}\right) \tag{98}$$

Later Kiryanov and Filinovskii [281] and Tong and associates [282] considered a reversible chemical reaction. Their solution simplifies to (98) if one assumes irreversibility and high rate of chemical reaction. Möller and Heckner [283] considered both the effect of the following chemical reaction and the incomplete reversibility of the charge-transfer step. Kabakchi and Filinovskii [284] also discussed the influence of following first-order chemical reactions of different rates on the electrode reaction.

DIMERIZATION REACTIONS

The influence of the following dimerization reactions on the $E_{1/2}$ potential of the current–potential dependence was considered in an approximate way by Galus and Adams [280]. A somewhat different equation was presented later

[4]:

$$E^k_{1/2} = E^0_f + \frac{RT}{3nF} \ln \left(\frac{0.863v^{1/3}}{D^{1/3}} \right) + \frac{RT}{3nF} \ln \left(\frac{k_1 C^0_{Ox}}{\omega} \right) \tag{99}$$

Such an equation was also derived by Bonnaterre and Cauquis [285].

The equation of the current–potential dependence is [4]

$$E = E^k_{1/2} + \frac{RT}{nF} \ln \left(\frac{i_l - i}{i_l^{1/3} i^{2/3}} \right) \tag{100}$$

When chemical equilibrium is not greatly shifted to dimer formation, the equation describing the $i-E$ curve is more complex. These problems were also discussed by Bonnaterre and Cauquis [285, 286].

Dimerization of the primary product for the electrode reaction and also the mechanisms DIM 2 and DIM 3 (see Section 2.3.1, "Dimerization Reactions") were considered by Nadjo and Savéant [181].

Elving and co-workers [287] studied an electrode reaction consisting of two successive one-electron transfer steps coupled with dimerization of an intermediate product.

APPLICATIONS TO ORGANIC SYSTEMS

Philip [288] investigated the oxidation of N,N,N',N'-tetramethyl-p-phenylenediamine and found that when $k_1/\omega^{1/2} > 10$ with typical values of D and v, the solution of Tong and associates [282] is identical to that of Galus and Adams [280]. Earlier, Tong and associates [282] studied the electrode reactions of compounds similar to that studied by Philip.

3.3.2   Charge Transfer With Cyclic Regeneration of the Reactant

REGENERATION OF THE REACTANT

The theoretical elaboration of the electrode reaction (62) occurring on the RDE was given by Koutecky and Levich [289]. Their equation for very fast chemical reactions is identical to that derived for SEV (64), because in both cases a very fast chemical reaction controls the mass-transport rate.

The validity of (64) is observed when $\mu \ll \delta$ (μ is the thickness of the diffusion layer) or, in an expanded form, when

$$\left(\frac{D}{k_1 C^0_A + k_2} \right)^{1/2} \ll 1.61 D^{1/3} v^{1/6} \omega^{-1/2} \tag{101}$$

In slower chemical reactions Haberland and Landsberg [290] solved the problem by

$$k^{i_l}/i_l = x \coth x \tag{102}$$

where $x = 1.61(v/D)^{1/6} k_l/\omega^{1/2}$ and i_l is the limiting current in the absence of a chemical reaction. Related problems were also considered by others [291–293].

CHARGE TRANSFER FOLLOWED BY DISPROPORTIONATION OF THE PRIMARY PRODUCT

The reaction formulated schematically by (68) was theoretically elaborated by Ulstrup [294]. Using a more accurate method Filinovskii and Pleskov [4] obtained similar results. Holub [295] presented a general theory of electrode reactions followed by disproportionation occurring on the RDE.

As is known, the ratio of kinetic to disproportionation-free limiting current changes from 1 to 2. Holub's solution covers this whole range of ratios, whereas Ulstrup's is valid only for ratios larger than 1.5.

Disproportionation reactions were also briefly considered by Bonnaterre and Cauquis [296], who discussed a more general scheme for such processes.

APPLICATIONS TO ORGANIC SYSTEMS

The reaction of *tert*-butylhydroperoxide with electrochemically produced ferrous ions was studied using RDE [292]. Also methanol was oxidized with octavalent osmium [297].

3.3.3 ECE Reactions

ECE PROCESS WITH FIRST-ORDER CHEMICAL REACTION

The theory of ECE reactions was presented by Adams and co-workers [298]. A more exact treatment was given by Karp [299] and Filinovskii [300]. The solution of Filinovskii, valid for $n_1 \neq n_2$ when Ox_2 is reduced at potentials more positive than the potentials of Ox reduction (71), has the following form:

$$i_l = 0.94 i_l^{k_1 = 0} \left\{ 1 + \frac{n_2}{n_1} \left[1 - \frac{(1 + \delta^2 k_1/1.9D)^{1/2}}{1 + \delta^2 k_1/D} \right] \right\} \tag{103}$$

In (103) $i_l^{k_1 = 0}$ is the limiting current of the reduction of Ox unaffected by the chemical step and δ is the thickness of the diffusion layer.

A more complicated ECE mechanism in which two charge transfers are separated by a dimerization reaction was detailed by Adams and co-workers [301].

APPLICATIONS TO ORGANIC SYSTEMS

Theoretical elaborations were applied [301] to the study of the anodic oxidation mechanism of several substituted triphenylamines, which proceeds by dimerization to tetraphenylbenzidines.

Adams and co-workers [302, 303] also showed that the ECE mechanism operates in the oxidation of some aromatic hydrocarbons, when 9,10-diphenylanthracene in acetonitrile was oxidized in the presence of nucleophiles.

3.3.4 First-Order Chemical Reactions Preceding the Charge-Transfer Step

The theory of the processes represented by (73) was formulated by Koutecky and Levich [289]. Their solution may be given by

$$\frac{k^{il}}{\omega^{1/2}} = \frac{i_l}{\omega^{1/2}} - \frac{D^{1/6} i_k}{1.61 v^{1/6} K(k_1 + k_2)^{1/2}} \tag{104}$$

This problem was also investigated by Dogonadze [304]. Equation (104) was applied to the study of dissociation and recombination of weak acids by Vielstich and Jahn [305, 306].

More accurate measurements after the application of two RDEs, connected synchronously in a bridge circuit, yielded [307] $k_1 = 9.1 \times 10^5$ s^{-1} and $k_2 = 5.2 \times 10$ L/mol·s for acetic acid.

The dissociation and recombination rate constants of trinitromethane in acetonitrile were found [308] to be equal, $k_1 = 4.5 \times 10^2$ s^{-1} and $k_2 = 8 \times 10^9$ L/mol·s.

3.4   Modifications of the RDE

Albery and co-workers [309] described the construction of the semitransparent RDE, which allows combination of photochemical and electrochemical experiments. The theory of the reaction occurring on the illumination of the transparent RDE for the process

$$A \xrightarrow{h\nu} B \xrightarrow{k} \text{Products} \tag{105}$$

was presented [310] and tested, taking the photochemical reaction

$$\tfrac{1}{2}C_2O_4^{2-} + Fe(C_2O_4)_3^{3-} \longrightarrow Fe(C_2O_4)_3^{4-} + CO_2 \tag{106}$$

as an example. In the dark there was no current at anodic potentials; however, it appeared upon illumination from the photogenerated Fe(II). A faster rotation rate yielded a smaller limiting current, since a lesser amount of Fe(II) was then produced. The cathodic current from Fe(III) reduction was smaller because reaction (106) occurred in the diffusion layer. In such studies a genuine steady state may be established and the rotation speed is a useful experimental variable. Experiment and theory agreed satisfactorily.

Albery and co-workers [311] also studied the theory of a transparent RDE for two photogenerated species reacting with second-order kinetics. The same group of workers [312, 313] used transparent RDE to study possible photogalvanic systems.

Miller and Bruckenstein [314] modulated sinusoidally the electrode rotation speed about a center value

$$\omega^{1/2} = \omega_0^{1/2} + \frac{\Delta\omega^{1/2}}{2} \sin 2\pi ft \tag{107}$$

where ω_0 and $\Delta\omega$ are the center angular velocity and the amplitude of introduced changes, respectively, and f is the modulation frequency. If $\Delta\omega^{1/2}/\omega_0^{1/2} \leqslant 0.1$, the diffusion layer at the electrode should be almost independent of time and no serious deviation from the hydrodynamic steady state occurs. Thus the modulated current may be described by expressing ω in (82) by (107). One then obtains

$$i_{\text{mod}} = 0.62nFAD^{2/3}\nu^{-1/6}C_{Ox}^0 \frac{\Delta\omega^{1/2}}{2} \sin 2\pi ft \tag{108}$$

or by dividing (108) by (82) one has

$$\frac{i_{mod}}{i_l} = \frac{\Delta\omega^{1/2}\sin 2\pi ft}{2\omega_0^{1/2}} \tag{109}$$

Using a synchronous phase-sensitive detector one may separate from the total current a peak-to-peak value Δi of the sinusoidal component of the current:

$$\Delta i = i_l \frac{\Delta\omega^{1/2}}{\omega_0^{1/2}} \tag{110}$$

Rotation speed modulation enables separation of the reactant current that depends on ω from currents that have a nondiffusional origin connected with the electrode reaction of the surface, charging of the double layer, and so forth.

This separation enables, according to Miller and Bruckenstein [315], increased sensitivity of the RDE method in analytical determinations, even up to 10^8 M.

An apparatus is described [316, 317] that permits automatic recording of the dependencies i versus $\omega^{1/2}$ (at constant potential) and E versus $\omega^{1/2}$ (at constant current). The linear dependence $1/i$ versus $1/\omega^{1/2}$ can also be automatically recorded [318], important when applying such plots in kinetic analysis.

Müller and Westfahl [319] introduced the rotating L-shaped electrode. This electrode was used to study the electrode kinetics of Fe^{3+}/Fe^{2+} and $Fe(CN)_6^{3-}/Fe(CN)_6^{4-}$ systems. The rate constant found using this electrode agreed satisfactorily with that obtained by conventional RDE for the Fe^{3+}/Fe^{2+} system.

An analysis of diffusional boundary-layer growth and the velocity distribution in a laminar boundary-layer flow near a rotating axisymmetric electrode were presented [320, 321].

3.5   Other Applications of the RDE

The rotating disk electrode was frequently used to study electrode reactions of inorganic substances [4, 221, 230] and when analyzing for trace amounts of metals [48, 50, 55, 322–328].

A well-controlled, high transport rate in the electrolytic accumulation step renders this electrode very useful in precise analysis. The determined metals may be either directly deposited onto the RDE surface, as in the determination of mercury [50, 328, 329], or codeposited with mercury [48, 55, 330, 331] to obtain lower residual currents.

The rotating twin electrode is sometimes used in a subtractive mode, as proposed by Sipos and associates [332] and as applied in other works [333, 334].

RDE voltammetry was also used to study the deposition of various metals [335–337], as well as anodic dissolution of metals and their corrosion. A detailed analysis of the dissolution of metals with the participation of oxygen and hydrogen ions was presented by Zembura and Fuliński [338]. The dissolution rate of metals in acids is usually proportional to acid concentration, and its dependence on the rotation rate square root illustrates that the dissolution process is controlled by the transport rate [339–345].

The influence of oxygen dissolved in copper on the kinetics and mechanism of copper dissolution in acidic solutions was studied by Zembura and Bugajski [346]. Dissolution of nonmetals was also studied. For instance, dissolution of terephtalic acid in aqueous amine solutions was studied by Krichevskii and Tsekhanskaya [347] and graphite in fused metal oxides by Boronenkov and associates [348]. The rotating disk composed of various materials placed in different media is especially well suited to studying disk material dissolution because of its well-controlled transport rate. It was also used to study metal ion sorption rates on silica and Dowex cation exchanger [349, 350].

Landsberg and Thiele [351] considered processes occurring on the RDE having very small inactive sites uniformly distributed over the surface, which may be applicable when explaining certain corrosion data.

3.6   Processes on the RDE Under Nonstationary Conditions

When the RDE potential is scanned so quickly that the thickness of the diffusion layer caused by this perturbation is smaller than that determined by (80), or if a current pulse is applied to the RDE, nonstationary conditions may result. Both situations have been theoretically discussed.

Fried and Elving [352], using the Nernst diffusion-layer concept, considered the dependence of the limiting current under fast potential scan. Later Girina and associates [353] considered this problem with stricter conditions. Results of these investigations enable prediction of conditions for the appearance of a maximum in the current–potential dependency.

The response of an irreversible [354] and quasi-reversible [355] reaction occurring on an RDE to a linear sweeping of the potential was analyzed with the aid of the Nernst diffusion model by Andricacos and Cheh. RDE polarization by a current pulse of constant intensity was considered by Levich [221] and Siver [356] and more rigorously examined by Hale [357] and Filinovskii and Kiryanov [358].

Chronopotentiometry using a RDE was theoretically considered by Buck and Keller [359] and also by Andricacos and Cheh [360] but with periodic current inputs. These results compared with those published earlier [358] favorably agreed with values of the instantaneous limiting-current densities.

3.7   Rotating Ring–Disk Electrodes

The rotating ring–disk electrode (RRDE), which together with the concentration distribution of the disk product at its surface is shown in Figure 3.17, was introduced by Frumkin and Nekrasov [361].

The disk electrode with radius r_1 situated in the center is concentrically surrounded by a ring made also from a conductor separated from the disk by a thin layer of insulator. Both disk and ring electrodes as well as an insulating layer with a total radius r_3 are placed in the same plane. In the well-working system the gap between both electrodes $r_2 - r_1$ should be very thin.

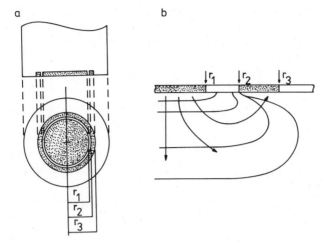

Figure 3.17 Schematic view of the ring-disk electrode (*left*) and concentration distribution at its surface (*right*).

From the mechanical viewpoint both electrodes form one system that rotates about a common axis that passes through the center of the disk. However, electrically both electrodes form two independent systems and have different independently applied potentials.

Let us assume that the soluble in solution product Red arising in the disk reaction

$$Ox + ne \longrightarrow Red \quad \text{disk electrode} \tag{111}$$

may react on the ring electrode if this electrode has the proper potential and may result in reoxidation of Red to Ox.

$$Red \longrightarrow Ox + ne \quad \text{ring electrode} \tag{112}$$

Since not all molecules (or ions) of Red may arrive at the surface of a ring electrode, the collection efficiency parameter N was introduced,

$$N = \frac{i_{ri} n_d}{i_d n_{ri}} \tag{113}$$

which determines the part of Red that is transported from the disk electrode to the ring and reacts at that electrode. N may be determined easily by experiment. In (113) i_{ri} and i_d are the current in the ring and disk electrode circuits, respectively, and n_d and n_{ri} are the numbers of electrons transferred in the elementary reaction occurring at the disk and ring electrodes; as was assumed in (111) and (112) n_d may be equal to n_{ri}.

The parameter N is important because its value may indicate adsorption of the product or its disappearance in a chemical reaction. In the latter case the kinetic collection efficiency N_k is lower than the parameter observed in the case of a transport-controlled process. N was calculated theoretically. With the RDE only the reactant transport along the y axis is important. However, for the RRDE, the radial component of the fluid velocity vector should also be considered.

Assuming the steady state exists, the transport equation to be solved is

$$V_r \frac{\partial C}{\partial r} + V_y \frac{\partial C}{\partial y} = D \frac{\partial^2 C}{\partial y^2} \tag{114}$$

$V_r = 0.51 r \omega^{3/2} v_y^{-1/2}$, while V_y has a more complex form. The solution of the problem presented, with the collection efficiency dependent on various parameters, was derived by several workers, using the proper boundary conditions for different parts of the RRDE.

The first theoretical treatment of that problem was presented by Ivanov and Levich [362]. It was only an approximate solution. Their N values were similar to those calculated on the basis of the approximate equation

$$N = [(r_3/r_1)^3 - (r_2/r_1)^3]^{2/3} \tag{115}$$

where r_1, r_2, and r_3 are shown in Figure 3.17.

Later Albery and Bruckenstein [363] derived the following expression for N:

$$N = 1 - F\left(\frac{\alpha}{\beta}\right) + \beta^{2/3}[1 - F(\alpha)] - [1 + \alpha + \beta]^{2/3}\left\{1 - F\left[\frac{\alpha}{\beta}(1 + \alpha + \beta)\right]\right\} \tag{116}$$

where $\alpha = (r_2/r_1)^3 - 1$, $\beta = (r_3/r_1)^3 - (r_2/r_1)^3$, and

$$F(x) = \frac{3^{1/2}}{4\pi} \ln\left[\frac{(1 + x^{1/3})^3}{1 + x}\right] + \frac{3}{2\pi} \arctan\left(\frac{2x^{1/3} - 1}{\sqrt{3}}\right) + \frac{1}{4}$$

When $x \to 0$, $F(x) \to 0$; for $x \to \infty$, $F(x) \to 1$.

The collection efficiency problem was also considered by Bruckenstein and Feldman [364], Matsuda [365], Tokuda and Matsuda [366], and Bard and Prater [367]. The latter authors used digital simulation when calculating N values. Comparison of solutions for N obtained by analytical and simulation methods has been discussed [368].

It follows from theoretical elaborations [362, 363] that N is independent of the rotation rate of the electrode system; however, N depends on the construction of RRDE. If r_3/r_2 is of the order of 1.4 and r_2/r_1 is between 1.02 and 1.10, N, as expected from the papers just cited, should be about 0.45. Collection efficiency experimentally determined [361, 363, 369–373] satisfactorily agreed with theoretical predictions.

In the experimental determination of N, rather reversibly reacting simple systems such as Fe^{3+}/Fe^{2+} [371], $Fe(CN)_6^{3-}/Fe(CN)_6^{4-}$ [372], or Cu^{2+}/Cu^+ [373] were selected to have good conditions for ring and disk reactions occurring with a 100% current efficiency. The product of the disk electrode should be chemically stable, not adsorbed on the disk, and well soluble in the solution.

The other parameter, the shielding efficiency S, was also introduced [373]. S is defined as the ratio of the ring electrode current observed when the disk electrode is at a potential on the limiting-current region for the ring electrode reaction i_{ri} to the current when the disk electrode circuit is open i_{ri}^0:

$$S = \frac{i_{ri}}{i_{ri}^0} \qquad (117)$$

It relates to the situation occurring when the concentration in the bulk of the substance that reacts at the ring electrode does not equal zero and the same electrode reaction occurs on both electrodes.

Since the concentration of the reactant at the ring surface is decreased because of its partial consumption by the disk reaction $i_{ri} < i_{ri}^0$.

The current i_{ri} is observed, although reactant is consumed at the disk electrode, because it diffuses to the ring from points normal to the disk, the gap region, and the ring electrode.

The decrease of the limiting current in the ring circuit will be [373]

$$i_{ri} = i_{ri}^0 - Ni_d \qquad (118)$$

The shielding efficiency is dependent only on the geometric parameters of RRDE.

To have better results, the electrode systems should have the lowest possible S. It is realized if RRDE has both a thin gap and a ring.

Using digital simulation Prater and co-workers [254] calculated the current densities at both the ring and disk surfaces. Although assumed to be uniform at the disk, simulated and experimental results show that current density is nonuniform at the ring.

3.8   Modification of Ring–Disk Electrodes

Miller and Visco [374] developed a variant of the RRDE in which the ring is divided into two half-rings. As it was shown by Miller [375] in studies of electrode reactions, such electrodes are particularly useful for investigating electrode process mechanisms occurring in several stages. Various potentials of both half-rings may be maintained. Figure 3.18 illustrates an experiment in which a disk electrode was anodized with a constant current of 448 $\mu A/cm^2$.

Anodic and cathodic currents present in the circuits of both half-rings focus

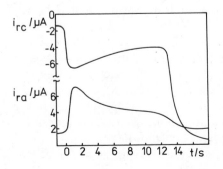

Figure 3.18 Split ring anodic (i_{ra}) and cathodic (i_{rc}) limiting currents during anodization of a copper disk in 1 M NaOH with a current density equal to 448 $\mu A/cm^2$; 2100 rpm. Potentials of the anodic and cathodic parts of the ring were equal, $+0.3$ V and -0.7 V, respectively.

on Cu(I) formation, which is oxidized and reduced at half-ring electrodes. This electrode was also applied to the study of the anodic behavior of Cu–Zn alloys in ammonia solutions [376]. Collection efficiency of an RRDE whose ring only partially served as an electrode was theoretically considered by Dikusar [377]; a disk electrode with two rings was also used [378, 379]. The two-ring concentric electrode was introduced in 1972 [380], and it was shown that such an electrode system is very useful for electrochemical reactions proceeding with the evolution of gaseous products. Filinovskii and associates [381] calculated the current on the second ring in two-ring electrodes. The two-ring electrode transport rate was considered by Dikusar [382].

Debrot and Heusler [383] described an RDE with an optically transparent stationary ring of quartz and showed that under typical conditions, concentrations of the order of $10^{-5} M$ may be detected by extinction measurement, with further sensitivity obtained by improved measurement techniques. Theoretical problems connected with using such electrodes were elaborated upon [384] with digital simulation.

Memming [385] used a rotating platinum-semiconductor ring–disk electrode to analyze photoelectrochemical processes.

3.9   Equipment Used in RRDE Voltammetry

Construction of the ring–disk electrode is described in the literature [230, 363, 386–388]. When preparing such electrodes, epoxy resin is usually used as a material that forms a nonconducting part of the electrode, including a gap between the ring and the disk. This material is inert and exhibits suitable resistance. When it is used, a gap of about 0.1 mm or smaller is obtained. A Teflon-coated material is too soft, since polishing the RRDE will eventually result in particles of conducting metals becoming imbedded in the Teflon-coated gap, thereby changing its nonconducting properties.

Demountable ring–disk electrodes were described by several authors [389–391].

The detailed preparation of a split-ring electrode was described by Miller [376].

Equipment used in RRDE voltammetry is manufactured by many producers. A comprehensive range of ring–disk electrodes and rotators with speeds up to 10^4 rpm are produced by Pine Instrument Company (USA). Both ring and disk electrodes of different dimensions, constructed from various materials, are available.

Tacussel, Metrohm, Pine Instrument, and Beckman also offer RRDEs made from various materials. Bruker also produces ring–disk electronics /E350/, a modular system providing potentiostatic control of both ring and disk electrodes, as well as a read-out of the currents.

Both ring and disk electrode circuits have to be completely electrically independent even though the distance separating both electrodes is very small. This cannot be obtained by using two ordinary potentiostats.

Bipotentiostats, providing simultaneous control of the potential of two electrodes, are very useful in RRDE voltammetry and were described earlier [373, 376]. The BIPAD bipotentiostat especially developed for this technique is produced by Tacussel.

3.10 Applications of the RRDE to the Study of Electrode Reaction Mechanisms

3.10.1 Diagnostic Criteria in Mechanism Elucidation

The RRDE system is very suitable to the study of the electrode reaction mechanisms. One may easily check whether the studied reaction is simple. For simple processes the following relation should be obeyed:

$$\frac{i_d n_{ri}}{i_{ri} n_d} N = 1 \tag{119}$$

All parameters in (119) were already described. In such a case N is independent of the disk electrode potential. Assuming $n_d = n_{ri}$, a simple case, the plot of i_{ri} versus i_d should be linear with a slope equal to the collection efficiency.

If the reaction occurring on the disk electrode is more complicated with an unstable primary product that decays because of the chemical reaction, then a slope of i_{ri} versus i_d dependence equals

$$\frac{i_{ri}}{i_d} = N_k \tag{120}$$

Obviously, $N_k < N$.

The RRDE method makes it possible to record the current–potential dependence for such unstable species. This dependence may also characterize the final products, because their presence may be monitored by separate steps on recorded curves.

Diagnostic criteria very useful in the elucidation of the mechanism of complicated reactions were presented by Bockris and associates [392].

The following reactions occurring on the disk electrode were considered:

$$Ox + (n_1 + n_2)e \rightarrow Red \tag{121}$$

$$Ox + n_1 e \rightarrow Red_1 \tag{122}$$

$$Red_1 + n_2 e \rightarrow Red \tag{123}$$

interrelated in the scheme

$$Ox + n_1 \rightarrow Red_1 + n_2 e \rightarrow Red$$
$$\downarrow \tag{124}$$
$$\text{to bulk}$$

where Red_1 is an intermediate.

On the ring electrode the oxidation reaction occurs:

$$Red_1 - n_1 e \rightarrow Ox \tag{125}$$

For scheme (124) the following equation was derived [392]:

$$\frac{i_d}{i_{ri}} = \frac{x+1}{N} + \frac{x+2\bar{k}}{N\omega^{1/2}} \tag{126}$$

where x is a constant dependent on the kinetics of reactions (121) and (122), while $\bar{k} = 1.61 D^{-1/2}\omega^{1/2}k$ with k denoting the rate constant of electroreduction of Red_1 to Red.

Bockris and associates [392] considered several typical cases. When it is assumed that only reaction (121) occurs, since no intermediate is produced, no ring current is observed.

If only reaction (122) proceeds and intermediate Red_1 is stable ($k=0$), then (126) simplifies to

$$\frac{i_d}{i_{ri}} = \frac{1}{N} \tag{127}$$

The ratio of disk and ring currents is independent of ω. Such independence of i_d/i_{ri} on ω is predicted also for the case when both reactions (121) and (122) occur with stable Red_1.

The following relation is then valid:

$$\frac{i_d}{i_{ri}} = \frac{x+1}{N} \tag{128}$$

If the electroreduction of Ox to Red proceeds only through intermediate Red_1 [reaction (121) is excluded], then

$$\frac{i_d}{i_{ri}} = \frac{1}{N} + \frac{2\bar{k}}{N\omega^{1/2}} \tag{129}$$

In this case the plots of i_d/i_{ri} versus $\omega^{-1/2}$ are represented by lines with slopes that are dependent on $\bar{k}$ and that intersect the current ratio axis at values of $1/N$.

When reactions (121)–(123) occur simultaneously, (126) cannot be simplified. The slope of the dependence of i_d/i_{ri} on $\omega^{-1/2}$ is dependent on potential. The dependencies of i_d/i_{ri} on $\omega^{-1/2}$ for all cases considered above are shown schematically in Figure 3.19.

Information about unstable intermediate products and their formation and decay rates are obtained by studying the dependence of their formation efficiency on potential of the disk electrode, bulk concentration of a reactant, and so forth.

Nekrasov [393] also discussed diagnostic criteria that permit intermediate identification and suggest the electrode reaction course. Such criteria were also presented in a review paper by Opekar and Beran [230].

3.10.2   Study of Adsorption

Theoretical bases of application of RRDE to the study of adsorption were developed by Bruckenstein and Napp [394]. Such measurements should be carried out in a nonstationary state. The transit time t_1 needed by the disk product to reach the inner radius of the ring electrode is described by [395]

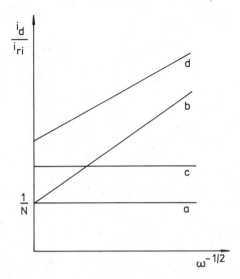

Figure 3.19   The dependence of i_d/i_{ri} on $\omega^{-1/2}$ in the case of a several step electrode reaction with different mechanisms: a—only reaction (122) occurs; b—intermediate undergoes further reaction (123); c—two parallel reactions, (121) and (122), occur; d—reactions (121), (122), and (123) occur parallelly.

$$t_1 = \frac{43.1}{f} \left(\frac{v}{D}\right)^{1/3} \left[\log\left(\frac{r_2}{r_1}\right)\right]^{2/3} \tag{130}$$

where f is the rotation rate of the electrode (in revolutions per minute) and other symbols were described earlier. The dependence of i_{ri} on time is determined by the adsorption of substrate and/or product of the disk electrode reaction. If this product is quickly and strongly adsorbed on the disk surface, then the current in the ring circuit will appear after a time longer than t_1 (Figure 3.20, Curve b). This elongation is caused by the adsorption of the product up to saturation of the disk surface. Consequently, the difference of the transit time in the absence and presence of adsorption may be used to determine the amount of adsorbed product. When the substrate is strongly adsorbed on the disk, then the current in the ring circuit appears at time t_1; but at the beginning of the i_{ri}–time dependence, a maximum is observed (Figure 3.20, Curve c). This maximum results from the additional transport to the ring of the substance that before the charge transfer was adsorbed on the disk surface.

3.10.3   Studies of Mechanism of Electrode Reactions of Organic Compounds

In recent years RRDE has frequently been used to study complicated organic reactions, thereby establishing the presence of unstable intermediates [393].

This method was used [396] to study products of an electrochemical reduc-

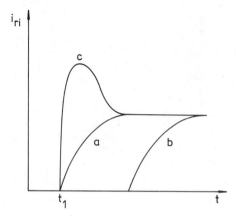

Figure 3.20 The ring current–time curves observed in the case when (a) both substrate and product are not adsorbed on the disk or adsorbed to the same extent; (b) there is a nonadsorbed substrate and fast and quantitative adsorption of the product; and (c) substrate is quickly adsorbed on the disk to a greater degree than product.

tion of benzil in strongly alkaline media; both *cis*- and *trans*-stilbendiolate were the concluded products. Ketolization rate constants of two isomers were also determined.

Neubert and Prater [397] studied the oxidation of *N,N*-dimethylaniline in strongly acidic aqueous media and suggest the deprotonation of the cation radicals formed in the first step with subsequent coupling of neutral radicals.

In a series of papers [398–400] Bard and co-workers studied the mechanisms of several electrohydrodimerization reactions.

The RRDE was also used in the study of the electrode reaction mechanism of (*E*)-1-phenyl-2-nitro-1-propene in DMF [401].

Nekrasov and co-workers [402] discussed the mechanism of the cathodic reduction of benzaldehyde, acetophenone, and their derivatives, based on the RRDE data, with free radical participation.

Nguyen Suan and Nekrasov studied the mechanism of the electroreduction of aromatic carbonyl compounds in alkaline [403] and in acidic nonbuffered solutions [404]. Podlibner and Nekrasov [405] have studied in detail the electroreduction of nitroferrocene in alkaline media and have shown the formation of an unstable intermediate, not described in the literature, which in the disproportionation reaction yields aminoferrocene, cyclopentadienyl anions, and $Fe(OH)_2$. The mechanism of electroreduction of 1,1,1-trinitroethane in alkaline media was also studied [406].

The following reactions were proposed:

$$H_3C—C(NO_2)_3 + 2e \rightarrow H_3C—C(NO_2)_2^- + NO_2^-$$

$$H_3C—C(NO_2)_2^- + e \rightarrow H_3C—C(NO_2)_2^{2-} \tag{131}$$

$$H_3C—C(NO_2)_2^{2-} + 5e + 5H_2O \rightarrow H_3C—C{=}NOH + 7OH^-$$
$$\underset{\text{NHOH}}{|}$$

The formation of the final and intermediate products was proved by the use of a RRDE. Similarly, the formation of intermediate anion and dianion during electroreduction of aromatic aldehydes and ketones in aqueous solutions was shown [403, 404, 407, 408]. The cathodic reduction of p-nitroaniline to p-phenylenediamine in acidic solutions was shown to proceed through an unstable intermediate [409].

3.10.4   Studies of Mechanism of Formation and Electroreduction of Two- and Three-Valent Inorganic Cations

As mentioned earlier the RRDE is also very useful when studying intermediates formed in the multielectron electrode reactions of inorganic ions and compounds. The formation of such intermediates is based on the current flowing in the ring circuit. This type of study was initiated by Nekrasov and Berezina [410], who detected Cu(I) on the platinum electrode in Na_2SO_4 solutions during electroreduction of Cu(II).

The formation of In^+ [374], Bi^+ [411], and Be^+ [412] was found in the course of the anodic oxidation of corresponding metals.

This electrode system was also applied to the spontaneous metal dissolution by interaction with high-valent ions of this metal. In general, a reaction of that type may proceed according to chemical and electrochemical mechanisms [413, 414]. Molodov and Yanov [414, 415] elaborated theoretically the criteria for the determination of the mechanism of such metal dissolution.

To distinguish the mechanisms operating in reactions of that type it was proposed [413, 415] that the dependence of the limiting current at a ring electrode and the potential of a disk on the rate of revolution of the electrode be studied.

Theoretical aspects of the application of the RRDE to determination of kinetic parameters of such reactions were also considered by Kiss, Farkas, and associates [416–419]. The RRDE was frequently applied to the study of the spontaneous dissolution of copper under the influence of Cu(II). In recent years this process in 5 M aqueous solution of $HClO_4$ [420] and in methanol with 0.5 M H_2SO_4 as a background electrolyte [413, 421, 422] was studied.

It was found that the reduction of Cu(II) to Cu(I) controls the overall rate of the process

$$Cu(II) + Cu \rightarrow Cu(I) \tag{132}$$

pointing to the validity of the electrochemical mechanism.

The RRDE was also used to study the dissolution of metals under the influence of oxidants such as oxygen [423].

In addition to the study of metals dissolution ([424–427] and earlier cited works) the anodic dissolution of such materials as titanium carbide was studied [428].

Nonconventional application of the RRDE has also been described; namely, the rate of development of a photographic film was studied by Fujishima and associates [429] using this electrode. A silver halide gelation emulsion film was used as a disk and Pt was used to construct the ring electrode. When hydroquinone was added as a developing agent, the reduction current at the ring electrode increased.

3.10.5   Studies of the Oxygen Electrochemical Reaction

One of the most intensely studied inorganic electrode reactions using the RRDE is the electroreduction of oxygen. This reaction was studied under various experimental conditions with a disk electrode constructed from various materials. The material of a ring electrode, which generally is used only as a sensor of intermediates, is not critical. It was found that the electroreduction of oxygen on the gold electrode in alkaline solutions occurs in two steps [430–432]

$$O_2 + H_2O + 2e \rightarrow HO_2^- + OH^- \tag{133}$$

$$HO_2^- + H_2O + 2e \rightarrow 3OH^- \tag{134}$$

In alkaline solutions the mechanism of oxygen electroreduction was also studied at platinum [433–440], nickel [441, 442], rhodium [443–445], pyrolytic graphite [446], palladium [447], and amalgamated gold electrodes [448]. In the latter work the surface-active compounds were present in the studied solutions.

In acid solutions the oxygen electrode reaction was investigated using platinum [435, 449, 450], rhodium [444], and gold [451] electrodes.

The electroreduction of oxygen in dimethylformamide is described in [451].

3.11   Electrode Reactions With Chemical Complications Occurring at the RRDE

Several electrode reaction schemes were considered theoretically. Charge transfer with a following first-order irreversible chemical reaction was detailed by Bruckenstein and Feldman [364] and later by Albery and Bruckenstein [452]. The equation given by the latter authors has the form

$$\frac{N}{N_k} = 1 + 1.28 \left(\frac{v}{D} \right)^{1/3} \frac{k_1}{\omega} \tag{135}$$

where N and N_k are collection efficiencies for transport-controlled and kinetically controlled processes, respectively. This equation is valid when the condition $k_1^{1/2} (0.51 \omega^{3/2} v^{-1/2} D^{1/2})^{2/3} < 0.5$ is satisfied. Reference [5] considers this problem further.

The theory of the charge-transfer process with a first-order, reversible, chemical reaction [452] and an irreversible, second-order, chemical reaction [453, 454] was also presented.

Prater and Bard [455] considered similar problems using digital simulation to calculate N_k and also outlined the theory of the catalytic process [456].

Albery and Drury [368] found that both analytical and numerical methods of solving kinetic problems give similar results. The analytical approach is useful when considering electrodes with both a thin-ring and a thin gap between the ring and the disk.

Puglisi and Bard [457] considered the second-order chemical process in which the primary product occurring on the disk electrode interacts with a substrate to yield either an electroinactive substance or a substance that may be oxidized or reduced immediately on the disk electrode. The same authors [458] studied electrohydrodimerization reactions using a RRDE.

Bard and co-workers have used [459, 460] this electrode in studies of electro-generated chemiluminescence.

Following Nekrasov [393] in the study of homogeneous kinetics by the RRDE method the limits of rate constants may be estimated for the first-order reactions:

$$3 \times 10^{-2} \, s^{-1} < k < 10^3 \, s^{-1}$$

which corresponds to the half-life $t_{1/2}$ in the limits $7 \times 10^{-4} \, s < t_{1/2} < 20 \, s$, and for reactions of the second order

$$5 \times 10^3 \, L/mol \cdot s < k < 1 \times 10^9 \, L/mol \cdot s$$

Using the RRDE the method of titration in the diffusion layer based on the reaction of the disk-generated titrant with the analyzed substance [461] was proposed. The excess of the titrant is monitored by the ring electrode. In this As(III) was titrated by generated bromine [461].

3.12   Flowing Solutions Along Two Stationary Working Electrodes

A concept similar to that of the RRDE was developed by Gerischer and co-workers [462], who prepared a narrow channel with one exchangeable wall equipped with two electrodes separated by an insulating film only a few micrometers thick.

A pump system passed the solution through this channel at a well-controlled rate. The product or intermediate produced at one electrode was transported to the second electrode, where it could be detected. This system is similar to that of the RRDE. Analogous to the collection efficiency described earlier, the parameter dependent on the distance between both electrodes has numerical values similar to those of N for the RRDE.

Theoretical considerations of the channel flow double-electrode system, comprising two neighboring rectangular electrodes similar to that described above [462], were presented by Tokuda and Matsuda [463]. Two such electrodes embedded in the wall surface of a channel are shown in Figure 3.21.

Solution is assumed to flow in the x direction. The channel width is assumed to be much larger than its height $2b$, so that channel flow may be considered two-dimensional. The general equation for the stationary current–potential curves of the process occurring on the detector electrode located downstream was derived.

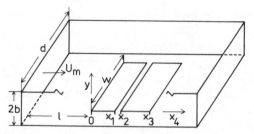

Figure 3.21 Schematic diagram of the cell with a double electrode in a channel flow: d, the width of the channel, is assumed to be much larger than its height $2b$. The origin of the x axis is the leading edge of the first generator electrode and x_1, x_2, and x_3 denote the back edge of the first electrode, the front, and the back edges of the detector electrode, respectively. W is the common width of the first and second electrodes. Flow of the solution takes place along the x direction with the rate U_m.

Matsuda and co-workers [464] presented theoretical equations for collection efficiency in the process,

$$A \pm n_1 e \longrightarrow B \quad \text{generator electrode}$$

$$B \xrightarrow{k} P \quad \text{solution} \tag{136}$$

$$B \pm n_2 e \longrightarrow Y \quad \text{detector electrode}$$

The authors, considering typical experimental conditions, conclude that the range of rate constants that can be determined by using their method is comparable to that of the RRDE, being equal to $1\,s^{-1} < k < 10^3\,s^{-1}$. The upper limit of k can be raised by making x_1, $x_2 - x_1$, a, and b smaller.

Still another arrangement was used by Albery and co-workers [465], in which solution flows down a tube past an electrode and then immediately through the cavity of an EPR spectrometer. The electrode consists of an annular ring that is flush with the wall of the tube. Thus, the entire apparatus is analogous to the RRDE except the detector electrode has been replaced by the EPR spectrometer [466].

Again, Albery and associates [466] theoretically considered the process in which the electrode reaction product dissociates into two radicals. The process consisting of the formation of free radicals that decompose in the bulk of the solution by second-order kinetics,

$$R^{\mp} + R^{\mp} \xrightarrow{k} \text{products} \tag{137}$$

has also been theoretically presented [467]. The convective diffusion equation was solved in this case by the simulation method. The results of the calculations were tested using the reduction of nitrobenzene as an example. Good agreement between theory and experiment was found.

The theory of EC reactions at channel and tubular electrodes was recently developed [468]. It was estimated that the first-order rate constant can be measured in the limits $10^{-2}\,s^{-1} < k < 10^2\,s^{-1}$.

3.13   Flow-Through Detectors

As mentioned earlier, in addition to electrodes moving in solutions, cells with stationary electrodes but moving solutions have been used. Examples of former systems were discussed in the preceding sections. This section presents even simpler systems, where only one electrode is placed in flowing solutions. Such systems have been investigated frequently in recent years because of their use as electrochemical detectors in continuous-flow analysis and in high-performance liquid chromatography. There is also the possibility of applying flow-through electrodes in electrochemical preparation of some compounds [469, 470], or in the direct electrowinning processes [471].

It is not the intention of the present author to cover this expanding field extensively. Only a brief outline of this problem is presented and is mostly related to electrochemical detection.

Chromatopolarography was introduced by Kemula [472] and this early method employed the dropping mercury electrode as a detector. The use of such an electrode in flowing systems has advantages, because the electrode surface is continuously renewed; however, the growth and fall of the mercury drop create an electrical noise.

To diminish these interferences, stationary mercury electrodes, such as a hanging mercury electrode, a mercury pool electrode, or a mercury film electrode with gold and platinum support, were used (for a review see [473, 474]).

These stationary electrodes suffer, however, from the greater influence of solution impurities that may be adsorbed onto the electrode surface during the experiment thereby changing its properties.

Using solid electrodes as voltammetric detectors has limitations similar to those of stationary mercury electrodes; in addition to adsorption, autoinhibition and passivation phenomena may occur; however, solid electrodes allow the application of the method to detection of substances that react with electrodes at positive potentials.

Electrodes made of platinum, vitreous carbon, carbon paste, and silicone-rubber-impregnated carbon were usually used [473–475]. Carbon paste electrodes are very useful because the electrode surface may be easily renewed.

Electrochemical detectors in liquid chromatography are used because of their selectivity, sensitivity, and linear response in a wide concentration range [476].

The comparative study of several detectors with DME [477] and solid electrodes [475] was carried out. In practice, different electrode arrangements are used.

Theoretical works on the description of the limiting currents observed with amperometric flow-through detectors were critically analyzed by Hanekamp and van Nieukerk [478]. They suggested the following general equation based on [221–223, 479–487]:

$$i_l = knC^0 D(Sc)^b d(Re)^a \tag{138}$$

where k is a dimensionless constant, d is characteristic of the electrode width, and C^0 and D are the concentration of a reactant and its diffusion coefficient, respectively. Values of a and b are determined by the specific quantities of the hydrodynamic conditions; Sc is the dimensionless Schmidt number

$$Sc = \frac{v}{D}$$

where v is a kinematic viscosity and Re is the Reynolds number

$$Re = \frac{\bar{V}}{r}$$

where $\bar{V}$ is a mean linear velocity of the fluid, l is the characteristic electrode length, and r is the radius of the electrode.

The values of the different parameters that appear in (138) for the different geometries and positions of the detector electrode are summarized in Table 3.2 [478].

Whether a detector is "parallel" or "opposite" depends on the direction of the mercury flow from the capillary—either parallel to or opposite to fluid flow. For these two detectors, $r_{Hg} = 0.026 \, (mt)^{1/3}$, where m is the mercury flow rate and t the drop time. The dropping mercury electrode was represented here as a flat plate, an assumption that may be justified at higher fluid velocities.

It follows from (138) that the current is dependent on the flow rate. This is not very convenient in experimental work, because to obtain reproducible results the flow rate should be precisely controlled and constant in time.

As a result, special detecting techniques were chosen so that the current response would not be influenced by the mobile phase flow.

The choice of these methods was based on the concept that the observed current will be independent of the flow rate if the thickness of the diffusion layer created at the electrode surface by the applied method is considerably lower than the thickness of the convective layer, which is controlled by the flow rate of the liquid.

Several methods, such as potential pulse techniques with a short pulse time, were applied that fulfilled this condition. Also, the RDE was used with an electrode revolving so quickly that the above-mentioned condition was fulfilled

Table 3.2   Parameters for Equation (138)

Type of Detector Electrode	k	a	b	d
Tube	8.0	1/3	1/3	$l^{1/3}r^{2/3}$
Thin-layer	0.8	1/2	1/3	r
Disk	3.3	1/2	1/3	r
Parallel	7.9	1/3	1/3	r_{Hg}
Opposite	3.7	1/2	1/2	r_{Hg}

[474]. The rapidly dropping mercury electrode was applied with polarographic detectors, and square-wave polarography was also used.

The detection of the substance was also based on changes in the differential capacity of the electrode when the analyzed substance is adsorbed on the electrode surface. This idea was developed and used in practice by Kemula and co-workers [474, 488, 489].

A detailed discussion of problems connected with electrochemical detection in flowing systems is beyond the scope of this chapter. The reader interested in such problems should consult review papers [473, 474, 490–492].

Acknowledgment

It is a pleasure to thank Professor Ralph N. Adams for his suggestion that I write this chapter in the new edition of the present series. I started the chapter while working in his laboratory during the summer semester of 1981. His advice and help at that time, and also much earlier when I worked with him for a longer time, are gratefully acknowledged.

References

1. S. Piekarski and R. N. Adams, "Voltammetry with Stationary and Rotated Electrodes," in A. Weissberger and B. W. Rossiter, Eds., *Physical Methods of Chemistry*, Vol. 1, Part 2A, 4th ed., Wiley-Interscience, New York, 1971.
2. R. N. Adams, *Electrochemistry at Solid Electrodes*, Dekker, New York, 1969.
3. Z. Galus, *Fundamentals of Electrochemical Analysis*, Horwood, Chichester, 1976.
4. Yu. V. Pleskov and V. Yu. Filinovskii, *Rotating Disk Electrode*, Nauka, Moscow, 1972 (in Russian).
5. W. J. Albery and H. L. Hitchman, *Ring–Disc Electrodes*, Clarendon, Oxford, 1971.
6. A. J. Bard and L. F. Faulkner, *Electrochemical Methods. Fundamentals and Applications*, Wiley, New York, 1980.
7. W. Kemula and Z. Kublik, *Anal. Chim. Acta*, **18**, 104 (1958); *Rocz. Chem.*, **32**, 941 (1958).
8. W. M. Peterson, *Am. Lab.*, **11(12)**, 69 (1979).
9. R. N. Adams, *Anal. Chem.*, **30**, 1576 (1958).
10. J. Tenygl, *Proc. J. Heyrovský Memorial Congress in Polarography*, Vol. 1, Prague, 1980, p. 51.
11. G. J. Hills and D. J. G. Ives, "The Calomel Electrode and Other Mercury–Mercurous Salt Electrodes," and G. J. Janz, "Silver–Silver Halide Electrodes," in D. J. G. Ives and G. J. Janz, Eds., *Reference Electrodes*, Academic, New York, 1961.
12. D. E. Smith, *Crit. Rev. Anal. Chem.*, **2**, 247 (1971).
13. Ch. Yarnitzky, "Experiments in Electronics Related to Electrochemistry," in E. Gileadi, E. Kirova Eisner, and J. Penciner, Eds., *Interfacial Electrochemistry. An Experimental Approach*, Addison-Wesley, Reading, MA, 1975.
14. R. R. Schroeder, "Operational Amplifier Instruments for Electrochemistry," in J. S. Mattson, H. B. Mark, Jr., and H. C. MacDonald, Jr., Eds., *Computers in Chemistry and Instrumentation*, Vol. 2, *Electrochemistry: Calculations, Simulation and Instrumentation*, Dekker, New York, 1973.
15. E. R. Brown, D. E. Smith, and G. L. Booman, *Anal. Chem.*, **40**, 1411 (1968).

16. D. Britz, *J. Electroanal. Chem.*, **88**, 309 (1978); J. E. Harrar and C. L. Pomernacki, *Anal. Chem.*, **45**, 57 (1973); D. T. Sawyer and J. L. Roberts, *Experimental Electrochemistry for Chemists* Wiley, New York, 1974, Chapter 5.

17. A. M. Bond, *Modern Polarographic Methods in Analytical Chemistry*, Dekker, New York, 1980, Chapter 2.

18. J. Crank, *The Mathematics of Diffusion*, Oxford University, Oxford, 1956, pp. 3 and 84.

19. A. Ševčik, *Collect. Czech. Chem. Commun.*, **13**, 349 (1948).

20. J. E. B. Randles, *Trans. Faraday Soc.*, **44**, 327 (1948).

21. W. H. Reinmuth, *Anal. Chem.*, **33**, 1793 (1961).

22. H. Matsuda, *Z. Elektrochem.*, **61**, 489 (1957).

23. Ya. P. Gokhstein, *Dokl. Akad. Nauk SSSR*, **126**, 598 (1959).

24. W. T. de Vries and E. van Dalen, *J. Electroanal. Chem.*, **6**, 490 (1963).

25. R. S. Nicholson and I. Shain, *Anal. Chem.*, **36**, 706 (1964).

26. C. Gumiński and Z. Galus, *Rocz. Chem.*, **39**, 1767 (1970).

27. J. W. Loveland and P. J. Elving, *J. Phys. Chem.*, **56**, 250 (1952).

28. J. W. Loveland and P. J. Elving, *J. Phys. Chem.*, **56**, 255 (1952).

29. E. Ahlberg and V. D. Parker, *Acta Chem. Scand. B*, **33**, 696 (1979).

30. S. P. Perone, *Anal. Chem.*, **38**, 1158 (1966).

31. D. Garreau and J. M. Savéant, *J. Electroanal. Chem.*, **50**, 1 (1974).

32. R. P. Frankenthal and I. Shain, *J. Am. Chem. Soc.*, **78**, 2969 (1956).

33. M. M. Nicholson, *J. Am. Chem. Soc.*, **76**, 2539 (1954).

34. P. T. Kissinger, J. B. Hart, R. N. Adams, *Brain Res.*, **55**, 209 (1973).

35. R. L. McCreery, R. Dreiling, and R. N. Adams, *Brain Res.*, **73**, 15 (1974).

36. M. A. Dayton, J. C. Brown, K. J. Stutts, and R. M. Wightman, *Anal. Chem.*, **52**, 946 (1980).

37. M. A. Dayton, A. G. Ewing, and R. M. Wightman, *Anal. Chem.*, **52**, 2392 (1980).

38. Z. Galus, J. Schenk, and R. N. Adams, *J. Electroanal. Chem.*, **135**, 1 (1982).

39. P. Delahay, *New Instrumental Methods in Electrochemistry*, Interscience, New York, 1954, Chapter 3.

40. Z. G. Soos and P. J. Lingane, *J. Phys. Chem.*, **68**, 3821 (1964).

41. G. P. Sato, M. Kakihana, H. Ikeuchi, and K. Tokuda, *J. Electroanal. Chem.*, **108**, 381 (1980).

42. K. B. Oldham, *J. Electroanal. Chem.*, **122**, 1 (1981).

43. K. Aoki and J. Osteryoung, *J. Electroanal. Chem.*, **122**, 19 (1981).

44. J. Koryta, J. Dvořák, and V. Boháckova, *Lehrbuch der Elektrochemie*, Springer, Wien, New York, 1975, Chapter 2.

45. A. T. Hubbard and F. C. Anson, "The Theory and Practice of Electrochemistry with Thin-Layer Cells," in A. J. Bard Ed., *Electroanalytical Chemistry*, Vol. 4, Dekker, New York, 1971.

46. A. T. Hubbard, *CRC Crit. Rev. Anal. Chem.*, **3**, 201 (1973).

47. T. R. Gilbert and D. N. Hume, *Anal. Chim. Acta*, **65**, 451 (1973); T. R. Copeland, J. H. Christie, R. A. Osteryoung, and R. K. Skogerboe, *Anal. Chem.*, **45**, 2171 (1973).

48. T. M. Florence, *J. Electroanal. Chem.*, **27**, 273 (1970).

49. Z. Stojek, B. Stepnik, and Z. Kublik, *J. Electroanal. Chem.*, **74**, 277 (1976).

50. M. Štulikova, *J. Electroanal. Chem.*, **48**, 33 (1973).

51. W. T. de Vries and E. van Dalen, *J. Electroanal. Chem.*, **8**, 366 (1964).

52. W. T. de Vries and E. van Dalen, *J. Electroanal. Chem.*, **12**, 189 (1966); **10**, 183 (1965).

53. W. T. de Vries and E. van Dalen, *J. Electroanal. Chem.*, **14**, 315 (1967); W. T. de Vries, *J. Electroanal. Chem.*, **9**, 448 (1965).

54. A. G. Stromberg, A. A. Zheltonozhko, and A. A. Kaplin, *Zh. Anal. Khim.*, **28**, 1045 (1973).
55. G. E. Batley and T. M. Florence, *J. Electroanal. Chem.*, **55**, 23 (1974).
56. Z. Stojek and Z. Kublik, *J. Electroanal. Chem.*, **105**, 247 (1979); **77**, 205 (1977).
57. R. F. Lane and A. T. Hubbard, *J. Phys. Chem.*, **77**, 1401 (1973).
58. R. W. Murray, *Acc. Chem. Res.*, **13**, 135 (1980); K. W. Willman and R. W. Murray, *J. Electroanal. Chem.*, **133**, 211 (1982). P. G. Pickup, C. R. Leidner, P. Denisevich, and R. W. Murray, *J. Electroanal. Chem.*, **164**, 39 (1984). C. R. Leidner, P. Denisevich, K. W. Willman, and R. W. Murray, *J. Electroanal. Chem.*, **164**, 63 (1984).
59. Y. P. Gokhstein and A. Y. Gokhstein, *Dokl. Akad. Nauk SSSR*, **128**, 985 (1959); "Multistage Electrochemical Reactions in Oscillographic Polarography," in I. S. Longmuir Ed., *Advances in Polarography*, Vol. 2, Pergamon, New York, 1960.
60. D. S. Polcyn and I. Shain, *Anal. Chem.*, **38**, 370 (1966).
61. W. F. Gutknecht and S. P. Perone, *Anal. Chem.*, **42**, 906 (1970).
62. H. Matsuda and Y. Ayabe, *Z. Elektrochem.*, **59**, 494 (1955).
63. Y. P. Gokhstein, *Dokl. Akad. Nauk SSSR*, **131**, 601 (1960).
64. W. H. Reinmuth, *Anal. Chem.*, **32**, 1891 (1960).
65. P. Delahay, *J. Am. Chem. Soc.*, **75**, 1190 (1953).
66. Y. P. Gokhstein and A. Y. Gokhstein, *Zh. Fiz. Khim.*, **34**, 1654 (1960).
67. R. D. de Mars and I. Shain, *J. Am. Chem. Soc.*, **81**, 2654 (1959).
68. R. S. Nicholson, *Anal. Chem.*, **37**, 1351 (1965).
69. P. K. Wrona, *Bull. Acad. Pol. Sci. Ser. Sci. Chem.*, **27**, 725 (1979).
70. C. P. Andrieux, L. Nadjo, and J. M. Savéant, *J. Electroanal. Chem.*, **26**, 147 (1970).
71. K. B. Oldham, *Anal. Chem.*, **41**, 1904 (1969).
72. J. C. Imbeaux and J. M. Savéant, *J. Electroanal. Chem.*, **44**, 169 (1973).
73. J. M. Savéant and D. Tessier, *J. Electroanal. Chem.*, **65**, 57 (1975).
74. M. E. Peover and J. S. Powell, *J. Electroanal. Chem.*, **20**, 427 (1969).
75. D. A. Corrigan and D. H. Evans, *J. Electroanal. Chem.*, **106**, 287 (1980).
76. K. B. Oldham and J. Spanier, *J. Electroanal. Chem.*, **26**, 331 (1970).
77. M. Goto and D. Ishi, *J. Electroanal. Chem.*, **61**, 361 (1975).
78. P. Dalrymple-Alford, M. Goto, and K. B. Oldham, *Anal. Chem.*, **49**, 1390 (1977).
79. A. M. Bond, *Anal. Chem.*, **52**, 1318 (1980).
80. G. C. Barker, "Some Possible Developments in ac Polarography," in I. S. Longmuir Ed., *Advances in Polarography*, Vol. 1, Pergamon, New York, 1960.
81. C. K. Mann, *Anal. Chem.*, **33**, 1484 (1961); **35**, 326 (1965); **36**, 2424 (1966).
82. R. Sh. Nigmatullin and M. R. Vyaselev, *Zh. Anal. Khim.*, **19**, 545 (1964).
83. J. H. Christie and P. J. Lingane, *J. Electroanal. Chem.*, **10**, 176 (1965).
84. D. R. Ferrier and R. R. Schroeder, *J. Electroanal. Chem.*, **45**, 343 (1973).
85. D. R. Ferrier, D. H. Chidester, and R. R. Schroeder, *J. Electroanal. Chem.*, **45**, 361 (1973).
86. R. C. Rooney, *J. Polarogr. Soc.*, **9**, 45 (1963).
87. L. Ja. Shekun, *Zh. Fiz. Khim.*, **36**, 455 (1962).
88. F. B. Stephens and J. E. Harrar, *Chem. Instrum.*, **1**, 169 (1968).
89. H. M. Davis and J. E. Seaborn, "A Differential Cathode-Ray Polarograph," in I. S. Longmuir Ed., *Advances in Polarography*, Vol. 1, Pergamon, New York, 1960.
90. H. M. Davis and H. I. Shalgosky, "The Performance of the Differential Cathode-Ray Polarograph," in I. S. Longmuir Ed., *Advances in Polarography*, Vol. 2, Pergamon, New York, 1960.
91. B. B. Damaskin, O. A. Petri, and V. V. Batrakov, *Adsorption of Organic Compounds on Electrodes*, Plenum, New York, 1971.

92. R. H. Wopschall and I. Shain, *Anal. Chem.*, **39**, 1514 (1967).
93. E. Laviron, *J. Electroanal. Chem.*, **52**, 355, 395 (1977).
94. S. W. Feldberg, "Digital Simulation of Electrochemical Surface Boundary Phenomena: Multiple Electron Transfer and Adsorption," in J. S. Mattson, H. B. Mark, Jr., and H. C. MacDonald, Jr., Eds., *Computers in Chemistry and Instrumentation*, Vol. 2, *Electrochemistry: Calculations, Simulation and Instrumentation*, Dekker, New York, 1973.
95. M. H. Hulbert and I. Shain, *Anal. Chem.*, **42**, 162 (1970).
96. F. C. Anson, *Anal. Chem.*, **36**, 932 (1964).
97. J. H. Christie, G. Lauer, R. A. Osteryoung, and F. C. Anson, *Anal. Chem.*, **35**, 1979 (1963).
98. F. C. Anson, *Anal. Chem.*, **38**, 54 (1966).
99. R. A. Osteryoung and F. C. Anson, *Anal. Chem.*, **36**, 975 (1964). R. W. Murray, *Acc. Chem. Res.*, **13**, 135 (1980). K. D. Snell and A. G. Keenan, *Chem. Soc. Rev.*, **8**, 259 (1979). W. R. Heinemann and P. T. Kissinger, *Anal. Chem.*, **50**, 166R (1978). W. C. Heinemann and P. T. Kissinger, *Anal. Chem.*, **52**, 138R (1980).
100. E. Laviron, *J. Electroanal. Chem.*, **112**, 1 (1980).
101. E. Laviron, *J. Electroanal. Chem.*, **112**, 11 (1980).
102. E. Laviron, *J. Electroanal. Chem.*, **122**, 37 (1981).
103. E. R. Brown and J. R. Sandifer, "Cyclic Voltammetry, ac Polarography, and Related Techniques," in B. W. Rossiter and J. F. Hamilton, Eds., *Physical Methods of Chemistry*, Vol. 2, Wiley, New York, 1985.
104. B. Alstad and V. D. Parker, *J. Electroanal. Chem.*, **112**, 163 (1980).
105. T. M. Krygowski, M. Lipsztajn, and Z. Galus, *J. Electroanal. Chem.*, **42**, 261 (1973).
106. M. Lipsztajn, T. M. Krygowski, and Z. Galus, *J. Electroanal. Chem.*, **49**, 17 (1974).
107. J. S. Jaworski and M. K. Kalinowski, *J. Electroanal. Chem.*, **76**, 301 (1977).
108. A. Kapturkiewicz and M. K. Kalinowski, *J. Phys. Chem.*, **82**, 1141 (1978).
109. M. Lipsztajn, T. M. Krygowski, E. Laren, and Z. Galus, *J. Electroanal. Chem.*, **54**, 313 (1974).
110. M. Lipsztajn, T. M. Krygowski, E. Laren, and Z. Galus, *J. Electroanal. Chem.*, **57**, 339 (1974).
111. M. Lipsztajn, M. Buchalik, and Z. Galus, *J. Electroanal. Chem.*, **105**, 341 (1979).
112. L. A. Avaca and A. Bewick, *J. Electroanal. Chem.*, **41**, 405 (1973).
113. O. Hammerich and V. D. Parker, *J. Am. Chem. Soc.*, **96**, 4289 (1974).
114. M. D. Ryan and D. H. Evans, *J. Electroanal. Chem.*, **67**, 333 (1976).
115. M. D. Ryan and D. H. Evans, *J. Electroanal. Chem.*, **121**, 881 (1974).
116. D. H. Evans and D. A. Griffith, *J. Electroanal. Chem.*, **136**, 149 (1982).
117. D. H. Evans and D. A. Griffith, *J. Electroanal. Chem.*, **134**, 301 (1982).
118. E. Ahlberg, B. Drews, and B. S. Jensen, *J. Electroanal. Chem.*, **87**, 141 (1978).
119. A. Demortier and A. J. Bard, *J. Am. Chem. Soc.*, **95**, 3495 (1973).
120. W. H. Smith and A. J. Bard, *J. Am. Chem. Soc.*, **97**, 5203 (1975).
121. I. Vartires, W. H. Smith, and A. J. Bard, *J. Electrochem. Soc.*, **122**, 894 (1975).
122. W. H. Smith and A. J. Bard, *J. Electroanal. Chem.*, **76**, 19 (1977).
123. R. D. Grypa and J. T. Maloy, *J. Electrochem. Soc.*, **122**, 377 (1975).
124. E. Ouziel and Ch. Yarnitzky, *J. Electroanal. Chem.*, **78**, 257 (1977).
125. M. R. Asirvatham and M. D. Hawley, *J. Electroanal. Chem.*, **53**, 293 (1974).
126. K. J. Borhani and M. D. Hawley, *J. Electroanal. Chem.*, **101**, 407 (1979).
127. V. N. Leibzon, A. S. Mendkovich, S. G. Majranovskii, T. A. Klimova, M. M. Krajushkin, S. S. Novikov, and V. V. Sevostjanova, *Electrokhimiya*, **12**, 1481 (1976).

128. V. D. Pokhodenko and E. P. Platonova, *Elektrokhimiya*, **10**, 789 (1974).
129. L. A. Tinker and A. J. Bard, *J. Electroanal. Chem.*, **133**, 275 (1982).
130. L. M. Doane and A. J. Fatiadi, *J. Electroanal. Chem.*, **135**, 193 (1982).
131. J. Robinson and R. A. Osteryoung, *J. Am. Chem. Soc.*, **101**, 323 (1979).
132. M. Delmar, P-C. Lacaze, J-Y. Dumousseau, and J-E. Dubois, *Electrochim. Acta*, **27**, 61 (1982).
133. A. F. Diaz, J. Crowley, J. Bargon, G. P. Gardini, and J. B. Torrance, *J. Electroanal. Chem.*, **121**, 355 (1981).
134. A. F. Diaz, A. Martinez, K. K. Kanazawa, and M. Salmon, *J. Electroanal. Chem.*, **130**, 181 (1981).
135. J. Madec and J. Courtot-Coupez, *J. Electroanal. Chem.*, **84**, 169 (1977).
136. B. S. Jensen and V. D. Parker, *J. Am. Chem. Soc.*, **97**, 5211 (1975).
137. F. Ammar, L. Nadjo, and J. M. Savéant, *J. Electroanal. Chem.*, **47**, 146 (1973).
138. K. Yasukouchi, I. Taniguchi, H. Yamaguchi, and K. Arakawa, *J. Electroanal. Chem.*, **121**, 231 (1981).
139. B. A. Olsen and D. H. Evans, *J. Electroanal. Chem.*, **136**, 139 (1982).
140. R. Raghavan and G. Dryhurst, *J. Electroanal. Chem.*, **129**, 189 (1981).
141. L. R. Sharma, A. K. Manchanda, G. Singh, and R. S. Verma, *Electrochim. Acta*, **27**, 223 (1982).
142. C. L. Schmidt, C. F. Kolpin, and H. S. Swofford Jr., *Anal. Chem.*, **53**, 41 (1981).
143. R. N. Goyal, A. Brajter-Toth, and G. Dryurst, *J. Electroanal. Chem.*, **133**, 287 (1982).
144. A. Anne and J. Moiroux, *J. Electroanal. Chem.*, **137**, 293 (1982).
145. G. Pilloni, G. Zotti, and S. Zecchin, *J. Electroanal. Chem.*, **125**, 129 (1981).
146. L. Roullier and E. Laviron, *J. Electronal. Chem.*, **134**, 181 (1982).
147. C. Degrand and L. L. Miller, *J. Electroanal. Chem.*, **132**, 163 (1982).
148. C. P. Andrieux, J. M. Savéant, and D. Tessier, *J. Electroanal. Chem.*, **63**, 429 (1975).
149. L. Nadjo, J. M. Savéant, and D. Tessier, *J. Electroanal. Chem.*, **52**, 403 (1974).
150. J. M. Savéant and D. Tessier, *J. Electroanal. Chem.*, **77**, 225 (1977).
151. E. Ahlberg and V. D. Parker, *J. Electroanal. Chem.*, **121**, 57 (1981).
152. B. Alstad and V. D. Parker, *J. Electroanal. Chem.*, **122**, 183 (1981); **133**, 33 (1982); **136**, 251 (1982).
153. Kh. Z. Brainina and M. B. Vydrevich, *J. Electroanal. Chem.*, **121**, 1 (1981).
154. F. Vydra, K. Štulik, and E. Juláková, *Electrochemical Stripping Analysis*, Horwood, Chichester, 1976.
155. R. N. Adams, *Anal. Chem.*, **48**, 1126A (1976).
156. D. Homolka and V. Mareček, *J. Electroanal. Chem.*, **112**, 91 (1980).
157. Z. Samec, V. Mareček, J. Koryta, and M. W. Khalil, *J. Electroanal. Chem.*, **83**, 393 (1977).
158. Z. Samec and V. Mareček, *J. Electroanal. Chem.*, **100**, 841 (1979).
159. A. Hofmanová, Le Q. Hung, and W. Khalil, *J. Electroanal. Chem.*, **135**, 257 (1982).
160. G. Geblewicz, Z. Koczorowski, and Z. Figaszewski, *Colloid and Surfaces*, **6**, 43 (1983).
161. Z. Koczorowski and G. Geblewicz, *J. Electroanal. Chem.*, **139**, 177 (1982).
162. Z. Samec, *J. Electroanal. Chem.*, **111**, 211 (1980); Z. Samec, D. Homolka, and V. Mareček, *J. Electroanal. Chem.*, **135**, 265 (1982).
163. H. W. Nürnberg and P. Valenta, *J. Electroanal. Chem.*, **57**, 125 (1974).
164. Y. M. Temerk, P. Valenta, and H. W. Nürnberg, *J. Electroanal. Chem.*, **131**, 265 (1982).
165. D. H. Evans, *J. Phys. Chem.*, **76**, 1160 (1972).
166. L. Nadjo and J. M. Savéant, *J. Electroanal. Chem.*, **48**, 113 (1973).

167. R. Caban and T. W. Chapman, *J. Electrochem. Soc.*, **123**, 1036 (1976).
168. Shi-Chern Yen and T. W. Chapman, *J. Electroanal. Chem.*, **135**, 305 (1982).
169. R. Speiser and A. Rieker, *J. Electroanal. Chem.*, **102**, 1 (1979).
170. E. Laviron, *J. Electroanal. Chem.*, **39**, 1 (1972).
171. E. Laviron, *J. Electroanal. Chem.*, **35**, 333 (1972).
172. A. Vallat, M. Person, and E. Laviron, *Electrochim. Acta*, **27**, 485 (1982).
173. J. M. Savéant and D. Tessier, *J. Electroanal. Chem.*, **61**, 251 (1975).
174. J. M. Savéant and E. Vianello, *C.R. Hebd. Seances Acad. Sci., Ser. C.*, **256**, 2597 (1963); **259**, 4017 (1964).
175. J. M. Savéant and E. Vianello, *Electrochim. Acta*, **12**, 1545 (1967).
176. R. S. Nicholson, *Anal. Chem.*, **37**, 667 (1965).
177. C. P. Andrieux, L. Nadjo, and J. M. Savéant, *J. Electroanal. Chem.*, **26**, 147 (1970); **42**, 223 (1973).
178. J. C. Imbeaux and J. M. Savéant, *J. Electroanal. Chem.*, **44**, 169 (1973).
179. E. Laviron, *Electrochim. Acta*, **16**, 409 (1971).
180. E. Laviron, *J. Electroanal. Chem.*, **34**, 463 (1972).
181. L. Nadjo and J. M. Savéant, *J. Electroanal. Chem.*, **44**, 327 (1973).
182. C. P. Andrieux and J. M. Savéant, *Bull. Soc. Chim. France*, 3281 (1972).
183. C. P. Andrieux and J. M. Savéant, *J. Electroanal. Chem.*, **33**, 453 (1971).
184. E. Lamy, L. Nadjo, and J. M. Savéant, *J. Electroanal. Chem.*, **50**, 141 (1974).
185. E. Lamy, L. Nadjo, and J. M. Savéant, *J. Electroanal. Chem.*, **42**, 189 (1973).
186. F. Ammar, C. P. Andrieux, and J. M. Savéant, *J. Electroanal. Chem.*, **53**, 407 (1974).
187. E. Laviron and Y. Mugnier, *J. Electroanal. Chem.*, **93**, 69, (1978).
188. L. Nadjo and J. M. Savéant, *J. Electroanal. Chem.*, **33**, 419 (1971).
189. F. Ammar, L. Nadjo, and J. M. Savéant, *J. Electroanal. Chem.*, **47**, 146 (1973).
190. A. Lasia, *J. Electroanal. Chem.*, **42**, 253 (1973).
191. A. Vallat and E. Laviron, *J. Electroanal. Chem.*, **74**, 309 (1976).
192. S. N. Frank, A. J. Bard, and A. Ledwith, *J. Electrochem. Soc.*, **122**, 898 (1975).
193. W. R. Fawcett and A. Lasia, *Can. J. Chem.*, **59**, 3256 (1981).
194. E. Laviron and A. Vallat, *J. Electroanal. Chem.*, **46**, 421 (1973).
195. J. M. Savéant and D. Tessier, *J. Phys. Chem.*, **82**, 1723 (1978).
196. J. M. Savéant and D. Tessier, *J. Electroanal. Chem.*, **64**, 143 (1975).
197. J. M. Savéant and D. Tessier, *J. Electroanal. Chem.*, **61**, 251 (1975).
198. V. D. Parker, *Acta Chem. Scand. B*, **34**, 359 (1980).
199. B. S. Jensen and V. D. Parker, *Acta Chem. Scand. B*, **30**, 749 (1976).
200. E. Laviron, *J. Electroanal. Chem.*, **42**, 415 (1973).
201. L. Nadjo and J. M. Savéant, *J. Electroanal. Chem.*, **30**, 41 (1971).
202. J. M. Savéant and E. Vianello, "Studies on Catalytic Currents in Oscillographic Polarography with a Linear Change of Voltage. Theory" (in French), in I. S. Longmuir, Ed., *Advances in Polarography*, Vol. 1, Pergamon, New York, 1960.
203. L. Rampazzo, *J. Electroanal. Chem.*, **14**, 117 (1967); *Ric. Sci.*, **36**, 998 (1966).
204. M. Mastragostino, L. Nadjo, and J. M. Savéant, *Electrochim. Acta*, **13**, 721 (1968).
205. L. Nadjo and J. M. Savéant, *J. Electroanal. Chem.*, **48**, 113 (1973).
206. M. L. Olmstead and R. S. Nicholson, *Anal. Chem.*, **41**, 862 (1969).
207. A. Vallat, M. Person, and E. Laviron, *J. Electroanal. Chem.*, **27**, 657 (1982).
208. P. J. Kudirka and R. S. Nicholson, *Anal. Chem.*, **44**, 1786 (1972).
209. M. Mastragostino and J. M. Savéant, *Electrochim. Acta*, **13**, 751 (1968).
210. L. Sipos, Lj. Jeftić, M. Branica, and Z. Galus, *J. Electroanal. Chem.*, **32**, 35 (1971).
211. B. Ćosović, Lj. Jeftić, M. Branica, and Z. Galus, *Croat. Chem. Acta*, **45**, 475 (1973).

212. J. M. Savéant and E. Vianello, *Electrochim. Acta*, **10**, 905 (1965).
213. R. S. Nicholson and I. Shain, *Anal. Chem.*, **37**, 178 (1965).
214. R. S. Nicholson and I. Shain, *Anal. Chem.*, **37**, 190 (1965).
215. C. Amatore and J. M. Savéant, *J. Electroanal. Chem.*, **85**, 27 (1977).
216. G. S. Alberts and I. Shain, *Anal. Chem.*, **35**, 1859 (1963).
217. D. W. Leedy and R. N. Adams, *J. Electroanal. Chem.*, **14**, 119 (1967).
218. A. Darchen and C. Monet, *J. Electroanal. Chem.*, **78**, 81 (1977).
219. J. M. Savéant and E. Vianello, *Electrochim. Acta*, **8**, 905 (1963).
220. Yu. M. Kargin, V. Zh. Kondranina, G. K. Budnikov, and N. A. Ulakhovich, *Izv. Akad. Nauk SSSR, Ser. Khim.*, 2436 (1971). G. K. Budnikov, T. V. Kalinina, N. A. Koren, and V. V. Kormachev, *Zh. Org. Khim.*, **41**, 2138 (1971).
221. V. G. Levich, *Physicochemical Hydrodynamics*, Prentice-Hall, Englewood Cliffs, NJ, 1962.
222. L. N. Klatt and W. J. Blaedel, *Anal. Chem.*, **40**, 512 (1968); **38**, 879 (1966). J. Jordan and R. A. Javick, *J. Am. Chem. Soc.*, **80**, 1264 (1958); *Electrochim. Acta*, **6**, 23 (1962). T. O. Oesterling and C. L. Olson, *Anal. Chem.*, **39**, 1543 (1967).
223. F. Strafelda and A. Kimla, *Collect. Czech. Chem. Commun.*, **28**, 1516 (1963); **30**, 3606 (1965).
224. K. W. Pratt, Jr. and D. C. Johnson, *Electrochim. Acta*, **27**, 1013 (1982).
225. W. Nernst, *Z. Phys. Chem. (Leipzig)*, **47**, 52 (1904).
226. W. Nernst and E. S. Merriam, *Z. Phys. Chem. (Leipzig)*, **53**, 235 (1905).
227. H. A. Laitinen and I. M. Kolthoff, *J. Phys. Chem.*, **45**, 1079 (1941).
228. W. J. Albery and S. Bruckenstein, *J. Electroanal. Chem.*, **144**, 105 (1983).
229. A. C. Riddiford, "The Rotating Disk System," in. P. Delahay, Ed., *Advances in Electrochemistry and Electrochemical Engineering*, Vol. 4, Interscience, New York, 1966.
230. F. Opekar and P. Beran, *J. Electroanal. Chem.*, **69**, 1 (1976).
231. S. Azim and A. C. Riddiford, *Anal. Chem.*, **34**, 1023 (1962).
232. K. F. Blurton and A. C. Riddiford, *J. Electroanal. Chem.*, **10**, 457 (1965).
233. K. B. Prater and R. N. Adams, *Anal. Chem.*, **38**, 153 (1966).
234. P. Kulesza, T. Jedral, and Z. Galus, *J. Electroanal. Chem.*, **109**, 141 (1980).
235. W. J. Albery and S. Bruckenstein, *Trans. Faraday Soc.*, **62**, 1920 (1966).
236. Z. Galus and R. N. Adams, *J. Phys. Chem.*, **67**, 866 (1963).
237. Z. Galus, C. Olson, H. Y. Lee, and R. N. Adams, *Anal. Chem.*, **34**, 164 (1962).
238. D. P. Gregory and A. C. Riddiford, *J. Chem. Soc.*, 3757 (1956).
239. V. G. Levich, *Acta Physicochim. URSS*, **17**, 257 (1942).
240. J. Newman, *J. Phys. Chem.*, **70**, 1327 (1966).
241. T. F. Kassner, *J. Electrochem. Soc.*, **114**, 689 (1967).
242. W. H. Smyrl and J. Newman, *J. Electrochem. Soc.*, **118**, 1079 (1971).
243. B. Miller and S. Bruckenstein, *Anal. Chem.*, **46**, 2033 (1974).
244. M. B. Bardin and A. N. Dikusar, *Elektrokhimiya*, **6**, 1147 (1970).
245. Yu. K. Delimarski and I. I. Penkalo, *Ukr. Khim. Zh., Russ. Ed.*, **36**, 1279 (1970).
246. C. M. Mohr and J. Newman, *J. Electrochem. Soc.*, **122**, 928 (1975).
247. I. Cornet, W. N. Lewis, and R. Kappesser, *Trans. Inst. Chem. Eng.*, **47**, T222 (1969).
248. M. H. Meklati and M. Daguenet, *J. Chim. Phys. Phys.-Chim. Biol.*, **70**, 1102 (1973).
249. M. Daguenet and J. Robert, *J. Chim. Phys. Phys.-Chim. Biol.*, **64**, 395 (1967).
250. N. Gregory, J. T. Stuart, and W. S. Walker, *Phil. Trans. R. Soc. London*, **A248**, 155 (1967).
251. H. E. Hintermann and E. Suter, *Rev. Sci. Instrum.*, **36**, 1610 (1965). C. A. Emery and

H. E. Hintermann, *Electrochim. Acta*, **13**, 127 (1968). M. Daguenet, I. Eelboin, and M. Froment, *C. R. Hebd. Seances Acad. Sci.*, **258**, 3694 (1964).

252. J. Newman, *J. Electrochem. Soc.*, **113**, 501 (1966); **113**, 1235 (1966).
253. V. Marathe and J. Newman, *J. Electrochem. Soc.*, **116**, 1704 (1969).
254. G. Neubert, E. Gorman, R. van Fleet, and K. B. Prater, *J. Electrochem. Soc.*, **119**, 677 (1972).
255. P. Delahay, *New Instrumental Methods in Electrochemistry*, Interscience, New York, 1954, Chapter 9.
256. D. Jahn and W. Vielstich, *J. Electrochem. Soc.*, **109**, 849 (1962).
257. J. E. B. Randles, *Can. J. Chem.*, **37**, 238 (1959).
258. J. Küta and E. Yeager, *J. Electroanal. Chem.*, **31**, 119 (1971).
259. J. Koryta, *Electrochim. Acta*, **6**, 67 (1962).
260. J. Małyszko, *Chimia*, **29**, 166 (1975).
261. J. Jordan and R. A. Javick, *Electrochim. Acta*, **6**, 23 (1962).
262. E. Laviron, *J. Electroanal. Chem.*, **124**, 19 (1981).
263. E. Laviron, *J. Electroanal. Chem.*, **131**, 61 (1982).
264. T. Ikeda, C. R. Leidner, and R. W. Murray, *J. Electroanal. Chem.*, **138**, 343 (1982). Kuo-Nan Kuo and R. W. Murray, *J. Electroanal. Chem.*, **131**, 37 (1982). P. Daum and R. W. Murray, *J. Phys. Chem.*, **85**, 389 (1981); R. D. Rocklin and R. W. Murray, *J. Phys. Chem.*, **85**, 2104 (1981). N. Oyama and F. C. Anson, *Anal. Chem.*, **52**, 1192 (1980).
265. E. A. Aikazyan and Yu. V. Pleskov, *Zh. Fiz. Khim.*, **31**, 205 (1957).
266. T. A. Miller, B. Lamb, K. Prater, J. K. Lee, and R. N. Adams, *Anal. Chem.*, **36**, 418 (1964).
267. F. Lohmann and W. Mehl, *Ber. Bunsenges. Phys. Chem.*, **71**, 493 (1967).
268. T. B. Denisova and M. Kh. Kishinevskii, *Tr. Kishinev. Politekh. Inst.*, **5**, 13 (1966).
269. N. G. Chovnykh and V. V. Vaschenko, *Zh. Fiz. Khim.*, **35**, 580 (1961).
270. J. E. L. Bowcott and B. A. Plunkett, *Electrochim. Acta*, **14**, 883 (1969).
271. S. Clarke and J. A. Harrison, *J. Electroanal. Chem.*, **36**, 109 (1972).
272. B. P. Nesterov and N. B. Korovin, *Elektrokhimiya*, **2**, 1296 (1966).
273. J. A. Harrison and Z. A. Khan, *J. Electroanal. Chem.*, **26**, 1 (1970).
274. J. A. Harrison and Z. A. Khan, *J. Electroanal. Chem.*, **28**, 131 (1970).
275. M. Fleischmann, K. Korinek, and D. Pletcher, *J. Electroanal. Chem.*, **34**, 499 (1972).
276. U. Eisner and Y. Zemer, *J. Electroanal. Chem.*, **38**, 381 (1972).
277. M. Březina, J. Koryta, T. Loučka, D. Maršikova, and J. Pradec, *J. Electroanal. Chem.*, **40**, 13 (1972).
278. M. Breitenbach and K. H. Heckner, *J. Electroanal. Chem.*, **29**, 309 (1971).
279. W. H. Tiedemann and D. N. Bennion, *J. Electrochem. Soc.*, **117**, 203 (1970).
280. Z. Galus and R. N. Adams, *J. Electroanal. Chem.*, **4**, 248 (1962).
281. V. A. Kiryanov and V. Yu. Filinovskii, *Lectures on Polarography*, Kiev, 1965, p. 42 through ref. [4].
282. L. K. J. Tong, Kai Liang, and W. R. Ruby, *J. Electroanal. Chem.*, **13**, 245 (1967).
283. D. Möller and K. H. Heckner, *J. Electroanal. Chem.*, **38**, 337 (1972).
284. S. A. Kabakchi and V. Yu. Filinovskii, *Elektrokhimiya*, **8**, 1428 (1972).
285. R. Bonnaterre and G. Cauquis, *J. Electroanal. Chem.*, **32**, 199 (1971).
286. R. Bonnaterre and G. Cauquis, *J. Electroanal. Chem.*, **32**, 215 (1971).
287. Z. Samec, W. T. Bresnahan, and P. J. Elving, *J. Electroanal. Chem.*, **133**, 1 (1982).
288. R. H. Philip, Jr., *J. Electroanal. Chem.*, **27**, 369 (1970).
289. J. Koutecky and V. G. Levich, *Zh. Fiz. Khim.*, **32**, 1565 (1958).

290. D. Haberland and R. Landsberg, *Ber. Bunsenges. Phys. Chem.*, **70**, 724 (1966).
291. P. Beran and S. Bruckenstein, *J. Phys. Chem.*, **72**, 3630 (1968).
292. F. Opekar and P. Beran, *J. Electroanal. Chem.*, **32**, 49 (1971).
293. Yu. S. Miliavskii, *Elektrokhimiya*, **10**, 449 (1974).
294. J. Ulstrup, *Electrochim. Acta*, **13**, 1717 (1968).
295. K. Holub, *J. Electroanal. Chem.*, **30**, 71 (1971).
296. R. Bonnaterre and G. Cauquis, *J. Electroanal. Chem.*, **31**, App. 15–18 (1971).
297. M. Fleischmann, D. Pletcher, and A. Rafinski, *J. Electroanal. Chem.*, **38**, 323 (1972).
298. P. A. Malachesky, L. S. Marcoux, and R. N. Adams, *J. Phys. Chem.*, **70**, 4068 (1966).
299. S. Karp, *J. Phys. Chem.*, **72**, 1082 (1968).
300. V. Yu. Filinovskii, *Elektrokhimiya*, **5**, 635 (1969).
301. L. S. Marcoux, R. N. Adams, and S. W. Feldberg, *J. Phys. Chem.*, **73**, 2611 (1969).
302. L. S. Marcoux, J. M. Fritsch, and R. N. Adams, *J. Am. Chem. Soc.*, **89**, 5766 (1967).
303. G. Manning, V. D. Parker, and R. N. Adams, *J. Am. Chem. Soc.*, **91**, 4584 (1969).
304. R. R. Dogonadze, *Zh. Fiz. Khim.*, **32**, 2437 (1958).
305. W. Vielstich, *Z. Anal. Chem.*, **173**, 84 (1960).
306. W. Vielstich and D. Jahn, *Z. Elektrochem.*, **64**, 43, 129 (1960).
307. W. J. Albery and R. P. Bell, *Proc. Chem. Soc. London*, 169 (1963).
308. V. A. Kokorekina, L. G. Feoktistov, V. Yu. Filinovskii, and S. A. Sevelev, *Elektrokhimiya*, **7**, 1196 (1971).
309. W. J. Albery, M. D. Archer, N. J. Field, and A. D. Turner, *Discuss. Faraday Soc.*, **56**, 28 (1973).
310. W. J. Albery, M. D. Archer, and R. G. Egdell, *J. Electroanal. Chem.*, **82**, 199 (1977).
311. W. J. Albery, W. R. Bowen, F. S. Fischer, and A. D. Turner, *J. Electroanal. Chem.*, **107**, 1 (1980).
312. W. J. Albery, W. R. Bowen, F. S. Fisher, and A. D. Turner, *J. Electroanal. Chem.*, **107**, 11 (1980).
313. W. J. Albery, P. N. Bartlett, W. R. Bowen, F. S. Fisher, and A. W. Foulds, *J. Electroanal. Chem.*, **107**, 23 (1980).
314. B. Miller and S. Bruckenstein, *J. Electrochem. Soc.*, **121**, 1558 (1974).
315. B. Miller and S. Bruckenstein, *Anal. Chem.*, **46**, 2026 (1974).
316. B. Miller, M. I. Bellavance, and S. Bruckenstein, *Anal. Chem.*, **44**, 1983 (1972).
317. B. Miller and S. Bruckenstein, *J. Electrochem. Soc.*, **117**, 1032 (1970).
318. K. J. Kretschmer, C. H. Hamann, and B. Fassbender, *J. Electroanal. Chem.*, **60**, 231 (1975).
319. L. Müller and M. Westfahl, *Z. Phys. Chem.* (*Leipzig*), **257**, 145 (1976).
320. R. Bachrun, A. Suwono, and M. Daguenet, *Electrochim. Acta*, **25**, 1561 (1980).
321. R. Bachrun and M. Daguenet, *J. Electroanal. Chem.*, **124**, 53 (1981).
322. F. Vydra and M. Štuliková, *J. Electroanal. Chem.*, **40**, 99 (1972).
323. M. Štuliková and F. Vydra, *J. Electroanal. Chem.*, **42**, 127 (1973).
324. M. Kopanica and F. Vydra, *J. Electroanal. Chem.*, **31**, 175 (1971).
325. M. Štuliková and F. Vydra, *J. Electroanal. Chem.*, **38**, 349 (1972).
326. T. M. Florence, *J. Electroanal. Chem.*, **49**, 255 (1974).
327. T. M. Florence and Y. J. Farrar, *J. Electroanal. Chem.*, **51**, 191 (1974).
328. R. W. Andrews, J. H. Larochelle, and D. C. Johnson, *Anal. Chem.*, **48**, 212 (1976).
329. I. Gustavsson and J. Golimowski, *Sci. Total Environ.*, **22**, 85 (1981).
330. J. Golimowski, P. Valenta, M. Stoeppler, and H. W. Nürnberg, *Talanta*, **26**, 649 (1979).
331. J. Golimowski, P. Valenta, and H. W. Nürnberg, *Z. Lebensm. Unters, Forsch.*, **168**, 353 (1979).

332. L. Sipos, T. Magjer, and M. Branica, *Croat. Chem. Acta*, **46**, 35 (1974).
333. L. Sipos, J. Golimowski, P. Valenta, and H. W. Nürnberg, *Fresenius Z. Anal. Chem.*, **298**, 1 (1979).
334. L. Sipos, P. Valenta, H. W. Nürnberg, and M. Branica, *J. Electroanal. Chem.*, **77**, 263 (1977).
335. N. Ibl and K. Schdegg, *J. Electrochem. Soc.*, **114**, 54 (1967).
336. M. Saloma and M. Holtan, *Acta Chem. Scand.*, **28**, 93 (1974).
337. J. A. Harrison, R. P. J. Hill, and J. Thompson, *J. Electroanal. Chem.*, **47**, 431 (1973).
338. Z. Zembura and A. Fuliński, *Electrochim. Acta*, **10**, 859 (1965).
339. R. D. Armstrong and G. M. Bulmann, *J. Electroanal. Chem.*, **25**, 121 (1970).
340. H. G. Feller, *Corros. Sci.*, **8**, 259 (1968).
341. Z. Zembura, *J. Electroanal. Chem.*, **46**, 243 (1973).
342. K. Heusler, *Z. Elektrochem.*, **65**, 192 (1961).
343. Z. Zembura and A. Maraszewska, *Rocz. Chem.*, **47**, 1503 (1973).
344. K. Heiz, *Werkst. und Korros.*, **15**, 63 (1964).
345. Z. Zembura and L. Burzyńska, *Corros. Sci.*, **17**, 871 (1977).
346. Z. Zembura and J. Bugajski, *Corros. Sci.*, **21**, 69 (1981).
347. I. R. Krichevskii and Ju. V. Tsekhanskaya, *Zh. Fiz. Khim.*, **30**, 2315 (1956); **33**, 2331 (1959).
348. V. N. Boronenkov, O. A. Esin, P. M. Shurigin, and B. A. Kuhtin, *Elektrokhimiya*, **1**, 1245 (1965).
349. F. Vydra, *J. Electroanal. Chem.*, **25**, App. 13 (1970); F. Vydra, *Collect. Czech. Chem. Commun.*, **37**, 123 (1972).
350. F. Vydra and M. Štuliková, *Collect. Czech. Chem. Commun.*, **38**, 2441 (1973).
351. R. Landsberg and R. Thiele, *Electrochim. Acta*, **11**, 1243 (1966). F. Scheller, S. Müller, R. Landsberg, and H-J. Spitzer, *J. Electroanal. Chem.*, **19**, 187 (1968). F. Scheller, R. Landsberg, and S. Müller, *J. Electroanal. Chem.*, **20**, 375 (1969). F. Scheller, R. Landsberg, and H. Wolf, *Z. Phys. Chem. (Leipzig)*, **243**, 345 (1970). H. Wolf and R. Landsberg, *J. Electroanal. Chem.*, **28**, 295 (1970).
352. I. Fried and P. J. Elving, *Anal. Chem.*, **37**, 464, 803 (1965).
353. G. P. Girina, V. Yu. Filinovskii, and L. G. Feoktistov, *Elektrokhimiya*, **3**, 941 (1967).
354. P. C. Andricacos and H. Y. Cheh, *J. Electroanal. Chem.*, **124**, 95 (1981).
355. G. C. Quintana, P. C. Andricacos, and H. Y. Cheh, *J. Electroanal. Chem.*, **144**, 77 (1983).
356. Yu. G. Siver, *Zh. Fiz. Khim.*, **34**, 577 (1960).
357. J. M. Hale, *J. Electroanal. Chem.*, **6**, 187 (1963); **8**, 332 (1964).
358. V. Yu. Filinovskii and V. A. Kiryanov, *Dokl. Akad. Nauk SSSR*, **156**, 1412 (1964).
359. R. P. Buck and H. E. Keller, *Anal. Chem.*, **35**, 400 (1963).
360. P. C. Andricacos and H. Y. Cheh, *J. Electroanal. Chem.*, **121**, 133 (1981).
361. A. N. Frumkin and L. N. Nekrasov, *Dokl. Akad. Nauk SSSR*, **126**, 115 (1959).
362. Yu. B. Ivanov and V. G. Levich, *Dokl. Akad. Nauk SSSR*, **126**, 1029 (1959).
363. W. J. Albery and S. Bruckenstein, *Trans. Faraday Soc.*, **62**, 1920 (1966).
364. S. Bruckenstein and G. A. Feldman, *J. Electroanal. Chem.*, **9**, 395 (1965).
365. H. Matsuda, *J. Electroanal. Chem.*, **16**, 153 (1968).
366. K. Tokuda and H. Matsuda, *J. Electroanal. Chem.*, **44**, 199 (1973).
367. A. J. Bard and K. B. Prater, *J. Electrochem. Soc.*, **117**, 207 (1970).
368. W. J. Albery and J. S. Drury, *J. Chem. Soc. Faraday Trans.*, **68**, 456 (1972).
369. S. Bruckenstein, *Elektrokhimiya*, **2**, 1085 (1966).
370. A. N. Frumkin, L. N. Nekrasov, V. G. Levich, and Y. B. Ivanov, *J. Electroanal. Chem.*, **1**, 84 (1959).

371. W. J. Albery, M. L. Hitchman, and J. Ulstrup, *Trans. Faraday Soc.*, **64**, 2831 (1968).
372. J. Margarit, G. Dabosi, and M. Levy, *Bull. Soc. Chim. Fr.*, 2096 (1972).
373. D. T. Napp, D. C. Johnson, and S. Bruckenstein, *Anal. Chem.*, **39**, 481 (1967).
374. B. Miller and R. E. Visco, *J. Electrochem. Soc.*, **115**, 251 (1968).
375. B. Miller, *J. Electrochem. Soc.*, **116**, 1675 (1969).
376. B. Miller, *J. Electrochem. Soc.*, **116**, 1117 (1969).
377. G. K. Dikusar, *Elektrokhimiya*, **11**, 1411 (1975).
378. K. E. Heusler and H. Schurig, *Z. Phys. Chem.* (*Frankfurt am Main*), **47**, 117 (1965).
379. G. V. Zhutaeva, V. S. Bagotskii, and N. A. Shumilova, *Elektrokhimiya*, **7**, 1707 (1971).
380. I. V. Kadija and V. M. Nakić, *J. Electroanal. Chem.*, **34**, 15 (1972); **35**, 177 (1972).
381. V. Yu. Filinovskii, I. V. Kadija, and B. Zh. Nikolich, *Elektrokhimiya*, **10**, 297 (1974).
382. G. K. Dikusar, *Elektrokhimiya*, **11**, 1413 (1975).
383. H. Debrot and K. E. Heusler, *Ber. Bunsenges. Phys. Chem.*, **81**, 1172 (1977).
384. R. Dörr and E. W. Grabner, *Ber. Bunsenges. Phys. Chem.*, **82**, 164 (1978).
385. R. Memming, *Ber. Bunsenges. Phys. Chem.*, **81**, 732 (1977).
386. B. Cavalier, C. Dezal, and J. Jacq, *Bull. Soc. Chim. Fr.*, 3210 (1966).
387. P. Beran and F. Opekar, *Chem. Listy*, **68**, 305 (1974).
388. R. H. Sonner, B. Miller, and R. E. Visco, *Anal. Chem.*, **41**, 1498 (1969).
389. G. V. Zhutaeva and N. A. Shumilova, *Elektrokhimiya*, **2**, 606 (1966).
390. A. N. Doronin, *Elektrokhimiya*, **4**, 1193 (1969).
391. G. W. Harrington, H. A. Laitinen, and V. Trendafilov, *Anal. Chem.*, **45**, 433 (1973).
392. A. Damjanović, M. A. Genshaw, and J. O'M. Bockris, *J. Chem. Phys.*, **45**, 4057 (1966).
393. L. N. Nekrasov, *Elektrokhimiya*, **11**, 851 (1975); "The Rotating Ring–Disk Electrode Method as a Tool of Study of the Kinetics and Mechanism of Electrode Reactions," in Ja. P. Stradins and S. G. Majranovskii, Eds., *Polarography Problems and Perspectives*, Zinatne, Riga, 1977 (in Russian).
394. S. Bruckenstein and D. T. Napp, *J. Am. Chem. Soc.*, **90**, 6303 (1968).
395. S. Bruckenstein and G. S. Feldman, *J. Electroanal. Chem.*, **9**, 395 (1965).
396. D. C. Johnson and R. R. Gaines, *Anal. Chem.*, **45**, 1670 (1973).
397. G. Neubert and K. B. Prater, *J. Electrochem. Soc.*, **121**, 745 (1974).
398. V. J. Puglisi and A. J. Bard, *J. Electrochem. Soc.*, **120**, 748 (1973).
399. Lun Shu, R. Yeh, and A. J. Bard, *J. Electrochem. Soc.*, **12**, 189 (1977).
400. J-M. Nigretto and A. J. Bard, *J. Electrochem. Soc.*, **123**, 1303 (1976).
401. R. Allensworth, J. W. Rogers, G. Ridge, and A. J. Bard, *J. Electrochem. Soc.*, **121**, 1412 (1974).
402. L. N. Nekrasov, L. N. Vykhodtseva, A. P. Korotkov, and L. P. Yureva, *Elektrokhimiya*, **13**, 735 (1977).
403. Nguyen Suan and L. N. Nekrasov, *Elektrokhimiya*, **9**, 1362 (1974).
404. Nguyen Suan and L. N. Nekrasov, *Elektrokhimiya*, **9**, 1752 (1974).
405. B. G. Podlibner and L. N. Nekrasov, *Elektrokhimiya*, **6**, 1155, 1580 (1970); **7**, 379 (1971).
406. I. P. Ryvkina, L. N. Nekrasov, V. A. Petrosjan, and V. I. Slovetskij, *Dokl. Akad. Nauk SSSR*, **220**, 1339 (1975); **222**, 617 (1975).
407. A. D. Korsun and L. N. Nekrasov, *Elektrokhimiya*, **5**, 212, (1969).
408. L. N. Nekrasov, N. N. Nefedova, and A. D. Korsun, *Elektrokhimiya*, **5**, 889 (1969).
409. L. N. Nekrasov, I. P. Ryvkina, and B. G. Podlibner, *Elektrokhimiya*, **8**, 1404 (1972).
410. L. N. Nekrasov, and N. P. Berezina, *Dokl. Akad. Nauk SSSR*, **142**, 858 (1962).
411. V. V. Gorodetskii, N. B. Shchelkanova, J. G. Goncharova, and V. V. Losiev, *Elektrokhimiya*, **12**, 1255 (1976).
412. J. Eckert and W. Forker, *Z. Phys. Chem.* (*Leipzig*), **253**, 153 (1973).

413. A. I. Molodov, L. A. Yanov, and V. V. Losiev, *Elektrokhimiya*, **12**, 513 (1976).
414. A. I. Molodov, *Elektrokhimiya*, **13**, 1625 (1977).
415. A. I. Molodov and L. A. Yanov, *Elektrokhimiya*, **12**, 513 (1978).
416. L. Kiss and J. Farkas, *Acta Chim. Acad. Sci. Hung.*, **96**, 127 (1977).
417. L. Kiss, J. Farkas, P. Kovacs, and L. Kozari, *Acta Chim. Acad. Sci. Hung.*, **97**, 399 (1978).
418. L. Kiss, J. Farkas, and I. Matrai, *Acta Chim. Acad. Sci. Hung.*, **100**, 135 (1979).
419. P. Joó, J. Farkas, and L. Kiss, *Acta Chim. Acad. Sci. Hung.*, **112**, 433 (1983).
420. A. I. Molodov, G. N. Markosjan, L. I. Lakh, and V. V. Losiev, *Elektrokhimiya*, **14**, 522 (1978).
421. A. I. Molodov, L. A. Yanov, and V. V. Losiev, *Zashch. Met.*, **12**, 578 (1976).
422. A. I. Molodov and L. A. Yanov, *Zashch. Met.*, **14**, 194 (1978).
423. B. Miller and M. I. Bellavance, *J. Electrochem. Soc.*, **119**, 1510 (1972).
424. R. D. Armstrong and I. Baurhoo, *J. Electroanal. Chem.*, **34**, 41 (1972); **40**, 325 (1972).
425. R. D. Armstrong, J. A. Harrison, H. R. Thirsk, and R. Whitfield, *J. Electrochem. Soc.*, **117**, 1003 (1970).
426. R. D. Armstrong and M. Henderson, *J. Electroanal. Chem.*, **26**, 381 (1970).
427. R. D. Armstrong, M. Henderson, and H. R. Thirsk, *J. Electroanal. Chem.*, **35**, 119 (1972).
428. R. D. Cowling and H. F. Hintermann, *J. Electrochem. Soc.*, **118**, 1912 (1971).
429. A. Fujishima, F. Karasawa, and H. Honda, *J. Electroanal. Chem.*, **134**, 187 (1982).
430. B. G. Podlibner and L. N. Nekrasov, *Elektrokhimiya*, **5**, 340, (1969).
431. M. R. Tarasevich, K. A. Radushkhina, V. Ju. Filinovskii, and R. Kh. Burshtein, *Elektrokhimiya*, **6**, 1522 (1970).
432. M. A. Genshaw, A. Damjanovic, and J. O'M. Bockris, *J. Electroanal. Chem.*, **15**, 163 (1967).
433. L. N. Nekrasov and L. Mjuller, *Dokl. Akad. Nauk SSSR*, **149**, 1107 (1963).
434. A. Damjanovic, M. A. Genshaw, and J. O'M. Bockris, *J. Electrochem. Soc.*, **114**, 1107 (1967).
435. A. Damjanovic, M. A. Genshaw, and J. O'M. Bockris, *J. Phys. Chem.*, **70**, 3761 (1966).
436. L. Mjuller and L. N. Nekrasov, *Zh. Fiz. Khim.*, **38**, 3028 (1964).
437. L. N. Nekrasov and L. Mjuller, *Dokl. Akad. Nauk SSSR*, **157**, 416 (1964).
438. L. N. Nekrasov and T. K. Zolotova, *Elektrokhimiya*, **4**, 864 (1968).
439. M. R. Tarasevich, R. Kh. Burshtein, and K. A. Radushkhina, *Elektrokhimiya*, **6**, 372 (1970).
440. M. R. Tarasevich and K. A. Radushkhina, *Elektrokhimiya*, **6**, 376 (1970).
441. Gu Lin-in, N. A. Shumilova, and V. S. Bagotskii, *Electrokhimiya*, **3**, 460 (1967).
442. G. P. Samojlov, E. I. Khrushcheva, N. A. Shumilova, and V. S. Bagotskii, *Elektrokhimiya*, **6**, 1347 (1970).
443. L. N. Nekrasov, E. I. Khrushcheva, N. A. Shumilova, and M. R. Tarasevich, *Elektrokhimiya*, **2**, 363 (1966).
444. M. A. Genshaw, A. Damjanovic, and J. O'M. Bockris, *J. Phys. Chem.*, **71**, 3722 (1967).
445. K. A. Radushkhina, M. R. Tarasevic, and R. Kh. Burshtein, *Elektrokhimiya*, **6**, 1352 (1970).
446. M. R. Tarasevich, F. Z. Sabirov, A. P. Mertsalova, and R. Kh. Burshtein, *Elektrokhimiya*, **4**, 432 (1968).
447. M. R. Tarasevich and V. S. Vilinskaya, *Elektrokhimiya*, **8**, 1489 (1972).
448. N. I. Dubrovina and L. N. Nekrasov, *Elektrokhimiya*, **8**, 1503 (1972).

449. L. Mjuller and L. N. Nekrasov, *Dokl. Akad. Nauk SSSR*, **154**, 437 (1964).
450. A. Damjanovic, M. A. Genshaw, and J. O'M. Bockris, *J. Electrochem. Soc.*, **114**, 466 (1967).
451. L. N. Nekrasov, L. A. Dukhanova, N. I. Dubrovina, and L. N. Vykhodtseva, *Elektrokhimiya*, **6**, 388 (1970).
452. W. J. Albery and S. Bruckenstein, *Trans. Faraday Soc.*, **62**, 1946 (1966). W. J. Albery and S. Bruckenstein, *Trans. Faraday Soc.*, **62**, 2598 (1966).
453. W. J. Albery and S. Bruckenstein, *Trans. Faraday Soc.*, **62**, 2584 (1966).
454. W. J. Albery, M. L. Hitchman, and J. Ulstrup, *Trans. Faraday Soc.*, **65**, 1101 (1969).
455. K. B. Prater and A. J. Bard, *J. Electrochem. Soc.*, **117**, 335 (1970).
456. K. B. Prater and A. J. Bard, *J. Electrochem. Soc.*, **117**, 1517 (1970).
457. V. J. Puglisi and A. J. Bard, *J. Electrochem. Soc.*, **119**, 833 (1972).
458. V. J. Puglisi and A. J. Bard, *J. Electrochem. Soc.*, **119**, 829 (1972).
459. J. T. Maloy, K. B. Prater, and A. J. Bard, *J. Am. Chem. Soc.*, **93**, 5959 (1971).
460. J. T. Maloy and A. J. Bard, *J. Am. Chem. Soc.*, **93**, 5968 (1971).
461. W. J. Albery and S. Bruckenstein, *Trans. Faraday Soc.*, **62**, 1938 (1966).
462. H. Gerischer, I. Mattes, and K. Braun, *J. Electroanal. Chem.*, **10**, 553 (1965).
463. K. Tokuda and H. Matsuda, *J. Electroanal. Chem.*, **52**, 421 (1974).
464. K. Aoki, K. Tokuda, and H. Matsuda, *J. Electroanal. Chem.*, **79**, 49 (1977).
465. W. J. Albery, B. A. Coles, A. M. Couper, and K. M. Garnett, *J. Chem. Soc. Chem. Commun.*, 198 (1974).
466. W. J. Albery, B. A. Coles, and A. M. Couper, *J. Electroanal. Chem.*, **65**, 901 (1975).
467. W. J. Albery, A. T. Chadwick, B. A. Coles, and N. A. Hampson, *J. Electroanal. Chem.*, **75**, 229 (1977).
468. B. A. Coles and R. G. Compton, *J. Electroanal. Chem.*, **127**, 37 (1981).
469. R. E. Sioda and W. Kemula, *Electrochim. Acta*, **17**, 1171 (1972).
470. R. E. Sioda, *Electrochim. Acta*, **19**, 57 (1974); *Chem. Eng.*, February 21, 57 (1983).
471. G. M. Cook, *Chem. Eng.*, February 21, 59 (1983).
472. W. Kemula, *Rocz. Chem.*, **26**, 281 (1952).
473. K. Štulik and V. Pacáková, *J. Electroanal. Chem.*, **129**, 1 (1981).
474. W. Kemula and W. Kutner, "Amperometric Flow-Through Detection in Liquid Chromatography," in E. Pungor and I. Buzas, Eds., *Modern Trends in Analytical Chemistry*, Akademiai Kiado, Budapest, 1984.
475. K. Štulik and V. Pacáková, *J. Chromatogr.*, **208**, 269 (1981).
476. P. T. Kissinger, *Anal. Chem.*, **49**, 447A (1977).
477. H. B. Hanekamp, P. Bos, U. A. Th. Brinkman, and R. W. Frei, *Fresenius Z. Anal. Chem.*, **297**, 404 (1979).
478. H. B. Hanekamp and H. J. van Nieuwkerk, *Anal. Chim. Acta*, **121**, 13 (1980).
479. A. Kimla and F. Strafelda, *Collect. Czech. Chem. Commun.*, **29**, 2913 (1964).
480. Y. Okinaka and I. M. Kolthoff, *J. Electroanal. Chem.*, **73**, 3326 (1977).
481. W. J. Blaedel, C. J. Olson, and L. R. Sharma, *Anal. Chem.*, **35**, 2100 (1963).
482. H. Matsuda, *J. Electroanal. Chem.*, **15**, 325 (1967).
483. K. Brunt and C. H. P. Bruins, *J. Chromatogr.*, **172**, 37 (1979).
484. G. Wranglen and O. Nilsson, *Electrochim. Acta*, **7**, 121 (1962).
485. S. L. Marchiano and A. J. Arvia, *Electrochim. Acta*, **12**, 801 (1967).
486. H. Matsuda, *J. Electroanal. Chem.*, **15**, 109 (1967).
487. J. Yamada and H. Matsuda, *J. Electroanal. Chem.*, **44**, 189 (1973).
488. W. Kemula, B. Behr, K. Chlebicka, and D. Sybilska, *Rocz. Chem.*, **39**, 1315 (1965).
489. W. Kemula and W. Kutner, *J. Chromatogr.*, **204**, 131 (1981).

490. H. B. Hanekamp, "Polarographic Continuous-Flow Detection," unpublished doctoral dissertation, Free University of Amsterdam, The Netherlands, 1981.
491. R. J. Rucki, *Talanta*, **27**, 147 (1980).
492. E. Pungor, Z. Feher, and M. Váradi, *CRC Crit. Rev. Anal. Chem.*, **9**, 97 (1980).

Chapter **4**

CYCLIC VOLTAMMETRY, AC POLAROGRAPHY, AND RELATED TECHNIQUES

Eric R. Brown and James R. Sandifer

1   INTRODUCTION

Cyclic voltammetry and alternating-current polarography are electrochemical techniques involving control of the potential of an electrode of the zeroth (inert) or first (dissolving) kind while monitoring the resultant current flow as the solution adjacent to the electrode is electrolyzed. Potential control results in several benefits relative to more passive or static techniques such as potentiometry (Chapter 2). If one simply measures the potential of an electrode dipping into an electrolytic solution relative to a suitable reference electrode, then only one datum of information is provided. The potential indicates the thermodynamic equilibrium state involving every redox couple in the solution. If the potential of the electrode is forced to vary from its equilibrium value, electrons will be transported between the electrode and the adjacent solution in an effort to readjust the concentrations of the various oxidized and reduced species to new values dictated by the applied potential. The term "adjacent solution" confines the electrolysis to a region typically a few microns from the electrode surface. However, if the potential is held at a particular value long enough, the entire solution will be electrolyzed such that the concentrations of the various electroactive species will adjust themselves to a new state of thermodynamic equilibrium characteristic of the applied potential. The transport of electrons between the electrode and the electrolytic solution constitutes a flow of current, which is measured in the external circuit used to control the potential in the first place. The current in this dynamic technique is proportional to the concentration of those electroactive species that are most perturbed at the applied potential.

Potential-control experiments therefore provide analytically useful information in the magnitude of the currents, which are proportional to concentrations, and in the potentials at which the currents flow, which are indicative of the formal redox potentials of the electroactive species involved.

1.1 Cyclic Voltammetry

1.1.1 Description

An experiment may be performed in which the potential of the electrode is varied linearly with time, linear sweep voltammetry, or, if the electrode is a dropping mercury electrode, polarography (Chapter 9), and the current is monitored at each point along the potential/time axis. If the rate of the potential sweep is extremely slow, or if the solution is confined to a very thin layer, then complete electrolysis of the sample can be accomplished during the time required for each infinitesimally small potential increment. The varying potential will cause the redox couples throughout the solution to vary through a range of thermodynamic states of equilibrium. If the sweep rate is increased, or if the solution is very thick, only the electroactive species near the surface of the electrode will be affected by the changing potential.

At potentials more than about 120 mV beyond the formal redox potential of a particular couple, that couple will convert itself almost entirely into one form or the other (oxidized or reduced) by accepting electrons from or donating electrons to the electrode. The resultant current then reaches a maximum, which depends upon the scan rate, as will be discussed in greater detail later. As the potential is scanned beyond the 120 mV range, the current decreases because additional electroactive species cannot diffuse into the region adjacent to the electrode rapidly enough to maintain the high current level that was previously established by the electroactive species already there. Many redox systems include several couples and their associated maxima. Ruthenium(III) bipyridyl, for example, can be reduced successively to three different products at three different potentials, each resulting in a maximum in the voltammogram caused by one-electron transfer processes.

Cyclic voltammetry takes the linear-scan voltammetric experiment described previously one step further. At a suitable potential beyond a given current maximum the direction of potential scan is reversed or "cycled." A second current maximum of opposite sign results, which corresponds to reelectrolysis of products formed before the scan reversal, as illustrated in Figure 4.1. The characteristics of this maximum give cyclic voltammetry one of its major advantages. If a homogeneous chemical reaction consumes the product formed during the initial sweep *within the time necessary to reverse the scan*, the maximum of opposite sign will be totally absent. The size, shape, and position of these maxima, as functions of scan rate and potential at scan reversal, can all serve as diagnostic criteria concerning the nature of the homogeneous chemical reaction.

The chemical reactions that occur in the solution adjacent to the electrode may obey any one of the countless kinetic schemes available in nature. Furthermore, they may occur between successive charge-transfer reactions at the electrode—so-called "ECE"—mechanisms. Cyclic voltammetry provides a powerful set of diagnostic criteria for the elucidation of diverse kinetic schemes. Expected responses may be calculated by mathematically combining Fick's laws of diffusion, the Nernst equation, and the kinetic description (rate and equilibrium constants) of the homogeneous reactions. Many of these responses are reviewed here along with various procedures for deriving others, either numerically or analytically.

1.1.2   Complications

It must be realized that the nature of the charge-transfer process at the electrode may limit or at least complicate the use of cyclic voltammetry as a

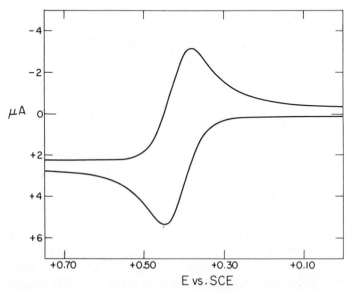

Figure 4.1 Reversible cyclic voltammetric curve. Experimental conditions: 1 mM ferrocene in acetonitrile, 0.1 M tetrabutylammonium perchlorate, Pt electrode, scan rate 10 V/min.

diagnostic tool. Basically, three problems associated with charge transfer arise. First, the Nernst equation, based on thermodynamic considerations, applies rigorously *only at zero current*. Surface concentrations may not obey the Nernst equation if there is a net flow of current. The electrode reaction is then said to be "irreversible" in an electrochemical sense that is analogous to thermodynamically irreversible processes. That is, the maximum useful work involved in the process is less than the decrease in Gibbs free energy.

Second, the solution adjacent to the electrode consists of charged ions (supporting electrolyte) and dipolar solvent molecules, which may orient and distribute themselves about the electrode in response to its potential and resultant electric field. Since these species are not discharged (oxidized or reduced) by the electrode, a net electrical charge on the solution side of the electrode–solution interface occurs that is exactly equal to the charge on the electrode but of opposite sign. This electrified interface serves as a potential-dependent capacitance, which will store charge in response to the varying potential applied during the cyclic voltammetry experiment. A current, not dependent upon the electrolysis of the electroactive species, flows to charge the capacitance. There are other examples of such currents and they are collectively referred to as "nonfaradaic."

Third, reactants or products of the charge-transfer (faradaic) reaction may adsorb to the electrode. Adsorption circumvents the diffusion process, thus causing massive currents that flow for short durations. The adsorbed species is thermodynamically different from the free species, causing a shift in the potential corresponding to the current maximum of the cyclic voltammogram. In addition, the adsorbed layer is both resistive, because it blocks the normal current flow of diffusing species, and capacitive, because it acts as a dielectric at the charged interface.

Complications arising from these various kinetic and electrodic effects can cause the cyclic voltammograms to be virtually uninterpretable. Mathematical descriptions of the current response may not be derivable as analytically useful equations, although numerical methods may still be used. Equations that can be derived often contain terms that cannot be evaluated, such as the value of the double-layer capacitance. It is then often convenient to resort to more specific techniques, which can discriminate between various effects on the basis of some parameter other than current, voltage, or time, the only parameters considered until now.

1.2   Ac Voltammetry

1.2.1   Description

One such technique is ac voltammetry (or polarography if a dropping mercury electrode is employed), in which a small-amplitude (< 10 mV) alternating voltage is added to the ramp excitation signal described above. The small-amplitude signal serves to "linearize" and therefore simplify the response of the system, whereas the large-amplitude ramp controls the ratio of oxidized-to-reduced species at the surface of the electrode. The alternating-current response represents the rate of interconversion between oxidized and reduced species at the applied frequency, which is high compared with the changing dc ramp voltage. Only species immediately adjacent to the electrode can respond on this time scale. Concentrations of electroactive species farther out into the bulk, which are responding to the dc signal, do not have time to diffuse to the surface of the electrode within the period of the alternating voltage. Rectification of the alternating current, followed by filtration, results in a direct current proportional to the alternating current. An example of an ac polarogram with the effect of drop growth is shown in Figure 4.2.

In essence, the time dependence of the system response has been converted to frequency dependence with the advantage that transients no longer need to be measured. Instead of measuring a series of events occurring within a given

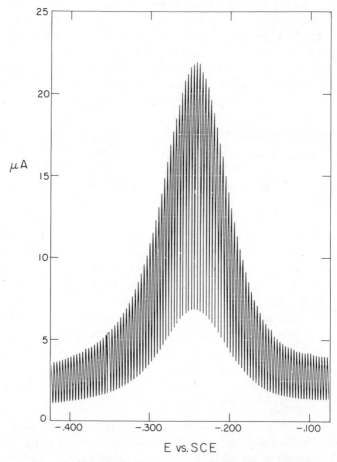

Figure 4.2   Reversible ac polarogram. Experimental conditions: 1 mM Fe(III) oxalate in 0.5 M K$_2$C$_2$O$_4$, DME electrode, applied potential is 10 mV p-p at 1.50 Hz, dc potential scan 60 mV/min.

interval of time, we are now measuring the repetitive occurrence of a single event at a specific frequency.

Frequency is a powerful variable. Faradaic or charge-transfer currents increase with the square root of frequency and for this reason high frequencies are preferable for analytical work. However, nonfaradaic charging currents increase directly with frequency. Therefore, one must not employ frequencies that are so high that the nonfaradaic processes become predominant. A time-domain analogy to this situation can be found in pulse polarography (Chapter 9), in which current measurements are made at some specific time after application of a voltage pulse (but still within the duration of the pulse). Nonfaradaic currents that decay rapidly are no longer flowing, but faradaic currents of analytical and mechanistic interest have not had a chance to decay.

In ac voltammetry, the measured quantities of interest are the amplitude of the sinusoidal current response, which occurs a specific number of times (cycles) per second, and the phase shift between the applied signal and the resultant current. In general, this current is not in phase with the applied voltage. In other words, maxima in the current response do not occur simultaneously with maxima in the voltage excitation. Currents that flow through capacitors, for example, lead the applied voltages that drive them by 90°. An applied sine-wave voltage, therefore, results in a cosine-wave current response. (The cosine leads the sine by 90° because the cosine of an angle equals the sine of that angle plus 90°.) A simple resistor provides current in phase with the excitation voltage. This phase difference (phase angle) between response and excitation provides an important parameter to separate the various faradaic and nonfaradaic contributions to the measured response. Sinusoidal-voltage excitation applied to electrochemical cells results in currents that may be represented as the sum of a sine wave and a cosine wave. The sine-wave part of the response is in phase with the sine-wave excitation and its amplitude is said to be the "real," or resistive, component of the current. The amplitude of the cosine wave, leading the sine-wave excitation by 90°, is said to be the "quadrature," or capacitive, component of the current, since 90° is one-fourth of a circle. This component is also referred to as the "imaginary" component because the total response may be represented alternatively as a complex number. The coefficient of the imaginary part of that number is the amplitude of the cosine. Resolution of the total response into real and quadrature components can be readily achieved instrumentally.

These components have great value because they serve to separate the current into resistive, capacitive, and diffusive components on the following basis: Current flow in a resistor is not shifted in phase and therefore appears in the amplitude of the sine wave, making no contribution to the amplitude of the cosine wave. Current flowing through a capacitor appears only in the amplitude of the cosine wave. Diffusion currents contain equal resistive and capacitive components at all frequencies, causing a 45° phase shift and making equal contribution to both waves. The phase shift and its dependence upon frequency serve to sort out these three contributions to the current.

1.2.2 Admittance Techniques

These methods involve the measurement of real and quadrature alternating currents driven by a small-amplitude alternating voltage covering several decades of frequencies at a constant dc potential. The admittance is the current divided by the voltage. Impedance techniques involve the measurement of real and quadrature alternating voltages driven by small-amplitude alternating currents. The impedance is the voltage divided by the current. Admittance and impedance are therefore reciprocals of each other, and it is immaterial which way the data are collected. Ac voltammetry is the special case of an admittance measurement made at a single frequency, without regard to phase, as the dc potential is varied linearly. This technique may be extended by monitoring only the inphase component of the current, eliminating some of the nonfaradaic contribution to the current response. A complete admittance (impedance) characterization would involve phase and amplitude measurements at all frequencies and at all applied voltages.

1.2.3 Nonlinear Response

That part of the electrochemical cell that consists of the electrode, the interface, and the adjacent solution is a nonlinear circuit element. Figure 4.1 shows this clearly; the response current does not vary linearly with the applied voltage over an extended voltage range. However, for a small voltage range at any point on the curve an approximation of linearity can be made. Thus impedance measurements, which employ small-amplitude excitations, result in more nearly linear responses than are obtained from large-amplitude techniques such as cyclic voltammetry. As a result, many of the processes that contribute to the impedances of electrochemical cells can be represented simply by arrangements of linear elements such as resistors and capacitors in analogous electrical networks referred to as "equivalent circuits." Such an equivalent-circuit model of electrochemical cells is extremely useful, since complete mathematical descriptions of total response can be rather complicated. Diffusive (faradaic) processes are represented in such circuits by an infinite series of resistors and capacitors called a transmission line, the impedance of which follows a square root of frequency dependence.

Faradaic processes are somewhat nonlinear even at small-amplitude applied alternating voltages. In other words, the current response is slightly distorted from a true sine wave. This distortion or nonlinearity results in current components that occur at frequencies higher than the applied voltage. These frequencies are integer multiples or harmonics of the fundamental applied frequency and provide another means of separating faradaic from nonfaradaic currents. The nonlinear faradaic impedance elements respond at harmonic frequencies whereas the linear nonfaradaic elements do not. In principle, if one monitors only the second harmonic frequency during an ac voltammetry experiment, one will see the contributions of only the faradaic processes. The experimental difficulty with this technique is that the second harmonic signals are generally rather small. Increasing the amplitude of the applied fundamental excitation voltage increases the higher harmonic responses because the non-linearity of the cell impedance is magnified at larger signal levels. In fact, the second harmonic signal increases as the square of the applied voltage, the third harmonic increases as the cube, and so on.

1.3   Model of Electrochemical Reactions

Understanding a process as complex as a series of reactions at an electrode surface requires some model of the process. Two alternatives are apparent in selecting a useful framework from which experimental data can be rationalized. The first of these is a mathematical model complete insofar as experimental capabilities allow its authenticity to be verified. To those well schooled in mathematical operations and their significance, this is perhaps the only model of real meaning. However, to the majority of chemists, such an approach leaves room for vast misunderstanding. Thus an intuitive model, which affords ready incorporation of new concepts, might be useful. Such a model was first presented in detail by Vlček [1, 2] in describing the nature of charge-transfer reactions of coordination compounds. However, this model is general enough to be valid for all types of charge-transfer reactions. It consists of seven sequential steps:

1. The reactant is brought to the vicinity of the electrode by mass transfer.
2. A chemical reaction occurs, which yields the species actually entering the inner part of the electrical double layer.
3. A structural rearrangement occurs yielding the species that takes part in the actual electron-transfer reaction.
4. The actual electron transfer takes place.
5. A structural reorganization occurs leading to the immediate product of charge transfer.
6. A chemical reaction of the immediate product of charge transfer occurs, which yields the species stable in solution.
7. The final product of charge transfer is removed to the bulk of the solution by mass transfer.

2   FUNDAMENTAL MATHEMATICAL CONCEPTS

2.1   Mass Transport

Consider an electrochemical cell in which the only mode of mass transport is diffusion to a planar electrode and in which there are no homogeneous kinetic complications. We begin with Fick's Second Law

$$\frac{\partial C_{Ox}}{\partial t} = D_{Ox} \frac{\partial^2 C_{Ox}}{\partial x^2} \tag{1}$$

where C_{Ox} is the time- and distance-dependent concentration of the oxidized species, D_{Ox} is its diffusion coefficient, x is distance, and t is time. Equations of this form can be dealt with far more easily in the "Laplace plane" than in the time domain, as will be demonstrated shortly. Transformation of (1) into the Laplace plane simply involves the integration [3]

$$\mathscr{L}(C_{Ox}) \equiv \bar{C}_{Ox} \equiv \int_0^\infty e^{-pt} C_{Ox} dt \tag{2}$$

where the transformation has occurred from the time domain to the Laplace (p) domain. Laplace transforms of many functions can be found in tables. They will be indicated in this chapter by writing a bar above the function. Transformation of (1) requires the Laplace transform of a derivative. It is a property of Laplace transforms that

$$\mathscr{L}\left(\frac{\partial C_{Ox}}{\partial t}\right) \equiv p\bar{C}_{Ox} - C_{Ox}^* \tag{3}$$

where C_{Ox}^* is the initial ($t=0$) concentration of oxidized species at the surface of the electrode and also its time-independent concentration in the bulk of the solution. Superscript $*$ will always indicate bulk concentrations in this chapter. Equation (3) follows from substitution of $\partial C_{Ox}/\partial t$ into (2), followed by integration by parts.

The Laplace transform of (1) is then the linear differential equation

$$p\bar{C}_{Ox} - C_{Ox}^* = D_{Ox} \frac{\partial^2 \bar{C}_{Ox}}{\partial x^2} \tag{4}$$

which can be integrated to the particular solution

$$\bar{C}_{Ox} = K_1 e^{\sqrt{p/D_{Ox}}x} + K_2 e^{-\sqrt{p/D_{Ox}}x} + \frac{C_{Ox}^*}{p} \tag{5}$$

A general solution can be found by using two boundary (distance) conditions to evaluate the independent parameters K_1 and K_2. One boundary condition is common to all the techniques considered in this chapter. The concentration of oxidized species far from the surface of the electrode does not vary with time— the semi-infinite, linear diffusion approximation. If $x=0$ corresponds to the surface of the electrode, it follows that $\bar{C}_{Ox} \to \bar{C}_{Ox}^*$ as $x \to \infty$. K_1 must therefore be equal to zero to prevent $\bar{C}_{Ox} \to \infty$. The Laplace transform of a constant is the constant divided by the Laplace variable p. Therefore, (5) becomes

$$\bar{C}_{Ox} = \bar{C}_{Ox}^*(1 + K_2 e^{-\sqrt{p/D_{Ox}}x}) \tag{6}$$

It is not our intention to measure the distance dependence of the concentration, as described by (6), but rather the current dependence upon applied potential. The relationship between current and concentration is given by Fick's First Law

$$i = +nFD_{Ox} \left.\frac{\partial C_{Ox}}{\partial x}\right|_{x=0} = -nFD_{Red} \left.\frac{\partial C_{Red}}{\partial x}\right|_{x=0} \tag{7}$$

which expresses the dependence of the flux of oxidized or reduced species (in moles/cm$^2 \cdot$s) at the surface of the electrode upon their concentration gradients. The number of equivalents of electrons transferred at the electrode per mole of species consumed, n, and F, the Faraday constant (coulombs per equivalent), have been included in the equation to convert the flux from moles/cm$^2 \cdot$s to coulombs/cm$^2 \cdot$s, which is the current density. The plus sign associated with the oxidized species indicates that it is diffusing toward the electrode

$$\left(\frac{\partial C_{Ox}}{\partial x} \bigg|_{x=0} \text{ is positive} \right)$$

while the reduced species is diffusing away

$$\left(\frac{\partial C_{Red}}{\partial x} \bigg|_{x=0} \text{ is negative} \right)$$

The Laplace transform of (7) is given by (8)

$$\bar{i} = +nFD_{Ox} \frac{\partial \bar{C}_{Ox}}{\partial x} \bigg|_{x=0} = -nFD_{Red} \frac{\partial \bar{C}_{Red}}{\partial x} \bigg|_{x=0} \tag{8}$$

Equation (6) may be evaluated at $x=0$ to yield

$$K_2 = \frac{\bar{C}_{Ox} - \bar{C}_{Ox}^*}{\bar{C}_{Ox}^*} = \frac{\overline{\Delta C_{Ox}}}{\bar{C}_{Ox}^*} \tag{9}$$

where $\overline{\Delta C_{Ox}}$ is the Laplace transform of the difference in concentration between species at the electrode surface and in the bulk of the solution. Likewise, it is the difference between the concentrations at the electrode surface before and after application of a perturbing potential. The derivative of (6) with respect to distance, evaluated at $x=0$, is

$$\frac{\partial \bar{C}_{Ox}}{\partial x} \bigg|_{x=0} = -\bar{C}_{Ox}^* K_2 \sqrt{\frac{p}{D_{Ox}}} \tag{10}$$

Equations (8), (9), and (10) may be combined to yield the expression

$$\bar{i} = -nF\sqrt{D_{Ox}}\{\sqrt{p}\}\{\overline{\Delta C_{Ox}}\} \tag{11}$$

and an equivalent expression written in terms of the reduced species

$$\bar{i} = nF\sqrt{D_{Red}}\{\sqrt{p}\}\{\overline{\Delta C_{Red}}\} \tag{12}$$

Equations (11) and (12) have opposite signs because, as considered here, the oxidized and reduced species diffuse in opposite directions with respect to the surface of the electrode, as formerly expressed in (7). These equations actually give the relationship between current and concentration at the electrode surface. They are of the form

$$\text{Response} = \text{System} \times \text{Excitation} \tag{13}$$

and were rigorously derived, but only for the case of diffusional mass transport with rapid electron transfer and no kinetic complications. The simple form of (13) *in the Laplace plane* is the justification for having transformed (1) in the first place. As will be shown later, transformation back into the time domain will result in rather complicated expressions. Expressions like (13) are generally applicable when kinetic complications arise except that different kinetic systems will have different system transforms, $\sqrt{p}$ in this case. Many system transforms have been tabulated and can be found in the literature [4–6]. We will return to the question of coupled homogeneous kinetics later. Equation (11) is often written

$$\overline{\Delta C_{Ox}} = -\frac{1}{nF\sqrt{D_{Ox}}}\{\bar{i}\}\left\{\frac{1}{\sqrt{p}}\right\} \tag{14}$$

which is composed of the product of two Laplace transforms. A property of Laplace transforms is that the inverse of such products is given by the convolution integral [3],

$$\mathscr{L}^{-1}\{f(p)g(p)\} = F(t) * G(t) = \int_0^t F(t-\tau)G(\tau)d\tau \tag{15}$$

In this equation f and g are the two Laplace transforms, F and G are their inverses, $*$ represents the convolution operation and τ is a dummy variable. Application of (15) to (14) results in

$$\Delta C_{Ox} = C_{Ox} - C_{Ox}^* = -\frac{1}{nF\sqrt{D_{Ox}}\sqrt{\pi}}\int_0^t \frac{i(\tau)}{(t-\tau)^{1/2}}\,d\tau \tag{16}$$

The analogous expression for the reduced species (12) is

$$\Delta C_{Red} = C_{Red} - C_{Red}^* = \frac{1}{nF\sqrt{D_{Red}}\sqrt{\pi}}\int_0^t \frac{i(\tau)}{(t-\tau)^{1/2}}\,d\tau \tag{17}$$

2.2   Coupled Chemical Reactions

As stated before, solution chemical reactions cause the system transforms to be different from (11) and (12). These transforms may be derived, as shown by the following example.

Assume that one of the redox species is in equilibrium with some other species that is inert with respect to charge-transfer reactions, with the electrode *at the potential of interest*. The equilibrium would typically be an acid–base equilibrium in a buffered solution (7). This kinetic situation can be expressed as

$$Y \underset{k_b}{\overset{k_f}{\rightleftharpoons}} Ox \underset{-e^-}{\overset{+e^-}{\rightleftharpoons}} Red \tag{18}$$

where Ox is the oxidized form of the redox couple and Red is the reduced form. The inert species is Y, the conjugate base of the acid Ox, and k_f and k_b are forward and backward rate constants. The time-dependent variations in the concentration of Y, Ox, and Red are given by the system of equations

$$\frac{\partial C_Y}{\partial t} = D_{Ox}\frac{\partial^2 C_Y}{\partial x^2} - k_f C_Y + k_b C_{Ox} \tag{19}$$

$$\frac{\partial C_{Ox}}{\partial t} = D_{Ox}\frac{\partial^2 C_{Ox}}{\partial x^2} + k_f C_Y - k_b C_{Ox} \tag{20}$$

$$\frac{\partial C_{Red}}{\partial t} = D_{Red}\frac{\partial^2 C_{Red}}{\partial x^2} \tag{21}$$

Notice that Y has been given the same diffusion coefficient as Ox (D_{Ox}), whereas the diffusion coefficient of Red (D_{Red}) is different. This is a necessary assumption to simplify the derivation and is generally reasonably good since Y and Ox are closely related in structure.

As before, the derivation progresses by taking Laplace transforms of these equations

$$p\bar{C}_Y - C_Y^* = D_{Ox}\frac{\partial^2 \bar{C}_Y}{\partial x^2} - k_f\bar{C}_Y + k_b\bar{C}_{Ox} \tag{22}$$

$$p\bar{C}_{Ox} - C_{Ox}^* = D_{Ox}\frac{\partial^2 \bar{C}_{Ox}}{\partial x^2} + k_f\bar{C}_Y - k_b\bar{C}_{Ox} \tag{23}$$

$$p\bar{C}_{Red} - C_{Red}^* = D_{Red}\frac{\partial^2 \bar{C}_{Red}}{\partial x^2} \tag{24}$$

Equation (24) can be integrated to a general solution by using the boundary condition expressed by (8)

$$\overline{\Delta C}_{Red} = \frac{\bar{i}}{nF\sqrt{D_{Red}}}\left\{\frac{1}{\sqrt{p}}\right\} \tag{25}$$

The result is the same as that found earlier, (12). Equations (22) and (23) each contain two dependent variables and cannot, therefore, be integrated separately. We proceed by taking linear combinations of these equations. The sum of (22) and (23) is

$$p(\bar{C}_{Ox} + \bar{C}_Y) - (C_{Ox}^* + C_Y^*) = D_{Ox}\frac{\partial^2(\bar{C}_{Ox} + \bar{C}_Y)}{\partial x^2} \tag{26}$$

Since Y does not undergo charge transfer at the surface of the electrode, it does not contribute to the current. It then follows that

$$\frac{\partial \bar{C}_Y}{\partial x}\bigg|_{x=0} = 0$$

and (8) can be used to find a general solution to (26)

$$\Delta(\bar{C}_{Ox} + \bar{C}_Y) = -\frac{\bar{i}}{nF\sqrt{D_{Ox}}}\left\{\frac{1}{\sqrt{p}}\right\} \tag{27}$$

Now multiply (22) by k_f and (23) by k_b. The difference between the two resulting expressions is

$$p(k_f\bar{C}_Y - k_b\bar{C}_{Ox}) - (k_f C_Y^* - k_b C_{Ox}^*) = D_{Ox}\frac{\partial^2(k_f\bar{C}_Y - k_b\bar{C}_{Ox})}{\partial x^2}$$

$$-(k_f + k_b)(k_f\bar{C}_Y - k_b\bar{C}_{Ox}) \tag{28}$$

Notice that the second term on the left-hand side of (28) is zero if the system is at equilibrium initially. Equation (28) may be integrated to a general solution

$$\Delta(k_f\bar{C}_Y - k_b\bar{C}_{Ox}) = \frac{\bar{i}}{nF\sqrt{D_{Ox}}}\left(\frac{k_b}{\sqrt{p + k_f + k_b}}\right) \tag{29}$$

by again employing the boundary condition expressed by (8). If (27) is multiplied by k_f, then (30a) results when (29) is subtracted

$$\overline{\Delta C_{Ox}} = -\frac{\overline{i}}{nF\sqrt{D_{Ox}}}\left\{\left(\frac{k_f}{k_f+k_b}\right)\frac{1}{\sqrt{p}}+\left(\frac{k_b}{k_f+k_b}\right)\frac{1}{\sqrt{p+k_f+k_b}}\right\} \qquad (30a)$$

Equation (30a) can be rewritten in terms of the equilibrium constant, $K = k_f/k_b$,

$$\overline{\Delta C_{Ox}} = -\frac{i}{nF\sqrt{D_{Ox}}}\left\{\left(\frac{K}{1+K}\right)\frac{1}{\sqrt{p}}+\frac{1}{1+K}\frac{1}{\sqrt{p+k}}\right\} \qquad (30b)$$

where $k = k_f + k_b$.

The system transform of the reaction (18) is then given by (25) and (30), for the responses of the concentrations of the reduced and oxidized species, respectively.

2.3   Relationship Between Concentration and Potential

Notice that the excitation transforms in (11), (12), (25), and (30) are in terms of $\overline{\Delta C_{Ox}}$ and $\overline{\Delta C_{Red}}$. Thus the equations are not written from the viewpoint of the experimenter, but rather from the point of view of the species in solution. Current will flow in response to changes in concentration at the surface of the electrode. As far as the experimenter is concerned, however, the excitation is the applied voltage. One must then be concerned with the relationship between surface concentration changes and applied voltage. This relationship can be expressed by (31) [5, 8]

$$i = nFAk_s[C_{Ox}e^{-(\alpha nF/RT)(E-E^0)} - C_{Red}e^{((1-\alpha)nF/RT)(E-E^0)}] \qquad (31)$$

where E is the time-dependent, applied potential and E^0 is the standard redox potential. Derivation of (31) is dependent upon the assumption that charge transfer between the electrode and the oxidized or reduced species occurs through an energy barrier whose height can be altered by the applied potential. The term k_s is the "heterogeneous charge-transfer rate constant," and although it has the units of velocity, it is nevertheless a measure of the height of this barrier at the standard redox potential. The symmetry of the barrier, with respect to the reaction coordinate, is indicated by the value of α, which is called the "transfer coefficient." If $\alpha = \frac{1}{2}$, the barrier is symmetric, and variations in applied potential can lower the barrier with equal ease for both oxidation and reduction. If $\alpha < \frac{1}{2}$, the barrier can be lowered more readily when $E > E^0$, and oxidation is favored. If $\alpha > \frac{1}{2}$, the barrier can be lowered more readily when $E < E^0$, and reduction is favored. In the limit that $i/k_s \to 0$, either because $i = 0$ or because $k_s \to \infty$, (31) reduces to the Nernst equation

$$E = E^0 + \frac{RT}{nF}\ln\left(\frac{C_{Ox}}{C_{Red}}\right) \qquad (32)$$

Equations (11) and (12) with $\sqrt{p}$ replaced by the appropriate system transform, (25) and (30) if homogeneous kinetics are involved, and (31) are central to virtually every electrochemical technique in which an applied current or an applied voltage is involved. These equations can be combined with appropriate time-dependent voltage expressions, E, to derive expected responses.

As an important example, consider what happens when $E \ll E^0$, in (32). Now C_{Ox}, the concentration of the oxidized species at the surface of the electrode, must approach zero. Then $\overline{\Delta C_{Ox}} = \overline{C}_{Ox} - \overline{C}_{Ox}^*$ becomes $-\overline{C}_{Ox}^*$, the initial or bulk concentration. Equation (11) then becomes

$$\overline{i} = nF\sqrt{D_{Ox}}\sqrt{p}\,\overline{C}_{Ox}^* \tag{33}$$

Since C_{Ox}^* is a constant, $\overline{C}_{Ox}^* = C_{Ox}^*/\sqrt{p}$ and (33) becomes

$$\overline{i} = nF\sqrt{D_{Ox}}\,C_{Ox}^*/\sqrt{p} \tag{34}$$

On inverse Laplace transformation,

$$i = nFC_{Ox}^*\sqrt{D_{Ox}}/\sqrt{\pi t} \tag{35}$$

This equation is referred to as the "Cottrell" equation and shows that purely diffusional currents obey $t^{-1/2}$ behavior. E has been made so large that all charge-transfer effects (electrode to species) have been overcome and the current flows under diffusion control. The applied potential is then said to be on the "diffusion plateau" of a voltammogram measured under stirred conditions.

2.4   Half-Wave Potential

The Nernst equation can be written in an experimentally more convenient form by defining the "half-wave" potential, $E_{1/2}$, and referencing E to this value rather than to E^0. At the half-wave potential, the total current flow in the cell is one-half the sum of the maximum possible anodic and cathodic faradaic currents. These are the diffusion-limited currents measured in voltammetry and polarography. Consequently, $E_{1/2}$ may be determined as shown in Figure 4.3.

Furthermore, the surface concentrations, C_{Ox} and C_{Red}, are one-half their maximum possible values at $E_{1/2}$. They may be calculated from (11) and (12) in the following way. These equations are combined and inversely transformed giving

$$\sqrt{D_{Ox}}(C_{Ox} - C_{Ox}^*) = -\sqrt{D_{Red}}(C_{Red} - C_{Red}^*) \tag{36}$$

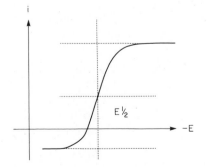

Figure 4.3   Example of measurement of half-wave potential.

The maximum cathodic current flows when $C_{Ox}=0$, and (36) becomes

$$(\sqrt{D_{Red}}C_{Red})_{max\,cat} = \sqrt{D_{Ox}}C^*_{Ox} + \sqrt{D_{Red}}C^*_{Red} \qquad (37)$$

At the maximum anodic current $C_{Red}=0$, and (36) becomes

$$(\sqrt{D_{Ox}}C_{Ox})_{max\,an} = \sqrt{D_{Ox}}C^*_{Ox} + \sqrt{D_{Red}}C^*_{Red} \qquad (38)$$

At the half-wave potential the surface concentrations are halfway between the two extremes expressed by (37) and (38). Therefore,

$$(\sqrt{D_{Ox}}C_{Ox})_{E_{1/2}} = [0 + (\sqrt{D_{Ox}}C_{Ox})_{max\,an}]/2$$
$$= [\sqrt{D_{Ox}}C^*_{Ox} + \sqrt{D_{Red}}C^*_{Red}]/2 \qquad (39)$$

and

$$(\sqrt{D_{Red}}C_{Red})_{E_{1/2}} = [(\sqrt{D_{Red}}C_{Red})_{max\,cat} + 0]/2$$
$$= [\sqrt{D_{Ox}}C^*_{Ox} + \sqrt{D_{Red}}C^*_{Red}]/2 \qquad (40)$$

The ratio of C_{Ox} to C_{Red} at $E_{1/2}$ is then equal to $\sqrt{D_{Red}}/\sqrt{D_{Ox}}$. Substitution of these parameters into the Nernst equation leads to

$$E_{1/2} = E^0 + \frac{RT}{nF} \ln\left(\frac{\sqrt{D_{Red}}}{\sqrt{D_{Ox}}}\right) \qquad (41)$$

Reference of E to $E_{1/2}$ then yields the desired alternative form of the Nernst equation

$$E = E_{1/2} + \frac{RT}{nF} \ln\left(\frac{\sqrt{D_{Ox}}C_{Ox}}{\sqrt{D_{Red}}C_{Red}}\right) \qquad (42)$$

We also introduce the parameter e^j, which will be needed in subsequent sections

$$e^j = \frac{\sqrt{D_{Ox}}C_{Ox}}{\sqrt{D_{Red}}C_{Red}} \qquad (43)$$

where

$$j = \frac{nF}{RT}(E - E_{1/2}) \qquad (44)$$

2.5 Digital Simulation

Mathematical descriptions of electrochemical systems involve integration of the appropriate differential equation, for example, (1), to obtain concentration profiles near the surface of the electrode and subsequently the fluxes of the various electroactive species. The current is proportional to flux, see (7). These equations are substituted into either the Nernst equation or the absolute rate expression (31) as boundary conditions along with the appropriate potential–time function to obtain ultimately an expression relating current to potential, or to time if potential is time independent. The resulting equations can be solved analytically only in a few cases; otherwise, numerical methods must be used.

In a numerical approach well documented in the mathematical literature the concentration profile near the electrode is calculated by discretizing Fick's law into finite intervals of time, Δt, and distance, Δx, as illustrated in Figure 4.4. At a planar electrode we can define volume elements as the electrode area times Δx. This finite difference approach, as it is called, is illustrated for Fick's Second Law and a planar electrode [10–12]:

$$\frac{\Delta C(x, t)}{\Delta t} = D \frac{\Delta^2 C(x, t)}{\Delta x^2} \tag{45}$$

In finite difference form (45) becomes

$$C(x, t+\Delta t) = C(x, t) + D\Delta t \left(\frac{C(x+\Delta x, t) - 2C(x, t) + C(x-\Delta x, t)}{\Delta x^2} \right) \tag{46}$$

As the time and distance elements become smaller, the accuracy of the simulation, that is, the concentration at any time and distance, increases at the expense of computational time. Eventually, round-off errors caused by finite word length start contributing to error as Δx and Δt are decreased, and error may actually increase [12]. One calculates the concentration at time $t+\Delta t$ in terms of the concentration at time t so the method is iterative. The new concentration at any volume element is determined by its old concentration and the old

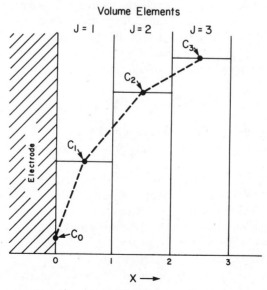

Figure 4.4 Schematic drawing of electrode surface and adjacent volume elements. Reprinted from [9], p. 188, by courtesy of Marcel Dekker, Inc.

concentration in the elements on either side of it. If iteration proceeds from time zero, then each new time involves a new iteration, which we number k. The distance, or volume elements using a unit electrode area, are also numbered from the electrode surface, j, so (46), after rearranging, becomes

$$C(j, k+1) = C(j, k) + \frac{D\Delta t}{\Delta x^2} [C(j+1, k) - 2C(j, k) + C(j-1, k)] \qquad (47)$$

The number of volume elements to be considered includes only those within the diffusion layer, which is approximately $(Dt)^{1/2}$ cm thick. In practice $6(Dt)^{1/2}$ is recommended [10], although smaller distances may be used with little loss in accuracy, particularly in cyclic voltammetry. As the experiment proceeds, more and more volume elements must be considered for each time iteration.

The concentration at the first element, bounded on one side by the electrode, cannot be calculated by (47). Rather, the surface concentration, related to the potential through the Nernst equation or the absolute rate expression, is used. At the first volume element

$$C(1, k+1) = C(1, k) + \frac{D\Delta t}{\Delta x^2} \left[C(2, k) - C(1, k) - \frac{Z(k)\Delta x}{D} \right] \qquad (48)$$

where $Z(k)$ is the flux term at the electrode

$$Z(k) = D[C(1, k) - C(0, k)]/0.5\Delta x \qquad (49)$$

The current is calculated from the flux at each iteration $i/nFA = Z(k)$ and the charge passed can be obtained by summing the current for each iteration.

Reilley and co-workers [13] and Sandifer and Buck [14] have improved this approach by considering some of the assumptions made in and errors resulting from use of (49). The assumption is that the flux at the electrode surface is approximated by the flux at $\Delta x/4$ from the surface, obtained from (49). If the electrode is placed at the center of the first volume element, the flux at the surface now becomes

$$Z(k) = D[C(2, k) - C(1, k)]/\Delta x \qquad (50)$$

and (48) is eliminated. The concentration in the first volume element (the electrode surface) is controlled by the boundary condition imposed by the experiment.

This method avoids "recycling" the flux term in the iterations as the concentration profile is calculated. Increased accuracy results and computation time, particularly in chronoamperometry, is decreased by a factor of 2 or 3 [15]. Although this method is described by Bard and Faulkner [11] for cyclic voltammetry, it requires more computation time than Feldberg's original method [10] to avoid a time displacement error [14].

An advantage to digital simulation of the electrochemical experiment occurs when coupled chemical reactions are treated. Normally the effects of the chemical reaction are added to the diffusion equation, which significantly complicates the subsequent derivation by Laplace transform techniques.

Second-order equations, in fact, cannot be solved this way because one cannot obtain the Laplace transform of the product of two independent variables inevitably appearing in such equations. With digital simulation, the changes in concentration caused by chemical reaction are treated after the changes caused by diffusion have been calculated. Since the time increments are so small, the errors created by treating the concentration changes sequentially instead of simultaneously are negligible.

Another advantage to digital-simulation techniques is the ease with which nonplanar electrode geometry can be treated. Feldberg [10], in his original monograph, gives the appropriate equations for spherical and cylindrical electrodes. These equations have been used by Osteryoung and co-workers [15] to simulate the technique of differential pulse polarography at the dropping mercury electrode. Ruzic and Feldberg [16] have demonstrated the importance of including sphericity in ac polarography and in second-harmonic ac polarography with amalgam formation. Other electrode configurations, such as the ring–disk [17] and rotating disk [18] involving hydrodynamic mass-transfer [10] and "active" semiconductor [19] electrodes, have also been simulated. Mass-transfer complications, such as adsorption [9, 20], have been treated by suitable modification to the diffusion equation. Polymer films with non-equivalent redox sites have also been studied for the cyclic voltammetric experiment [21].

Because of the power of digital simulation, several studies have been devoted to improving the computation time and accuracy involved in doing the calculations. One improvement in computational time can be achieved by discretizing distance (volume) or time or both in a nonlinear fashion. Joslin and Pletcher [22], and later Feldberg [23], improved computation time from tenfold to a few percent by smoothly increasing the distance between each distance element as one moves away from the electrode surface. The improvement depends on the exact function used for doing this. Osteryoung and co-workers [15] used a variable-time element to achieve a fivefold decrease in computation time. A combination of variable-time with variable-volume elements has been employed by Seeber and Stefani [24] to achieve ten- to thirtyfold decreases in computer time for cyclic voltammetry and chronoamperometry.

Magno, Bontempelli, and Perosa [25] have recently compared several of these procedures for simulations of cyclic voltammetry. They found that expansion of the time grid at long times along with expansion of the volume elements offered no improvement in computation time over expansion of the volume elements alone. Using an exponentially expanding volume element function [23], they showed a time improvement of almost a factor of 20 compared with using a fixed volume element for an EC_{irr} mechanism. More modest improvements in calculation time occurred for uncomplicated electron-transfer reactions. Lasia [26] has recently discussed the accuracy of various simulations of cyclic voltammetry and chronoamperometry using both explicit and implicit finite difference methods.

The time for simulation computations becomes prohibitive only for systems involving rapid, coupled chemical reactions. The chemical-reaction layer then becomes much smaller than the diffusion or charge-transfer reaction-layer thickness, requiring very small space and time increments for accurate computation. By treating the chemical reaction as being either at steady state (near the electrode) or at equilibrium (far from the electrode), Ruzic and Feldberg [27] have created an equivalent heterogeneous reaction parameter to describe both the charge-transfer and the homogeneous chemical reaction. Accuracy was preserved with significant improvements in computational time using a reversible, preceding reaction as an example. Magno and co-workers [28] have simulated cyclic voltammetric curves for several complex mechanisms including fast reversible and irreversible coupled kinetics. Limitations in the finite difference simulation for these mechanisms are described and methods of avoiding these problems are given.

Another numerical technique for solving the differential equations describing electrochemical systems has been used, orthogonal collocation. The concentration profile, $C(x, t)$, is substituted by a function of Legendre polynomials. Coefficients are selected to satisfy the mass-transport equation, including coupled chemical reactions, at selected points, the collocation points. As few as seven points are required to accurately describe the current function in cyclic voltammetry [29]. Taking first and second derivatives of $C(x, t)$ with respect to t at these collocation points yields another set of polynomial equations. The coefficients can be evaluated, after application of required boundary conditions, using standard computer subroutines. This technique has been described by Whiting and Carr [30] for chronoamperometry for the ECE mechanism and for disproportionation, requiring about 5 s of computation time. Speiser, Pons, and Rieker [29, 31] have used the technique for cyclic voltammetry. Yen and Chapman [32] have provided a general set of boundary-value collocation functions for use with cyclic voltammetry with reversible charge transfer that can include a following irreversible chemical reaction.

3   CYCLIC VOLTAMMETRY

Cyclic voltammetry is probably the most widely used of the modern electroanalytical techniques. The easily recognizable response of current maxima as the potential is cycled makes this technique without peer for preliminary studies of unknown electrochemical systems.

The current function, peak current divided by the square root of the scan rate, can be measured as a function of scan rate to unambiguously determine the mechanism of electrochemical reactions in many cases. The presence of adsorption, multiple charge-transfer reactions, and coupled chemical reactions can be determined easily with a minimum of experimental effort.

Conversely, the mathematical difficulty of obtaining closed-form analytical expressions relating current, potential, and concentration makes this technique difficult to use quantitatively. In addition, the current peaks are quite broad, making graphical information inexact. The experimental difficulty of measuring return peak current base lines also limits the quantitative value of the technique. These difficulties create large uncertainty in analyzed parameters such as charge transfer and coupled chemical-reaction-rate constants. Half-wave potentials can only be determined accurately under special conditions. Nevertheless, considerable effort has gone into calculating the cyclic voltammetric response of many different mechanisms. Working curves have been obtained in many cases as aids in analyzing experimental data quantitatively. This section describes such work and illustrates the utility of the technique as applied to several electrochemical systems.

3.1   Derivation of the Current Function

In the linear-sweep or cyclic voltammetric experiment the potential–time function is given by (51)

$$E = \begin{cases} E_i - vt & \text{for } 0 \leqslant t \leqslant t_s \\ E_i - 2vt_s + vt & \text{for } t \geqslant t_s \end{cases} \tag{51}$$

where E_i is the initial potential, v is the sweep rate in volts per second, and t_s is the time when the potential scan is reversed. Generally the sweep rate is the same in both directions, but not necessarily. Asymmetric potential scans with faster return rates have been suggested by Savéant [33] to simplify analysis in complex kinetic schemes.

The concentration expressions derived earlier in the form of the convolution integrals, (16) and (17), are used in the Nernst equation for reversible charge transfer (42), to obtain an integral equation relating current density, potential, and time

$$\int_0^t \frac{i \, d\tau}{nFC_{Ox}^*(\pi D_{Ox}\{t-\tau\})^{1/2}} = \frac{1}{1+e^j} \tag{52}$$

where j is given in terms of the half-wave potential, as used in Section 2, (44),

$$j = \frac{nF}{RT}(E - E_{1/2}) \tag{53}$$

and E is given by (51) above. Since our goal is to obtain an expression for current density as a function only of potential, not time, a transformation of variables is made using (54)

$$at = \frac{nFvt}{RT} \tag{54}$$

such that combining (51), (53), and (54) yields (55)

$$j(at) = \begin{cases} \dfrac{nF}{RT}(E_i - E_{1/2}) - at & 0 \leqslant t \leqslant t_s \\ \dfrac{nF}{RT}(E_i - E_{1/2}) + at - 2at_s & t \geqslant t_s \end{cases} \tag{55}$$

Making the following variable changes in (52)

$$\tau = z/a \tag{56}$$

and

$$\chi(at) = \frac{i}{nFC_{Ox}^*(\pi a D_{Ox})^{1/2}} \tag{57}$$

yields the dimensionless integral expression

$$\int_0^{at} \frac{\chi(z)dz}{(at-z)^{1/2}} = \frac{1}{1 + e^{j(at)}} \tag{58}$$

The solution of (58) gives values of $\chi(at)$ as a function of at and ultimately furnishes current-density values as a function of potential, according to

$$i = nFC_{Ox}^*(\pi D_{Ox}a)^{1/2}\chi(at) \tag{59}$$

Equation (58) has been solved analytically [34, 37], but a numerical solution is more generally applicable to cyclic voltammetric experimentation. Nicholson and Shain [38] have used this technique to calculate $\chi(at)$ to an accuracy of ± 0.001 for a single cyclic scan as a function of potential. Digital-simulation techniques [10] have also been used to obtain values of $\pi^{1/2}\chi(at)$ as a function of potential for substitution in (59).

The solution of (59) for the peak current of the forward potential scan, where $\pi^{1/2}\chi(at) = 0.4463$, is given by (60)

$$i_p = 2.69 \times 10^5 n^{3/2} D_{Ox}^{1/2} C_{Ox}^* v^{1/2} \tag{60}$$

where i_p is given in A/cm^2 and concentrations are given in moles/cm^3. Equation (60) is often called the Randles–Ševčik equation after the two investigators who first derived it independently [36, 39]. A good discussion of the historical development of this equation has been given by Adams [40].

The peak potential is $28.5/n$ mV negative of $E_{1/2}$ or

$$E_p = E_{1/2} - 28.5/n \, mV \tag{61}$$

Because the peak is quite broad, it is often easier to measure the half-peak potential, which is related to the half-wave potential by (62)

$$E_{p/2} = E_{1/2} + \frac{28.0}{n} \, mV \tag{62}$$

These equations are for a reduction reaction. For oxidation the signs of the numerical terms are reversed.

On the reverse scan or reoxidation response, the position of the peak depends on vt_s, the switching potential [37, 38]. As this potential moves toward more negative values, the position of the anodic (reoxidation) peak becomes constant at $29.5/n$ mV anodic of the half-wave potential. With the switching potential more than $100/n$ mV cathodic of the reduction peak, the separation of the two peaks will be $58/n$ mV and independent of the rate of potential scan. This is a commonly used criterion of reversibility. In fact, for a reversible wave the polarographic half-wave potential is just halfway between the two peak potentials and can be measured within ± 5 mV from (63). This equation is valid as long as the ratio of $k_s/v^{1/2}$ is greater than 0.02, where k_s is the charge-transfer rate constant, (31), and v is the scan rate in volts per second.

$$E_{1/2} = \frac{E_p^{Red} + E_p^{Ox}}{2} \tag{63}$$

According to (59) the current at any potential increases linearly with the square root of the potential scan rate. Normally, the peak current is chosen for experimental use because it is easier to measure. For convenience, the peak current divided by the square root of scan rate, the "current function," is examined as a function of scan rate. If (59) is obeyed exactly, no variation should occur in the current function with changes in the rate of potential scan.

The peak current for the return wave has the same value as the forward peak current when the base line is properly obtained. This base line is not zero current but is the decaying current line as the potential sweeps past the forward peak. To determine this decay accurately, the forward reduction current must be measured or calculated past the switching potential. Polcyn and Shain [41] and Eggins and Smith [42] have used an estimation technique to calculate the curve

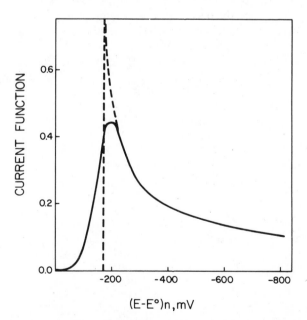

Figure 4.5 Comparison of descending branch of stationary electrode voltammogram to potentiostatic current–time curve. (——) voltammogram, (- - -) Cottrell equation. E^0 marks potential (time) where an equivalent potential step occurs to match the voltammogram. Reprinted from [41], by courtesy of *Analytical Chemistry*.

assuming $t^{-1/2}$ or diffusion-controlled decay. The method requires matching the diffusion-controlled current decay [Cottrell equation (25) in Section 2] to the cyclic voltammetric current past the peak potential. The variable parameter is t_0, the time corresponding to the initiation of the equivalent potential step required to match the latter part of the voltammetric curve, as shown in Figure 4.5. The current match is given by (64)

$$\sqrt{\pi}\chi(at) = \left(\frac{RT}{\pi n F}\right)^{1/2} \frac{1}{\{v(t-t_0)\}^{1/2}} \tag{64}$$

where v is the scan rate in volts per second. Using (51), (64) can be written

$$\sqrt{\pi}\chi(at) = \left(\frac{RT}{\pi n F}\right)^{1/2} \frac{1}{(E-E_0)^{1/2}} \tag{65}$$

where E_0 is the potential at time t_0, expressed with respect to the peak potential E_p or the half-wave potential $E_{1/2}$. The value of t_0 or E_0 depends on how much of the voltammetric curve is matched. To improve the match Polcyn and Shain [41] used another parameter β, which is less than 1, in the numerator of (66)

$$\sqrt{\pi}\chi(at) = \left(\frac{RT}{\pi n F}\right)^{1/2} \frac{\beta}{(E-E_0)^{1/2}} \tag{66}$$

Values of β and E_0 [41, 42] are given in Table 4.1 for matching different parts of the voltammetric decay of a reversible electron-transfer reaction. These values can be used to calculate the decay current at long times for use as the base line of the peak current in the reverse cycle. The method has also been used to estimate diffusion coefficients from CV data [42].

Table 4.1   Values of E_0 and β to Calculate Base Line Current Function Beyond the Switching Potential

Valid Potential Range $(mV)^a$	$E_0(mV)^b$	β
20–830	3.28	0.987
50–830	15.11	0.986
100–830	17.30	0.985
50–300	16.31	0.977
100–300	20.84	0.970
170–970	25.0	0.993
270–970	19.6	0.995
370–970	16.7	0.996

^{a}Potential range past the peak potential.
^{b}Value of E_0 is measured with respect to $E_{1/2}$. It is positive of $E_{1/2}$ for reductions and negative of $E_{1/2}$ for oxidations.

An empirically obtained alternative has been suggested by Nicholson [43], which does not require additional measurements or calculations. The current ratio can be calculated using the uncorrected reverse peak current, measured from the zero current line and the current at the switching potential, $(i_{sp})_0$, using (67)

$$\frac{i_p^a}{i_p^c} = \frac{(i_{ap})_0}{i_{cp}} + \frac{0.485(i_{sp})_0}{i_{cp}} + 0.086 \tag{67}$$

This method of obtaining the current ratio is illustrated in Figure 4.6 and is applicable to any cyclic voltammetric curve.

Theoretical cyclic voltammograms for reversible electrochemical reactions of nonunity reaction order have been studied by Shuman [44],

$$mOx + ne^- \rightleftharpoons mRed$$

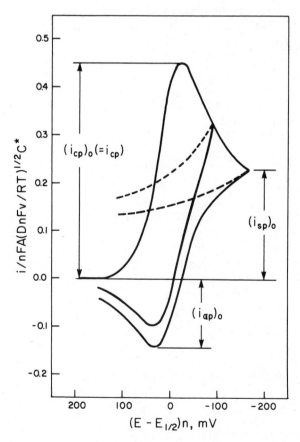

Figure 4.6 Cyclic voltammogram showing parameters necessary to calculate peak current ratio according to (67). Reprinted from [43], by courtesy of *Analytical Chemistry*.

Table 4.2   Diagnostic Criteria for Cyclic Voltammetry
and Reversible Charge Transfer

$$Ox + ne^- \rightleftharpoons Red$$

Properties of the potential of the response:

E_p is independent of v

$E_p^c - E_p^a = 59/n$ mV at 25°C and is independent of v

Properties of the current function:

$i/v^{1/2}$ is independent of v

Properties of the anodic-to-cathodic current ratio:

i_p^a/i_p^c is unity and independent of v

Other:

Wave shape is independent of v

As the reaction order increases, m/n becomes larger, the response becomes less peak-shaped and peak current becomes lower. The current parameter, $\pi^{1/2}\chi(at)$, decreases from 0.4463 for $m=1$ to 0.353 for $m=2$ to 0.303 for $m=3$. The peak potential also shifts cathodically of $E_{1/2}$ as m increases ($E_p - E_{1/2} = 36.0/n$ mV for $m=2$ and $49.8/n$ mV for $m=3$). Experimental data for a typical system of nonunity reaction order, the dissolution of mercury in cyanide media, were in good agreement with the theoretical treatment [44]. Thus, extension of the theory to all reaction orders of common interest appears possible.

Table 4.2 summarizes the properties of cyclic voltammetric data for reversible charge transfer.

3.2   Quasi-Reversible Charge Transfer

This section describes the mechanism in which the current is controlled by a mixture of diffusion and charge-transfer kinetics. Because diffusion plays a part in controlling the surface concentrations of the redox couple, these expressions are the same as those used for the reversible case, but are now substituted into the absolute rate expression (31) instead of the Nernst equation. The result is an equation that relates the current to the charge-transfer rate, where k_s is the charge-transfer rate constant at the standard potential, E^0, and α is the charge-transfer coefficient of the reduction step.

The resultant expression, in dimensionless parameters, is given by

$$\frac{\chi(at)}{\psi} e^{\alpha j(at)} = 1 - (1 + e^{j(at)}) \int_0^{at} \frac{\chi(z)dz}{(at-z)^{1/2}} \tag{68}$$

Equation (68) has been made dimensionless by the substitutions [43]

$$\psi = \frac{k_s}{(\pi a D)^{1/2}} \tag{69}$$

where

$$D = D_{Ox}^{1-\alpha} D_{Red}^{\alpha} \tag{70}$$

Expressions for at and $\chi(at)$ are given by (54) and (57), respectively. If (68) can be solved for $\chi(at)$, the cyclic voltammetric current can be obtained from (59) as a function of at, which is related to potential.

Nicholson [45] has presented (68) in a slightly different form and has obtained a numerical solution for different values of k_s and α. The dimensionless parameter ψ contains both k_s and α, and Nicholson has shown that when $\psi \geqslant 7$, the solution of $\chi(at)$ becomes equal to the Nernstian case discussed previously. When $\psi < 0.001$, the solution of (68) is the same as for the irreversible case discussed in the next section. The region between $\psi = 0.001$ and $\psi = 7$ is that of the so-called quasi-reversible mechanism of interest here.

Through several calculations of theoretical curves, Nicholson has found that in the region of $0.3 \leqslant \alpha \leqslant 0.7$, the separation of anodic and cathodic peaks is almost independent of α, being mainly a function of k_s. This approximation becomes less accurate, however, as ψ becomes smaller. At $\psi = 0.5$ there is only a 5% variation in peak separation as α is varied from 0.3 to 0.7. The peak separation at this point is $105/n$ mV, which is significantly different from the reversible peak separation of $59/n$ mV.

The exact peak separation as a function of ψ is given in the working curve in Figure 4.7. The data used to construct this curve are given by Nicholson [45] and by others [31]. The useful working range of Figure 4.7 for determination of k_s is $0.5 \leqslant \psi \leqslant 5.0$. Since ψ is also a function of scan rate, this experimental parameter can be varied to make ψ fall within the required range. If the scan rate is slow

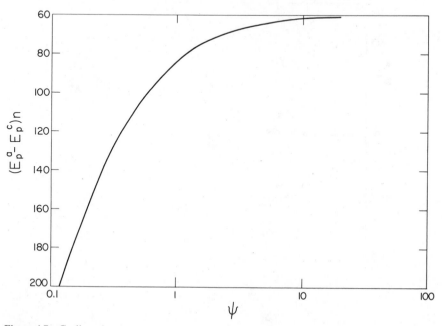

Figure 4.7 Cyclic voltammetric working curve for quasi-reversible reaction. Plot of $n(E_p^a - E_p^c)$ against ψ (69).

enough, the peak separation approaches reversible behavior, and if it is very fast, irreversible behavior is approached.

Experimentally, a scan rate is chosen to obtain a peak separation somewhere between $120/n$ and $60/n$ mV. From Figure 4.7 the corresponding value of ψ is determined. Equation (69) is then used to determine k_s. The method has recently been extended to temperature-dependent studies of k_s [46]. To use this method rigorously D_{Ox}, D_{Red}, and α must be known, but except for α values very different from 0.5 and D_{Red} very different from D_{Ox}, the approximation $D_{Ox}^{1/2} = D_{Red}^{1/2} = D^{1/2}$ can be used.

The value of α can only be estimated by matching experimental curves with theoretical curves for various values of α. For large values of α, the reduction peak is much sharper and the current larger than that of the corresponding anodic peak. The opposite is true for small values of α.

Examples of the shape of the cyclic voltammetric response for $\psi = 0.5$ and various values of α are calculated in Figure 4.8. The cyclic voltammetric response for quasi-reversible charge transfer is summarized in Table 4.3.

Nicholson [45] has used this technique to study the reduction of Cd(II) at a mercury electrode in 1 M Na$_2$SO$_4$. He used published values of D_{Ox} and D_{Red} for this system and an estimated value of $\alpha = 0.25$ obtained by comparing various theoretically calculated curves for shape and symmetry with experi-

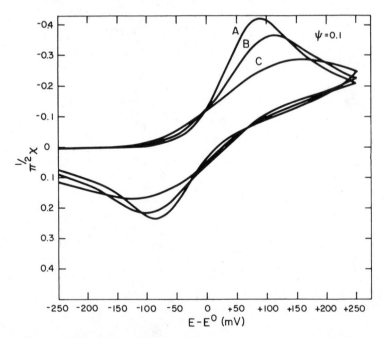

Figure 4.8 Example of cyclic voltammetric curves for $\psi = 0.1$ and various α values: (A) $\alpha = 0.3$, (B) $\alpha = 0.5$, (C) $\alpha = 0.7$.

Table 4.3   Diagnostic Criteria for Cyclic Voltammetry and
Quasi-Reversible Charge Transfer

$$Ox + ne^- \overset{k_{s,\alpha}}{\rightleftharpoons} Red$$

Properties of the potential of the response:

E_p shifts with v

$E_p^c - E_p^a$ may approach $60/n$ mV at low v but increases as v
increases

Properties of the current function:

$i_p/v^{1/2}$ is virtually independent of v

Properties of the anodic-to-cathodic current ratio:

i_p^a/i_p^c equal to unity only for $\alpha = 0.5$

Others:

The response visually broadens as v is increased

---

mental curves. The value of $k_s = 0.24$ cm/s remains fairly constant with scan rate variations from 48 to 120 V/s.

Three different studies of quinone reduction on solid electrodes in aprotic media illustrate the difficulties of charge-transfer studies. Rate constants varied from 0.052 cm/s to 0.12 cm/s [47] for benzoquinone on platinum in DMF, whereas 0.24 cm/s was obtained in acetonitrile [48]. Similar differences were obtained for a gold electrode. Variations in electrode preparation could explain these differences.

Klingler and Kochi [49] have recently obtained an expression for k_s assuming initially that the electron-transfer reaction is irreversible (see next section). This can be achieved experimentally by using fast scan rates:

$$k_s = 2.18(D\alpha a)^{1/2} \exp\left(\frac{\alpha^2 nF}{RT}\{E_p^c - E_p^a\}\right) \tag{71}$$

As the scan rate increases, the apparent value of k_s also increases until the increasing peak separation counterbalances the scan-rate effect. The charge-transfer rate constant for reduction of $K_3Fe(CN)_6$ in 1 M KCl at platinum was studied using (71). Alpha (α) was obtained from the shape of the cyclic voltammetric wave assuming irreversibility [see (77) in the next section]. Using (71), $k_s = (8.7 \pm 0.8) \times 10^{-2}$ cm/s, when $\alpha = 0.25$, in good agreement with earlier studies of this reaction [50]. Scan rates in excess of 100 V/s were used in this work. The value obtained using Nicholson's working curve (Figure 4.7) was $k_s = (2.3 \pm 0.8) \times 10^{-2}$ cm/s, much smaller than values reported previously. Klingler and Kochi [49] attributed this difference to use of $\alpha = 0.5$ in deriving Figure 4.7 instead of the actual value of $\alpha = 0.25$.

3.3   Irreversible Charge Transfer

An irreversible electrochemical reaction is one for which the charge-transfer rate constant is small. Thus the potential at which a significant extent of reaction occurs may be quite cathodic of the standard potential. At such potentials the reverse charge-transfer process does not occur. If the absolute rate expression (31) is then employed to describe the current, the second term on the right-hand side becomes zero. Thus, for an irreversible reaction, the absolute rate expression can be simplified to (72)

$$\frac{i(t)}{nF} = C_{Ox} k_s \exp\left(\frac{-\alpha nF}{RT} \{E - E^0\}\right) \tag{72}$$

If the reaction is irreversible, the peak separation is so large that there is often no current observed on the return potential sweep in cyclic voltammetry.

By using the same boundary conditions as employed for the reversible case, with the exception that the concentration of the product species is assumed to be zero, an expression for the surface concentration of the reactant species is obtained. The current is then related to the charge-transfer rate by (72) using the value for C_{Ox} from (16). The appropriate value of E is substituted into (72) as well, and the resultant integral equation is solved.

If the product of a reversible charge-transfer reaction is destroyed by a rapid chemical reaction such that no reverse reaction can proceed, the overall reaction sequence will be reduced to a response having the same qualitative form as that of the irreversible charge-transfer reaction and will yield the same response if $\alpha = 1$. Such reactions can be treated by the equations derived for the irreversible system. However, it should be emphasized that slow charge transfer and rapid charge transfer coupled to a very rapid and irreversible chemical reaction are not conceptually equivalent processes. Thus, care should be exercised in terming a reaction irreversible in the absence of knowledge of the overall reaction.

The appropriate integral equation describing the response of the irreversible reaction for the cyclic voltammetric technique is given by (73)

$$\frac{\chi(at)}{\psi} e^{\alpha j(at)} = 1 - \int_0^{at} \frac{\chi(z)dz}{(at-z)^{1/2}} \tag{73}$$

where now $j(at)$ is given by (74) for the single potential scan

$$j(at) = \frac{nF}{RT}(E_i - vt - E^0) \tag{74}$$

Equation (73) has been made dimensionless by using the definition for $\chi(at)$ in (57) and the definition of ψ in (75)

$$\psi = \frac{k_s}{(\pi D_{Ox} a)^{1/2}} \tag{75}$$

where a is given by (54).

These equations differ slightly from those given by Nicholson and Shain [38] because no distinction is made between the total number of electrons consumed per mole and the number consumed in the rate-determining, charge-transfer step, as these two quantities are assumed to be the same.

Delahay [51] and Matsuda and Ayabe [34] have given a numerical solution for (73) but the data of Nicholson and Shain [38] are more accurate, being given to within ± 0.001.

From the potential scale given by Nicholson and Shain [38] in their tabulation, the peak potential can be calculated for an irreversible response.

$$E_p = E^0 - \frac{RT}{\alpha nF} [0.780 + \ln(D_{Ox}\alpha a)^{1/2} - \ln k_s] \tag{76}$$

There is about a $30/\alpha n$ mV cathodic shift in peak potential for every tenfold increase in scan rate, and this criterion can be used to characterize an irreversible response. An example of this shift is shown in Figure 4.9, in which the current is plotted for various scan rates. The same shift per tenfold scan-rate increase applies to the half-peak potential also [38]. Since this parameter is frequently measured more accurately, it can be used as well to characterize the irreversible response. Thus the shape of the response is independent of scan rate and the separation of E_p and $E_{p/2}$ is given by (77)

$$E_p - E_{p/2} = \frac{-1.857RT}{\alpha nF} \tag{77}$$

An alternative expression to (76) can be obtained relating peak current to peak potential, from which the parameters k_s and α can be found [35, 38].

$$i_p = 0.227nFC^*_{Ox}k_s \exp\left(\frac{-\alpha nF}{RT}\{E_p - E^0\}\right) \tag{78}$$

A plot of log i_p versus $E_p - E^0$ (or $E_{p/2} - E^0$) for different scan rates yields α from the slope and k_s from the intercept provided E^0 is known.

Reinmuth [35] has shown that at the foot of the response the current is independent of scan rate (see Figure 4.9) and can be related to the potential and

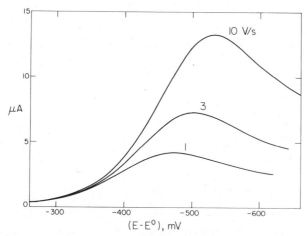

Figure 4.9 Theoretical irreversible cyclic voltammetric waves. $k_s = 10^{-6}$ cm/s, $\alpha = 0.5$, $n = 1$, $D_{Ox} = D_{Red} = 10^{-6}$ cm^2/s, area $= 0.02$ cm^2, $C^*_{Ox} = 10^{-6}$ M, $E^0 = 0.0$ V, scan rate as shown.

the initial potential, E_i. The initial potential is chosen at the foot of the response where no current flows

$$i = nFC_{Ox}^* k_s \exp\left[\frac{-\alpha nF}{RT}(E - E_i)\right] \tag{79}$$

Nicholson and Shain [38] have shown that this equation is valid only for current values less than 0.1 i_p, but (79) does offer a convenient way to obtain k_s and α, even when the standard potential E^0 for the reaction is unknown.

By the way of summary, Figure 4.10 shows the effect of the charge-transfer rate constant on the cyclic voltammetric response as the rate is varied from essentially reversible (diffusion-controlled) behavior to irreversible behavior. Table 4.4 summarizes the important characteristics of cyclic voltammetric waves for irreversible charge transfer.

Sundholm [52] has studied the irreversible reduction of $PtCl_4^{2-}$ on mercury in 1 M NaClO$_4$ and has shown that $i_p/v^{1/2}$ is independent of scan rate while the peak shifts cathodically as the scan rate increases. Knowing the value of E^0 for this reduction, Sundholm has reported values of $\alpha = 0.23$ and $k_s = 9 \times 10^{-6}$ cm/s, an essentially irreversible reaction. Taylor and Humffray [53] have measured the rate constants of several reduction reactions of metal ions on a glassy carbon electrode. Rate constants of 10^{-4} cm/s and less, with α values quite different from 0.5, were obtained reliably.

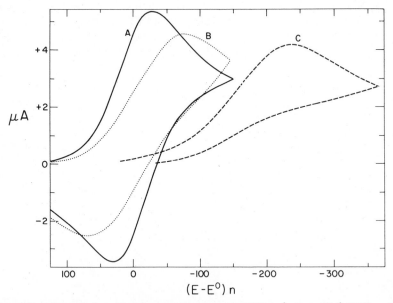

Figure 4.10 Theoretical cyclic voltammetric curves. $A = 0.02$ cm^2, $C_{Ox} = 10^{-6}$ M, $D_{Ox} = D_{Red} = 10^{-6}$ cm^2/s, scan rate $= 1$ V/s, $E^0 = 0$. Curve A—reversible curve; Curve B—quasi-reversible, $k_s = 0.03$, $\alpha = 0.5$; Curve C—irreversible, $k_s = 10^{-4}$ cm/s, $\alpha = 0.5$.

Table 4.4   Diagnostic Criteria for Cyclic Voltammetry and
Irreversible Charge Transfer

$$Ox + ne^- \xrightarrow{k_s, \alpha} Red$$

Properties of the potential of the response:

   E_p shifts cathodically by $30/\alpha n$ mV per tenfold increase in v

Properties of the current function:

   $i_p/v^{1/2}$ is constant with v

Other:

   There is no current on the reverse scan;

   the wave shape is determined by α, (77), and is independent of v

---

3.4   Coupled Chemical Reactions

If all electrochemical reactions were merely simple charge-transfer reactions, as discussed so far, the characterization of reaction mechanisms would not be difficult. However, many charge-transfer reactions are coupled to homogeneous chemical reactions and to other charge-transfer reactions in a bewildering variety of mechanisms. The presence of coupled chemical reactions is observed by the perturbations they cause to the uncomplicated charge-transfer processes already described. There are two objectives of studies of electrochemical reactions that include coupled chemical reactions: (a) a characterization of the overall electrochemical mechanism and nature of the charge-transfer process, and (b) evaluation of the kinetic parameters of the coupled chemical reactions. The real strength of cyclic voltammetry lies in fulfilling the first objective— characterization of the overall mechanism.

Consider, for example, a reversible redox reaction in which the product of charge transfer undergoes a reversible chemical reaction:

$$Ox + ne \rightleftharpoons Red$$

$$Red \underset{k_b}{\overset{k_f}{\rightleftharpoons}} Z$$

In cyclic voltammetry the current response is quite different from the uncomplicated reversible case. The magnitude of the reverse current response is less than that observed in the absence of the chemical reaction. The degree to which the current is decreased depends on the time scale of the experiment relative to the rate of the forward chemical reaction, k_f. Expanding the range of potential examined in the cyclic experiment can determine whether the products of the chemical reaction are themselves electroactive. The ratio of forward to backward current response as a function of solution concentration can often determine the stoichiometry of the chemical reaction. A reaction second-order in the reduced species, Red, will show a decrease in relative response of the reverse cycle at the same scan rate, as concentration is increased.

Differences of this type lead to rather severe complications in the complete theoretical description of the processes even as they afford easily observed qualitative information. For this reason cyclic voltammetry is often supplemented by other electrochemical techniques to study complex kinetic processes in detail. The following sections focus on those mechanisms that have received sufficient attention, both theoretical and experimental, to make cyclic voltammetry of practical use in their study.

3.4.1   Preceding Chemical Reaction

This reaction sequence is discussed on the basis of the equations

$$Z \underset{k_b}{\overset{k_f}{\rightleftharpoons}} Ox$$

$$Ox + ne^- \rightleftharpoons Red$$

with the chemical reaction assumed to be first order. The forward chemical reaction is that which produces the electrochemically reactive species, and the equilibrium constant is defined as

$$K = k_f/k_b$$

With a preceding chemical reaction, the magnitude of the cathodic current response is expected to be sensitive to the rate of the chemical reaction. If the rate of the chemical reaction is slow compared with the time scale of the experiment, no significant conversion of the electroinactive form into the active form can take place. Under such conditions the cathodic response has the characteristics of the reversible case but the current magnitude is determined by the equilibrium concentration of the electroactive species. If a firm assessment of n has been reached, the lower current is of significant diagnostic value. At the opposite extreme in rate of the chemical reaction, the current has the characteristics of the reversible case with variations in position on the potential axis.

Under conditions intermediate to the above cases, Nicholson and Shain [38] have shown through calculation of a large number of theoretical response curves that a working curve very nearly fits the equation

$$\frac{i_k}{i_d} = \frac{1}{1.02 + 0.471(a/k)^{1/2}K^{-1}} \tag{80}$$

where $k = k_f + k_b$, i_k is the observed peak current, and i_d is the diffusion-controlled peak current expected in the absence of the chemical complications. The i_d value is calculated or determined from an experiment on a time scale slow with respect to the rate of the chemical reaction.

The anodic current response is not quite so sensitive to variations in the kinetic parameters as the cathodic response and therefore offers little additional data. However, for cases in which the i_d value cannot be obtained, the ratio of i_a to i_c can be used for kinetic measurements through a working curve of i_a/i_c as a function of $(a/k)^{1/2}K^{-1}$. The anodic response is a function of the switching potential and such working curves must be calculated for a particular value of the switching potential.

The variation of i_a/i_c as a function of scan rate is quite useful for diagnostic purposes, approaching a value of unity at slow scan rates and increasing at higher scan rates.

The position of the response on the potential axis is also determined by the kinetic parameters. Working curves have been presented [38] for this behavior that afford kinetic measurements, but the potential shift does not seem to offer the advantages of the current measurements. The potential behavior offers diagnostic data for differentiation from the similar case with irreversible charge transfer only at low values of the kinetic parameter, that is, when the response approaches reversible behavior. Thus the utility of the potential variations appears trivial in light of the other quite clear diagnostic information.

Figure 4.11 shows theoretically calculated voltammograms for the cathodic scan only, to illustrate the general shape of the response. The curve becomes very flat as the scan rate increases and the peak current is much less than that obtained for the corresponding reversible case.

Bailey and co-workers [54] have applied this treatment qualitatively to an estimation of the equilibrium constant for the second protonation of phenazine (P)

$$PH^+ + H^+ \rightleftharpoons PH_2^{2+}$$

through the electrochemical reaction

$$PH_2^{2+} + e^- \rightleftharpoons PH_2^+$$

Values of $K(k)^{1/2}$ were obtained from experimental data for i_k/i_d, the value of i_d being determined from the limiting value of the peak current at lower scan rates.

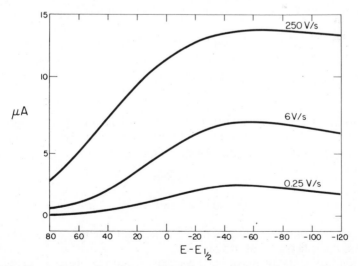

Figure 4.11 Theoretical cyclic voltammograms for chemical reaction preceding a reversible charge transfer. $n=1$, $A=0.02$ cm^2, $C_{Ox}^*=10^{-6}$ M, $D_{Ox}=D_{Red}=10^{-6}$ cm^2/s, $K=1$, $k=100$ s^{-1}, scan rate shown on curve.

By estimation of the upper limit of the forward reaction, an estimate of the equilibrium constant was obtained.

The chemical reaction may precede an irreversible charge-transfer reaction. This mechanism describes a number of reactions occurring in aqueous solution and has received considerable study with dc polarography (see Chapter 9). With an irreversible charge-transfer reaction, only variations in the cathodic current and in the position of the response on the potential axis are available for use. Nicholson and Shain [38] have treated this case, and the correlations below are from their work.

If the time scale of the experiment is great relative to the rate of the chemical reaction, that is, if k/b is small ($b = \alpha a$), the response is that of the uncomplicated irreversible charge transfer with the exception of a displacement on the potential axis. If k/b is large and $(b/k)^{1/2}K^{-1}$ is also large, the response obtained is no longer peak-shaped and both the potential of the response and the magnitude of the current are independent of b. Under these conditions the current is directly proportional to $K(k)^{1/2}$.

At intermediate values of $(b/k)^{1/2}K^{-1}$ with k/b large, numerical solution of the integral equation for values of the kinetic parameters leads to the empirical equation

$$\frac{i_k}{i_d} = \frac{1}{1.02 + 0.531(b/k)^{1/2}K^{-1}} \tag{81}$$

where again i_k is the observed current and i_d is the current expected in the absence of the chemical reaction.

Since other correlations are not available, the variation of the potential of the response with variation in the kinetic parameter assumes greater importance. This shift decreases to a relative independence of $(b/k)^{1/2}K^{-1}$ for small values of the kinetic parameter.

Qualitatively, this case is rather easy to note with the lack of anodic response, the decrease in current function $(i_p/v^{1/2})$ with an increase in scan rate, and the general nature of the irreversible charge-transfer case.

3.4.2 Following Chemical Reaction

Treatment of this mechanism can be divided into two categories: (a) the reaction is reversible and (b) the reaction is essentially irreversible, where the equilibrium between Red and Z lies far to the right.

$$Ox + ne^- \rightleftharpoons Red$$

$$Red \underset{k_b}{\overset{k_f}{\rightleftharpoons}} Z \qquad K = k_b/k_f$$

Experimental conditions can often be adjusted to reduce the mechanism to these less complicated limiting cases.

With a following chemical reaction, the magnitude of the cathodic current response does not vary significantly and is thus of little value for either qualitative or quantitative use. Conversely, relatively large changes in the magnitude of the anodic current are expected at an appropriate balance of the chemical kinetic and experimental parameters. Therefore the variations in anodic current, or the ratio of anodic-to-cathodic current as a function of the time scale of the experiment, serve as a prime variable for both qualitative and quantitative measurements.

The values of the equilibrium constant and the rate constants determine the applicability of the treatment that follows. If the rate constant for the forward chemical reaction is small and the equilibrium constant is large, that is, equilibrium shifted toward the reactant, the reaction system reduces to the reversible case. As the rate constant for the forward reaction increases, treatment of the system as charge transfer with a reversible chemical reaction becomes applicable. If the equilibrium constant is small, the system reduces to the irreversible following reaction with the exception of the position of the response in potential. Thus, instances in which the rate constant of the chemical reaction is significant and the equilibrium constant is not excessively small are the most important.

Illustrative cyclic voltammetric responses for cases of interest are shown in Figure 4.12. At high values of the rate of potential scan relative to the chemical

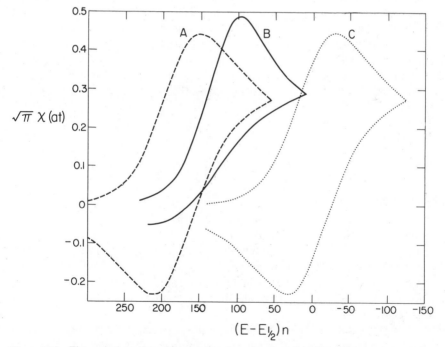

Figure 4.12 Theoretical cyclic voltammetric curves for reversible following chemical reaction. Curve A—fast chemical reaction; wave has reversible shape with potential shifted by $(RT/nF)\ln(1+K)/K$, where $K = 10^{-3}$; Curve B—intermediate chemical reaction rate, $K = 10^{-3}$, $k/a = 10^{4}$; Curve C—slow chemical reaction. Essentially reversible behavior.

kinetics, the reversible response is obtained at the expected potential. With sufficiently rapid kinetics, a reversible response is observed, but is shifted in potential by an amount reflecting the magnitude of the equilibrium constant. The intermediate cases of balance of the kinetic parameter and time scale of the experiment yield a response that is qualitatively different from the reversible case and appears at a potential other than the reversible potential.

For these cases the ratio of anodic-to-cathodic current can serve as a useful parameter for quantitative measurements. However, the magnitude of the anodic current is a function of the switching potential. A working curve of the ratio of anodic-to-cathodic current as a function of $(a/k)^{1/2}K^{-1}$ is employed, but such curves must be calculated with an arbitrarily selected switching potential [38]. The ratio i_p^a/i_p^c approaches unity as K increases and/or $(a/k)^{1/2}$ decreases. The ratio decreases at higher values of $(a/k)^{1/2}$ and/or as K decreases. If the equilibrium constant is known, kinetic data can thus be calculated from a single current–potential curve.

The ratio of anodic-to-cathodic peak current is useful for diagnostic purposes. At low values of the scan rate, the ratio approaches unity. A decrease in the ratio i_p^a/i_p^c is noted as the scan rate is increased. This behavior alone almost serves to differentiate this case from all others.

The behavior of the cathodic peak potential over the range of kinetic parameters for a reversible following chemical reaction is somewhat more complex. Under conditions in which the following chemical reaction is in complete equilibrium with the system at all times and the observed response has the appearance of the reversible system in other respects, the response is shifted anodically on the potential axis by the quantity $(RT/nF)\ln[(1+k)K^{-1}]$. When the rate of the chemical reaction is slow, the response occurs at the reversible potential.

In the cases in which k/a is large and the kinetic parameter $(a/k)^{1/2}K^{-1}$ small, the equation

$$E_p = E_{1/2} - \frac{RT}{nF}\left[0.780 + \ln\left(\frac{a}{k}\right)^{1/2} - \ln(1+K)\right] \tag{82}$$

is obeyed [38]. Thus under this relationship the peak potential shifts more cathodically by $60/n$ mV for a tenfold increase in $(a/k)^{1/2}K^{-1}$.

Intermediate values of $(a/k)^{1/2}K^{-1}$ lead to an intermediate variation in the position of the response on the potential axis. At large values of the parameter, the peak potential is independent of scan rate, that is, reversible plus $\ln[(1+K)/K]$ term. A working curve can be constructed from published data [38] relating the potential to the kinetic parameter. Both $E_{1/2}$ and K must be known for use of this relationship.

Table 4.5 summarizes the cyclic voltammetric response for charge transfers coupled to reversible chemical reactions assuming rapid charge transfer.

Irreversible following chemical reactions produce at most a 10% increase in the magnitude of the cathodic peak current compared with the reversible charge-transfer mechanism in the absence of a chemical reaction. Thus, this parameter

Table 4.5   Diagnostic Criteria for Cyclic Voltammetry With a Preceding or Following Reversible Chemical Reaction

---

Preceding Reaction:

$$Z \underset{k_b}{\overset{k_f}{\rightleftharpoons}} Ox + ne^- \rightleftharpoons Red; \quad K = k_f/k_b$$

Properties of the potential of the response:

   E_p shifts anodically with an increase in v

Properties of the current function:

   $i_p/v^{1/2}$ decreases as v increases

Properties of the anodic-to-cathodic current ratio:

   i_p^a/i_p^c is generally greater than unity and increases as v increases with a value of 1 approached at lower values of v

Others:

   A response similar to the reversible case, but lower in magnitude, is obtained when the chemical kinetics are slow and K is moderate in value

Following Reaction:

$$Ox + ne^- \rightleftharpoons Red \underset{k_b}{\overset{k_f}{\rightleftharpoons}} Z; \quad K = k_b/k_f$$

Properties of the potential of the response:

   E_p shifts cathodically with an increase in v by an amount that approaches $60/n$ per tenfold increase in v if k is large and K is small; intermediate values of K and k lead to a shift of a lesser magnitude

Properties of the current function:

   $i_p/v^{1/2}$ virtually constant with v

Properties of the anodic-to-cathodic current ratio:

   i_p^a/i_p^c decreases from unity as v increases

Others:

   If K is small and the chemical kinetics are rapid, a response typical of reversible charge transfer, except for shifts in potential, will be noted

---

has no real use for either qualitative characterization or quantitative measurement. The magnitude of the anodic response, however, is influenced markedly by the coupled chemical reaction except at very low values of the rate constant relative to the scan rate (low values of k_f/a), where the response is essentially that of the reversible case. At large values of the rate constant, the anodic response is essentially that of the irreversible charge transfer and little information concerning the chemical reaction is obtained. At intermediate values of k_f/a, however, the magnitude of the anodic current response provides the basis for excellent measurements of the kinetics of the chemical reaction.

   Through calculation of a large number of theoretical response curves varying both k_f/a and the switching potential E_s, Nicholson and Shain [38] have shown that the problem of variation of the anodic current with switching potential can be circumvented by use of the parameter $k_f\tau$, where τ is the time in seconds required to scan from the $E_{1/2}$ to the switching potential. A working curve of

i_p^a/i_p^c as a function of log $k_f\tau$ (Figure 4.13) can be constructed, which is useful throughout the range in which the response can be distinguished from the reversible case ($k_f\tau = 0.02$) or for which an anodic response can be measured ($k_f\tau = 1.6$). Of course, $E_{1/2}$ for the reaction must be known accurately to use this relationship. If $E_{1/2}$ is not available from an independent measurement, one can obtain a value from the limiting behavior at rapid scan rates, assuming that the charge-transfer process is fast.

The variation in anodic response as a function of the rate of potential scan is also quite useful for qualitative purposes. As the scan rate is increased, i_p^a/i_p^c increases toward unity (reversible case) as a limit. Since other mechanisms in which the anodic current can be measured have a behavior markedly different from that of this case, the variation of i_p^a/i_p^c with scan rate is a particularly valuable diagnostic tool. However, the magnitude of the cathodic current as a function of scan rate must be monitored to avoid confusion with the ECE cases.

The variation in peak potential with the kinetic parameters for the irreversible following reaction is also of value. At higher values of k_f/a, the relationship

$$E_p = E^0 + \frac{RT}{2nF}\left[\ln\left(\frac{k_f}{a}\right) - 1.56\right] \tag{83}$$

has been obtained [38], which describes a cathodic shift of $30/n$ mV for a tenfold decrease in k_f/a.

For quantitative characterization, the variation of E_p is also useful if a suf-

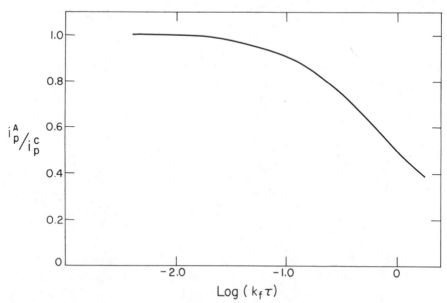

Figure 4.13 Cyclic voltammetric working curve of peak current ratio versus the kinetic parameter $k_f\tau$ for an irreversible following chemical reaction.

Table 4.6 Diagnostic Criteria for Cyclic Voltammetry With an Irreversible Chemical Reaction Following Charge Transfer

$$Ox + ne^- \rightleftharpoons Red \xrightarrow{k_f} Z$$

Properties of the potential of the response:

E_p shifts cathodically by $30/n$ mV for low v with lesser shift at higher values of v

Properties of the current function:

$i_p/v^{1/2}$ is independent of v

Properties of the current ratio:

i_p^a/i_p^c increases toward unity as v increases

---

ficiently wide range of scan rates can be employed [55]. At low values of the scan rate, E_p shifts cathodically by $30/n$ with a tenfold increase in the scan rate, (83). As the scan rate is increased, lesser magnitudes of this shift are observed until E_p becomes constant with scan rate, as in the reversible case. These parameters are summarized in Table 4.6.

Kuempel and Schaap [56] have studied the rate of ligand exchange between cadmium ion and calcium ethylenediaminetetraacetate by cyclic voltammetry. Cadmium amalgam is oxidized to form Cd(II) in calcium–EDTA solutions. The rate of formation of the Cd–EDTA complex is very rapid and the equilibrium lies far to the right (favoring the complex). Under these conditions the working curve of Figure 4.13 can be used. A pseudo-first-order rate constant of 10 s^{-1}, with a maximum uncertainty of $\pm 8\%$, was obtained. Multiplying this rate by the EDTA concentration yields a second-order rate constant of $(2.6 \pm 0.7) \times 10^8$ $M^{-1} \text{s}^{-1}$, in agreement with the value of $(6.1 \pm 0.7) \times 10^8$ reported by Aylward and Hayes [57] using ac polarography and faradaic impedance. A rate of 2.3×10^9 $M^{-1} \text{s}^{-1}$ was obtained by Matsuda and Tamamushi [58] using ac polarography by reducing the complex at the electrode surface. This work is discussed in the section on ac polarography.

The rate constant for the ligand-exchange reaction between pyridine and cyanopyridine hemochrome yielding pyridine hemochrome has also been evaluated by Davis and Orleron [59], using cyclic voltammetry. The more rapid reaction of pyridine with the cyanide hemochrome ($k_f = 20 \text{ s}^{-1}$), compared with 0.5 s^{-1} for the cyanopyridine hemochrome reaction, could also be examined by this technique. It thus appears that the cyclic voltammetric examination of ligand-exchange kinetics is broadly applicable.

Stapelfeldt and Perone [60] have applied cyclic voltammetry and the working curve in Figure 4.13 to a study of the reduction of benzil. At pH values greater than 11, benzil is reduced to the stilbene diolate, which undergoes an irreversible rearrangement

$$\phi\text{—C—C—}\phi + 2e^- + H^+ \rightleftharpoons \phi\text{—C=C—}\phi$$
$$\underset{O\ \ O}{\overset{||\ \ ||}{}} \qquad\qquad \underset{HO\ \ O^-}{\overset{|\ \ |}{}}$$

to the benzoin anion

$$\phi\text{—C}\text{=}\text{C}\text{—}\phi \xrightarrow{k_f} \phi\text{—CH}\text{—C}\text{—}\phi$$
$$\quad\ \ \overset{|}{\text{HO}}\ \ \overset{|}{\text{O}^-} \qquad\qquad \overset{|}{\text{O}^-}\ \ \overset{\|}{\text{O}}$$

with a rate constant of approximately $2\ \text{s}^{-1}$.

Perone and Kretlow [61] have also evaluated the rate constant for the hydration of the oxidation product of ascorbic acid by means of cyclic voltammetric data. The rate constant of $1.4 \times 10^3\ \text{s}^{-1}$ they obtained was in excellent agreement with the value from potential step measurements.

The preceding treatment of irreversible following reactions has assumed the charge-transfer reaction is fast, that is, diffusion controlled. The effects of slow charge transfer, always with $\alpha = 0.5$, on shifts in peak potential with scan rate have been considered by Evans [62] and by Nadjo and Savéant [63] using digital simulation. A general condition was obtained [62] in which the charge-transfer rate constant is fast enough or the chemical reaction slow enough to yield potential shifts in agreement with the "reversible" theory already described. The condition is given by

$$k_f \leqslant k_s^2/100\pi D \tag{84}$$

At scan rates below $k_f/a = 1$, the potential shift is controlled by the chemical reaction. At higher scan rates the potential shift, which may be small if k_s is large, is controlled by the charge-transfer rate constant. If the inequality of (84) is not obeyed, then peak potentials shift with scan rate in response to both rate parameters. Since the shifts are additive, peak potential shifts may vary from $30\ \text{mV/decade}$ to $60\ \text{mV/decade}$ of scan rate depending on the ratio of k_s^2/k_f [63]. In such cases peak current ratios as a function of scan rate would still identify the following chemical reaction mechanism, but the working curve of Figure 4.13 would no longer be valid. Examples of published work where slow charge transfer may have been a complicating factor were mentioned [62].

3.4.3   Catalytic Regeneration of Reactant

The catalytic mechanism refers to the case in which the product of an electrode reaction undergoes a homogeneous, irreversible, first-order or psuedo-first-order chemical reaction to regenerate the starting material. This scheme is represented by

$$\text{Ox} + ne^- \underset{k_c}{\rightleftharpoons} \text{Red} + \text{Z}$$

Generally, Red must react with some reagent Z in solution to regenerate Ox. By keeping this reagent in large excess, a pseudo-first-order reaction rate constant k_c is determined, from which the second-order rate constant k_2 can be determined using the known concentration of the reagent C_Z

$$k_2 = k_c(C_Z)^{-1}$$

The more general case of second-order catalysis with varying degrees of reversibility of the homogeneous reaction have been described using digital simulation to calculate the cyclic voltammetric response [64, 65].

The specific catalytic mechanism above has been treated theoretically by Savéant and Vianello [66] for the single-sweep technique and by Nicholson and Shain [38] for the cyclic experiment, assuming the charge transfer is reversible. Savéant and Vianello [67] also considered the case of quasi-reversible charge transfer and presented the integral equation for this case. Their subsequent discussion, however, dealt only with the case of reversible charge transfer.

Numerical values of $\chi(at)$ have been given [38, 66] for various values of the kinetic parameter k_c/a. These calculations show that as the term k_c/a becomes small (rapid scan rates or slow kinetics), the current approaches that for the simple reversible case. As k_c/a becomes larger the current decreases but is always larger than for the reversible case. Ultimately, the current becomes constant for increasing values of k_c/a, and the peak of the response disappears. The response assumes a sigmoidal shape, and the limiting current at cathodic potentials is given by (85) [66]. This is the case of pure catalytic control of the current [67].

$$i_{lim} = nFC^*_{Ox}(D_{Ox}k_c)^{1/2} \tag{85}$$

The half-wave potential now is also the reversible half-wave potential for the redox couple being studied, and the current is independent of scan rate. An expression for the current at any point along the wave is given by

$$i = \frac{nFC^*_{Ox}(D_{Ox}k_c)^{1/2}}{1+e^j} \tag{86}$$

where j is given by

$$j = \frac{nF}{RT}(E - E_{1/2}) \tag{87}$$

Savéant and Vianello [67] have given the corresponding pure catalytic current expression for a quasi-reversible charge transfer:

$$i = \frac{nFC^*_{Ox}(D_{Ox}k_c)^{1/2}}{1+e^j+[(D_{Ox}k_c)^{1/2}/k_s]e^{\alpha j}} \tag{88}$$

The current is still independent of scan rate and yields the same limiting current at cathodic potentials given by (85), but the half-wave potential is cathodic of the reversible $E_{1/2}$ and dependent on the kinetic parameters of the system.

The ratio of the limiting catalytic current to the diffusion current in the absence of the catalytic reaction can be used to determine the catalytic rate constant k_c. Combining (85) and (59) yields (89), since $\pi^{1/2}\chi(at) = 0.446$ at the peak of the diffusion-controlled response [65]

$$\frac{(i_\infty)_c}{(i_p)_d} = \frac{k_c^{1/2}}{0.446a^{1/2}} \tag{89}$$

Nicholson and Shain [38] have examined $(i_\infty)_c/(i_p)_d$ as a function of the kinetic parameter for the general case of mixed diffusion and catalytic current. The linear portion of the curve ($k_c/a > 1$) is where (85), (86), and (89) are applicable.

The case of mixed catalytic and diffusion control is similar to the reversible case in that the anodic peak current is equal to the cathodic peak current when measured using the cathodic response as a base line. However, in contrast to the reversible response, the potential of the cathodic peak is not constant with scan rate but shifts anodically by about $60/n$ mV for a tenfold increase in scan rate. Moreover, the value of the peak current, or plateau current if there is no peak, for the catalytic case is always larger than that of the corresponding reversible current, as mentioned above. Because the peak potential does shift anodically as the scan rate increases, a measure of the catalytic rate can be obtained by plotting the peak potential against the kinetic parameter k_c/a. Because the half-peak potential is easier to measure, it is the recommended parameter to use. $E_{p/2}$ varies between the value of the reversible case, $E_{p/2} - E_{1/2} = 28/n$ mV, for small k_c/a, and that of the pure catalytic case, $E_{p/2} - E_{1/2} = 0$, when k_c/a becomes large. Experimentally, the scan rate is adjusted so that there is still a peak in the response, $E_{p/2}$ is measured for various values of $\log(k_c/a)$, and results are compared with a working curve to obtain k_c [38].

The wave shape also begins to change markedly going from essentially diffusion control to catalytic control of the current. Examples of this are shown in Figure 4.14. Diagnostic criteria for the catalytic mechanism are summarized in Table 4.7.

The catalytic reaction following an irreversible charge-transfer reaction has also been treated by Nicholson and Shain [38]. The behavior is similar to that of the reversible charge-transfer case, but the kinetic parameter is now $k_c/\alpha a$. Many of the same criteria can be used to diagnose the catalytic reaction.

In the case of complete catalytic control, a closed-form solution for the wave is obtained similar to (86):

$$i = \frac{nFC_{Ox}^*(D_{Ox}k_c)^{1/2}}{1 + [(\alpha D_{Ox}k_c)^{1/2}/k_s]e^{\alpha j}} \tag{90}$$

At very cathodic potentials the limiting current expressed by (85) is obtained and a plot of $(i_\infty)_c/(i_v)$ as a function of the kinetic parameters can be used to obtain k_c. The linear portion of the plot for the irreversible case is now given by

$$\frac{(i_\infty)_c}{(i_p)} = \frac{k_c^{1/2}}{0.496(\alpha a)^{1/2}} \tag{91}$$

In the region of partial charge-transfer control and catalytic control, the potential of the wave shifts cathodically as the kinetic parameter $k_c/\alpha a$ is increased. A measure of the catalytic rate could be obtained from this shift in the same manner as for the reversible case.

Savéant and Vianello [67] have reported an experimental study of three catalytic reactions coupled to a reversible charge-transfer reaction. They used

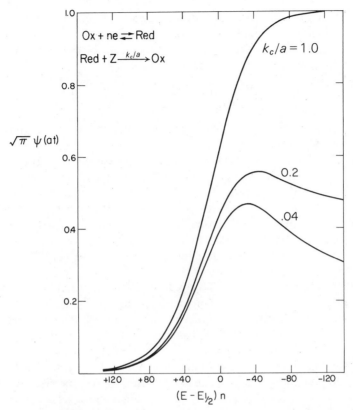

Figure 4.14 Cyclic voltammetric curves for the catalytic regeneration mechanism for various values of the kinetic parameter k_c/a.

Table 4.7 Diagnostic Criteria for Cyclic Voltammetry With Catalytic Regeneration

$$Ox + ne \rightleftharpoons Red + Z$$
$$\underset{k_c}{\overline{\qquad\qquad}}$$

Properties of the potential of the response:
  E_p shifts anodically by a maximum of 60/n mV per tenfold decrease in v; no dependence of E_p on v is observed for large or small values of k_c/a

Properties of the current function:
  $i_p/v^{1/2}$ increases at low values of v and becomes independent of v at high values

Properties of the current ratio:
  i_p^a/i_p^c is unity

Others:
  As k_c/a becomes large the response approaches a sigmoid shape; i_p^c is always larger than the reversible response

the method of calculating the pseudo-first-order rate constant according to (89) for various scan rates and various concentrations of the catalytic regenerating agent. To calculate the second-order rate constant k_2, (89) can be rewritten to include the concentration of the catalytic species Z.

$$\frac{(i_\infty)_c}{(i_p)_d} = \frac{(C_Z k_2)^{1/2}}{0.446 a^{1/2}} \tag{92}$$

The scan rates used for these studies ranged from 0.1 to 1.0 V/s, and the regenerating agent was present in thirtyfold excess or more.

The reversibility of the systems was checked at scan rates up to 10 V/s, and peak separations (anodic and cathodic) in the absence of any catalytic agent were about $60/n$ mV at all scan rates used in the subsequent catalytic studies.

The three systems studied were the reaction of H_2O_2 with Fe(II)–EDTA in acetate buffer at pH 4.7, the reaction of NH_2OH with Ti(III) in 0.2 M oxalate, and the reaction of NH_2OH with Ti(III) in 2 M H_3PO_3. The rate constant for the reaction of the Ti(III)–oxalate complex was found to be 42 $M^{-1} s^{-1}$, in good agreement with other studies [67].

Another example of the irreversible pseudo-first-order catalytic mechanism, the reoxidation of ferrous triethanolamine complex by hydroxylamine in alkaline solution, was studied by Polcyn and Shain [68]. Peak current ratios, compared to the uncomplicated diffusion-controlled case with no hydroxylamine present, were measured at various scan rates and concentrations of added hydroxylamine. Using a working curve of i_c/i_d [38], values of k_c/a were obtained and plotted against $1/a$. The slope of the straight line yields k_c at each hydroxylamine concentration. The second-order rate constant $k_2 = k_c/$[hydroxylamine] was calculated as 229 ± 5 $M^{-1} s^{-1}$ for three different concentrations, in agreement with other studies [68]. A second method at one concentration was used by matching the whole experimental voltammogram to (86). The result is shown in Figure 4.15. The value of k_c/a was 11.1 at 0.098 M hydroxylamine yielding $k_2 = 217$ $M^{-1} s^{-1}$.

Murthy and Reddy [69] have made use of (92) to study the reaction of Fe(III) in 0.1 M H_2SO_4 with leucomethylene blue generated at a platinum wire electrode in a cyclic voltammetric experiment. The average second-order rate constant of $5.5 \pm 0.5 \times 10^3$ $M^{-1} s^{-1}$ yields purely catalytic curves (see the top curve in Figure 4.14) at $[Fe(III)] = 10^{-2}$ M and scan rates of less than 100 mV/s. The value of the rate constant is obtained by assuming two ferric ions are required to oxidize leucomethylene blue.

3.4.4   Dimerization Reactions

The irreversible, second-order, following reaction or dimerization has received significant theoretical and experimental study.

$$Ox + ne^- \rightleftharpoons Red$$

$$2Red \xrightarrow{k_2} Z$$

Treatment of the cyclic voltammetric response with respect to potential shifts

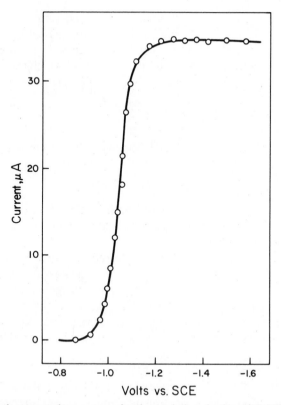

Figure 4.15 Single-sweep voltammogram for the catalytic reduction of Fe(III)–triethanolamine complex with hydroxylamine. Line—theoretical line; points—experimental. $v = 0.051$ V/s, $k_c/a = 11.1$, hydroxylamine = 0.098 M. Reprinted from [68], by Courtesy of *Analytical Chemistry*.

and peak current ratios has been presented by Olmstead and co-workers [70] and by Nadjo and Saveant [63]. A combination of integral equation procedures and finite difference simulations was used to solve the appropriate nonlinear differential equations.

As with the first-order, following reaction, the peak current of the initial scan is not overly sensitive to the kinetic parameter, $\psi = k_2 C^*_{Ox}/a$. The maximum increase in peak current is about 20% over the uncomplicated reversible mechanism [71]. The peak potential shifts anodically as scan rate increases but the sensitivity is less than for a first-order reaction, being $20/n$ mV per decade decrease in scan rate [63, 72].

$$E_p = E^0 + \frac{RT}{3nF}\left(\ln\frac{k_2 C^*_{Ox}}{a} - 3.12\right) \qquad (93)$$

Compare (93) with (83). The important variable in (93) is the concentration, which will also cause peak potential shifts of 20 mV/log C^*_{Ox} [63] at constant scan rate. The influence of peak potential on initial concentration is an important diagnostic, indicative of second-order, following chemical reactions.

The shift in potential with scan rate can be used to determine the dimerization rate constant if scan rates can be achieved that are fast enough to exceed the chemical reaction rate. The peak potential at high scan rates becomes independent of potential (reversible case). The scan rate at which this peak potential intersects the line given by (93) can be used to calculate k_2 within $\pm 50\%$ [72].

$$k_2 = 0.8 v_i C_{Ox}^* F/RT \tag{94}$$

The scan rate at the intersection is v_i in V/s. An important consideration to using this method is whether charge transfer becomes rate limiting at high scan rates. Evans [62] has shown that

$$k_2 C_{Ox}^* \leqslant k_s^2/100\pi D \tag{95}$$

is a general condition that must be obeyed in dimerization reactions for such criteria as (94) to be valid, although Nadjo and Savéant [63] claim this limit is overly restrictive.

The most sensitive parameter is still the return wave, so the peak current ratio is a measure of the dimerization reaction. Olmstead and co-workers [70] have presented a working curve of i_p^a/i_p^c plotted against the kinetic term, $\log(k_2 C_{Ox}^* \tau)$, where τ is the time to scan between $E_{1/2}$ and the switching potential. The curve shifts somewhat on the $\log(k_2 C_{Ox}^* \tau)$ axis, depending on the switching potential. This can be accounted for by using a new axis variable, w, such that

$$\log w = \log(k_2 C_{Ox}^* \tau) + 0.034(a\tau - 4) \tag{96}$$

The working curve of i_p^a/i_p^c plotted against $\log w$ in (96) is shown in Figure 4.16. Diagnostic criteria for the dimerization mechanism are summarized in Table 4.8.

More complex dimerization schemes, applicable to organic radical chemistry, have been considered [72]. These include radical–radical ion coupling and ion–ion coupling. These schemes also involve the electrode activity of the dimer formed in solution, producing multiple waves on a CV scan. Experimental examples have been reported [73].

Two examples of a reversible dimerization reaction have appeared. The first

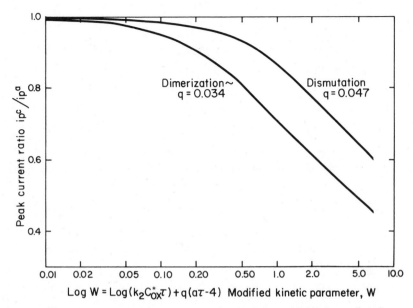

Figure 4.16 Working curve for dimerization and dismutation mechanism for cyclic voltammetry. Peak current ratio plotted against the modified kinetic parameter W, given by (96) for dimerization and (97) for dismutation.

Table 4.8   Diagnostic Criteria for Cyclic Voltammetry With a Following, Irreversible Dimerization Reaction

$$Ox + ne^- \rightleftharpoons Red$$
$$2Red \xrightarrow{k_d} Z$$

Properties of the potential of the response:
   E_p shifts cathodically by $20/n$ mV per tenfold increase in v and per tenfold decrease in initial concentration, C_{Ox}^*

Properties of the current function:
   $i_p^c/v^{1/2}$ decreases a maximum of 20% from low to high v

Properties of the current ratio:
   i_p^a/i_p^c increases with v and decreases as C_{Ox}^* increases

is a study of the oxidation of 2,6-di-*tert*-butyl-4-ethylphenoxide anion in acetonitrile [74]. The radical, formed reversibly at -0.3 V versus SCE in the first oxidation, couples reversibly to form another electroactive species reducible at -2.0 V. As the scan rate increases, the dimer has less time to decompose on the reverse scan, so the first wave becomes irreversible at the same time the wave at -2.0 V increases. The curves as a function of scan rate are shown in Figure 4.17.

A similar mechanism [75] occurs during the oxidation of 2,5-dithiacyclooctane (DTCO) in actonitrile with one additional complication. The cation radical formed in the first step is oxidized at a potential 20 mV more negative than the first oxidation. The combined wave becomes less reversible as the scan rate

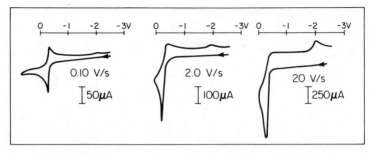

Figure 4.17 Cyclic voltammograms of 4.14 mM 2,6-di-*tert*-butyl-4-ethylphenoxide in acetonitrile with 0.1 M tetraethylammonium perchlorate on a glassy carbon electrode. Scan rates shown on figure [74].

increases or as the concentration increases. The dimer reduction wave occurs at more negative potentials at high scan rates. The mechanism can be summarized as follows:

$$DTCO \rightleftharpoons DTCO \cdot^+ + e^- \qquad E_1 = 0.335 \text{ V versus Ag/AgNO}_3$$

$$DTCO \cdot^+ \rightleftharpoons DTCO^{+2} + e^- \qquad E_2 = 0.315 \text{ V}$$

$$2DTCO \cdot^+ \rightleftharpoons (DTCO)_2^{2+} \qquad K_D = 5000 \ M^{-1} \ s^{-1}$$

$$(DTCO)_2^{2+} + 2e^- \rightleftharpoons 2DTCO \qquad E_p = -0.6 \text{ V}$$

3.4.5   Disproportionation

This reaction is a special type of regeneration mechanism in which the product of the electrode reaction undergoes a second-order cross-redox reaction to reform one molecule of the starting material.

$$A + ne^- \rightleftharpoons B$$

$$2B \xrightarrow{2k_d} A + C$$

A paradox, first considered by Feldberg [76], exists, because the reaction implies that the electrode reaction $B + n_2 e^- \rightarrow C$ should occur more easily than the reduction of A to B. In addition, the oxidation of C back to B should also occur at the electrode, the potential depending on the reversibility of the homogeneous disproportionation reaction. The complexities of these additional electrode reactions have been explored at length by Savéant and coworkers and will be described in more detail in the section on the ECE mechanism. Only the single-electrode reaction followed by irreversible dismutation is discussed here. This mechanism has been studied theoretically at length by Mastragostino, Nadjo, and Savéant [77] for the single-sweep experiment and by Olmstead and Nicholson [78] for the cyclic experiment. Nadjo and Savéant [63] have also dealt briefly with the effect of slow charge transfer on the response of the system. The kinetic parameter, $\psi = 2k_d C_A/a$, is similar to that of the dimerization reaction. The parameter of importance that distinguishes the two mechanisms is the change in peak current as the scan rate changes. Because of the regeneration, the peak current function increases as the scan rate decreases, going from 0.446 to 1.054 as the response shifts from complete diffusion control to complete kinetic control [77]. No peak occurs at the largest values of ψ, making differentiation from the pseudo-first-order catalytic case difficult. The current function versus potential for complete kinetic control has been tabulated [78]. As the kinetic parameter increases (slow scan rates or fast kinetics) the peak potential becomes more positive by 20 mV per decade as the scan rate decreases in the limit of complete kinetic control [77]. The concentration also causes a 20 mV per decade shift as it is increased. The shape of the whole response can be seen from tabulations of the potential dependence of current function as the kinetic parameter is varied. The return peak current response is quite sensitive to the kinetics of dismutation, and ratios of peak current response, i_p^a/i_p^c, have been calculated by Olmstead and Nicholson [78]. Since the values change slightly with switching potential, a parameter w,

$$\log w = \log(k_2 C_A \tau) + 0.047(a\tau - 4) \tag{97}$$

related to the kinetic parameter, is used as in the dimerization case, (96). In (97) τ is the time to scan between $E_{1/2}$ and the switching potential. Values of i_p^a/i_p^c plotted against $\log w$ are given in Figure 4.16 for both dimerization and dismutation. Table 4.9 summarizes the diagnostic criteria for this mechanism.

Farnia and co-workers [79] have presented an elegant study of the irreversible disproportionation of triphenyl ethylene radical anion in dimethylformamide (DMF). They fit their measured current ratios for the first reduction wave to the working curve of i_p^a/i_p^c versus $\log w$ (Figure 4.16) to obtain the rate constant as a function of temperature. The values were in good agreement with the rate measured by other means. Although the peak potential shifts were not used for quantitative measurements, the results as a function of scan rate were in accord with the theoretical predictions of Nadjo and Savéant [63] over a three-hundredfold range of scan rates.

Another radical that undergoes irreversible dismutation is formed by oxidation of 1-phenyl-3-pyrazolidinone over a wide pH range.

$$2R \xrightarrow{2k_d} P + Z$$

The rate constant, $2k_d$, is 320 $M^{-1} s^{-1}$ at 25°C, measured by following the loss of radical spectrally [80]. The cyclic voltammetric response of this system at a hanging mercury electrode at pH 11.5 is shown in Figure 4.18 at several scan rates. Digital simulation using the rate constant above yields the points in Figure 4.18. Correlation of the simulations to the experiment is made by using a diffusion coefficient of 8.6×10^{-6} cm^2/s and an electrode area of 0.032 cm^2. The

Table 4.9   Diagnostic Criteria for Cyclic Voltammetry With a Following
Disproportionation Reaction

$$Ox + ne^- \rightleftharpoons Red$$
$$2Red \xrightarrow{k_d} Ox + Z$$

Properties of the potential of the response:
   E_p shifts cathodically by $20/n$ mV per tenfold increase in v and per tenfold decrease
   in C^*_{Ox}

Properties of the current function:
   $i^c_p/v^{1/2}$ decreases by more than a factor of 2 as v increases

Properties of the current ratio:
   i^a_p/i^c_p increases with v and decreases as C^*_{Ox} increases

Other:
   Sigmoidal current shape at lowest v; current at low v is larger than for reversible
   reaction

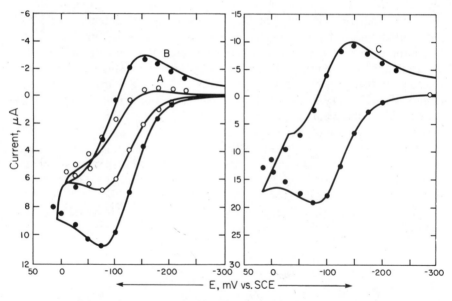

Figure 4.18   Cyclic voltammetric curve of 1-phenyl-3-pyrazolidinone on a hanging mercury electrode at pH 11.5. Lines—experimental, Points—calculated for $k_d = 320\ M^{-1}\ s^{-1}$ $D_{Ox} = D_{Red} =$
8.6×10^{-6} cm²/s, concentration $= 2$ mM, $A = 0.032$ cm². Scan rate: Curve A, 10 mV/s; Curve B,
20 mV/s; Curve C, 100 mV/s.

agreement is quite good over the whole forward part of the wave. Deviations
on the return wave are caused by proximity of mercury dissolution and the
influence of spherical diffusion [78].

3.5   Sequential Electron-Transfer Reactions

Reactions in which more than one electron is reversibly added or removed
from a species are quite common in both organic and inorganic electrochemistry.

These reactions can be divided into two types: (a) The second electron is added more easily ($E_2^0 > E_1^0$) and (b) the second electron is added with more difficulty ($E_2^0 < E_1^0$). Obviously the possibility exists that $E_1^0 = E_2^0$, a special case that can sometimes be studied experimentally with organic redox reactions where E_1^0 and E_2^0 are pH dependent. These examples are really a type of ECE mechanism where the intervening chemical step is a rapid, reversible protonation reaction.

The two extremes are easily discerned using cyclic voltammetry [41]. When the second electron transfer is much easier, $E_0^2 - E_0^1 > 180$ mV for reduction, the response is simply that of a two-electron reaction. Peak separation is 30 mV and the peak current response, $\pi^{1/2}\chi_p(at)$, is $2^{3/2}$ of that for a one-electron reaction; see (57). Other techniques such as exhaustive coulometry, may be required to determine the n value unambiguously in the absence of reliable diffusion coefficient data. The other limiting case, $E_0^2 - E_0^1 > -180$ mV, is also observed easily because the response is that of two essentially independent, reversible, one-electron reductions with peak separations of each step about 60 mV [41].

Uncertainty arises when the two half-wave potentials are not well separated, because only one peak is observed in each scan direction, but the separation is larger than 30 mV. Calculations have been made for several peak separations ranging from +200 mV to −200 mV [81]. The results are plotted in Figure 4.19 and listed in Table 4.10 for both the observed peak separation and the peak

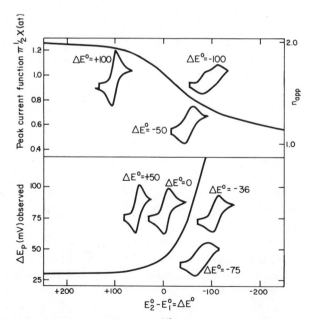

Figure 4.19 Calculated peak current function, $\pi^{1/2}\chi(at)_p$, and observed $E_p^a - E_p^c$ as a function of $(E_2^0 - E_1^0) = \Delta E^0$ for EE mechanism with reversible charge transfer. Examples of cyclic voltammogram shapes are shown for several ΔE^0 values.

Table 4.10 Potential Parameters for EE Mechanism and Reduction[a]

$\Delta E = E_2^0 - E_1^0$	$\Delta E_p = E_p^c - E_p^a$ [b]	$E_p^c - E_{p/2}$ [c]
200	29.8	29
100	30.8	31
60	34.0	33
40	37.9	34
20	37.0	37
0	42.2	41.5
−10	45.2	44
−20	49.3	48
−30	54.9	54
−35.61	58.5	57.0
−40	62.7	59.6
−50	72.4	68.3
−60	85.2	78.1
−70	98.9	89.7
−80	113	101.8
−90	126	113.1
−100	140	125.4
−110	152	137.7
−120	164	149.5
−140	188	172.7
−160	210	193.2
−200	252	236.2

[a]Values in millivolts.
[b]For two distinct waves E_p^c is the most negative forward wave and E_p^a the most positive return wave. Switching potential is 250 mV beyond $E_{1/2} = (E_2^0 + E_1^0)/2$.
[c]The first nine values from [82]. Others from [81].

current response, $\pi^{1/2}\chi_p(at)$, as a function of $E_2^0 - E_1^0$ for two consecutive reduction reactions. The shape of the whole cyclic voltammogram is also illustrated for some specific cases.

The observed peak separation ΔE ranges from 30 mV, the two-electron limit, to larger values. By the time $E_2^0 - E_1^0$ exceeds about −75 mV, a shoulder is observed in the cyclic voltammogram, as Figure 4.19 shows. The current response of the peak decreases from a two-electron reaction response to a one-electron response as ΔE^0 becomes more negative. For the case of $E_2^0 = E_1^0$ the peak response is more than twice the one-electron response, but less than a two-electron response [41]. Theoretical peak separation is 42 mV.

The peak-separation method is subject to considerable error unless the switching potential is considerably past the second peak. Experimentally, this may not be possible as solvent or electrode decomposition may interfere. The difficulty can be alleviated by analyzing only the forward scan, measuring the parameter $E_p - E_{p/2}$. Richardson and Taube [81] have extended the earlier analysis [82] of this parameter to greater values of $E_2^0 - E_1^0$. The values of $E_p - E_{p/2}$ as a function of $E_2^0 - E_1^0$ are also included in Table 4.10. The cyclic voltammetric method has been compared to pulse polarography for measurement of $\Delta E_{1/2} = E_2^0 - E_1^0$ for reduction of several binuclear ruthenium(III) ammine complexes [81]. Values of $\Delta E_{1/2}$ ranged from -50 to $-150\,\text{mV}$ indicating the second reduction was more difficult in all cases.

An example of a sequential electron transfer with both reductions occurring at the same potential has been reported [83]. The compound is a binuclear copper(II) complex, which is reversibly reduced in DMF.

The single wave centered about $-0.47\,\text{V}$ versus SCE has all the criteria of a reversible reaction, but peak separation is $42\,\text{mV}$ for scan rates between 10 and $200\,\text{mV/s}$. No coulometric results were reported, but peak currents were larger than expected for a one-electron reduction. The peak separation and current magnitudes conform to a sequential two-electron reduction with $E_1^0 = E_2^0$.

Another special case exists when $E_2^0 - E_1^0 = -RT/F \ln 4$ or $-35.6\,\text{mV}$ at 25°C. This splitting can result in a molecule containing two identical, non-interacting redox centers where no significant solvent reorganization occurs during the charge-transfer reaction [84]. Electron addition follows simple statistics.

This spacing results in a voltammetric wave with $60\,\text{mV}$ peak separation, but a current twice as large is expected for a one-electron reaction. Such a reaction has been observed by Bard and co-workers [85] for the reduction of 9,10-anthrylbis(styryl ketone), ABSK, in acetonitrile on platinum.

ABSK

Peak separation is $55 \pm 5\,\text{mV}$ for scan rates of 50–$500\,\text{mV/s}$ with the peak current function and i_p^c/i_p^a independent of scan rate. Only exhaustive coulometry of the parent (reduction) and the product (oxidation) showed this to be an overall two-electron reaction. Apparently, steric factors cause extensive rotation of the styryl keto group out of the anthracene plane, effectively insulating the two redox centers completely. Diradicals of alkyl p-nitrophenyl compounds have also been reported by Ammar and Savéant [84] with two potential steps separated by only 36 mV. The phenyl groups must be separated by two or three methylene groups for effective insulation.

The influence of slow charge-transfer kinetics on the EE mechanism has been described by Ryan [86] for the cyclic voltammetric experiment. Because the electrode reaction occurs at potentials quite positive of E_1 when E_2 is more positive than E_1, severe demands are made on the rate of the first electron transfer. The apparent rate constant will be less than that for either of the two steps. When the first electron-transfer step is rate limiting, the apparent rate constant k_s' is given by (98)

$$k_s' = k_{s1} \exp(-\alpha_1 F\Delta E/2RT) \tag{98}$$

where $\Delta E = E_2^0 - E_1^0$ [86]. Similarly, if the second electron transfer is rate limiting

$$k_s' = k_{s2} \exp[-(1-\alpha_2)F\Delta E/2RT] \tag{99}$$

For $\Delta E > 180$ mV, the theory for quasi-reversible electron transfer [43] can be used to measure k_s'.

A complication of slow charge transfer at the electrode is that charge transfer of the intermediate species in solution can become important. This is the second-order disproportionation reaction already discussed. The exact conditions of ΔE and charge-transfer parameters where disproportionation becomes important have been outlined by Ryan [86]. Obviously slow charge transfer increases the significance of the disproportionation step. A systematic approach to studying this mechanism has been given by Ryan [86] when only one wave is observed in the cyclic experiment.

First, the presence of the EE mechanism is obtained, usually by double-potential step chronoamperometry or chronocoulometry (Chapter 6). Second, the presence of disproportionation is determined by changing concentration and noting changes in wave shape and position of the peaks. Third, the presence of slow charge transfer is determined by peak-potential variations with scan rate. With peak-potential separations between 30 and 150 mV one of the charge-transfer rates is slow. If the first step is slow, peak current ratios, i_p^a/i_p^c, will be greater than 1, and if the second charge transfer is slow the ratio is always less than 1. Finally the parameters E^0, k_s, and α for each charge transfer can be determined using the equations presented by Ryan [86].

The reduction of benzil (Bn) in 0.1 M barium perchlorate with DMF solvent provides an example of a quasi-reversible sequential reduction mechanism [86, 87] where the disproportionation reaction is not important.

Bn

Digital simulation was used to match experimental curves to theoretical calculations.

The reaction is not really an uncomplicated EE mechanism because a fast ion-pairing equilibrium occurs between the two electron-transfer steps. It is fast enough, however, not to alter the kinetic response of the system, but only shifts the apparent potentials of both electron-transfer reactions.

The second electron-transfer reaction, reduction of the ion pair $Ba^{2+}-Bn^{\overline{\cdot}}$, is essentially irreversible but occurs about 200 mV more positive than the first reduction of benzil, which forms the radical anion. The overall mechanism and constants involved are given below for the hanging mercury-drop electrode.

$$Bn + e^- \rightleftharpoons Bn^{\overline{\cdot}} \qquad\qquad E_1 = -1.530 \text{ V versus SRE*}$$

$$k_s = 0.1 \text{ cm/s} \qquad\qquad \alpha = 0.7$$

$$Bn^{\overline{\cdot}} + Ba^{2+} \rightleftharpoons Ba^{2+} - Bn^{\overline{\cdot}} \qquad K = 1000 \pm 100 \ M^{-1}$$

$$Ba^{+2} - Bn^{\overline{\cdot}} + e^- \rightleftharpoons BaBn \qquad E_2 = -1.315 \text{ V versus SRE*}$$

$$k_s = 5 \times 10^{-4} \text{ cm/s} \qquad\qquad \alpha = 0.66$$

Other solvents and cations show the same overall mechanism, but rate constants, potentials, and ion-paring constants are quite different. For example, in acetonitrile with $NaClO_4$ the reduction is a completely reversible sequential reduction with E_2 about 80 mV more negative than E_1 [87].

Smith and Bard [88] have shown that the reduction of cyclooctatetraene (COT) in liquid ammonia on gold with 0.1 M KI electrolyte at $-38°C$ proceeds by a sequential EE mechanism with slow charge transfer of the first step. Peak separation at 500 mV/s was 440 mV. The intermediate radical anion forms in very low concentration indicating E_2 more positive than E_1.

Digital simulation assuming a simple EE mechanism matches experiment closely when $E_2 - E_1 = 220$ mV, $k_{s1} = 1.58 \times 10^{-4}$, and $k_{s2} = 1.26 \times 10^{-2}$ cm/s at $-38°C$. Values of $\alpha_1 = 0.4$ and $\alpha_2 = 0.5$ were also used in the simulation. Ion pairing of the intermediate radical anion with K^+ also occurs in this system. Although the constants are probably not accurate, the good fit of the simulation to experiment shows that overall two-electron reduction proceeds without the intervention of following chemical reactions [88].

*SRE is a silver wire electrode in 0.01 M AgNO$_3$ and 0.10 M TBAP in DMF [87].

3.6   The ECE Mechanism

The preceding treatment of two sequential charge-transfer reactions is expanded to include the case of an intervening chemical reaction.

$$A + ne^- \rightleftharpoons B \qquad E_1^0$$

$$B \underset{k_b}{\overset{k_f}{\rightleftharpoons}} C \qquad K = k_b/k_f, \; k = k_f + k_b$$

$$C + n_2e^- \rightleftharpoons D \qquad E_2^0$$

Electron-transfer reactions of organic compounds proceed sequentially by one-electron steps where the intermediate species must undergo a homogeneous chemical reaction, for example, ionization following oxidation or protonation following reduction, before the second electron transfer can occur. Such reactions are often fast and chemically reversible and may be difficult to measure electrochemically. Other reactions such as hydrolysis or solvolysis, nucleophilic or electrophilic addition, dimerization, and disproportionation are often irreversible and occur on a measurable time scale. These reactions form products that are also electrochemically active in the potential range of the first electron transfer. Such reductions can occur more easily than the first step, $E_2^0 > E_1^0$, or with greater difficulty, $E_2^0 < E_1^0$.

The first theoretical treatment of the ECE reaction using cyclic voltammetry was made by Nicholson and Shain [89] for the irreversible chemical reaction:

$$A + ne^- \rightleftharpoons B$$

$$B \xrightarrow{k_f} C$$

$$C + ne^- \rightleftharpoons D$$

For our discussion, this mechanism can be divided into two cases, depending on the potential of the second step with respect to the first. The first case, conceptually the easiest to visualize, occurs when the second reduction is more difficult than the first, that is, when $\Delta E = E_2^0 - E_1^0$ is negative. Under these conditions the first charge-transfer process has the properties of the irreversible EC mechanism discussed earlier. Thus the quantitative relationships as well as the diagnostic criteria, Table 4.4, for this case can be used. At small values of the chemical rate constant with respect to the time scale of the experiment (k_f/a is small), the response approaches reversible behavior in all results.

As the value of k_f/a increases, the second process begins to appear as a broad, relatively ill-defined response. This response increases to the limiting case of two fully developed cathodic responses at large values of k_f/a. The relative magnitude of the two current responses can be employed for quantitative measurements through data presented by Nicholson and Shain [89] for a working curve of i_p^c (A to B)/i_p^c (C to D) as a function of k_f/a.

The nature of the total anodic response is also a function of the rate of the chemical reaction. With the second charge transfer reversible and uncomplicated by chemical kinetics, the anodic response is simply that of a reversible charge transfer. However, at intermediate values of k_f/a, the magnitude of the anodic response for the second charge transfer is a function of both the separation in the two processes and the switching potential. With the quantitative relationships noted above, this anodic process offers little of value for other than qualitative purposes. At smaller values of the kinetic parameter, the anodic response of the first process can be employed for quantitative purposes. Simulated cyclic current–potential curves are shown in Figure 4.20 to illustrate the qualitative nature of the response as a function of k_f/a.

The variations in potential for these coupled processes are of no great value for quantitative use in view of the excellent properties of the current measurements. As the rate constant for the chemical reaction increases, the variation in potential of the first process can be employed for qualitative purposes to provide assurance that the mechanism is that treated.

Differentiation between an ECE mechanism and an uncomplicated, multistep charge-transfer reaction is also of interest. As discussed earlier, if the potentials of the two processes are sufficiently separated ($> 100/n$ mV), the two processes behave independently as uncomplicated charge-transfer systems in all respects. Excellent agreement with the theory was obtained by Polcyn and Shain [41] for the two-step reduction of copper(II) in an ammonium chloride–ammonium hydroxide medium. Thus, examination of the composite response

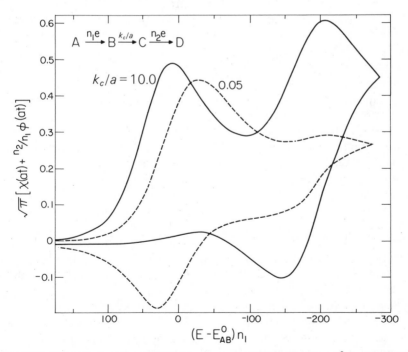

Figure 4.20 Theoretical cyclic voltammetric curves for ECE mechanism. E_2^0 is cathodic of E_1^0 and the chemical reaction is irreversible. Values of k_f/a are shown.

as a function of scan rate affords ready differentiation of the simple charge-transfer system from the ECE mechanism wherein the response is dependent on the time scale of the experiment.

As the separation in E_1^0 and E_2^0 decreases, the nature of the response becomes more complicated. Two responses may be observed if the kinetic parameter is large, but the nature of the current–potential curve becomes less recognizable as the kinetic parameter decreases.

Calculations [89] for the case in which the separation in potential is zero provide insight into the quantitative aspects of the response. The response for the product couple is likely to occur at potentials cathodic of the reversible value since the concentration of C is determined by the production of B (and the chemical reaction) and the concentration of B does not reach a limiting value at E_2^0. The magnitude of this effect also depends on the rate constant of the chemical reaction. At the appropriate value of the rate constant of the chemical reaction, two peaks may actually be observed since the product couple may be shifted cathodically and the original couple shifted anodically by the following chemical reaction.

The nature of the anodic response is also a function of the two processes. At lower values of the kinetic parameter, the anodic response is primarily that of the original couple, whereas at high values the response is that of the product couple. The anodic response is a compound function of the switching potential as a result of the potential shifts in the two processes, and thus it is difficult to employ quantitatively. More importantly, the anodic current cannot be measured accurately because of the difficulty in establishing the cathodic base line.

Quantitative measurements for an ECE mechanism that does not involve couples separated markedly from each other in potential appear to be impractical by any means other than computer-aided comparison of experimental and theoretically calculated responses.

With the product couple more readily reduced than the parent couple, only a single cathodic response is observed. If the rate constant for the chemical reaction is rapid, essentially all the initial species, A, will be converted to the final product, D, of the mechanism. As the potential is returned to anodic values, an anodic response is observed at more anodic potential as D is converted to C. A subsequent cathodic scan from the anodic limit reveals the reduction portion of the product couple, $C \rightarrow D$.

This reaction sequence is very common for the oxidation of organic compounds as noted by Adams and co-workers [40, 90]. The oxidation of triphenylamine studied by these workers [90] may be considered a classic example of this reaction. A typical cyclic voltammetric response for the oxidation of triphenylamine is illustrated in Figure 4.21. As discussed above, essentially no cathodic $(B \rightarrow A)$ response for the initial oxidation is observed, but a redox system $(C \rightleftharpoons D)$ at more cathodic potentials is apparent. The product redox couple results from reactions of N,N,N',N'-tetraphenylbenzidine produced in a bimolecular coupling of the triphenylamine cation radicals.

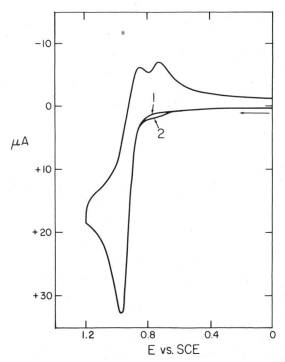

Figure 4.21   Cyclic voltammetric curve of triphenylamine in acetonitrile. Supporting electrolyte—
0.1 M TBAP, 5×10^{-4} M triphenylamine. Scan rate $= 18$ mV/s. Numbers refer to first and second
scan. Large pyrolytic graphite electrode.

The peak current function $i_p/v^{1/2}$ for the initial oxidation is large at low scan
rates and decreases to a constant value at higher scan rates as shown in Figure
4.22.

The appearance of the reaction for the product couple as well as the magnitude of the initial cathodic current as a function of scan rate can readily provide
a qualitative estimate of the rate constant of the chemical reaction. However,
quantitative measurements by cyclic voltammetry for this case are difficult
except by comparison with theoretically calculated response curves.

Oxidation of methoxyphenols in aqueous acid solutions provides another
example of the irreversible ECE mechanism [91]. The two-electron oxidation
product undergoes hydrolytic cleavage to form a quinone and methanol.

For an ECE process, the current magnitude is a function of $n_1 + n_2$ rather
than the $n^{3/2}$ function of the Randles–Ševčik equation. Nelson [92] has suggested that some ECE processes may be differentiated from a direct multielectron transfer by the current magnitude since the current for a direct two-
electron transfer is 2.83 times that for a one-electron transfer, whereas the current for the ECE reaction with n_1 and $n_2 = 1$ is only twice that for a one-electron
transfer.

When the intervening chemical reaction is considered reversible in the ECE

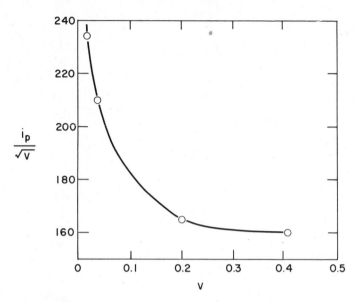

Figure 4.22   Current function behavior of oxidation peak for triphenylamine as a function of scan rate. Experimental conditions: same as Figure 4.21.

mechanism, the reaction scheme becomes much more complicated, particularly when the second electron transfer occurs more easily than the first, $E_2^0 \gg E_1^0$. The overall cyclic voltammetric response for the first sweep can be broken into kinetic zones. These zones are described by two parameters, $K = k_b/k_f$ and $\lambda = k/a$, where $k = k_f + k_b$, as shown in Figure 4.23. The nomenclature used by Savéant and co-workers [77, 93] has been retained.

The zone described by DO is the diffusion-controlled zone, where essentially no contribution by the second electron transfer occurs because λ is too small or K is very large (B predominates). A reversible one-electron reaction is observed centered about E_1^0. At large values of K and λ a second wave appears at more negative potentials than the first, centered at $E_2' = E_2^0 - 30/n \log K$. Zone DE depicts a reversible two-electron reaction centered about

$$\frac{E_1^0 + E_2^0}{2} - \frac{RT}{2nF} \ln K \tag{100}$$

because the equilibrium constant is too small to shift E_2' very much. The intermediate diffusional zone, DI, occurs when the chemical step is still very fast, but the equilibrium lies more toward B. When this occurs the apparent potential of the second wave approaches that of the first and the wave shape and height are intermediate between those of a one-electron and two-electron wave. Zone transitions from DE to DI to DO (as log K increases at large λ) are exactly those described earlier for the simple EE mechanism. In those examples, the second electron-transfer was shifted by ion pairing of the radical anion with metal

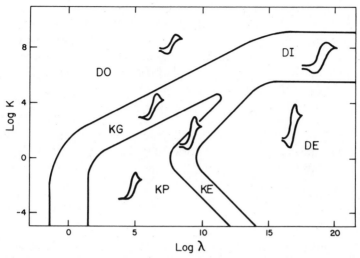

Figure 4.23   Kinetic zones for ECE mechanism with E_2^0 positive of E_1^0. See text for explanation of terminology. Reprinted from [93], by courtesy of Elsevier Press.

cations, but the simple EE mechanism is valid, because the kinetic parameter is so large.

The kinetic zones in Figure 4.23 are in three parts. The region KG is controlled by both diffusion and kinetics under conditions where the concentration of C is essentially zero [77]. The bottom of this zone where λ is small is essentially that for the irreversible reaction of B to C treated by Nicholson and Shain [89] and described earlier. Zone KE describes the situation of fast, but still rate-limiting, kinetics where the peak current approaches a two-electron height and width. The pure kinetic zone, KP, is controlled by the rate of the chemical reaction B→C. Peak currents are constant with scan rate, v, as the peak shifts anodically 30 mV/log v.

In electrochemical studies of organic compounds, the product of the first electron transfer is a radical, cationic if produced by oxidation and anionic if produced by reduction. These species can undergo proton loss or gain as the intervening electrochemical step producing a second electroactive species. Other reactions of the radical can occur, notably dimerization. Dimerization can occur concurrent with ionization or can follow the proton loss/gain. It can also occur with the neutral radical and the charged radical. Obviously the pH of the solution will control which of these reactions is rate limiting. Savéant and co-workers have studied this complication extensively both theoretically [77, 94] and experimentally [95–97]. The reduction of some carbonyl compounds in alcoholic aqueous solution occurs in two sequential steps, gradually shifting to a single two-electron wave as the pH decreases. Such reactions have the form of the classic ECE mechanism with alcohols as the main product [95]. In other cases dimerization obviously occurs based on product isolation. Because these rates, based on peak-potential shifts, are pH dependent, the dimerization involves a protonated ketyl radical [95]. Oxidation of 1,2-enediamines [97] and reduction of imines [96] give evidence of a straightforward ECE mechanism although dimeric diamines are found in the electrolysis products.

A second complication to ECE reduction mechanisms when $E_2 \gg E_1$ is the possibility of direct solution electron transfer

$$B + C \rightleftharpoons A + D$$

called either disproportionation or half-regeneration. If kinetics are fast, the overall reaction approaches a straightforward two-electron reaction depending on the equilibrium constant of the solution reaction. The equilibrium constant is determined by the potential difference, $E_2^0 - E_1^0$, in millivolts.

$$\log K_{eq} = \frac{E_2^0 - E_1^0}{59.1} \tag{101}$$

This mechanism exists in aqueous solution for the oxidation of hydroquinones, p-aminophenols, and ascorbic acid at mercury, except in very alkaline solution. Only when the solution kinetics are slow can this complication be studied. Savéant and co-workers [94] have studied this reaction sequence theoretically and shown how changes in rate constants, concentrations, and sweep rate can shift the system from ECE to disproportionation. Peak-potential shifts with scan rate vary from 20 mV/decade for rate control by the half-regeneration reaction to 30 mV/decade for reaction control by the intervening chemical reaction B→C.

The reduction wave of benzophenone to the corresponding alcohol shows 20 mV/log v dependence and [OH$^-$] dependence, indicating the reaction proceeds through the solution electron-transfer disproportionation between the ketyl radical anion and the protonated anion [95]. The rate constant is estimated to be $10^6 \ M^{-1} \ s^{-1}$.

More recently Amatore and Savéant [98] have discussed whether it is even possible to distinguish an ECE mechanism when disproportionation is significant, that is, $E_2 \gg E_1$ for reduction. They conclude that in the absence of other experimental variables, such as pH, the time scale of the experiment must be below 0.1 μs, probably outside the range of electrochemical kinetic techniques. At times longer than this, disproportionation is always the kinetically important reaction.

3.7   Adsorption

Supply of reactant to the electrode surface may be governed by an adsorption process as well as by diffusion, severely complicating the overall electrochemical process to be studied. Although adjustment of experimental conditions can reduce the adsorption complications, recognition of the presence of adsorption is mandatory to adequate use of cyclic voltammetry. One must know which electroactive species is adsorbed and whether the rate of adsorption is fast (equilibrium conditions) or slow compared with the time scale of the experiment. Diagnostic criteria are presented below for the investigation of most situations.

For equilibrium conditions, an adsorption isotherm is used to describe the process. For the situation in which no interaction among adsorbed species occurs, the Langmuir isotherm is used frequently, as given by

$$\Gamma = \frac{\Gamma^* \beta C}{1 + \beta C} \tag{102}$$

where Γ is the surface concentration, Γ^* the saturation value of the surface concentration, C the solution concentration, and β the proportionality constant. It should be noted that this isotherm assumes that a limiting surface coverage will be reached. The free energy of adsorption is then given by

$$\Delta G = -RT \ln \beta \tag{103}$$

with the value of β dependent on a number of experimental conditions including the particular electrode material, the solvent system, and the presence of other adsorbable materials. The "strength" of the adsorption is embodied in the value of the free energy and, through Γ^*, the extent to which material is adsorbed. Thus, "strong" adsorption denotes both a greater free energy and a greater extent of adsorption.

The kinetics of adsorption may be important, particularly at low concentrations. Cyclic voltammetry is not well suited to studies of slow adsorption processes unless one can use slowly renewable electrode surfaces, such as a long-drop-time mercury electrode. Feldberg [9] has used digital simulation techniques to study the effects of slow adsorption of both reactant and product on the current response in cyclic voltammetry. For this discussion adsorption–desorption is assumed to be at equilibrium.

Processes involving strong adsorption of the product of charge transfer have been treated by Wopschall and Shain [99] for the reduction of methylene blue. A typical cyclic voltammetric response involving strong adsorption of the product of charge transfer is illustrated in Figure 4.24. Under conditions in which both adsorption and diffusion control are significant, a response prior to the diffusion-controlled response is obtained for the reduction to the adsorbed state. The reverse potential scan also contains a response related to the adsorption process, in this case following the diffusion-controlled response. Observation of both the cathodic and anodic responses is mandatory for characterization of the process as the sequence

$$Ox + ne^- \rightleftharpoons Red$$

$$Red \rightleftharpoons Red_{(ads)}$$

The separation between peak potentials of the adsorption- and diffusion-controlled responses is a function of the free energy of adsorption. As the energy

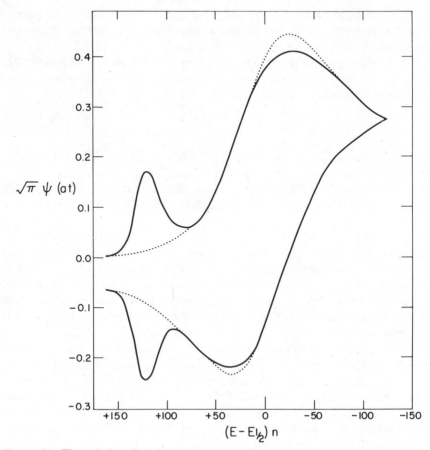

Figure 4.24 Theoretical cyclic voltammetric wave with strong adsorption of product. Dotted curve shows normal reversible wave.

of adsorption increases, the separation in the peak potential of the two responses ΔE_p increases. However, this separation is also a function of the bulk concentration of reactant and cannot be used directly to calculate the free energy of adsorption. The reader is referred to the original literature [99] for a discussion of the measurements required to relate ΔE_p to ΔG. The shape of the adsorption-controlled response is a function of the potential dependency of the isotherm. If adsorption increases as the potential is made more negative, there is a sharper current response than would otherwise occur if adsorption were potential independent. The sharpness of the response is also concentration dependent. The width increases at lower concentrations because diffusion cannot maintain surface excesses in complete equilibrium with the solution concentration.

The solution concentration also governs the relative importance of adsorption and diffusion processes when strong adsorption of product occurs. At very low concentrations, reduction to the adsorbed state is the primary process. As the concentration increases, the relative height of the adsorption response with respect to the diffusion-controlled response decreases, Increasing the concentration shifts the potential of the adsorption process toward positive values by approximately $60/n$ mV/decade increase in concentration, if the isotherm is potential independent.

At very slow scan rates, the adsorption response may not appear. As the scan rate is increased, the adsorption response increases relative to the diffusion-controlled response. At very rapid scan rates, only the adsorption response is noted. This behavior illustrates the point that although the peak current is an increasing function of the scan rate, the total coulombs involved in the experiment decrease with an increase in scan rate. At very rapid scan rates, the total

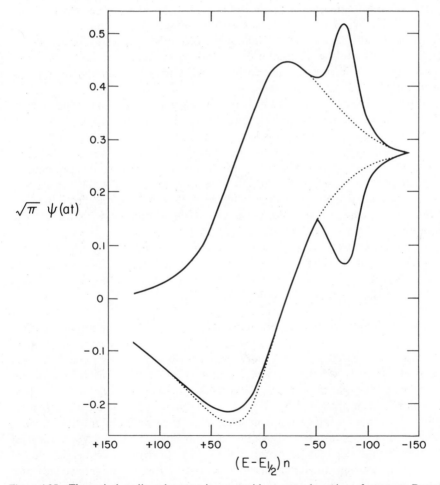

Figure 4.25 Theoretical cyclic voltammetric wave with strong adsorption of reactant. Dotted curve shows normal reversible wave.

coulombs transferred become equal to (or less than) the amount that can be transferred to adsorbed product.

Processes involving strong adsorption of the reactant produce a current response past the diffusion-controlled process as illustrated in Figure 4.25. Appearance of adsorption-controlled response for both directions of the scan implies reversible adsorption–desorption, that is,

$$Ox \rightleftharpoons Ox_{(ads)}$$

$$Ox_{(ads)} + ne^- \rightleftharpoons Red$$

The variation of the adsorption response with experimental conditions parallels that for strong adsorption of product.

The relative magnitudes of the adsorption and diffusion responses are a function of concentration, with the adsorption response predominant at lower concentrations. Integration of the total area under the adsorption response (coulombs) provides convenient access to surface coverage values. The limiting value of the integrated area as the concentration is increased is a convenient means to estimate the maximum or saturation surface coverage. As the bulk concentration is increased, the relative magnitude of the adsorption response decreases.

When the electroactive species is only weakly adsorbed, the cyclic voltammetric response is not markedly different from that of the uncomplicated case. The free energy of adsorption is low, so the potential difference for reduction of the solution species and the adsorbed species is too small to show a separate response. However, the magnitude of the response reflects the presence of adsorption as illustrated in Figure 4.26. The peak potentials for the forward and return scan are also closer together. Differentiating this response from that of a multielectron charge-transfer reaction is possible by studying the scan rate and concentration dependence of the current.

At fast scan rates, more of the total charge is passed through the adsorbed reactant, causing the current function, $i_p/v^{1/2}$, to increase with increasing scan rates [99]. Since this behavior is unique, a significant increase in the current function at faster scan rates is a strong indication of the presence of weak adsorption.

The concentration of reactant may also provide diagnostic information for the detection of the presence of weak adsorption. As the concentration of reactant is increased, a greater portion of the total response occurs through diffusion control. A limiting behavior at high concentrations is that of uncomplicated charge transfer. The theoretical response as a function of concentration has been presented by Wopschall and Shain [99]. These authors have emphasized that the full variation in behavior from a predominantly adsorption-controlled to a predominantly diffusion-controlled reaction involves several orders of magnitude in concentration, which is frequently beyond experimental practicality. Perhaps the best experimental compromise for diagnostic purposes is to use the variation in response with scan rate and confirm the behavior at

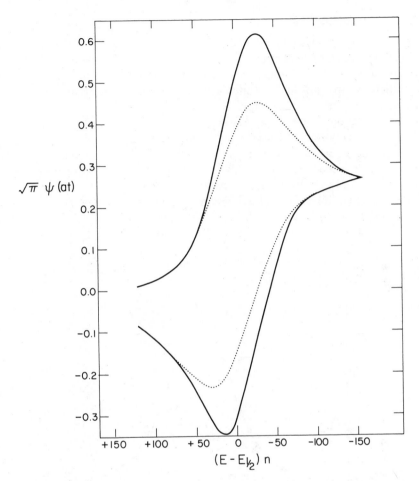

Figure 4.26 Theoretical cyclic voltammetric wave with weak adsorption of reactant. Dotted curve shows normal reversible wave.

several concentrations. Use of cyclic voltammetry specifically for the study of weak adsorption appears not to be fruitful, particularly in comparison with other techniques.

The above treatment has also assumed the absence of any complicating chemical reactions. Unfortunately, many electrochemical reactions of organic compounds involve both coupled chemical reactions and adsorption. The manner in which these two processes reflect their presence on each other has been presented by Wopschall and Shain [100] for the case of a following, irreversible chemical reaction with weak adsorption of the reactant. A semi-empirical method was developed by these authors and applied successfully to the reduction of azobenzene.

3.8   Thin Surface Films

The electrochemistry of surface layers adsorbed or precipitated on the electrode surface is of great interest, both for analytical studies and for electrode-mechanism studies. Stripping analysis, in which metal ions are concentrated by reduction into thin films of mercury on platinum or graphite and subsequently reoxidized, is an example of an analytical method. Studies of oxide layers or other precipitated salts also fall under the purview of thin-layer techniques. The response at chemically modified electrodes [101] in which electroactive species are irreversibly bonded to the electrode, for example, by silanization, can be described mathematically by the same thin-layer equations. Historically, these equations were first derived for, and applied to, electrochemical cells in which a thin layer of solution is constrained next to the electrode by some mechanical means, minimizing the effects of mass transfer.

The equation describing the cyclic voltammogram of a reversible charge-

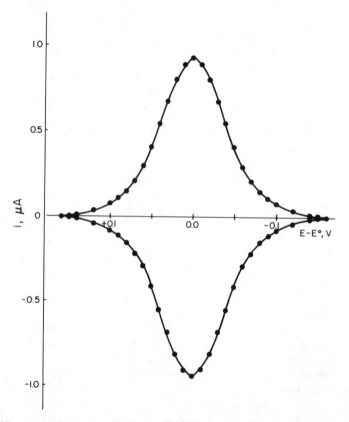

Figure 4.27 Calculated cyclic current–potential curve for reduction in a thin-layer cell. $v =$ 1 mV/s, volume = 1 μL, concentration = 1 mM [102]. Reprinted from [102], p. 132, by courtesy of Marcel Dekker, Inc.

transfer reaction at slow scan rates in a thin cell has been derived by Hubbard and Anson [102]

$$i = \frac{nFalC^*_{Ox}e^j}{(1+e^j)^2} \tag{104}$$

where a is the scan rate, nFv/RT, and j is the potential expression, (53). The thin-cell thickness is l in cm. Concentration is C^*_{Ox} in mM, and the current is expressed in $\mu A/cm^2$. The peak current occurs at the formal potential of the redox reaction and is given by

$$i_p = \frac{nFalC^*_{Ox}}{4} \tag{105}$$

On the return sweep, the reoxidation current is the mirror image of the forward current reflected about the zero-current axis as shown in Figure 4.27. The curves are symmetrical about the E^0 potential axis also, although this may be difficult to determine experimentally except at very slow scan rates. The width of the peak at half-height is $90/n$ mV [102].

If the thin layer of reactant is obtained by adsorption to the electrode surface,

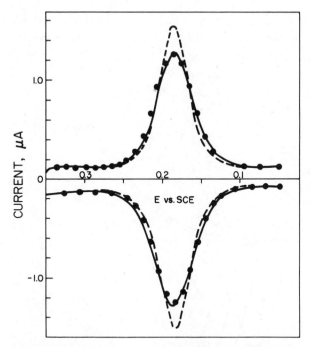

Figure 4.28 Cyclic voltammogram for 1.9×10^{-10} mol/cm^2 of 9,10-phenanthrenequinone irreversibly adsorbed on the basal plane of a pyrolytic graphite electrode in 1 M HClO$_4$. $v = 50$ mV/s, (——) experimental voltammogram. (----) calculated voltammogram according to (106), (●) points calculated assuming destabilizing interaction of species. Reprinted from [104], by courtesy of *Analytical Chemistry*.

the concentration terms in (104) and (105) are modified somewhat [103]

$$i = \frac{nFa\Gamma_{Ox}^*(b_{Ox}/b_{Red})e^j}{(1+(b_{Ox}/b_{Red})e^j)^2} \qquad (106)$$

The terms, $b_{Ox} = \beta_{Ox}\Gamma_{Ox}$ and $b_{Red} = \beta_{Red}\Gamma_{Red}$, are the Langmuir adsorption coefficients for the oxidized and reduced forms, respectively. The peak current becomes

$$i_p = \frac{nFa\Gamma_{Ox}^*}{4} \qquad (107)$$

In the thin-layer cell VC_{Ox}^* moles of material react and in the adsorbed layer the equivalent quantity is $A\Gamma_{Ox}^*$, where A is the electrode area in square centimeters.

The position of the adsorption peak with respect to the formal potential of the redox couple in solution depends on the relative strength of adsorption of reduced and oxidized species as discussed in the last section. The surface standard potential, where the peak occurs, is given by [103]

$$E_s^0 = E^0 + \frac{RT}{nF} \ln\left(\frac{b_{Red}}{b_{Ox}}\right) \qquad (108)$$

The preceding discussion applies only to ideal surfaces, assuming no interaction among species adsorbed. Brown and Anson [104] have studied several systems, of which Figure 4.28 is an example, where interactions occur between

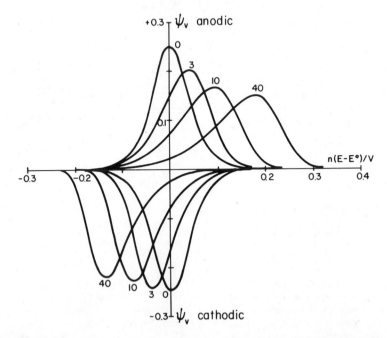

Figure 4.29 Cyclic voltammograms of surface films with slow charge-transfer control. $\alpha = 0.6$; the values of vnF/RTk_s are shown on the curve. Reprinted from [105], by courtesy of Elsevier Press.

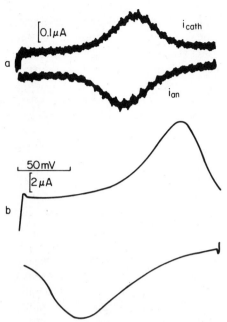

Figure 4.30 Cyclic voltammograms of surface-adsorbed benzo[c]cinnoline. Starting potential, -0.65 V, concentration in solution $= 3 \times 10^{-6}$ M. Scan rate: Curve a, 0.4 V/s; Curve b, 6.67 V/s. Reprinted from [106], by courtesy of Elsevier Press.

the reduced and oxidized species to destabilize the adsorbed reactants on the pyrolytic graphite electrode [104].

Irreversible and quasi-reversible charge-transfer reactions have been studied in adsorbed layers and thin-layer cells [102, 103]. The peaks become less symmetric and the peak potentials start separating as the charge-transfer rate decreases. Laviron [103] has given equations that describe the peak separation in the cyclic experiment in terms of k_s and α. Calculated curves are shown in Figure 4.29 for various values of a/k_s with $\alpha = 0.6$ [105]. Experimentally, benzo[c]cinnoline adsorbed on mercury at pH 8.70 gave the curves shown in Figure 4.30 at two scan rates [106]. The values of $\alpha = 0.53$ and $k_s = 38.9$ s^{-1} were measured for this two-electron reduction.

Chemical reactions coupled to surface electrochemical reactions or occurring in thin-layer cells have been described theoretically and experimentally by Laviron [103]. These include the EC mechanism of which the benzidine rearrangement following reduction of azobenzene on the mercury electrode surface is an example. Other mechanisms have included the ECE reaction in a thin layer and regeneration mechanisms in thin layers or adsorbed films.

4 ADMITTANCE AND IMPEDANCE TECHNIQUES

The voltage excitation used in cyclic voltammetry is a triangular wave with a peak-to-peak amplitude that can be several volts but with a frequency generally below 100 Hz. By comparison, the admittance technique relies on a sinusoidal excitation that has a peak-to-peak amplitude of only a few millivolts but a frequency of up to 100 kHz or more. Cyclic voltammetry is then a large-amplitude, low-frequency technique whereas the admittance method is a small-amplitude, high-frequency technique.

Small applied voltages have the obvious disadvantage that they will result in small current responses. However, this is a problem only at low frequencies. Just as peak currents are dependent upon scan rate in cyclic voltammetry, alternating currents are dependent upon frequency in admittance measurements. The high frequencies employed in admittance measurements result in much larger, and therefore more analytically useful, responses than are seen in cyclic voltammetry. Furthermore, the small-amplitude excitations enable one to linearize the absolute rate equation (31) or the Nernst equation (32) and therefore avoid the cumbersome convolution expressions (58) encountered in cyclic voltammetry. The mathematics are then enormously simplified and the data may be treated more analytically than was possible with cyclic voltammetry.

The admittance technique does not have the great diagnostic capability possessed by cyclic voltammetry regarding homogeneous kinetics. However, its ability to characterize interfacial processes is probably unparalleled among the various electrochemical methods. This is especially true of fast processes. The analytical utility of admittance measurements, in the form of ac voltammetry (Section 5), rivals that of pulse polarography (Chapter 9).

The impedance of a cell is simply the reciprocal of its admittance. Measurement of current responses to voltage excitations (admittances) are often easier to make than voltage responses to current excitations (impedances). However, the data, measured at different frequencies, may then be plotted as impedances.

For small applied voltages it is convenient to express (11), the current response expression, in terms of voltage rather than concentration. A small voltage change, ΔE, will cause changes in C_{Ox} and C_{Red} (ΔC_{Ox} and ΔC_{Red}, respectively) according to (109)

$$E + \Delta E = E^0 + \frac{RT}{nF} \ln \left(\frac{C_{Ox} + \Delta C_{Ox}}{C_{Red} + \Delta C_{Red}} \right) \qquad (109)$$

If (32), the Nernst equation, is subtracted from (109), one obtains

$$\Delta E = \frac{RT}{nF} \left\{ \ln \left(\frac{C_{Ox} + \Delta C_{Ox}}{C_{Ox}} \right) - \ln \left(\frac{C_{Red} + \Delta C_{Red}}{C_{Red}} \right) \right\} \qquad (110)$$

Since ΔE is small (< 5 mV), $\Delta C_{Ox}/C_{Ox}$ and $\Delta C_{Red}/C_{Red}$ will be $\ll 1$ and (110) may be linearized by truncating the series,

$$\ln(1 + \chi) = \chi - \frac{\chi^2}{2} + \frac{\chi^3}{3} - \frac{\chi^4}{4} + \cdots \qquad (111)$$

where $\chi = \Delta C_{Ox}/C_{Ox}$, or $\Delta C_{Red}/C_{Red}$ as needed, after the first term. Therefore

$$\Delta E \simeq \frac{RT}{nF} \left\{ \frac{\Delta C_{Ox}}{C_{Ox}} - \frac{\Delta C_{Red}}{C_{Red}} \right\} \qquad (112)$$

A relationship between ΔC_{Ox} and ΔC_{Red} can be obtained by comparison of (11) and (12), from which it follows that

$$\sqrt{D_{Ox}}\Delta C_{Ox} = -\sqrt{D_{Red}}\Delta C_{Red} \tag{113}$$

The Laplace transform of (112) now becomes

$$\overline{\Delta E} = \frac{RT}{nF}\left(\frac{1}{\sqrt{D_{Ox}}C_{Ox}} + \frac{1}{\sqrt{D_{Red}}C_{Red}}\right)\sqrt{D_{Ox}}\,\overline{\Delta C}_{Ox} \tag{114}$$

Substitution of (114) into (11) yields

$$\bar{i} = \frac{-\dfrac{n^2F^2}{RT}\{\sqrt{p}\}\{\overline{\Delta E}\}}{\left\{\dfrac{1}{\sqrt{D_{Ox}}C_{Ox}} + \dfrac{1}{\sqrt{D_{Red}}C_{Red}}\right\}} \tag{115}$$

This is the desired relationship, in the Laplace domain, between current and voltage. It can be written more conveniently by defining

$$\sigma = \frac{RT}{\sqrt{2}n^2F^2}\left(\frac{1}{\sqrt{D_{Ox}}C_{Ox}} + \frac{1}{\sqrt{D_{Red}}C_{Red}}\right) \tag{116}$$

The equation then becomes

$$\bar{i} = -\frac{1}{\sqrt{2}\sigma}\{\sqrt{p}\}\{\overline{\Delta E}\} \tag{117}$$

Assume that an excitation of the form

$$\Delta E = \Delta E_{dc} + \Delta E_{ac}\sin\omega t \tag{118}$$

is applied to the cell. ΔE_{dc} and ΔE_{ac} are constant voltages. According to tables of Laplace transforms, (118) may be transformed into

$$\overline{\Delta E} = \frac{\Delta E_{dc}}{p} + \frac{\omega\Delta E_{ac}}{\omega^2 + p^2} \tag{119}$$

which, when substituted into (117), leads to

$$\bar{i} = -\frac{1}{\sqrt{2}\sigma}\left\{\frac{\Delta E_{dc}}{\sqrt{p}} + \frac{\omega\sqrt{p}}{p^2 + \omega^2}\Delta E_{ac}\right\} \tag{120}$$

Inverse transformation, using the convolution integral (15), results in

$$i = -\frac{1}{\sqrt{2}\sigma}\left\{\frac{\Delta E_{dc}}{\sqrt{\pi t}} + \omega\Delta E_{ac}\int_0^t \frac{\cos\omega(t-\tau)}{\sqrt{\pi\tau}}d\tau\right\} \tag{121}$$

The resultant current flow consists of a $1/\sqrt{t}$ transient term and a convolution integral. The transient term accounts for whatever current must flow to change the average surface concentration of oxidized species from a ratio that satisfies the Nernst equation with E volts applied to a ratio that satisfies the Nernst equation with $E + \Delta E_{dc}$ volts applied. An instrument that measures only recurring signals will not monitor the transient term. Therefore, it does not require further consideration. Additional transient terms will arise from evaluation of the convolution integral in (121) and, for the same reason, they are also of no concern. They may be avoided by taking the limit of the integral as $t \to \infty$. In this way, only the "steady-state" solution is found, which is the quantity measured experimentally. With the trigonometric identity

$$\cos \omega(t - \tau) = \sin \omega t \sin \omega \tau + \cos \omega t \cos \omega \tau \qquad (122)$$

Equation (121) becomes

$$\Delta i_{ac} = - \frac{\omega \Delta E_{ac}}{\sqrt{2}\sigma} \left\{ \sin \omega t \int_0^\infty \frac{\sin \omega \tau}{\sqrt{\pi \tau}} \, d\tau + \cos \omega t \int_0^\infty \frac{\cos \omega \tau}{\sqrt{\pi \tau}} \, d\tau \right\} \qquad (123)$$

where Δi_{ac} is the sinusoidal part of the total current. The integrals may be easily evaluated. They both equal $1/\sqrt{2\omega}$ and the final result is

$$\Delta i_{ac} = - \frac{\Delta E_{ac}}{2\sigma} \sqrt{\omega} \{ \sin \omega t + \cos \omega t \} \qquad (124)$$

We can now define admittance, Y, as the ratio $|\Delta i_{ac}/\Delta E_{ac}|$ and (124) becomes

$$Y = \frac{\sqrt{\omega}}{2\sigma} (\sin \omega t + \cos \omega t) \qquad (125)$$

The sine and cosine functions show that Y is composed of real (inphase) and quadrature (out-of-phase) components, since these functions are quadrature to each other, as discussed in the Introduction. This quadrature relationship is most often, and more conveniently, expressed by writing Y as a complex number,

$$Y = \frac{\sqrt{\omega}}{2\sigma} (1 + j) \qquad (126)$$

where $j = \sqrt{-1}$; j is quadrature to unity in the same sense that the cosine is quadrature to the sine, as may be seen from Figure 4.31. A vector has been drawn in an arbitrary direction and represented as the unit vector. A "j" vector

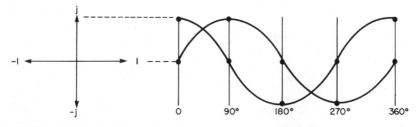

Figure 4.31 Relationship between the sine and cosine functions and rotating unit and j vectors. The projections of these vectors define sine and cosine waves as they rotate in a counterclockwise direction.

has then been drawn perpendicular to it, but sharing the same origin. It is apparent that multiplication of the unit vector by j rotates it $90°$ (quadrature) in a counterclockwise direction. A second multiplication by j rotates the vector an additional $90°$ since $\sqrt{-1} \cdot \sqrt{-1} = -1$. A third multiplication by j rotates the vector $90°$ more, to $-j$, and a fourth multiplication rotates it back to its initial orientation. As also shown in Figure 4.31, the rotation of the unit vector generates a sine wave when its reflection on the perpendicular axis is plotted against angle of rotation. Similarly, rotation of the j vector generates a cosine wave.

Y, as expressed in (126), is a vector and may be mapped onto Figure 4.31, as shown in Figure 4.32. In that case Y_R, the real component of the vector, is oriented in the direction of unity and Y_Q, the quadrature component of the vector, is oriented in the j direction. Rotation of Y in a counterclockwise direction at constant angular velocity (frequency) now generates (125)—the sum of sine and cosine waves—when the reflection of Y on the perpendicular axis is plotted versus angle of rotation. Since angle of rotation is directly proportional to time, it follows that a higher angular velocity will generate a sinusoidal wave with identically higher frequency.

This vector representation of Y is extremely useful. From the figure, the phase angle, ϕ, by which Y leads the applied voltage (which may be represented as a vector pointing in the "Y_R" direction of the figure) is simply arctan (Y_Q/Y_R). The amplitude (modulus) of Y is $\sqrt{Y_R^2 + Y_Q^2} = \sqrt{\omega}/\sqrt{2\sigma}$ by the Pythagorean theorem. As will be shown later, Y, Y_R, Y_Q, or ϕ can all be measured independently.

We can also define the Laplace transform of the admittance, $\bar{Y}$, as $|\overline{\Delta i/\Delta E}|$. Equation (117) now becomes

$$\bar{Y} = \frac{1}{\sqrt{2\sigma}} \sqrt{p} \tag{127}$$

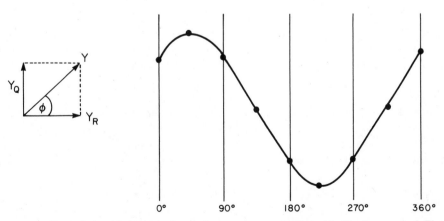

Figure 4.32 Relationship between the admittance vector and a sinusoid with phase angle ϕ. The sinusoid is the projection of Y as it rotates in a counterclockwise direction.

$\bar{i}$ has been replaced by $\overline{\Delta i}$ in (117) because the current responses are expected to be small. Note that if p is replaced by $j\omega$, (127) leads directly to (126) since $\sqrt{j} = (1+j)/\sqrt{2}$ by DeMoivre's theorem [107]

$$Z^{1/n} = r^{1/n} \left[\cos \left(\frac{\theta}{n} + 2\pi m \right) + j \sin \left(\frac{\theta}{n} + 2\pi m \right) \right]$$

$$m = 0, 1, 2, \ldots, n-1 \qquad (128)$$

Z is any complex number, n is an integer, θ is the angle between Z and the real axis, and r is the modulus (the square root of the sum of the squares of the two components) of Z.

Substitution of $j\omega$ for p amounts to a Fourier transformation, since when Y is expressed in the form of (125), it is actually a Fourier series truncated after the first term. Any single-valued, continuous, and recurring function may be expanded into a Fourier series of the form

$$\frac{1}{2} a_0 + \sum_{n=1}^{\infty} (a_n \cos nx + b_n \sin nx) \qquad (129)$$

By analogy with (125), $a_0 = 0$, $a_1 = b_1 = \sqrt{\omega}/2\sigma$, and all other a_n's and b_n's with $n > 1$ are zero. Equation (125) does not have terms with $n > 1$ (terms which represent harmonic frequencies) as a direct result of the linearization of (110). The equivalency of (125) and (126) has already been explained.

The preceding discussion demonstrates that the admittance of an electrochemical system can easily be derived from its system transform. One must simply perform a Fourier transformation by substituting $j\omega$ for p in the system transform. Additional manipulations with DeMoivre's theorem (128) will then be required to get the final result into the form of (126). Cases in which kinetic complications are involved may also be treated in this way, as will be shown later (Section 4.4).

4.1   Measurement Principles

Currents proportional to Y_R and Y_Q can be measured experimentally in the same way that Fourier coefficients are determined mathematically in textbook discussions of the Fourier series. In general, the total current flow will be

$$\Delta i_{ac} = \Delta i_R \sin \omega t + \Delta i_Q \cos \omega t \qquad (130)$$

where $Y_R = \Delta i_R / \Delta E_{ac}$ and $Y_Q = \Delta i_Q / \Delta E_{ac}$. To determine Δi_R we simply multiply (130) by a filtered sample of the input signal, (118) with $\Delta E_{dc} = 0$, and integrate across one complete cycle. The limits of the integration do not matter as long as they cover a complete cycle. The result is

$$\int_{\text{one cycle}} \Delta i_{ac} \Delta E \, dt = \Delta i_R \Delta E_{ac} \int_{\text{one cycle}} \sin^2 \omega t \, dt + \Delta i_Q \Delta E_{ac}$$

$$\int_{\text{one cycle}} \sin \omega t \cos \omega t \, dt \qquad (131)$$

Substitution of $\theta = \omega t$ into (131) results in

$$\int_{\text{one cycle}} \Delta i_{\text{ac}} \Delta E \, dt = \frac{\Delta i_R \Delta E_{\text{ac}}}{\omega} \int_0^{2\pi} \sin^2 \theta \, d\theta + \frac{\Delta i_Q \Delta E_{\text{ac}}}{\omega} \int_0^{2\pi} \sin \theta \cos \theta \, d\theta$$

(132)

where the limits of integration have been set at 0 and 2π, consistent with the allowed values of θ. The second term on the right-hand side of (132) integrates to zero, whereas the first term integrates to $\pi \Delta i_R \Delta E_{\text{ac}}/\omega$. Division by $\tau/2\Delta E_{\text{ac}}$, where $\tau = 1/f = 2\pi/\omega$, results in Δi_R. This process can be accomplished at frequencies below about 1 Hz with a commercially available, multiplier integrated circuit, a sample of the applied signal, and a gated integrator set to the period of interest. One must then make the final division manually. At higher frequencies the output of the multiplier can be filtered. Then integration and division steps are not necessary since filtration may be thought of as a continuous averaging process of integration followed by division. A stable output is therefore produced.

A similar process is used to determine i_Q, in which case (130) is multiplied by a sample of the applied signal shifted $90°$

$$\Delta E_Q = \Delta E_{\text{ac}} \cos \omega t$$

(133)

Equation (132) then becomes

$$\int_{\text{one cycle}} \Delta i_{\text{ac}} \Delta E \, dt = \frac{\Delta i_R \Delta E_{\text{ac}}}{\omega} \int_0^{2\pi} \sin \theta \cos \theta \, d\theta + \frac{\Delta i_Q \Delta E_{\text{ac}}}{\omega} \int_0^{2\pi} \cos^2 \theta \, d\theta$$

(134)

Now the first term integrates to zero and the second to $\Delta i_Q \Delta E_{\text{ac}} \tau/2$. Proceeding as before, Δi_Q may be determined.

Resolution of the total current into its real and quadrature components can also be accomplished with a square wave, for which the Fourier series representation is

$$\Delta E = \frac{4\Delta E_{\text{ac}}}{\pi} \sum_{n=0}^{\infty} \frac{\sin(2n+1)\omega t}{(2n+1)}$$

(135)

Only the first two terms of this series are needed to illustrate the use of this function

$$\Delta E = \frac{4\Delta E_{\text{ac}}}{\pi} (\sin \omega t + \tfrac{1}{3} \sin 3\omega t)$$

(136)

Equation (132) now becomes

$$\int_{\text{one cycle}} \Delta i_{\text{ac}} \Delta E \, dt = \frac{4\Delta E_{\text{ac}}}{\pi \omega} \left\{ \Delta i_R \int_0^{2\pi} \sin^2 \theta \, d\theta + \frac{\Delta i_R}{3} \int_0^{2\pi} \sin \theta \sin 3\theta \, d\theta \right.$$

$$\left. + \Delta i_Q \int_0^{2\pi} \sin \theta \cos \theta \, d\theta + \frac{\Delta i_Q}{3} \int_0^{2\pi} \cos \theta \sin 3\theta \, d\theta \right\}$$

(137)

Tables of integrals show that all these terms integrate to zero except the first one.

Therefore

$$\int_{\text{one cycle}} \Delta i_{\text{ac}} \Delta E \, dt = \frac{4\Delta E_{\text{ac}}}{\omega} \Delta i_R = \left(\frac{2\Delta E_{\text{ac}} \tau}{\pi} \right) \Delta i_R \qquad (138)$$

Similarly, $(2\Delta E_{\text{ac}} \tau/\pi)\Delta i_Q$ may be obtained by multiplication of the signal by a square wave shifted $90°$ followed by integration over one cycle. As before, one multiplies by the suitably phase-shifted square wave and filters the result to obtain a value proportional to Δi_R or Δi_Q.

The successful use of square waves to accomplish resolution illustrates a great advantage of this approach. Notice that the second and fourth terms in (137) integrate to zero *because they contain a frequency term* (3θ) *different from the fundamental* (θ). Experimentally, such terms can arise from noise, and it is satisfying to know that this technique (cross correlation) not only picks out the desired component of phase but also discriminates against all other signals which do not have the desired frequency. It provides an automatic noise filter. Unfortunately, errors can result from coherent odd-harmonic terms which may appear in Δi_{ac} (130) since these components will couple with the odd-harmonic term of the square wave (135).

Multiplication by a square wave can be accomplished without resort to multiplier circuits. Instead one can use two switches, such as field effect transistors, which are alternately turned on and off. The signal to be resolved, (130), is first split into two parts. The first part is fed into the input of one of the switches while the second part is inverted $180°$ and fed into the input of the other switch. The switches are then opened (turned off) and closed (turned on) by a timing circuit in accordance with whether Δi_R or Δi_Q is being resolved. The outputs of the two switches are then added together and integrated or filtered depending upon the frequency.

The procedure outlined above is most applicable to analog instrumentation. Speed advantages accrue from use of Hadamard, rather than Fourier, transform when data are processed by microcomputer, however. This speed advantage derives because multiplications by sine waves are not required for Hadamard transformation. Further discussion can be found in [108].

4.2   Quasi-Reversible Charge Transfer

Removal of the reversible charge-transfer restriction imposed in the preceding discussion requires use of a linearized form of (31), rather than the linearized Nernst equation (112). Equation (31) interrelates four parameters: current, voltage, and the concentrations of the oxidized and reduced species at the surface of the electrode. Any small change in one of these parameters causes small changes in each of the other three. Therefore linearization of the equation begins with replacing each of these parameters with differentials, as was done before in the linearization of the Nernst equation; i is replaced by $i + \Delta i$, E by $E + \Delta E$, C_{Ox} by $C_{\text{Ox}} + \Delta C_{\text{Ox}}$, and C_{Red} by $C_{\text{Red}} + \Delta C_{\text{Red}}$. The resultant expression is then simplified using the series expansion

$$e^{\chi} = 1 + \chi + \frac{\chi^2}{2!} + \frac{\chi^3}{3!} + \frac{\chi^4}{4!} + \cdots \tag{139}$$

truncated after the second term. In (139), $\chi = nF\Delta E/RT$ and the truncation error is not too great provided ΔE is less than about 5 mV. Subtraction of (31) then provides, after Laplace transformation,

$$\overline{\Delta i} = -nFk_s \left\{ \left[\frac{\alpha nF}{RT} C_{Ox}e^{-\alpha nF(E-E^0)/RT} + \frac{(1-\alpha)nF}{RT} C_{Red}e^{(1-\alpha)nF(E-E^0)/RT} \right] \overline{\Delta E} \right.$$

$$\left. - \left[e^{-\alpha nF(E-E^0)/RT}\overline{\Delta C_{Ox}} \right] + \left[e^{(1-\alpha)nF(E-E^0)/RT}\overline{\Delta C_{Red}} \right] \right\} \tag{140}$$

The first term on the right-hand side of (140) has the form of Ohm's law: current equals voltage divided by resistance. It therefore follows that a resistance, referred to as the "charge-transfer resistance," R_{CT}, may be defined:

$$R_{CT} = \frac{RT/n^2F^2k_s}{\{\alpha C_{Ox}e^{-\alpha nF(E-E^0)/RT} + (1-\alpha)C_{Red}e^{(1-\alpha)nF(E-E^0)/RT}\}} \tag{141}$$

Equation (140) can now be combined with (141) and expressions for $\overline{\Delta C_{Ox}}$ and $\overline{\Delta C_{Red}}$ obtained from (11) and (12) to yield

$$-\overline{\Delta i} = \frac{\overline{\Delta E}}{R_{CT}} + k_s \frac{\overline{\Delta i}}{\sqrt{p}} \left\{ \frac{e^{-\alpha nF(E-E^0)/RT}}{\sqrt{D_{Ox}}} + \frac{e^{(1-\alpha)nF(E-E^0)/RT}}{\sqrt{D_{Red}}} \right\} \tag{142}$$

$\overline{\Delta i}$ has replaced $\overline{i}$ in (11) and (12), since the surface concentrations are changing as a result of small changes in the magnitude of a large current. We are interested only in these small changes. We also define σ as

$$\sigma = \frac{k_s}{\sqrt{2}} \left\{ \frac{e^{-\alpha nF(E-E^0)/RT}}{\sqrt{D_{Ox}}} + \frac{e^{(1-\alpha)nF(E-E^0)/RT}}{\sqrt{D_{Red}}} \right\} R_{CT} \tag{143}$$

from which it follows that (140) rearranges to

$$-\overline{\Delta E} = \left(R_{CT} + \sqrt{2}\frac{\sigma}{\sqrt{p}} \right) \overline{\Delta i} \tag{144}$$

which is of the form, response = system × excitation, with the excitation being $\overline{\Delta i}$. The Laplace transform of the impedance, $\overline{Z}$ may now be defined analogously to $\overline{Y}$, that is, $\overline{Z} = \overline{\Delta E}/\overline{\Delta i}$. Equation (144) now becomes

$$-\overline{Z} = R_{CT} + \sqrt{2}\frac{\sigma}{\sqrt{p}} \tag{145}$$

$j\omega$ substitution, involving DeMoivre's theorem, leads to the impedance, Z,

$$-Z = R_{CT} + \frac{\sigma}{\sqrt{\omega}}(1-j) \tag{146}$$

Z is a vector, analogous to Y (126), and has real (Z_R) and quadrature (Z_Q) components that are analogous to Y_R and Y_Q.

Before proceeding further some comments need to be made concerning the minus sign associated with Z in (146). Since Z is a vector, the minus sign means that Z points in the -1 direction (see Figure 4.31) instead of the $+1$ direction. In other words the impedance is 180° out of phase with the current excitation. This is merely an artifact of the way current was defined in (7). It is customary to simply ignore the minus sign. However, one should not be disconcerted by the observation that currents that should be in phase with voltages sometimes appear to be 180° out of phase, depending upon how one arranges the experiment.

Impedance measurements are normally made at the equilibrium potential, the applied potential at which there is no flow of direct current. In that case the Nernst equation is obeyed. It then follows that

$$e^{-\alpha n F(E-E^0)/RT} = \left(\frac{C_{Ox}}{C_{Red}}\right)^{-\alpha} \tag{147}$$

and

$$e^{(1-\alpha)nF(E-E^0)/RT} = \left(\frac{C_{Ox}}{C_{Red}}\right)^{1-\alpha} \tag{148}$$

Substitution of (147) and (148) into the expression for the charge-transfer resistance (141) and σ (143) leads to

$$R_{CT} = \frac{RT}{n^2 F^2 k_s} C_{Ox}^{(1-\alpha)} C_{Red}^{\alpha} \tag{149}$$

and

$$\sigma = \frac{RT}{\sqrt{2}n^2 F^2} \left\{ \frac{1}{C_{Ox}\sqrt{D_{Ox}}} + \frac{1}{C_{Red}\sqrt{D_{Red}}} \right\} \tag{150}$$

Notice that (150) and (116) are the same.

4.3 Equivalent Circuits

In electrical circuits, elements of impedance add together when placed in series with each other. Therefore an "equivalent circuit" representation of (146) is given in Figure 4.33 where the resistor has the value R_{CT} and W, the "Warburg" impedance, has the value $(\sigma/\sqrt{\omega})(1-j)$.

The double-layer capacitance provides a capacitive path for current flow, which does not occur through charge transfer or diffusion. Capacitive (charging) currents flow in parallel to the network in Figure 4.33, so the equivalent circuit becomes that given in Figure 4.34. Current is driven through the double-layer capacitance by time-dependent voltage according to [109]

$$\Delta i_c = C_{dl} \frac{d\Delta E}{dt} \tag{151}$$

Figure 4.33 Equivalent-circuit representation of the Warburg diffusional impedance with associated charge-transfer resistance.

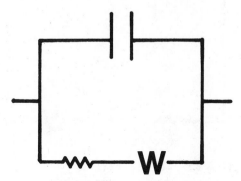

Figure 4.34 Equivalent-circuit representation of the Warburg diffusional impedance with associated charge-transfer resistance and double-layer capacitance.

where Δi_c is the small differential of capacitive current driven by the small voltage drop, ΔE, across the circuit. Laplace transformation leads to

$$\overline{\Delta i_c} = C_{dl}(p\overline{\Delta E} - E_0) \tag{152}$$

E_0 is the initial voltage drop across the circuit and can only lead to unwanted transient terms. It may therefore be ignored. Substitution of $j\omega$ for p leads to

$$Y_c = j\omega C_{dl} \tag{153}$$

where Y_c is the admittance of the capacitor. The impedance of the capacitor, Z_c, is $1/Y_c = -j/\omega C_{dl}$, and is commonly referred to as the "capacitive reactance."

Another element usually included in the equivalent circuit of an electrochemical cell is the "series" resistance, R_S, which represents the sum of all series resistances in the cell. R_S includes line and contact resistances, which arise because of opposition to electron flow, but the major portion of R_s is simply caused by the difficulty associated with moving ions within the bulk solution between the electrodes. It is often referred to as the "solution resistance" for this very good reason. The equivalent circuit (Randles equivalent circuit) is given in Figure 4.35.

The various elements of impedance shown in the Randles equivalent circuit may be added together as though they were resistors. R_{CT} and W add directly

$$Z_B = R_{CT} + W \tag{154}$$

where Z_B is the impedance of the lower branch. While the impedances of elements in series add directly, the admittances of elements in parallel add directly. Therefore the admittance of the right-hand side of the circuit becomes

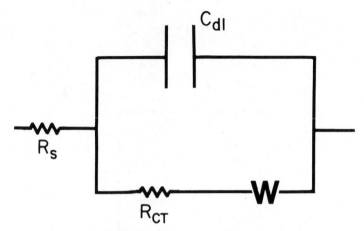

Figure 4.35 The Randles equivalent circuit.

$1/Z_B + Y_c$. The impedance of the circuit then becomes

$$Z = R_S + (Z_B^{-1} + Y_c)^{-1} \tag{155}$$

when R_S is added to the impedance of the right-hand side. Substitution of appropriate terms for these parameters leads to

$$Z_R = R_S + \frac{(R_{CT} + \sigma/\sqrt{\omega})}{(1 + \sqrt{\omega}C_{dl}\sigma)^2 + \omega^2 C_{dl}^2 (R_{CT} + \sigma/\sqrt{\omega})^2} \tag{156}$$

and

$$Z_Q = \frac{\omega C_{dl}(R_{CT} + \sigma/\sqrt{\omega})^2 + (\sigma/\sqrt{\omega})(1 + \sqrt{\omega}C_{dl}\sigma)}{(1 + \sqrt{\omega}C_{dl}\sigma)^2 + \omega^2 C_{dl}^2 (R_{CT} + \sigma/\sqrt{\omega})^2} \tag{157}$$

where $Z = Z_R - jZ_Q$.

These equations may appear formidable; however, they can be easily used to interpret experimental data. Two conditions are required. First, consider frequencies so high that $\sigma/\sqrt{\omega} \ll R_{CT}$ and second, allow that $\sqrt{\omega}C_{dl}\sigma \ll 1$. These equations reduce to

$$Z_R = R_S + \frac{R_{CT}}{1 + (\omega R_{CT} C_{dl})^2} \tag{158}$$

$$Z_Q = \frac{\omega R_{CT}^2 C_{dl}}{1 + (\omega R_{CT} C_{dl})^2} \tag{159}$$

The first of these conditions simply means that diffusional processes, represented by the value of σ, require more time to occur than charge-transfer processes, represented by the value of R_{CT}. Frequencies are then considered that are so high that diffusion does not occur at all and only charge-transfer effects are seen. The second condition is a statement to the effect that these two processes are well resolved in time or frequency.

Equations (158) and (159) constitute a set of equations parametric in frequency. They can be combined into a single equation by eliminating the parameter ω. It follows from (158) that

$$1+(\omega R_{CT}C_{dl})^2 = \frac{R_{CT}}{Z_R - R_S} \tag{160}$$

which, when substituted into (159), yields (161) on rearrangement.

$$\omega = \frac{Z_Q}{(Z_R - R_S)R_{CT}C_{dl}} \tag{161}$$

Substitution of (161) into (158) leads to

$$Z_R - R_S = \frac{R_{CT}}{1 + \left(\dfrac{Z_Q}{Z_R - R_S}\right)^2} \tag{162}$$

which may be rearranged to

$$\left(Z_R - R_S - \frac{R_{CT}}{2}\right)^2 + Z_Q^2 = \frac{R_{CT}^2}{4} \tag{163}$$

This is the equation of a circle with center on the real axis and diameter R_{CT}. The circle intersects the real axis at R_S in the limit of infinite frequency; therefore, both of these parameters may be read directly from a plot of Z_Q versus Z_R. Such plots are "Argand plots" but are also known as Cole–Cole, Nyquist, or complex-plane plots. The Argand representation of a Randles equivalent circuit is shown in Figure 4.36. Such plots can also be used to determine the value of the double-layer capacitance because the frequency corresponding to the maximum value of Z_Q occurs when $\omega R_{CT}C_{dl} = 1$.

Notice that (163) is no longer obeyed at low frequencies. Then $\sigma/\sqrt{\omega} \gg R_{CT}$ and (156) and (157) reduce to

$$Z_R = R_S + R_{CT} + \sigma/\sqrt{\omega} \tag{164}$$

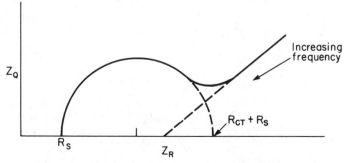

Figure 4.36   Argand representation of the impedance of the Randles equivalent circuit.

and

$$Z_Q = \sigma/\sqrt{\omega} + 2\sigma^2 C_{dl} \tag{165}$$

Therefore,

$$Z_R = R_S + R_{CT} - 2\sigma^2 C_{dl} + Z_Q \tag{166}$$

It follows that σ can be determined from the intercept of the linear portion of the plot with the real axis. Obviously, Argand plots can be extremely informative.

4.4 Coupled Chemical Reactions

The preceding treatment can be extended further to cover cases involving homogeneous kinetics. Substitution of (147), (148), and (149) into (140) and rearrangement results in

$$-\overline{\Delta E} = R_{CT}\overline{\Delta i} - \frac{RT}{nF} \left\{ \frac{\overline{\Delta C}_{Ox}}{C_{Ox}} - \frac{\overline{\Delta C}_{Red}}{C_{Red}} \right\} \tag{167}$$

By comparison with the Laplace transform of (112) (the linearized Nernst equation), this equation shows that a small current variation about zero average current causes a small potential variation about the equilibrium potential, which is composed of two parts. The first part is the "iR" drop across the activation resistance, R_{CT}, and represents the energy needed to move a coulomb of charge through the barrier mentioned earlier. Remember that a volt is a joule per coulomb. The second part is the "Nernst potential" needed to change the concentrations of Ox and Red by $\overline{\Delta C}_{Ox}$ and $\overline{\Delta C}_{Red}$, respectively. Notice that the $\overline{\Delta E}$ which appears in the Laplace transform of (112) is not the same $\overline{\Delta E}$ that appears in (167). The former was derived from the Nernst equation, whereas the latter was derived from the absolute rate expression (31) and therefore "contains" the Nernst equation in addition to the charge-transfer resistance.

Use of (167) to derive expected impedance responses in cases involving kinetic complications may be illustrated with the scheme

$$Y \underset{k_b}{\overset{k_f}{\rightleftharpoons}} Ox \underset{-e^-}{\overset{+e^-}{\rightleftharpoons}} Red \tag{168}$$

for which the transforms for $\overline{\Delta C}_{Ox}$ and $\overline{\Delta C}_{Red}$ were derived earlier and are given by (25) and (30). We replace $\overline{i}$ by $\overline{\Delta i}$ in (25) and (30) and substitute into (167). Since $\overline{Z} = \overline{\Delta E}/\overline{\Delta i}$, it follows that for this scheme

$$\overline{Z} = R_{CT} + \frac{\sqrt{2}K\sigma_{Ox}}{1+K} \left\{ \frac{1}{\sqrt{p}} \right\} + \frac{\sqrt{2}\sigma_{Ox}}{1+K} \left\{ \frac{1}{\sqrt{p+k}} \right\} + \sqrt{2}\sigma_{Red} \left\{ \frac{1}{\sqrt{p}} \right\} \tag{169}$$

where $K = k_f/k_b$, $k = k_f + k_b$, $\sigma_{Ox} = RT/\sqrt{2}n^2 F^2 \sqrt{D_{Ox}}C_{Ox}$, and $\sigma_{Red} = RT/\sqrt{2}n^2 F^2 \sqrt{D_{Red}}C_{Red}$. Fourier transformation by $j\omega$ substitution results in

$$Z = R_{CT} + \sqrt{2} \left\{ \left(\frac{K\sigma_{Ox}}{1+K} + \sigma_{Red} \right) \frac{1}{\sqrt{j\omega}} + \frac{\sigma_{Ox}}{1+K} \frac{1}{\sqrt{k+j\omega}} \right\} \tag{170}$$

DeMoivre's theorem (128) may now be used to transform this result into the desired form. In (128), $n=2$ for both $(j\omega)^{1/2}$ and $(k+j\omega)^{1/2}$, therefore $m=0$ and 1. Since $\theta+2\pi=\theta$, it follows that both values of m lead to the same root in each of these expressions. In the case of $(j\omega)^{1/2}$ we must evaluate

$$(j\omega)^{1/2}=\omega^{1/2}\left\{\cos\frac{\theta}{2}+j\sin\frac{\theta}{2}\right\} \tag{171}$$

Since the ω vector points only in the j direction, which is perpendicular to unity (see Figure 4.31), it follows that $\theta=90°$, so (171) becomes

$$(j\omega)^{1/2}=\left(\frac{\omega}{2}\right)^{1/2}(1+j) \tag{172}$$

a result that was given earlier. The reciprocal of (172) is

$$(j\omega)^{-1/2}=\frac{1-j}{\sqrt{2\omega}} \tag{173}$$

Evaluation of $(k+j\omega)^{1/2}$ is a little more complicated. The relationships between k, ω, r, and θ are shown in Figure 4.37, which shows that θ changes with ω. From this figure, $\cos\theta=k/r$. Since

$$\cos\frac{\theta}{2}=\sqrt{\frac{1+\cos\theta}{2}}$$

and

$$\sin\frac{\theta}{2}=\sqrt{\frac{1-\cos\theta}{2}}$$

(trigonometric identities) it follows that

$$(k+j\omega)^{1/2}=\frac{1}{\sqrt{2}}[\sqrt{r+k}+j\sqrt{r-k}] \tag{174}$$

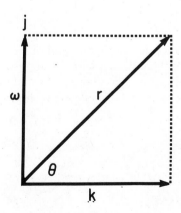

Figure 4.37 Vectorial relationship between the frequency ω and the rate constant k.

Therefore,

$$(k+j\omega)^{-1/2} = \frac{1}{\sqrt{2r}} \{(r+k)^{1/2} - j(r-k)^{1/2}\} \tag{175}$$

From Figure 4.37, $r = \sqrt{\omega^2 + k^2}$, so

$$(k+j\omega)^{-1/2} = \frac{1}{\sqrt{2}} \left\{ \left(\frac{\sqrt{\omega^2+k^2}+k}{\omega^2+k^2} \right)^{1/2} - j \left(\frac{\sqrt{\omega^2+k^2}-k}{\omega^2+k^2} \right)^{1/2} \right\} \tag{176}$$

Substitution of (173) and (176) into (170) yields the final result, $Z = Z_R - jZ_Q$, where

$$Z_R = R_{CT} + \left(\frac{K\sigma_{Ox}}{1+K} + \sigma_{Red} \right) \frac{1}{\sqrt{\omega}} + \frac{\sigma_{Ox}}{1+K} \left(\frac{\sqrt{\omega^2+k^2}+k}{\omega^2+k^2} \right)^{1/2} \tag{177}$$

and

$$Z_Q = \left(\frac{K\sigma_{Ox}}{1+K} + \sigma_{Red} \right) \frac{1}{\sqrt{\omega}} + \frac{\sigma_{Ox}}{1+K} \left(\frac{\sqrt{\omega^2+k^2}-k}{\omega^2+k^2} \right)^{1/2} \tag{178}$$

These equations appear more formidable than they are. At frequencies so high that $\omega \gg k$, they reduce to the simple Warburg expression (146) and may be combined to

$$Z_R = R_{CT} + Z_Q \tag{179}$$

At frequencies so low that $\omega \ll k$,

$$Z_R = R_{CT} + \left(\frac{K\sigma_{Ox}}{1+K} + \sigma_{Red} \right) \frac{1}{\sqrt{\omega}} + \frac{\sigma_{Ox}}{1+K} \sqrt{\frac{2}{k}} \tag{180}$$

and

$$Z_Q = \left(\frac{K\sigma_{Ox}}{1+K} + \sigma_{Red} \right) \frac{1}{\sqrt{\omega}} \tag{181}$$

which may be combined to

$$Z_R = R_{CT} + Z_Q + \frac{\sigma_{Ox}}{1+K} \sqrt{\frac{2}{k}} \tag{182}$$

Again an expression of the Warburg form is obtained. An Argand plot of (178) versus (177) is shown in Figure 4.38. Notice that the intercept with the abscissa at the limit of infinite frequency provides the value of R_{CT}, as in the case with no kinetic complications, and that the difference between the two intercepts shown provides the value

$$\Delta Z_R = \frac{\sigma_{Ox}}{1+K} \sqrt{\frac{2}{k}} \tag{183}$$

When $\omega = k$

$$Z_R = R_{CT} + Z_Q + 0.4551 \Delta Z_R \tag{184}$$

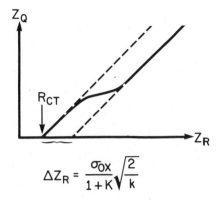

$$\Delta Z_R = \frac{\sigma_{OX}}{1+K}\sqrt{\frac{2}{k}}$$

Figure 4.38 Argand representation of the impedance associated with a typical kinetic scheme—a CE mechanism.

Therefore, k may be estimated from the frequency corresponding to the point midway between the two linear extrapolations.

Lelievre and Plichon [110] have used Argand representations of impedance data to study the base hydrolysis of N,N-dimethylquinonediimine formed by oxidation of N,N-dimethyl-p-phenylenediamine at a rotating platinum disk electrode—an example of an EC mechanism.

4.5 Applications

Before proceeding to situations in which impedances, or components thereof, are measured as functions of applied voltage, let us first review some applications of impedance measurements performed at the equilibrium potential. Such applications are truly legion, and we cannot hope to cover all of them here. The following examples are intended only to be illustrative of the power and scope of the technique.

Randles [111] and Ershler [112] first introduced impedance as a technique for measuring rapid charge-transfer rate constants, k_s, in 1947, although the diffusional element, W, of the Randles equivalent circuit (Figure 4.35) was contributed by Warburg [113] almost five decades earlier. Randles studied dilute-amalgam electrodes in his work. He plotted Z_R and Z_Q versus $1/\sqrt{\omega}$ in accordance with (146) and determined k_s from the difference between the parallel lines. Such plots, referred to as "Randles plots," are still commonly used. Sluyters [114] used the Argand-plot representation of the Randles equivalent circuit in 1960. However, such plots had been used for some time in the treatment of dielectric-constant data [115] and were referred to as Cole–Cole plots. Although dielectric data are more closely associated with the admittance plane, impedance plots are also sometimes referred to by this name. In addition, Euler and Dehmelt [116] had used plots of Z_Q versus Z_R in their studies of batteries, another important area of application for impedance measurements even to the present.

For example, Figure 4.39 is the Argand representation of a Leclanché cell [117], which may be compared with Figure 4.36. Notice, however, that the

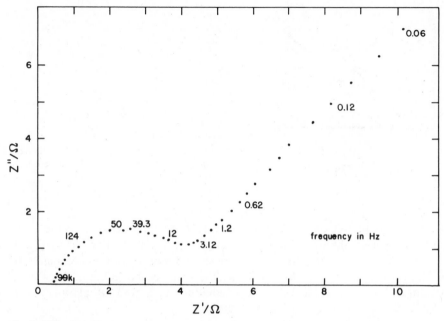

Figure 4.39 Argand representation of the impedance of a Leclanché cell. Ever Ready SP11.23C. Reprinted from [117], by courtesy of Chapman and Hall.

high-frequency semicircle is slightly depressed—its center below the real axis. This phenomenon will be discussed later. Impedance measurements have been found useful in battery technology for the measurements of states of charge and various quality-control parameters, or for attributing voltage losses to the different battery components on the basis of the frequency domains within which their effects are significant.

Sluyters and Oomen [118] continued the study of amalgam electrodes and generated the results shown in Figures 4.40 and 4.41 for the Hg/Hg_2^{2+} and $Zn(Hg)/Zn^{2+}$ systems, respectively. Impedances are given in ohm·cm^2 and frequencies are given in kilohertz. These figures are often cited as verification of (156) and (157). Additional work by Sluyters and others during the 1960s is summarized in an excellent chapter in Bard's series on electroanalytical chemistry [119]. More recent applications include homogeneous kinetic complications [120], but these impedance measurements are generally made at potentials other than the equilibrium potential and therefore will not be discussed until later.

Buck [121] has reviewed the application of impedance techniques to the study of ion-selective electrodes. Mertens, Van den Winkel, and Vereecken have used impedance measurements in their mechanistic studies of lanthanum fluoride crystal electrodes [122] and Gavach, Seta, and Henry have used them to investigate ionic transfer across aqueous–liquid membrane interfaces [123]. The high bulk resistances and geometrical capacitances ($\varepsilon\varepsilon_0 A/d$) of many ion-

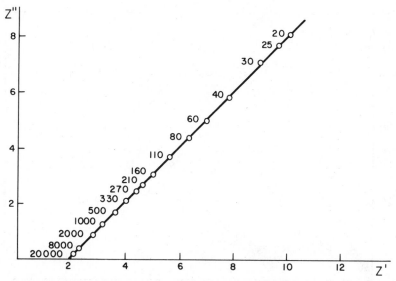

Figure 4.40 Argand representation of the impedance of the Hg/Hg_2^{2+} system. Reprinted from [118], by courtesy of Koninklijke Nederlandse Chemische Vereniging.

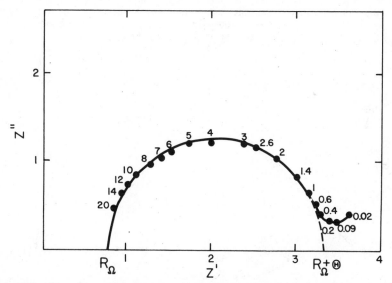

Figure 4.41 Argand representation of the impedance of the $Zn(Hg)/Zn^{2+}$ system. Reprinted from [118], by courtesy of Koninklijke Nederlandse Chemische Vereniging.

selective electrodes might be expected to preclude measurement of interfacial impedances. However, as shown in Figure 4.42, the interfacial impedances of pH glass electrodes are quite comparable to the bulk parameters and can be measured easily [124].

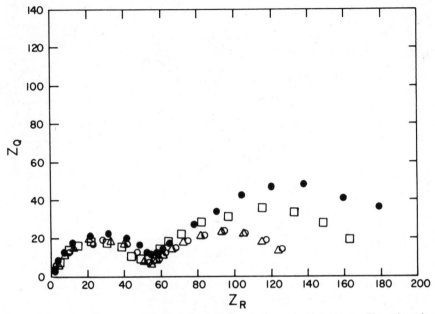

Figure 4.42 Argand representation of the impedance of a general-purpose, pH-sensing glass electrode with different surface treatments. The invariant high-frequency arcs are caused by the bulk properties of the glass itself; the low-frequency arcs are due to the variable interfacial impedances expressed in $M\Omega$. Reprinted from [124], by courtesy of Elsevier Press.

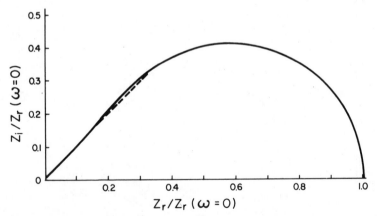

Figure 4.43 Argand representation of a surface film or finite diffusion layer. Also referred to as a "finite Warburg." Reprinted from [118, 124], by courtesy of Elsevier Press.

One is often concerned with the diffusional impedances of surface films in the study of ion-selective electrodes, and the theoretical impedances of such films may be represented as shown in Figure 4.43 [125]. A similar figure appears in Sluyters-Rehbach and Sluyter's chapter [119] but describes convective diffusion rather than diffusion through a film. An analogy is therefore clearly drawn between a film and the laminar diffusion layer caused at electrode surfaces by the convective flow of solutions. Derivation of the mathematical representation of Figure 4.43 requires that finite, rather than semi-infinite, boundary conditions be used. One then arrives at a more complicated form of (6). This "thin-layer" situation is beyond the scope of this chapter except to point out the form of the response.

Notice in Figure 4.42 that the arcs are depressed in a manner similar to the battery results shown in Figure 4.39. This phenomenon also appears in the Argand plots of the admittances and impedances of solid electrolytes, as shown for β-alumina [126] in Figure 4.44. Applications to the study of solid electrolytes have followed from earlier work by Bauerle [127], who studied the zirconia–yttria system.

There are actually two departures from ideal behavior in Figure 4.44. The first is the lowered, high-frequency semicircle and the second is the greater-than-unity slope in the linear, low-frequency region. Both departures can be described mathematically by introducing a new, frequency-dependent, empirical element Z_E, of the form

$$Z_E = R(j\omega\tau)^{-\alpha} \tag{185}$$

where τ is a normalizing time constant. Application of DeMoivre's theorem

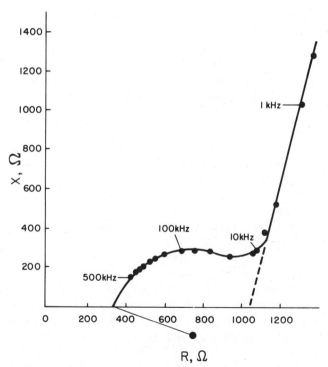

Figure 4.44 Argand representation of the impedance of undoped β-alumina at 3°C. Reprinted with permission from [126], Pergamon Press, Ltd.

(128) shows that (185) can be written

$$Z_E = \frac{\cos(\alpha\pi/2) - j\,\sin(\alpha\pi/2)}{(\omega\tau)^\alpha}\, R \tag{186}$$

therefore

$$Z_E = Z_{R,E} - jZ_{Q,E} \tag{187}$$

where

$$Z_{R,E} = \frac{\cos(\alpha\pi/2)}{(\omega\tau)^\alpha}\, R \tag{188}$$

and

$$Z_{Q,E} = \frac{\sin(\alpha\pi/2)}{(\omega\tau)^\alpha}\, R \tag{189}$$

It follows that an Argand plot of (185) is a straight line, which makes an angle of $\alpha\pi/2$ with the real axis (slope $= \tan \alpha\pi/2$). If Z_E is placed in parallel with a resistor with value R, one obtains

$$Z = \frac{R}{1 + (j\omega\tau)^\alpha} \tag{190}$$

The equations that describe a semicircle in an Argand plot (158) and (159) may be combined and rearranged to

$$Z = \frac{R}{1 + j\omega\tau} \tag{191}$$

where R_S has been set equal to zero, all subscripts have been dropped to show that we are now discussing a general, empirical case, and $\tau = RC$. Comparison of (190) and (191) reveals that the experimentally observable difference lies in the value of α. Equation (191) is a specific instance of the more generally applicable (190), in which $\alpha = 1$. This value of α transforms Z_E into a capacitive reactance [see (153)]. One should be aware that many authors prefer to write $1 - \alpha$ in place of the α presented here. It can be shown that (190) describes a semicircle with center below the real axis, depressed by the angle $\alpha\pi/2$ as measured according to the construction shown in Figure 4.45.

The exact meaning of α has been somewhat elusive and many explanations of $\alpha < 1$ have been proposed [128, 129]. Of course, if $\alpha = \frac{1}{2}$, one has the simple Warburg case; however, values both greater and less than $\frac{1}{2}$ are not uncommon.

One possible cause of $\alpha < \frac{1}{2}$ that is of particular concern to us here is attributed to surface roughness. Notice in (6) that the concentration of electroactive species varies exponentially from the surface of the electrode to the bulk of the solution with a factor $e^{-x/\delta}$ where $\delta = \sqrt{D/p}$. δ may be defined as the "diffusion-layer thickness." Its frequency dependence may be approximated by $j\omega$ substitution, which leads to

$$\delta = \sqrt{\frac{D}{j\omega}} \tag{192}$$

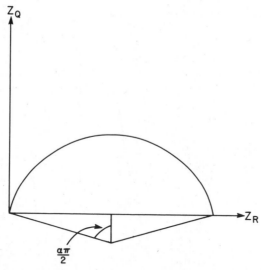

Figure 4.45   Argand impedance representation showing empirical α behavior.

The magnitude of δ is then its modulus, which is simply $\sqrt{D/\omega}$. Obviously the reaction layer becomes thinner as the frequency is increased. If it becomes comparable to the roughness of the surface, there will be an effective increase in surface area, as illustrated in Figure 4.46.

As shown rigorously by de Levie [130], surface roughness effectively shifts the phase angle of the Warburg impedance from 45° $(\alpha = \frac{1}{2})$ to 22.5° $(\alpha = \frac{1}{4})$. Analogously, double-layer capacitive reactances are shifted from 90° $(\alpha = 1)$ to 45° $(\alpha = \frac{1}{2})$ and may be mistakenly identified as Warburg components. Candy and co-workers [131] have demonstrated that this effect can be used to characterize porous electrodes in terms of pore size, number, and effective area. Some of their data are shown in Figure 4.47. Notice that the high-frequency linear region is caused by the double-layer capacitance and not the diffusional Warburg impedance since the solution contains no electroactive species. This case may be similar to the β-alumina situation seen in Figure 4.44, which was explained in terms of grain boundary resistance and capacitance [126].

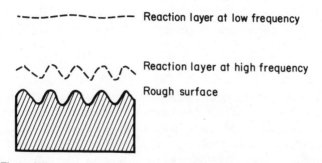

Figure 4.46   Relationship between diffusion layer and surface roughness.

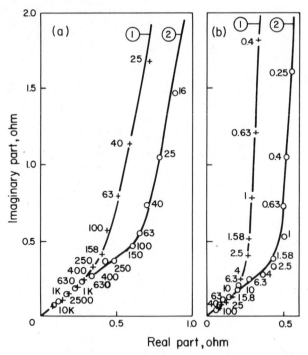

Figure 4.47 Argand representations of the impedances of porous electrodes. (a) Gold powder electrode; (b) Raney gold electrode. Reprinted with permission from [131], Pergamon Press, Ltd.

The impedances of variously shaped pores can be calculated numerically using the procedure outlined by Keiser and co-workers [132].

Until now, the question of adsorption has been addressed only implicitly by way of the double-layer capacitance, the value of which depends upon adsorbed species that may or may not be electrochemically active [133]. Impedance measurements can be used to study electrode processes in which a variety of different adsorption mechanisms may be active. For example, the electroactive species might diffuse to the surface of the electrode and adsorb to it with an accompanying partial transfer of charge. The remainder of the total charge transferred will then come from the adsorbed intermediate so formed. The resultant product might now desorb and diffuse away. This process may be represented as

$$Ox \xrightarrow{i_1} Int \xrightarrow{i_2} Red \tag{193}$$

where i_1 and i_2 are the components of the total current that result in the formation of the intermediate, Int, and the reduced species, Red, respectively. Armstrong and Henderson [134] have treated this situation theoretically and we shall follow a limiting case similar to their derivation to illustrate the effects of adsorption in a general way and also to show how one can approach such problems theoretically. An assessment of the power of impedance techniques for the study of adsorption is beyond the scope of this chapter. However, one should be aware of the effects of adsorption so that they will be recognized if encountered.

We begin with (193) but restrict our discussion to cases that do not involve diffusion, or perhaps more particularly to frequencies so high that diffusion does not have time to occur. Only species immediately adjacent to the surface of the electrode are affected. We also define $v_1 = i_1/n_1 F$ and $v_2 = i_2/n_2 F$ as the fluxes, in mol/cm²·s, that are the equivalents of the currents i_1 and i_2. Then the rate of accumulation of the surface excess of intermediate, Γ, is given by

$$\frac{d\Gamma}{dt} = v_1 - v_2 \tag{194}$$

Since impedance measurements involve only small changes in Γ, (194) may be written

$$\frac{d\Delta\Gamma}{dt} = v_1 - v_2 \tag{195}$$

which on Laplace transformation becomes

$$p\overline{\Delta\Gamma} = \bar{v}_1 - \bar{v}_2 \tag{196}$$

The rates of adsorption and desorption, $\bar{v}_1$ and $\bar{v}_2$, depend upon the amount of intermediate already adsorbed and upon the applied potential. These dependencies can be expressed in a Taylor series

$$\bar{v}_1 = \left(\frac{\partial v_1}{\partial \Gamma}\right)_E \overline{\Delta\Gamma} + \left(\frac{\partial v_1}{\partial E}\right)_\Gamma \overline{\Delta E} \tag{197}$$

and

$$\bar{v}_2 = \left(\frac{\partial v_2}{\partial \Gamma}\right)_E \overline{\Delta\Gamma} + \left(\frac{\partial v_2}{\partial E}\right)_\Gamma \overline{\Delta E} \tag{198}$$

The total change in current caused by the adsorption–desorption process is given by

$$\overline{\Delta i} = F(n_1\bar{v}_1 + n_2\bar{v}_2) \tag{199}$$

Equations (195) through (199) can be combined to yield

$$\bar{Y} = \frac{1}{R_\infty} + \frac{1}{R_0}\left(\frac{1}{1 + p\tau}\right) \tag{200}$$

where

$$\bar{Y} = \overline{\Delta i}/\overline{\Delta E} \tag{201}$$

$$\tau = \frac{1}{\left[\left(\frac{\partial v_1}{\partial \Gamma}\right)_E - \left(\frac{\partial v_2}{\partial \Gamma}\right)_E\right]} \tag{202}$$

$$\frac{1}{R_\infty} = F\left[n_1 \left(\frac{\partial v_1}{\partial E}\right)_\Gamma + n_2 \left(\frac{\partial v_2}{\partial E}\right)_\Gamma \right] \tag{203}$$

and

$$\frac{1}{R_0} = \tau F\left[n_1 \left(\frac{\partial v_1}{\partial \Gamma}\right)_E + n_2 \left(\frac{\partial v_2}{\partial \Gamma}\right)_E \right]\left[\left(\frac{\partial v_1}{\partial E}\right)_\Gamma - \left(\frac{\partial v_2}{\partial E}\right)_\Gamma \right] \tag{204}$$

Equation (200) becomes, by Fourier transformation ($j\omega$ substitution),

$$Y = \frac{1}{R_\infty} + \frac{1}{R_0} \frac{1}{1 + j\omega\tau} \tag{205}$$

The important thing to notice about (205) is that since the various partial derivatives of v_1 and v_2 can be either positive or negative, R_0 and R_∞ can also be either positive or negative. (τ must always be positive if a steady-state condition can be achieved.) Negative resistances make sense only if one realizes that these are differential quantities. In fact, the nature of the impedance technique itself is such that only differential quantities—R, C, Y, and Z—can be measured. Small changes in current, voltage, concentration, and so forth are involved and these occur around large values that are essentially at equilibrium. This fact led to our original definition of admittance in terms of differentials [see (125) and (201)]. A negative resistance can therefore refer to a region of the current–voltage curve in which an increase in voltage causes a decrease in current, without regard to sign. It does not mean that current flows in a direction opposite to that dictated by the applied voltage. Differential capacitances, analogously defined, can also be negative.

The experimental consequences of negative resistances and capacitances are "inductive" effects. In electronics, an inductor, L, generates a voltage drop proportional to the rate of change of the current that flows through it [109].

$$\Delta E = L\frac{d\Delta i}{dt} \tag{206}$$

Laplace transformation, followed by $j\omega$ substitution, leads to the inductive admittance

$$Y_L = -\frac{j}{\omega L} \tag{207}$$

which is totally quadrature and, by comparison with (153), 180° out of phase (because of the minus sign) with the capacitive admittance. In the impedance plane one then sees data points below the real axis. Results similar to those reported by Keddam and co-workers [135, 136] for iron dissolution, shown in Figure 4.48, can be obtained. Interpretation of these results involves the inclusion of three adsorbed reaction-intermediate species in the hypothesized mechanism.

A further consequence of negative resistance values is that they allow sustained, steady-state, current oscillations at some applied voltages. Obviously

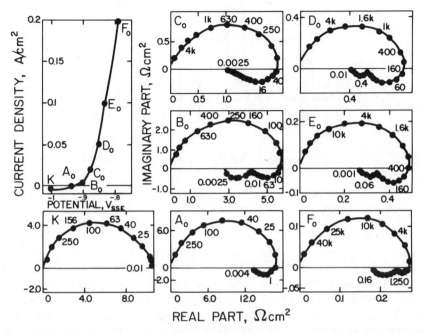

Figure 4.48 Argand representations of the impedance of an iron electrode in pH 0 solution. Each graph refers to a corresponding point on the polarization curve. Frequencies shown in Hz [135]. Reprinted with permission from the publisher, The Electrochemical Society, Inc.

such effects are beyond the scope of this chapter. We wish only to point out that they exist. The interested reader is referred to the work of de Levie [137, 138], who introduced negative charge-transfer resistances, which do not follow from the rate expression (31). However, the rate expression is not sacrosanct.

If diffusional effects are included in the treatment of adsorption processes, one is led to the equivalent circuit shown in Figure 4.49 [139], where C_1 is approximately equal to the double-layer capacitance in the absence of adsorption and W_1 is the corresponding diffusional impedance. C_2 and W_2 are terms that result from adsorption. Their values depend upon the particular mechanism involved. Notice that the charge-transfer resistance is conspicuously absent from this equivalent circuit. Slow charge-transfer effects can be included in the general analysis but the resultant expressions are too complicated to be of concern to us here [140].

Further information can be found in the treatment by D.D. Macdonald and McKubre [141] and in a review by R. D. Armstrong and co-workers [142].

By way of comparison, impedance and admittance techniques have a major advantage over other electrochemical techniques in that the measurements are made at steady state (zero direct current) or at pseudo-steady state (small changes in direct current on the experimental time scale). It then becomes far easier to separate the faradaic contributions to the total current from other contribu-

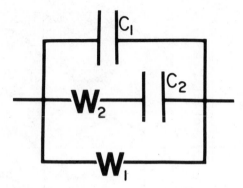

Figure 4.49 Equivalent circuit for which the effects of adsorption are included in C_2 and W_2. C_1 is approximately equal to the double-layer capacitance and W_1 is the diffusional or Warburg impedance. W_2 has the same $1/\sqrt{\omega}$ frequency dependence as W_1. Reprinted from [139], by courtesy of Elsevier Press.

tions, such as double-layer charging, since the potential dependence of these contributions is avoided. Other techniques, such as cyclic voltammetry, greatly perturb the electrochemical system with large applied voltages. They therefore incorporate nonfaradaic potential-dependent contributions into the net response. In the case of double-layer charging, for example, a single value of C_{dl} at a given applied potential can be assigned to the cell when impedance techniques are considered. However, this value will change with applied voltage and its voltage functionality would have to be convoluted into the analysis if cyclic voltammetric responses are considered. This great strength of the impedance technique can also be its great weakness. One deals only with differential values of resistance and capacitance. Integral values tend to be a problem. For example, if one simply wishes to know how much adsorption occurs at the surface of an electrode without regard to mechanism, one is probably better off using an integral technique such as chronocoulometry (see Chapter 6).

Each of the applications discussed above concerns the measurement of impedances that are dominated either by surface phenomena or by events that occur within the bulk of the solution. This is a direct result of using metallic electrodes to make the measurements. Use of semiconductor electrodes introduces additional resistors and capacitors into the equivalent circuit description of the cell. These elements result from inefficient movement of charge within the electrode, establishment of a space-charge region (diffuse double layer) on the electrode side of the electrode–solution interface, and charging and discharging of "surface states." Surface states are interfacial sites at which an electron can have an energy which is forbidden within the bulk of the electrode, that is, an energy level which is within the band gap of the semiconductor. Impedance measurements can be effectively used to characterize these effects [142], and the principles involved are not substantially different from those already discussed.

4.6   Effects of Changing Potential

Impedance measurements are not confined to the equilibrium potential. Potential dependence appears in the forms of (141) and (143), the equations for charge-transfer resistance and Warburg coefficient, respectively. Substitution of $e^j = \sqrt{D_{Ox}}C_{Ox}/\sqrt{D_{Red}}C_{Red}$ (43), where $j = (nF/RT)(E - E_{1/2})$, into these equations leads to the alternative forms

$$R_{CT} = \frac{RT}{n^2 F^2 k_s (\sqrt{D_{Ox}/D_{Red}})^\alpha C_{Ox}^*} \left(\frac{1 + e^j}{e^{j(1-\alpha)}} \right) \qquad (208)$$

and

$$\frac{1}{2\sigma} = \frac{n^2 F^2 \sqrt{D_{Ox}} C_{Ox}^*}{\sqrt{2} RT} \left[\frac{1}{4 \cosh^2(j/2)} \right] \qquad (209)$$

$[\cosh(j/2) = (e^{j/2} + e^{-j/2})/2]$. C_{Ox}^* has been introduced by resorting to (36) with C_{Red}^* equal to zero. With the trigonometric identity

$$\sin(\theta_1 + \theta_2) = \sin\theta_1 \cos\theta_2 + \sin\theta_2 \cos\theta_1 \qquad (210)$$

where $\theta_1 = \omega t$ and $\theta_2 = \pi/4$, (124) becomes

$$I(\omega t) = \frac{n^2 F^2 \sqrt{D_{Ox}} C_{Ox}^*}{4RT \cosh^2(j/2)} \Delta E_{ac} \sqrt{\omega} \sin(\omega t + \pi/4) \qquad (211)$$

which is the potential-dependent form of the Warburg admittance most commonly used in ac voltammetry, where $-\Delta i_{ac}$ is usually redefined $I(\omega t)$.

Commonality between the impedance and ac voltammetry techniques can be further illustrated with (177) and (178), the equations for Z_R and Z_Q that describe the impedance characteristics of the slow charge-transfer case with preceding equilibrium, as described earlier. In ac voltammetry it is customary to plot the cotangent of the phase angle, ϕ, against either $\sqrt{\omega}$ or $(E - E_{1/2})$. Since $\cot\phi = Z_R/Z_Q$, it follows from these equations that

$$\cot\phi = 1 + \frac{R_{CT}}{\sigma} \sqrt{\omega} \qquad (212)$$

when $\omega \gg k$ and

$$\cot\phi = 1 + \frac{R_{CT}(1 + K) + \sigma_{Ox}\sqrt{2/k}}{K\sigma + \sigma_{Red}} \sqrt{\omega} \qquad (213)$$

when $\omega \ll k$; $\sigma = \sigma_{Ox} + \sigma_{Red}$.

Notice in the derivation of (169) that $\sigma_{Ox}/\sigma_{Red} = \sqrt{D_{Ox}}C_{Ox}/\sqrt{D_{Red}}C_{Red} = e^j$. Combining this expression with (177) and (178) results in

$$\cot\phi = \frac{R_{CT}\dfrac{(1+K)}{\sigma_{Ox}}\sqrt{\omega} + K + (1+K)e^j + \left[\dfrac{\sqrt{1+g^2}+g}{1+g^2} \right]^{1/2}}{K + (1+K)e^j + \left[\dfrac{\sqrt{1+g^2}-g}{1+g^2} \right]^{1/2}} \qquad (214)$$

where $g = k/\omega$. If charge transfer is very rapid, $R_{CT} = 0$ and (214) provides the value of cot ϕ over the entire frequency and potential range. The equation in this form will be discussed further in Section 5 on ac voltammetry. If charge transfer is slow, the potential dependence of R_{CT} and σ_{Ox} must be included.

It follows from (36) that

$$\sqrt{D_{Ox}}C_{Ox} + \sqrt{D_{Red}}C_{Red} = \sqrt{D_{Ox}}C_{Ox}^* + \sqrt{D_{Red}}C_{Red}^* \tag{215}$$

which combines with $\sqrt{D_{Ox}}C_{Ox}/\sqrt{D_{Red}}C_{Red} = e^j$ to yield

$$\sqrt{D_{Ox}}C_{Ox} = (\sqrt{D_{Ox}}C_{Ox}^* + \sqrt{D_{Red}}C_{Red}^*)e^j/(1+e^j) \tag{216}$$

Substitution into the expression for σ_{Ox} [see (169)] leads to

$$\sigma_{Ox} = \frac{RT}{\sqrt{2}n^2F^2} \frac{(1+e^j)}{(\sqrt{D_{Ox}}C_{Ox}^* + \sqrt{D_{Red}}C_{Red}^*)e^j} \tag{217}$$

Division of (208) by (217) and multiplication by $1 + K$ provides

$$\frac{R_{CT}}{\sigma_{Ox}}(1+K) = \frac{\sqrt{2}}{k_s}(1+K)\left(\sqrt{\frac{D_{Red}}{D_{Ox}}}\right)^\alpha \left(\frac{\sqrt{D_{Ox}}C_{Ox}^* + \sqrt{D_{Red}}C_{Red}^*}{C_{Ox}^*}\right)e^{\alpha j} \tag{218}$$

which may be inserted directly into (214) to complete the derivation of the potential and frequency dependence of cot ϕ.

It should be emphasized that all the above potential-dependent expressions were derived with the assumption that the Nernst equation, as embodied in the e^j expression, is obeyed on the dc time scale. This can be true in the case of the R_{CT} equation (218) only if the potential is varied slowly enough to keep the direct current nearly zero. Otherwise the results will be dependent on sweep rate and these equations will no longer be obeyed.

The importance of the diffusion-layer thickness has been discussed previously with regard to surface roughness. It should be emphasized that it is also important with regard to sweep rate. The ramp voltage applied in ac voltammetry defines a diffusion-layer thickness analogous to that defined by the alternating voltage applied in the admittance technique. The equations derived above apply only if the diffusion layer defined by the ramp, within which concentrations vary linearly, is much larger than the diffusion layer defined by the alternating voltage, within which the concentration varies sinusoidally. This condition is equivalent to saying that the layer of solution adjacent to the electrode is in a state of equilibrium with respect to the large-amplitude signal (the Nernst equation applies) and is perturbed only by the concentration waves caused by the ac signal. These waves dampen to insignificance without ever experiencing a significant nonzero average concentration gradient within this layer, even though the average concentration is much different from that of the bulk of the solution. The experimental test of this condition is that the results be independent of sweep rate.

The preceding discussion illustrates clearly the relationship between impedance (or admittance) and ac voltammetry, which could easily be referred to as admittance with potential sweep. However, we have confined ourselves to planar stationary electrodes. The dropping mercury electrode, used in ac polarography, presents additional complications.

5 AC VOLTAMMETRY

Ac voltammetry refers to the technique of linearly scanning the dc potential, as in linear-sweep voltammetry but at the same time applying a small-amplitude sine wave. The frequency of the sine wave is high enough that the dc potential is essentially constant for several cycles. In ac polarography (dropping mercury electrode) the scan rate is the same as used in dc polarography, that is, 0.5–10 mV/s depending on drop time. At a stationary electrode the scan rate can be an order of magnitude more rapid as long as the sine-wave frequency is high.

The different time scales of the two processes allow considerable simplification in deriving the current–potential expression for various mechanisms. One can consider the dc process as creating the surface concentrations, which are then perturbed slightly by the ac process. As long as the dc diffusion-layer thickness is large compared to the distance of the ac concentration perturbations, the ac signal "sees" a concentration of electroactive species at the electrode that is constant on the ac time scale.

The validity of applying ac polarographic theory to stationary electrodes using fast dc programs has received considerable attention. Bond, O'Halloran, Ruzic, and Smith [144, 145] extended the earlier studies of Unterkofler and Shain [146] to show that as long as the scan rate was slow compared to the applied ac frequency, both reversible and quasi-reversible systems obeyed the equations derived for polarography [147]. They ascribed a lower limit of 128 ac cycles/dc cyclic sweep, which is equivalent to an applied ac frequency of 16 Hz using a 50 mV/s dc sweep over a potential range of ± 200 mV [145]. An analytical solution for the total alternating and direct current response was obtained by Henderson and Gordon [148], who suggested an even lower limit of 20 ac cycles per sweep to separate the direct current response from the periodic ac response accurately. They pointed out, however, that commonly used alternating current detectors probably would require up to 200 ac cycles/dc cyclic sweep for accurate separation, nearly the same as the limit obtained by Bond and co-workers [144].

Tokuda and Matsuda [149] have studied the theory of ac voltammetry at rotating disk electrodes for small-amplitude sinusoidal, rectangular, and triangular applied potentials. They conclude that as long as $\omega^{1/2} > 3\sqrt{\pi D}/\delta$, where δ is the convective diffusion-layer thickness governed by the rotation rate, the ac polarographic equations [147] for reversible and quasi-reversible systems are valid [149].

Sophisticated potential functions, both dc and ac, have been applied to the cell to elicit the desired response, that is, ac current and its phase-angle relation to the applied ac potential. Instead of the slow dc ramp employed conventionally, normal pulse polarographic wave forms have been used [150, 151]. Rapid ramps, up to 0.6 V/s, on slowly growing mercury drops [152, 153] and cyclic programs on stationary electrodes [142, 145] have been employed. The ac voltage need not be a sine wave; small-amplitude square waves [154], triangular waves [155], and digitized sine waves [156] have been used. Smith and co-workers [157–164] have developed a sophisticated, computer-controlled system, which applies a range of frequencies consisting of odd harmonics of the lowest frequency with random phase-angle relations among them. Even random or pseudo-random white noise has been used as an excitation source [151]. Fast Fourier Transform (FFT) analysis of frequencies up to 125 kHz provides faradaic admittance data up to 60 kHz. The dc signal is a staircase, also generated by computer. Data at several frequencies are obtained simultaneously in time scales of a few seconds although significantly more time is required for data processing [164].

In all these methods the ac response can be described by the reversible or quasi-reversible mechanism derived by Smith [147] in the absence of coupled chemical reactions. The cyclic ac voltammetric response for slow charge transfer on the dc time scale, where the alternating current on the return dc-potential sweep is shifted from the response on the forward sweep, is a special case and is discussed later.

5.1   Reversible Response

To obtain an alternating current expression as a function of applied dc potential we can use the linearization method outlined in Section 4 on ac-admittance and impedance techniques.

The resultant current-density expression in Section 4 (211) for reversible charge transfer is repeated here:

$$I(\omega t) = \frac{n^2 F^2 C_{Ox}^* \Delta E_{ac}(\omega D_{Ox})^{1/2}}{4RT \cosh^2(j/2)} \sin(\omega t + \pi/4) \tag{219}$$

The phase-angle relationship between the applied potential and the resultant current is given by $\pi/4$, that is, the current leads the potential by $\pi/4$ radians or $45°$ for a diffusional process at all dc potentials.

To obtain (219), the "dc" surface concentration must be obtained as pointed out by Smith [147], who used a more rigorous method to obtain (219). The same result can be obtained using the diffusion-layer-thickness concept as described by Sluyters-Rehbach and Sluyters [119] for the reversible case.

Equation (219) has been obtained by assuming only planar diffusion. It is interesting that this same expression is obtained by assuming diffusion to an expanding plane. This fact applies to the reversible case only and is a manifesta-tion of the time dependence of the dc process. There is no change in the type of time dependence of the dc process between the expanding plane model and the planar model, so the time dependence of the ac process remains the same, as does the expression for the current magnitude. For other reaction mechanisms, the current expression is different for the models, but again it is attributable to changes in the dc process. This is discussed in more detail in later sections.

Referring to (219), we note that when the alternating current is plotted against dc potential, the shape of the wave is given by $1/\cosh^2(j/2)$, since j is the expression of dc potential

$$j = \frac{nF}{RT}(E_{dc} - E_{1/2}) \qquad (220)$$

When j equals zero, $1/\cosh^2(j/2)$ has a maximum value of 1; therefore, there is a peak in $I(\omega t)$ at $E_{dc} = E_{1/2}$.

It can also be shown from (219) that the width of the ac wave at the half-peak height is $90/n$ mV at 25°C. Substituting for the definition of j, one can show further that

$$E_{dc} = E_{1/2} \pm \frac{2RT}{nF} \ln\left[\left(\frac{I_p}{I}\right)^{1/2} + \left(\frac{I_p - I}{I}\right)^{1/2}\right] \qquad (221)$$

The term I_p is the peak current of the ac wave and I is the value at any potential E_{dc}. Thus, one obtains a linear plot of E_{dc} versus $\log\{[I_p/I]^{1/2} + [(I_p - I)/I]^{1/2}\}$ with a slope of $\pm 118/n$ mV at 25°C [147]. The slope is positive when E_{dc} is positive of $E_{1/2}$ and negative when E_{dc} is negative of $E_{1/2}$.

Historically, the experimental parameters of interest with respect to phase-angle measurement are the resistive and capacitive components of the faradaic impedance, discussed in Section 4. The technique of ac polarography, however, yields the same experimental information in a slightly different way. Any combination of two of the four possible parameters—total current, resistive current, capacitive current, and phase angle—yields the other two. The theory of ac polarography uses the total current and phase-angle information for diagnostic study, whereas faradaic-impedance theory makes use of the capacitive and resistive components of the total admittance.

Equation (219) also shows that $I(\omega t)$ is proportional to $\omega^{1/2}$ and has a zero intercept at $\omega = 0$. Table 4.11 summarizes the properties of the reversible ac polarogram useful for diagnostic purposes.

It should be noted that all of the above criteria regarding the shape of the wave are valid only for small amplitudes of applied potential. The value of the

Table 4.11   Diagnostic Criteria for ac Polarography with Reversible Charge Transfer

---

Properties of the potential of the response:
   E_p is independent of ω; half-peak width is $90/n$ mV at 25°C and independent of ω

Properties of the alternating current:
   $I(\omega t)$ is proportional to $\omega^{1/2}$ with a zero intercept

Properties of the phase angle:
   ϕ is 45° and constant with respect to both ω and E_{dc}

Others:
   Linear plot of E_{dc} versus $\log\left[\left(\frac{I_p}{I}\right)^{1/2} + \left(\frac{I_p - I}{I}\right)^{1/2}\right]$
   with a slope of $118/n$ mV at 25°C

---

applied potential should be less than $16/n$ mV peak to peak. If this criterion is not followed, there will be a significant contribution to the total alternating current by the higher harmonic components. As Smith [147] has shown, currents of all harmonic frequencies flow in the cell, but these higher harmonic components are negligible if ΔE is small.

Many electrochemical systems that show reversible behavior with other techniques deviate from diffusion control on time scales accessible with ac polarography. Three systems with very fast charge-transfer rates can be used to study reversible responses into the kilohertz frequency range. They are $Cd^{2+}/Cd(Hg)$ in $1\,M$ KCl or $1\,M$ $NaClO_4$ and tris-oxalato iron in $1\,M$ $K_2C_2O_4 + 0.05\,M$ $H_2C_2O_4$ [165]. The charge-transfer rate constants have all been measured [166] for these readily available, easy-to-use systems using faradaic rectification methods and ac polarography [167].

5.2   Quasi-Reversible Charge Transfer

By far the major applications of fundamental and second-harmonic ac polarography have been to the measurement of charge-transfer rate parameters, k_s and α, for many electrochemical reactions at both dropping-mercury and stationary electrodes such as platinum. The range of k_s values measurable with commercially available instruments spans the range from irreversible reactions, $k_s \leqslant 10^{-6}$ cm/s, to fast charge transfer, $k_s > 1$ cm/s.

The quasi-reversible response can be obtained as outlined earlier for the reversible case, except the surface concentrations are obtained from the absolute rate expression, (31) in Section 2. The current-density expression is given by (222)–(229) using Smith's [147] notation, assuming only C_{Ox} initially present in solution. If both redox species are present, the concentration term becomes $C_{Ox}^* + C_{Red}^*$.

$$I(\omega t) = I_{rev} F(\lambda t^{1/2}) G(\omega^{1/2}\lambda^{-1})\sin(\omega t + \phi) \tag{222}$$

where

$$I_{rev} = \frac{n^2 F^2 C_{Ox}^* (\omega D_{Ox})^{1/2}\Delta E_{ac}}{4RT\,\cosh^2(j/2)} \tag{223}$$

$$F(\lambda t^{1/2}) = 1 + \frac{(\alpha e^{-j} - \beta)\psi(\lambda t^{1/2})D^{1/2}}{k_s e^{-\alpha j}} \tag{224}$$

$$G(\omega^{1/2}\lambda^{-1}) = \left\{\frac{2}{1 + [1 + (2\omega)^{1/2}/\lambda]^2}\right\}^{1/2} \tag{225}$$

$$\phi = \cot^{-1}\left[1 + \frac{(2\omega)^{1/2}}{\lambda}\right] \tag{226}$$

$$\lambda = \frac{k_s}{D^{1/2}}(e^{-\alpha j} + e^{\beta j}) \tag{227}$$

$$D = D_{Ox}^{\beta} D_{Red}^{\alpha} \tag{228}$$

$$\psi(\lambda t^{1/2}) = \frac{k_s e^{-\alpha j}}{D^{1/2}} \exp(\lambda^2 t) \operatorname{erfc}(\lambda t^{1/2}) \tag{229}$$

The expression for $\psi(\lambda t^{1/2})$ is the direct current expression at constant potential. The term I_{rev} is the reversible current magnitude (219).

The expression for $F(\lambda t^{1/2})$ gives the deviation from reversibility attributable to effects of the charge-transfer process on the dc response. When $\lambda t^{1/2} > 50$, the dc process is essentially reversible, $F(\lambda t^{1/2})$ is unity [147], and the drop-time dependence of (222) is eliminated. For the normal polarographic drop life of 5–10 s, the dc process is reversible for $k_s > 10^{-2}$ cm/s.

The expression $G(\omega^{1/2}\lambda^{-1})$ is a measure of the deviation from reversibility of the ac process. This term is unity when $(2\omega)^{1/2}\lambda^{-1} \ll 1$ [147]. For any given value of rate constant, the deviations from reversibility become greater as the frequency is increased. Thus, as faster charge-transfer rates are studied, higher frequencies must be used to obtain reliable results.

The deviations discussed above are actually surface-concentration deviations. If the charge transfer is very fast, essentially thermodynamic (reversible) surface concentrations are observed. As the charge-transfer rate becomes slower, it takes time to reach these equilibrium values, hence the time dependence of the $F(\lambda t^{1/2})$ term for the dc process.

The time dependence of the ac process is embodied in the frequency ω. As the frequency increases there is less time for the concentration to relax to the equilibrium value before a new cycle begins. Thus, there is a lag, which decreases the current magnitude as expressed by $G(\omega^{1/2}\lambda^{-1})$. This lag also causes a phase shift between the applied ac potential and the resulting current as seen by (226). The ratio $(2\omega)^{1/2}\lambda^{-1}$ again expresses this deviation from reversible behavior. The reversible case has a constant phase shift of $\pi/4$ and $\cot(\pi/4) = 1$. In the quasi-reversible case there is an additional shift represented by the ratio $(2\omega)^{1/2}\lambda^{-1}$ attributable to the finite rate of the charge-transfer process (see also the discussion in Section 4 on faradaic impedance). As the charge-transfer rate constant decreases, this deviation in phase angle from $\pi/4$ manifests itself at lower and lower frequencies.

The expression for $\psi(\lambda t^{1/2})$ has been derived assuming planar diffusion. The corresponding expression assuming diffusion to an expanding plane also yields a time-dependent term [168, 169]:

$$\psi(\lambda t^{1/2})_{\text{ep}} = \left(\frac{7}{3\pi}\right)^{1/2} \frac{(1.61 + \lambda t^{1/2})}{(1.13 + \lambda t^{1/2})^2} \tag{230}$$

It should be mentioned again that no time dependence is observed for the quasi-reversible case when the dc process is reversible, that is, $k_s > 10^{-2}$ cm/s. One concludes that the time dependence of the ac polarographic wave is caused only by nonequilibrium of the dc process [146].

With these thoughts in mind, several criteria can be developed that describe the quasi-reversible system and differentiate it from other mechanisms. For any value of k_s, $F(\lambda t^{1/2})$ becomes unity when $\alpha e^{-j} - \beta = 0$, as can be seen from

(224), even though the dc process may be non-Nernstian. This occurs at only one potential, E_{dc}^*.

$$E_{dc}^* = E_{1/2} + \left(\frac{RT}{nF}\right) \ln\left(\frac{\alpha}{1-\alpha}\right) \tag{231}$$

This "crossover" potential has been used to measure the charge-transfer coefficient by measuring the alternating current versus dc potential as a function of drop life [170]. For a reduction process, current positive of this potential increases as the drop life increases, and current negative of it decreases as the drop life increases. At E_{dc}^* the current is independent of drop life.

Hung and Smith [170] calculated α based on the "crossover" potential for the reduction of V(III) to V(II) in $1\,M$ H_2SO_4. Their reported α value of 0.46 ± 0.03 is in excellent agreement with the value reported by Randles [171] for this system. Based on their experiments with other systems, Hung and Smith state that this method is probably useful only for systems in which k_s is between 10^{-4} cm/s and 5×10^{-3} cm/s [170]. Larger values do not show sufficient time dependence, and smaller values yield currents too small for accurate measurements.

Because the phase angle is independent of drop life, yet depends on the charge-transfer rate, it offers an excellent means of measuring k_s. The cot ϕ at any point along the wave is linear with $\omega^{1/2}$ with an intercept of unity and a slope of $\sqrt{2}/\lambda$. The charge-transfer rate constant can be determined from λ if the diffusion coefficients and α are known. Examples of a plot of the phase angle at $E_{1/2}$ versus $\omega^{1/2}$ for three different values of k_s are given in Figure 4.50 [167].

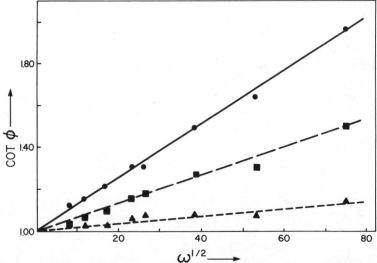

Figure 4.50   Cot ϕ–$\omega^{1/2}$ data for cadmium, chromium cyanide, and iron oxalate systems. Systems: ($\bullet$) 3.00×10^{-3} M Cd^{3+} in $1.00\,M$ Na$_2$SO$_4$, 25°C; ($\blacksquare$) 2.00×10^{-3} M Cr(CN)$_6^{3-}$ in $1.00\,M$ KCN, 25°C; ($\blacktriangle$) 3.00×10^{-3} M Fe(C$_2$O$_4$)$_3^{3-}$ in $0.50\,M$ K$_2$C$_2$O$_4$, 25°C. Reprinted from [167], by courtesy of *analytical Chemistry*.

At any frequency, cot ϕ varies with the dc potential with a maximum value reached at $(E_{dc})_{max}$

$$(E_{dc})_{max} = E_{1/2} + \frac{RT}{nF} \ln \left(\frac{\alpha}{1-\alpha} \right) \tag{232}$$

Note that this is the same potential as the crossover potential and is frequency independent. The value of cot ϕ at $(E_{dc})_{max}$ is given by (233)

$$\cot \phi_{max} = 1 + \frac{(2\omega D)^{1/2}}{k_s[(\alpha/\beta)^{-\alpha} + (\alpha/\beta)^{\beta}]} \tag{233}$$

Both k_s and α can be obtained by using these two equations [146, 172], but the calculated value of k_s is dependent on the value of α obtained from either (231) or (232). In (233) the value of D can be obtained by determining the diffusion coefficients D_{Ox} and D_{Red} from some other technique, perhaps dc polarography, if k_s is not too small. Figure 4.51 shows the plot of cot ϕ versus $(E_{dc} - E_{1/2})$ for one of the values of k_s given in Figure 4.50 at $\omega = 5300$ rad/s.

The values of k_s for the systems illustrated in Figure 4.50 were determined from the dependence of cot ϕ_{max} as a function of $\omega^{1/2}$ and the potential of the maximum using (232) and (233). The values given in Table 4.12 are the averages from determinations at eight frequencies and agree with previously published results [167].

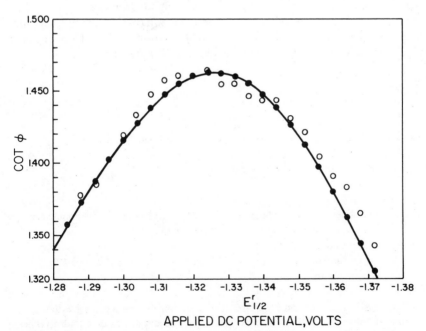

Figure 4.51 Typical cot ϕ–E_{dc} results with chromium cyanide system. System: $2.00 \times 10^{-3} M$ Cr(CN)$_6^{3-}$ in $1.00 M$ KNC, 25°C. ($\bigcirc$) Experimental; ($\bullet$) theory for $\alpha = 0.59$ and $k_s = 0.28$ cm/s (solid curve drawn through these points). Reprinted from [167], by courtesy of *Analytical Chemistry*.

Table 4.12   Heterogeneous Charge-Transfer Rate Parameters Obtained by ac Polarography in the Noncoherent Wave Frequency Multiplex Mode [167]

System	k_s(cm/s)[a]	α
1. $Cd^{2+}/Cd(Hg)$ in 1.00 M Na_2SO_4 at 25°C	0.15 ± 0.01	0.30 ± 0.03
2. $Cr(CN)_3^{5-}/Cr(CN)_6^{4-}$ in 1.00 M KCN at 25°C	0.27 ± 0.02	0.59 ± 0.03
3. $Fe(C_2O_4)_3^{3-}/Fe(C_2O_4)_3^{4-}$ in 0.500 M $K_2C_2O_4$ at 25°C	1.2 ± 0.03	$>0.5^b$

[a]Rate parameters calculated from peak cot ϕ measurements give average value from measurements at eight frequencies; uncertainties = average deviations.
[b]Cot ϕ peak too shallow for accurate estimate of peak potential.

Expressions for the peak-current magnitude and the position of the peak cannot be obtained except for the high-frequency limit, and these are given elsewhere [173]. At low frequencies the position of the peak approaches $E_{1/2}$ [as $G(\omega^{1/2}\lambda^{-1})$ approaches unity]. The peak current approaches linearity with $\omega^{1/2}$.

As the frequency decreases, the experimental parameters all approach the behavior of the reversible case since there is more time for equilibrium to occur. Of course, for very slow charge-transfer reactions, this behavior is not reached until the frequency becomes so low that the steady-state approximation used in deriving (222) is no longer valid [174]. Table 4.13 summarizes the observable criteria for the quasi-reversible reaction using ac polarography.

For rapid charge-transfer rate constants, the alternating current maximum occurs at $E = E_{1/2}$ ($\alpha = 0.5$). The slope of cot ϕ versus $\omega^{1/2}$, from (226) and (227), is $\sqrt{2}/\lambda$ and λ at $E_{1/2}$ is $2k_s/D^{1/2}$. Thus the slope is independent of α.

$$\text{Slope} = \sqrt{\frac{D}{2}} \frac{1}{k_s} \tag{234}$$

Table 4.13   Quasi-Reversible Diagnostic Criteria in ac Polarography

Properties of the potential of the response:
   E_p varies with ω if $\alpha \neq 0.5$

Properties of the alternating current:
   $I(\omega t)$ is nonlinear with $\omega^{1/2}$; linearity with $\omega^{1/2}$ may be noted at low frequencies and a limiting amplitude may be noted at high frequencies

Properties of the phase angle:
   cot ϕ varies linearly with $\omega^{1/2}$, approaching unity at low ω; cot ϕ versus E_{dc} exhibits a maximum

Others:
   The wave is dependent on drop time, the alternating current increases with drop time at potentials anodic of a "crossover" potential and decreases with drop time cathodic of this potential.

Using this equation Creason and Smith [158] reevaluated the iron oxalate system in $1\,M\ K_2C_2O_4$ plus $0.05\,M\ H_2C_2O_4$ using Fast Fourier Transform admittance measurements at $E_{1/2}$ using fourteen simultaneously applied frequencies. The least-squares straight line of cot ϕ versus $\omega^{1/2}$, covering frequencies from 39.9 to 985 Hz, was $1.06\pm0.04\times10^{-3}$ from ensemble averaging of 100 measurements. The intercept of 1.003 ± 0.002 compares with the theoretical value of unity, (226). Using $D_{Ox}=D_{Red}=D=4.94\times10^{-6}$ cm^2/s, $k_s=1.48$ ±0.05 cm/s, in agreement with other results [166, 167] (see also Table 4.12). Cot ϕ versus E_{dc} did not vary enough in this frequency range to obtain α with any accuracy using (226). The results of other studies of rapid charge-transfer reactions in nonaqueous media have been summarized by Smith [165].

Bond and co-workers [175] have studied the reduction of several transition metal complexes of tris-acetylacetonate (acac) and its sulfur analog on platinum and mercury using (232) to determine α. They also used the high-frequency limit of the peak current to obtain k_s. The high-frequency limit expression for $I(\omega t)_{peak}$, assuming the dc process is reversible, is given by (235) [146, 173].

$$I(\omega t)_{peak}=\frac{n^2F^2C_{Ox}^*\Delta Ek_s\beta^\beta\alpha^\alpha}{RT}\left(\frac{D_{Ox}}{D_{Red}}\right)^{\alpha/2}\tag{235}$$

For α values near 0.5 and for $D_{Ox}\simeq D_{Red}$, (235) becomes

$$I(\omega t)_{peak}=\frac{n^2F^2C_{Ox}^*\Delta Ek_s}{2RT}\tag{236}$$

A plot of $I(\omega t)_{peak}$ versus $\omega^{1/2}$ for these complexes is given in Figure 4.52, which shows the current becoming independent of frequency. The simplified current expression (236) was used because $\alpha\simeq0.5$ for these complexes. Values of k_s were 0.13 cm/s and 0.023 cm/s for Fe(acac)$_3$ and Cr(acac)$_3$, respectively.

Cyclic ac voltammetry on a stationary electrode provides another way to measure k_s and α for uncomplicated charge-transfer reactions. For completely reversible reactions on the dc-process time scale (sweep rate), the ac responses for the forward and return sweeps are superimposable [143, 145] unless amalgam formation occurs. The methods discussed previously can be used to analyze the charge-transfer rate parameters. When the process becomes slow enough that the dc surface concentrations are no longer at equilibrium, the ac response for the forward and reverse processes separates, that is, hysteresis with respect to the dc-potential axis is observed [144]. The alternating peak currents in the two directions differ also, as shown in Figure 4.53. The crossover potential where the forward and back currents are equal can be used to determine α.

$$E_{CO}=E_{1/2}+\frac{RT}{nF}\ln\left(\frac{\alpha}{1-\alpha}\right)\tag{237}$$

Notice this is the same expression as (231), which describes the potential where $I(\omega t)$ is independent of drop life for a non-Nernstian dc process, in conventional ac polarography.

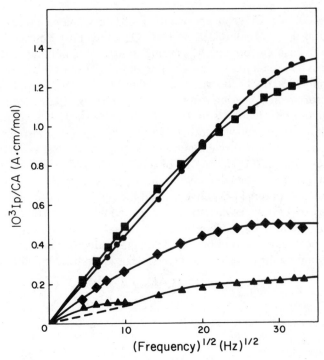

Figure 4.52 Ac frequency dependence of the first reduction waves of the chelates: ●, Cr(SacSac)₃; ■, Fe(acac)₃; ◆, Mn(acac)₃; ▲, Cr(acac)₃. Reprinted from [175], by courtesy of *Inorganic Chemistry*.

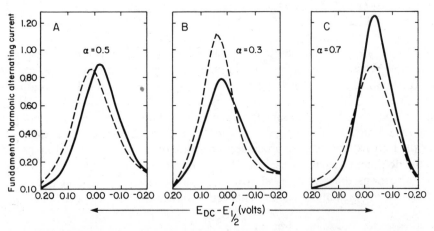

Figure 4.53 Predicted fundamental harmonic ac cyclic voltammograms with non-Nernstian dc behavior: illustration of effect of α. $v = 50$ mV/s, $\omega = 500 \times 2\pi$ s^{-1}, $\Delta E = 5.00$ mV, $k_s = 4.4 \times 10^{-3}$ cm/s, $\alpha = 0.50$ (A), 0.30 (B), 0.70 (C). (——) = forward (cathodic) scan; (- - - -) = reverse (anodic) scan. Alternating-current units $= RT(\omega t)/n^2 F^2 A(2\omega D_{Ox})^{1/2} C_{Ox} \Delta E$ (normalized total fundamental harmonic faradaic current; planar diffusion). Reprinted from [144], by courtesy of *Analytical Chemistry*.

The peak current ratio, R_p, where

$$R_p = \frac{\text{forward-scan peak current}}{\text{reverse-scan peak current}} \tag{238}$$

goes through a minimum as k_s is varied, having a value at large k_s of unity and a near-unity value for small $k_s \simeq 10^{-5}$ cm/s. Since the ratio is also dependent on α, it is apparently not a useful analytical parameter [143]. The peak separation of the forward and backward scan is related to k_s, but the dependence is also a function of α and, to some extent, frequency and scan rate. Nevertheless, with α obtained from (237) the working curves of Bond and co-workers [144] can be used to obtain k_s from the peak separation. Figure 4.54 is an example of ac cyclic voltammograms of Mn(acac)$_3$ in acetone on platinum at four scan rates, showing the constancy of E_{CO}. The peak separation varies with scan rate and was used to obtain $k_s = (1.5 \pm 0.2) \times 10^{-3}$ cm/s [144].

Difficulties with this method of analysis result when amalgam formation occurs [144]. The product of electron transfer accumulates in the drop or mercury thin film, increasing the return-current magnitude and making it scan-rate dependent. The crossover potential, E_{CO}, is no longer a constant, but varies with dc scan rate. O'Halloran and Blutstein [176] have studied the reduction of Mn(II) to Mn(Hg) in 0.5 M NaCl at a hanging mercury drop. The forward and reverse peak current ratios vary with scan rate and no crossover potential is observed. The system is further complicated by a very asymmetric charge transfer, $\alpha = 0.9$, and rate parameters were obtained only by matching both the

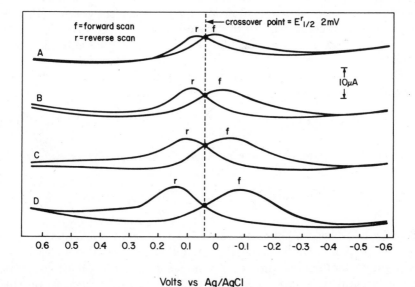

Figure 4.54 Fundamental harmonic ac cyclic voltammograms of 1 mM Mn(acac)$_3$ on Pt in 0.1 M TEAP/acetone. Applied 10 mV p-p sine wave at 100 Hz. Dc scan: (A) 20 mV/s, (B) 50 mV/s, (C) 100 mV/s, (D) 200 mV/s. Reprinted from [144], by courtesy of *Analytical Chemistry*.

dc cyclic voltammograms and the ac cyclic voltammograms using digital simulation [144, 176]. Best fits over a scan range of 2 mV/s to 5 Mv/s ($\omega = 400$ Hz) were obtained with $k_s = 4.0 \times 10^{-4}$ cm/s and $\alpha = 0.9$.

A recent extension of ac polarography for studies of quasi-reversible electron-transfer reactions has been made by Pons and co-workers [177]. Reflectance of monochromatic light impinging on the electrode is monitored instead of the alternating current. They have shown that the reflectance measured at the frequency of the applied potential is 90° out of phase with the alternating current so that the previously derived dependence of cot ϕ on frequency [(226) and (227)] or dc potential [(232) and (233)] used in alternating current studies is applicable. The advantage of using sinusoidally modulated ac reflectance spectroscopy is that the nonfaradaic double-layer response in the current measurement is eliminated in the spectroscopic measurement. The technique was tested on the ferricyanide/ferrocyanide couple in KCl at a platinum electrode by measuring the slope of cot ϕ as a function of frequency according to (234) to obtain k_s after mathematical correction for uncompensated solution resistance.

5.3   Irreversible Charge Transfer

Very slow charge-transfer reactions have received little experimental attention using ac polarographic techniques because the response is very small, that is, the charge-transfer resistance, (208) in Section 4, is very large. Timmer, Sluyters-Rehbach, and Sluyters have shown, both theoretically and experimentally [178], that such a signal does exist for an irreversible reaction caused by slow charge transfer. Smith and McCord [179] independently showed theoretically that a signal could exist for this reaction and extended their work to fast following chemical reactions, including the catalytic regeneration of starting material.

The theoretical results show that the peak current depends only on the transfer coefficient α, but not on k_s.

$$I_p(\omega t) = \left(\frac{7}{3\pi t}\right)^{1/2} \frac{\alpha n^2 F^2 C_{Ox}^* D_{Ox}^{1/2} \Delta E_{ac}}{RT} \tag{239}$$

Similarly, the dc potential of the peak is related to the half-wave potential of the irreversible dc polarogram [179]

$$(E_{dc})_{peak} = (E_{1/2})_{irr} - \frac{RT}{2\alpha n F} \ln Q \tag{240}$$

where $Q = 1.349(2\omega t)^{1/2}$. Thus it appears that no information regarding the value of k_s can be obtained from ac polarography that cannot be obtained from dc polarography for the irreversible case, although it may be possible to determine αn more accurately. Thus k_s only affects the dc potential of the ac wave. These diagnostic criteria are summarized in Table 4.14.

Sluyters and co-workers have studied the reduction of Eu(III) to Eu(II) in $1 M$ NaClO$_4$ solution using faradaic impedance measurements and ac polaro-

Table 4.14   Diagnostic Criteria for Irreversible Charge Transfer in ac Polarography

$$Ox + ne^- \xrightarrow{k_s} Red$$

Properties of the potential of the response:

E_p shifts cathodically by $15/\alpha\eta$ mV for a tenfold increase in ω

Properties of the alternating current:

I_p is independent of ω at high frequencies, independent of k_s at all frequencies, and proportional to $t^{-1/2}$ (drop time)

Properties of the phase angle:

ϕ is much less than $45°$

Others:

The ac response is very small and, in some cases, may not be observable

---

graphy [178]. Using a value of $\alpha = 0.41$ and $k_s = (2.9 \pm 0.2) \times 10^{-4}$ cm/s, they were able to match their experimental ac polarograms to the theoretically calculated ones between 320 and 2000 Hz. The rate parameters were calculated from faradaic impedance measurements (see Section 4). These rate parameters agree well with results of other workers [178, 180].

5.4   Coupled Chemical Reactions

Unlike cyclic voltammetry, ac voltammetry does not offer the immediate qualitative characterization of mechanisms in which homogeneous chemical reactions are coupled to the charge-transfer step. A peak-shaped, alternating current response still occurs although the magnitude and symmetry may vary. The phase angle as a function of dc potential and frequency is a good qualitative parameter, but requires considerable experimental effort to obtain unless one has the capability of doing Fast Fourier Transform analysis. Even so, nonfaradaic effects such as double-layer charging and uncompensated iR potential drop must be eliminated before useful measurements are obtained.

The problem is also difficult theoretically. As mentioned earlier, the current expression for the dc process must be obtained before the alternating current expression is complete. In the past this problem has severely limited accurate solutions to a closed-form, alternating current expression. Recently, Ruzic, Smith, and Feldberg [181] and Ruzic and Smith [182] have attacked the problem using digital simulation to solve the dc surface concentrations, upon which the ac signal makes its perturbation. The ac boundary-value problem is treated separately [183]. Separation of the two processes applies only to small-amplitude ac potentials. With the dc concentrations in hand, analytical expressions for the ac process can be obtained for coupled chemical mechanisms of any complexity. The usual assumptions regarding equality of diffusion coefficients and fast charge transfer can even be waived. A summary of several reaction schemes and predictions of theoretical ac voltammetric response has been presented on the basis of the above theoretical approach [184].

With these experimental and theoretical complexities in mind, one can see why nonexperts have been reluctant to embrace ac voltammetry as the method of choice for studying coupled chemical reaction mechanisms. Nevertheless, examples of such studies do exist in the literature and some of them will be described.

5.4.1   Preceding Chemical Reaction

The preceding coupled chemical reaction is given by the reaction scheme

$$Y \underset{k_b}{\overset{k_f}{\rightleftharpoons}} Ox + ne \underset{}{\overset{ks,\alpha}{\rightleftharpoons}} Red$$

where $K = k_f/k_b$ and $k = k_f + k_b$.

The general case of a chemical reaction preceding a quasi-reversible charge transfer has been treated theoretically at some length by McCord and Smith [185] using the expanding-plane model. A discussion of the phase-angle expressions has also been given earlier for this mechanism [186].

When the preceding chemical reaction is very fast, a diffusion-controlled wave results because the chemical reaction has no influence on the flux of the electroactive species. In this case, however, the potential of the peak is shifted negative for reduction, depending on the value of the equilibrium constant K

$$E_p = E_{1/2} + \frac{RT}{nF} \ln\left(\frac{K}{1+K}\right) \tag{241}$$

If the equilibrium constant is very large, there is no thermodynamic potential shift, as can be seen from (241). This means that Ox is the predominant species in solution. The current magnitude is also controlled by the sum of the bulk concentration of C_Y, the electroinactive species, and C_{Ox}, the electroactive species.

As the equilibrium constant becomes smaller and Y starts to become the predominant species in solution, the peak potential shifts to more negative potentials. This shift can be used to measure complexation constants just as is done using dc polarography. Bond [187, 188] and Bond and Waugh [189] have used ac polarography to measure several metal–fluoride complexes in acid media using short drop times. These results have been summarized by Smith [165].

The ac polarographic kinetic parameter of interest is k/ω, where k is the sum of the rate constants of the forward and reverse chemical reactions. As k/ω becomes smaller (decreasing rate constant or increasing frequency), two effects are observed. The alternating polarographic current begins to decrease and the wave shifts to more anodic potentials. This shift is similar to that observed in the cyclic experiment as the scan rate increases. The anodic shift of E_p as the frequency is increased is a good diagnostic criterion for the preceding reaction mechanism. The peak-current values are very sensitive to the rate parameters of the chemical reaction and become more so as the frequency is increased.

The phase angle also has properties unique to the preceding chemical reaction. At a given frequency, cot ϕ plotted against dc potential has a sigmoidal shape with a limiting value at cathodic potentials. Examples of this behavior are shown in Figure 4.55. Experimentally, this may not be easily

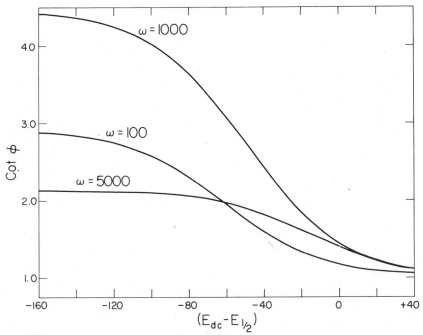

Figure 4.55   Theoretical cot ϕ versus E_{dc} for a preceding chemical reaction. $K = 0.10$, $k = 5000 \text{ s}^{-1}$, reversible charge transfer.

observed because the current becomes rather small at potentials cathodic of the peak current. The shape of cot ϕ versus dc potential is the mirror image of that of the following reaction where cot ϕ values reach a limiting value at potentials anodic to the half-wave potentials [146, 186].

This limiting value of the cotangent, cot ϕ^{lim}, goes through a maximum as the kinetic parameter k/ω is varied. The limit occurs at different values of k/ω, depending on the equilibrium constant, so it may be necessary to cover a frequency range of two orders of magnitude or more to observe this phenomenon.

An assumption implicit in the above discussion is that the charge-transfer reaction remains essentially reversible at the frequency necessary to determine the cot ϕ^{lim}_{max}. If this is not true, phase shifts associated with the charge-transfer rate complicate the analysis by increasing cot ϕ near $E_{1/2}$.

It is possible to obtain useful kinetic parameters even when slow charge transfer does occur. Hawkridge and Bauer [190] studied the reduction of Cu(II) at several pH values and measured cot ϕ versus E_{dc} at several frequencies. An example is shown in Figure 4.56 for 100 Hz measurements at three pH values where the characteristic maximum in cot ϕ near $E^{dc}_{1/2}$ indicates slow charge transfer. At negative potentials the increasing values of cot ϕ are indicative of control by the preceding reaction, assumed to be dehydration of $Cu(H_2O)_6^{2+}$ [190]. A plateau in cot ϕ versus E_{dc}, as shown in Figure 4.55, is not observed here.

The phase angle at the current peak, plotted against the square root of

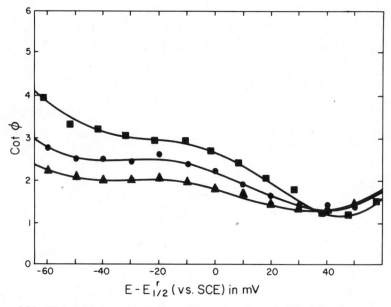

Figure 4.56 Plot of cot ϕ versus $E - E_{1/2}$. Parameter values: Cu(II) = 1.00 mM, 1 M LiNO$_3$, $T = 298$ K, $f = 100$ Hz, $\Delta E = 10$ mV (rms). ▲, pH 5.75; ●, pH 3.40; ■, pH 1.00. Reprinted from [190], by courtesy of *Analytical Chemistry*.

frequency, serves as a diagnostic criterion for coupled chemical reactions. A maximum in the plot is indicative of either a preceding or a following chemical reaction coupled to the charge-transfer process [147]. An example of this is shown in Figure 4.57 for the same reduction of Cu(II) to Cu(Hg) at three pH values [190]. The frequency of the maximum is an indication of the rate constant k_f, although it also depends on the equilibrium constant K. The dependence of cot ϕ on E_{dc} is used to determine whether a preceding or a following reaction is operative.

If the equilibrium constant is known, the exact dependence of cot ϕ on frequency and dc potential can be calculated:

$$\cot \phi = \frac{K + (1+K)e^j + \left[\dfrac{(1+g^2)^{1/2} + g}{1+g^2}\right]^{1/2}}{K + (1+K)e^j + \left[\dfrac{(1+g^2)^{1/2} - g}{1+g^2}\right]^{1/2}} \tag{242}$$

$$g = k/\omega \tag{243}$$

This result is similar to the expression for cot ϕ obtained in Section 4 (214), which also includes a term caused by slow charge transfer.

Matsuda and Tamamushi [191] have used ac polarography and (242) to

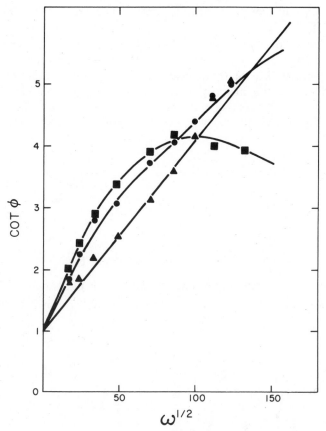

Figure 4.57 Plot of cot ϕ versus $\omega^{1/2}$. Parameter values: Cu(II) = 1.00 mM, $T = 298$ K, 1 M LiNO$_3$. ▲, pH 5.75; ●, pH 3.40; ■, pH 1.00. Reprinted from [190], by courtesy of *Analytical Chemistry*.

study the reduction of Cd(II) in EDTA, which is preceded by dissociation of the Cd–EDTA complex.

$$Cd(EDTA)^{2-} + H^+ \underset{k_b}{\overset{k_f}{\rightleftharpoons}} Cd^{2+} + HEDTA^{3-}$$

$$Cd^{2+} + 2e^- \overset{DME}{\rightleftharpoons} Cd(Hg)$$

Theoretical curves of cot ϕ versus $\omega^{1/2}$ were calculated for various values of k/ω at the ac peak potential, which was slightly negative of $E_{1/2}^{dc}$, but relatively constant from 70 to 2000 Hz. The curve best matched experiment for $k = 1000$, using $K = 10^{-3.67}$ [191]. From the concentration of free ligand, HEDTA^{3-}, the appropriate second-order rate constant, $k_b/[\text{HEDTA}^{3-}]$, was calculated to be $2.3 \times 10^9 \ M^{-1} \ \text{s}^{-1}$, in agreement with results of other studies (see Table I of [191]).

Another case of the preceding mechanism has been studied using ac techniques [192]. The reaction is the reduction of Cd(II) in nitrilotriacetate (NTA) as a function of pH. The rate parameters were obtained by matching the expanding-plane theoretical model [185] to the alternating current response at several frequencies. The interesting feature of this work is that the complex itself is also reduced at the mercury electrode in an irreversible charge-transfer reaction at potentials negative of the free Cd(II) reduction, which is completely reversible. The overall mechanism is outlined below.

Wave I

$$Cd(NTA)^- + H^+ \underset{k_b}{\overset{k_f}{\rightleftharpoons}} Cd^{2+} + HNTA^{2-}$$

$$Cd^{2+} + 2e^- \rightleftharpoons Cd(Hg)$$

Wave II

$$Cd(NTA)^- + H^+ + 2e^- \overset{k_{s,\alpha}}{\rightleftharpoons} Cd(Hg) + HNTA^{2-}$$

An example of the match of experiment and theory is illustrated in Figure 4.58 for reduction at pH 4.28 at four frequencies. The dots are experimental; the

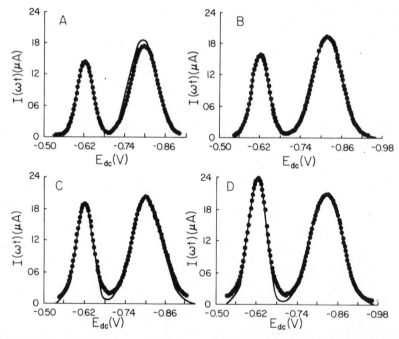

Figure 4.58 Fundamental harmonic ac polarographic results with CdNTA system at pH 4.28. (●) Experimental points at (A) 39 Hz, (B) 105 Hz, (C) 228 Hz, (D) 448 Hz; (——) theoretically calculated response with expanding-plane model. Reprinted from [192], by courtesy of *Analytical Chemistry*.

line is theoretical using parameters given in Figure 4.58. The more negative wave is the irreversible reduction of the complex itself (Wave II) since $k_s = 1.5 \times 10^{-7}$ cm/s and $\alpha = 0.63$. Notice that the current magnitude is independent of frequency as described in the last section (see Table 4.14). A measure of cot ϕ as a function of $\omega^{1/2}$ at the current peak of the first wave shows a maximum at low frequency, decreasing toward unity at high frequencies. The charge-transfer rate constant is unknown, but must be very large for Cd(II) reduction in NTA.

Other parameters can also be used to characterize the preceding reaction mechanism. The drop-time dependence of the ac polarographic wave offers an excellent means of determining whether or not such a mechanism is operative. The current increases as the drop time increases and the peak potential becomes more cathodic [185]. This positive time dependence is exactly opposite that observed for a following chemical reaction. Diagnostic criteria for a preceding chemical reaction are summarized in Table 4.15.

The variations in current magnitude as a function of frequency can be used to determine rate parameters if some estimate of the diffusion current, for example, in the absence of complexing species, can be obtained. The quantity $i_p/i_{p,d}$, where i_p is the observed peak current and $i_{p,d}$ is the calculated diffusion-controlled peak current, is plotted against log $\omega^{-1/2}$. The shape of the curve can be compared to that of the working curve shown in Figure 5 of [185] to obtain an estimate of K. The separation of the experimental curve from the working curve along the x axis can be used to obtain a measure of k. If K is known from some other source, this method can yield fairly good results. A large range of frequencies must be used for best results.

5.4.2   Following Chemical Reaction

In many ways the following reversible chemical reaction has the same effect on the ac polarographic response as does the preceding reversible reaction. The maximum effects on current magnitude and cot ϕ occur at potentials positive, instead of negative, of $E_{1/2}$. A theoretical discussion of the current response [193] and the phase-angle behavior [186] has been given for this mechanism coupled to a quasi-reversible charge-transfer reaction using the expanding-plane diffusion model. Only the case of reversible charge transfer is

Table 4.15   Diagnostic Criteria for a Preceding Chemical Reaction Using ac Polarography

---

Properties of the potential of the response:
   E_p shifts anodically as ω is increased and cathodically as the drop time is increased
Properties of the alternating current:
   $I(\omega t)$ increases by less than $\omega^{1/2}$ as ω increases; $I(\omega t)$ increases with drop time
Properties of the phase angle:
   cot ϕ versus E_{dc} has a sigmoidal shape with a limiting value at cathodic potentials
   and a value of unity at anodic potentials

---

considered here.

$$Ox + ne^- \rightleftharpoons Red$$

$$Red \underset{k_b}{\overset{k_f}{\rightleftharpoons}} Z \qquad K = k_b/k_f \quad \text{and} \quad k = k_f + k_b$$

The equilibrium constant is defined as $[R]/[Z]$ so that ac polarographic working curves for the preceding and following mechanism are the same.

When the chemical rate is sufficient to maintain equilibrium, the following expression for the peak potential of the ac wave is obtained [194]:

$$E_p = E_{1/2} + \frac{RT}{nF} \ln\left(\frac{1+K}{K}\right) \tag{244}$$

which is analogous to the preceding reaction case (241).

As the equilibrium shifts toward formation of product from Red, K becomes small and the potential of the wave shifts positive from $E_{1/2}$. When Red predominates, K is large and there is little shift in the peak potential from $E_{1/2}$.

At the opposite extreme in chemical kinetics, when the chemical rate is very slow, the peak potential is the reversible half-wave potential $E_{1/2}$. As the kinetic parameter k/ω is varied from large values (large rate constant or low frequency) to smaller values, the peak potential shifts cathodically, in a manner similar to the shift in peak potential of the cyclic voltammetric response as a function of the balance of the chemical kinetics and the time scale of the experiment. This cathodic shift in the peak potential as the frequency is increased can be used to indicate the possible presence of a following chemical reaction.

Another means to test for a chemical reaction involves monitoring the peak current as a function of frequency. The term $i_p/i_{p,d}$ is plotted as a function of $\omega^{1/2}$. A reversible charge transfer with no coupled chemical reactions has a value of $i_p/i_{p,d}$ of unity over the whole frequency range. In the presence of either a following or a preceding reaction, the term $i_p/i_{p,d}$ decreases as the frequency increases [185, 193]. If the chemical rate is very rapid and the equilibrium favors the electroactive species, there may be little change in the ratio over a moderate frequency range, but the value of $i_p/i_{p,d}$ will be less than unity. As the rate of the chemical reaction goes to zero, the system approaches diffusion control and the ratio again approaches unity. Note that this test indicates the presence of a coupled chemical reaction but says nothing about the type of reaction. Further information from either shifts in the peak potential with frequency or phase-angle data must be obtained to characterize the nature of the chemical reaction.

The effects of a following chemical reaction on phase-angle data show up most strongly on the anodic side of the wave. The variation in cot ϕ as a function of E_{dc} is shown in Figure 4.59 for typical values of K and k. The sigmoidal curve reaches a limiting value of cot ϕ at anodic potentials, denoted as cot ϕ^{lim}. Unfortunately, this behavior may be experimentally difficult to detect because of the low current values at these potentials. This cot ϕ versus E_{dc} behavior is the mirror image of the behavior of a system with a preceding chemical reaction (discussed earlier) about the value $E_{dc} = E_{1/2}$.

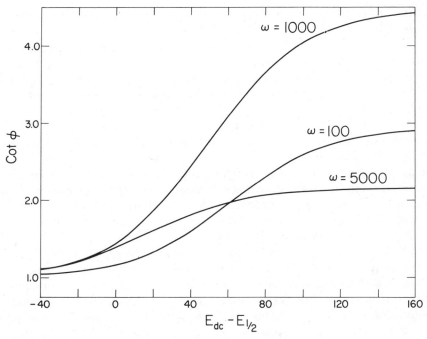

Figure 4.59   Theoretical cot ϕ versus E_{dc} for a following chemical reaction. $K = 0.10$, $k = 5000$ s^{-1}, reversible charge transfer.

Since it may be difficult to measure cot ϕ^{lim}, McCord and Smith [185] have suggested using the value of cot ϕ at the peak potential. Working curves may actually be calculated at any potential by using the expression

$$\cot \phi = \frac{K + (1+K)e^{-j} + \left[\dfrac{(1+g^2)^{1/2} + g}{1+g^2} \right]^{1/2}}{K + (1+K)e^{-j} + \left[\dfrac{(1+g^2)^{1/2} - g}{1+g^2} \right]^{1/2}} \tag{245}$$

where g is k/ω.

Note that this is the same expression as for the preceding reaction case (242) with e^{j} replaced by e^{-j}, again illustrating the mirror-image response of these two mechanisms.

In addition to the shift in peak potential with changes in frequency and the behavior of cot ϕ, a following chemical reaction may be characterized by the negative time dependence of the alternating current [193]. This means that as the drop time increases, the alternating current decreases. The peak potential also shifts to more anodic potentials as the drop time increases.

Other mechanisms can cause a negative drop-time-dependent ac polarographic wave, so this behavior is not by itself a sufficient diagnostic criterion. However, coupled with the effects previously mentioned, it affords a fairly reliable indication of this mechanism. The diagnostic criteria for reversible charge transfer followed by a reversible chemical reaction are summarized in Table 4.16.

Several simplifications arise when the chemical reaction can be considered irreversible, proceeding in only one direction. McCord and co-workers [193] have shown that this situation arises when $K(kt)^{1/2} \lesssim 10^{-3}$. For example, if $K = 10^{-4}$ and $t = 5$ s, the chemical reaction is essentially irreversible if $k_f > 20$ s^{-1}.

The previous discussion of drop-time dependence and potential shifts is generally applicable to the irreversible following reaction also. The behavior of cot ϕ with potential is also similar although the value of cot ϕ^{lim} may be quite large. The main difference lies in the variation of cot ϕ^{lim} with the chemical rate constant k_f. In the reversible chemical reaction, cot ϕ reaches a maximum value as k is varied, the value of k depending on the equilibrium constant K. With an irreversible chemical reaction, cot ϕ increases without limit as k_f increases.

Aylward and co-workers [195] have obtained a working curve for the case in which the chemical reaction rate is much less than the applied frequency. This curve relates the ratio $i_p/i_{p,d}$ to the dimensionless parameter $k_f t$, where t is the drop time. McCord and co-workers [193] obtained similar curves for the planar-diffusion model and the expanding-plane model. These working curves are illustrated in Figure 4.60, taken from [193], and show the importance of using the more exact expanding-plane model for this type of analysis.

Aylward and Hayes [196] have used this technique to measure the rate of formation of Cd–EDTA complex by the reaction scheme

$$Cd(Hg) \rightleftharpoons Cd^{2+} + 2e^-$$

$$Cd^{2+} + HY^{3-} \xrightarrow{k_f} CdHY^-$$

where Y^{4-} is the EDTA tetravalent anion. The equilibrium constant for the second reaction is small, $K = 10^{-3.67}$ [191], and the pseudo-first-order rate constant is kept small by having a low concentration of HY^{3-}. Under these experimental conditions, the constant $K(k_f t)^{1/2} \approx 10^{-3}$ and the reaction is essentially irreversible. Aylward and Hayes measured $i_p/i_{p,d}$ by faradaic im-

Table 4.16 Diagnostic Criteria for the Reversible Following Reaction Using ac Polarography

---

Properties of the potential of the response:
  E_p shifts cathodically as ω is increased; E_p shifts anodically as the drop time is increased

Properties of the alternating current:
  $I(\omega t)$ increases by less than $\omega^{1/2}$ as ω is increased if k is not large and K is not small; $I(\omega t)$ decreases as the drop time is increased

Properties of the phase angle:
  cot ϕ has a sigmoidal shape with a limiting value at anodic potentials and a value of unity at cathodic potentials

---

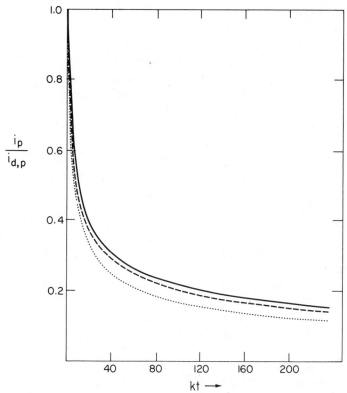

Figure 4.60 Ac polarographic working curve for irreversible following chemical reaction with reversible charge transfer. (———) Expanding-plane theory. (– – –) Aylward, Hayes, and Tamamushi theory [195]. (· · ·) Stationary-plane theory. Reprinted from Elsevier Publishing Co., [193].

pedance techniques at varied frequencies, drop lives, and concentrations of HY^{3-} and their value of $(6.1 \pm 0.7) \times 10^8\ M^{-1} s^{-1}$ can be compared with $23 \times 10^8\ M^{-1} s^{-1}$, obtained by Matsuda and Tamamushi [191] as described earlier.

Because of the unique behavior of $\cot \phi$ for the irreversible reaction, it can be used to determine the rate of the following reaction if the reversible half-wave potential is known. The appropriate expression for the cotangent of the phase angle under these conditions is given by (246), which is obtained from (245) when $K = 0$. A working curve of $\cot \phi$ versus $g = k_f/\omega$ can be generated from (246) for some convenient potential, and the experimental $\cot \phi$ at this potential yields k_f since the frequency is known. By varying the applied frequency, a series of values of k_f is obtained to determine the uncertainty of the measurements.

$$\cot \phi = \frac{1 + e^j \left[\dfrac{(1+g^2)^{1/2} + g}{1+g^2} \right]^{1/2}}{1 + e^j \left[\dfrac{(1+g^2)^{1/2} - g}{1+g^2} \right]^{1/2}} \tag{246}$$

A more general approach to data analysis, in which $E_{1/2}$ need not be known nor the special condition applied that $k_f \ll \omega$, arises through relating the direct polarographic current to the alternating polarographic current [193]. A series of working curves is constructed by calculating the ratio $i_{ac}/(nF\Delta E/RT)i_{dc}$ versus the potential $(E_{dc} - E_{1/2})$ for several values of the kinetic parameter k_f/ω. Experimental curves of the current ratio as a function of E_{dc} are compared with the theoretical curves. The shape of the theoretical curve that best matches that of the experimental curve yields k_f, and the position of the experimental curve on the potential axis yields $E_{1/2}$.

The cot ϕ parameter can also be used in this manner to obtain k_f and $E_{1/2}$. Theoretical curves using (246) can be constructed of cot ϕ versus $(E_{dc} - E_{1/2})$ for various values of k_f/ω. The shape of the experimental curve of cot ϕ versus E_{dc} yields k_f, and the position yields $E_{1/2}$.

The reduction of cyclooctatetraene in DMF was studied by Huebert and Smith [197] using fundamental- and second-harmonic ac polarography. The second reduction step of the anion radical to the dianion is followed by protonation in an essentially irreversible reaction. A plot of cot ϕ at the current peak versus $\omega^{1/2}$ showed the characteristic increase in cot ϕ at low frequencies associated with a coupled chemical reaction. The increase was small, going from unity to about 1.5. In addition the system is complicated by slow charge transfer, so cot ϕ also increases linearly with $\omega^{1/2}$ at high frequencies instead of falling back toward unity. The rate constant of the following reaction was actually determined from the second-harmonic polarographic data. The rate was variable depending on amounts of residual water in the solution.

5.4.3   Catalytic Regeneration

The catalytic mechanism that involves regeneration of the starting material, usually in a pseudo-first-order mechanism, has received considerable theoretical [147, 179, 198] and experimental [199, 200] study using ac polarography. The general current expression for $I(\omega t)$, based on the expanding-plane model, is fairly complex and is presented here only for a special case. Earlier work concentrated on the expression for cot ϕ, which is independent of electrode geometry and can be used to determine the catalytic rate constant, k_c, expressed in s^{-1}.

$$Ox + ne \underset{k_c}{\overset{k_{s,\alpha}}{\rightleftharpoons}} Red$$

The expression for cot ϕ as a function of k_c and ω is best expanded to include the effects of slow charge transfer because the regenerative scheme places greater demands on the rate of charge transfer than does diffusion.

$$\cot \phi = \frac{\dfrac{(2\omega)^{1/2}}{\lambda} + \left[\dfrac{(1+g^2)^{1/2} + g}{1+g^2} \right]^{1/2}}{\left[\dfrac{(1+g^2)^{1/2} - g}{1+g^2} \right]^{1/2}} \tag{247}$$

λ is given by (227) or by $2k_s/D^{1/2}$ at $E_{dc}=E_{1/2}$ and $g=k_c/\omega$. Only when the charge-transfer rate constant k_s is large enough that $\lambda \gg k_c^{1/2}$, does (247) become independent of k_s. A feature of (247) is that no dc-potential term appears except in λ, that is, the value of cot ϕ is constant across the whole ac polarogram for rapid charge transfer (i.e., large values of λ).

If charge transfer is fast, (247) is simplified considerably.

$$\cot \phi = (1+g^2)^{1/2} + g \tag{248}$$

which, at very large values of k_c or low frequency ($g^2 \gg 1$), becomes cot $\phi = 2k_c/\omega$. Thus as the frequency decreases, cot ϕ becomes very large, going to infinity as $\omega \to 0$. This is an excellent criterion for this mechanism.

Examples of the dependence of cot ϕ on frequency are shown in Figure 4.61 for several values of the catalytic rate constant with and without charge-transfer control, using $k_s = 0.1$ cm/s, not an unduly small value.

The large value of cot ϕ at low frequencies in the catalytic case means that the equivalent faradaic impedance becomes purely resistive. The total amount of Red formed at the electrode surface is immediately reoxidized to Ox, causing a very steep concentration gradient. Diffusion, from which the capacitive component of the faradaic impedance is obtained, now plays a minimal role in mass transport (see Section 4).

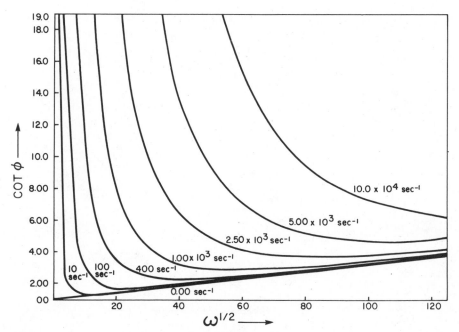

Figure 4.61 Frequency dependence of cot ϕ with catalytic reaction at $E_{dc}=E_{1/2}$ with varying k_c. $k_s = 0.100$ cm/s, $\alpha = 0.500$, $D = 1.00 \times 10^{-5}$ cm^2/s, $n=1$, $T=25°C$. Reprinted from [186], by courtesy of *Analytical Chemistry*.

Under reversible charge-transfer conditions, a working curve of cot ϕ plotted against the kinetic parameter k_c/ω can be used to obtain the rate constant k_c. This curve is shown in Figure 4.62. The linear portion of the curve represents essentially complete kinetic control of the wave. Note that this region occurs for large values of k_c/ω, which means that experiments can be conducted at fairly low frequencies where charge transfer has a minimal effect on the phase angle.

There is a lower limit to the pseudo-first-order kinetic rate that can be measured by this technique set by the lowest frequency that can be obtained conveniently. The steady-state assumptions [199] made in deriving (247), as well as experimental considerations, set this lower limit at about 10 Hz; that is, it is difficult to study reactions in which $k_c < 30\,\text{s}^{-1}$. Experimentally it is also difficult to measure accurately values of cot ϕ greater than 10.

If $k_c \gg \omega$, the problem can be avoided by measuring the current amplitude instead. Assuming fast catalytic kinetics, the current expression simplifies to that given by (249) [179, 200].

$$I(\omega t) = I_{\text{rev}} k_c^{1/2} F(\lambda_k)\sin \omega t \tag{249}$$

where

$$F(\lambda_k) = \frac{\lambda_k[\lambda_k + \alpha(1 + e^{-j})]}{(1 + \lambda_k)^2} \tag{250}$$

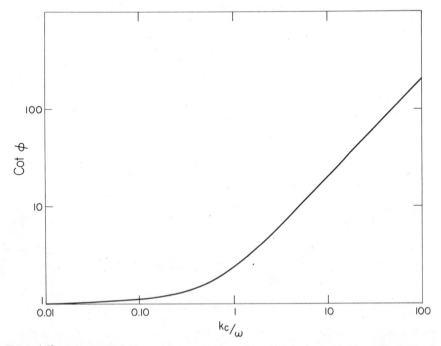

Figure 4.62 Ac polarographic working curve of cot ϕ versus k_c/ω for catalytic regeneration with reversible charge transfer.

and

$$\lambda_k = \lambda / k_c^{1/2} \tag{251}$$

The expression I_{rev} is given by (223), the reversible ac polarographic response. The term $F(\lambda_k)$ becomes unity if charge transfer is fast, that is, $\lambda \gg k_c^{1/2}$.

The current is now a direct measure of k_c and is considerably larger than the current in the absence of the catalytic reaction. A plot of $I(\omega t)$ as a function of concentration of catalytic reagent should yield a straight line, from which the second-order rate constant can be calculated. If charge transfer starts to become rate limiting, which it may at high catalyst concentrations, the wave will be broadened with respect to the reversible case, and a plot of $I(\omega t)$ versus concentration would curve downward from a straight line. The potential of the $I(\omega t)$ peak also shifts to negative potentials [200].

Smith [199] has tested (247) experimentally by determining k_c for a system using dc polarographic data and obtaining k_s and α from ac polarographic data. Phase-angle data were then obtained using the catalytic system and the values were compared with theory and with the values of k_c, k_s, and α previously obtained. The experimental plots of cot ϕ against frequency and dc potential at several frequencies were compared with the theoretical calculations, with excellent agreement.

The system Smith used was the Ti(IV)/Ti(III) couple in 0.2 M $K_2C_2O_4$, with $KClO_3$ as the catalytic regenerating agent. Charge-transfer data were obtained in the absence of ClO_3^- ion and then a fiftyfold excess of ClO_3^- ion was added to obtain phase-angle data on the pseudo-first-order catalytic system. A value of $k_s = 0.046$ cm/s with $\alpha = 0.35$ was obtained by ac polarography and the second-order catalytic rate of 4.25×10^3 M^{-1} s^{-1} was obtained from dc polarography at three chlorate concentrations. Data for cot ϕ using ac polarography and these constants gave excellent agreement with theoretical curves. Because of the somewhat low value of k_s, it might be difficult to study this system by using Figure 4.62 and the assumption of charge-transfer reversibility.

The catalytic regeneration of Fe(TEA)$^-$, the triethanolamine complex of Fe(III), by chlorite anion in alkaline solution was also studied by Bullock and Smith [200]. The charge-transfer rate constant is somewhat faster, $k_s = 0.21$ cm/s and $\alpha = 0.5$, as measured in the absence of catalyst. In 0.01 M ClO_2^-, $k_c = 320$ s^{-1}, obtained from dc polarography. Using these rate parameters the fundamental- and second-harmonic current response at 448 Hz ($\omega = 2810$) and 896 Hz matched the calculated curves very well. The rate parameter k_c was not reevaluated from the ac polarographic data.

5.4.4   Disproportionation and Dimerization

The application of ac polarography to the study of second-order chemical reactions, for example, dismutation and dimerization, coupled to charge-transfer reactions has been presented by Hayes and co-workers [183]. The ac and dc processes are treated separately by assuming a steady-state concentration of reacting species at the electrode surface on the ac time scale. This concentration is obtained using numerical techniques or digital simulation of the dc process [181]. The important kinetic parameter on the ac time scale is given by (252)

$$g = \frac{2k_D C(0, t)}{\omega} \tag{252}$$

where $C(0, t)$ is the time-dependent dc surface concentration and k_D is the dismutation or dimerization rate constant. The surface-concentration term in (252) means that the observable parameters such as $\cot \phi$ and $I(\omega t)$ are now dependent on drop time and concentration in ways unique to these two mechanisms. An added complication, particularly for phase-angle measurements, is that the electrode-geometry model chosen to represent the dropping mercury electrode (DME) has a significant impact on the values calculated for $\cot \phi$. Consequently, the value of k_D obtained by matching experiment to theory will depend on the model chosen.

The expressions for $\cot \phi$ as a function of the kinetic parameter g are given in (253) for the two mechanisms, primarily for comparison with the expressions for the simple, irreversible, following reaction, (246), and the catalytic regeneration reaction, (247).

Dismutation

$$\cot \phi = \frac{\dfrac{(2\omega)^{1/2}}{\lambda}(1+e^j) + \frac{1}{2} + (\frac{1}{2}+e^j)\left[\dfrac{(1+g^2)^{1/2}+g}{1+g^2}\right]^{1/2}}{\frac{1}{2}+(\frac{1}{2}+e^j)\left[\dfrac{(1+g^2)^{1/2}-g}{1+g^2}\right]^{1/2}} \tag{253a}$$

Dimerization

$$\cot \phi = \frac{\dfrac{(2\omega)^{1/2}}{\lambda}(1+e^j) + 1 + e^j\left[\dfrac{(1+g^2)^{1/2}+g}{1+g^2}\right]^{1/2}}{1+e^j\left[\dfrac{(1+g^2)^{1/2}-g}{1+g^2}\right]^{1/2}} \tag{253b}$$

The frequency dependence of $\cot \phi$ on $\omega^{1/2}$ for these two mechanisms is much like that for the catalytic regeneration step. Chemical kinetics cause an increase in $\cot \phi$ only at low frequencies, whereas slow charge transfer increases $\cot \phi$ at higher frequencies. The exact value of $\cot \phi$ at any frequency is determined by the initial concentration and the drop time since these parameters control $C(0, t)$ in (252).

Cot ϕ as a function of dc potential across the polarographic wave exhibits a peak even for reversible charge transfer. This effect is apparently unique to these mechanisms. Of course the exact value of $\cot \phi$ at the peak and even the potential of the peak varies with drop time and initial concentration.

The value of peak current as a function of $\omega^{1/2}$ is strongly dependent on the homogeneous kinetics and is not linear with $\omega^{1/2}$. The current is also not linear with initial concentration.

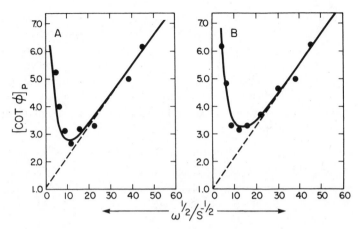

Figure 4.63 Peak cot ϕ versus $\omega^{1/2}$ profiles with UO_2^{2+}/UO_2^+ couple. ($\bullet$) Experimental points for (A) 10 mM UO_2^{2+} and (B) 20 mM UO_2^{2+} using 10 mV p-p sine wave; (———) theoretical profile for $\alpha = 0.5$, $k_s = 0.025$ cm/s, $k_D = 1 \times 10^4$ M^{-1} s^{-1}, $t = 3.50$ s. Expanding-sphere model. Reprinted from [183], by courtesy of Elsevier Press.

Because the responses of cot ϕ and $I(\omega t)$ depend so much on experimental parameters, it is not possible to construct general working curves for fitting experimental data to theory. Nevertheless, Hayes, and co-workers [183] have applied ac polarography to the study of two second-order, coupled chemical reactions. One of these is the disproportionation of U(V) following reduction of U(VI) in strongly acid solution, 6 M $HClO_4$. The other reaction is the dimerization of the ketyl radical following reduction of benzaldehyde in alkaline aqueous ethanol medium. Rate parameters were obtained by fitting experimental values of cot ϕ and $I(\omega t)$ versus $\omega^{1/2}$ to the theoretical expressions obtained using the exact expanding-sphere model for the DME.

An example of the fit is shown in Figure 4.63 for cot ϕ versus $\omega^{1/2}$ for the uranium disproportionation, measured at two concentrations. Large values of cot ϕ at low frequency are caused by the chemical reaction, $g \gg 1$ in (253). The linear variation of cot ϕ versus $\omega^{1/2}$ at higher frequencies is caused by the $(2\omega)^{1/2}/\lambda$ term in (253), indicating that slow charge transfer is the rate-limiting step at these frequencies, and the expression for cot ϕ reduces to that given earlier for slow charge transfer (226). This appears to be the only reported study of the charge-transfer kinetics for this system. Values for k_D, the dismutation rate constant, are consistent with the results of many other studies for this system, once acidity and ionic strength are taken into consideration [183].

5.5 Multiple-Electron Transfer With Coupled Reactions

Complex reaction schemes involving sequential electron transfers, both with and without intervening homogeneous chemical reactions, have received considerable theoretical study. Hung and Smith have evaluated the response of multiple-electron steps for both the reversible [201] and quasi-reversible [202] mechanisms using the expanding-plane model when the waves are close together. If $E_1^0 \gg E_2^0$ for a reduction mechanism, two separate, independent waves are observed as expected. When $E_1^0 \ll E_2^0$, a single two-electron wave is observed. No experimental studies of these mechanisms by ac voltammetry have been reported. Spherical diffusion with amalgam formation has also been studied theoretically [203] for these mechanisms.

Theoretical studies of the ECE mechanism have been presented for three types of intervening chemical reactions:

1. The chemical reaction is reversible and first-order [204].
2. The reaction involves catalytic regeneration of starting material [205].
3. The reaction is second-order involving irreversible reproportionation of the product of the first electron-transfer reaction [182].

$$A + e^- \rightleftharpoons B$$

$$B + e^- \rightleftharpoons C$$

$$A + C \rightarrow 2B$$

None of these mechanisms has been studied experimentally.

A mechanism that has been studied in detail theoretically and experimentally involves two independent electrode reactions coupled by a homogeneous redox reaction [191].

$$A + e^- \rightleftharpoons B \qquad E_1^0$$

$$C + e^- \rightleftharpoons D \qquad E_2^0$$

$$A + D \underset{k_2}{\overset{k_1}{\rightleftharpoons}} B + C$$

Experimentally this reaction mechanism is useful for studying the kinetics of the homogeneous reaction only when the charge-transfer rate of the second step is slow ($k_s < 10^{-5}$ cm/s) and the waves are well separated [181]. Under these conditions, the height of the second wave is a measure of the forward rate constant k_1 when $k_1 \gg k_2$. Figure 4.64 illustrates the current ratio, R, of this second wave as a function of dc potential for several values of $k = k_1 + k_2$. The current ratio is given by

$$R = \frac{I(\omega t)_k}{I(\omega t)_{k=0}} \tag{254}$$

and can be measured by scanning the potential of the second wave in the presence and absence ($k = 0$) of the first redox system, that is, species A. For large values of k the ratio reaches a constant value in Figure 4.64 which is related to the initial concentration of A and C in the system. For equal diffusion currents this ratio, $R_{max} = [A]/[C] + 1$. An alternate set of working curves of k versus $\log(R_{max} - R)$ is shown in Figure 4.65 for the plateau value of R, curve B, and for the value of R measured at the dc polarographic half-wave potential, $E_{1/2}$, which is obtained with only C present in the solution, curve A. The value of R is independent of frequency and applied ac potential [181].

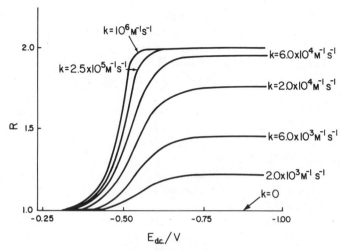

Figure 4.64   Predicted ac polarographic kinetic/nonkinetic current ratio (R) at second wave with cross-redox mechanism and $k_1 \gg k_2$. Parameter values: $T = 298$ K, $C_A^* = C_C^* = 1.00 \times 10^{-3}$ M, $D_A = D_B = D_C = D_D = 4 \times 10^{-6}$ cm²/s, $t = 4.00$ s, $\omega = 250$ s⁻¹, $E_1^0 = 1.00$ V, $E_{1/2} = -0.50$ V (when $k = 0$), $k_1 = \infty$, $k_2 \leqslant 10^{-6}$ cm/s, $\alpha_2 = 0.50$, k values shown in figure ($k = k_1 + k_2 \simeq k_1$), expanding-plane model. Reprinted from [181], by courtesy of Elsevier Press.

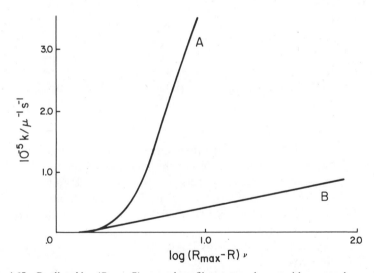

Figure 4.65   Predicted $\log (R_{\max} - R)$ versus k profiles at second wave with cross-redox mechanism and $k_1 \gg k_2$. Parameter values: same as Figure 4.64. (A) Profile at dc half-wave potential (where E_1 is value observed when chemical reaction is nonexistent, i.e., $C_A^* = 0$). (B) Profile at plateau of R versus E_{dc} plot. Reprinted from [181], by courtesy of Elsevier Press

Smith and co-workers [206, 207] have tested this mechanism using the iron–chromium system in 1 M perchlorate at pH 1 where the more easily reduced couple A→B is $Fe^{3+} \to Fe^{2+}$ and the second couple C→D is $Cr^{3+} \to Cr^{2+}$. Because the phase angle is not a significant variable, only the inphase current is measured. This effectively removes the double-layer charging current from the measurement. The ratio on the plateau was, in fact, independent of frequency from 300 to 7000 rad/s and also independent of applied potential up to 120 mV peak-to-peak. Large amplitudes are desirable because the irreversible reduction of Cr^{3+} results in small alternating currents, but the current is proportional to ΔE; see (239).

The value of k_1 where the current ratio in the plateau becomes independent of k is determined only by the time scale of the dc process, that is, the drop time, and the concentration of the species in solution. Rapid-drop experiments using mechanically controlled drop times can increase the upper values of k_1 accessible to measurement by this technique, but lower concentrations (below 1 mM) are difficult to use because of the small currents obtained from irreversible charge-transfer reactions.

Matusinovic and Smith [208, 209] have used this technique to measure the Eu(II) reduction of several Co(III) pentammine complexes in acid perchlorate solution. Drop times ranged from 1 to 10 s and concentrations from 1 to 10 mM. Second-order rate constants from $4\,M^{-1}\,s^{-1}$ to $2 \times 10^4\,M^{-1}\,s^{-1}$ were obtained. Optimum results are obtained for reactions with half times from 10^{-2} to 1 s. Concentrations can be varied to put the reaction within this time slot.

5.6   Adsorption

Ac polarography has not been used to study adsorption processes of electroactive species although complex theoretical treatment has been available for some time. The use of electroactive surface layers formed by strong irreversible adsorption, chemical bonding to the electrode, and polymeric coating of the electrode has increased the interest in ac polarography as a technique for their study.

5.6.1   Surface Layers

Both Laviron [105, 210, 211] and, independently, Kakutani and Senda [212] have simplified their treatment to consider only the case of irreversible adsorption and charge transfer occurring between the adsorbed molecules. Mass transport to and from solution is ignored. Both Langmuir adsorption, with no interaction between surface species, and Frumkin adsorption, which considers such interactions, have been used.

For reversible charge transfer and Langmuir adsorption, the equation for the ac voltammetric wave is

$$I(\omega t) = \frac{n^2 F^2 \omega \Delta E \Gamma^*_{Ox}(b_{Ox}/b_{Red})e^j}{RT[1+(b_{Ox}/b_{Red})e^j]^2} \tag{255}$$

Comparison of the cyclic voltammetric current for the same conditions, (106) in Section 3, reveals that they are identical except that the scan rate v has been replaced by $\omega \Delta E$. The peak potential occurs at E^0_s given by (108) and the half-peak width is $90/n$ mV. The peak shape with respect to dc potential is exactly the same as the diffusion-controlled ac voltammetric wave, but the potential is shifted by $(RT/nF)\ln(b_{Red}/b_{Ox})$ because of the adsorption coefficients b_{Ox} and b_{Red} [see (108)].

If interaction among the adsorbed species occurs, the waves are sharpened for attractive interactions and broadened for repulsive interactions just as in cyclic voltammetry. Theoretical curves are given in Figure 4.66 for various values of the interaction parameter for both the cyclic dc and the ac response [105]. Only the positive-going peak is observed in the ac voltammetric experiment.

For the reversible case, the phase angle between the applied potential and the current given by (255) is 90°, so $\cot \phi = 0$ and there is no inphase component of the surface redox process [211, 212].

The mechanism of slow charge transfer has also been considered by Laviron [211] and by Kakutani and Senda [212] for both Langmuir and Frumkin adsorption isotherms. The phase angle ϕ depends on the charge-transfer rate constant and α, where $\beta = 1 - \alpha$.

$$\cot \phi = \frac{\omega}{k_s}\left(\frac{b_{Ox}}{b_{Red}}e^{-\alpha j} + \frac{b_{Ox}}{b_{Red}}e^{-\beta j}\right)^{-1} \tag{256}$$

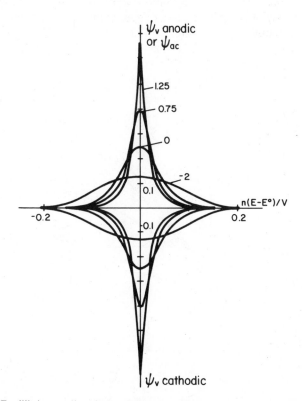

Figure 4.66 Equilibrium cyclic voltammetric peaks and ac peaks at 25°C in the case of a Frumkin isotherm. The value of the interaction parameter is shown on each curve. Reprinted from [105], by courtesy of Elsevier Press.

The phase angle goes through a maximum with potential such that

$$\cot \phi_{\max} = \frac{\omega}{k_s}\left[\left(\frac{\alpha}{\beta}\right)^{-\alpha} + \left(\frac{\alpha}{\beta}\right)^{\beta}\right]^{-1} \tag{257}$$

and the potential at $\cot \phi_{\max}$

$$E_{\cot \phi_{\max}} = E_0^s + \frac{RT}{nF}\ln\left(\frac{\alpha}{\beta}\right) \tag{258}$$

As the applied frequency increases, the phase angle ϕ decreases, and the current at any potential approaches a maximum value at very high frequencies, as shown in Figure 4.67 for the current peak calculated for several values of k_s in s^{-1}. The ac voltammogram deviated from the reversible shape since ϕ is potential dependent. It becomes broader and asymmetric if α is significantly different from 0.5.

If $\alpha = 0.5$, $\cot \phi_{\max}$ occurs at the peak potential and is linear with ω. The value of k_s can be determined from the slope of $\cot \phi$ versus ω since the term in brackets in (257) is equal to 1 and the slope of $\cot \phi_{\max}$ versus ω becomes $1/k_s$. Laviron [211] has used this method to measure the charge-transfer rate constant for three species irreversibly adsorbed to a dropping mercury electrode. The values of $\cot \phi$ versus frequency are plotted in Figure 4.68. The slopes of the straight

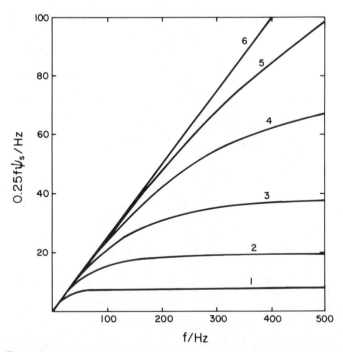

Figure 4.67 Variations of $f\psi_s$ as a function of f for $\alpha=0.5$. k_s: (1) 100, (2) 250, (3) 500, (4) 1000, (5) 2000, (6) 10,000 s^{-1}. Reprinted from [211], by courtesy of Elsevier Press.

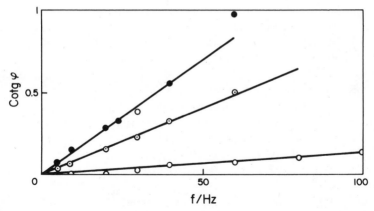

Figure 4.68 Cot ϕ as a function of the frequency. ($\bigcirc$) Azobenzene, pH 1, $C = 3.5 \times 10^{-9}$ mol/cm^3; ($\odot$) phenazine, pH 6.8, $C = 4 \times 10^{-9}$ mol/cm^3; ($\bullet$) benzo[c]cinnoline, pH 11.5, $C = 10^{-9}$ mol/cm^3; $\tau = 5$ s, $m = 1.62 \times 10^{-3}$ g/s. Reprinted from [211], by courtesy of Elsevier Press.

lines yield $k_s = 2300$ s^{-1} for azobenzene, 380 s^{-1} for phenazine, and 220 s^{-1} for benzo[c]cinnoline [211]. Peak separations from dc cyclic voltammograms confirm these values. Independent measurements confirm that α values are indeed approximately 0.5.

5.6.2 Tensammetry

In addition to ac polarographic waves observed for electroactive species, another type of wave can be observed when a nonelectroactive species is adsorbed or desorbed at the electrode surface. This wave has been called a tensammetric wave [213] and occurs when the potential-dependent adsorption–

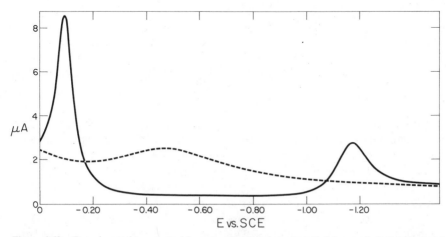

Figure 4.69 Experimental tensammetric wave of isobutyl alcohol. Solid curve is 0.42 M isobutyl alcohol in 0.5 M KNO$_3$ supporting electrolyte. Dashed curve is 0.5 M KNO$_3$ only. $\Delta E = 5$ mV at 80 Hz. The drop oscillations are not shown.

desorption process results in a change in the double-layer capacitance. This process gives rise to a peak-shaped ac polarographic wave over the dc potential range where it occurs. Many organic surfactants, both charged and uncharged, exhibit this behavior even though they do not partake in any faradaic reactions. An example of a tensammetric wave is shown in Figure 4.69 for the adsorption and desorption of isobutanol on mercury in 0.5 M KNO$_3$.

A detailed discussion of experimental observations of this phenomenon is beyond the scope of this chapter and the interested reader is referred to the monograph of Breyer and Bauer [173], in which this process is discussed more thoroughly. A qualitative examination by cyclic voltammetry provides ready differentiation of tensammetric waves from waves for adsorption-controlled faradaic processes.

Canterford [214] has shown how tensammetric waves can be used for analysis of surface-active species that do not undergo a faradaic reaction at the electrode surface.

5.7   Nonlinear Faradaic Response

Although the equivalent circuit of a faradaic response has been represented as a combination of linear circuit elements (resistors and capacitors) in Section 4, the faradaic impedance is really nonlinear, as discussed in Section 1. Linear circuit elements are justified because ΔE_{ac} is small, allowing the series expansions for the potential function, (111) or (139), to be truncated after the linear term. Higher harmonic signals result from inclusion of these higher-order potential terms. For the application of a sinusoidal potential of frequency ω, one can measure current response at higher multiples of this frequency, namely, 2ω, 3ω, and so on. These currents are called the second harmonic, third harmonic, and so on. Smith [147] has presented expressions for harmonic currents up to the fifth, and measurements have been made [215] up to the fourth harmonic for reversible electron transfer.

The current response for the reversible second-harmonic current is given by (259).

$$I(2\omega t) = \frac{n^3 F^3 C_{Ox}^* (2\omega D_{Ox})^{1/2} \Delta E^2 \, \sinh(j/2)}{16 R^2 T^2 \cosh^3(j/2)} \, \sin(2\omega t - \pi/4) \tag{259}$$

The dc-potential dependence is given by the hyperbolic sine and cosine functions of $j/2$. Two peaks are observed, one positive, the other negative, if phase-sensitive detection is used. The current is zero at $E_{dc} = E_{1/2}$ [$j = 0$ and $\sinh(0) = 0$], and the current magnitude at other potentials is proportional to the square of the applied sine-wave amplitude, ΔE. The two peaks are separated by $68/n$ mV and are symmetric about $E_{1/2}$ [147]. Three peaks are observed for the third harmonic, four peaks for the fourth harmonic, and so on [215].

Only the second-harmonic response has been studied extensively for both analytical and mechanistic studies. This work is described in more detail later.

In addition to the higher-harmonic currents, a direct current is also generated, proportional to the amplitude of the applied frequency. It has dc-potential dependence equal to that of the second-harmonic current. If this current is blocked by applying an opposing voltage or by using a high-impedance measuring circuit, the blocking voltage as a function of dc potential is termed the "faradaic rectification polarogram." The technique both at equilibrium dc potentials and as a function of applied dc potential has been extensively developed theoretically [174, 216, 217] and experimentally [218–220] to the measurement of k_s and α for quasi-reversible systems. An example of rectification polarograms for $Fe(III)(C_2O_4)_3^{3-}$ at four frequencies is shown in Figure 4.70. The lines are calculated and matched to the experimental ΔE values using $\alpha = 0.16$ and $k_s = 1.29$ cm/s, which agrees with the ac polarographic results described earlier [167] for this system. Interested readers are referred to the original literature mentioned above for a more detailed discussion of the technique.

Another manifestation of nonlinear faradaic response is probed by a technique called intermodulation polarography. Two periodically varying potentials, usually sinusoidal, of different frequencies, are applied to the cell. Current response occurs at both the applied frequencies and the second harmonic of each applied frequency. In addition, a current response occurs at the sum and the difference of the two applied frequencies. Intermodulation polarography usually measures the current of the difference frequency.

The intermodulation current-density polarogram response is given by (260) [221] for a planar electrode.

$$I(\omega_1 - \omega_2)t = \frac{n^3 F^3 C_{Ox}^*(\omega_1 - \omega_2)^{1/2} D_{Ox}^{1/2} \Delta E^2 \sinh(j/2)}{8R^2 T^2 \cosh^3(j/2)} \sin\left[(\omega_1 - \omega_2)t + \frac{3\pi}{4}\right]$$

$$(260)$$

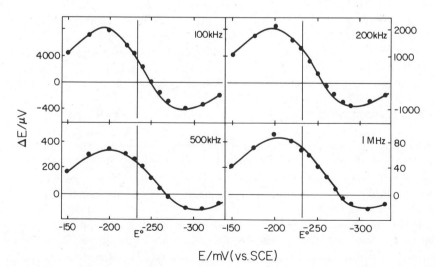

E/mV(vs.SCE)

Figure 4.70 Rectification polarograms of 1 mM Fe(III) in 1.0 M $K_2Ox + 0.05$ M Ox. Reprinted from [218], by courtesy of Elsevier Press.

This current is larger than the second harmonic current by $\sqrt{2}$ measured at the same frequency and has the same dc-potential dependence. The phase angle, however, differs by $\pi/2$. Bond and co-workers [222] and Sluyters and co-workers [223] have presented instrumentation to measure phase-selective intermodulation polarograms and have given examples of the response, verifying (260).

Despite the several manifestations of nonlinear faradaic response of an electrochemical cell, only second-harmonic ac polarography and voltammetry have been studied extensively. The technique embodies the advantage of all the other nonlinear techniques in that efficient separation of the faradaic and nonfaradaic current response occurs. Both the uncompensated cell resistance and the double-layer capacitance are essentially linear elements and contribute little or no current at harmonic frequencies (see Section 7). This property has led to extensive investigations of second-harmonic polarography for analytical and mechanistic studies.

McCord and Smith [224] presented a detailed theoretical study of second-harmonic polarography for the quasi-reversible mechanism. The equations are very complex, particularly if the dc process is also slow and if spherical correction effects are considered. Working curves were constructed relating the peak current (compared with the reversible current) to the dimensionless parameter, $\sqrt{2\omega D/k_s}$, for various α values. Similar curves relating the peak potential separation to $\sqrt{2\omega D/k_s}$ for various α values were presented. The working curves were used to obtain approximate values of k_s and α experimentally [225]. Refining these values required matching the whole current versus dc-potential wave to calculated values via a trial-and-error process. The method was extended [226] to an amalgam-forming system, in which the dc process was also slow, with good results. The method obviously requires extensive computer facilities to carry out the theoretical calculations.

Phase-angle measurements have also proven useful in second-harmonic ac polarography for both mechanistic [227] and analytical studies [228]. Both the resistive (R or inphase) and quadrature (Q or 90° phase) components of the current can be measured to obtain $\cot \phi$, which is the ratio $I(2\omega t)_Q/I(2\omega t)_R$. An example of these two currents is given in Figure 4.71 for reduction of Cd^{2+} in 1.0 M Na_2SO_4 at a mercury electrode.

Examination of (259) reveals that for a reversible system the phase angle is $-\pi/4$ or $-45°$ for the first peak of the second-harmonic polarogram. At $E_{1/2}$ the sign of the term $\sinh(j/2)$ changes from plus to minus and the phase angle changes by $\pi/2$ or 180°. For a quasi-reversible system the change is much less abrupt with potential and is larger, particularly at higher frequencies. Matching experimental phase-angle polarograms to theory can also be used to obtain the parameters k_s and α.

If the charge-transfer step is reversible on the time scale of the second-harmonic frequency, the current for the resistive and capacitive signals goes through zero at $E = E_{1/2}$ (259). Bond and co-workers [228] demonstrated this for Cd^{2+} reduction in 1 M KCl up to 1000 Hz for the resistive-current signal.

Theoretical studies of second-harmonic polarography have been extended to

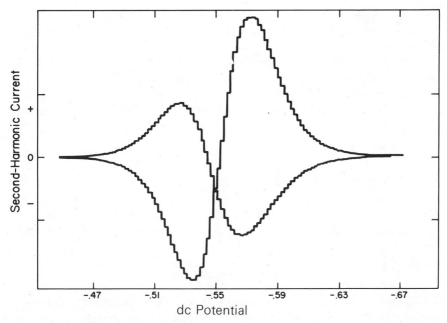

Figure 4.71 Inphase and quadrature second-harmonic ac polarograms for $Cd^{2+}/Cd(Hg)$ in 1.0 M Na_2SO_4. Dc scan rate 50 mV/s, 10 mV peak-to-peak sine wave at 1000 Hz. Current measured by sample-and-hold circuitry. Polarogram with largest signal excursion is quadrature component. Reprinted from [227], by courtesy of *Analytical Chemistry*.

mechanisms involving coupled chemical reactions. These include preceding reactions [229], following reactions [230], and catalytic regeneration [200]. No rate parameters have been measured by this technique although calculated second-harmonic polarograms were matched to experimental results for several coupled-reaction mechanisms [192, 200, 231].

Second-harmonic polarography seems to offer a unique advantage in measuring the thermodynamically reversible potential of a redox couple even in the presence of fast, coupled chemical reactions. For a fast, following, irreversible reaction, the inphase and quadrature currents will intersect at zero current at $E_{1/2}$ as long as $k < 3\omega$ [230]. The current signals need not be symmetrical, but the two currents must intersect at zero current [232]. The requirement does not seem unduly restrictive since measurements at 1000 Hz could be used to obtain $E_{1/2}$ even when the following reaction occurs at 2×10^4 s^{-1}. The other requirement is that the charge-transfer rate constant be fast compared with the measuring frequency.

Breslow and co-workers [233, 234] have used this technique to measure thermodynamic potentials for the oxidation of hydrocarbons on platinum. Ahlberg and Parker [235] have evaluated the technique and suggest using the quadrature-signal intersection at zero current to measure $E_{1/2}$ because the resistive current may be distorted by small amounts of uncompensated resistance and adsorption. They measured $E_{1/2}$ for diphenylanthracene in aceto-nitrile with various levels of pyridine such that the pseudo-first-order following reaction was as high as 10^3 s^{-1}.

Charge-transfer reactions coupled to reversible chemical reactions are also amenable to measurement of $E_{1/2}$ by second-harmonic polarography [229, 230]. The frequency requirement now is that $k < 0.2\omega$, slightly more stringent than for irreversible reactions. No experimental test of this limit has been reported.

6   RELATED TECHNIQUES

6.1   Laplace Transformation

In Section 2 system transforms were derived in the Laplace domain and then Fourier transformed into the frequency domain or inverse-Laplace transformed into the time domain. These transformations are ones of increasing complexity, however. Consider the kinetic scheme given by (18),

$$Y \underset{k_b}{\overset{k_f}{\rightleftharpoons}} Ox \underset{-e^-}{\overset{+e^-}{\rightleftharpoons}} Red$$

for example, for which the system transform with respect to the oxidized species is given by (30). If a potential $E \ll E^0$ is applied to the electrode at time zero, $C_{Ox} = 0$ and (30) may be rearranged to

$$\frac{nFC^*\sqrt{D}}{i\sqrt{p}} = 1 + \frac{1}{K}\frac{\sqrt{p}}{\sqrt{p+k}} \tag{261}$$

where C^* is the sum of the initial concentrations of Ox and Y, and $D = D_{Ox}$. Inverse transformation of this equation back into the time domain results in

$$\frac{i(1-K^2)}{nFC^*\sqrt{DK}}$$

$$= \frac{e^{-kt}-K}{\sqrt{\pi t}} + \sqrt{\frac{k}{1-K^2}}\, e^{K^2kt/(1-K_2)}\left\{\mathrm{erf}\left(\sqrt{\frac{kt}{1-K^2}}\right) - \mathrm{erf}\left(\sqrt{\frac{K^2kt}{1-K^2}}\right)\right\} \tag{262}$$

where erf is the error function, so named because it is the area under the bell-shaped, standard normal- or Gaussian-distribution error curve and is given by the expression

$$\mathrm{erf}(x) = \frac{2}{\sqrt{\pi}}\int_0^x e^{-y^2}\,dy \tag{263}$$

Even in the usual situation that $k \ll 1$, (262) still has the rather complicated form

$$\frac{i}{nFC^*\sqrt{D}} = \sqrt{k_f K}\, e^{k_f Kt}\, \mathrm{erfc}(\sqrt{k_f Kt}) \tag{264}$$

where erfc is the complementary error function, $(1 - \mathrm{erf})$. By comparison, the

Laplace transform of (264) is

$$\frac{nFC^*\sqrt{D}}{\bar{i}\sqrt{p}} = 1 + \frac{\sqrt{p}}{\sqrt{k_f K}} \tag{265}$$

It should be apparent that interpretation of data will be somewhat easier if one can avoid the inverse transformation. This is not overwhelmingly difficult. One simply feeds the experimentally determined current response into a computer and numerically evaluates the Laplace transform of the current $\bar{i}$ from the equation

$$\bar{i} = \int_0^\infty e^{-pt} i \, dt \tag{266}$$

with several values of p. Equation (266) is analogous to (2). A plot of $nFC^*\sqrt{D}/\bar{i}\sqrt{p}$ against $\sqrt{p}$ then provides a straight line with slope $1/\sqrt{k_f K}$, according to (265).

Equation (264) may be written in dimensionless form

$$i/i_d = \sqrt{\pi}\lambda e^{\lambda^2} \operatorname{erf}(\lambda) \tag{267}$$

where i_d, given by the Cottrell equation (35), is the function which the current would follow in the absence of kinetic complications, and $\lambda = \sqrt{k_f K t}$. Time-domain data would be interpreted by comparing values of i/i_d to a working curve of (267) to obtain values of $\sqrt{k_f K}$ at different times during the current decay. A "best fit" average would then be calculated.

Several advantages accrue from the Laplace transform technique, especially when one transforms both the current response and the voltage step itself. Armed with both $\bar{i}$ and $\bar{E}$, a function $\bar{Z} = \bar{E}/\bar{i}$, termed the "transient impedance," may be defined. If $j\omega$ is substituted into (266) and its counterpart for $\bar{E}$, equations of the form

$$\bar{i}(j\omega) = \int_0^\infty i \cos(\omega t) dt - j \int_0^\infty i \sin(\omega t) dt \tag{268}$$

result from which real and quadrature transient impedances may be calculated. Equivalent circuit representations of these functions may then be formulated and analyzed as discussed in the impedance section to account for nonfaradaic effects. One should be aware, however, that the transient impedances may or may not be equivalent to the impedances discussed earlier, since one might be derived from large-amplitude signals, and would therefore be integral, while the other is derived from small-amplitude signals and is therefore differential.

Since the voltage excitation is transformed, its exact time dependence is not important. It simply must be transformable. Oscillations or "ringing" in the voltage step would alter the form of the time response of the current and therefore invalidate (264). However, the transient impedances would not be affected because they are ratios. In this way, frequencies as high as 10^8 rad/s can be achieved.

For further details of this method the reader is referred to [236–241].

6.2   Fourier Transformation

Computerized data acquisition facilitates Laplace transformation, resulting in the benefits already discussed. One can more easily Fourier transform the acquired transient data, however, by using the Fast Fourier Transform (FFT) algorithm devised by Cooley and Tukey [242]. Use of the transformed data to calculate transient impedances or admittances then follows analogously. The Fourier transform of the current response to a potential excitation will be a set of Fourier coefficients of the inphase (real) and out-of-phase (quadrature) sine and cosine terms in (129). At each value of ω the Fourier transform of the current can then be expressed in terms of the resistive (R) and quadrature (Q) components

$$\bar{i}(j\omega) = i_R(\omega) - ji_Q(\omega) \tag{269}$$

We write $\bar{i}(j\omega)$ to show the correspondence between the Fourier transform and the $j\omega$-substituted Laplace transform. Fast Fourier Transformation of the potential excitation results in

$$\bar{E}(j\omega) = E_R(\omega) - jE_Q(\omega) \tag{270}$$

From these equations it follows that

$$\bar{Y}(j\omega) = \frac{\bar{i}(j\omega)}{\bar{E}(j\omega)} = Y_R(\omega) + jY_Q(\omega) \tag{271}$$

where

$$Y_R = \frac{i_R E_R + i_Q E_Q}{E_R^2 + E_Q^2} \tag{272}$$

and

$$Y_Q = \frac{i_R E_Q - i_Q E_R}{E_R^2 + E_Q^2} \tag{273}$$

Equation (271) is identical with expressions derived in Section 4. However, it was previously derived from assumed sinusoidal potential excitations. It is apparent from the derivation given here that any excitation, which is Fourier transformable, may be used to generate the current transient data for subsequent mathematical analysis. Smith's group discussed this technique in a series of pioneering papers [160, 243, 244] and studied various transformable voltage excitations extensively [244]. They found that an excitation consisting of a set of equal-amplitude, odd-harmonic sinusoids with frequencies greater than a fundamental but with random phases was preferable to other possibilities [161, 244]. They have used this excitation in applications of the FFT technique to ac polarography and have shown that admittance spectra, covering frequencies up to 125 kHz [161], can be obtained within the lifetime of a single drop [164], even when the dc component of the voltage excitation is a pulse [151] (pulse polarography).

6.3   Convolution or Semi-Integral Techniques

Another "excitation-independent" technique involves the convolution integral that appears in (16). Note that the derivation of this equation required only one assumption—that diffusion obeys Fick's laws with no kinetic complications. It was then necessary to assume a mathematical relationship between applied voltage and surface concentrations of electroactive species to proceed to the various current-response equations characteristic of the different experimental techniques. Such mathematical relationships may be avoided, however, by simply evaluating the integral in (16) numerically with the current response from any potential excitation. Computerized data acquisition facilitates this step. Equation (16) then becomes

$$C_{Ox} = C_{Ox}^* - \frac{m(t)}{nF\sqrt{D_{Ox}}} \tag{274}$$

where $m(t)$ is the value of the integral

$$m(t) = \frac{1}{\sqrt{\pi}} \int_0^t \frac{i(\tau)}{(t-\tau)^{1/2}} \, d\tau \tag{275}$$

If the time dependence of the voltage excitation is of such a form that $E \ll E^0$ as $t \to \infty$, it follows that $C_{Ox} \to 0$ and (274) becomes

$$C_{Ox}^* = \frac{m(\infty)}{nF\sqrt{D_{Ox}}} \tag{276}$$

Equations (274) and (276) combine to

$$C_{Ox} = \frac{m(\infty) - m(t)}{nF\sqrt{D_{Ox}}} \tag{277}$$

Similarly, the expression,

$$C_{Red} = \frac{m(t)}{nF\sqrt{D_{Red}}} \tag{278}$$

results from (17) if the concentration of the reduced species is initially zero. The convolution integral therefore provides the concentrations of oxidized and reduced species at the surface of the electrode regardless of the exact mathematical relationship between concentration and potential.

Comparison of the convolution integral with the Riemann–Liouville definition of the "semi-integral" [245] reveals that they are one in the same. Therefore, m is the semi-integral of the current, as written

$$m = \frac{d^{-1/2}i}{d^{-1/2}t} \tag{279}$$

The convolution [72, 246] and semi-integral [247, 248] approaches are therefore identical; however, the Savéant school refers to the convolution integral with the letter "I," whereas the Oldham school refers to the semi-integral with the letter "m."

Substitution of (277) and (278) into the Nernst equation results in

$$E(t) = E_{1/2} + \frac{RT}{nF} \ln \left[\frac{m(\infty) - m(t)}{m(t)} \right] \tag{280}$$

If the applied voltage varies linearly from an initial potential E_0 with a rate of v V/s,

$$E(t) = E_0 - vt \tag{281}$$

then time is no longer explicit in (280) and it may be written [249]

$$E = E_{1/2} + \frac{RT}{nF} \ln \left[\frac{m_c - m}{m} \right] \tag{282}$$

where m_c is the cathodic limit of $m(\infty)$ when $E \ll E^0$. A parameter, m_a, which is the anodic limit of $m(\infty)$ when $E \gg E^0$, can be introduced to account for the more general case in which both oxidized and reduced species are present initially. The expression then becomes

$$E = E_{1/2} + \frac{RT}{nF} \ln \left(\frac{m_c - m}{m - m_a} \right) \tag{283}$$

Equation (282) can be written explicitly as a function of m

$$m = \tfrac{1}{2} m_c \left\{ 1 - \tanh \left[\frac{nF}{2RT} (E - E_{1/2}) \right] \right\} \tag{284}$$

and this is the equation of a reversible "neopolarogram" [249], a graph of the semi-integral versus the applied voltage. In the most usual experimental arrangement, the potential is swept linearly (281); however, (284) holds regardless of the time dependence of E. However, $m_c = nFC_{Ox}^* \sqrt{vD_{Ox}}$, and a similar expression can be written for m_a.

Equations can also be derived that describe the shapes of irreversible and quasi-reversible neopolarograms [250] on the assumption that the rate expression (31) is valid. However, the fact that m provides the correct surface concentrations of oxidized and reduced species regardless of the exact form of the mathematical relationship between concentration and potential provides a useful means of studying this relationship. A general form may be written [251]

$$i = nFk(E)[C_{Ox} - C_{Red} e^{(nF/RT)(E - E^0)}] \tag{285}$$

where $k(E)$ is a potential-dependent rate constant for the reduction of the oxidized species. A similar expression may be written for oxidation. Substitution of (277) and (278) into this expression leads to

$$\ln[k(E)] = \ln \sqrt{D_{Ox}} - \frac{1}{i} \ln \left\{ m_c - m \left[1 + \sqrt{\frac{D_{Ox}}{D_{Red}}} \, e^{(nF/RT)(E - E^0)} \right] \right\} \tag{286}$$

which may be used to calculate $k(E)$ as a function of E using experimental values of m taken from a neopolarogram. A plot of $\ln[k(E)]$ versus E should yield a straight line with a slope of $\alpha nF/RT$ if (31), the rate expression, is obeyed. Savéant and Tessier found significant nonlinearity when they applied this technique to their studies of t-nitrobutane in aprotic solvents [251].

One advantage the semi-integral method has over the other excitation-independent techniques is that, although algorithms are available for the computer calculation of semi-integrals [252, 253], it is also possible to obtain direct readouts from analog instrumentation [254].

An additional advantage that results from the independence of m and the applied potential is that the data are easily corrected for uncompensated resistance and double-layer capacitance. The applied potential may be corrected to

$$E_{corr} = E + iR_u \qquad (287)$$

where R_u is the uncompensated resistance of the solution (R_s). The measured current may be corrected for double-layer charging using (288) [252]

$$i_f = i - C_{dl}v + R_u C_{dl}\left(\frac{di}{dt}\right) \qquad (288)$$

where i_f is the faradaic current, before evaluation of the convolution integral.

A further advantage of the convolution or semi-integral technique is that treatment of data for electrode processes with coupled chemical reactions becomes somewhat simplified [246] relative to the more conventional linear-sweep methods.

Savéant and Tessier [255] have taken advantage of these features to study fast charge-transfer reactions with associated fast homogeneous kinetics (reduction of benzaldehyde in alkaline ethanol) at sweep rates up to 2300 V/s. Their experimental evaluation of responses obtained at such high sweep rates appears in reference [256].

6.4 Semidifferential Electroanalysis

Analytical use may be made of m_c since it is proportional to C_{Ox}^*. However, sensitivity and resolution can be enhanced by recording the first derivative of m as a function of E rather than of m itself. We define e as the first derivative of m and therefore the "semiderivative" of i [257].

$$e = \frac{dm}{dt} = \frac{d}{dt}\left(\frac{d^{-1/2}i}{dt^{-1/2}}\right) = \frac{d^{1/2}i}{dt^{1/2}} \qquad (289)$$

Since $dm/dE = (dm/dt)(dt/dE)$, according to the chain rule of differentiation, (284) becomes [258],

$$e = \frac{nFm_c}{4RT}\,\text{sech}^2\left[\frac{nF}{2RT}(E - E_{1/2})\right] \qquad (290)$$

where sech is the hyperbolic secant function and

$$\text{sech}\,x = (\cosh x)^{-1} = \left(\frac{e^x + e^{-x}}{2}\right)^{-1} \qquad (291)$$

Equation (290) describes the symmetrical peak shape of a reversible derivative neopolarogram. Corresponding expressions have been derived to describe the irreversible and quasi-irreversible situations [258].

The simple form of (290) is in stark contrast to the "nonanalytical" form of the corresponding cyclic voltammetric wave. Therefore, further enhancement of the sensitivity and resolving power of semidifferential electroanalysis, relative to linear-sweep voltammetry [259], by computerized curve-fitting procedures is greatly facilitated [260].

Further differentiation of (289) results in

$$e' = \frac{d}{dt}\left(\frac{d^{1/2}i}{dt^{1/2}}\right) = \frac{d^{1.5}i}{dt^{1.5}} \tag{292}$$

and

$$e'' = \frac{d}{dt}\left(\frac{d^{1.5}i}{dt^{1.5}}\right) = \frac{d^{2.5}i}{dt^{2.5}} \tag{293}$$

The properties of the 2.5-order derivative (293) have been investigated experimentally [261]. It was found that further enhancement of sensitivity and resolution is achieved by this higher order of differentiation.

7   EXPERIMENTAL TECHNIQUES

7.1   Instrumentation

Although this chapter is concerned mainly with cyclic voltammetry and alternating-current techniques, the need for other techniques in mechanistic studies must be recognized. No one technique, used exclusively, is as powerful as the judicious use of several. Accordingly, instrument manufacturers have sought to produce reasonably priced, accurate, but above all, versatile, electrochemical instrumentation. Those with no background in electronics can purchase completely assembled instruments and use the manufacturer's literature to obtain acceptable performance.

For those with some background in electronic instrumentation, laboratory-assembled, modular instruments may be satisfactory. Such "black boxes" may be interconnected to construct potentiostats, function generators, or signal-conditioning devices. Experienced electrochemists can build their own instruments, and many detailed circuit diagrams have been published, together with operational explanations. It is generally not too difficult to adapt these to the instrumentation requirements of a particular application.

Electrochemical instruments may be conveniently divided into three basic units, consistent with the "response = system × excitation" approach introduced earlier in this chapter. The first unit is the signal generator, which produces the input potential excitation. The second unit is the potentiostat, which applies the excitation to the system, the electrochemical cell. The third unit, the signal-conditioning device, processes the response of the cell to which it is interfaced through a current-to-voltage transducer. These units are shown in Figure 4.72. A fourth element, a computer, has also been included. In many cases its presence is optional; in others it is mandatory. It may control the signal generator or the signal-conditioning device or it can *be* either of these devices. The ever-present recorder is also included in the figure.

Note that the cell is connected to the potentiostat through three electrodes. Of these, the "indicator" electrode is the one of most immediate interest. All the phenomena discussed in this chapter occur at the surface of this electrode. In most cases the electrode is connected to earth ground through a low-input-impedance current-measuring device. The low impedance of this device ensures that the leads from the indicator electrode are at virtual ground. Indicator electrodes are also referred to as "working" electrodes.

Current is driven through the cell by voltage applied at the "auxiliary" or "counter" electrode. The magnitude of the current is determined by the magnitude of this applied voltage divided by the sum of auxiliary and indicator electrode interfacial impedances plus solution resistance.

The "reference" electrode is connected to the potentiostat through a high-impedance voltage-measuring device. The high impedance of this device ensures that no current flows through the reference electrode, the main advantage of the three-electrode system. The potential of the reference electrode would otherwise be perturbed by current flow. Placement of the tip of the reference electrode as close as possible to the indicator electrode minimizes the voltage-drop contributions from the auxiliary electrode and the solution resistance. Since no current flows through the reference electrode its impedance is inconsequential, provided it is not so high as to cause instrumental instability.

The potentiostat itself is a "balancing" instrument such that the voltage applied across the cell (auxiliary to indicator) will be whatever is required to cause the voltage drop across the interface of the indicator electrode (reference to indicator) to equal the value assigned by the signal generator. The usual

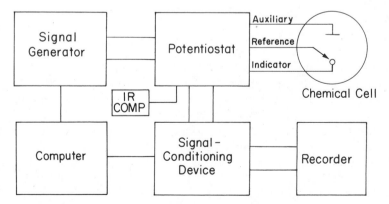

Figure 4.72 Generalized block diagram of electrochemical instrument and cell.

electrical convention would be to measure the voltage of the indicator electrode relative to the reference electrode. However, since the indicator is held at virtual ground potential, the actual voltage measured is that of the reference electrode relative to the indicator. This arrangement leads to the electrochemical sign convention mentioned earlier.

7.2 Signal Generation

It is convenient to have one instrument, such as the Hewlett Packard Model 3325A synthesizer/function generator, which can generate both triangular and sinusoidal potential excitations. However, separate instruments can also be used since most commercial potentiostats can add the outputs of at least two signal generators at their inputs. A popular signal generator is the EG&G PARC 175 Universal Programmer, which has the important feature that the high and low limits of the amplitude of its triangle wave, as well as the scan rate, can be adjusted independently.

Sine-wave generators should be chosen with some care. Sometimes sine waves are formed by "shaping" triangle waves with diode circuits. Shaping causes a 2–3% distortion in the wave, which appears mainly at the second- and third-harmonic frequencies. Measurements made at the fundamental frequency with these generators are not affected by these distortions, but a second-harmonic content of 0.2% or greater in the excitation signal cannot be tolerated for the measurement of second-harmonic faradaic responses.

As mentioned before, the signal generator might be a computer, as in the example of staircase voltammetry [262]. Digital computers cannot generate a continuously varying analog signal such as the triangle wave. Instead, they must generate successive voltages in a stepwise manner, resulting in a staircase-shaped voltage/time profile. Staircase voltammograms must then be deconvoluted [263] to make them comparable to their analog-determined counterparts. Thus, staircase voltammetry is an example of computer use as both input and output signal-conditioning devices.

7.3 Signal Conditioning—Digital Processing

Several examples of digital (computer) processing of output signals have already been cited with regard to Laplace [236–241] and Fourier [242–244] transformations as well as convolution [246, 251, 253, 254], semi-integral [245, 247–250], and semiderivative [257–261] techniques. In these examples the output of the current-to-voltage transducer is "digitized" through an analog-to-digital converter and stored as a set of binary numbers in the memory of the computer for subsequent transformation. Use of the computer as a data-acquisition device as well as a "number cruncher" has the enormous advantage of speed, coupled with accuracy. Analog data-acquisition devices, such as X-Y recorders or storage oscilloscopes, are limited by response time, in the former case, or resolution, in the latter case.

7.4   Signal Conditioning—Analog Processing

Digital computers have found their way into almost every recently designed electrochemical instrument of any sophistication. Thus we have digital ac polarography [264], in which a computer-generated sine wave is added to the computer-generated staircase, mentioned earlier [262], to produce the excitation; the response of the cell is acquired by the same computer, which is then used further to process the data. In addition, the computer may be used to initiate and/or terminate the experiment.

One gets the impression that the experimenter is only needed to mix solutions. However, even with their great power, digital computers have not completely displaced analog signal-processing devices from electrochemistry. Analog devices are relatively simple and inexpensive to build or they may be purchased for immediate use. On the other hand, digital computers are still expensive (at least relative to homemade analog devices) and they require more than just "plug-in" interfacing. They also require programming, which can represent a sizable expenditure of time and effort. Note, however, that computer programs that can write other computer programs are becoming available [265].

Analog lock-in detectors for applications in impedance and admittance measurements are described in references [266–269]. These are "home-built," inexpensive instruments, which can measure inphase and out-of-phase or quadrature signals at selected frequencies. Commercially available instruments with similar capabilities can be purchased from companies such as Hewlett Packard, Ithaco, or EG&E PARC. Kojima and Bard [270] have described the use of an EG&G PARC model HR-8 lock-in amplifier in determining rates of rapid electrode reactions.

Computerized, automatic instruments have been described by Armstrong and co-workers [271] and by Sluyter's group [272] for measuring impedances or admittances. One of the most sophisticated of all the computerized instruments is that described by Huebert and Smith [167], which was designed for the rapid measurement of admittance during ac-polarography experiments. These instruments combine the best of both the analog and the digital worlds.

7.5   Correction of Nonfaradaic Effects

The goal of electrochemical experimentation is to probe the faradaic process occurring at the working electrode. Unfortunately, the solution resistance between the reference and working electrodes and the double-layer capacitance associated with the working electrode–solution interface, interfere with accurate measurement of the faradaic admittance/impedance. The relationship of the impedances of these elements to the faradaic impedance is given by the equivalent circuit diagram in Figure 4.35 (Section 4), the Randles equivalent circuit. In impedance measurements over a large frequency range, the values of the solution resistance and double-layer capacitance can be obtained directly from

Argand plots as described in Section 4. Similarly, the Fourier transform technique of Huebert and Smith [167], for admittance measurements with simultaneous application of a wide range of frequencies in ac voltammetry, can be used to measure these parameters directly. Computer analysis of the raw data can then correct for their effects at intermediate frequencies to obtain accurate values for the faradaic admittance. In the other techniques of interest here, cyclic voltammetry and ac voltammetry at a single applied frequency, the nonfaradaic correction must be made manually after the experiment or instrumentally during the experiment.

The effect of solution resistance, and the corresponding iR drop in the voltage applied to the working electrode, is very difficult to correct after the experiment. The iR voltage drop opposes both the applied dc and ac potentials. In cyclic voltammetry the potential drop broadens the wave, mimicking the effect of slow charge transfer, as described by Nicholson [71]. In ac polarography the effect is even more insidious. The iR drop decreases the current response, particularly in second-harmonic polarography where $I(2\omega t)$ is proportional to ΔE^2. The iR drop also alters the apparent phase relation between the applied signal and the observed current, creating errors in measurement of cot ϕ. Although corrections can be made after the experiment, once the cell resistance is known, the computations are not trivial.

The best way to make the correction is to add back a signal equal to the iR drop during the experiment. This is accomplished by taking a portion of the output of the current-to-voltage transducer in Figure 4.72 and adding it back into the potentiostat. As the current increases so does the signal fed back to the potentiostat. The proportion of the current-transducer signal fed back depends on the actual value of the cell resistance. This and other iR compensation techniques have been reviewed by Britz [273].

Because this correction involves positive feedback, the possibility exists that the potentiostat can lose control of the cell potential and go into oscillation. The onset of oscillation has been suggested as an indication of complete compensation [274]. Various circuit modifications can be made to the potentiostat to prevent oscillation even for 100% compensation. These include capacitance bypass between the auxiliary and reference electrodes [274] and capacitance damping of the feedback signal itself [273]. Dual reference electrodes, one of which has very small, high-frequency impedance, will also aid in potentiostat stability [273]. With such damping, even overcompensation can occur, that is, more signal can be fed back than is necessary. This means accurate knowledge of cell resistance is required for proper compensation, and one cannot rely on the onset of oscillation to indicate adequate compensation. Britz [273] has also reviewed methods of measuring the cell resistance. Obviously, damping techniques will limit accurate response of the potentiostat at high frequency so the tendency is to minimize the value of capacitance used for damping and tolerate a small amount of uncompensated resistance to maintain stable instrument operation [273].

The double-layer charging current is proportional to the rate of change of potential (151) and so is proportional to the scan rate in cyclic voltammetry and the applied frequency in ac techniques. At high scan rates or high frequencies, the charging current can become larger than the faradaic current, obscuring the response of interest. Correction can be accomplished after the experiment by measuring the double-layer current in the absence of the faradaic species and subtracting the values at each potential from the total response. In cyclic voltammetry the correction is easily accomplished, but in ac voltammetry the method involves vectorial subtraction; that is, both the amplitude and the phase of the signal must be taken into account. Such corrections are tedious, although computer data acquisition and analysis simplify the correction.

The problems of vectorial subtraction in ac polarography can be ignored if the quadrature current is measured instead of the total current. Proper instrumental iR-drop compensation must also be employed. The double-layer current is inphase with the quadrature signal so its magnitude can be subtracted directly from the total quadrature current. The resistive-current signal requires no correction because there is no charging-current component at this phase angle. These two signals can be used to obtain the phase angle and its cotangent.

$$\cot \phi = \frac{I(\omega t) \text{ quadrature current} - I(\omega t) \text{ charging current}}{I(\omega t) \text{ resistive current}}$$

An alternative method involves the use of two potentiostats and two cells, only one of which contains the electroactive species [274]. By synchronizing the drop growth and fall mechanically, the double-layer current can be subtracted using analog techniques. Again proper iR compensation is required in both cells.

References

1. A. A. Vlček, *Progress in Polarography*, Vol. 1, Interscience, New York, 1962, p. 269.
2. A. A. Vlček, *Progress in Inorganic Chemistry*, Vol. 5, Interscience, New York, 1963, p. 211.
3. A. L. Nelson, K. W. Folley, and M. Coral, *Differential Equations*, D. C. Heath and Company, Boston, 1964, or any other standard text on this subject.
4. J. W. Ashley, Jr. and C. N. Reilley, *J. Electroanal. Chem.*, 7, 253 (1964).
5. C. N. Reilley, in I. M. Kolthoff and P. J. Elving, Eds., *Treatise on Analytical Chemistry*, Vol. 4, Wiley, New York, 1966, p. 2109.
6. C. N. Reilley and R. W. Murray, in I. M. Kolthoff and P. J. Elving, Eds., *Treatise on Analytical Chemistry*, Vol. 4, Wiley, New York, 1966, p. 2163.
7. J. Koutecký and R. Brdička, *Collect. Czech. Chem. Commun.*, 12, 337 (1947).
8. P. Delahay, *Double Layer and Electrode Kinetics*, Interscience, New York, 1966.
9. S. W. Feldberg, in J. S. Mattson, H. B. Mark, and H. C. McDonald, Eds., *Electrochemistry—Calculations, Simulations, and Instrumentation* (*Computers in Chemistry and Instrumentation*), Vol. 2, Dekker, New York, 1972, Chapter 7.
10. S. W. Feldberg, "Digital Simulation: A General Method," in A. J. Bard, Ed., *Electroanalytical Chemistry*, Vol. 3, Dekker, New York, 1969.
11. A. J. Bard and L. R. Faulkner, *Electrochemical Methods—Fundamentals and Applications*, Wiley, New York, 1980, Appendix B.

12. D. Britz, *Digital Simulation in Electrochemistry*, Springer-Verlag, Berlin, 1981.
13. M. K. Hanafey, R. L. Scott, T. H. Ridgeway, and C. N. Reilley, *Anal. Chem.*, **50**, 116 (1978).
14. J. R. Sandifer and R. P. Buck, *J. Electroanal. Chem.*, **49**, 161 (1974).
15. J. W. Dillard, J. A. Turner, and R. A. Osteryoung, *Anal. Chem.*, **49**, 1246 (1977).
16. I. Ruzic and S. W. Feldberg, *J. Electroanal. Chem.*, **63**, 1 (1975).
17. K. B. Prater, in J. S. Mattson, H. B. Mark, and H. C. McDonald, Eds., *Electrochemistry—Calculations, Simulations, and Instrumentation* (*Computers in Chemistry and Instrumentation*), Vol. 2, Dekker, New York, 1972, Chapter 8.
18. S. W. Feldberg, *J. Electroanal. Chem.*, **109**, 69 (1980).
19. D. Laser and A. J. Bard, *J. Electrochem. Soc.*, **123**, 1828, 1837 (1976).
20. R. D. Armstrong and M. F. Bell, *J. Electroanal. Chem.*, **58**, 419 (1975).
21. P. J. Peerce and A. J. Bard, *J. Electroanal. Chem.*, **114**, 89 (1980).
22. T. Joslin and D. Pletcher, *J. Electroanal. Chem.*, **49**, 171 (1974).
23. S. W. Feldberg, *J. Electroanal. Chem.*, **127**, 1 (1981).
24. R. Seeber and S. Stefani, *Anal. Chem.*, **53**, 1011 (1981).
25. F. Magno, G. Bontempelli, and D. Perosa, *Anal. Chim. Acta*, **147**, 65 (1983).
26. A. Lasia, *J. Electroanal. Chem.*, **146**, 397 (1983).
27. I. Ruzic and S. W. Feldberg, *J. Electroanal. Chem.*, **50**, 153 (1974).
28. F. Magno, G. Bontempelli, and M. Andreuzzi-Sedea, *Anal. Chim. Acta*, **140**, 65 (1982).
29. B. Speiser and A. Rieker, *J. Electroanal. Chem.*, **102**, 1 (1979).
30. L. F. Whiting and P. W. Carr, *J. Electroanal. Chem.*, **81**, 1 (1977).
31. B. Speiser, *J. Electroanal. Chem.*, **110**, 69 (1980); B. Speiser, S. Pons, and A. Rieker, *Electrochim. Acta*, **27**, 1171 (1982).
32. S. Yen and T. W. Chapman, *J. Electroanal. Chem.*, **135**, 305 (1982).
33. J. M. Savéant, *Electrochim. Acta*, **12**, 999 (1967).
34. H. Matsuda and Y. Ayabe, *Z. Elektrochem.*, **59**, 494 (1955); A. Y. Gokhshtein and Y. P. Gokhshtein, *Dokl. Akad. Nauk SSSR*, **131**, 601 (1960).
35. W. H. Reinmuth, *Anal. Chem.*, **32**, 1891 (1960).
36. A. Ševčik, *Collect. Czech. Chem. Commun.*, **13**, 349 (1948).
37. J. C. Myland and K. B. Oldham, *J. Electroanal. Chem.*, **153**, 43 (1983); K. B. Oldham, *J. Electroanal. Chem.*, **105**, 373 (1979).
38. R. S. Nicholson and I. Shain, *Anal. Chem.*, **36**, 706 (1964).
39. J. E. B. Randles, *Trans. Faraday Soc.*, **44**, 327 (1948).
40. R. Adams, *Electrochemistry at Solid Electrodes*, Dekker, New York, 1969.
41. D. S. Polcyn and I. Shain, *Anal. Chem.*, **38**, 370 (1966).
42. B. R. Eggins and N. H. Smith, *Anal. Chem.*, **51**, 2282 (1979).
43. R. S. Nicholson, *Anal. Chem.*, **37**, 1406 (1965).
44. M. S. Shuman, *Anal. Chem.*, **41**, 142 (1969).
45. R. S. Nicholson, *Anal. Chem.*, **37**, 1351 (1965).
46. J. E. J. Schmitz and J. G. M. van der Linden, *Anal. Chem.*, **54**, 1879 (1982).
47. A. Capon and R. Parsons, *J. Electroanal. Chem.*, **46**, 215 (1973); T. W. Rosanske and D. H. Evans, *J. Electroanal. Chem.*, **72**, 277 (1976).
48. R. Samuelsson and M. Sharp, *Electrochim. Acta*, **23**, 315 (1978).
49. R. J. Klingler and J. K. Kochi, *J. Phys. Chem.*, **85**, 1731 (1981).
50. N. Tanaka and R. Tamamushi, *Electrochim. Acta*, **9**, 963 (1964).
51. P. Delahay, *J. Am. Chem. Soc.*, **75**, 1190 (1953).
52. G. Sundholm, *Electrochim. Acta*, **13**, 2111 (1968).
53. R. J. Taylor and A. A. Humffray, *J. Electroanal. Chem.*, **42**, 347 (1973).
54. D. N. Bailey, D. M. Hercules, and D. K. Roe, *J. Electrochem. Soc.*, **116**, 190 (1969).

55.   J. M. Savéant and E. Vianello, *Electrochim. Acta*, **12**, 629 (1967).
56.   J. R. Kuempel and W. B. Schaap, *Inorg. Chem.*, **7**, 2435 (1968).
57.   G. H. Aylward and J. W. Hayes, *Anal. Chem.*, **37**, 195 (1965).
58.   K. Matsuda and R. Tamamushi, *Bull. Chem. Soc. Japan*, **41**, 1563 (1968).
59.   D. G. Davis and D. T. Orleron, *Anal. Chem.*, **38**, 179 (1966).
60.   H. E. Stapelfeldt and S. P. Perone, *Anal. Chem.*, **41**, 623 (1969).
61.   S. P. Perone and W. T. Kretlow, *Anal. Chem.*, **38**, 1761 (1966).
62.   D. H. Evans, *J. Phys. Chem.*, **76**, 1160 (1972).
63.   L. Nadjo and J. M. Savéant, *J. Electroanal. Chem.*, **48**, 113 (1973).
64.   C. P. Andrieux, J. M. Dumas-Bouchiat, and J. M. Savéant, *J. Electroanal. Chem.*, **87**, 39 (1978); C. P. Andrieux, C. Blocman, J. M. Dumas-Bouchiat, F. M'Halla, and J. M. Savéant, *J. Electroanal. Chem.*, **113**, 19 (1980)
65.   D. M. Dimarco, P. A. Forshey, and T. Kuwana, in J. S. Miller, Ed., *Chemically Modified Surfaces in Catalysis and Electrocatalysis*, ACS Symposium Series 192, Washington, D.C., 1982, Chapter 6.
66.   J. M. Savéant and E. Vianello, in I. S. Langmiur, Ed., *Advances in Polarography*, Vol. 1, Pergamon, New York, 1960.
67.   J. M. Savéant and E. Vianello, *Electrochim. Acta*, **10**, 905 (1965).
68.   D. S. Polcyn and I. Shain, *Anal. Chem.*, **38**, 376 (1966).
69.   A. S. N. Murthy and K. S. Reddy, *Electrochim. Acta*, **28**, 1677 (1983).
70.   M. L. Olmstead, R. G. Hamilton, and R. S. Nicholson, *Anal. Chem.*, **41**, 260 (1969).
71.   R. S. Nicholson, *Anal. Chem.*, **37**, 667 (1965).
72.   C. P. Andrieux, L. Nadjo, and J. M. Savéant, *J. Electroanal. Chem.*, **26**, 147 (1970).
73.   L. Nadjo and J. M. Savéant, *J. Electroanal. Chem.*, **33**, 419 (1971); C. P. Andrieux and J. M. Savéant, *J. Electroanal. Chem.*, **33**, 453 (1971).
74.   J. A. Richards and D. H. Evans, *J. Electroanal. Chem.*, **81**, 171 (1977); D. H. Evans, *Acc. Chem. Res.*, **10**, 313 (1977).
75.   G. S. Wilson, D. D. Swanson, J. T. Klug, R. S. Glass, M. D. Ryan, and W. K. Musher, *J. Am. Chem. Soc.*, **101**, 1040 (1979); M. D. Ryan, D. D. Swanson, R. S. Glass, and G. S. Wilson, *J. Phys. Chem.*, **85**, 1069 (1981).
76.   S. Feldberg, *J. Phys. Chem.*, **73**, 1238 (1969).
77.   M. Mastragostino, L. Nadjo, and J. M. Savéant, *Electrochim. Acta*, **13**, 721 (1968).
78.   M. L. Olmstead and R. S. Nicholson, *Anal. Chem.*, **41**, 862 (1969).
79.   G. Farnia, F. Maran, G. Sandona, and M. G. Severin, *J. Chem. Soc. Perkin II*, 1153 (1982).
80.   A. Castellan, F. Masetti, U. Mazzucato, and E. Vianello, *J. Phot. Sci.*, **14**, 164 (1966).
81.   D. E. Richardson and H. Taube, *Inorg. Chem.*, **20**, 1278 (1981).
82.   R. L. Myers and I. Shain, *Anal. Chem.*, **41**, 980 (1969).
83.   D. E. Fenton, R. R. Schroeder, and R. L. Lintvedt, *J. Am. Chem. Soc.*, **100**, 1931 (1978).
84.   F. Ammar and J. M. Savéant, *J. Electroanal. Chem.*, **47**, 115 (1973).
85.   T. H. Teherani, L. A. Tinker, and A. J. Bard, *J. Electroanal. Chem.*, **90**, 117 (1978).
86.   M. D. Ryan, *J. Electrochem. Soc.*, **125**, 547 (1978).
87.   M. D. Ryan and D. H. Evans, *J. Electroanal. Chem.*, **67**, 333 (1976).
88.   W. H. Smith and A. J. Bard, *J. Electroanal. Chem.*, **76**, 19 (1977).
89.   R. S. Nicholson and I. Shain, *Anal. Chem.*, **37**, 178 (1965).
90.   E. T. Seo, R. F. Nelson, J. M. Fritsch, L. S. Marcoux, D. W. Leedy, and R. N. Adams, *J. Am. Chem. Soc.*, **88**, 3498 (1966).
91.   M. Petek, S. Bruckenstein, B. Feinberg, and R. N. Adams, *J. Electroanal. Chem.*, **42**, 397 (1973).
92.   R. F. Nelson, *J. Electroanal. Chem.*, **18**, 329 (1968).

93. J. M. Savéant, C. P. Andrieux, and L. Nadjo, *J. Electroanal. Chem.*, **41**, 137 (1973).
94. C. P. Andrieux, L. Nadjo, and J. M. Savéant, *J. Electroanal. Chem.*, **42**, 223 (1973); C. Amatore and J. M. Savéant, *J. Electroanal. Chem.*, **85**, 27 (1977).
95. L. Nadjo and J. M. Savéant, *J. Electroanal. Chem.*, **47**, 146 (1973).
96. C. P. Andrieux and J. M. Savéant, *J. Electroanal. Chem.*, **33**, 453 (1971).
97. C. P. Andrieux and J. M. Savéant, *J. Electroanal. Chem.*, **28**, 339 (1970).
98. C. Amatore and J. M. Savéant, *J. Electroanal. Chem.*, **86**, 227 (1978).
99. R. H. Wopschall and I. Shain, *Anal. Chem.*, **39**, 1514, 1527 (1967).
100. R. H. Wopschall and I. Shain, *Anal. Chem.*, **39**, 1535 (1967).
101. R. W. Murray, *Acc. Chem. Res.*, **13**, 135 (1980).
102. A. T. Hubbard and F. C. Anson, "The Theory and Practice of Electro-Chemistry With Thin Layer Cells," in A. J. Bard, Ed., *Electroanalytical Chemistry*, Vol. 4, Dekker, New York, 1970, Chapter 2.
103. E. Laviron, "Voltammetric Methods for the Study of Adsorbed Species," in A. J. Bard, Ed., *Electroanalytical Chemistry*, Vol. 12, Dekker, New York, 1982, Chapter 2.
104. A. P. Brown and F. C. Anson, *Anal. Chem.*, **49**, 1589 (1977).
105. E. Laviron, *J. Electroanal. Chem.*, **100**, 263 (1979).
106. E. Laviron, *J. Electroanal. Chem.*, **101**, 19 (1979).
107. G. B. Thomas, Jr., *Calculus and Analytic Geometry*, Addison-Wesley, Reading, MA, 1972.
108. P. F. Seelig and R. de Levie, *Anal. Chem.*, **52**, 1506 (1980); C. C. Chang and R. de Levie, *Anal. Chem.*, **55**, 356 (1983); D. E. Smith, in A. G. Marshall, Ed., *Fourier, Hadamard and Hilbert Transforms in Chemistry*, Plenum, New York, 1982, Chapter 15.
109. P. Lorrain and D. R. Carson, *Electromagnetism, Principles and Applications*, W. H. Freeman, San Francisco, CA, 1979.
110. D. Lelievre and V. Plichon, *Electrochim. Acta*, **23**, 725 (1978).
111. J. E. B. Randles, *Disc. Faraday Soc.*, **1**, 11 (1947).
112. H. Ershler, *Disc. Faraday Soc.*, **1**, 269 (1947).
113. E. Warburg, *Ann. Phys. (Leipzig)*, **67**, 493 (1899); **6**, 125 (1901).
114. J. H. Sluyters, *Rec. Trav. Chim. Pays-Bas*, **79**, 1092 (1960).
115. K. S. Cole and R. H. Cole, *J. Chem. Phys.*, **9**, 341 (1941).
116. J. Euler and K. Dehmelt, *Z. Elektrochem.*, **61**, 1200 (1957).
117. S. A. G. R. Karunathilaka, N. A. Hampson, R. Leek, and T. J. Sinclair, *J. Appl. Electrochem.*, **10**, 357 (1980).
118. J. H. Sluyters and J. J. C. Oomen, *Rec. Trav. Chim. Pays-Bas*, **79**, 1101 (1960).
119. M. Sluyters-Rehbach and J. H. Sluyters, in A. J. Bard, Ed., *Electroanalytical Chemistry*, Vol. 4, Dekker, New York, 1970, p. 1.
120. C. P. M. Bongenaar, A. G. Remijnse, M. Sluyters-Rehbach, and J. H. Sluyters, *J. Electroanal. Chem.*, **111**, 139 (1980).
121. R. P. Buck, *HSI, Hung. Sci. Instrum.*, **49**, 7 (1980).
122. J. Mertens, P. Van den Winkel, and J. Vereecken, *J. Electroanal. Chem.*, **85**, 277 (1977).
123. C. Gavach, P. Seta, and F. Henry, *Bioelectrochem. Bioenerg.*, **1**, 329 (1974).
124. J. R. Sandifer and R. P. Buck, *J. Electroanal. Chem.*, **56**, 385 (1974).
125. R. P. Buck, *J. Electroanal. Chem.*, **18**, 381 (1968).
126. J. H. Kennedy, J. R. Akridge, and M. Kleitz, *Electrochim. Acta*, **24**, 781 (1979).
127. J. E. Bauerle, *J. Phys. Chem. Solids*, **30**, 2657 (1969).
128. H. J. deBruin and A. D. Franklin, *J. Electroanal. Chem.*, **118**, 405 (1981).
129. J. R. Macdonald, *J. Phys. Chem.*, **61**, 3977 (1974).
130. R. de Levie, in P. Delahay, Ed., *Advances in Electrochemistry and Electrochemical Engineering*, Vol. 6, Interscience, New York, 1967.

131. J-P. Candy, P. Fouilloux, M. Keddam, and H. Takenouti, *Electrochim. Acta*, **26**, 1029 (1981).

132. H. Keiser, K. D. Beccu, and M. A. Gutjahr, *Electrochim. Acta*, **21**, 539 (1976).

133. P. Delahay, *Double Layer and Electrode Kinetics*, Interscience, New York, 1965.

134. R. D. Armstrong and M. Henderson, *J. Electroanal. Chem.*, **39**, 81 (1972).

135. M. Keddam, O. R. Mattos, and H. Takenouti, *J. Electrochem. Soc.*, **128**, 257 (1981).

136. M. Keddam, O. R. Mattos, and H. Takenouti, *J. Electrochem. Soc.*, **128**, 266 (1981).

137. R. de Levie, *J. Electroanal. Chem.*, **25**, 257 (1970).

138. R. de Levie and L. Pospišil, *J. Electroanal. Chem.*, **22**, 277 (1969).

139. M. Sluyters-Rehbach and J. H. Sluyters, *J. Electroanal. Chem.*, **65**, 831 (1975).

140. P. Delahay, *J. Electroanal. Chem.*, **19**, 61 (1968).

141. D. D. Macdonald and M. C. H. McKubre, in J. O'M. Bockris, B. E. Conway, and R. E. White, Eds., *Modern Aspects of Electrochemistry*, No. 14, Plenum, New York, 1982.

142. R. D. Armstrong, M. F. Bell, and A. A. Metcalfe, in H. R. Thirsk, Ed., *Electrochemistry*, Vol. 6, The Chemical Society, London, 1978.

143. J. F. McCann and S. P. S. Badwal, *J. Electrochem. Soc.*, **129**, 551 (1982); M. P. Dare-Edwards, A. Hamnett, and P. R. Trevellick, *J. Chem. Soc., Faraday Trans. I*, **79**, 2111 (1983).

144. A. M. Bond, R. J. O'Halloran, I. Ruzic, and D. E. Smith, *Anal. Chem.*, **48**, 872 (1976); **50**, 216 (1978).

145. A. M. Bond, R. J. O'Halloran, I. Ruzic, and D. E. Smith, *J. Electroanal. Chem.*, **90**, 381 (1978).

146. W. L. Unterkofler and I. Shain, *Anal. Chem.*, **37**, 218 (1965).

147. D. E. Smith, in A. J. Bard, Ed., *Electroanalytical Chemistry*, Vol. 1, Dekker, New York, 1966, p. 1.

148. D. Henderson and J. G. Gordon II, *J. Electroanal. Chem.*, **108**, 129 (1980).

149. K. Tokuda and H. Matsuda, *J. Electroanal. Chem.*, **82**, 157 (1977); **90**, 149 (1978).

150. A. M. Bond and B. S. Grabaric, *J. Electroanal. Chem.*, **87**, 251 (1978).

151. J. W. Hayes and D. E. Smith, *J. Electroanal. Chem.*, **114**, 283 (1980).

152. H. Blutstein and A. M. Bond, *Anal. Chem.*, **46**, 1934 (1974).

153. C. I. Mooring and H. L. Kies, *J. Electroanal. Chem.*, **78**, 219 (1977); *Anal. Chim. Acta*, **94**, 135 (1977).

154. K. Okamoto, *Rev. Polarogr. (Japan)*, **11**, 225 (1964); **12**, 40, 50 (1964).

155. J. H. Sluyters, J. S. M. Breukel, and M. Sluyters-Rebach, *J. Electroanal. Chem.*, **31**, 201 (1971).

156. J. E. Anderson and A. M. Bond, *Anal. Chem.*, **53**, 1394 (1981).

157. D. E. Smith, *Anal. Chem.*, **48**, 517A (1976).

158. S. C. Creason and D. E. Smith, *Anal. Chem.*, **45**, 2401 (1973).

159. S. C. Creason and D. E. Smith, *J. Electroanal. Chem.*, **47**, 9 (1973).

160. S. C. Creason and D. E. Smith, *J. Electroanal. Chem.*, **36**, App. I (1972).

161. R. J. Schwall, A. M. Bond, R. J. Loyd, J. G. Larsen, and D. E. Smith, *Anal. Chem.*, **49**, 1797 (1977).

162. R. J. Schwall, A. M. Bond, and D. E. Smith, *Anal. Chem.*, **49**, 1805 (1977).

163. R. J. Schwall, A. M. Bond, and D. E. Smith, *J. Electroanal. Chem.*, **85**, 217, 231 (1977).

164. R. J. O'Halloran, J. C. Schaar, and D. E. Smith, *Anal. Chem.*, **50**, 1073 (1978); R. J. O'Halloran and D. E. Smith, *Anal. Chem.*, **50**, 1391 (1978).

165. D. E. Smith, in Louis Meites, Ed., *Critical Reviews of Analytical Chemistry*, Vol. 2, Chemical Rubber Company, Cleveland, 1971, p. 247; M. Grzeszczuk and D. E. Smith, *J. Electroanal. Chem.*, **157**, 205 (1983).

166. R. deLeeuwe, M. Sluyters-Rehbach, and J. H. Sluyters, *Electrochim. Acta*, **14**, 1183 (1969); J. Struijs, M. Sluyters-Rehbach, and J. H. Sluyters, *J. Electroanal. Chem.*, **146**, 263 (1983).

167. B. J. Huebert and D. E. Smith, *Anal. Chem.*, **44**, 1179 (1972).

168. H. Matsuda, *Z. Elektrochem.*, **62**, 977 (1958).

169. D. E. Smith, T. G. McCord, and H. L. Hung, *Anal. Chem.*, **39**, 1149 (1967).

170. H. L. Hung and D. E. Smith, *Anal. Chem.*, **36**, 922 (1964).

171. J. E. B. Randles, *Can. J. Chem.*, **37**, 238 (1959).

172. E. R. Brown, H. L. Hung, T. G. McCord, D. E. Smith, and G. L. Booman, *Anal. Chem.*, **40**, 1424 (1968).

173. B. Breyer and H. H. Bauer, in P. J. Elving and I. M. Kolthoff, Eds., *Alternating Current Polarography and Tensammetry*, Vol. 13 of Treatise on Analytical Chemistry, Interscience, New York, 1963.

174. P. Delahay, in P. Delahay and C. W. Tobias, Eds., *Advances in Electrochemistry and Electrochemical Engineering*, Vol. 1, Interscience, New York, 1961, Chapter 5.

175. A. M. Bond, R. L. Martin, and A. F. Masters, *Inorg. Chem.*, **14**, 1432 (1975).

176. R. J. O'Halloran and H. Blutstein, *J. Electroanal. Chem.*, **125**, 261 (1981).

177. A. S. Hinman, J. F. McAleer, and S. Pons, *J. Electroanal. Chem.*, **154**, 45 (1983).

178. B. Timmer, M. Sluyters-Rehbach, and J. H. Sluyters, *J. Electroanal. Chem.*, **14**, 169, 181 (1967).

179. D. E. Smith and T. G. McCord, *Anal. Chem.*, **40**, 474 (1968).

180. L. Gierst and R. Cornelissen, *Collect. Czech. Chem. Commun.*, **25**, 3004 (1960).

181. I. Ruzic, D. E. Smith, and S. W. Feldberg, *J. Electroanal. Chem.*, **52**, 157 (1974).

182. I. Ruzic and D. E. Smith, *J. Electroanal. Chem.*, **58**, 145 (1975).

183. J. W. Hayes, I. Ruzic, D. E. Smith, J. R. Delmastro, and G. L. Booman, *J. Electroanal. Chem.*, **51**, 245, 269 (1974).

184. I. Ruzic, R. J. Schwall, and D. E. Smith, *Croat. Chem. Acta*, **48**, 651 (1976).

185. T. G. McCord and D. E. Smith, *Anal. Chem.*, **41**, 116 (1969).

186. D. E. Smith, *Anal. Chem.*, **35**, 602 (1963).

187. A. M. Bond, *J. Electroanal. Chem.*, **20**, 109, 223 (1969); **23**, 269, 277 (1969); **28**, 443 (1970).

188. A. M. Bond, *J. Electrochem. Soc.*, **117**, 1145 (1970); *Anal. Chim. Acta*, **53**, 159 (1971).

189. A. M. Bond and A. B. Waugh, *Electrochim. Acta*, **15**, 1471 (1970).

190. F. M. Hawkridge and H. H. Bauer, *Anal. Chem.*, **44**, 364 (1972).

191. K. Matsuda and R. Tamamushi, *Bull. Chem. Soc. Japan*, **41**, 1563 (1968).

192. K. R. Bullock and D. E. Smith, *Anal. Chem.*, **46**, 1069 (1974).

193. T. G. McCord, H. L. Hung, and D. E. Smith, *J. Electroanal. Chem.*, **21**, 5 (1969).

194. J. Heyrovsky and J. Kuta, *Principles of Polarography*, Academic, New York, 1966.

195. G. H. Aylward, J. W. Hayes, and R. Tamamushi, in J. A. Friend and F. Gutman, Eds., *Proceedings of the First Australian Conference on Electrochemistry 1963*, Pergamon, Oxford, 1964, pp. 323–331.

196. G. H. Aylward and J. W. Hayes, *Anal. Chem.*, **37**, 195 (1965).

197. B. J. Huebert and D. E. Smith, *J. Electroanal. Chem.*, **31**, 333 (1971).

198. J. H. Sluyters and M. Sluyters-Rehbach, *J. Electroanal. Chem.*, **23**, 457 (1969); **26**, 237 (1970).

199. D. E. Smith, *Anal. Chem.*, **35**, 610 (1963).

200. K. R. Bullock and D. E. Smith, *Anal. Chem.*, **46**, 1567 (1974).

201. H. L. Hung and D. E. Smith, *J. Electroanal. Chem.*, **11**, 237 (1966).

202. H. L. Hung and D. E. Smith, *J. Electroanal. Chem.*, **11**, 425 (1966).

203. I. Ruzic and D. E. Smith, *J. Electroanal. Chem.*, **57**, 129 (1974).
204. I. Ruzic, H. R. Sobel, and D. E. Smith, *Electroanal. Chem.*, **65**, 21 (1975).
205. D. E. Smith and H. R. Sobel, *Anal. Chem.*, **42**, 1018 (1970).
206. R. J. Schwall and D. E. Smith, *J. Electroanal. Chem.*, **94**, 227 (1978).
207. R. J. Schwall, I. Ruzic, and D. E. Smith, *J. Electroanal. Chem.*, **60**, 117 (1975).
208. T. Matusinovic and D. E. Smith, *J. Electroanal. Chem.*, **98**, 133 (1979).
209. T. Matusinovic and D. E. Smith, *Inorg. Chem.*, **20**, 3121 (1981).
210. E. Laviron, *J. Electroanal. Chem.*, **105**, 25 (1979).
211. E. Laviron, *J. Electroanal. Chem.*, **97**, 135 (1979).
212. T. Kakutani and M. Senda, *Bull. Chem. Soc. Japan*, **52**, 3236 (1979).
213. B. Breyer and S. Hacobian, *Aust. J. Sci. Res., Ser. A*, **5**, 500 (1952).
214. D. R. Canterford, *J. Electroanal. Chem.*, **118**, 395 (1981).
215. A. M. Bond, in *Modern Polarographic Methods in Analytical Chemistry*, Dekker, New York, 1980, pp. 295–297.
216. H. P. Agarwal, in A. J. Bard, Ed., *Electroanalytical Chemistry*, Vol. 7, Dekker, New York, 1974, pp. 161–271.
217. F. van der Pol, M. Sluyters-Rehbach, and J. H. Sluyters, *J. Electroanal. Chem.*, **40**, 209 (1972).
218. F. van der Pol, M. Sluyters-Rehbach, and J. H. Sluyters, *J. Electroanal. Chem.*, **45**, 377 (1973).
219. F. van der Pol, M. Sluyters-Rehbach, and J. H. Sluyters, *J. Electroanal. Chem.*, **58**, 177 (1975).
220. H. P. Agarwal and P. Jain, *Electrochim. Acta*, **26**, 621 (1981).
221. T. G. McCord, E. R. Brown, and D. E. Smith, *Anal. Chem.*, **38**, 1615 (1966).
222. H. Blutstein, A. M. Bond, and A. Norris, *Anal. Chem.*, **48**, 1975 (1976).
223. J. H. Wolsink, M. Sluyters-Rehbach, and J. H. Sluyters, *J. Electroanal. Chem.*, **117**, 213 (1981).
224. T. G. McCord and D. E. Smith, *Anal. Chem.*, **40**, 289 (1968).
225. T. G. McCord and D. E. Smith, *Anal. Chem.*, **41**, 131 (1969).
226. T. G. McCord and D. E. Smith, *Anal. Chem.*, **42**, 126 (1970).
227. D. E. Glover and D. E. Smith, *Anal. Chem.*, **43**, 775 (1971).
228. H. Blutstein, A. M. Bond, and A. Norris, *Anal. Chem.*, **46**, 1754 (1974).
229. T. G. McCord and D. E. Smith, *J. Electroanal. Chem.*, **26**, 61 (1970).
230. T. G. McCord and D. E. Smith, *Anal. Chem.*, **41**, 1423 (1969).
231. T. G. McCord and D. E. Smith, *Anal. Chem.*, **42**, 2 (1970).
232. A. M. Bond and D. E. Smith, *Anal. Chem.*, **46**, 1946 (1974).
233. M. R. Wasielewski and R. Breslow, *J. Am. Chem. Soc.*, **98**, 4222 (1976); J. L. Grant and R. Breslow, *J. Am. Chem. Soc.*, **99**, 7745 (1977).
234. B. Jaun, J. Schwarz, and R. Breslow, *J. Am. Chem. Soc.*, **102**, 5741 (1980).
235. E. Ahlberg and V. D. Parker, *Acta Chem. Scand. B*, **34**, 91 (1980).
236. A. A. Pilla, *J. Electrochem. Soc.*, **117**, 467 (1970).
237. A. Rajakumar and P. R. Krishnaswamy, *Ind. Eng. Chem., Process Des. Dev.*, **14**, 250 (1975).
238. A. A. Pilla, *J. Electrochem. Soc.*, **118**, 1295 (1971).
239. A. A. Pilla and G. S. Margules, *J. Electrochem. Soc.*, **124**, 1697 (1977).
240. A. J. Bard and L. Faulkner, *Electrochemical Methods*, Wiley, New York, 1980, Chapter 9.
241. K. Doblhofer and A. A. Pilla, *J. Electroanal. Chem.*, **39**, 91 (1972).
242. J. W. Cooley and J. W. Tukey, *Math. Comp.*, **19**, 297 (1965).

243. S. C. Creason and D. E. Smith, *J. Electroanal. Chem.*, **40**, App. 1 (1972).
244. S. C. Creason, J. W. Hayes, and D. E. Smith, *J. Electroanal. Chem.*, **47**, 9 (1973).
245. K. B. Oldham and J. Spanier, *The Fractional Calculus*, Academic, New York, 1974.
246. J. C. Imbeaux and J. M. Savéant, *J. Electroanal. Chem.*, **44**, 169 (1973).
247. K. B. Oldham, *Anal. Chem.*, **41**, 1904 (1969).
248. K. B. Oldham, *Anal. Chem.*, **44**, 196 (1972).
249. M. Goto and K. B. Oldham, *Anal. Chem.*, **45**, 2043 (1973).
250. M. Goto and K. B. Oldham, *Anal. Chem.*, **48**, 1671 (1976).
251. J. M. Savéant and D. Tessier, *J. Electroanal. Chem.*, **65**, 57 (1975).
252. A. J. Bard and L. Faulkner, *Electrochemical Methods*, Wiley, New York, 1980, Chapter 6.
253. K. B. Oldham, *J. Electroanal. Chem.*, **121**, 341 (1981).
254. K. B. Oldham, *J. Electroanal. Chem.*, **45**, 39 (1973).
255. J. M. Savéant and D. Tessier, *J. Phys. Chem.*, **82**, 1723 (1978).
256. J. M. Savéant and D. Tessier, *J. Electroanal. Chem.*, **77**, 225 (1977).
257. M. Goto and D. Ishii, *J. Electroanal. Chem.*, **61**, 361 (1975).
258. P. Dalrymple-Alford, M. Goto, and K. B. Oldham, *J. Electroanal. Chem.*, **85**, 1 (1977).
259. M. Goto, M. Kato, and D. Ishii, *Anal. Chim. Acta*, **126**, 95 (1981).
260. J. J. Toman and S. D. Brown, *Anal. Chem.*, **53**, 1497 (1981).
261. M. Goto, T. Hirano, and D. Ishii, *Bull. Chem. Soc. Japan*, **51**, 470 (1978).
262. M. D. Ryan, *J. Electroanal. Chem.*, **79**, 105 (1977).
263. H. L. Suprenant, T. H. Ridgway, and C. N. Reilley, *J. Electroanal. Chem.*, **75**, 125 (1977).
264. J. E. Anderson and A. M. Bond, *Anal. Chem.*, **53**, 1394 (1981).
265. G. H. Morrison, *Anal. Chem.*, **53**, 2161 (1981).
266. R. de Levie and A. A. Husovsky, *J. Electroanal. Chem.*, **20**, 181 (1969).
267. A. J. Bentz, J. R. Sandifer, and R. P. Buck, *Anal. Chem.*, **46**, 543 (1974).
268. D. E. Mathis and R. P. Buck, *Anal. Chem.*, **48**, 2033 (1976).
269. H. Blutstein, A. M. Bond, and A. Norris, *J. Electroanal. Chem.*, **89**, 75 (1978).
270. H. Kojima and A. J. Bard, *J. Electroanal. Chem.*, **63**, 117 (1975).
271. R. D. Armstrong, M. F. Bell, and A. A. Metcalfe, *J. Electroanal. Chem.*, **77**, 287 (1977).
272. C. P. M. Bongenaar, M. Sluyters-Rehbach, and J. H. Sluyters, *J. Electroanal. Chem.*, **109**, 23 (1980).
273. D. Britz, *J. Electroanal. Chem.*, **88**, 309 (1978).
274. E. R. Brown, T. G. McCord, D. E. Smith, and D. D. DeFord, *Anal. Chem.*, **38**, 1119 (1966).

Chapter **5**

CONTROLLED-POTENTIAL ELECTROLYSIS AND COULOMETRY

Louis Meites

1   INTRODUCTION

Many studies have been made of the reactions that occur when solutions of reducible or oxidizable substances are electrolyzed. Some were made to identify the products that are formed or to elucidate the mechanisms by which the reactions occur. Others were made to obtain information that would be useful in devising new procedures of separation, synthesis, or analysis. Still others were made to reveal similarities or differences between the behaviors of closely related substances, or to uncover correlations between those behaviors and their chemical constitutions or molecular structures.

Because these studies have had such diverse aims, and because they have extended over many decades, they have employed three related but distinctly different techniques. One was constant-voltage electrolysis, which has been authoritatively reviewed by Swann [1]. An approximately constant, and usually rather large, electromotive force (emf) is applied across a cell whose electrodes have large areas. A large current flows, and substantial amounts of product are obtained rapidly. Many large-scale syntheses are still performed in this way. The advantage of the technique is that it requires only relatively simple and inexpensive apparatus. Its disadvantage is that it is unselective: different reductions and oxidations often occur simultaneously, giving rise to mixtures of products in proportions that depend on the experimental conditions and usually vary as an electrolysis proceeds. It is difficult to reproduce the results, and nearly impossible to obtain fundamental information.

Voltammetry, along with a number of other techniques related to it, is a newer approach. Voltammetry at dropping mercury electrodes (polarography) is described by Vlček, et al. in Chapter 9; voltammetry at other indicator electrodes is described by Geiger and Hawley in Chapter 1, as well as in other chapters in this volume. Most voltammetric indicator electrodes have very small areas. Hence the currents flowing through them are small, and it is easy to measure or control their potentials, which are important because they govern the rates and extents of the electron-transfer processes that occur at the surfaces of the electrodes. By proper selection of the potential, it is often possible to isolate and study one particular half-reaction to the exclusion of others: to reduce or oxidize a single constituent of a mixture, or to effect just one of a number of possible steps. Voltammetry is very useful in studying fundamental electrochemistry because of its selectivity; but because the indicator electrodes used in voltammetry, and the currents that flow through them, are so small they cannot be used to reduce or oxidize significant amounts of material within a reasonable length of time.

Controlled-potential electrolysis combines the valuable features of constant-voltage electrolysis and voltammetry. As in constant-voltage electrolysis, the processes of interest occur at the surface of a large electrode, and therefore give rise to large currents that cause them to proceed rapidly and to reach completion quickly. As in voltammetry, the potential of that electrode is controlled, and therefore the selectivity of controlled-potential electrolysis is comparable to that of voltammetry. It is the combination of speed with selectivity that entitles controlled-potential electrolysis to a place among the most versatile and useful electrochemical techniques.

In controlled-potential coulometry the current–time curve is integrated under conditions such that only a single substance is reduced or oxidized at the surface of the electrode whose potential is controlled, and the electrolysis and integration are prolonged until virtually all of that substance has been consumed. The total quantity of electricity Q_∞ (in coulombs) consumed in reducing or oxidizing N^0 moles of material is then given by Faraday's law:

$$Q_\infty = \int_0^\infty i\, dt = nFN^0 \tag{1}$$

where i is the current (in amperes) t seconds after the start of the electrolysis, n is the number of electrons transferred to or from each molecule or ion of the electroactive substance, and F is the number of coulombs per faraday. Integrating the current with respect to time makes it possible to evaluate n if N^0 is known, or to evaluate N^0 if n is known. The former possibility is helpful in elucidating the course and product(s) of the half-reaction; the latter one is the basis of many analytical procedures.

As (1) implies, the current varies with time during an electrolysis at controlled potential. This is because any substance, whether present initially or formed as an intermediate, that can undergo reduction or oxidation is converted as time goes on into others that cannot. With a simple half-reaction, in which the starting material is converted into a single product, either directly or in a series of steps of which all are much faster than the mass-transfer process, the current decays exponentially with time. Other kinds of dependences on time arise from the occurrence of slow steps, which may be either chemical or electrochemical. There are two kinds of mechanisms: consecutive and branched. A consecutive mechanism, in which each substance can react in only one way, yields a single product and, normally, an integral value of n. A branched one, in which at least one intermediate reacts in two or more different ways at comparable rates, gives a mixture of products and, normally, a nonintegral value of n. If the competing reactions have different orders, the proportions of the products and the value of n depend on the initial concentration of the starting material. Combining coulometric data with information on current–time curves, and with identification and determination of the products obtained, makes it possible to elucidate the mechanism of the process and to evaluate the rate constants of the slow steps.

Like controlled-potential electrolysis, in which a substantial volume of a stirred solution is electrolyzed under conditions that are hydrodynamically rather ill-defined, the electrolysis of a thin layer of a stationary solution confined between two electrodes is based on ideas closely similar to those of voltammetry. However, the two techniques differ so much in apparatus and methodology that the latter is represented here only by a few references [2–5] for the convenience of those who may wish to draw comparisons and contrasts.

2 PRINCIPLES OF ELECTROLYSIS: THE BEHAVIORS OF SIMPLE HALF-REACTIONS

An electrolytic reaction is brought about by applying an emf across two electrodes in contact with a solution. Electrons flow through the external circuit from the anode to the cathode. At the interface between the cathode and the solution in contact with it, electrons are transferred from the cathode to some constituent of the solution, which is reduced. Sometimes that constituent is the solvent, and the solvated electron or free radical that is formed initially may diffuse for some distance away from the surface of the electrode before it reacts with a molecule or ion of a solute, but such details of the electron-transfer process will not concern us here. Conversely, at the surface of the anode, some constituent of the solution loses electrons to the anode and is oxidized. Electricity is carried through the solution by the migration of cations towards the negative electrode (which is often, but not necessarily, the cathode) and of anions toward the positive electrode (which may or may not be the anode).

The substance of interest reacts at the surface of one electrode, which is called the "working electrode," regardless of whether it is the cathode or the anode. Other substances react at the other electrode, which is called the "auxiliary (or counter) electrode." Their behaviors are of no concern as long as they, or the products of the reactions that they undergo, do not interact with the substances present, or formed, in the solution surrounding the working electrode. To prevent such interactions, the working and auxiliary electrodes are usually located in different compartments separated by a porous diaphragm.

The total or applied emf V that is imposed across these two electrodes is distributed in accordance with the equation

$$V = E_a - E_c + iR \tag{2}$$

where i is the current (in amperes), R is the resistance (in ohms), and E_a and E_c are the potentials of the anode and cathode, respectively, referred to the same reference electrode. It does not matter what reference electrode is chosen, but for work with aqueous solutions the saturated calomel electrode (SCE) is the nearly universal choice. The factors that affect i will be discussed below; among them, the concentration of the substance of interest is especially important because it varies as an electrolysis proceeds. The value of R depends on the concentration and mobility of each of the ionic species present, and on the areas of the electrodes, the distance between them, and the thickness and porosity of the diaphragm or membrane that may be used to separate the solutions in contact with the electrodes. Both the designs of cells and the selections of supporting electrolytes are largely governed by the necessity of minimizing the iR drop and its variation across the surface of the working electrode.

The potential of each electrode may be regarded as the sum of a reversible potential and an overpotential. The former depends on the standard potential of the half-reaction taking place at the electrode and on the activities, at the surface of the electrode, of the substances involved in that half-reaction. It is described by the Nernst equation and is the potential that the electrode would have if there were no current flowing through the cell, so that the bulk of the solution had the same composition as the layer at the surface of the electrode, or if all the chemical and electrochemical processes involved in the half-reaction were extremely fast. Some half-reactions are indeed very fast, but the great majority involve at least one slow step and consequently exhibit irreversible or non-Nernstian behavior. The overpotential reflects the occurrence of a slow step, and its value is affected by changing any experimental variable that influences the rate of that step. Such variables include the efficiency of stirring, the temperature, the composition (and especially the pH and ionic strength) of the solution, and the nature and, with solid working electrodes, even the prior history of the surface of the electrode. The overpotentials differ for different half-reactions, and vary in different and unpredictable ways as the experimental conditions are changed. Even changes that seem very minor may have profound effects on the yields, and even on the identities, of the products obtained. Many striking examples are cited by Swann [1], Allen [6], Zuman [7], and the authors of Weinberg's *Technique of Organic Electrosynthesis* [8].

The importance of the potential of the working electrode was originally recognized by Haber [9, 10] and can be illustrated by considering the electrolytic reduction of p-benzoquinone. Figure 5.1 is a polarogram of an aqueous solution that contained approximately 0.001 M p-benzoquinone in a supporting electrolyte consisting of 0.1 M potassium dihydrogen phosphate, 0.1 M potassium monohydrogen phosphate, and 1 M potassium nitrate, and having a pH of 6.8. It was obtained with a conventional dropping mercury electrode [11, 12] as the indicator electrode and a saturated calomel electrode as the reference electrode. Both the current and the resistance of the cell were so small that the iR term in (2) was always negligible, so that the applied emf at each point was equal to the difference between the potentials of the two electrodes.

The polarogram may be divided into five regions. In the first, the potential of the dropping electrode is more positive than about $+0.2$ V versus SCE, and some anodic process occurs at its surface. Under the conditions of Figure 5.1 mercury is oxidized to a sparingly soluble mercury(I) phosphate; under other conditions some constituent of the supporting electrolyte may be oxidized, but in any event this region is devoid of interest in controlled-potential electrolysis. So is the second region, which in Figure 5.1 extends from about $+0.2$ to about $+0.1$ V versus SCE. In this region the potential of the dropping electrode is neither sufficiently positive to effect the oxidation of mercury, nor sufficiently negative to effect the reduction of the quinone, at an appreciable rate; and the very small currents that flow at the dropping electrode are due chiefly to charging

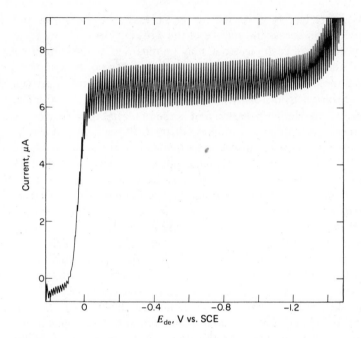

Figure 5.1 Polarogram of 0.001 *M* *p*-benzoquinine in 0.1 *M* potassium dihydrogen phosphate–0.1 *M* potassium monohydrogen phosphate–1 *M* potassium nitrate, pH 6.8, secured with a conventional dropping mercury electrode. Reprinted, with permission, from L. Meites, "Controlled-Potential Electrolysis," in A. Weissberger and B. W. Rossiter, Eds., *Physical Methods of Chemistry*, Vol. 1, Part IIA, 4th ed., Wiley-Interscience, 1971.

of the electrical double layers around the growing drops. The third region extends from about $+0.1$ to about -0.1 V versus SCE. It contains the rising portion of a wave that corresponds to the half-reaction $Q + 2H^+ + 2e = H_2Q$, where Q represents the quinone and H_2Q the corresponding hydroquinone. Since this half-reaction behaves reversibly [13], the potential E of the dropping electrode conforms to the equation

$$E = E^0 - \frac{RT}{nF} \ln \left(\frac{a_{H_2Q}}{a_{H^+}^2 a_Q} \right)$$

$$= E^{0'} - \frac{RT}{nF} \ln \left(\frac{[H_2Q]}{[H^+]^2[Q]} \right)$$

(3)

where E^0 is the standard and $E^{0'}$ the formal potential of the couple, and the activities denoted by a and the concentrations denoted by square brackets are those in the thin layer of solution immediately adjacent to the surface of the electrode. Since both $E^{0'}$ and, because the solution is well buffered, $[H^+]$ are constant, the ratio $[H_2Q]/[Q]$ must vary as the potential changes. It increases as the potential becomes more negative. At a potential near -0.1 V it is so large that nearly every molecule of the quinone is reduced on reaching the electrode surface, and a comparatively large current is consumed. The fourth region extends from about -0.1 to about -1.2 V versus SCE. Over this entire range the potential is so much more negative than $E^{0\prime}$ that $[H_2Q]/[Q]$ must be extremely large. Because $[Q]$ must therefore be extremely small, molecules of the quinone cannot survive at the surface of the electrode, but are reduced as rapidly as they reach it by diffusion or any other mechanism. The rate at which they do so is independent of the potential of the electrode, and hence the current that is consumed in reducing them is also independent of potential. It is known as the "limiting current" and is proportional to the concentration of the quinone in the bulk of the solution. The fifth and final region begins at about -1.2 V versus SCE and is the region in which some other substance begins to undergo reduction. Under these conditions the half-reaction

$$H_2PO_4^- + e = HPO_4^{2-} + \tfrac{1}{2}H_2$$

occurs along with reduction of the quinone.

A polarogram of generally similar shape would be obtained with another electroactive substance whose reduction was irreversible or non-Nernstian. There would again be five regions, of which all but the third would have the same significance. In the third region, although the current would again increase as the potential became more negative, it would do so for a different reason. With a non-Nernstian couple it is the rate, rather than the equilibrium, of the process that is significant. In the simplest case the starting material Ox is converted into the product Red in a single electron-transfer step, but this involves an energy of activation so large that equilibrium is not attained during a measurement of the current. The half-reaction Ox $+ ne \rightarrow$ Red is characterized by the value of $k_{s,h}$, a standard heterogeneous rate constant that applies to both the forward (cathodic) and the reverse (anodic) process at the standard potential of the couple. If that value is very small, which is what the term "irreversible" implies, only a very small current results from the slow reduction of Ox at the standard potential of the couple. As the potential of the indicator electrode becomes more negative, the cathodic current i_c changes in accordance with the equation [14–16]

$$i_c = -nFAk_{s,h}c_{Ox}^0 \exp[-\alpha n_a F(E - E^0)/RT] \tag{4}$$

where i_c is expressed in amperes; the negative sign reflects a recent International Union of Pure and Applied Chemistry (IUPAC) prescription [17] that cathodic currents should be regarded as negative; n is the number of electrons consumed in reducing each ion or molecule of Ox; F, R, and T are the number of coulombs per faraday, the gas constant, and the temperature (in kelvins), respectively; A is the area of the electrode (cm^2); and c_{Ox}^0 is the concentration (mol/cm^3) of Ox at the surface of the electrode. The transfer coefficient α is a parameter whose value is considered to reflect the symmetry of the activation-energy barrier for the half-reaction and depends on the experimental conditions, including the potential of the electrode. Its value must lie between 0 and 1, and is usually between about 0.3 and 0.7. The parameter n_a is the number of electrons involved in the rate-controlling step (of which there is assumed to be only one); its value must be integral, cannot exceed that of n, and is usually equal to either 1 or 2. The reader is advised to study Parsons' comments [16] on the significance of these parameters as an antidote to this transcription of ideas that have permeated the electroanalytical literature for several decades.

Although the value of i_c is small for an irreversible half-reaction when $E = E^0$, it increases as E becomes more negative because of the exponential term in (4). For small values of i_c the increase is exponential, but appreciable values of i_c reflect the reduction of Ox at appreciable rates, and are accompanied by a decrease of c_{Ox}^0 below the bulk concentration. If the potential is much more negative than E^0, c_{Ox}^0 approaches zero and the current reaches a limiting value that is controlled by the rate at which ions or molecules of Ox are transported from the bulk of the solution to the surface of the electrode, just as though the couple were reversible. The general shape of the current–potential curve does not depend on whether the couple is reversible or not, but reversible and irreversible couples do behave quite differently in electrolyses that are carried out in the third region. The differences are described near the end of this section.

Figure 5.2 is another current–potential curve for the solution of p-benzoquinone that gave the polarogram shown in Figure 5.1. It was obtained by using a large pool of mercury as the indicator electrode in place of the dropping mercury electrode, and employing an efficient propeller-type stirrer to agitate the interface between the electrode and the solution. The general shapes of the two curves are very much the same: the same five regions can be identified

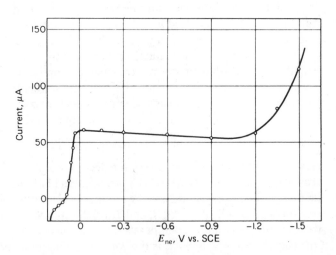

Figure 5.2 Current–potential curve of the solution of Figure 5.1 at a stirred mercury-pool electrode having an area of 40 cm². The small decrease of the current between 0 and 0.9 V versus SCE is attributable to the partial electrolytic depletion of the solution while the measurements were being made. Reprinted, with permission, from L. Meites, "Controlled-Potential Electrolysis," in A. Weissberger and B. W. Rossiter, Eds., *Physical Methods of Chemistry*, Vol. 1, Part IIA, 4th ed., Wiley-Interscience, 1971.

without difficulty in Figure 5.2, and they occur at very nearly the same potentials as in Figure 5.1. However, the limiting current, which is only about 6 μA in Figure 5.1, is roughly 60 mA in Figure 5.2. This is due partly to the much larger area of the pool electrode, and partly to the fact that stirring brings molecules of the quinone to the surface of the pool much more rapidly than diffusion brings them to the surface of the dropping electrode. The reduction of the quinone is about 10^4 times as fast at the pool electrode as at the dropping electrode—so fast, indeed, that the concentration of the quinone in the bulk of the solution decreased appreciably during the minute or so that was needed to obtain the data plotted in Figure 5.2. Because large electrodes in stirred solutions yield high limiting currents, half-reactions proceed rapidly at their surfaces and reach completion within reasonable lengths of time, and this is the fundamental reason why controlled-potential electrolyses are practical and useful.

Most controlled-potential electrolyses are performed either to obtain the product(s) of the half-reaction that occurs at the surface of the working electrode or to find the quantity of electricity consumed in that half-reaction for one of the purposes described in connection with (1). For either purpose they are usually performed in the fourth region, on the "plateau" of the wave. That choice maximizes the speed and completeness of the desired reaction while providing reasonable freedom from interference by other half-reactions that occur at more negative potentials. In the third region, on the rising portion of the wave, the reduction is slower and cannot be made complete if the half-reaction is reversible. In the fifth region some other half-reaction takes place in addition to the desired one. This is always deleterious in coulometric experiments because the unwanted half-reaction consumes electricity, and is deleterious even in synthetic work if that half-reaction involves the further reduction of the desired product.

Electrolyses may, however, be carried out in other regions for other purposes. A solution of p-benzoquinone might be electrolyzed in the second region, below the foot of the wave, to destroy impurities that are more easily reducible before determining the quinone polarographically or in some other way, to effect the coulometric determination of some more easily reducible substance without reducing any of the quinone at the same time, or to oxidize any of the hydroquinone that might be present. Electrolyses in the third region show how the potential affects the extent of reduction and provide values of the formal potentials of reversible half-reactions. They are especially useful in elucidating mechanisms like

$$Ox + n_1 e \rightarrow I$$
$$I \rightarrow P$$
$$I + n_2 e \rightarrow Red$$

where the relative proportions of the products P and Red can be changed by varying the potential of the working electrode because this affects the rate constant for the reduction of the intermediate I.

Evidently the potential of the working electrode is of great importance. To choose it rationally, one must have a reasonable amount of information about the electrochemical behavior of the system. As Figures 5.1 and 5.2 suggest, polarographic data provide a useful starting point for work with mercury-pool electrodes, and voltammetric data obtained with platinum microelectrodes serve the same purpose for work with large platinum electrodes in stirred solutions. A chemist who wanted to assay a sample of p-benzoquinone by controlled-potential coulometry could tell from Figure 5.1 that a preliminary electrolysis with a stirred mercury pool at a potential near $+0.15$ V versus SCE in this supporting electrolyte would remove any more easily reducible constituents without affecting the quinone, and that the quinone could then be rapidly and quantitatively reduced by a second electrolysis at any potential between about -0.1 and -1.2 V versus SCE. If the sample were known to contain no impurity that is more difficultly reducible than the quinone, any potential between these limits might be chosen. One of, say, -0.2 V would be preferable to one of, say, -1.1 V because, although the rate of reduction of dihydrogen phosphate ion at the latter potential is too small to yield an increase of current that can be discerned on the scale of Figure 5.1, it is not completely negligible. In accordance with (4), it decreases as the potential becomes less negative; at -0.2 V versus SCE it would be small enough to ignore in any electrolysis involving more than a microgram of the quinone.

Very large critical compilations of polarographic and voltammetric data [18, 19] are available to guide such choices, but they must be used with care because the correspondence is far less close for irreversible processes than for reversible ones. The polarographic half-wave potential $E_{1/2,\mathrm{DME}}$, defined as the potential at which the maximum current attributable to the substance of interest is half as large as on the plateau, varies with the drop time τ at the half-wave potential [20, 21], and the rapidity of mass transfer in a stirred solution renders the slowness of the rate-determining electron-transfer step even more pronounced. For an irreversible process the theoretical value of the difference, $\Delta E_{1/2}$, between $E_{1/2,\mathrm{DME}}$ and the voltammetric half-wave potential $E_{1/2,\mathrm{v}}$ at a large electrode in a stirred solution is given by [15, 22]

$$\Delta E_{1/2} = \frac{0.05916}{\alpha n_a} \log \left(\frac{1.349(D_{\mathrm{Ox}}\tau)^{1/2}}{\delta} \right) \qquad (5a)$$

at 25°C: D_{Ox} is the diffusion coefficient of Ox $(\mathrm{cm}^2/\mathrm{s})$, and δ is a distance (cm) equivalent to the thickness of the diffusion–turbulence-damping region in the layer of the phase containing Ox (which is normally the solution, but may be the amalgam when the process of interest is the oxidation of a metal dissolved in mercury) that is adjacent to the surface of the electrode. The value of δ is familiarly called the "thickness of the Nernst diffusion layer." If D_{Ox}, τ, and δ have the typical values 6×10^{-6} cm^2/s, 4 s, and 0.003 cm, respectively, (5a) becomes

$$\Delta E_{1/2} = \frac{0.014}{\alpha n_a} \qquad (5b)$$

or roughly 30 mV if αn_a has the common value of 0.5. Differences ten or twenty times as large are not uncommon. Karp and Meites [22] found the half-wave potential of hydrogen peroxide in a neutral phosphate buffer to be -0.75 V versus SCE at a dropping mercury electrode but -1.31 V versus SCE at a large stirred mercury pool; Streuli and Cooke [23] found the corresponding figures for 1-nitroso-2-naphthol in an ammoniacal buffer to be -0.27 and -0.80 V versus SCE, though the difference for the same compound does not exceed 0.1 V in an acetate buffer of pH 5.5 [18, 24]. Hence, as was first pointed out by Charlot, Badoz-Lambling, and Trémillon [25] and Rechnitz [26], and as has been stressed by many others since, polarographic and voltammetric data provide useful indications of what can be done, and how and within what range of conditions it can be done, but they do not provide a completely reliable guide to the optimum conditions for doing it.

This is because there are important differences between the experimental conditions associated with the two techniques. The most important ones are in:

1.   The concentration of the electroactive substance, which is most often between 0.01 and 1 mM (10^{-5}–10^{-3} M) in voltammetry, but is rarely below 1 mM in controlled-potential electrolysis, and is sometimes as high as 25 or even 100 mM, especially in preparative work. This is a matter of custom rather than necessity; there is no real difficulty in carrying out controlled-potential electrolyses with solutions as dilute as 0.001 mM under typical conditions, though to be sure it would often be unwise to use such dilute solutions when the purpose is to obtain and characterize the product. But it is no less unwise to attempt to correlate the values of n and the identities and proportions of the products obtained in polarography at a low concentration with those obtained in controlled-potential electrolysis at a much higher one, for if the mechanism involves any second- or higher-order process the correlation will be far from straightforward.

2.   The rates of mass transfer between the bulk of the solution and the surface of the electrode, which are at least an order of magnitude higher in controlled-potential electrolysis than in polarography. An intermediate that can either undergo further reduction or participate in some homogeneous chemical reaction is much more likely to do the latter in controlled-potential electrolysis, because it is so much more rapidly carried away from the surface of the electrode. The difficulty of reconciling polarographic data for many substances with the results of controlled-potential electrolyses in non-aqueous, and particularly aprotic, solvents often reflects the greater opportunity that stirring provides for side reactions of free-radical intermediates.

3.   The effects of adsorption, which are far more pronounced in controlled-potential electrolysis. The rate of adsorption is finite, and in voltammetry with a hanging mercury-drop electrode the scan rate may be so high that the entire voltammogram is obtained in a fraction of the time that would be required to reach equilibrium in the adsorption of an intermediate or product. In polarography under the same conditions, adsorption must begin afresh as each new

drop is born, but may become substantially complete long before the drop has attained maturity, and may therefore be much more prominent than it is with a hanging drop. At a pool electrode, adsorption may be prominent during almost all of the electrolysis.

For such reasons it is naive to think that what has been said here about the p-benzoquinone–p-benzohydroquinone couple has simple parallels in the behaviors of other, more complex, systems. In selecting the potential at which a controlled-potential electrolysis is to be carried out, data obtained with dropping electrodes and other microelectrodes are fully reliable only for reversible couples. When irreversibility is marked—that is, when slow steps are prominent—it is much safer to rely on a current–potential curve obtained, under exactly the same conditions, with the large electrode that will be used in the electrolysis. The composition of the solution, the concentration of the electroactive substance, and the efficiency of stirring are the most important of the conditions that should be duplicated. It is easy to obtain such a curve manually by adjusting the potentiostat to give each of a number of potentials, say 0.1–0.3 V apart, depending on which of the five regions is being observed, and measuring the current at each potential. The measurements must be made quickly to avoid excessive electrolytic depletion of the solution: the negative slope of the plateau in Figure 5.2 is an artifact attributable to the time that was expended in obtaining the four points at potentials between -0.05 and -1.2 V versus SCE. Some potentiostats have bridges driven by synchronous motors to permit automatic recording of current–potential curves, which can also be recorded with an ordinary polarograph if the concentration of the electroactive substance is sufficiently low—say, 0.01 mM or less—and if the resistance of the cell is not too high. In reporting such curves, or the values of the half-wave potentials and limiting currents obtained from them, it is essential to specify the material and form of the working electrode (for example, mercury pool or platinum gauze), its approximate area, and the value of the mass-transfer constant s_{Ox} either for the electroactive substance being studied or, if slow chemical steps are known or thought to be involved in the half-reaction, for a reference or "pilot" substance, such as cadmium ion, under experimental conditions that are as nearly as possible identical.

The rate of a controlled-potential electrolysis is interesting in analytical and synthetic work, and is crucial in mechanistic studies. Assuming that the half-reaction $Ox + ne \rightarrow Red$ occurs in a single step and that the rate of reoxidation of Red is negligible—which is true on the plateau of the wave no matter whether the half-reaction is reversible or not, and is also true on the rising part of a totally irreversible wave—Karp and Meites [22] obtained the equation

$$i = \frac{s_{Ox} nFVAk_{Ox} c_{Ox}^{b,0}}{Ak_{Ox} + Vs_{Ox}} \exp\left(\frac{-s_{Ox} Ak_{Ox} t}{Ak_{Ox} + Vs_{Ox}}\right) \qquad (6)$$

where $c_{Ox}^{b,0}$ is the concentration (mol/cm^3) of Ox in the bulk of the solution at the start of the electrolysis; k_{Ox} is the heterogeneous rate constant (cm/s) of the electron-transfer process and is given by

$$k_{Ox} = k_{s,h} \exp\left[\frac{-\alpha n_a F(E - E^0)}{RT}\right] \tag{7}$$

[cf. (4)]; and s_{Ox} is the mass-transfer constant (s^{-1}) of Ox, defined by

$$s_{Ox} = D_{Ox} A / V\delta \tag{8}$$

in which A is the area (cm^2) of the interface between the electrode and the solution, and V is the volume (cm^3) of the phase containing Ox. It is convenient to define an electrolytic rate constant s_{Ox}^*, whose value depends on the potential of the working electrode, by writing

$$s_{Ox}^* = \frac{s_{Ox} A k_{Ox}}{A k_{Ox} + V s_{Ox}} \tag{9}$$

Combining this with (6) yields

$$i = s_{Ox}^* n F V c_{Ox}^{b,0} \exp(-s_{Ox}^* t) \tag{10}$$

as the equation for the current–time curve obtained in this simple case.

There are two extreme situations. On the plateau of the wave, k_{Ox} is very much larger than $V s_{Ox}/A$; then s_{Ox}^* and s_{Ox} are identical, and (10) becomes

$$i = s_{Ox} n F V c_{Ox}^{b,0} \exp(-s_{Ox} t) \tag{11}$$

which is equivalent to an expression first obtained by Lingane [27]. Near the foot of a totally irreversible wave, however, k_{Ox} is much smaller than $V s_{Ox}/A$; then (9) and (10) become, respectively,

$$s_{Ox}^* = \frac{A k_{Ox}}{V} \quad \left(= \frac{k_{Ox}\delta}{D_{Ox}}\right) \tag{12}$$

and

$$i = n F A k_{Ox} c_{Ox}^{b,0} \exp\left(\frac{-A k_{Ox} t}{V}\right) \tag{13}$$

According to (13), a plot of ln i against t has a slope that is equal to $-A k_{Ox}/V (= -s_{Ox}^*)$; according to (7), a plot of ln s_{Ox}^* against the potential E of the working electrode has a slope that is equal to $-\alpha n_a F/RT$, or $-\alpha n_a/25.69$ mV^{-1} at 25°C. Alternatively, the dependence of s_{Ox}^* on potential over the rising part of a totally irreversible wave can be written in the form [22, 28]

$$E = E^0 + \frac{RT}{\alpha n_a F} \ln\left(\frac{A k_{s,h}}{V s_{Ox}}\right) - \frac{RT}{\alpha n_a F} \ln\left(\frac{s_{Ox}^*}{s_{Ox} - s_{Ox}^*}\right) \tag{14}$$

so that a plot of E against ln $s_{Ox}^*/(s_{Ox} - s_{Ox}^*)$, which is analogous to the conventional polarographic log plot, is linear and has a slope equal to $-25.69/\alpha n_a$ mV at 25°C. Either of these procedures serves for evaluating the important parameter αn_a, although the one based on (14) is superior because it is not restricted to small values of k_{Ox} (and hence to the foot of the wave). There are some processes, such as the reduction of hydrogen ion [28], for which αn_a has the same values at dropping electrodes as at large stirred mercury pools, but there are others, such as the reduction of chromium(VI) in solutions of sodium hydroxide [29], for which the differences are substantial.

In any event, (10) asserts that the current decays exponentially with time. This is one of the characteristic features of the behavior of the one-step mechanism contemplated here. (Others are listed in Section 6.1.) We may write either

$$\log i = \log i^0 - 0.434 \, s_{Ox}^* t \tag{15a}$$

or

$$\ln i = \ln i^0 - s_{Ox}^* t \tag{15b}$$

where i^0 is the current at the start of the electrolysis and is given by

$$i^0 = s_{Ox}^* nFVc_{Ox}^{b,0} \tag{16}$$

A typical current–time curve, together with the corresponding plot of $\log i$ against t, is shown in Figure 5.3.

The variation of current with time has important consequences. One is that, in view of (2), it necessitates continuous readjustment of the emf V that must be

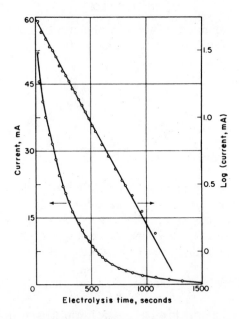

Figure 5.3 Plots of the current and of the logarithm of the current against time during an electrolysis of 100 cm^3 of the solution of Figure 5.1 and 5.2 with a stirred mercury-pool cathode having an area of 40 cm^2 and maintained at a potential of -0.5 V versus SCE. Reprinted, with permission, from L. Meites, "Controlled-Potential Electrolysis," in A. Weissberger and B. W. Rossiter, Eds., *Physical Methods of Chemistry*, Vol. 1, Part IIA, 4th ed., Wiley-Interscience, 1971.

applied to the cell in order to keep the potential of the working electrode at a constant value. The half-reaction that occurs at a platinum or graphite auxiliary electrode is usually the reduction of hydrogen ion or the solvent if the working electrode is the anode; or the oxidation of the solvent, chloride ion, or some depolarizer (such as hydrazine) that is present in large excess if the working electrode is the cathode. Current–potential curves for all these processes show that the potential of the auxiliary electrode is unlikely to vary more than a few tenths of a volt over the range of current densities likely to be involved in a typical electrolysis. Hence the difference $E_a - E_c$ between the potentials of the two electrodes is approximately constant in any one electrolysis; it is unlikely to exceed 5 V, and indeed is rarely above 2 V. However, if the initial current i^0 is even as large as 1 A and if the resistance of the cell is 100 ohms, the iR drop will be equal to 100 V at the start of the electrolysis and will decrease toward zero as the electrolysis proceeds. It is primarily to compensate for variations of the iR drop that the applied emf must be varied, and it was the difficulty of effecting the variation manually in such a way as to prevent excessive deviations of the potential of the working electrode away from the desired value that made the invention of the potentiostat [30] essential before controlled-potential electrolysis could begin to find its rightful place among the techniques that are useful in practical work.

Another consequence of the variation of current with time is that attempts to employ (1) entail integrations, with respect to time, of currents that vary over several orders of magnitude. The design of potentiostats and current integrators long claimed an excessive fraction of the energies of those concerned with the field, and for some years the papers describing new instruments probably outnumbered those describing new applications. The currently popular solutions to these problems are described briefly in Sections 3.1 and 5.2.

Equation (10) makes it possible to estimate the time that will be required to achieve any desired degree of completion when the half-reaction occurs·by the simple mechanism considered here. Suppose, for example, that the solution of Figures 5.1 and 5.2 is to be electrolyzed with a stirred mercury-pool electrode at a potential of -0.15 V versus SCE. An estimate of the value of s_{Ox}^* (which is identical with s_{Ox} because that potential lies on the plateau of the wave) could be obtained from (16). According to Figure 5.2, the current at the start of the electrolysis would be approximately 6×10^{-2} A; since $n = 2$ F/mol, $F = 9.65 \times 10^4$ C/F (A·s/F), $V = 50$ cm^3, and $c_{Ox}^{b,0} = 1 \times 10^{-6}$ mol/cm^3, s_{Ox}^* would be approximately equal to 6×10^{-3} s^{-1}. This is a fairly typical value for the conditions under which Figure 5.2 was obtained: 50 cm^3 of an aqueous solution at 25°C contained in a working-electrode compartment that was made from a 250-cm^3 Erlenmeyer flask and that also contained 30 cm^3 of mercury (giving a surface area of about 40 cm^2) efficiently stirred by a two-bladed, propeller-type stirrer rotated in the interface between the mercury and the solution. With the aid of (14) it can easily be shown that the current, and hence the concentration of quinone remaining, will decrease to a tenth of its initial value after 384 $(= 2.303/s_{Ox}^*)$ s, to a hundredth of its initial value after 768 s, and so on. About

25 minutes should suffice to reduce 99.99% of the starting material. As is true of any first-order process—and this electrolysis is a first-order process because both the concentration of the quinone and the rate at which it decreases are proportional to the current—the time required to reach any particular degree of completion does not depend on the initial concentration of the starting material. It does, however, depend on the factors that affect s_{Ox}^* in accordance with (8) and (9). These include:

1. The potential of the working electrode, because k_{Ox} increases as the plateau is approached;
2. The area of the interface between the electrode and the solution and
3. The volume of the phase containing Ox, so that one goal in designing cells for controlled-potential electrolysis is to minimize the ratio V/A, which is equal to the average depth of the phase containing the starting material;
4. The efficiency of stirring, which is related to the thickness of the Nernst diffusion layer; and
5. The temperature and viscosity of the phase containing Ox, because these affect the value of D_{Ox}, unless the electrolysis is conducted at a potential near the foot of the wave.

The temperature at which a controlled-potential electrolysis is carried out and the viscosity of the supporting electrolyte are generally governed by other considerations, but attention to the rest of these experimental variables is essential in most work.

Controlled-potential electrolyses are occasionally described as time-consuming. The reader can evaluate this assertion on the basis of the information given in the preceding paragraph. Since a potentiostat renders it unnecessary for the operator to pay any attention whatever to the electrolysis while it is in progress, he or she is needed only to prepare the solution, fill the cell, deaerate the solution, and empty and clean the cell when the electrolysis is complete. Those to whom instrument time is of exceptional importance can use ultrasonic agitation [31], or extremely efficient mechanical agitation [32] of the interface between the working electrode and the solution, to achieve values of s_{Ox} that are an order of magnitude higher than the one quoted above. However, an electrolysis may be greatly prolonged by:

1. Failure to maximize the stirring efficiency.
2. Deliberately selecting a potential near the foot of the wave that is of interest (to reveal, or exaggerate the effect of, a suspected slow chemical side reaction of an intermediate).
3. The occurrence of a very slow chemical step preceding the last electron-transfer step.

In any controlled-potential electrolysis there is a limit to the degree of completion that can eventually be reached. It is set by the Nernst equation: if N_{Ox}^0 is the number of moles of p-benzoquinone that are present at the start of the electrolysis under discussion, while f is the fraction of it that is reduced at equilibrium, (3) becomes, at 25°C,

$$E = E^{0\prime} - 0.02958 \log \left(\frac{f N_{Ox}^0 / V}{[H^+]^2 [(1-f) N_{Ox}^0 / V]} \right)$$

$$= E^{0\prime} - 0.05916 \, \text{pcH} - 0.02958 \log \left(\frac{f}{1-f} \right)$$

(17)

The formal potential of the couple is $+0.696$ V versus the normal hydrogen electrode (NHE), or $+0.455$ V versus SCE, at an ionic strength of $1\,M$ and at 25°C [33]. Introducing this value, together with $E = -0.15$ V versus SCE and the approximation pcH $= 6.8$ (where pcH denotes the negative decadic logarithm of the concentration of hydrogen ion and 6.8 is the measured pH-value) into (17) yields $\log[f/(1-f)] = 6.86$. Hence f must be virtually equal to 1 and $(1-f)$, the fraction remaining unreduced, is equal to 1.4×10^{-7}, which corresponds to $f = 0.99999986$. Similarly, it can be calculated that $f = 0.9999932$ at -0.10 V versus SCE, or 0.99967 at -0.05 V versus SCE. The first of these values is so close to 1 that electrolysis at a potential more negative than -0.15 V versus SCE would not be worth considering unless an enormously large excess of the quinone had to be even more nearly completely reduced prior to the detection or determination of some more difficultly reducible constituent of the solution. The last value is certainly high enough to be considered quantitative, but electrolysis at any potential less negative than -0.05 V versus SCE should not be considered for any purpose except the very specialized one described in Section 4.3. Virtually identical values could have been obtained by taking the half-wave potential of the polarographic or voltammetric wave to be equal to the formal potential *in the medium used*—that is, to $E^{0\prime} - 0.05916 \, \text{pcH}$. This is not exact (because it ignores the difference between the diffusion coefficients of the oxidized and reduced forms) and is permissible only if the wave is known to be reversible, but has the advantage that a value of the half-wave potential is far more likely to be available than one of the formal potential.

If the reduction of Ox is irreversible, it will not proceed at an appreciable rate unless the potential of the working electrode is considerably more negative than the formal potential in the medium used. However, the Nernst equation still governs the equilibrium that is eventually established. These things mean that a totally irreversible half-reaction will proceed to completion if it occurs at all, and that complete reduction can be attained even at a potential near the foot of the voltammogram. Figure 5.4 shows two different cases. Curve a represents the reduction of the hexaaquachromium(III) ion, and Curve b represents the oxidation of the hexaaquachromium(II) ion produced by that reduction [34]. These curves show that it is possible to reduce this species of chromium(III) quantitatively even at a potential as positive as -0.8 V versus SCE. At that potential the reduction would be extremely slow because the current is much smaller than the limiting current, which must be approximately equal to 350 mA even though the final rise of current due to the reduction of hydrogen ion is so

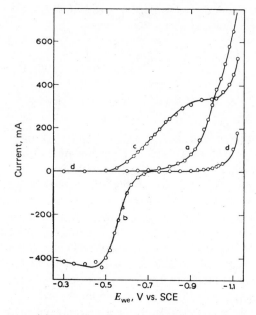

Figure 5.4 Current–potential curves obtained with a stirred mercury-pool electrode having an area of 40 cm^2 in 6 M hydrochloric acid solutions containing (a) 14 mM hexaaquachromium(III); (b) 14 mM chromium(II), prepared by the controlled-potential electroreduction of solution (a) at -1.10 V versus SCE; and (c) 14 mM tetraaquadichlorochromium(III), prepared by the controlled-potential electrooxidation of solution (b) at -0.40 V versus SCE. Curve (d) represents the residual current obtained with 6 M hydrochloric acid alone. Reprinted, with permission, from L. Meites, "Controlled-Potential Electrolysis," in A. Weissberger and B. W. Rossiter, Eds., *Physical Methods of Chemistry*, Vol. 1, Part IIA, 4th ed., Wiley-Interscience, 1971.

close to the chromium wave that there is only a bare suggestion of an inflection around that current. Nevertheless, such a reduction would eventually approach completion because the rate of reoxidation of chromium(II) at -0.8 V is entirely insignificant.

Under the conditions of Figure 5.4 the electrolytic oxidation of the hexa-aquachromium(II) ion yields the tetraaquadichlorochromium(III) ion as the predominating product [35, 36], and Curves *b* and *c* therefore represent the behavior of the couple

$$Cr(OH_2)_4Cl_2^+ + 2H_2O + e = Cr(OH_2)_6^{2+} + 2Cl^-$$

which is much more nearly reversible than the couple corresponding to Curve *a*. The reduction of the dichlorochromium(III) species will proceed quantitatively, and with reasonable speed, at any potential more negative than about -0.8 V versus SCE, and the oxidation of chromium(II) will be quantitative at any potential more positive than about -0.45 V, but either process will be incomplete and will yield a mixture of the two oxidation states if it is carried out at any intermediate potential. If one wished to reduce the $Cr(OH_2)_4Cl_2^+$ ion quantitatively with the least possible interference from the reduction of hydrogen ion (the process responsible for the final current rise, which appears to begin at about -0.9 V, on Curve d), it would be better to perform the electrolysis at -0.8 V than on the plateau (at, say, -0.95 V), even though it would proceed less rapidly.

Electrolysis on the rising part of an irreversible wave often serves to bring about a desired half-reaction while minimizing the extent of another that occurs with only slightly less facility. This has the corollary that, if an irreversible cathodic process is responsible for the final current rise, it will proceed at a finite rate even at a potential so positive that no increase of current due to that process can be detected on inspecting the voltammogram: even if the $Cr(OH_2)_4Cl_2^+$ ion is reduced at -0.8 V, as was recommended above, a small current due to the reduction of hydrogen ion will flow throughout the electrolysis, and will continue to flow after the reduction of the chromium is complete. Section 5 will discuss the phenomenon further and will show how the necessary correction can be applied in controlled-potential coulometry, but there are implications for synthetic work as well. Suppose that a substance gives two irreversible waves and that it is desired to prepare the product of the half-reaction that is responsible for the first one. Especially if the waves are fairly close together, it may be dangerous to attempt an electrolysis at a potential on the plateau of the first wave, for the second half-reaction may occur at a significant rate at such a potential, and its intermediates and products may react with those of the desired process to yield final products having strange and misleading natures.

3   APPARATUS FOR CONTROLLED-POTENTIAL ELECTROLYSIS

3.1   Electrical Apparatus

The preceding section showed that controlled-potential electrolysis involves monitoring the difference of potential between the working electrode and the reference electrode, and maintaining that difference at a nearly constant value by adjusting the emf imposed across the working and auxiliary electrodes. The variation that can be tolerated depends on the current–potential curve for the system and on the purpose of the electrolysis. Under the conditions of Figure 5.2, fluctuations as large as ± 0.8 V around a nominal working-electrode potential of -0.75 V versus SCE could be tolerated if the purpose was simply to obtain a quantitative yield of the hydroquinone. At an instant when the potential lies at the positive extreme of this range, the rate of reduction of the quinone would be smaller than it is on the plateau and, because the half-reaction is reversible under these conditions, there might even be some reoxidation of the hydroquinone that had been formed during the earlier stages, but these things would merely prolong the electrolysis a little. At an instant when the potential lies at the negative extreme, a large fraction of the current would be consumed in reducing dihydrogen phosphate ions but, because hydrogen has no effect on any of the substances involved in the half-reaction, its formation would merely increase the electric bill a little. If the purpose were to determine the quinone by measuring the quantity of electricity consumed in reducing it, the reduction of dihydrogen phosphate ion would lead to a positive error, and its rate would have to be minimized by avoiding excursions to potentials more negative than about -0.8 V versus SCE, or an even less negative value if the concentration of quinone were very low. In either event the allowable range would be further decreased by the presence of anything else that could be reduced. Sometimes the wave that is of interest is so closely followed by another one, which may be due either to another constituent of the solution or to a subsequent step in the reduction or oxidation of the substance of interest, that control to ± 0.05 V may be essential. In coulometric work under conditions such that the solvent or supporting electrolyte is oxidized or reduced at an appreciable rate, measurement of and correction for the continuous faradaic background current (Section 5.1) will be difficult and unsatisfactory unless the potential is controlled within ± 10 mV. For some specialized purposes, including the evaluation of a formal potential in the fashion described by Section 4.3, it may even be necessary to achieve control within ± 1 mV.

Especially with mercury-pool working electrodes, the precision of control is usually less important than the speed with which control is regained after a disturbance. So that the durations of electrolyses will be as short as possible, it is common to use a propeller-type stirrer partly immersed in the mercury and rotated at such a rate that droplets of mercury are almost, but not quite, detached from the surface and thrown through the solution. Rapid fluctuations of the surface area are easy to see, and they persist even if the stirrer is rotated much less rapidly. Suppose that a mercury pool has exactly the desired potential at some instant, and that its area decreases suddenly in the next instant. The decrease will be accompanied by an instantaneous change of the efficiency of stirring and consequently of the values of s_{Ox}, i, and iR. Unless the potentiostat responds instantaneously, the change in the value of iR will cause the value of $E_a - E_c$ to undergo an equal but opposite change. The potential of the working electrode is driven away from the desired value until the potentiostat reasserts control.

Other important characteristics of a potentiostat are the emf V and the power Vi that it can supply to the electrolytic cell. Because, as was shown above, the iR term in (2) usually overwhelms the difference of potential $E_a - E_c$, the latter will be ignored in this discussion. Under typical conditions the limiting current is roughly equal to $50nc_{Ox}$ mA at a mercury pool, where c_{Ox} is the concentration of the electroactive species in *mmol/dm³* and n is the number of electrons involved in reducing or oxidizing each mole of it. A 10 mM solution of a substance that undergoes a two-electron reduction may be expected to yield an initial current of about 1 A in an electrolysis performed on the plateau of its wave. The resistance R of the cell will depend partly on its design and partly on the nature and concentration c of the supporting electrolyte with which it is filled. For cells like the one shown in Figure 5.8, a crude but useful approximation is

$$R = \frac{3 \times 10^4}{\Lambda c} \tag{18}$$

where c is the concentration and Λ is the equivalent conductance of the supporting electrolyte. With $1\,M$ potassium chloride (for which $\Lambda = 150\,cm^2/$ ohm·equivalent), this yields $R = 200$ ohms, so that an applied emf of about 200 V, and consequently a power of about 200 W, would be needed at the outset. Few commercial potentiostats can supply 200 W to the cell, or can provide 200 V at any current; even if they could, drastic measures would have to be taken to prevent the solution from boiling in the face of such a power dissipation. This is one reason for using relatively concentrated supporting electrolytes whenever possible; it is an even more compelling reason for avoiding high concentrations of the electroactive substance. In the electrolysis being considered, the initial current would be only about 0.1 A, and the power required would be only about 20 W, if the initial concentration were 1 mM rather than 10 mM; these figures are well within the capabilities of commercially available potentiostats. If, for some reason, the concentration of potassium chloride could not exceed 0.1 M, it would be best to keep the initial concentration of the electroactive substance down to about 0.1 mM. Of course there is some elasticity in these figures, but the problems are greatly exacerbated in non-aqueous solvents, in which relatively high resistances are not uncommon.

The danger is that the potentiostat may be unable to keep the working electrode at the desired potential throughout the electrolysis. Suppose that one wants to perform a reduction on the plateau of a wave, but that the limiting current is initially so large that the value of V in (2) exceeds the maximum available. The working electrode will assume a potential on the rising part of the wave. As the electrolysis proceeds, the concentration of the electroactive substance and the limiting current will decrease, the current that can be forced through the cell will become equal to a larger fraction of the limiting current, and the potential of the working electrode will drift toward the desired value. The current will be nearly constant until the desired value is attained; only then will it begin to decrease in the expected way. Figure 5.5 shows several current–time curves that illustrate these phenomena.

Sometimes such behavior is innocuous, but sometimes it is not. If there are no possible side reactions, as in the reduction of cadmium(II) to cadmium amalgam in an aqueous medium, the fact that part of the electrolysis is carried out on the rising portion of the wave can have no effect except to prolong the electrolysis. On the other hand, the reduction of benzaldehyde at -1.5 V versus SCE (which lies on the plateau of its wave) at pH 3 gives a quantitative yield of benzyl alcohol by a mechanism that can be abbreviated as

$$\phi CHO + H^+ + e \rightarrow \phi \dot{C}HOH$$
$$\phi \dot{C}HOH + H^+ + e \rightarrow \phi CH_2OH$$

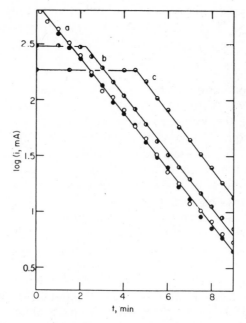

Figure 5.5 Effect of a limited voltage output on the current–time curve in controlled-potential electrolysis. The data were secured in oxidations of vanadium(II) to vanadium(III) at a mercury anode in 4 M hydrochloric acid. The cell resistance was 18.8 ohms; higher values were secured by connecting a resistance box in series with the cell. The nominal control potential was -0.30 V versus SCE, but the potentiostat was adjusted to provide a maximum output voltage of only 5.5 V. The total resistance was (a) 18.8 (open circles) or 25 ohms (solid circles), (b) 45 ohms, and (c) 75 ohms. Reprinted, with permission, from L. Meites, "Controlled-Potential Electrolysis," in A. Weissberger and B. W. Rossiter, Eds., *Physical Methods of Chemistry*, Vol. 1, Part IIA, 4th ed., Wiley-Interscience, 1971.

At a potential on the rising portion of the wave, however, the intermediate can no longer be reduced as rapidly as it is formed at the surface of the electrode. Instead, part of it dimerizes to yield hydrobenzoin:

$$2\phi\dot{C}HOH \longrightarrow \begin{array}{c} \phi CHOH \\ | \\ \phi CHOH \end{array}$$

A mixture of products is obtained, the yield of benzyl alcohol is less than quantitative, and the value of n obtained coulometrically is not only nonintegral but also dependent on the experimental conditions. The literature contains many examples of the confusion and error that can result.

The principle of operation of a potentiostat can be gleaned from Figure 5.6.

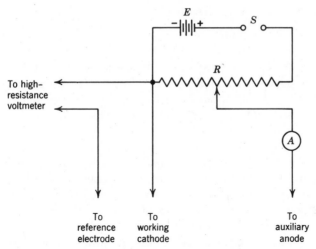

Figure 5.6 Simple manual circuit for controlled-potential electrolysis. *E* is a storage battery or other dc source free from appreciable ac ripple; *S* is a single-pole–single-throw on-off switch; *a* is a milliammeter; and *R* is a heavy-duty rheostat having a resistance of about 10 ohms. Reprinted, with permission, from L. Meites, "Controlled-Potential Electrolysis," in A. Weissberger and B. W. Rossiter, Eds., *Physical Methods of Chemistry*, Vol. 1, Part IIA, 4th ed., Wiley-Interscience, 1971.

The difference between the potentials of the working and reference electrodes is observed by means of a digital or other voltmeter, and the emf applied across the working and auxiliary electrodes is adjusted, by means of a rheostat acting as a voltage divider, in such a way as to keep the observed difference of potential constant and equal to the desired value. An ammeter connected between the auxiliary electrode and the rheostat will provide a visual indication of the progress of the electrolysis. One could of course replace the ammeter with a standard resistor across which there is connected either a strip-chart recording potentiometer (to provide a graphical record of the current–time curve) or an A/D converter (to permit storing current–time data in the core memory of a computer). Coulometric measurements can be made by connecting a current integrator in series with, or in place of, the ammeter.

As was mentioned above, human beings are scarcely capable of the speed and precision that is needed to obtain acceptable control with this simple circuit under any but the most favorable conditions. The principles, design, construction, and use of electronic potentiostats, which can maintain such control automatically even under adverse conditions, have been well and thoroughly reviewed by Fisher in 1974 [37]. Analog instruments have reigned supreme for many years and have undergone little if any further improvement since Fisher's review was written. However, recent developments in other areas of scientific instrumentation make it safe to predict that they will give way to digital [38–41] or hybrid microprocessor-controlled analog potentiostats [42–46] in the not too far distant future.

3.2   Cells and Reference Electrodes

Nearly every user of controlled-potential electrolysis forms his or her own ideas and prejudices about the features that render a cell useful and convenient, and no good purpose would be served by discussing all the cells that have been described. This subsection therefore emphasizes the functions that cells must serve and the principal difficulties that arise in designing them, and it gives only brief descriptions of a very few cells that have proven to be especially suitable for a number of different purposes.

Figure 5.7 shows the very simple cell employed by Lingane [47] for controlled-potential electroseparations and electrogravimetric determinations of metal ions by deposition onto a platinum cathode. Concentric cylinders of platinum gauze serve as the working and auxiliary electrodes, a saturated calomel electrode is used as the reference electrode, and the solution is stirred by means of a magnetic stirrer and stirring bar.

Because they are simple, inexpensive, and easy to use, such single-compartment cells are very convenient for the purposes mentioned, but they have two interrelated drawbacks that render them unsuitable for most other purposes. One is that the product of the half-reaction occurring at one electrode may be electroactive at the other. Suppose that a dilute aqueous solution of a metal-ion sulfate, nitrate, or perchlorate is electrolyzed with a platinum anode. Oxygen is

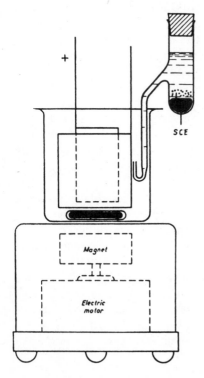

Figure 5.7   Cell for controlled-potential separations and determinations of metals with a platinum working electrode. Reprinted, with permission, from L. Meites, "Controlled-Potential Electrolysis," in A. Weissberger and B. W. Rossiter, Eds., *Physical Methods of Chemistry*, Vol. 1, Part IIA, 4th ed., Wiley-Interscience, 1971.

evolved, and convection and diffusion bring some of it to the surface of the cathode, where it is reduced (to hydrogen peroxide, water, or a mixture of the two, depending on the potential of the cathode and the material from which it is made). A current, arising from the oxidation of water at one electrode and the reduction of oxygen at the other, will continue to flow long after all of the metal ion has been removed from the solution. Similarly, in a solution containing a high concentration of chloride ion, chlorine is formed at the anode and reduced back to chloride ion at the cathode. The electroactive substance itself may give rise to a cyclic process if the half-reaction occurring at the working electrode yields a soluble product: the reduction of iron(III) at a working cathode might yield iron(II), which would be reoxidized at the anode. All such processes eventually reach a steady state, in which the intermediate is consumed at one electrode just as rapidly as it is formed at the other. Noncyclic processes are also possible: in the electrolysis of a solution of nitrobenzene it is possible to establish conditions under which the nitrobenzene is reduced to phenylhydroxylamine at one electrode while the latter is oxidized to nitrosobenzene at the other. The nitrobenzene and phenylhydroxylamine would eventually be exhausted and the current would decay toward zero, but the composition of the resulting solution would have little obvious bearing on the nature of the half-reaction that had occurred at the working electrode. Coulometric measurements are impossible in all such circumstances.

Equal or greater confusion may arise when an intermediate formed at the working electrode can react with a product or intermediate formed at the auxiliary electrode. Such reactions are especially to be feared in work with nonaqueous solutions, in which anionic radicals formed at one electrode are sure to react with cationic ones formed at the other if they are allowed to mix.

In the controlled-potential electrodeposition of a metal these problems are unimportant, but for almost any other purpose it is better to use a cell in which the solutions surrounding the cathode and anode are separated by a diaphragm. Such a cell also makes it possible to exclude oxygen from the solution in the working-electrode compartment, which is essential in coulometric work and is a wise precaution in most other applications. Unless the potential of the working electrode is so positive that oxygen cannot be reduced at an appreciable rate, it is customary to pass a rapid stream of oxygen-free nitrogen or argon through that solution by way of a sintered-glass gas-dispersion cylinder, both for some time before the electrolysis is begun and throughout its whole duration. The following subsection includes a discussion of the meaning of "oxygen-free" in this connection.

One widely useful double-diaphragm cell is shown in Figure 5.8 [48]. The solution in the working-electrode compartment contains the substance that is of interest; the other two compartments usually contain the supporting electrolyte alone. If the solution contains chloride ion at a high concentration and if the working electrode is the cathode, it is advisable to add a little hydrazine [49] to the auxiliary-electrode compartment to prevent anodic attack on the auxiliary electrode. The level of the solution in the auxiliary-electrode com-

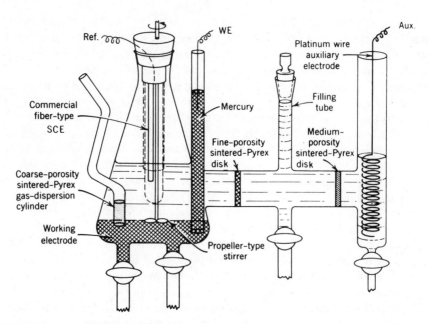

Figure 5.8 Double-diaphragm cell for controlled-potential electrolyses with a mercury working electrode. Reprinted with permission, from L. Meites, "Controlled-Potential Electrolysis," in A. Weissberger and B. W. Rossiter, Eds., *Physical Methods of Chemistry*, Vol. 1, Part IIA, 4th ed., Wiley-Interscience, 1971.

partment should be slightly higher than that of the solution in the working-electrode compartment to avoid loss of the electroactive substance by bulk flow of the solution out of the latter compartment. The reference electrode may be a commercial saturated calomel electrode of the type sold for use with pH meters, provided that neither potassium chloride nor any other substance will precipitate in the liquid junction when the electrode is brought into contact with the solution being electrolyzed. Precipitation leads to an increase of the resistance between the working and reference electrodes, and eventually to inability of the potentiostat to maintain control. The danger is least with electrodes in which the junction is made by an asbestos fiber; with palladium-junction electrodes it is so great that these should not be used. If the solution in the working-electrode compartment contains silver ion or a high concentration of a non-aqueous solvent, of hydrochloric acid, or of perchlorate ion, it is better to use a silver–silver chloride electrode of the type shown in Figure 5.9 [50]. This has the advantage that its compartments can be filled with a solution that will minimize the danger of precipitation. If, for example, the solution being electrolyzed contains much perchlorate ion, the potassium chloride normally present in the electrode and bridge compartments of the reference electrode can be replaced with sodium chloride. Similarly, 50% ethanol saturated with lithium chloride might be used in work with solutions containing 50% ethanol. Some solutions contain ions, such as cyanide and sulfide, that would

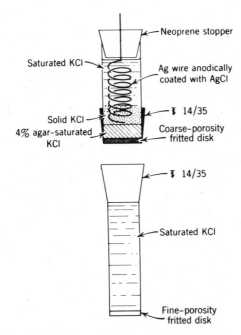

Figure 5.9 Silver–silver chloride saturated potassium chloride reference electrode and bridge. The lower (bridge) compartment serves to prevent contamination of the solution in the upper (electrode) compartment by constituents of the solution being electrolyzed, even in a very prolonged electrolysis. The solution in the bridge compartment is discarded and replaced after each electrolysis. Reprinted, with permission, from L. Meites, "Controlled-Potential Electrolysis," in A. Weissberger and B. W. Rossiter, Eds., *Physical Methods of Chemistry*, Vol. 1, Part IIA, 4th ed., Wiley-Interscience, 1971.

affect the potential of the reference electrode if they came into contact with it, and this can be prevented, even in very long experiments, by filling the bridge compartment with saturated potassium chloride. Whatever its design, the reference electrode should be placed in the cell in such a way that the liquid junction is made as near the surface of the working electrode as possible, for some fraction of the iR drop between the auxiliary and working electrodes occurs between the tip of the reference electrode and the surface of the working electrode in most cells and causes a corresponding emf to appear in the control circuit. The danger is especially great in work with solvents having low dielectric constants, in which high conductivities cannot be attained.

The blades of a propeller-type stirrer should be about half immersed in the mercury, and the stirring motor should usually be adjusted to a speed that is just barely insufficient to throw an occasional droplet of mercury up into the solution. These dispositions are important because they govern the value of s_{Ox} and thus the rate of the overall process. Under properly arranged conditions s_{Ox} is typically equal to about $6 \times 10^{-3} \, s^{-1}$ with 50 cm^3 of solution in this cell, so that 99.9% completion will be attained within about 20 min if the system exhibits simple behavior. Although (8) is by no means completely reliable—for example, doubling the area of the working electrode without changing the stirring arrangements does not double the value of s_{Ox}, as the equation predicts that it should, because stirring is less efficient around the periphery of the electrode than in the immediate vicinity of the stirrer—it does accurately describe the relation between s_{Ox} and V [51], and therefore s_{Ox} may be increased to about 0.01 s^{-1} by using only 35 cm^3 of solution. A four-bladed paddle-type stirrer gives values of s_{Ox} about 50% higher when used in the same way [26], while values as high as 0.1 s^{-1} should be attainable with ultrasonic agitation and in other ways [31, 52–55].

Cells like the one shown in Figure 5.8 suffer from two disadvantages that are important in certain kinds of work. One, which was mentioned in the preceding paragraph, is that the efficiency of stirring is not constant over the entire surface of the working electrode. Hence the thickness of the Nernst diffusion layer is greater near the walls of the cell than in its center, and the length of time that is required for an ion or molecule of any product of an electron-transfer step to traverse the diffusion layer depends on where it is formed. If the mechanism is branched and includes a second- or pseudo-second-order step, as does the one

$$Ox + n_1 e \rightarrow I$$
$$2I \rightarrow P$$
$$I + n_2 e \rightarrow Red$$

the fraction of the I that undergoes dimerization increases with the time that is spent in the diffusion layer, where, during most of the electrolysis, its concentration is higher than in the bulk of the solution. The proportions of P and Red in the final product depend on the overall efficiency of stirring, and are not the same at different points on the surface of the electrode [56].

The other disadvantage of such cells is that different points on the surface of the working electrode are at different distances from the auxiliary electrode. The surface of the working electrode is equipotential, but there is an iR drop across the working-electrode compartment through the layer of solution in contact with the surface of the electrode. Consequently, there is a variation of the difference of potential between a point on the surface of the electrode and an adjacent point in the solution, and it is this difference of potential that is generally known as the "potential" of the electrode. If the working electrode is the cathode, the potential will be less negative at the left-hand side of the pool, and more negative at its right-hand side, than it is in the immediate vicinity of the tip of the reference electrode. The resulting phenomena were first discussed by Booman and Holbrook [57] and by Harrar and Shain [58], and Harrar's summary of them [59] is both illuminating and thorough. The variation of potential increases with both the iR term in (2) and the ratio of the distances between the auxiliary electrode and the farthest and nearest points on the surface of the working electrode; and the effect of the variation becomes more important as the permissible range of potentials becomes narrower. If the nominal control potential is on the plateau of the voltammetric wave but very near the final current rise, reduction of the supporting electrolyte may occur at an appreciable rate at the portion of the working electrode having the most negative potential near the beginning of the electrolysis, when the iR drop through the cell is large. As the electrolysis proceeds and the iR drop decreases, the potentials at points nearest the auxiliary electrode become less negative, and the rate of reduction of the supporting electrolyte in that region decreases to the value that corresponds to the nominal control potential. An excess of electricity is consumed, and an accurate correction for the error cannot be made. Conversely, if the nominal control potential is on the plateau but near the rising part of the wave, the potentials at points farthest from the auxiliary electrode may lie on the rising part of the wave for some time after the beginning of the electrolysis, and approach the desired value on the plateau as the iR drop is decreased by depletion of the starting material. If the half-reaction can follow only a single course, there will be no error in the quantity of electricity consumed, although the rate constant s_{Ox}^{*} will vary with time and a plot of $\ln i$ against t will not be accurately linear [60]. However, electrolysis on the rising part of a wave does not always give the same product as electrolysis on the plateau: on the rising portion, an electroactive intermediate can escape from the surface of the working electrode and undergo some side reaction in the bulk of the solution, while on the plateau it cannot escape because it will be further reduced or oxidized as rapidly as it is formed. There will again be an error in the quantity of electricity that is consumed, correction for the error will again be impossible, and the interpretation of the results will be complicated by the fact that a mixture of products will be obtained. The least favorable situation is of course that in which one is attempting to perform an electrolysis on a plateau so short that the iR drop causes part of the surface of the electrode to have a potential on the rising part of the wave that is of interest, and another part to have a potential that encroaches on the unwanted wave that follows.

These problems can be minimized by keeping the iR drop as low as possible, but can be avoided only by means of a cell in which the working electrode is everywhere equidistant from the auxiliary electrode. This is much easier to do with solid working electrodes than with mercury pools. Booman and Holbrook [57] designed a cell in which the center of the pool is occupied by a solid stirrer, so that electrolysis proceeds only at the surrounding ring-shaped area, and in which a flat circle of platinum wire is placed in a toroidal auxiliary-electrode compartment located directly above the ring. A further advantage of this arrangement is that it gives rise to less fluctuation of the electrode area, and therefore to less noisy current–time curves, than a propeller-type stirrer does. Jones, Shults, and Dale [61] placed a platinum-wire auxiliary electrode in an unfired Vycor tube above the mercury pool but a little to one side of the shaft of the stirrer, which was located along the vertical axis of the cell. Karp [32] combined the features of these cells by bringing the shaft of a stirrer, like that used by Booman and Holbrook, up through the bottom of the cell, which made it possible to place an auxiliary-electrode compartment like that used by Jones,

Shults, and Dale directly above the center of the stirrer. Ordinary magnetic stirring bars are difficult to center on the surface of a mercury pool, and one that wanders away from the center is likely to stick on, and may even break, the tip of the reference electrode. Nevertheless, Shia [60], following Moinet and Peltier [62], was able to use a "floating" stirring bar (e.g., Markson Science, Inc., Del Mar, California, Catalog No. A-3759) together with a large spiral of platinum wire mounted directly above the center of the pool and separated from it by a fine-porosity fritted disk, and found that this arrangement gave both reasonably high values of s_{Ox} and reasonably low noise levels in the measured currents. The systems characteristics of all these cells have been studied by Harrar and Pomernacki [63], whose paper should be consulted for details.

3.3   Auxiliary Equipment

For the sake of simplicity, many controlled-potential electrolyses can be performed in cells open to the air. Even at a potential where oxygen can be reduced at the working electrode, its presence may have no effect on the completeness of a separation or on the course and yield of an electrosynthesis. When this is true, and when the effects of its reduction on coulometric data and current–time curves are unimportant, there is no need to exclude it. Nevertheless, it cannot be presumed to be innocuous at potentials where it is not reduced, for it is thermodynamically a much more powerful oxidizing agent than its voltammetric properties suggest. It can therefore be reduced at an appreciable rate by an intermediate or product even though its direct electrolytic reduction is immeasurably slow. For example, coulometric data indicate that 2 faradays are consumed in oxidizing 1 mole of anthracene dissolved in acetonitrile [64–68]. However, bianthrone, whose formation corresponds to a three-electron process, is obtained as the product, apparently as the result of a reaction between atmospheric oxygen and the substance formed by the electrooxidation. In general, it is unsafe to allow oxygen to come into contact with the solution, either during electrolysis itself or during the subsequent work-up, unless its presence is shown to have no deleterious effects.

Most experimenters remove dissolved oxygen by merely bubbling "prepurified" or "Seaford-grade" nitrogen through the solution, and seem to assume that the resulting concentration of dissolved oxygen is too low to have any significant effect. Whether or not this is justifiable depends on what one means by "significant." If the gas contains 10 ppm of oxygen, the equilibrium concentration of dissolved oxygen will be about 10^{-8} M. Assume that 60 cm^3 of a solution containing this concentration of dissolved oxygen is electrolyzed with a mercury cathode at a potential on the plateau of the second oxygen wave, where $n=4$ for the reduction of oxygen to water. The typical value of s_{Ox} cited above can be used to estimate that a steady current of about 1.2 μA will flow because of the reduction of oxygen; if the electrolysis proceeds for a total of 2000 s, about 25 nF of electricity will be consumed in that reduction. If the original solution were 1 mM in a substance Ox which undergoes the half-reaction $Ox + 2e = Red$, the total initial current would be about 60 mA and the total quantity of electricity consumed would be 120 μF. The current consumed by oxygen would produce no noticeable upward concavity in a plot of log i against t until after the total current had decayed through several decades, and the corresponding consumption of electricity would lead to a relative error of only 0.02% in the coulometric determination of Ox. If the initial concentration of Ox were 1 μM, the corresponding coulometric error would be 20%; it might just be possible to apply a correction accurate enough to avert tragedy. A worse danger is that the reaction between oxygen and the reduced form Red may be rapid enough to consume most of the oxygen that is passed into the solution; if the rate of flow of gas were, say, 300 cm^3/min in the above electrolysis, a total of 40 μmol of oxygen would enter the cell during 2000 s. If the stoichiometry of the reaction were $2Red + O_2 = 2Ox + 2H_2O$, the coulometric error would be about 60% with the 1 mM solution and several orders of magnitude with the 1 μM one if the above assumptions could be taken at face value. As the current would decay only to a value that would be very little smaller than the initial current for the 1 μM solution, the difficulty would be even worse than these calculations indicate.

In such circumstances it is necessary to remove even the trace of oxygen present in prepurified nitrogen. This may be done by passing the gas either over hot copper or through a solution of chromium(II) chloride. Passage through a 50 cm Vycor or quartz tube, electrically heated to 450–500°C and filled with copper turnings, decreases the oxygen content of a stream of nitrogen to an undetectably low value. The surface of the copper must be regenerated occasionally by passing tank hydrogen through the column for about 10 min; convenience and safety are best combined by doing this whenever the tank of nitrogen is changed.

Solutions of chromium(II) for removing oxygen from gas streams may be prepared by placing about 25 g of heavily amalgamated 20-mesh zinc in a 250-cm^3 Corning 31760 gas-washing bottle having a coarse- or medium-porosity fritted disk, adding about 150 cm^3 of 2 M hydrochloric acid containing 0.05 M potassium dichromate, stoppering tightly, and letting the mixture stand for 1 or 2 days. During this time the dichromate is reduced to chromium(II) ion by the zinc, and the solution is ready for use as soon as the characteristic blue color of that ion has appeared. A single wash bottle of this sort suffices for all purposes except coulometry on the microgram or submicrogram scale, for which two or three should be used. A similar wash bottle filled with water should follow the last solution of chromium(II) chloride to trap droplets containing chromium(II) chloride that might otherwise be sprayed over into the cell. Solutions of chromium(II) chloride deteriorate slowly on standing, turning brownish and then depositing precipitates, but may easily be regenerated by adding 10–20 cm^3 of concentrated hydrochloric acid. Solutions of vanadium(II) chloride serve the same purpose but are slightly less efficient; solutions of vanadium(II) sulfate [69] have been very widely adopted, but are less suitable because they evolve hydrogen sulfide slowly on standing.

The gas train should be constructed entirely from Tygon tubing and glass; rubber, polyethylene, and other porous materials are unsuitable.

Procedures for purifying mercury for use in polarography and other electrochemical techniques have been described by several authors [11, 70, 71]. Used mercury may be returned to the supplier for reprocessing or may be distilled to remove metals such as platinum and gold, which defy chemical separation. However, the first course is too expensive, and the second is too hazardous and time-consuming, to be adopted in a laboratory where much controlled-potential electrolysis is done and where 5–10 kg of mercury may be used each day. Of the simpler procedures that have been suggested, the writer naturally prefers his own [70], in which dissolved base metals are oxidized, first by efficient agitation of the dry mercury in air and then by prolonged aspiration of air through the mercury in contact with 1 M perchloric acid. This should be done routinely if base metals are deposited into the mercury. When they are not, as is true in much work with organic compounds, it probably suffices to wash used mercury with solvents that will dissolve those compounds and also the ionic constituents of the supporting electrolytes, then dry it thoroughly and pinhole it once or twice. Occasional more thorough purification is advisable to prevent the accumulation of excessive amounts of electroactive metals (such as lead, zinc, and the alkalies and alkaline earths), whose ions are present as trace impurities in most of the reagents from which supporting electrolytes are prepared. If a scum forms on the surface of mercury after it has been purified, the purification was grossly insufficient. It has been reported that mercury becomes contaminated with surface-active impurities on standing in polyethylene containers, and purification and storage should therefore be done in glass.

Most controlled-potential electrolyses are performed at ambient temperature. In work at other temperatures, in large-scale synthetic work with a diaphragm cell (where there is danger of excessive electrical heating), or in a kinetic study requiring close control of the temperature, it is advantageous to use a water-jacketed cell in conjunction with an external thermostat. Rechnitz [26] described a cell suitable for these purposes.

3.4   General Technique

Controlled-potential electrolyses are generally carried out in the following fashion. The directions refer explicitly to the use of the double-diaphragm cell shown in Figure 5.8; the modifications necessary for work with other cells are obvious.

One first prepares about 200 cm^3 of the supporting electrolyte and, in 50–100 cm^3 of that solution, dissolves an appropriate amount of the substance that is to be examined. The optimum concentration of that substance is usually between 10^{-4} and 10^{-3} M. If the solution is much more dilute, isolation and identification of the product are sure to be difficult; if it is much more concentrated, the iR drop through the cell may be so large that the potentiostat cannot bring the working electrode to the desired potential at the start of the electrolysis. Controlled-potential electrolyses are often carried out in mixtures of alcohols and other non-aqueous solvents with water; solutions in such solvents should be buffered, but the solubilities of many salts in them are so low that substantial buffer capacities cannot be secured. Appreciable variations of pH may then occur. In the bulk of the solution, these can be prevented by using an autotitrator to add acid or base to the solution in the working-electrode compartment at a rate that will just compensate for the consumption of hydrogen or hydroxide ion by the electrolytic reaction [72]. However, even this expedient cannot suffice to maintain a constant pH in the layer of solution that is immediately adjacent to the surface of the working electrode, for when hydrogen ions are produced or consumed in the half-reaction the difference between their concentrations in that layer and in the bulk of the solution depends on the current density and therefore cannot remain constant throughout an electrolysis.

Then the central compartment of the cell is filled with the supporting electrolyte and is tightly stoppered, and the auxiliary-electrode compartment is filled with the same solution to a level that will be just above the final level of the solution in the working-electrode compartment. A suitable volume of the solution containing the substance to be examined is added to the working-electrode compartment, the stirrer is started, and the solution is deaerated by passing a rapid stream of oxygen-free gas through it for 5–10 min. Meanwhile the potentiostat is turned on, allowed to warm up, and adjusted so that it will provide the desired control potential. This generally must be deduced from polarograms or other voltammetric curves, which must have been obtained under conditions—temperature, pH, nature of the buffer, ionic strength, solvent, and so on—as nearly as possible identical to those that will prevail during the electrolysis.

When the deaeration is judged to be complete, about 30 cm^3 of mercury is added to the working-electrode compartment. To ensure that the platinum lead-in wire is completely covered with mercury, it is often advisable to tilt the cell slightly. Deaeration is continued to remove the air that entered the cell during the addition of the mercury, and the position and speed of the stirrer are meanwhile adjusted as described above. Finally, the potentiostat is connected to the cell and the electrolysis is allowed to proceed without further attention.

Trouble often arises, especially (but not only) in coulometric work, from the presence of impurities in the solvents and reagents from which supporting electrolytes are prepared. The most ubiquitous of these are peroxides in such solvents as dioxane and tetrahydrofuran; electroactive heavy-metal ions, such as lead(II) and zinc(II), in inorganic salts; and alkali-metal ions in tetraalkylammonium salts. All these can usually be eliminated by pre-electrolyzing the supporting electrolyte alone *at the same potential* as will be used for the subsequent reduction or oxidation of the substance that is of interest. This is done by the procedure just described, and is continued until the current flowing through the cell has decreased to a negligible or constant value. Then the potentiostat is disconnected from the cell, a concentrated stock solution of the substance to be examined is added to the working-electrode compartment, the mixture is allowed to stand for a few minutes while dissolved oxygen is removed from it by the continued flow of the stream of gas, and finally the potentiostat is reconnected to the cell to begin the actual electrolysis.

Minor deviations from this procedure are occasionally necessary or desirable. If one wants to remove traces of, say, sodium ion from a supporting electrolyte prior to a reduction at a potential close to the half-wave potential of sodium ion, it is clearly advantageous to perform the pre-electrolysis at a slightly more negative potential, where the removal of sodium ion will be nearly complete. The resulting sodium amalgam should be removed from the cell and replaced with fresh mercury before beginning the subsequent electrolysis, for otherwise it will react with the substance being examined or with oxygen dissolved in the stock solution of that substance.

The current during a pre-electrolysis may not decrease to a negligible value. It certainly cannot do so if the potential of the working electrode corresponds to a point on either the initial or the final current rise on a voltammogram of the solution. The high steady current then obtained on pre-electrolysis reflects the rate of the oxidation of mercury (or platinum) or of the reduction of water, hydrogen ion, or some other major constituent of the solution. This kind of "background" current is discussed further in Section 5.1. If it is very large, it is likely to be fatal in any coulometric work except the very crudest. It can be decreased only by changing the conditions (solvent, pH, supporting electrolyte, and so on) in such a way that the wave in question becomes better separated from the initial or final current rise.

In this connection it may be pointed out that the final current rise at a large mercury electrode appears to begin at a potential that is considerably less negative than the one at which it begins on a polarogram of the same solution. At a dropping electrode the comparatively large "charging" or "capacity" current, which is consumed in charging the electrical double layers surrounding successive drops up to these relatively negative potentials, masks the exponential increase of the faradaic current until the latter has become quite large. With a large electrode, however, the charging current decreases to zero within a very short time after the beginning of an electrolysis, and the faradaic current can therefore be detected even at relatively positive potentials.

When a single-compartment cell is used, the current during a pre-electrolysis may decay only to a fairly high constant value for a quite different reason. Suppose that such a cell consists of a mercury cathode and a silver anode immersed in a solution in which silver(I) has an appreciable solubility. As the pre-electrolysis proceeds, oxygen and other impurities (such as peroxides and traces of metal ions) are reduced at the cathode, and at the same time some dissolved silver(I) is formed at the anode. As this reaches the surface of the cathode, it is reduced; eventually a steady state is reached in which silver ions enter the solution at the anode just as fast as they are deposited at the cathode. A steady current flows: it obeys (10), with $c_{Ox}^{b,0}$ equal to the steady-state concentration of dissolved silver(I), and is thus equal to 1 mA if that concentration is even as large as $2 \times 10^{-5} M$. A similar cyclic process, in which oxygen is formed at the anode and re-reduced at the cathode, can occur if an internal platinum auxiliary anode is used. Such phenomena were described in detail by Kruse [73] and can be avoided only by using diaphragm cells.

An increase of current during an electrolysis may result from electrical heating of the solution, particularly during an interval when the current would be nearly constant if the temperature were nearly constant. Such circumstances arise when the solution is saturated with the starting material and contains an excess of the suspended solid, when the current is resistance-limited, or when the consumption of starting material is nearly counterbalanced by the formation of another electroactive substance by a homogeneous reaction. They can also arise [74, 75] (a) from the formation of an electroactive substance by a homogeneous reaction between inert ones [76]; (b) from the formation of a substance that yields its limiting current at the potential where the electrolysis is carried out by a homogeneous reaction between a product and a starting material that yields only a fraction of its limiting current at that potential [50, 77]; or (c) from the increase of the rate of reduction of water or hydrogen ion that accompanies the deposition of a noble metal such as platinum and that reflects the decrease of the overpotential for the evolution of hydrogen [78]. All of these are fairly unusual, however, and most of them are fairly easy to diagnose. By far the most common situation is that in which the current decreases continuously as the electrolysis proceeds. It may decrease to the value obtained in the pre-electrolysis at the same potential, although it often decays to an appreciably higher value that remains essentially constant despite very prolonged further electrolysis. In either event the electrolysis is complete when the steady final current is attained. The mercury can then be drained out of the working-electrode compartment and the solution removed for whatever working-up may be appropriate. The other compartments of the cell should then be emptied and their contents discarded, and the cell should be washed very thoroughly, with special attention to the fritted disks. It should then be filled with water (or another solvent) or with the supporting electrolyte to be used in the next experiment: the fritted disks must not be allowed to dry out. The reference electrode should be allowed to stand in a solution of potassium chloride or some other appropriate electrolyte.

4 APPLICATIONS OF CONTROLLED-POTENTIAL ELECTROLYSIS

Controlled-potential electrolysis can be of value to the chemist in several ways. One is in purifying materials and removing electroactive constituents from solutions before undertaking other work with them. It is advantageous for these purposes because it makes possible the nearly complete removal of oxidizable or reducible impurities without contaminating the solutions with excess reagents, traces of resins, or other foreign materials. Another is in synthesizing the products of half-reactions, where it is advantageous because of its extreme selectivity and reproducibility. A third is in studying the mechanisms and behaviors of half-reactions, where its outstanding (but by no means its only) advantage is that it permits the identification and investigation of chemical reactions that are coupled with the electron-transfer process and that are too slow for convenient study by other techniques.

4.1   Purifications and Separations

Controlled-potential electrolysis is very much the most efficacious and convenient technique for ridding solutions of substances, such as the ions of the alkali and alkaline earth metals and of many heavy metals as well as alkynes, thiols, and a number of other organic compounds, which can be electrolytically converted into insoluble solids or substances soluble in mercury. A more specialized but related application is to the removal of minute traces of electroactive metals from mercury. The technique has been employed to prepare supporting electrolytes and mercury for use in polarography, stripping analysis, and other sensitive electrochemical techniques, as well as in absorption spectroscopy. For all of these purposes, the starting material has only to be subjected to a sufficiently prolonged electrolysis at a potential on the plateau of a voltammetric wave known to correspond to a half-reaction yielding a suitable product.

Extreme degrees of purification are readily achieved. For a reversible process such as the deposition of copper from V_s cm^3 of solution into V_a cm^3 of amalgam, the numbers N of moles of copper in the two phases at equilibrium are given by

$$E = E_{1/2} - \frac{RT}{nF} \ln \left(\frac{N_a/V_a}{N_s/V_s} \right) \tag{19}$$

Suppose that an alkaline tartrate solution, in which the polarographic half-wave potential $E_{1/2}$ of copper(II) is equal to -0.52 V versus SCE, is to be freed from copper by an electrolysis in which $V_s = 3 V_a$, which is a fairly typical value of this ratio. At 25°C, if the electrolysis is carried out at -0.72 V versus SCE, N_a/N_s will be equal to 1.94×10^6 at equilibrium, so that the concentration of copper(II) remaining in the solution will be only 5×10^{-7} times its initial concentration; it would of course be even smaller if the potential were more negative. An infinitely prolonged electrolysis would be required to produce this result, but one for $9.21/s_{Ox}^*$ ($= 4/0.434 \, s_{Ox}^*$) s would suffice to remove 99.99% of the copper that was present initially. If any cadmium happened to be present as well, only an insignificant fraction [approximately 0.001%, according to (19)] of it would be reduced, for this potential is appreciably more positive than the half-wave potential of cadmium(II), which is equal to -0.856 V versus SCE in this medium.

The degree of completion that could be attained by a very prolonged electrolysis on the plateau of a wave exceeds that computed from (19) if the wave is irreversible. Here too the extent of the purification achievable in practice depends chiefly on the value of s_{Ox} and the patience of the operator.

The technique has been used [79] to remove zinc from sodium hydroxide, nickel and zinc together from ammoniacal ammonium citrate, iron from a weakly acidic citrate solution, and alkali and alkaline earth metals from tetramethylammonium hydroxide; to remove traces of zinc and other electroactive metals from mercury by anodic stripping [80]; to remove electroactive impurities from a solution of potassium chloride before using it as a supporting electrolyte in anodic stripping voltammetry with a hanging-mercury-drop electrode [81]; and to prepare, by three successive electrolyses in different media at different potentials, a solution of copper(II) that was entirely free from any detectable trace of any other heavy metal [82].

Separations are based on the same ideas and have usually been followed by polarographic or other electrometric analyses of the electrolyzed solutions. Although the determination of a trace of a more difficultly reducible constituent of a solution containing a large excess of a more easily reducible one is a classic source of trouble in polarography, Lingane was able to determine $10^{-4}\%$ of zinc in cadmium and its salts after a controlled-potential electrodeposition of cadmium from an ammoniacal ammonium chloride solution [83], and as little as $10^{-5}\%$ of nickel and zinc have been determined in copper and its salts [82]. Garn and Halline [84] reduced nitrocellulose by controlled-potential electrolysis prior to determining phthalate ion polarographically; as this example indicates, physical separation is by no means essential. One might determine fumaric acid polarographically in the presence of excess maleic acid after reducing the latter at a potential and in a solution in which it is reducible while fumaric acid is not. The interference of acetaldehyde in the polarographic determination of acetone as acetone imine in a strongly ammoniacal solution [85] could be eliminated by reducing the acetaldehyde to ethanol in a weakly alkaline solution before adding ammonia.

4.2   Electrosyntheses

The first modern controlled-potential electrosynthesis was carried out by Lingane, Swain, and Fields [86]. These authors wished to prepare 9-(o-iodophenyl)dihydroacridine **(II)**, which they were unable to secure from 9-(o-iodophenyl)acridine **(I)** by chemical reduction because they could find no chemical reducing agent that was not either too weak to effect any reduction at all or so strong as to yield 9-phenyldihydroacridine **(III)** in a single step.

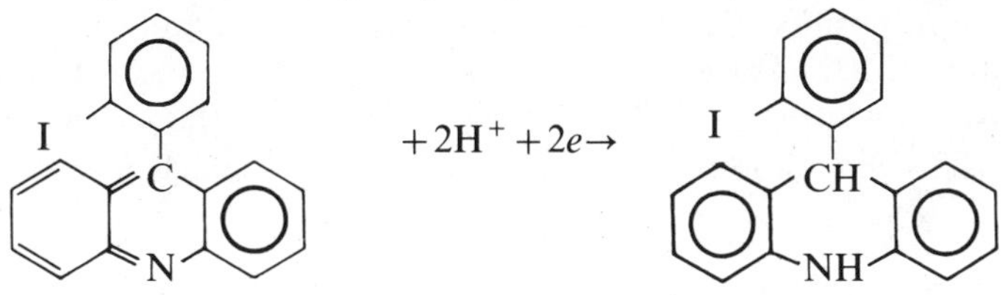

$$+2H^+ +2e\rightarrow \qquad (20a)$$

(I)   **(II)**

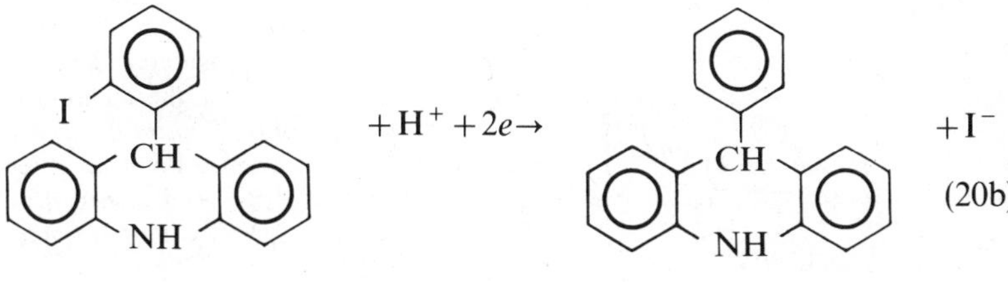

$$+H^+ +2e\rightarrow \qquad +I^- \qquad (20b)$$

(II)   **(III)**

A typical polarogram of 9-(o-iodophenyl)acridine in 90% ethanol containing 0.1 M potassium hydroxide and 0.5 M potassium acetate is shown in Figure 5.10. The first wave, for which $E_{1/2} = -1.32$ V versus SCE, corresponds to the desired reaction [20a)]; the second, for which $E_{1/2} = -1.62$ V versus SCE, corresponds to reaction (20b). Solutions containing 0.8 g of (I) in 500 cm^3 of this supporting electrolyte were electrolyzed at a working-electrode potential of -1.39 V versus SCE. This potential lies on the rising portion of the first wave [and probably, in view of (5), even nearer its foot on the stirred-pool voltammogram than on the polarogram in Figure 5.10]; it was chosen, in preference to a more negative potential on the plateau, to minimize the formation of (III). Had the first wave been reversible, electrolysis on its rising portion would have led to an equilibrium mixture containing an appreciable fraction of unreduced (I), but the wave is actually irreversible, and therefore the electrolysis proceeded to completion. It had to be performed in a diaphragm cell and in the absence of air because of the sensitivity of (II) to oxidation, either anodically or by dissolved oxygen. After the current had attained a small constant value, the product was isolated by extraction into chloroform. An average yield of 90% was obtained, the loss being attributed to incomplete extraction (or, perhaps, to oxidation by air during the recovery process) rather than to incomplete reduction.

It is the selectivity of the technique that is especially noteworthy. The difficulty of chemical reduction depends largely on the width of the range of potentials over which an acceptable yield of the desired product can be secured. In the reduction of p-benzoquinone under the conditions of Figure 5.1, any potential more negative than about -0.05 V versus SCE would suffice. Many reducing agents give potentials in this range. When a reducing agent Red reacts with the quinone, its oxidized form Ox is produced along with the hydroquinone. The ratio, [Red]/[Ox], of the concentrations of these two species decreases as the reduction proceeds. In accordance with the Nernst equation, the potential becomes more positive. The variation is not very important in this reaction because the reduction of the quinone proceeds quantitatively to the hydroquinone over an extremely wide range of potentials. In the reduction of 9-(o-iodophenyl)acridine, however, the allowable range is much narrower. If the

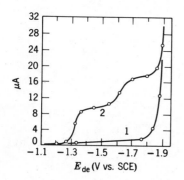

Figure 5.10 Polarograms of (1) air-free 0.1 M potassium hydroxide–0.5 M potassium acetate in 90% ethanol and (2) the same solution after the addition of 2.22 mM 9-(o-iodophenyl)acridine. Reprinted, with permission, from L. Meites, "Controlled-Potential Electrolysis," in A. Weissberger and B. W. Rossiter, Eds., *Physical Methods of Chemistry*, Vol. 1, Part IIA, 4th ed., Wiley-Interscience, 1971.

potential is more positive than about -1.28 V versus SCE, the reduction will not occur at an appreciable rate; if it is more negative than about -1.5 V, the yield will decrease and the product will be more or less contaminated with **(III)**. Even though a particular reducing agent may give a potential inside the acceptable range at the start of the reaction, its depletion and the accumulation of its oxidized form cause the potential to drift toward more positive values, and the drift may be so extensive that the reaction stops before it is complete. This argument is a little oversimplified because a chemical reduction need not occur in the same range of potentials as the corresponding electrochemical reduction, and the difference between the potentials needed to cause reactions (20a) and (20b) to occur at appreciable rates may be either wider or narrower for homogeneous reactions with chemical reducing agents than it is for heterogeneous reactions at the surfaces of electrodes. Nevertheless, it is common experience that many selective reductions or oxidations are easily effected by controlled-potential electrolysis although they are difficult or impossible with chemical reagents.

At least as important as the selectivity of the technique is the fact that it does not involve contaminating the solution with chemical reducing or oxidizing agents and the products of their reactions with the substance that is of interest. Solutions of selenium(-II) and tellurium(-II) [87], tungsten(III) and (V) [88], manganese(I) [89], and the lower oxidation states of rhenium [90], ruthenium [91], and osmium [92] have all been prepared in this way for examination by other electrochemical techniques in which foreign reducible and oxidizable substances might interfere. Similar considerations are responsible for the utility of controlled-potential electrolysis in synthesizing some medically important chelates of technetium in 10^{-9} M solutions [93].

Polarography is almost always used to guide the preliminary selection of conditions for controlled-potential electrosynthesis. Indeed, the most common use of controlled-potential electrosynthesis has been in preparing the products of polarographic half-reactions as an aid in the interpretation of polarographic waves. Electrolyses are customarily carried out at potentials near the midpoints of plateaus of polarographic waves and in the same supporting electrolytes in which these waves have been obtained. Citations of about 80 examples were collected more than a decade ago [75], and many hundreds of others have appeared since.

As was said in Section 2, however, the relationship between polarography and the processes occurring at large stirred mercury pools is by no means straightforward. In addition to the differences mentioned in Section 2, there is a difference between the time scales. In the "ECE" (electrochemical–chemical–electrochemical) mechanism (Case 5b, Section 6.6)

$$Ox + n_1 e \rightarrow I$$

$$I \rightarrow J \tag{21}$$

$$J + n_2 e \rightarrow Red$$

the chemical step cannot be detected by polarography or voltammetry unless it is fast enough to proceed to an appreciable extent during the few seconds that constitute the lifetime of a single drop or the duration of a single experiment. In a controlled-potential electrolysis the formation and subsequent reduction of J could not escape notice unless the intervening chemical step were very slow indeed. For example, Laitinen and Kneip [94] found that the polarographic behavior of p-N,N-dimethylaminoazobenzene in alkaline solutions corresponds to the two-electron process

$$C_6H_5N\!\!=\!\!NC_6H_4N(CH_3)_2 + 2H^+ + 2e \rightarrow C_6H_5NHNHC_6H_4N(CH_3)_2$$

whereas controlled-potential electrolysis yields a mixture of aniline and p-(N,N-dimethylamino)aniline resulting from a further two-electron reduction of the hydrazo compound. Similarly, the reduction of α-furildioxime in alkaline media consumes six electrons per molecule in polarography but eight in controlled-potential electrolysis under identical conditions [95]; the oxidation of 1,2-dimethylhydrazine at platinum electrodes in acidic aqueous solutions involves two to three electrons per molecule according to chronopotentiometry but involves six in controlled-potential electrolysis [96, 97]; and the oxidation of hypoxanthine involves four electrons according to thin-layer electrochemistry but involves six according to controlled-potential electrolysis [98]. The list might be extended almost indefinitely, and the same considerations apply to small-scale electrolyses with voltammetric indicator electrodes and large-scale electrolyses with working electrodes made from platinum or any other material.

4.3   Electrochemical Studies

There are many ways in which controlled-potential electrolysis can supplement the information that other techniques provide about the course, mechanism, and electrochemical behavior of a half-reaction. Many of these are discussed in other sections; this section emphasizes the phenomena that are observed in relatively simple situations. It will serve as a basis for discussions, in Sections 6 and 7, of the complications that can arise.

On discovering that a substance is electroactive, the electrochemist's first task is to identify the product(s) of its reduction or oxidation. The use of controlled-potential electrolysis in preparing those products has been described in the preceding subsection, and ways of identifying and determining them by other techniques are reviewed in Section 8.

Next one wants to know whether the process is reversible or not. The term "reversible" has no absolute significance: whether or not the half-reaction $Ox + ne = Red$ displays Nernstian behavior depends on the technique by which it is studied and, more fundamentally, on the value of $k_{s,h}$, the concentrations of Ox and Red, and the rate at which the flow of current forces it to occur. The couple $VO_2^+ + 2H^+ + e = VO^{2+} + H_2O(l)$ is reversible in potentiometry as long as its exchange current is large compared to the rates of the other processes, such as the oxidation of water and the reduction of oxygen or hydrogen ion, that occur at the zero-current potential. In voltammetry it behaves irreversibly because the value of $k_{s,h}$ is fairly small, and displacing the equilibrium at a finite rate reveals that it is not restored instantaneously. Similarly, the couple $V^{3+} + e = V^{2+}$ is reversible in polarography but not in linear-sweep voltammetry at high sweep rates, because the value of $k_{s,h}$ is high enough that equilibrium is closely approached during the life of a drop at a dropping electrode but is not high enough to produce the same result during a fast sweep. On the other hand, if the mechanism involves a deactivation of Red by reaction with some other substance to yield an inert product P

$$Ox + ne \rightarrow Red$$

$$Red + Z \rightarrow P$$

the process may appear to be reversible in linear-sweep or triangular-wave voltammetry because the duration of the experiment may be too short to permit an appreciable fraction of the Red to react, but may behave irreversibly toward a slower technique. Such behavior is common in "aprotic" solvents, where Red is often a free radical and Z may be either the trace of water present as an impurity or the very weakly (C-)acidic solvent itself. In this frame of reference, controlled-potential electrolysis provides tests of reversibility that cover a wide gamut. One, which can take either of two slightly different forms [99–101], is based on (17). In one form, a solution containing N_{Ox}^0 moles of the oxidized form Ox is electrolyzed at a potential E_{we} on the rising part of its wave, and a measurement is made of the quantity of electricity $Q_{\infty,r}$ that is consumed in reaching equilibrium. In the equilibrium mixture

$$[Ox] = \frac{N_{Ox}^0 - Q_{\infty,r}/nF}{V}$$

and

$$[Red] = \frac{Q_{\infty,r}}{nFV}$$

so that

$$E^{0\prime} = E_{we} + 2.303 \frac{RT}{nF} \log \left(\frac{Q_{\infty,r}}{nFN_{Ox}^0 - Q_{\infty,r}} \right) \tag{22}$$

The measurements should of course be repeated at a number of different values of E_{we}. Very high precision is required of the potentiostat because $Q_{\infty,r}$ is very sensitive to fluctuations of E_{we}. In the other form, which is simpler and more convenient, the entire electrolysis is performed at a potential on the plateau of the wave. When some suitable fraction—perhaps one-tenth—of the Ox has been reduced, the potentiostat is disconnected from the cell, a note is made of the quantity Q of electricity that has accumulated, and the zero-current difference of potential between the working and reference electrodes is measured with a potentiometer or, more conveniently, with a digital voltmeter having a high input impedance. The electrolysis is resumed and allowed to proceed a little farther, and the measurement is repeated. After the desired number of points has been obtained, the electrolysis is allowed to proceed to completion to provide a value of Q_∞ for comparison with the expected value, which is equal to nFN_{Ox}^0 in accordance with (1). The values of Q and E_{we} at each point are combined with the equation

$$E^{0\prime} = E_{we} + 2.303 \frac{RT}{nF} \log\left(\frac{Q}{Q_\infty - Q}\right) \tag{23}$$

which is exactly analogous to (22).

If the process is reversible, the value of $E^{0\prime}$ obtained in this way is constant and independent of E_{we}. If it is totally irreversible, the electrolysis proceeds to completion even at a potential on the rising part of the wave. If there is a slow inactivation of Red, the values of $E^{0\prime}$ will drift slowly with time.

Another criterion is based on work by Rogers and Merritt [99], Meites [75], and Bard and Santhanam [102]. In electrolyses conducted at different potentials on the rising part of a reversible wave, plotting $\ln i$ against time gives straight lines that are nearly parallel for the simple reversible process $Ox + ne = Red$; they would be exactly parallel if the mass-transfer constants s_{Ox} and s_{Red} had identical values. If a reversible charge-transfer step is followed by inactivation of Red, the plots are concave downward, and this is the best way of detecting and studying a moderately slow inactivation step. If the process is totally irreversible (as it may be if Red is rapidly inactivated), the plots have different slopes $(= -s_{Ox}^*)$: the value of αn_a can be calculated from (14), and the values of i^0 obtained from their intercepts on the ordinate axis are proportional to their slopes, in accordance with (10) and (16).

A voltammogram obtained at the end of an electrolysis can be compared with a similar one obtained at the beginning in the fashion illustrated by Figure 5.4. If Red undergoes any reaction that is too slow to reach completion during an electrolysis, its limiting current will continue to change with time after the electrolysis is complete, and this is the best way of studying extremely slow subsequent reactions.

Other ways of identifying chemical processes that consume Red are outlined in Section 6.3, and Sections 6 and 7 also describe ways of studying many other kinds of chemical reactions that may be coupled with electrochemical ones.

5   COULOMETRY AT CONTROLLED POTENTIAL

5.1   Theory

Coulometry at controlled potential involves the integration, with respect to time, of the current that flows during a controlled-potential electrolysis. The quantity of electricity consumed increases as the electrolysis proceeds and approaches the limit described by (1):

$$Q_\infty = \int_0^\infty i\, dt = nFN^0 \tag{1}$$

where N^0 denotes the number of moles of starting material. On the simplest level there are three requirements that must be satisfied in order to evaluate either n or N^0 correctly from a measured value of Q_∞ and a known value of the other:

1. The electrolysis must be carried virtually to completion.
2. Only a single substance may be reduced or oxidized at the working electrode.
3. For analytical purposes, every ion or molecule of the starting material must react in the same way, so that its reduction or oxidation involves the same number of electrons.

The first of these requirements is easily satisfied by prolonging the measurement until a sufficiently large fraction—say, 99.9%—of the starting material has been reduced or oxidized. If the half-reaction occurs in a single step, so that the current–time curve is described by (10), this point will be reached when the current has decayed to 0.1% of its initial value. As will shortly appear, not all current–time curves obey (10), but there are few mechanisms in which at least 99.9% of the starting material will not be reduced or oxidized by the time the total current drops to 0.1% of its initial value. Electrolyses involving couples that are reversible or nearly so, and that are conducted at potentials where the rates of the backward reactions are appreciable, form nearly the only exception.

The second requirement is less simple. Of course one should avoid potentials so positive that any constituent of the supporting electrolyte, or the working electrode itself, is oxidized. Nor would one ever deliberately carry out such a measurement in the presence of a foreign substance that gives a wave preceding or coinciding with the one corresponding to the process of interest. If such a substance is present in the sample, it can usually be removed or rendered innocuous by a pre-electrolysis under properly chosen conditions before the reduction or oxidation of the substance of interest is begun.

No matter how much care is taken in such matters, however, it is never possible to avoid the consumption of some charge by extraneous processes. It is therefore essential to understand the natures and manifestations of these processes, the magnitudes of the errors to which they give rise, and the ways in which corrections for them can be made. The following discussion is based on that of Meites and Moros [50], who identified six components of the total quantity of electricity consumed. Those that are most important here are:

1. Q_∞, which is defined by (1) and whose evaluation is the purpose of any coulometric experiment.

2. Q_c, the "charging" quantity of electricity, which is consumed in charging the electrical double layer at the surface of the working electrode up to the potential employed in the electrolysis. The value of Q_c is given by

$$Q_c = \kappa A (E_{max} - E) \tag{24}$$

where κ is the differential capacity (in $\mu F/cm^2$) of the double layer; A is the area (in cm^2) of the interface between the working electrode and the solution; E_{max} is the potential of the electrocapillary maximum, which depends on the material from which the working electrode is made and on the composition of the supporting electrolyte; E is the potential of the working electrode, against the same reference electrode as E_{max}; and Q_c is expressed in microcoulombs (1 μC = 0.01036 nF). For a mercury electrode in an aqueous solution at a potential more negative than E_{max}, which in many supporting electrolytes is in the vicinity of -0.6 V versus SCE, a useful first approximation can be made by taking the value of κ to be independent of E and equal to about 20 $\mu F/cm^2$. Hence, if $A = 40$ cm^2 and $E = -1.1$ V versus SCE, which are typical values, the value of Q_c is only about 4 nF. This is negligible in any experiment involving more than a few micromoles of starting material. In work with smaller amounts, Q_c should be evaluated by the procedure described by Meites and Moros, which is based on the fact that charging of the double layer reaches completion within a very short interval (whose duration depends on the speed of response of the potentiostat) after the start of the electrolysis. Correction for it is made by subtracting it from the total quantity of electricity consumed in the electrolysis of the substance of interest. No one has yet faced the problem that must arise when the value of κ is altered by adsorption or deposition of the product of the electrolysis onto the surface of the working electrode, in which event this procedure will give an erroneous value of Q_c.

3. $Q_{f,i}$, the "faradaic impurity" quantity of electricity, which results from the reduction or oxidation of impurities in the supporting electrolyte, the sample, or the working electrode. This is best eliminated by prior electrolytic (Section 4.1) or other purification.

4. $Q_{f,c}$, the "continuous faradaic" quantity of electricity, which arises from the continuous slow reduction or oxidation of some major constituent of the supporting electrolyte, such as water or hydrogen ion. The rates of such processes, and hence the currents attributable to them, are described by (7) and (13), and are never equal to zero, though they may be negligibly small at potentials sufficiently far removed from either the initial or the final current rise. Only an insignificant fraction of the water or hydrogen ion (or, in a buffered solution, the acidic component of the buffer) could be consumed during any practical electrolysis, and therefore the continuous faradaic current $i_{f,c}$ remains constant. This is in contrast to the current corresponding to $Q_{f,i}$, which decays toward zero as the electrolysis proceeds. A correction for $Q_{f,c}$ should be based on two measurements of that current: one at the end of an electrolysis of the supporting electrolyte alone, and the other at the end of an electrolysis of a solution containing the substance that is being studied or determined. The electrolyses must be carried out under identical conditions and must be prolonged until constant steady-state currents have been attained, and the values of these currents should be identical within a reasonable estimate of the precision attained in measuring them. If they are, they may be averaged to obtain a value of $i_{f,c}$. Subsequent electrolyses of solutions containing the substance of interest are prolonged until the steady state has been attained, and the total quantity of electricity Q that has been consumed is measured at some later instant, t s after the start of the electrolysis. The product $i_{f,c}t$ is subtracted from Q. Often, however, the two

steady-state currents are not identical, and correction in this or any other simple manner is then certain to give an erroneous result.

Either for analytical purposes or because one wishes to minimize the extent of a competing second-order side reaction or the variation of potential across the surface of a working electrode when the cell has a relatively high resistance, it is sometimes necessary or desirable to work with extremely dilute solutions of the substance being determined or studied. By exercising scrupulous care, several investigators [63, 103–106] have succeeded in securing useful results with as little as a few micrograms of material, corresponding to an average of about 0.1 μF of electricity. In one exceptionally favorable case [79], an accuracy and a precision of a few percent were secured with less than 0.1 μg of material, corresponding to 2 nF of electricity. It is certainly safe to say that work can be done routinely with solutions as dilute as 2×10^{-6} M. When as much as 50 μF of electricity is involved, an accuracy and a precision of 0.1% should be attainable under any ordinary conditions; this quantity of electricity will be consumed in a one-electron reduction or oxidation of the electroactive material contained in 50 cm^3 of a 1 mM solution. Few other techniques provide comparable accuracy, precision, and selectivity.

5.2   Apparatus and Techniques for Current Integration

Coulometry at controlled potential was first suggested by Hickling [30] and was developed by Lingane [107, 108]. Many different ways of integrating the current with respect to time have been devised and tested, but only a few have survived. They fall into two groups, which reflect the most common uses of the technique:

1.   In evaluations of n, most workers are content with an accuracy and precision of a few percent, on the grounds that, if one is trying to decide whether a half-reaction involves one electron or two, a relative error of even 5% in an experimental value of n can produce neither error nor doubt. The writer thinks this to be ill-judged. There are substances whose reductions or oxidations involve many more than one or two electrons, and for which imprecise coulometric data may be useless at best, and misleading at worst. There are others whose reductions or oxidations involve mechanistic complexities that can be diagnosed and studied with the aid of current–time curves obtained with precise apparatus and careful technique, but about which useful information cannot be gotten in ways that barely serve to reveal whether n is equal to 1 or to 2. How far these considerations have affected practical work may be ascertained from the result of an examination of 227 values of n obtained by controlled-potential coulometry and cited in one recent volume of a large collection of electrochemical data [18]. Fifty-six percent of them had merely been said to be integral, the experimental values having presumably been so poor that the original authors were unwilling to expose their defects in print. Of the others, 16% lay within 1% of an integer or were accompanied by relative mean or standard deviations that were equal to, or smaller than, 1%; while 41% seemed

to have accuracies and precisions that were poorer than 5%. Although some of these categories must have included systems in which chemical side reactions consumed electroactive intermediates and led to authentically nonintegral values of n, it is clear that high accuracy and precision are not considered to be indispensable by those who are chiefly interested in such work.

2.   In analytical applications, however, the highest attainable accuracy and precision are regarded as being essential, and figures on the order of $\pm 0.1\%$ are routinely sought.

Current integrators can be divided into two corresponding groups: those for which the accuracy and precision are no better than 1%, and those for which they are better than 0.1%.

Of the devices in the first group, the strip-chart recording potentiometer with a built-in integrator is the most convenient and most versatile, and has the further advantage of providing a record of the current–time curve if one is wanted. If the current at the start of the electrolysis will not be too high (say, 1 A or less), it suffices to connect the input terminals of the recorder across a precision 1-ohm resistor—or, for greater flexibility, a resistance box calibrated at several different settings—in series with the auxiliary electrode. Currents exceeding 1 A should be avoided for reasons that were given above, but could be handled by interposing a precision voltage divider between the resistor and the recorder. Moderately heavy damping is usually desirable to remove as much as possible of the noise that arises from phenomena to be discussed below. Lingane and Jones [109] were the first to use a ball-and-disk integrator for controlled-potential coulometry, at a time when integrating recorders were not yet commercially available. A Solion cell [110] is a cheaper and simpler alternative that should be satisfactory if the current–time curve is not wanted and if the initial current is not too small. Gas coulometers, titration coulometers, and other kinds of coulometers have largely disappeared from practical work because they are ill-suited to integrating small currents and are also inconvenient to use. Their sole advantages are their cheapness and simplicity. Those to whom these virtues are attractive should consult Lingane's descriptions of these devices [111].

If the current obeys (10)—that is, if it decays exponentially toward zero— the value of Q_∞ may be obtained directly from a plot of $\log i$ or $\ln i$ against time, as was first pointed out by MacNevin and Baker [112]. Equation (1) may be rewritten to give

$$Q_\infty = nFN^0_{Ox} = nFVc^{b,0}_{Ox} = \frac{s^*_{Ox}nFVc^{b,0}_{Ox}}{s^*_{Ox}} = \frac{i^0}{s^*_{Ox}} \tag{25}$$

As is shown by (15), the value of i^0 may be obtained from the intercept, and that of s^*_{Ox} may be obtained from the slope, of either plot. MacNevin and Baker used working electrodes made of platinum gauze and were able to obtain values of Q_∞ having accuracies and precisions of about $\pm 2\%$. With mercury pools the situation is less favorable, for fluctuations of the efficiency of stirring and of the area of the electrode lead to fluctuations of the current that is due to the reduction or oxidation of the substance being studied [113]. Because the instantaneous value of dA/dt is always finite, they also lead to the continuous flow of a current that corresponds to charging and discharging of the electrical double layer. On the average, this current becomes equal to zero within a very short time after the beginning of the electrolysis, but the interval over which the total current is measured may be so short that a true average cannot be obtained. These phenomena account for the noise that is always observed on recorded current–time curves. Shia and Meites [113], who made measurements of current with a digital multimeter having an integration period of 100 ms, described the standard error s_i of a single measurement by the equation

$$s_i = Z_1 + Z_2 i \tag{26}$$

The value of Z_1 reflects the fluctuations of charging current and is proportional both to the integral capacity of the double layer at the potential E that is employed and to the difference between E and the potential of the electrocapillary maximum. The value of Z_2 reflects the variations of the faradaic current due to the substance that is of interest. Shia and Meites reported values of Z_1 that ranged from 10 to 50 μA, and values of Z_2 that were close to 0.013; both of these must depend on the design and rate of rotation of the stirrer.

The precision of a value of Q_∞ obtained by this technique will obviously depend on the values of Z_1 and Z_2, becoming worse as they increase. The results obtained by MacNevin and Baker would be difficult to duplicate with mercury pools because both Z_1 and Z_2 must be higher for mercury pools than for solid electrodes. On the other hand, much of the precision inherent in the data is sacrificed by employing a graphical technique. On the basis of some realistic assumptions regarding the number of measurements of current that are made and the way in which they are distributed along the time axis, Shia and Meites concluded that values of Q_∞ having relative precisions of about 0.5% could be attained on evaluating i^0 and s_{Ox}^* by nonlinear regression onto (15b) [114]. Very much better results can be secured by using the measured currents to compute values of the current integral Q_t at each of the experimental points, and then performing nonlinear regression onto the equation

$$Q_t = Q_\infty [1 - \exp(-s_{Ox}^* t)] \tag{27}$$

which is the integrated form of (15a).

The chief advantage of the MacNevin–Baker technique is that it makes it unnecessary to prolong an electrolysis to the bitter end; its chief disadvantage is that it is fatally easy to employ even if the current does not decay exponentially toward zero. Because the experimenter must know how the system behaves before applying this technique, it is probably best suited to routine analyses, and should be restricted to those that do not require the highest attainable accuracy and precision. More accurate and precise variants of it are described below.

In applications that do demand high accuracy and precision, two kinds of analog integrators have superseded all others. In one [52, 61, 115–117], a chopper-stabilized operational-amplifier circuit is used to amplify the iR drop across a standard resistor in series with the auxiliary electrode, and the amplified output is used to charge a precision capacitor. The capacitor is discharged before the electrolysis is begun; the emf across it at any later instant is proportional to the current integral and can be measured with a digital voltmeter. In the other [118–121], the iR drop is presented to the input of a voltage-to-frequency converter, which emits a train of pulses whose frequency at any instant is proportional to the current. The number of pulses is counted and is proportional to the current integral. Special provisions are needed if the sign of the current may change during an electrolysis. Routine analyses of a great many similar samples would be facilitated by the automatic controlled-potential coulometric analyzer, complete from sample changer to print-out device, designed [122] and further improved [123] by Phillips, Milner, et al.

An instrument of either type has the advantage of providing direct readings of the current integral, and makes it possible to evaluate a current integral with a precision of the order of $\pm 0.02\%$. The exceptional user whose requirements are even more stringent can satisfy them, though at the cost of much added complexity of manipulation, by employing a second cell containing an exactly known amount of some reference substance whose controlled-potential electro-reduction is known to be free from complications. This amount must be sufficient to consume more, but only very slightly more, electricity than the reduction or oxidation of the substance being determined, and the conditions must be so arranged that the standard substance always yields a limiting current larger than the current due to the substance being determined. The two cells are connected in series and an electrolysis is performed by controlling the potential of the working electrode of the cell containing the unknown substance. The current is not integrated during this electrolysis. When it is complete, there will still be some of the standard substance remaining in the other cell, and its amount is determined by controlled-potential coulometry. Any error in integrating the current affects only the determination of the excess of the standard substance, and extreme accuracy and precision can be attained [124, 125].

Digital integration was first employed by Shia and Meites [60, 113], who gave a comprehensive discussion of the factors that govern its accuracy and precision. An integrating digital voltmeter, or an integrating analog-to-digital converter [126], is used to obtain a digital representation of the iR drop across a standard resistor in series with the cell. Successive pairs of values of this emf, obtained at intervals of Δt s, are combined by trapezoidal integration and added to obtain values of the current integral Q_t at intervals of $\Delta t'$ s. The standard error of any one value of Q_t—say, the kth one, $Q_{t,k}$—is given by

$$s_{Q_{t,k}} = \Delta t \left[\sum_{l=1}^{k} \left\{ \left[Z_1 + Z_2 \left(\frac{\Delta Q_t}{\Delta t'} \right)_l \right]^2 \left(\frac{\Delta t'}{\Delta t} \right)_l \right\} \right]^{1/2}$$

where Z_1 and Z_2 refer to (26). The values of Z_1 and Z_2 depend on the conditions under which the electrolysis is carried out, and so does the rate of decay of

$\Delta Q_t/\Delta t'$, which is equal to the average value of the current during the interval that elapses between two successive instants at which values of Q_t are stored. Figure 5.11 shows how the absolute and relative standard errors of a single value of Q_t vary during an electrolysis under typical conditions. As the electrolysis nears completion the relative standard error approaches a value of about 0.05%. The chief advantage of the technique is that the data are easily subjected to nonlinear regression analysis once they have been stored in core memory; applying this to the data of Figure 5.11, and employing the same values of Z_1 and Z_2 to minimize the quantity

$$\frac{(Q_{t,k,\text{meas}} - Q_{t,k,\text{calc}})^2}{s_{Q_{t,k}}^2}$$

by regression onto (27) was shown to be capable of giving values of Q_∞ having relative standard errors smaller than 0.01%.

An earlier paragraph described the technique suggested by MacNevin and Baker for evaluating Q_∞ from data that do not extend to the completion of an electrolysis. Several others are possible if a direct-reading current integrator is available. Like the MacNevin–Baker technique, they all assume that the current–time curve obeys (11), so that the current decays exponentially toward zero. One [48] involves measuring the counting rate (which is proportional to the current) of the integrator at two different times during the electrolysis. Suppose that R_1 is the counting rate at an instant when $Q_1 \mu F$ has passed through the cell, and that R_2 is the counting rate at some later instant when

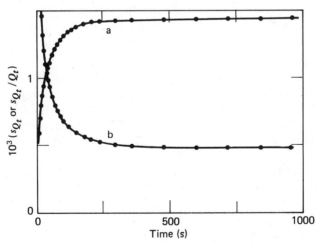

Figure 5.11   Effect of time on (a) the standard error of Q_t and (b) the relative standard error of Q_t, in an electrolysis for which $Z_1 = 10 \mu A$, $Z_2 = 0.01$, $i^0 = 30$ mA, and $s_0 = 0.01$ s^{-1}, so that $Q_\infty = 3$ C ($= 31 \mu F$). It is assumed that the current obeys (11), although the inclusion of a continuous faradaic background current equal to $30 \mu A$ (0.1% of the initial current) would make no visible difference on this scale. Reproduced, with permission, from G. A. Shia and L. Meites, *J. Electroanal. Chem.*, **87**, 369 (1978).

$Q_2 \mu F$ has been consumed. At each instant, both the current and the quantity of electricity that remains to be consumed are proportional to the concentration of the electroactive substance not yet reduced or oxidized, and

$$\frac{Q_\infty - Q_2}{Q_\infty - Q_1} = \frac{R_2}{R_1} \tag{28}$$

Harrar, et al. [127] employed a modification of this equation as the basis of "predictive coulometry," in which on-line predictions of the value of Q_∞ are updated continuously as new data are acquired, and in which the electrolysis is terminated when a number of successive predicted values agree within some predetermined figure. Instantaneous values of the current should not be used in place of R_1 and R_2 because they are afflicted by noise due to stirring. Much better results are obtained by writing

$$R_j = \frac{Q_{f,j} - Q_{0,j}}{\Delta t} \tag{29}$$

where Δt is the duration of the jth interval and $Q_{0,j}$, Q_j, and $Q_{f,j}$ are the quantities of electricity accumulated at its beginning, midpoint, and end, respectively. The length of the interval is important. If it is too short, the values of R obtained from (29) will be insufficiently precise; if it is too long, they will misrepresent the currents at the midpoints of the intervals. These objections are overcome in a second technique [128], in which the readings of the integrator are recorded at three equally spaced times. If they are successively Q_1, Q_2, and Q_3, then

$$Q_\infty = Q_1 + \frac{(Q_2 - Q_1 Q_2)^2}{2Q_2 - (Q_1 + Q_3)} \tag{30}$$

from which values accurate to $\pm 0.2\%$ can be secured if Q_1 is at least equal to $0.9 Q_\infty$ and if $Q_2 - Q_1$ is at least equal to $0.2(Q_\infty - Q_1)$. The accuracy decreases as either Q_1 or $(Q_2 - Q_1)$ decreases, but values accurate to $\pm 0.5\%$ are usually obtained even if $Q_1 = 0.5 Q_\infty$. Hence reasonable accuracy can be secured when the electrolysis is as little as 70% complete (which requires only $1.2/s_{Ox}^*$ s, or 200 s if $s_{Ox}^* = 6 \times 10^{-3}$ s^{-1}). A third technique is based on the fact, which is implicit in the sentence preceding (28), that the quantity Q_R of electricity remaining to be accumulated at any instant is proportional to the current at that instant, so that

$$Q_R (= Q_\infty - Q_t) = ki \tag{31a}$$

This can be rearranged to give

$$Q_t = Q_\infty - ki \tag{31b}$$

which shows that a plot of Q_t versus i is linear and has an intercept equal to Q_∞. Hanamura [129–131] constructed a device that yields such plots automatically, but a similar extrapolation can of course be made no matter how the values of Q_t and i are obtained. Linear regression is more attractive than graphical extrapolation, but unweighted linear regression onto (31b) as it stands will give the wrong result because it traduces the fundamental assumption that the random errors in the independent variable are insignificant in comparison with those in the dependent variable. Consequently it is simpler to effect regression onto the rearranged equation

$$i = \frac{Q_\infty}{k} - \left(\frac{1}{k}\right) Q_t \tag{32}$$

Comparing (32) and (25) shows that $k = 1/s_{Ox}^*$. It is not known what accuracy and precision can be obtained in this way.

Holland, Weiss, and Pietri [132] analyzed solutions of plutonium(III) by terminating the electrolysis and integration when oxidation to plutonium(IV) was 99.6–99.8% complete, then measuring the potential of an indicator electrode immersed in the solution. By combining that potential with the known formal potential of the plutonium(IV)–plutonium(III) couple and with the Nernst equation, they were able to calculate the fraction of the plutonium(III) that had escaped oxidation, and to apply an appropriate correction to the quantity of electricity consumed. Although it takes only about 60% as long to attain 99.7% completion as it does to attain 99.99% completion, they achieved excellent results. However, the procedure can be used only with couples that behave reversibly. Even for such couples it is inferior in principle (because of the logarithmic relationship embodied in the Nernst equation) to using amperometry or spectrophotometry (or any other technique in which the signal is proportional to concentration) to determine the concentration of starting material that has survived the electrolysis, which can be done regardless of the reversibility of the couple. That is the function of the measurements of i in the techniques described in the preceding paragraph, which of course become more precise as the last measured value of Q approaches Q_∞ more closely.

All these techniques can cope with a finite continuous faradaic background current if its value is known in advance and is unaffected by the product of the half-reaction that occurs at the working electrode; none can cope with a nonexponential decay of the current. The requirements are restrictive, and failure to keep them in mind may be very dangerous.

To the analyst confronted with a half-reaction that is free from complications, such as side reactions that consume electroactive material and catalytic processes that regenerate it, controlled-potential coulometry has two very great advantages over almost every other analytical technique. One is that it is extraordinarily precise, rapid, and selective. The other is that it yields a result whose relationship to the quantity of the substance being determined is given definitely by Faraday's law instead of having to be evaluated with the aid of standard solutions. To be sure, there is a constant of proportionality, n, that must be known in controlled-potential coulometric analysis, but *if the system exhibits simple behavior* it is a constant whose value does not vary continuously with small changes in the conditions of measurement, as do the constants of proportionality involved in other instrumental techniques of analysis. The other side of this coin is that the technique provides an extremely sensitive way of detecting the occurrence of any chemical complication in the overall process.

5.3   Evaluations of n

Controlled-potential coulometry is the standard technique for evaluating n, the number of electrons taking part in a half-reaction for each ion or molecule of the electroactive substance. A solution containing a known amount of that substance is subjected to controlled-potential electrolysis, Q_∞ is evaluated with the aid of one of the instruments or techniques described in the preceding subsection, and the value of n is calculated from (1). Citations of a great many examples may be found in the review by Perrin [133], in the monographs edited by Baizer [134] and Weinberg [8], and in compilations of electrochemical data [18, 19].

In principle, all that is required is to perform an electrolysis under the conditions described in Section 3.3 while evaluating the current integral in one of the ways described in Section 5.2. This is so simple that the present subsection will concentrate on the difficulties that may arise and on the precautions that should be taken.

It is useful to distinguish between two related quantities. One, denoted by the symbol n, is the number of electrons associated with each molecule or ion of the electroactive substance in the balanced chemical equation for the half-reaction. The other, denoted by the symbol n_{app} (where the subscript signifies "apparent"), is the number of faradays associated with each mole of the electroactive substance in a controlled-potential coulometric experiment. For a system that exhibits simple behavior, the two are identical: the equation for the half-reaction that describes the oxidation of chromium(II) under the conditions of Figure 5.4 is

$$Cr(OH_2)_6^{2+} + 2Cl^- = Cr(OH_2)_4Cl_2^+ + 2H_2O + e$$

so that the value of n is 1, while fifteen coulometric experiments with initial concentrations of chromium(II) ranging from about 0.033 to 17 mM [34] gave the mean value $n_{app} = 1.000_0 \pm 0.001_0$ (mean deviation). However, if the charge-transfer step is accompanied by chemical side reactions, the two may differ considerably: the half-reaction for the reduction of molybdenum(VI) at about -0.5 V versus SCE in a supporting electrolyte containing about 0.1 M sulfuric acid and 0.1 M sodium sulfate may be written

$$MoO_4^{2-} + 8H^+ + 3e = Mo^{3+} + 4H_2O$$

so that $n = 3$, but if some nitrate ion is present it will be reduced by the molybdenum(III). The current will not decrease to zero until all of the nitrate ion has been reduced, and the value of n_{app} will exceed 3 to an extent that depends on the ratio of the initial concentrations of nitrate ion and molybdenum(VI), and that may be extremely large. Only by coincidence will an integral value be obtained. In general, a nonintegral value of n_{app} betokens the occurrence of some chemical complication.

Since evaluations of n_{app} are performed only with systems whose behaviors are not yet very well understood, shortcuts are inadvisable because the experimenter cannot be sure that they are justified. Section 5.2 described several techniques by which the value of Q_∞ may be calculated if it is known, or can safely be assumed, that (15a) and (15b) are obeyed; these should not be used in evaluations of n_{app}. Actual integration of the current–time curve is essential, and should be prolonged until the current has attained a steady constant value. That value should be compared with the one that flows in an electrolysis of the supporting electrolyte alone under exactly the same conditions. If they are not identical, it is hazardous to attempt any correction unless both are very much smaller than the initial current that flowed during the reduction or oxidation of the substance being studied, in which event the correction is hardly worth applying. If the correction is appreciable, it should be based on knowledge of how the steady-state current varies with the concentration of the product of the half-reaction.

The interpretation of a value of n_{app} may not be completely straightforward even if the value is integral. There may be two or more different half-reactions for which the values of n are the same. For example, the fact that n_{app} is nearly equal to 1 for N-hydroxypyridine-2-thione ($=$RSH) at a mercury anode [135] might signify that the half-reaction is either

$$RSH + Hg \rightarrow RSHg + H^+ + e$$

or

$$2RSH \rightarrow RSSR + 2H^+ + 2e$$

Such an ambiguity can be resolved by isolating and identifying the product, as can the one that arises when the initial product is transformed into another by hydrolysis or some other reaction. The latter is exemplified by the reduction of cyclohexyl nitrate [136], in which a nitrite ester is formed in the charge-transfer step and then undergoes hydrolysis to yield cyclohexanol as the product of a two-electron reduction.

Situations in which the values of n_{app} are not integral are more interesting than those in which they are. The reduction of benzyldimethylanilinium bromide gives values of n_{app} ranging from 2 to 1.4 [137, 138]: the addition of the first electron gives dimethylaniline and the benzyl radical

$$C_6H_5CH_2\overset{+}{N}(CH_3)_2C_6H_5 + e = (CH_3)_2NC_6H_5 + C_6H_5\overset{\cdot}{C}H_2$$

of which the latter can be further reduced to toluene by the overall half-reaction

$$C_6H_5\overset{\cdot}{C}H_2 + H^+ + e = C_6H_5CH_3$$

This pathway predominates, so that $n_{app} = 2$, if the initial concentration of benzyldimethylanilinium bromide is sufficiently low. As that concentration increases, there is a corresponding increase of the concentration of benzyl radical at any instant, the second-order dimerization

$$2C_6H_5\overset{\cdot}{C}H_2 = C_6H_5CH_2CH_2CH_3$$

becomes more prominent, and the value of n_{app} decreases toward 1.

Generally similar behavior is observed in many systems, of which only a few can be listed here. They include the reduction of oxalic acid, which yields mixtures of glycolic acid and glyoxylic acid [139–141]; the reductions of many carbonyl compounds, which yield mixtures of carbinol and pinacol by the general scheme

$$R_2CO \xrightarrow{+H^+ + e} R_2\dot{C}OH \begin{cases} \xrightarrow{\text{dimerization}} \begin{matrix} R_2COH \\ | \\ R_2COH \end{matrix} \\ \xrightarrow{+H^+ + e} R_2CHOH \end{cases}$$

of which many examples are given in [6]; the oxidation of hydrazine in aqueous solutions, which yields nitrogen almost exclusively at low concentrations but increasing proportions of ammonia at higher ones by a mechanism that can be abbreviated as

$$N_2H_4 \xrightarrow{-2H^+ - 2e} N_2H_2 \begin{cases} \xrightarrow{\text{dimerization}} N_4H_4 \xrightarrow{+OH^-} NH_3 + N_3^- + H_2O \\ \xrightarrow{-2H^+ - 2e} N_2 \end{cases}$$

[142]; and the reduction of picric acid. Using a solution that contained 0.4 mM picric acid and 0.1 M hydrochloric acid, Lingane [107] found $n_{app} = 17.07$, and concluded that the half-reaction yields bis(3,5-diamino-4-hydroxyphenyl)-hydrazine

which might of course undergo the benzidine rearrangement. Both Bergman and James [143] and Meites and Meites [144] obtained the same value of n_{app} under the same conditions, but the latter authors showed that the value depends on both the concentration of hydrochloric acid and the initial concentration of picric acid in the fashions shown in Figure 5.12. If the initial concentration of picric acid was very low, the mean value of n_{app} was found to be $18.00 \pm 0.01_6$, in exact agreement with the expected value for reduction to s-triaminophenol. At higher concentrations, however, the value decreases, and may become considerably smaller than 17. Certainly the value $n_{app} = 17$ is an artifact, and it can even be inferred from these curves that values as small as 12 might be obtained under some conditions.

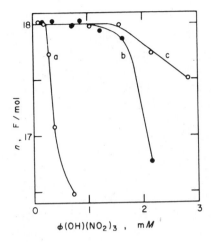

Figure 5.12 Effect of the initial concentration of picric acid on the value of n_{app} obtained on reduction at -0.40 V versus SCE in (a) 0.1, (b) 1, and (c) 3 M hydrochloric acid. Reprinted, with permission, from L. Meites, "Controlled-Potential Electrolysis," in A. Weissberger and B. W. Rossiter, Eds., *Physical Methods of Chemistry*, Vol. 1, Part IIA, 4th ed., Wiley-Interscience, 1971.

In all these cases the rate constant of the second-order side reaction could be evaluated with the aid of coulometric data obtained over a wide range of initial concentrations of the electroactive material [145]. This is discussed further in Section 6. Here it suffices to stress that a single nonintegral value of n_{app} is uninformative, and that data showing how that value is affected by changing the experimental conditions are far more revealing. One should at least investigate the effect of changing the initial concentration of the electroactive material: if this has no effect on the value of n_{app}, the side reaction must be pseudo-first-order, rather than pseudo-second-order as in the examples cited in the preceding paragraph. Other variables that deserve investigation are the efficiency of stirring (i.e., the value of s_{Ox}^{*}), the length of time for which the solution is allowed to stand before the electrolysis is begun, the concentration of the supporting electrolyte, and the temperature.

Even an integral value of n_{app} is suspect if it corresponds to an unlikely product or to one that cannot be isolated from the solution after the electrolysis is complete. An example is the value of 3 obtained for the reduction of glyoxal [146], in which the formation of erythritol is improbable. Any mechanism of the general form

$$Ox + n_1 e \rightarrow I$$

$$I \rightarrow P \tag{33}$$

$$I + n_2 e \rightarrow Red$$

in which some intermediate I is partially consumed in giving rise to an inert product P, will yield a value of n_{app} that lies between n_1 and $(n_1 + n_2)$, and that might therefore be fortuitously equal to 3 if $n_1 = n_2 = 2$. Section 6 lists several such mechanisms, along with others of different types that can behave similarly, and shows how they can be distinguished.

5.4   Analytical Applications

Controlled-potential coulometry is a valuable analytical technique because of its selectivity, accuracy, and precision, and also because many prior separations and, in inorganic analysis, adjustments of oxidation state can be effected in a controlled-potential electrolytic cell before the coulometric measurement is begun. Although a listing of the analytical methods based on the technique would be out of place here, a brief outline will be given of the considerations that should enter into the design of any analytical method.

If one wants to analyze a sample of which the substance to be determined is the only electroactive constituent, it is best to select a supporting electrolyte in which that substance yields a well-defined wave at a voltammetric indicator electrode. A medium that gives rise to an ill-defined wave may also be suitable, but chemical side reactions and other processes that affect the total consumption of electricity are common for such waves, and special care should therefore be taken to ascertain that good results are obtained with known samples. A suitable volume of the supporting electrolyte is deaerated and pre-electrolyzed, in the general fashion described in Section 3.4. Usually the pre-electrolysis is performed at a potential that lies on the plateau of the wave, although one on the rising part of the wave may be selected instead if the rate of reduction or oxidation of the supporting electrolyte is inconveniently large on the plateau and if the half-reaction of interest is irreversible; see Figure 5.4 and the discussion that accompanies it. When the current has decreased to a small constant value, the potentiostat is disconnected from the cell, a suitable small volume of a solution containing the substance to be determined is added to the working-electrode compartment, and the resulting mixture is deaerated very thoroughly. The current integrator is connected in series with the auxiliary electrode, the potentiostat is reconnected to the cell, and the electrolysis is resumed and carried to completion. It is advisable to check the procedure with known amounts of the substance being determined, and to make sure that the result is not affected by reasonable changes of the efficiency of stirring, the potential of the working electrode, and other experimental variables.

If the product obtained in this way could be reoxidized or re-reduced at some other potential, one should consider determining it rather than the starting material. A solution of vanadium(IV) could be standardized by measuring the quantity of electricity that is consumed when a known aliquot of it, in, say, 0.1 M sulfuric acid as the supporting electrolyte, is electrolyzed at a potential where the overall half-reaction is $VO^{2+} + 2H^+ + 2e = V^{2+} + H_2O$. However, the rate of reduction of hydrogen ion is high at any such potential, and it is difficult to apply a correction because that rate varies with the concentration of vanadium(II) and is therefore not constant during the electrolysis. Far better results could be achieved by measuring the quantity of electricity consumed in reoxidizing the vanadium(II) to vanadium(III), for this occurs at a much less negative potential and without interference from the induced reduction of hydrogen ion [50].

Prior separations are effected in much the same way. To determine zinc(II) in a sample that also contains copper(II), the copper must first be rendered innocuous because it is reduced at more positive potentials than zinc(II). In an ammoniacal ammonium citrate medium, the polarographic half-wave potentials are -0.165 and -0.340 V versus SCE for the reduction of copper(II) successively to copper(I) and copper amalgam, and -1.236 V versus SCE for the reduction of zinc(II) to zinc amalgam. Hence one could determine zinc by the following steps:

1. Deaeration of the supporting electrolyte, accompanied by pre-electrolysis at, say, -1.45 V versus SCE to remove traces of impurities, such as nickel(II) and cobalt(II), that can be reduced at that potential.

2. Addition of a portion of the sample and controlled-potential deposition of the copper at, say, -1.0 V versus SCE.

3. Removal and replacement of the mercury.

4. Controlled-potential deposition of the zinc at -1.45 V versus SCE.

5. Measurement of the quantity of electricity consumed in stripping the zinc from the amalgam at, say, -0.5 V versus SCE.

The third step is necessary if the concentrations of copper(II) and zinc(II) are high, because the intermetallic compound CuZn [147–150] would then precipitate in the mixed amalgam, and its redissolution is so slow that the precipitated zinc would not be oxidized in Step 5. If the concentrations of both copper and zinc in the amalgam were sure to be so small that precipitation could not occur, one could omit the third step and combine the second and fourth into a single electrolysis at -1.45 V. Similarly, the first step is intended to eliminate the possibility of forming NiZn, CoZn, or another compound of the same sort, and might likewise be omitted if the concentrations of zinc and the other metals are very low.

The analysis most frequently performed by controlled-potential coulometry is probably that of reactor fuel solutions and other materials for uranium [59, 103]. A solution containing uranium(VI) in 0.5–$1.0\,M$ sulfuric acid as the supporting electrolyte is pre-electrolyzed with a mercury-pool working electrode at a potential (usually $+0.085$ V versus SCE) where iron(III) and other strong oxidizing agents are reduced, but where uranium(VI) is not. Then the quantity of electricity is measured in a second electrolysis at a potential (usually -0.325 V versus SCE) where uranium(VI) is reduced to uranium(IV). The only substances that interfere are those, including copper(II), that are reduced at the latter potential but not at the former one. If only a moderate amount of copper is present, it can be determined by measuring the quantity of electricity consumed in a third electrolysis at $+0.085$ V [151], and that quantity can be subtracted from the one measured at -0.325 V to evaluate the quantity of electricity consumed in reducing uranium(VI). There are also procedures in which interfering substances are removed by extracting with methyl isobutyl ketone [152, 153], triisooctylamine [154], or tri-n-octylphosphine oxide [155, 156), followed by back-extraction of the uranium into an aqueous solution, and other ways of effecting preliminary separations can be used as the experimenter's resources and preferences dictate.

The accuracy and precision of an analytical method depend on the integrator that is used, on the magnitudes of the "background" corrections for Q_c and $Q_{f,i}$, on the care that is taken to eliminate electroactive impurities, and on the quantity of material being determined. Control charts reproduced by Fisher [37] show that the determination of uranium in reactor fuel solutions and similar materials can be carried out routinely with a relative standard error on the order of 0.1%, and the literature contains many examples of the achievement of accuracy and precision on the order of 0.02% by a single operator with a single instrument. These are generally at or above the 0.1-mF level and with systems in which the continuous faradaic current is small—say, 0.2%, or less, of the initial current due to the substance being determined—and in which the supporting electrolyte can be rigorously purified by pre-electrolysis before the sample is added. The accuracy and precision become poorer if correction must be made for an appreciable continuous faradaic current, for reasons that can be discerned by inspecting Figure 5.13. No matter how the extrapolation is made, its uncertainty becomes larger as the value of $i_{f,c}$ increases and as that of s_{Ox}^* decreases.

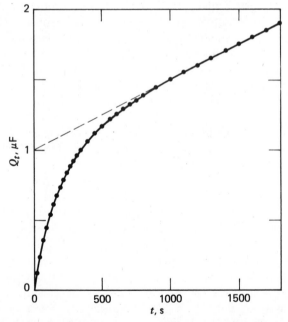

Figure 5.13 Principle of correcting for $i_{f,c}$ algebraically or graphically. The circles represent the dependence of Q_t on time during an electrolysis in which (34) was obeyed, with $Q_{corr} = 1.000$ mF, $s = 0.005$ s^{-1}, and $i_{f,c} = 48.2$ mA ($= 5 \times 10^{-4}$ mF/s). The dashed line represents a calculation of Q_{corr}, the quantity of electricity that corresponds to the substance being determined, by correcting for the current that continues to flow after the reduction or oxidation of that substance is complete.

Shia and Meites [113] found that the relative precision that can be attained in evaluating Q_∞ by nonlinear regression onto the three-parameter equation

$$Q_t = Q_\infty[1 - \exp(-s^*_{Ox}t)] + i_{f,c}t \tag{34}$$

is several times as large as that with which it can be evaluated by a similar procedure when $i_{f,c}$ is negligible and the data conform to the two-parameter equation

$$Q_t = Q_\infty[1 - \exp(-s^*_{Ox}t)] \tag{27}$$

which can be obtained from (15b) and (25), but that it is independent of the value of $i_{f,c}$ unless this is so large that it affects the relative magnitudes of the terms on the right-hand side of (26).

6   STUDIES OF REACTION MECHANISMS

6.1   Introduction

The half-reaction $Ox + ne \rightarrow Red$ that takes place at the working electrode may occur in a single step or in a series of steps of which the first involves electron transfer and is rate-determining; it may occur in a series of steps of which two or more are comparably slow; or it may involve the formation of intermediates that can undergo side reactions leading to other products. These three situations give rise to different kinds of behavior. The first and simplest may be identified by the following criteria:

1. R is the only product obtained.
2.   The quantity Q_∞/FN^0_{Ox} is equal to the integral value of n.
3.   A plot of $\log i$ versus t is truly linear.
4.   All these things are true regardless of variations of the efficiency of stirring, the initial concentration of Ox, or (as long as the rate of the reoxidation of Red remains negligible) the potential of the working electrode.

Although the second and third of these criteria are in the forms that most readers will probably find most convenient, the writer believes them to be grossly inferior to the single criterion that the deviation plot [114] obtained from nonlinear regression onto (27) shows no evidence of nonrandom deviations.

Spectroscopic [157–160], titrimetric [161–164], chromatographic [165–168], and many other techniques have been used to determine the products of controlled-potential electrolysis and to ascertain whether mixtures have been obtained. The second criterion is straightforward in the light of Section 5 above, but the third is not, because small but significant deviations from linearity are difficult to detect, and also because the noise that afflicts measurements of the current may be undesirably large. Uchiyama, et al. [169] regarded plots of i versus Q_t [99, 131] as superior to plots of $\log i$ versus t on the ground that the former are less affected by the residual current, but the writer does not agree. Continuous recording of the current always gives a noisy curve, but does so for the reasons discussed in connection with (26) rather than because of any phenomenon associated with the "residual current," and plots of i or log i versus any other variable differ chiefly in the portions of them in which the noise is most prominent. In the writer's opinion the best procedure is to estimate the average current during each of a number of successive intervals of duration $2\,\Delta t$, using a variant of (29):

$$i_{t,\text{est}} = \frac{Q_{t+\Delta t} - Q_{t-\Delta t}}{2\,\Delta t} \tag{35}$$

as was first done by Ficker and Meites [170]. It can be shown [171] that the ratio of this estimate to the true value is given by

$$\frac{i_{t,\text{est}}}{i_{t,\text{true}}} = \frac{\sinh(s_{\text{Ox}}^*\,\Delta t)}{s_{\text{Ox}}^*\,\Delta t} \tag{36}$$

if the current does decay exponentially with time, and hence that the relative error is smaller than 0.2% if Δt does not exceed $0.1/s_{\text{Ox}}^*$. If s_{Ox}^* has the typical value $6 \times 10^{-3}\,\text{s}^{-1}$, this corresponds to an interval of 15 s or less. Plots versus time of the resulting values of log $i_{t,\text{est}}$ are shown in Figure 5.14 for two electroactive substances.

Both of these plots appear to be satisfactorily linear, but in fact one of them is not. An even more sensitive test is needed. A suitable one may be based on the quantity Q_R defined by (31a). Combining that definition with (27) yields

$$s_{\text{Ox}}^* = -\frac{1}{t}\ln\!\left(\frac{Q_R}{Q_\infty}\right) \tag{37}$$

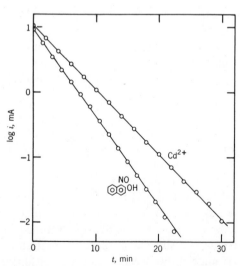

Figure 5.14 Plots of log $i_{t,\text{est}}$ against t for the controlled-potential electroreductions of cadmium(II) and 1-nitroso-2-naphthol [172] at mercury electrodes in acetate buffers at potentials on the plateaus of their respective waves. Reprinted, with permission, from L. Meites, "Controlled-Potential Electrolysis," in A. Weissberger and B. W. Rossiter, Eds., *Physical Methods of Chemistry*, Vol. 1, Part IIA, 4th ed., Wiley-Interscience, 1971.

Values of Q_R can be computed from those of Q_t and Q_∞, or can be measured or recorded [75] if a direct-reading current integrator is available. Harrar, et al. [127] devised ways, based on (37), of obtaining plots of s^*_{Ox} versus t directly. Even higher sensitivity may be obtained by writing (37) both for a time t_1 when $Q_R = Q_{R,1}$ and for a later time t_2 when $Q_R = Q_{R,2}$. The resulting expressions can be combined to give

$$s^*_{Ox} = \frac{1}{t_2 - t_1} \ln\left(\frac{Q_{R,1}}{Q_{R,2}}\right) \qquad (38)$$

The value of s^*_{Ox} will be constant if, but only if, the current–time curve is accurately described by (10). Figure 5.15 shows plots of s^*_{Ox} versus time corresponding to the curves in Figure 5.14. An essentially horizontal straight line is obtained for the reduction of cadmium(II), which is free from complications under these conditions: the small drift at values of t exceeding 20 min merely reflects the small uncertainty in the value of Q_∞ used in evaluating $Q_{R,1}$ and $Q_{R,2}$. The reduction of 1-nitroso-2-naphthol, however, occurs by an ECE mechanism (Case 5b, Section 6.6) [24, 172–175], which can probably be represented by the equations

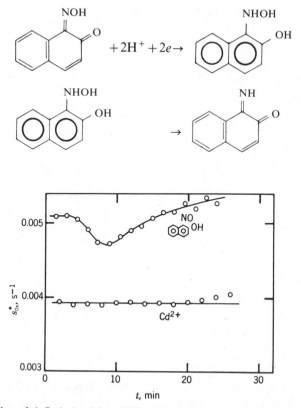

Figure 5.15 Plots of s^*_{Ox} [calculated from (39)] against time for the data of Figure 5.14. Reprinted, with permission, from L. Meites, "Controlled-Potential Electrolysis," in A. Weissberger and B. W. Rossiter, Eds., *Physical Methods of Chemistry*, Vol. 1, Part IIA, 4th ed., Wiley-Interscience, 1971.

Although the loss of water is too rapid to produce any obvious deviation from linearity in the plot of Figure 5.14, the occurrence of some chemical complication is unmistakably revealed by Figure 5.15.

The second of the above possibilities comprises mechanisms that involve two or more steps, but in which each substance can react in only one way. A typical mechanism of this sort is

$$Y \rightarrow Ox$$

$$Ox + ne \rightarrow Red$$

in which the current is governed by the kinetics of the prior transformation of the electroinactive precursor Y into Ox as well as by the factors that govern the rate of the charge-transfer step in the simple case, and another is the ECE mechanism mentioned several times above. Here:

1. Red is again the only product obtained.
2. The quantity Q_∞/FN^0 (which is equal to n_{app}) is again equal to the integral value of n, although N^0 must now be taken as the sum of the numbers of moles of Ox and Y present initially. When these things are true, the yield of product and the coulometric data are alike unaffected by the occurrence of a prior or intervening chemical step.
3. However, the value of s_{Ox}^* computed from (37) or (38) is not constant throughout the electrolysis (so that a plot of log i versus t is not linear).
4. The shape of a plot of either s_{Ox}^* or log i versus t is affected by variations of the efficiency of stirring, of the potential of the working electrode on the rising part of the wave if the wave is irreversible, of the initial concentration of the starting material if a pseudo-second- or higher-order rate-determining step is involved, and of pH if hydrogen ions are involved in that step.

The third of the above possibilities includes "branched" mechanisms such as

$$Ox + n_1 e \rightarrow I$$

$$2I \rightarrow D$$

$$I + n_2 e \rightarrow Red$$

of which several examples were cited in Section 5.3. For any such mechanism:

1. The yield of Red is less than 100%.
2. The value of n_{app} will differ from that corresponding to reduction to Red and will usually be nonintegral.
3. A plot of log i versus t may or may not be linear.

4. The value of n_{app} and the shape of a nonlinear plot of $\log i$ versus t will be affected by variations of the efficiency of stirring, of the potential of the working electrode, of the concentration of starting material if the competing step is a pseudo-second- or higher-order one, and of pH if it involves hydrogen ions.

The behaviors of many specific mechanisms of these general types have been worked out in detail and reviewed elsewhere [75, 102]. The following discussion includes very brief descriptions of the cases that have been investigated, with emphasis on differential diagnosis. Frequent reference is made to procedures for evaluating the rate constants of homogeneous reactions. As the peculiar advantage of these techniques is that they are applicable to reactions that involve unstable substances and are therefore difficult to study in other ways, there are few situations in which the values thus obtained have been confirmed by other techniques: just enough [77, 176–178] to show, when taken together with the nature of the theoretical foundation, that these procedures rest on a firm and trustworthy footing.

Such comparisons would undoubtedly be much more numerous if equations describing the dependences of i and Q_t on time were less complex than they are. Most evaluations of rate constants have been made by such expedients as constructing a family of "working curves" for a number of arbitrarily chosen values of a dimensionless parameter, such as k/s_{Ox}^*, where k is the pseudo-first-order rate constant of a chemical step. Similar experimental plots are matched with the working curves to find the value of the dimensionless parameter that gives the best match, and that value is combined with estimates of the experimental variables (such as s_{Ox}^* in the example just given) to obtain a value of k. Apart from being imprecise and cumbersome, such a procedure cannot succeed unless the values of all the other parameters are either known or devoid of appreciable effect on the working curves.

The alternative is to evaluate the desired rate constant directly, together with all the other parameters, by nonlinear regression. This has been shown [24] to succeed with the ECE mechanism (Case 5b below), for which the equation for the current–time curve is strikingly ill-suited to any other technique for handling data. Evaluations of rate constants, and comparisons of the results with those obtained in other ways, may be expected to become much more common in the future.

6.2 Catalytic Processes

CASE 1a. REGENERATION OF ELECTROACTIVE MATERIAL BY A HOMOGENEOUS REACTION FOLLOWING ELECTION TRANSFER

The mechanism

$$Ox + ne \rightarrow Red$$
$$Red + Z \rightarrow Ox \tag{39}$$

where Z may be hydrogen ion, water, or some other major constituent of the solution, gives rise to plots of log i (corrected, if necessary, for the continuous faradaic current obtained with the supporting electrolyte alone) versus t that are concave upward and have a finite horizontal asymptote. This asymptote represents a steady state in which Ox is reduced at the surface of the working electrode just as rapidly as it is produced by the homogeneous reaction. The catalytic current at the steady state is proportional both to c_{Red} and to c_Z, where each c is the steady-state concentration of the species denoted by the subscript; but it is essentially independent of the potential of the working electrode. The order and rate constant of the homogeneous step are easily obtained. If that step is first-order with respect to Red, its rate constant (or, more properly, its pseudo-first-order rate constant kc_Z) can also be evaluated by extrapolating the final linear portion of a plot of Q_t versus t back to $t = 0$; the intercept is less than the expected coulometric result nFN_{Ox}^0.

The foregoing description applies to the ordinary situation in which the product kc_Z is fairly small, so that the steady-state current is much smaller than the current at the start of the electrolysis. At the other extreme, kc_Z may be so large that almost all of the Red produced at the electrode surface is reoxidized before it can escape from the diffusion layer into the bulk of the solution, and in this event the current is abnormally large and virtually constant. This situation arises in the reduction of molybdenum(V) in saturated hydrazine dihydrochloride [179]; other examples could be constructed in profusion, but a detailed treatment is neither available nor likely to be of much interest [50, 75, 176, 178, 180–185].

CASE 1b.   CATALYTIC REDUCTION BY A CHARGE-TRANSFER STEP INVOLVING THE PRODUCT RED

The mechanism

$$Ox + ne \rightarrow Red$$
$$Red + H^+ \rightarrow RedH^+ \quad \quad (40)$$
$$RedH^+ + e \rightarrow Red + \tfrac{1}{2}H_2$$

involves the reduction of hydrogen ion (or some other major constituent of the supporting electrolyte) catalyzed by Red. It closely resembles Case 1a but differs from it in two respects:

1. The steady-state catalytic current depends on the potential of the working electrode.
2. If the rate of protonation of Red is high, as is usually true, the steady-state current will be proportional to the activity of hydrogen ion only over the range of pH-values in which most of the Red remains unprotonated; it will be independent of the activity of hydrogen ion in the range where most of the Red is in the protonated form [50, 75, 77].

CASE 1c. CATALYTIC REDUCTION BY A CHARGE-TRANSFER STEP INVOLVING THE PRODUCT OX

The related mechanism

$$Ox + ne \rightarrow Red$$
$$Ox + H^+ \rightarrow OxH^+ \tag{41}$$
$$OxH^+ + e \rightarrow Ox + \tfrac{1}{2}H_2$$

has been described as reduction of hydrogen ion induced by the reduction of Ox. It can be imagined to involve two alternative modes of decomposition of the product obtained when the first electron is transferred to an OxH^+ ion or molecule (or to a hydrogen ion or a molecule of water acting as a bridge between the electrode and an ion or molecule of Ox) [75, 77, 101, 158, 186].

6.3   Other Subsequent and Prior Reactions of Red or Ox

CASE 2a.   INACTIVATION OF RED BY A HOMOGENEOUS REACTION FOLLOWING ELECTRON TRANSFER

The mechanism

$$Ox + ne \longrightarrow Red$$
$$Red + Z \xrightarrow{k} P \tag{42}$$

in which Z is some major constituent of the solution and P is an electrolytically inert product, was discussed in another connection in Section 5. If the electrolysis is conducted at a potential where Red is not reoxidized, a plot of $\log i$ versus t is linear and n_{app} is equal to n. The occurrence of the inactivation step can be detected by isolating and identifying the products. Its pseudo-first-order rate constant kc_Z can be evaluated by following the time dependence of the concentration of Red or P, by reversal coulometry, or by analysis of current–time data obtained in an electrolysis on the rising portion of the polarographic wave if the Ox–Red couple is reversible. Other ways of identifying and studying this mechanism were described in Section 4.3 [178, 187].

CASE 2b.   INACTIVATION OF RED BY PARALLEL PSEUDO-FIRST-ORDER REACTIONS

The mechanism

$$Ox + ne \longrightarrow Red$$
$$Red + Z_1 \xrightarrow{k_1} P_1 \tag{43}$$
$$Red + Z_2 \xrightarrow{k_2} P_2$$

gives rise to behavior similar to that encountered in Case 2a. Determining the yields of P_1 and P_2 provides a value of the ratio $k_1 c_{Z_1}/k_2 c_{Z_2}$ of the pseudo-first-order rate constants; their sum can be evaluated in any of the ways applicable in Case 2a [178].

CASE 2c.   SUBSEQUENT INACTIVATION STEPS

Amatore and Savéant [188] examined the relative yields of the products obtained from the mechanism

$$Ox + e \rightarrow Red$$
$$Red \rightarrow I$$
$$2I \rightarrow D \qquad\qquad (44)$$
$$I \rightarrow P$$

CASE 2d.   FORMATION OF OX BY A HOMOGENEOUS PRIOR REACTION

The mechanism

$$Y \underset{k_b}{\overset{k_f}{\rightleftharpoons}} Ox$$
$$Ox + ne \xrightarrow{\hspace{1.5cm}} Red \qquad\qquad (45)$$

which is classically associated with the kinetic current in polarography, was mentioned above. The shape of the current–time curve depends on the relative values of the rate constants k_f and k_b, which pertain to the transformations of Y into Ox and of Ox into Y, respectively, and s_{Ox}^*. At one extreme, if both k_f/k_b and k_f/s_{Ox}^* are large, a plot of $\log i$ versus t is so nearly linear that the occurrence of the prior reaction may escape detection; at the other, if both of these quantities are small, a steady state, in which Ox is reduced just as rapidly as it is produced by the prior reaction, may be reached almost immediately after the beginning of the electrolysis. An intermediate case of special importance is that in which k_f is very much smaller than s_{Ox}^* but is of the same order of magnitude as k_b, so that the initial equilibrium mixture contains comparable concentrations of Y and Ox. By analogy with Case 4a below, one can dissect the resulting plot of $\log i$ versus t into two linear segments, of which the first can be ascribed to the reduction of Ox that was present initially and the second to the reduction of Ox that is formed from Y as the electrolysis proceeds. From the slopes and intercepts of these segments it is possible to evaluate the rate constants (whose ratio is equal to the equilibrium constant of the prior reaction) and also the initial concentrations of Y and Ox [74, 170, 189–191].

CASE 2e.   INACTIVATION OF OX BY A HOMOGENEOUS PSEUDO-FIRST-ORDER REACTION

The mechanism

$$Ox + ne \rightarrow Red$$
$$Ox + Z \rightarrow Y \qquad\qquad (46)$$

corresponds to Case 2c with $k_f = 0$ but differs from it chemically: what is contemplated here is the addition of Ox to a supporting electrolyte containing a large excess of some substance Z that inactivates it, followed by the commencement of the electrolysis at a time, t_1 seconds later, at which the inactivation reaction is still far from complete. A plot of $\log i$ versus t is linear, but Q_∞ is smaller than nFN_{Ox}^0. Coulometric data, which should include information on the effect of varying t_1, permit the evaluation of the rate constant k_b for the reaction of Ox with Z [189].

CASE 2f.   INACTIVATION OF OX BY HOMOGENEOUS REACTIONS OF OTHER TYPES

If (a) the initial concentration of Z in Case 2d above is not very much larger than that of Ox, or (b) Ox is inactivated by dimerization or some other second-order process of the form

$$2Ox \rightarrow Y$$

Q_∞ will again be smaller than nFN_{Ox}^0, but a plot of $\log i$ versus t will not be exactly linear. No example of either case is known [189].

6.4   Reactions of Ox and Red with Products or Intermediates

CASE 3a.   REACTION OF OX WITH RED TO YIELD AN INERT PRODUCT

If k is the second-order rate constant for the reaction of Ox with Red in the simplest "father–son" mechanism [192]

$$Ox + ne \longrightarrow Red$$
$$Ox + Red \xrightarrow{k} P$$

(47)

the value of n_{app} approaches n as kc_{Ox}^0/s_{Ox}^* approaches zero, and approaches $n/2$ as kc_{Ox}^0/s_{Ox}^* increases without limit. At either of these extremes a plot of $\log i$ versus t is virtually linear. The first extreme is trivial; the second can generally be diagnosed either by identifying the product or from the value of n_{app}, which may be half-integral. Of more interest is the intermediate situation, in which the final mixture contains appreciable concentrations of both Red and P, in which n_{app} lies between $n/2$ and n and can be used to evaluate kc_{Ox}^0/s_{Ox}^*, and in which the plot of $\log i$ versus t is not linear. This can often be achieved by varying the initial concentration of Ox, since Red is formed almost exclusively if kc_{Ox}^0/s_{Ox}^* is less than about 0.03, while P is formed almost exclusively if kc_{Ox}^0/s_{Ox}^* exceeds about 30. One may also vary s_{Ox}^* by changing the efficiency of stirring or the potential of the working electrode, and it may sometimes be possible to vary k by changing the composition of the supporting electrolyte. In reversal coulometry (where a certain fraction of the starting material Ox is reduced to Red by the flow of a quantity of electricity denoted by Q_f, and the applied potential is reset to a value at which Red is reoxidized to Ox, and a second quantity of electricity Q_b^0 is accumulated during an exhaustive electrolysis without changing the rate of mass transfer) the value of Q_b^0/Q_f varies slightly with that of the dimensionless parameter s/kc_{Ox}^0 if the latter is small, and k can be evaluated from working plots given by Yeh and Bard [193]; also [157, 178, 194].

CASE 3b.   THE EFFECT OF A COMPETING REACTION THAT INACTIVATES RED

The mechanism

$$Ox + ne \longrightarrow Red$$
$$Ox + Red \xrightarrow{k_1} P_1 \tag{48}$$
$$Red + Z \xrightarrow{k_2} P_2$$

differs from Case 3a in that it includes a pseudo-first-order inactivation of Red by reaction with some species Z present in large excess, and gives rise to generally similar behavior. The range of variation of n_{app} is decreased by the occurrence of the inactivation reaction, and the value of n_{app} varies with the concentration of Z [178].

CASE 3c.   REACTION OF OX WITH AN INTERMEDIATE TO YIELD AN INERT PRODUCT

The mechanism

$$Ox + n_1 e \longrightarrow I$$
$$Ox + I \xrightarrow{k} P \tag{49}$$
$$I + n_2 e \longrightarrow Red$$

is a variant of Case 3a, the limits of n_{app} now being $(n_1 + n_2)$ as $k c_{Ox}^0 / s_{Ox}^*$ approaches zero and $n_1/2$ as $k c_{Ox}^0 / s_{Ox}^*$ increases without limit. It should be noted that the homogeneous reaction cannot occur if the electrolysis is performed at a potential on the plateau of the wave, where I is reduced as rapidly as it is formed, for the concentration of Ox then cannot be appreciable at the surface of the electrode and the concentration of I cannot be appreciable anywhere. Electrolysis on the rising portion of the wave does, however, allow some of the I to survive long enough to participate in the homogeneous reaction unless it is much more easily reduced than Ox [32, 75].

CASE 3d.   REACTION OF OX WITH RED TO YIELD AN ELECTROACTIVE PRODUCT

If the values of s^* for Ox and I are comparable, the mechanism

$$Ox + n_1 e \longrightarrow Red$$
$$Ox + Red \xrightarrow{k} I \tag{50}$$
$$I + n_2 e \longrightarrow Red$$

can be identified only by following the dependence on time of the concentration of Ox or I. Much the same thing must be true in reversal coulometry, for Yeh and Bard [193] found Q_b^0 / Q_f to be nearly independent of the fraction of the Ox that is reduced in the forward electrolysis in the closely related mechanism

$$Ox + n_1 e \longrightarrow I$$
$$I + Ox \xrightarrow{k} J \tag{51}$$
$$J + n_2 e \longrightarrow Red$$

However, the overpotential for the reduction of Ox may so far exceed that for the reduction of I that a potential lying near the foot of the wave of Ox lies on or near the plateau of the wave of I. Then the current rises at the start of the electrolysis, passes through a maximum, and finally decays toward zero. The rate constant of the homogeneous reaction can be evaluated from data that include the maximum value of the current [77, 186].

CASE 3e. REACTION OF AN INTERMEDIATE WITH RED TO YIELD
AN ELECTROACTIVE PRODUCT

If the rate constant k for the homogeneous reaction in the mechanism

$$Ox + n_1 e \longrightarrow I$$
$$I + Red \xrightarrow{k} J \qquad (52)$$
$$J + n_2 e \longrightarrow Red$$

is very small, the current may decay nearly to zero as the reduction of Ox to I approaches completion, and then remain small for some time while the concentration of Red is slowly increased by the reactions that follow. As more and more Red accumulates, the rate of the homogeneous reaction increases and the current rises in proportion to the concentration of J. After passing through a maximum when most of the I has been consumed, the current decreases again as the remaining J is reduced.

Only one example is available as yet, and its classification is rather uncertain. It arose in the reduction of 1,3-diphenyl-1,3-propanedione in dimethyl sulfoxide and was interpreted by assuming the validity of the rate equation

$$- \frac{dc_I}{dt} = k c_{Red}$$

according to which the homogeneous reaction is zeroth order in I. However, the values of k thus computed were so strongly dependent on c_{Ox}^0 that the assumption must be regarded as dubious until further evidence is adduced [76].

6.5 Parallel Reductions

CASE 4a. REDUCTIONS OF TWO CHEMICALLY UNRELATED SPECIES

In the absence of any chemical interactions among the various substances involved, the controlled-potential electroreduction of a mixture of Ox_1 and Ox_2

$$Ox_1 + n_1 e \rightarrow Red_1$$
$$Ox_2 + n_2 e \rightarrow Red_2 \qquad (53)$$

gives rise to a plot of log i versus t that can be dissected into linear segments if the values of s^* for Ox_1 and Ox_2 are sufficiently different. The plot becomes linear at long times, after virtually all of the more rapidly reduced component (assumed to be Ox_1) has been consumed. The slope of this linear portion is equal to $-0.434 s_2^*$, and its intercept at zero time, log i_2^0, is easily obtained by extrapolation. The quantity of electricity required to reduce all the Ox_2 initially present is given by

$$Q_2 = \frac{i_2^0}{s_2^*}$$

Values of i_2 at various times during the first part of the electrolysis may be obtained from the same extrapolation. Subtracting these from the corresponding total measured currents gives values of i_1, and a plot of log i_1 versus t is linear; its slope and intercept at zero time may be combined in the above fashion to obtain a value of Q_1, or one may write simply

$$Q_1 = Q_\infty - Q_2$$

where Q_∞ is the total quantity of electricity consumed by the mixture. These considerations have been employed in analyzing some simple mixtures [195]. This kind of kinetic analysis may prove useful in dealing with mixtures of substances that yield overlapping irreversible voltammetric waves. It is generally necessary (and possible) to select a working-electrode potential that lies near the plateau of one of these waves but near the foot of the other, so that the values of s^* for the two electroactive substances will be very different [75, 196–198].

CASE 4b.   REDUCTIONS OF OX ALONG PARALLEL PATHS

The mechanism

$$Ox + n_1 e \rightarrow Red_1$$
$$Ox + n_2 e \rightarrow Red_2$$

(54)

gives rise to a strictly linear plot of log i versus t and a value of n_{app} intermediate between n_1 and n_2. As the heterogeneous rate constants for the two electron-transfer processes are likely to vary with potential in different ways, the value of n_{app} will usually be potential-dependent. Isolation and identification of the products are essential for proper diagnosis.

The related mechanism

$$Ox + H^+ = OxH^+ \quad \text{(fast)}$$
$$Ox + n_1 e \rightarrow Red_1$$
$$OxH^+ + n_2 e \rightarrow Red_2$$

(55)

behaves similarly, although the value of n_{app} will vary with the pH in a manner that depends on whether n_1/n_2 is larger or smaller than 1. Some of the characteristics of Case 2c may appear as the pH is increased and the rate of protonation decreases [199].

6.6   Other Mechanisms

CASE 5a.   CONSECUTIVE REDUCTIONS WITHOUT AN INTERVENING CHEMICAL STEP

The mechanism

$$Ox + n_1 e \rightarrow I$$
$$I + n_2 e \rightarrow Red$$

(56)

gives $n_{app} = (n_1 + n_2)$ at any potential, but the shape of a plot of $\log i$ versus t depends on the heterogeneous rate constant k_1 for the reduction of I. At a potential near the foot of the wave of I, where this rate constant is small, much of the I can escape into the bulk of the solution, and then its reduction will continue slowly for a long time after the last of the Ox has been consumed. A plot of $\log i$ versus t then has the shape described above in connection with Case 4a, and can be dissected in the manner prescribed there. The ratio Q_1/Q_2 of the resulting current integrals is very nearly equal to n_1/n_2 in this region (but not at potentials nearer the plateau of the wave of I), and the dependence on potential of the slope of the more slowly decaying segment can be used to evaluate αn_a for the reduction of I. As the potential becomes more negative, the plot of $\log i$ versus t becomes more nearly linear, the slope of its first portion remaining constant while that of the second one increases, and on the plateau of the wave of I the observed behavior is that of the simple overall mechanism [22, 32, 74, 75, 194, 200]

$$Ox + (n_1 + n_2)e \rightarrow Red$$

CASE 5b.   THE ECE MECHANISM

The mechanism

$$Ox + n_1 e \longrightarrow I$$

$$I \xrightarrow{k} J \tag{57}$$

$$J + n_2 e \rightarrow Red$$

often called the "ECE" ("electrochemical–chemical–electrochemical") mechanism, gives $n_{app} = (n_1 + n_2)$. As was explained in Section 4.2, this value of n_{app} exceeds the overall n-value deduced from polarographic data if k is so small that the transformation of I into J, and the subsequent further reduction of J, do not proceed almost to completion during the life of a drop at a dropping electrode. For any given common value of the diffusion coefficients of the substances involved, the shape of a plot of $\log i$ versus t depends on the value of the ratio k/s^*, where $s^* = s_{Ox}^* = s_J^*$, and nearly (but not quite) uniquely characteristic shapes are obtained if the value of k/s^* is neither very large nor very small. Since, even apart from the possibility of varying k by altering the composition of the supporting electrolyte, values of s^* ranging over about four orders of magnitude can be obtained without great difficulty by adjusting the efficiency of stirring and the area of the working electrode, curves having different ones of these characteristic shapes can be secured in successive electrolyses as an aid to diagnosis. It may further be noted that, if an electrolysis is allowed to proceed for some (not too long) time and is then interrupted, the transformation of I into J will continue after the flow of current has ceased, and will lead to an increase in the total concentration of reducible material. Consequently, the current that flows immediately after the electrolysis is resumed will exceed the current that flowed immediately before it was interrupted.

For a long time the ECE mechanism was the only one known to give rise to curves having these shapes, but at least two others are now recognized. One [201, 202] occurs in Case 4a when Red_1 is a solid whose presence on the surface of the electrode decreases the rate at which electrons can be transferred between the working electrode and the solution. Adsorption should be capable of producing similar effects. The other

$$YOx_3 = Y + 3Ox \quad \text{(fast)}$$
$$Ox + n_1 e \rightarrow Red$$

(58)

is one in which the electron-transfer step is preceded by a fast homogeneous equilibrium in which both Y and the compound YOx_3 are electrolytically inert; this is believed to account for the current–time curves obtained in the anodic stripping of zinc from platinum–zinc amalgams [78, 191].

Despite these complications, controlled-potential electrolysis is of great value in the diagnosis of ECE mechanisms and in evaluations of k for the chemical steps they involve. This is especially true when these steps are too slow for convenient study by faster techniques such as chronopotentiometry.

The preceding sentence does *not* mean that controlled-potential electrolysis is useful only for the detection and study of slow steps. Evaluations of the rate constant for the chemical step that intervenes between the two charge-transfer steps in the reduction of p-nitrosophenol, using such comparatively fast techniques as chronopotentiometry, chronoamperometry, and chronocoulometry, have given values in the vicinity of $0.4 \, s^{-1}$ over the range of pH-values from about 1.8 to 5.5 [203–207]. Evaluating the rate constant for the corresponding step in the reduction of 1-nitroso-2-naphthol, using current–time data obtained in controlled-potential electrolysis with a large stirred mercury pool, gave $1.1 \pm 0.1_5 \, s^{-1}$ at pH 5.5 [24]; and it was estimated that a value an order of magnitude higher might be attainable in the face of typical noise levels.

Yeh and Bard [193] considered the behavior toward reversal coulometry of the closely related mechanism

$$Ox + n_1 e \longrightarrow I$$
$$I + Z \xrightarrow{\ k\ } J$$
$$J + n_2 e \longrightarrow Red$$

(59)

(which differs from the standard ECE mechanism only in that another constituent Z, present at a large constant concentration, is implicated in the transformation of I into J), and found the ratio Q_b^0/Q_f to be a function of the dimensionless parameter s^*/kc_Z. (For a first-order chemical transformation there is of course a similar dependence on the corresponding parameter s^*/k.) If $Q_f/n_1 F V c_{Ox}^{b,0} = 0.999$—that is, if the reduction to I is virtually complete—the variation is most sensitive when the value of s^*/kc_Z lies between 10 and 20 [32, 74, 77, 95, 208–211].

CASE 5c.   ECE-LIKE MECHANISMS INVOLVING REDOX REACTIONS BETWEEN THE
INTERMEDIATES

Two different kinds of behavior can be envisioned for a system that undergoes reduction by the ECE mechanism:

1.   If J is reduced at a more negative potential than Ox there should, in principle, be two waves on a voltammogram of Ox. On the plateau of the first wave, Ox is reduced to I and some I is transformed into J, but because J cannot be reduced at an appreciable rate the current must be indistinguishable from that for an n_1-electron reduction. At a more negative potential the rate of reduction of J will be higher, and the current must increase to an extent that depends on the rate at which J is formed from I. The latter rate may be so small that the second wave is imperceptible on the voltammogram, but controlled-potential electrolysis, even at a potential preceding the start of that wave, must nevertheless yield J as the eventual product and must give $Q_\infty = n_1 F V c_{Ox}^{b,0}$. At such a potential the data will conform to Case 2a (or some variant thereof), whereas they will conform to Case 5a at a potential sufficiently negative to fall on the rising part or plateau of the second wave. This behavior is illustrated by the reduction of heptaphenyltropylium bromide at -1.4 V versus SCE from solutions in acetonitrile containing 0.1 M tetra-n-butylammonium perchlorate [212], where the inactivation step of Case 2a is thought to involve dimerization followed by rearrangement:

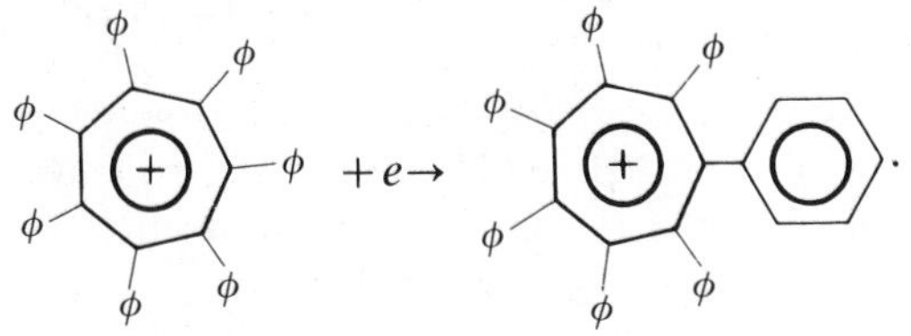

(I)

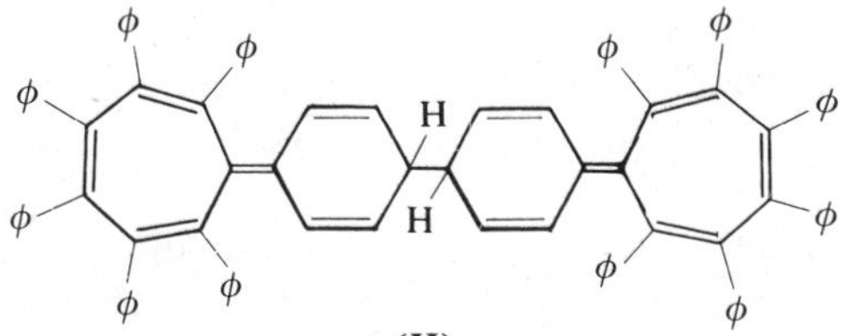

(II)

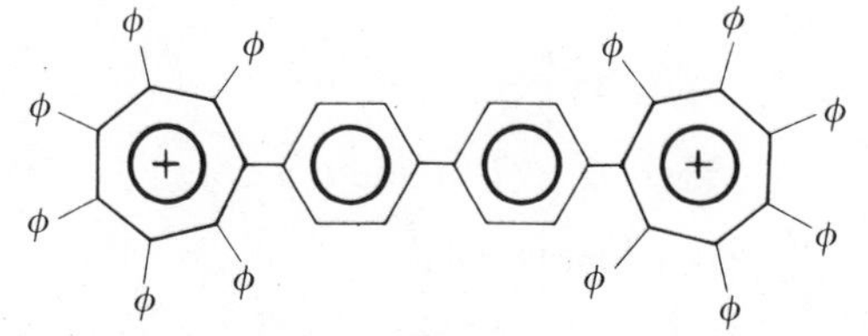

(III)

At the end of the electrolysis the solution gives a wave at -1.9 V. Electrolysis of the starting material at that potential would display the behavior of the ECE mechanism with complications arising from the complexity of this conversion of I ($=$ **I**) into J ($=$ **III**).

2. If, however, J is reduced at a less negative potential than Ox, there can, in principle, be only one wave on the voltammogram, for there is no potential at which J can be formed but can survive for long enough to escape from the surface of the electrode. An example arises in the reduction of picolinic acid at -1.3 V versus SCE from an aqueous acetate buffer [213], where the mechanism may, for the present purpose, be represented by

(pyridine)–COOH $+2H^+ + 2e \rightarrow$ (pyridine)–CH(OH)$_2$

(pyridine)–CH(OH)$_2$ $\rightarrow$ (pyridine)–CHO + H$_2$O

2 (pyridine)–CHO $+2H^+ + 2e \rightarrow$ (pyridine)–CHOH–CHOH–(pyridine)

Picolinic aldehyde, for which $E_{1/2} = -0.59$ V versus SCE (and which is therefore more easily reducible than the acid), can be found in the solution during the course of the electrolysis, although its maximum concentration is only about 4% of the initial concentration of starting material.

There is much evidence that the second of these kinds of behavior is the more common. Hence, if both of the charge-transfer steps are assumed to be reversible for simplicity, the Ox–I couple usually has a more negative formal potential than the J–Red couple, which means that the reaction I $+$ J $=$ Ox $+$ Red is usually spontaneous. On this basis two general mechanisms have been conceived and described at some length. They are often called "DISP" (for "disproportionation") mechanisms, although the terminology is objectionable: disproportionation is the redox analog of autoprotolysis, and does not occur in these mechanisms. One of them, generally called the "DISP1" mechanism [214–216], is represented by the equations

$$Ox + n_1 e \longrightarrow I$$
$$I \underset{k_b}{\overset{k_f}{\rightleftharpoons}} J \quad \text{(rate-determining)} \tag{60}$$
$$I + J = Ox + Red$$

The other, the "DISP2" mechanism [217–219], is represented by the equations

$$\text{Ox} + n_1 e \rightarrow \text{I}$$
$$\text{I} = \text{J} \quad \text{(fast equilibrium)} \tag{61}$$
$$\text{I} + \text{J} \rightarrow \text{Ox} + \text{Red} \quad \text{(rate-determining)}$$

Most of the early investigations and comparisons of these mechanisms concentrated on their behaviors toward chronoamperometry and similar techniques, and led to the conclusions that the DISP2 mechanism can be easily distinguished from either the ECE or the DISP1 mechanism, but that discrimination between the ECE and DISP1 mechanisms is much more difficult [220]. Amatore and Savéant compared the behaviors of the ECE and DISP1 mechanisms directly [221], and expressed the view that the DISP1 mechanism is overwhelmingly the more common of the two and that it may be impossible to prove, from any data obtained during an electrolysis, that a half-reaction is following the ECE mechanism rather than the DISP1 mechanism [222].

Much more might be said about the concepts that are involved, but in view of the space available here the reader must instead be referred to the literature cited above and to an important series of papers by Amatore and Savéant. These authors examined [223] the effect of the rate constants k_1 and k_2 on the ratio of the amounts of Red formed in the chemical and electrochemical pathways of the general mechanism

$$\text{Ox} + e \longrightarrow \text{I}$$
$$\text{I} \xrightarrow{k_1} \text{J}$$
$$\text{J} + e \longrightarrow \text{Red} \tag{62}$$
$$\text{I} + \text{J} \xrightarrow{k_2} \text{Ox} + \text{Red}$$

and also discussed the behaviors of several related schemes, usually with an emphasis on the relative proportions of the different products envisioned:

1.
$$\text{Ox} + e \longrightarrow \text{I}$$
$$2\text{I} \xrightarrow{k_1} \text{D}$$
$$\text{I} \xrightarrow{k_2} \text{J} \tag{63}$$
$$\begin{cases} \text{J} + e \rightarrow \text{Red,} \quad \text{or} \\ \text{I} + \text{J} = \text{Ox} + \text{Red} \end{cases}$$

[224] (see also Case 5f below).

2. The same mechanism with the additional steps [225]:

$$\text{Ox} + \text{I} \xrightarrow{k_3} \text{K}$$
$$\text{K} + e \longrightarrow \text{P} \tag{64}$$
$$\text{K} + \text{I} = \text{Ox} + \text{P}$$

3. First- or pseudo-first-order inactivation of J [226] (compare Case 5e):

$$Ox + e \longrightarrow I$$
$$I \xrightarrow{k_1} J$$
$$J \xrightarrow{k_2} P \tag{65}$$
$$J + e \rightarrow Red$$
$$I + J = Ox + Red$$

which is thought to be involved in many instances of reductive cleavage [227] as well as in electrochemically induced aromatic nucleophilic substitution [228].

4. The branched scheme

$$Ox + e \rightarrow I$$
$$I \xrightarrow{k_1} J$$
$$J \xrightarrow{k_2} K$$
$$J + e \longrightarrow Red \tag{66}$$
$$I + J = Ox + Red$$
$$K - e \longrightarrow P$$
$$Ox + K = I + P$$

[229], which is associated with the reductions of aromatic halides in the presence of soft nucleophiles [228, 230] (and in particular with the reduction of 2-chloroquinoline from solutions in liquid ammonia in the presence of benzene-thiolate [229]), and with reductions of aromatic halides in the presence of alkoxides as proton acceptors [231].

5. A scheme resembling the (Electrochemical–Chemical–Electrochemical–Chemical–Electrochemical) "ECECE" mechanism (Case 5d):

$$Ox + e \longrightarrow I$$
$$I \xrightarrow{k_1} J$$
$$J + e \longrightarrow Red_1$$
$$I + J = Ox + Red_1 \tag{67}$$
$$J \xrightarrow{k_2} K + L$$
$$L + e \longrightarrow Red_2$$
$$I + L = Ox + Red_2$$

[232], which is thought to be involved in reductions of aromatic halides in protogenic solvents [233].

CASE 5d.   THE ECECE MECHANISM

The mechanism

$$Ox + n_1 e \longrightarrow I$$
$$I \xrightarrow{k_1} J$$
$$J + n_2 e \longrightarrow K \qquad (68)$$
$$K \xrightarrow{k_2} L$$
$$L + n_3 e \longrightarrow Red$$

gives $n_{app} = (n_1 + n_2 + n_3)$ and plots of log i versus t that are more complex than those for Case 5b [74].

CASE 5e.   THE ECE MECHANISM WITH A COMPETING PSEUDO-FIRST-ORDER
INACTIVATION OF I

The mechanism

$$Ox + n_1 e \longrightarrow I$$
$$I \xrightarrow{k_1} J$$
$$I \xrightarrow{k_2} P \qquad (69)$$
$$J + n_2 e \longrightarrow Red$$

gives plots of log i versus t that strongly resemble those for Case 5b but, because some of the intermediate is converted into the electrolytically inert product P, the value of n_{app} is smaller than $(n_1 + n_2)$. Ways of evaluating the rate constants for the two chemical steps have been devised [210, 234].

CASE 5f.   THE ECE MECHANISM WITH A COMPETING PSEUDO-SECOND-ORDER
INACTIVATION OF I OR J

Both Bard and Mayell [210] and Meites [145] attempted to elucidate the behavior of the mechanism

$$Ox + n_1 e \longrightarrow I$$
$$2I \xrightarrow{k} D \qquad (70)$$
$$I + n_2 e \longrightarrow Red$$

by assuming (a) that $s^* = s^*_{Ox} = s^*_I$ (i.e., that the electrolysis is performed at a potential where both Ox and I yield their limiting currents, and also that the mass-transfer constants s_{Ox} and s_I have identical values) and (b) that the current at any instant is given by an equation of the form

$$i = s^* n_1 F V c^b_{Ox} + s^* n_2 F V c^b_I$$

where c^b is the bulk concentration, at the instant in question, of the species denoted by the subscript. Bard and Mayell gave plots of n_{app} against log $c^{b,0}_{Ox}$ for various values of the ratio s^*/k; Meites gave an empirical equation that could be used to compute the product $kc_{Ox}^{b;0}$ from values of n_{app} and s^* and discussed the problem of evaluating s^* from a plot of $\log i$ versus t.

Of course these attempts yielded similar results. An increase of either k or the initial concentration of Ox tends to increase the fraction of the I that is consumed by the pseudo-second-order dimerization, and consequently the value of n_{app} varies from $(n_1 + n_2)$ when $kc_{Ox}^{b;0}$ is very small to n_1 when $kc_{Ox}^{b;0}$ is very large. Unfortunately, the two assumptions are mutually exclusive: if I yields its limiting current at the potential employed, no significant amount of it can escape from the surface of the electrode into the bulk of the solution. Hence the bulk concentration of I must always be virtually zero, so that both the extent of dimerization in the bulk of the solution and the current resulting from the reduction of I brought to the working-electrode surface from the bulk of the solution must also be virtually zero, and it is on the extents of these reactions in the bulk of the solution that the computations are based. It has been argued [75] that these authors have in fact portrayed the behavior of the related mechanism

$$Ox + n_1 e \longrightarrow I$$
$$I \xrightarrow{k_1} J$$
$$2J \xrightarrow{k_2} D$$
$$J + n_2 e \longrightarrow Red$$

(71)

in which the first- or pseudo-first-order transformation of I into J is just slow enough to cause most of the I to be swept away from the electrode surface into the bulk of the solution before it is converted into J, but is not so slow as to be otherwise rate-determining. According to Klatt's discussion of this mechanism [234], it yields current–time curves so closely resembling those for the ECE mechanism that the two mechanisms are virtually indistinguishable on this basis. Those curves are governed by the values of the dimensionless parameters k/s^* and $k_2 c_{Ox}^{b;0}/s^*$, where k_2 is the second- or pseudo-second-order rate constant for the dimerization of J. However, this mechanism can be distinguished from the ECE mechanism by virtue of the fact that it gives $n_{app} < (n_1 + n_2)$ if the rate of dimerization is appreciable, and Klatt described a procedure for estimating the values of both the above parameters: it consisted of combining current–time data with the approximation $s^* = i^0/n_1 FV c_{Ox}^{b;0}$. Amatore and Savéant [188] considered the "disproportionation" step $I + J = Ox + Red$ in this scheme, and expressed the belief that with this modification it accounts for the reductions of aromatic carbonyl compounds in neutral media, and possibly also for the reductions of aromatic thiocarbonates.

Bard and Mayell [210] also discussed the similar mechanism

$$Ox + n_1 e \longrightarrow I$$
$$I \xrightarrow{k_1} J$$
$$2I \xrightarrow{k_2} D$$
$$J + n_2 e \longrightarrow Red$$

(72)

which gives rise to very similar behavior.

It is very difficult to obtain an exact treatment of the case that does not involve the intervening pseudo-first-order transformation, and such a treatment may be difficult to employ because the efficiency of stirring varies from one point on the surface of the working electrode to another and cannot be averaged as it can when only first-order processes are involved. Rangarajan [235] has made a promising approach to the complexities of the situation, and much remains to be said in this area [107, 142, 144, 236–244].

CASE 5g.   REGENERATION OF OX BY DISPROPORTIONATION OF AN INTERMEDIATE

The mechanism

$$Ox + ne \longrightarrow I$$
$$2I \xrightarrow{k} Ox + Red$$

(73)

of which the classical example is the reduction of uranium(VI) in an acidic solution, has been investigated by constructing working plots that show how i/i^0 depends on $s_{Ox}^* t$ and $kc_{Ox}^{b,0}/s_{Ox}^*$, where k is the second- or pseudo-second-order rate constant for the disproportionation. The best results are obtained under such conditions that $kc_{Ox}^{b,0}/s_{Ox}^*$ is of the order of magnitude of 1. Experimental data gave a value of 139 ± 7 dm^6/mol^2 s for the *third*-order rate constant ($= kc_{H^+}$) that governs the rate of disproportionation of uranium(V) in solutions containing perchloric acid and sodium perchlorate, and this is in good agreement with other values secured by a wide variety of techniques [245].

6.7   Summary

Almost all of the work that has been reviewed in this section was done between about 1958 and 1970. During that period it became clear that controlled-potential electrolysis could play an important role in elucidating the mechanisms of half-reactions. There were several groups that devoted their efforts to obtaining and scrutinizing the data—such as current–time curves and plots of n_{app} against the initial concentration of starting material—that the technique can provide. For some systems the data conformed to previously recognized patterns of behavior. For others they did not, and a number of the cases that have been listed arose from attempts to envision mechanisms that might account for the data obtained in studying new compounds.

There were several reasons for the near-cessation of such work that occurred around 1970. One was that the next important step, which would have involved developing precise techniques for evaluating the rate constants of the chemical steps in mechanisms of different kinds, could not be taken with the data-handling techniques that were available at the time. Another was that the criteria that had been developed for distinguishing among different mechanisms were mis-applied by some authors, who reached false and misleading interpretations as a result, and apparently seemed excessively subtle and difficult to others, who failed to draw conclusions that were clearly indicated by the data they had obtained. A third was the widespread inertia in reacting to the growing body of evidence that, as was mentioned in Section 2, half-reactions do not necessarily follow the same course under the very different conditions associated with different techniques: when controlled-potential electrolysis and voltammetry pointed to different mechanisms, it was often simply presumed that the former was in error.

The first of these problems has been solved and is now only a matter of history. The major contribution to a solution to the second is a scheme, devised by Shia [60, 246] and combining nonlinear regression analysis with automated classification of deviation patterns, for identifying any of these mechanisms on the basis of data on the variation of Q_t with time during an electrolysis. The third remains as a matter of philosophy. When controlled-potential electrolysis and polarography, triangular-wave voltammetry, or any other technique, suggest that the half-reaction follows different courses, the writer thinks it irresponsible to dismiss the results of either technique because they do not agree with those obtained with the other. A simple example arises in the reduction of α-mercaptopyruvic acid. At a dropping mercury electrode, reduction of the carbonyl group predominates at pH values near 5, but at a mercury pool cleavage of the C—S bond accounts for about 40% of the current [247]. In any such situation controlled-potential electrolysis will give current–time curves, products, and other information that will not be easy to reconcile with the results and interpretations obtained from electrochemical or most other techniques, but no explanation of the behavior can be complete unless it accounts for both kinds of observations. In the present state of our knowledge the explanation may often be speculative or obscure, and any discrepancies may have to be left as a puzzle to be solved at some future time. This is neither a sin nor a reflection on the quality of the work that has been done: experience has shown that our ability to solve puzzles does increase as time goes by. It is quite another thing to pretend that puzzles do not exist, or to avoid performing experiments that may uncover them.

7   COMBINATIONS OF CONTROLLED-POTENTIAL ELECTROLYSIS WITH OTHER TECHNIQUES

There are several ways in which controlled-potential electrolysis can be combined with other techniques to provide information about half-reactions. They can be divided, rather arbitrarily, into two categories: those in which the other technique is used to identify or study the product that is formed in a controlled-potential electrolysis, and those in which it is used to follow the progress of a reaction that occurs during a controlled-potential electrolysis, or the appearance and decay of an intermediate that is formed.

The products obtained from controlled-potential electrolysis have been identified by many different techniques, sometimes singly and sometimes in combination. Among these are elemental analysis; measurements of melting point; gas–liquid and liquid–liquid chromatography; mass spectrometry; and infrared, visible–ultraviolet, nuclear magnetic resonance, and electron paramagnetic resonance spectroscopy. Solid products formed on the surfaces of platinum, gold, and other solid working electrodes have been identified and studied by X-ray techniques, including electron-probe microanalysis, and by electron spectroscopy for chemical analysis (ESCA) and many other spectroscopic techniques for analyzing solid films. Electron paramagnetic resonance spectroscopy has long been invaluable for the identification and characterization of free radicals, which are often formed in electrolyses, particularly in "aprotic" solvents. Several excellent reviews [248–250] are available and should be consulted for information about experimental techniques and summaries of the understanding that has been gained through their use.

Solutions that are well suited to controlled-potential electrolysis are generally well suited to examination by other electrometric techniques, and identifications of electroactive products by polarography and techniques related to it are therefore very common. For example, the reduction of trichloroacetate ion in an ammoniacal solution can be shown to yield dichloroacetate ion by comparing its polarographic characteristics with those of authentic dichloroacetate ion [251], and the reduction of β-N-triethylaminoacrolein at a potential on the plateau of its first wave yields a product whose polarographic behavior is identical with that of acrolein [252]. Figure 5.4 showed a similar but slightly more complex example.

When another technique is used to monitor the appearance or disappearance of a product, starting material, or intermediate in a controlled-potential electrolysis, observations and measurements may be made either discontinuously (by interrupting the electrolysis periodically while the desired data are obtained) or continuously (by carrying out the electrolysis without interruption). Continuous observation is preferable for two reasons. One is that it entails less manipulation while the electrolysis is in progress, and therefore offers less opportunity for contamination by traces of air, water vapor, and other substances that might react with the constituents of the solution. The other is that it can provide data on the time dependence of the concentration of an intermediate or product in a form that is well adapted to elucidations of mechanism and evaluations of rate constants by nonlinear regression. If the electrolysis is interrupted occasionally, the equations needed for these purposes become discontinuous and much more difficult to manipulate. Usually it is best and simplest to circulate solution from the electrolytic cell into a separate cell which may, for example, be placed in the cell compartment of an ultraviolet–visible spectrophotometer or in the cavity of an electron paramagnetic resonance spectrometer. Alternatively, the solution may be passed through a tubular voltammetric indicator electrode, or into a cell designed for voltammetric examinations of flowing solutions; or a vibrating or rotating dropping electrode may be placed in the electrolytic cell itself. All these provide assurance that the data pertain to homogeneous solutions—which is not true if the entire contents of the electrolytic cell are examined, for example, by mounting that cell in the cavity of an electron paramagnetic resonance (EPR) spectrometer. However, they entail the disadvantage that a very short-lived intermediate may decompose during the time required for it to reach the external cell, although it might not do so to an appreciable extent before reaching a transducer located in the solution being electrolyzed.

Discontinuous observation is useful only with stable intermediates and products, and is well suited to the use of polarography and other voltammetric techniques that involve the diffusion of an electroactive substance from a quiet solution to a stationary electrode. Typically it entails interrupting the electrolysis after it has proceeded for an appropriate time—perhaps until about 20% of the starting material has been consumed—and removing a portion of the solution from the working-electrode compartment to an external spectrophotometric or polarographic cell, examining it, returning it to the electrolytic cell, and resuming the electrolysis. This was first done by Lingane and Small [88], and many others have followed their lead. Polarographic measurements are simplified, and oxidation by air is avoided, by providing the stopper of the working-electrode compartment of the electrolytic cell with a vertical tube through which a dropping mercury electrode can be inserted into the solution [253]. Zuman [254] stressed the importance of recording the entire polarogram, rather than merely monitoring the current at a fixed potential, so that new waves can be detected and so that variations of relative wave heights can be observed, and for similar reasons it is better to obtain a complete absorption spectrum than to monitor the absorbance at a single wavelength.

When a solution whose polarogram (or voltammogram, chronopotentiogram, etc.) consists of a single cathodic wave is subjected to controlled-potential electroreduction at a potential on the plateau of that wave, the polarogram may change in any of four ways:

1. The height of the original wave decreases in proportion to the quantity of electricity that has flowed through the cell, and no new wave appears on the polarogram. This behavior is observed with the simple mechanism corresponding to the equation $Ox + ne \rightarrow Red$, in Cases 1a–2b above, and in Case 2c if the rate of transformation of Y into Ox is high (if it is not, the height of the wave will increase for some time after each interruption of the electrolysis), but not with most others unless the parameters they involve have values that cause the deviations from proportionality to be too small for convenient detection.

2. The height of the original wave decreases, but does not do so in proportion to Q_t, and no new wave appears. This behavior is observed in Cases 2c–2e under certain conditions, and is characteristic of Cases 3a–5e.

3. The height of the original wave decreases, and a new wave appears at a more negative potential. This behavior is observed in Cases 3d, 3e, and 5a–5e if the half-wave potential of the electroactive intermediate (I or J) is more negative than the potential at which the electrolysis is carried out. Sometimes, however, the intermediate can be reduced at a more positive potential than the starting material, and in that event the behavior described in the next paragraph is observed instead.

4.   The height of the original wave decreases and a new wave appears at a more positive potential. This resembles the behavior described in the preceding paragraph, differing from it only in that the electroactive intermediate here is more readily reducible than the starting material.

Examples of all of these are known. A different situation is that in which the starting material gives two voltammetric waves. Then electrolysis at a potential on the rising part or plateau of the first wave may not affect the height of the second wave, or may cause it to either decrease or increase.

1.   If the height of the second wave remains the same, the mechanism can be described by the equations of Case 5a:

$$Ox + n_1 e \rightarrow I \tag{56}$$

$$I + n_2 e \rightarrow Red$$

With an organic compound these can take two different forms. The compound may contain only a single electroactive group, which undergoes stepwise reduction. If that group is represented by Ox_1 and the rest of the molecule by M, one has

$$Ox_1 - M + n_1 e \rightarrow I_1 - M \tag{74}$$

$$I_1 - M + n_2 e \rightarrow Red_1 - M$$

On the other hand, the molecule may contain two electroactive groups, Ox_1 and Ox_2, which are again reduced at different potentials. Then one has

$$Ox_1 - M - Ox_2 + n_1 e \rightarrow Red_1 - M - Ox_2 \tag{75}$$

$$Red_1 - M - Ox_2 + n_2 e \rightarrow Red_1 - M - Red_2$$

2.   If the height of the second wave decreases as the electrolysis proceeds, it may do so in either of two ways:

a.   The ratio i_1/i_2 of the heights of the two waves remains constant in Case 4b, or when the product I of the first half-reaction undergoes a reaction that deactivates it, and that is too slow to occur to a substantial extent during the life of one drop, but is fast enough to approach completion at every instant during the electrolysis:

$$Ox + n_1 e \rightarrow I$$

$$I \rightarrow P \tag{76}$$

$$I + n_2 e \rightarrow Red$$

In this situation I does not accumulate in the solution, and the second wave on the polarogram reflects the reduction of I formed at the surface of the drop.

b.   The ratio i_1/i_2 decreases in a variant of Case 2d:

$$Ox + n_1 e \rightarrow I$$

$$Ox + Z \rightarrow I \tag{77}$$

$$I + n_2 e \rightarrow Red$$

in which I can be produced either by the electrochemical reduction of Ox or by a homogeneous reaction that inactivates Ox. Much the same behavior occurs in Case 4a if both of the electroactive substances are in equilibrium with a common precursor Y and if both of the equilibria are very slow, and similar variants of many other cases can be imagined.

3. The most likely explanation of an increase of the height of the second wave is that the mechanism conforms to Case 5b, that J is more difficultly reducible than Ox, and that the transformation of I into J is sufficiently slow that it occurs only incompletely during the life of a drop. Another possibility is represented by the equations

$$Ox + n_1 e \rightarrow I$$

$$I + n_2 e \rightarrow Red_1 \tag{78}$$

$$I \rightarrow J \quad (slow)$$

$$J + n_3 e \rightarrow Red_2$$

in which I and J are reduced at about the same potential (which must be more negative than the potential at which the electrolysis is carried out) but with $n_3 > n_2$. A similar mechanism, with $n_3 < n_2$, would account for a decrease in the height of the second wave that was accompanied by a decrease of the ratio i_1/i_2.

Other techniques have also been used to monitor the courses of controlled-potential electrolyses. The writer [29] employed measurements of pH in studying the reduction of aluminum(III) from "neutral" unbuffered solutions. A half-reaction of the type $M^{n+} + ne = M(Hg)$ can be followed by integrating the appropriate peak on the α or γ spectrum of a solution spiked with a radioisotope of the ion, and the concentration–time curves for different ions can be followed separately during their simultaneous reductions [255]. There are many other possibilities.

8   CONCLUSION

Since the last edition of this volume was published over a decade ago, there has been a dramatic increase in the frequency with which controlled-potential electrolysis has been used to study the behaviors of organic systems. In conjunction with new techniques for identifying and determining products and for handling the data it provides, it has become invaluable for the elucidation of processes that have often proven to be more complex than they had previously been thought to be. Although its most frequent use during this decade has been in carrying out half-reactions on a scale large enough to provide identifiable amounts of their products, it has many other capabilities that should reach full fruition in the next decade.

Acknowledgments

This work was supported in part by grant number CHE-8026035 from the National Science Foundation, and in part by an appointment as Visiting

Research Scientist at the University of Pisa under the sponeorship of the Consiglio Nazionale delle Richerche.

References

1. S. Swann, Jr., "Electrolytic Reactions," in A. Weissberger, Ed., *Techniques of Organic Chemistry*, Vol. 2, Interscience, New York, 1948.
2. C. N. Reilley, *Rev. Pure Appl. Chem.*, **18**, 137 (1968).
3. A. T. Hubbard and F. C. Anson, in A. J. Bard, Ed., *Electroanalytical Chemistry*, Vol. 4, Dekker, New York, 1970, pp. 129–214.
4. W. R. Heineman, T. P. DeAngelis, and J. F. Goelz, *Anal. Chem.*, **47**, 1364 (1975).
5. A. T. Hubbard, *CRC Crit. Rev. Anal. Chem.*, **3**, 201 (1973).
6. M. J. Allen, *Organic Electrode Processes*, Reinhold, New York, 1958.
7. P. Zuman, in M. M. Baizer, Ed., *Organic Electrochemistry*, Dekker, New York, 1982.
8. N. L. Weinberg, Ed., *Technique of Organic Electrosynthesis*, in A. Weissberger, Ed., *Techniques of Chemistry*, Vol. 5, Wiley-Interscience, New York, 1974.
9. F. Haber, *Z. Elektrochem.*, **4**, 506 (1898).
10. F. Haber, *Z. Elektrochem.*, **5**, 77 (1899).
11. A. A. Vlček, J. Volke, L. Pospíšil, and R. Kalvoda, "Polarography," in B. W. Rossiter and J. F. Hamilton, Eds., *Physical Methods of Chemistry*, Wiley-Interscience, New York, 5th ed., 1985.
12. L. Meites, *Polarographic Techniques*, 2nd ed., Interscience, New York, 1965.
13. O. H. Müller, *Chem. Rev.*, **24**, 95 (1939).
14. S. Glasstone, K. J. Laidler, and H. Eyring, *The Theory of Rate Processes*, McGraw-Hill, New York, 1941.
15. P. Delahay, *New Instrumental Methods in Electrochemistry*, Interscience, New York, 1954.
16. R. Parsons, *Pure Appl. Chem.*, **52**, 233 (1979).
17. J. Jordan and L. Meites, *Pure Appl. Chem.*, **45**, 133 (1976).
18. L. Meites, P. Zuman, et al., *CRC Handbook Series in Organic Electrochemistry*, CRC Press, Boca Raton, FL: Vols. I and II, 1977; Vol. III, 1978; Vol. IV, 1979; Vol. V, 1982; Vol. VI, 1983.
19. L. Meites, P. Zuman, et al., *CRC Handbook Series in Inorganic Electrochemistry*, CRC Press, Boca Raton, FL: Vol. I, 1980; Vol. II, 1981; Vol. III, 1983; Vol. IV, 1984; Vol. V, in press; Vol. VI, in preparation.
20. J. Koutecký, *Collect. Czech. Chem. Commun.*, **18**, 597 (1953).
21. L. Meites and Y. Israel, *J. Am. Chem. Soc.*, **83**, 4903 (1961).
22. S. Karp and L. Meites, *J. Electroanal. Chem.*, **17**, 253 (1948).
23. C. A. Streuli and W. D. Cooke, *Anal. Chem.*, **26**, 963 (1954).
24. A. J. Dombroski, L. Meites, and K. I. Rose, *J. Electroanal. Chem.*, **137**, 67 (1982).
25. H. Charlot, J. Badoz-Lambling, and B. Trémillon, *Electrochemical Reactions*, Elsevier, Amsterdam, 1962.
26. G. A. Rechnitz, *Controlled-Potential Analysis*, Macmillan, New York, 1963.
27. J. J. Lingane, *Electroanalytical Chemistry*, Interscience, New York, 1953, pp. 192–195.
28. L. Meites, *J. Electroanal. Chem.*, **7**, 337 (1964).
29. L. Meites, unpublished results (1967).
30. A. Hickling, *Trans. Faraday Soc.*, **38**, 27 (1942).
31. A. J. Bard, *Anal. Chem.*, **35**, 1125 (1963).
32. S. Karp, Ph.D. Thesis, Polytechnic Institute of Brooklyn, New York, 1967.

33. L. Meites, "Standard and Formal Potentials," in L. Meites, Ed., *Handbook of Analytical Chemistry*, McGraw-Hill, New York, 1963, p. 5–11.
34. L. Meites, *Anal. Chim. Acta*, **18**, 364 (1958).
35. J. G. Jones and F. C. Anson, *Anal. Chem.*, **36**, 1137 (1964).
36. H. N. Ostensen, B. S. in Chem. Thesis, Polytechnic Institute of Brooklyn, New York, 1963.
37. D. J. Fisher, "Advances in Instrumentation for DC Polarography and Coulometry," in C. N. Reilley and R. W. Murray, Eds., *Advances in Analytical Chemistry and Instrumentation*, Vol. 10, Wiley, London, 1974, Chapter 1, pp. 1–158.
38. W. W. Goldsworthy and R. G. Clem, *Anal. Chem.*, **43**, 1718 (1971).
39. W. W. Goldsworthy and R. G. Clem, *Anal. Chem.*, **44**, 1360 (1972).
40. W. R. White, *Anal. Lett.*, **5**, 875 (1972).
41. C. L. Pomernacki and J. E. Harrar, *Anal. Chem.*, **47**, 1894 (1975).
42. G. Lauer and R. A. Osteryoung, *Anal. Chem.*, **40** (10), 30A (1968).
43. S. P. Perone, D. O. Jones, and W. F. Gutknecht, *Anal. Chem.*, **41**, 1153 (1969).
44. T. Kugo, Y. Umezawa, and S. Fujiwara, *Chem. Instrum.*, **2**, 189 (1969).
45. D. E. Smith, *CRC Crit. Rev. Anal. Chem.*, **2**, 246 (1971).
46. L. Ramaley, *Chem. Instrum.*, **6**, 119 (1975).
47. J. J. Lingane, *Anal. Chim. Acta*, **2**, 592 (1948).
48. L. Meites, *Anal. Chem.*, **27**, 1116 (1955).
49. J. J. Lingane and S. L. Jones, *Anal. Chem.*, **23**, 1804 (1951).
50. L. Meites and S. A. Moros, *Anal. Chem.*, **31**, 23 (1959).
51. L. Meites, unpublished results (1959).
52. G. L. Booman, *Anal. Chem.*, **29**, 213 (1957).
53. G. C. Goode and J. Herrington, *Anal. Chim. Acta*, **33**, 413 (1965).
54. R. G. Clem, F. Jakob, D. H. Anderberg, and L. D. Ornelas, *Anal. Chem.*, **43**, 1398 (1971).
55. R. G. Clem, *Anal. Chem.*, **43**, 1853 (1971).
56. R. I. Gelb, personal communication (1963).
57. G. L. Booman and W. B. Holbrook, *Anal. Chem.*, **35**, 1793 (1963).
58. J. E. Harrar and I. Shain, *Anal. Chem.*, **38**, 1148 (1966).
59. J. E. Harrar, "Techniques, Apparatus, and Analytical Applications of Controlled-Potential Coulometry," in A. J. Bard, Ed., *Electroanalytical Chemistry*, Vol. 8, Dekker, New York, 1974, pp. 1–167.
60. G. A. Shia, Ph.D. Thesis, Clarkson College of Technology, 1982.
61. H. C. Jones, W. D. Shults, and J. N. Dale, *Anal. Chem.*, **37**, 690 (1965).
62. C. Moinet and D. Peltier, *Bull. Soc. Chim. Fr.*, **1969**, 690.
63. J. E. Harrar and C. L. Pomernacki, *Anal. Chem.*, **45**, 57 (1973).
64. K. E. Friend and W. E. Ohnesorge, *J. Org. Chem.*, **28**, 2435 (1963).
65. E. J. Majeski, J. D. Stuart, and W. E. Ohnesorge, *J. Am. Chem. Soc.*, **90**, 633 (1968).
66. V. D. Parker, *Chem. Commun.*, 1131 (1969).
67. V. D. Parker, *Acta Chem. Scand.*, **24**, 3162 (1970).
68. L. Eberson and V. D. Parker, *Chem. Commun.*, 1290 (1970).
69. L. Meites and T. Meites, *Anal. Chem.*, **20**, 984 (1948).
70. L. Meites, *Polarographic Techniques*, 2nd ed., Interscience, New York, 1965, p. 83.
71. J. Heyrovský and P. Zuman, *Practical Polarography*, Academic, New York, 1968, p. 8.
72. J. W. Collat and J. J. Lingane, *J. Am. Chem. Soc.*, **76**, 4214 (1954).
73. J. M. Kruse, *Anal. Chem.*, **31**, 1854 (1959).
74. R. I. Gelb and L. Meites, *J. Phys. Chem.*, **68**, 630 (1964).

75. L. Meites, *Pure Appl. Chem.*, **18**, 35 (1969).
76. R. C. Buchta and D. H. Evans, *Anal. Chem.*, **40**, 2181 (1968).
77. Y. Israel and L. Meites, *J. Electroanal. Chem.*, **8**, 99 (1964).
78. R. S. Rodgers and L. Meites, unpublished results (1968).
79. L. Meites, *Anal. Chem.*, **27**, 416 (1955).
80. L. Meites, *Anal. Chim. Acta*, **20**, 456 (1959).
81. R. D. DeMars and I. Shain, *Anal. Chem.*, **29**, 1825 (1957).
82. L. Meites, *Anal. Chem.*, **27**, 977 (1955).
83. J. J. Lingane, *Electroanalytical Chemistry*, Interscience, New York, 1953, pp. 430–432.
84. P. D. Garn and E. W. Halline, *Anal. Chem.*, **27**, 1563 (1955).
85. P. Zuman, *Collect. Czech. Chem. Commun.*, **15**, 839 (1950).
86. J. J. Lingane, C. G. Swain, and M. Fields, *J. Am. Chem. Soc.*, **65**, 1348 (1943).
87. J. J. Lingane and L. W. Niedrach, *J. Am. Chem. Soc.*, **71**, 196 (1949).
88. J. J. Lingane and L. A. Small, *J. Am. Chem. Soc.*, **71**, 973 (1949).
89. S. A. Moros and L. Meites, *J. Electroanal. Chem.*, **5**, 90 (1963).
90. C. L. Rulfs and P. J. Elving, *J. Am. Chem. Soc.*, **73**, 3284 (1951).
91. G. A. Rechnitz, *Inorg. Chem.*, **1**, 953 (1962).
92. R. E. Cover and L. Meites, *J. Am. Chem. Soc.*, **83**, 4706 (1961).
93. C. D. Russel, *Int. J. Appl. Radiat. Isot.*, **28**, 241 (1977).
94. H. A. Laitinen and T. J. Kneip, *J. Am. Chem. Soc.*, **78**, 736 (1956).
95. R. I. Gelb and L. Meites, *J. Phys. Chem.*, **68**, 2599 (1964).
96. D. M. King and A. J. Bard, *Anal. Chem.*, **36**, 2351 (1964).
97. D. M. King and A. J. Bard, *Anal. Chem.*, **87**, 419 (1965).
98. A. C. Conway, R. N. Goyal, and G. Dryhurst, *J. Electroanal. Chem.*, **123**, 243 (1981).
99. L. B. Rogers and C. Merritt, Jr., *J. Electrochem. Soc.*, **100**, 131 (1953).
100. L. Meites, unpublished results (1954).
101. S. A. Moros, Ph.D. Thesis, Polytechnic Institute of Brooklyn, New York, 1961.
102. A. J. Bard and K. S. V. Santhanam, "Application of Controlled-Potential Coulometry to the Study of Electrode Reactions," in A. J. Bard, Ed., *Electroanalytical Chemistry*, Vol. 4, Dekker, New York, 1970, pp. 215–315.
103. G. L. Booman, W. B. Holbrook, and J. E. Rein, *Anal. Chem.*, **29**, 210 (1957).
104. L. Meites, *Anal. Chim. Acta*, **18**, 364 (1958).
105. M. R. Lindbeck and H. Freund, *Anal. Chem.*, **37**, 1647 (1965).
106. M. T. Kelley, W. L. Belew, G. V. Pierce, W. D. Shults, H. C. Jones, and D. J. Fisher, *Microchem. J.*, **10**, 315 (1966).
107. J. J. Lingane, *J. Am. Chem. Soc.*, **67**, 1916 (1945).
108. J. J. Lingane, *Anal. Chim. Acta*, **2**, 584 (1948).
109. J. J. Lingane and S. L. Jones, *Anal. Chem.*, **22**, 1220 (1957).
110. J. J. Banewicz, G. R. Argue, and R. F. Stewart, U.S. Patent 3,275,903, Sept. 27, 1966.
111. J. J. Lingane, *Electroanalytical Chemistry*, Interscience, New York, 1953, pp. 452–459.
112. W. M. MacNevin and B. B. Baker, *Anal. Chem.*, **24**, 986 (1952).
113. G. A. Shia and L. Meites, *J. Electroanal. Chem.*, **87**, 369 (1978).
114. L. Meites, *CRC Crit. Rev. Anal. Chem.*, **8**, 1 (1979).
115. M. T. Kelley, H. C. Jones, and D. J. Fisher, *Anal. Chem.*, **31**, 488 (1959).
116. M. T. Kelley, H. C. Jones, and D. J. Fisher, *Anal. Chem.*, **31**, 956 (1959).
117. M. T. Kelley, H. C. Jones, and D. J. Fisher, *Talanta*, **6**, 185 (1960).
118. R. Ammann and J. Desbarres, *Bull. Soc. Chim. Fr.*, **1962**, 1012.
119. R. Ammann and J. Desbarres, *J. Electroanal. Chem.*, **4**, 121 (1962).
120. E. N. Wise, *Anal. Chem.*, **34**, 1181 (1962).

121. A. J. Bard and E. Solon, *Anal. Chem.*, **34**, 1181 (1962).
122. G. Phillips and G. W. C. Milner, *Analyst*, **94**, 833 (1969).
123. G. Phillips, D. A. Newton, and J. D. Wilson, *J. Electroanal. Chem.*, **75**, 77 (1977).
124. G. A. Rechnitz and K. Srinivasan, *Anal. Chem.*, **36**, 2417 (1964).
125. G. C. Goode and J. Herrington, *Anal. Chim. Acta*, **38**, 369 (1967).
126. F. E. Woodard, W. S. Woodward, and C. N. Reilley, *Anal. Chem.*, **53**, 1251A (1981).
127. F. B. Stephens, F. Jakob, L. P. Rigdon, and J. E. Harrar, *Anal. Chem.*, **42**, 964 (1970).
128. L. Meites, *Anal. Chem.*, **31**, 1285 (1959).
129. S. Hanamura, *Talanta*, **2**, 278 (1959).
130. S. Hanamura, *Talanta*, **3**, 14 (1960).
131. S. Hanamura, *Talanta*, **9**, 901 (1962).
132. M. K. Holland, J. R. Weiss, and C. E. Pietri, *Anal. Chem.*, **50**, 236 (1978).
133. C. L. Perrin, "Mechanisms of Organic Polarography," in S. G. Cohen, A. Streitwieser, and R. W. Taft, Eds., *Progress in Physical Organic Chemistry*, Vol. 3, Interscience, New York, 1965, pp. 165–316.
134. M. M. Baizer, Ed., *Organic Electrochemistry*, Dekker, New York, 1982.
135. A. F. Krivis and E. S. Gazda, *Anal. Chem.*, **41**, 212 (1969).
136. F. Kaufman, H. Cook, and S. Davis, *J. Am. Chem. Soc.*, **74**, 4997 (1952).
137. P. J. Elving, I. Rosenthal, and M. K. Kramer, *J. Am. Chem. Soc.*, **73**, 1717 (1951).
138. I. Rosenthal, C-S. Tang, and P. J. Elving, *J. Am. Chem. Soc.*, **74**, 6112 (1952).
139. Tafel and Friedrichs, *Ber.*, **37**, 3187 (1904).
140. Baur, *Z. Elektrochem.*, **25**, 102 (1919).
141. Mohrschultz, *Z. Elektrochem.*, **32**, 434 (1916).
142. S. Karp and L. Meites, *J. Am. Chem. Soc.*, **84**, 906 (1962).
143. I. Bergman and J. C. James, *Trans. Faraday Soc.*, **50**, 60 (1954).
144. L. Meites and T. Meites, *Anal. Chem.*, **28**, 103 (1956).
145. L. Meites, *J. Electroanal. Chem.*, **5**, 270 (1963).
146. P. J. Elving and C. E. Bennett, *J. Am. Chem. Soc.*, **76**, 1412 (1954).
147. A. G. Stromberg and V. Gorodovykh, *Zh. Neorg. Khim.*, **8**, 2355 (1963).
148. A. I. Zebreva, *Tr. Inst. Khim. Nauk, Akad. Nauk Kaz. SSR*, **15**, 54 (1967).
149. Z. Galus, *CRC Crit. Rev. Anal. Chem.*, **4**, 359 (1974).
150. R. S. Rodgers and L. Meites, *J. Electroanal. Chem.*, **125**, 167 (1981).
151. A. D. Horton, L. G. Farrar, B. B. Hobbs, and W. D. Shults, *Proceedings of the 2nd Conference on Analytical Chemistry in Nuclear Technology*, USAEC Rept. TID-7568 (Part 2), Oak Ridge, Tennessee, 1959, pp. 96–98.
152. G. L. Booman and W. B. Holbrook, *Anal. Chem.*, **31**, 10 (1959).
153. W. J. Maeck, G. L. Booman, M. C. Elliot, and J. E. Rein, *Anal. Chem.*, **30**, 1902 (1958).
154. F. L. Moore, *Anal. Chem.*, **30**, 908 (1958).
155. W. D. Shults and L. B. Dunlap, *Anal. Chim. Acta*, **29**, 254 (1963).
156. W. D. Shults and L. B. Dunlap, *Anal. Chem.* **35**, 921 (1963).
157. D. H. Geske, *J. Phys. Chem.*, **63**, 1062 (1959).
158. M. Spritzer and L. Meites, *Anal. Chim. Acta*, **26**, 58 (1962).
159. J. L. Sadler and A. J. Bard, *J. Am. Chem. Soc.*, **90**, 1979 (1968).
160. L. C. Portis, V. V. Bhat, and C. K. Mann, *J. Org. Chem.*, **35**, 2175 (1970).
161. P. Zuman, *Collect. Czech. Chem. Commun.*, **25**, 3245 (1960).
162. R. F. Dapo and C. K. Mann, *Anal. Chem.*, **35**, 677 (1963).
163. O. Manoušek, O. Exner, and P. Zuman, *Collect. Czech. Chem. Commun.*, **33**, 3988 (1968).
164. T. Osa, A. Yildiz, and T. Kuwana, *J. Am. Chem. Soc.*, **91**, 3994 (1969).

165. P. Zuman, O. Manoušek, and V. Horák, *Collect. Czech. Chem. Commun.*, **29**, 2906 (1964).
166. D. Barnes, R. Belcher, and P. Zuman, *Talanta*, **14**, 1197 (1967).
167. V. D. Parker and B. E. Burgert, *Tetrahedron Lett.*, 2411 (1968).
168. K. Nyborg, *Acta Chem. Scand.*, **24**, 1609 (1970).
169. S. Uchiyama, K. Nozaki, and G. Muto, *J. Electroanal. Chem.*, **79**, 413 (1977).
170. H. K. Ficker and L. Meites, *Anal. Chim. Acta*, **26**, 172 (1962).
171. H. N. Ostensen and L. Meites, unpublished results (1963).
172. K. I. Rose, B.S. in Chem. Thesis, Polytechnic Institute of Brooklyn, New York, 1966.
173. R. M. Elofson and J. G. Atkinson, *Can. J. Chem.*, **34**, 4 (1956).
174. H. P. Cleghorn, *J. Chem. Soc. B*, 1387 (1970).
175. J. Bonastre, A. Castetbon, and P. Mericam, *Bull. Soc. Chim. Fr.*, **1977**, 1099.
176. S. A. Moros and L. Meites, *J. Electroanal. Chem.*, **5**, 103 (1963).
177. J. T. Lundquist, Jr. and R. S. Nicholson, *J. Electroanal. Chem.*, **16**, 445 (1968).
178. D. H. Geske and A. J. Bard, *J. Phys. Chem.*, **63**, 1057 (1959).
179. L. Meites, unpublished results (1955).
180. J. Badoz-Lambling, *J. Electroanal. Chem.*, **1**, 44 (1959/60).
181. G. A. Rechnitz and H. A. Laitinen, *Anal. Chem.*, **33**, 1473 (1961).
182. G. A. Rechnitz and J. E. McClure, *Talanta*, **10**, 417 (1963).
183. G. A. Rechnitz and J. E. McClure, *Talanta*, **12**, 153 (1965).
184. G. A. Rechnitz and J. E. McClure, *Anal. Chem.*, **36**, 2265 (1964).
185. J. A. Page and E. J. Zinser, *Talanta*, **12**, 1051 (1965).
186. J. G. McCullough and L. Meites, *J. Electroanal. Chem.*, **19**, 111 (1968).
187. A. J. Bard and S. V. Tatwawadi, *J. Phys. Chem.*, **68**, 2676 (1964).
188. C. Amatore and J. M. Savéant, *J. Electroanal. Chem.*, **125**, 1 (1981).
189. A. J. Bard and E. Solon, *J. Phys. Chem.*, **67**, 2326 (1963).
190. R. S. Rodgers and L. Meites, *J. Phys. Chem.*, **73**, 4348 (1969).
191. R. S. Rodgers and L. Meites, *J. Electroanal. Chem.*, **125**, 167 (1981).
192. T. Wasa and P. J. Elving, *J. Electroanal. Chem.*, **91**, 249 (1978).
193. L. R. Yeh and A. J. Bard, *J. Electroanal. Chem.*, **81**, 319 (1977).
194. J. G. McCullough, Ph.D. Thesis, Polytechnic Institute of Brooklyn, New York, 1967.
195. L. Meites, *Pure Appl. Chem.*, **18**, 65 (1969).
196. G. L. Booman and W. B. Holbrook, *Anal. Chem.*, **37**, 795 (1965).
197. I. Shain, J. E. Harrar, and G. L. Booman, *Anal. Chem.*, **37**, 1768 (1965).
198. D. T. Pence and G. L. Booman, *Anal. Chem.*, **38**, 1112 (1966).
199. H. Lund, *Acta Chem. Scand.*, **13**, 249 (1959).
200. J. G. Mason, *J. Electroanal. Chem.*, **11**, 462 (1966).
201. J. E. Harrar and L. P. Rigdon, *Anal. Chem.*, **41**, 758 (1969).
202. C. O. Huber and L. Lemmert, *Anal. Chem.*, **38**, 128 (1966).
203. G. S. Alberts and I. Shain, *Anal. Chem.*, **35**, 1859 (1963).
204. R. S. Nicholson and I. Shain, *Anal. Chem.*, **37**, 190 (1965).
205. R. S. Nicholson, J. M. Wilson, and M. C. Olmstead, *Anal. Chem.*, **38**, 542 (1966).
206. H. B. Herman and A. J. Bard, *J. Phys. Chem.*, **70**, 396 (1966).
207. H. B. Herman and H. N. Blount, *J. Phys. Chem.*, **73**, 1406 (1969).
208. G. Costa, P. Rozzo, and P. Batti, *Ann. Chim. (Rome)*, **45**, 387 (1955).
209. G. Costa and A. Puxeddu, *Ric. Sci.*, **27**, 894 (1957).
210. A. J. Bard and J. S. Mayell, *J. Phys. Chem.*, **66**, 2173 (1962).
211. J. S. Mayell and A. J. Bard, *J. Am. Chem. Soc.*, **85**, 421 (1963).
212. J. M. Lael, T. Teherani, and A. J. Bard, *J. Electroanal. Chem.*, **91**, 275 (1978).

213. O. R. Brown, J. A. Harrison, and K. S. Sastry, *J. Electroanal. Chem.*, **58**, 387 (1975).
214. D. Hawley and S. W. Feldberg, *J. Phys. Chem.*, **70**, 3459 (1966).
215. L. Nadjo and J. M. Savéant, *J. Electroanal Chem.*, **33**, 419 (1971).
216. D. H. Evans, T. W. Rosanske, and P. J. Jimenez, *J. Electroanal. Chem.*, **51**, 449 (1974).
217. G. L. Booman and D. T. Pence, *Anal. Chem.*, **37**, 1367 (1965).
218. M. Mastragostino, L. Nadjo, and J. M. Savéant, *Electrochim. Acta*, **13**, 721 (1968).
219. S. W. Feldberg, *J. Phys. Chem.*, **73**, 1238 (1969).
220. C. Amatore and J. M. Savéant, *J. Electroanal. Chem.*, **102**, 21 (1979).
221. C. Amatore and J. M. Savéant, *J. Electroanal. Chem.*, **85**, 27 (1977).
222. C. Amatore and J. M. Savéant, *J. Electroanal. Chem.*, **86**, 227 (1978).
223. C. Amatore and J. M. Savéant, *J. Electroanal. Chem.*, **123**, 189 (1981).
224. C. Amatore and J. M. Savéant, *J. Electroanal. Chem.*, **125**, 23 (1981).
225. C. Amatore and J. M. Savéant, *J. Electroanal. Chem.*, **126**, 1 (1981).
226. C. Amatore and J. M. Savéant, *J. Electroanal. Chem.*, **123**, 203 (1981).
227. J. Grimshaw, R. J. Haslett, and J. Trocha-Grimshaw, *J. Chem. Soc. Perkin Trans. I*, 2448 (1977).
228. J. M. Savéant, *Acc. Chem. Res.*, **13**, 823 (1980).
229. C. Amatore, J. Pinson, J. M. Savéant, and A. Thiébault, *J. Electroanal. Chem.*, **123**, 231 (1981).
230. C. Amatore, J. Chaussard, J. Pinson, J. M. Savéant, and A. Thiébault, *J. Am. Chem. Soc.*, **101**, 6012 (1979).
231. C. Amatore, J. Badoz-Lambling, C. Bonnel-Huyghe, J. Pinson, J. M. Savéant, and A. Thiébault, unpublished work cited in ref. 229.
232. C. Amatore, F. M'Halla, and J. M. Savéant, *J. Electroanal. Chem.*, **123**, 219 (1981).
233. F. M'Halla, J. Pinson, and J. M. Savéant, *J. Am. Chem. Soc.*, **102**, 4120 (1980).
234. L. N. Klatt, *J. Electroanal. Chem.*, **55**, 161 (1974); cf. also L. N. Klatt and R. L. Rouseff, *J. Am. Chem. Soc.*, **94**, 7295 (1971) and *J. Electroanal. Chem.*, **41**, 411 (1973).
235. S. Rangarajan, personal communication (1968).
236. P. J. Elving and J. T. Leone, *J. Am. Chem. Soc.*, **80**, 1021 (1958).
237. M. J. Allen and A. H. Corwin, *J. Am. Chem. Soc.*, **72**, 114 (1950).
238. M. J. Allen and A. H. Corwin, *J. Am. Chem. Soc.*, **72**, 117 (1950).
239. M. J. Allen, *J. Am. Chem. Soc.*, **72**, 3797 (1950).
240. M. J. Allen, *J. Org. Chem.*, **15**, 435 (1950).
241. M. J. Allen, *J. Am. Chem. Soc.*, **73**, 3503 (1951).
242. H. A. Levine and M. J. Allen, *J. Chem. Soc.*, 254 (1952).
243. M. J. Allen, J. E. Fearn, and H. A. Levine, *J. Chem. Soc.*, 2220 (1952).
244. M. J. Allen, *Proc. Intern. Comm. Electrochem. Thermodynam. Kinet.*, VI Meeting, 481 (1955).
245. J. Mocák and D. I. Bustin, *J. Electroanal. Chem.*, **79**, 307 (1977).
246. L. Meites and G. A. Shia, "Automatic Elucidation of Reaction Mechanisms in Stirred-Pool Controlled-Potential Chronocoulometry," in B. R. Kowalski, Ed., *Chemometrics: Theory and Application*, ACS Symposium Series 52, American Chemical Society, Washington, DC, 1977.
247. M. B. Fleury, J. Tohier, and P. Zuman, *J. Electroanal. Chem.*, in press.
248. R. N. Adams, *J. Electroanal. Chem.*, **8**, 151 (1964).
249. D. H. Geske, in A. Streitwieser, Jr. and R. W. Taft, Eds., *Progress in Physical Organic Chemistry*, Vol. 4, Interscience, New York, 1967, p. 125.
250. B. Kastening, in P. Zuman and L. Meites with I. M. Kolthoff, Eds., *Progress in Polarography*, Vol. 3, Interscience, New York, 1972, p. 195.

251. T. Meites and L. Meites, *Anal. Chem.*, **27**, 1531 (1955).
252. P. Zuman and V. Horák, *Collect. Czech. Chem. Commun.*, **26**, 176 (1961).
253. H. A. Catherino and L. Meites, *Anal. Chim. Acta*, **23**, 57 (1960).
254. P. Zuman, *The Elucidation of Organic Electrode Processes*, L. Meites, Ed., Academic, New York, 1969, pp. 89–90.
255. K. Samhoun and F. David, *J. Electroanal. Chem.*, **106**, 161 (1980).

Chapter **6**

CHRONOAMPEROMETRY, CHRONOCOULOMETRY, AND CHRONOPOTENTIOMETRY

Royce W. Murray

4   THE PRACTICE OF CHRONOAMPEROMETRY, CHRONOCOULOMETRY,
     AND CHRONOPOTENTIOMETRY

1   INTRODUCTION

This chapter reviews three electrochemical techniques involving application of either a current or potential step excitation to an electrochemical cell and measurement of some transient response related to electrolysis of a component of that cell. These techniques, chronopotentiometry, chronoamperometry, and chronocoulometry, are collectively herein referred to as the "chronomethods." In each the *solution in the cell compartment containing the working electrode is unstirred* and contains, in addition to the electroactive sample component, a large excess of an electroinactive salt, the *supporting electrolyte*. These conditions ensure that mass transport of sample component to the working electrode surface occurs through *diffusional* motion.

The sample component in our discussions is represented as an oxidized substance Ox, and its electrolysis, consuming n electrons per mole, as

$$Ox + ne \rightleftharpoons Red \tag{1}$$

Species Ox and product Red are solution-soluble unless otherwise noted. Their initial concentrations in the cell are denoted as C_{Ox}^b and C_{Red}^b, their instantaneous concentrations at distance x from the electrode surface as $C_{Ox(x,t)}$, and their diffusion coefficients as D_{Ox} and D_{Red}, respectively. Our discussions of this reduction process are transposable to the reverse (oxidation) reaction with no essential differences.

1.1   Elements of Chronopotentiometric, Chronoamperometric, and Chronocoulometric Experiments

The conceptually simplest chronopotentiometric experiment involves applying a step current excitation (A in Figure 6.1) to the working electrode and measuring its subsequent time-dependent potential relative to the potential of a reference electrode in the cell (B in Figure 6.1). A cathodic current step causes first a change in working electrode potential, at a rate limited by the capacitive charging of the working electrode's double layer, to a sufficiently negative potential to drive reaction (1). The potential then changes more slowly as Ox is reduced, until the supply of Ox at the electrode surface is exhausted $[C_{Ox(0,t)} \rightarrow 0]$, at which time the *transition time* (τ), the current efficiency for reaction (1) drops below unity and the potential rises to that of another electrode reaction (which for the moment we assume is uninteresting). The parameters of

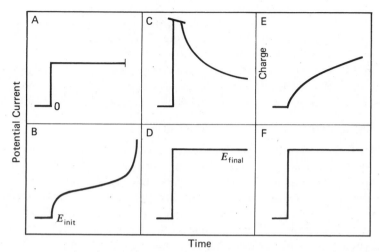

Figure 6.1   Excitations and responses of chronomethods: Chronopotentiometry, Curves A (excitation and B (response); chronoamperometry, Curves C (response) and D (excitation); chronocoulometry, Curves E (response) and F (excitation).

interest in the potential–time response curve (the *chronopotentiogram*) for reaction (1) are the value of τ, the shape of the E–t curve, and their dependencies on the value of the impressed cathodic current i.

In a chronoamperometric experiment, a potential step is applied to the working electrode (D in Figure 6.1) and the resulting current–time response (C in Figure 6.1) is measured. In the simplest case, the initial working electrode potential (E_{init}) is sufficiently positive that reaction (1) does not proceed, and the potential attained by the step E_{final} is sufficiently negative to drive the surface concentration of reactant Ox immediately to zero $[C_{Ox(0,t)} \to 0]$. The initially large cathodic current rapidly decays as the solution nearest the electrode surface becomes depleted of Ox. The current–time (i–t) response, or *chronoamperogram*, is interesting with respect to its shape and magnitude of current at any given time and to its dependency on the value of E_{final} [in cases insufficiently negative to drive $C_{Ox(0,t)}$ to zero].

The chronocoulometric experiment is operationally identical to chronoamperometry except that one integrates the current attributable to reaction (1) and records charge versus time (E and F in Figure 6.1). While the charge–time measurement is in principle no more informative than the current–time measurement, in practice Q–t data are sometimes more easily interpreted. As in chronoamperometry, parameters of interest in the Q–t curve, or *chronocoulogram*, are shape, amplitude, and dependency on E_{final}.

Useful variants of these experiments exist, notably ones in which a *reverse step excitation* follows the initial "forward" one. In such cases reaction (1) is driven in the oxidation direction following application of the reverse step. For chronopotentiometry, chronoamperometry, and chronocoulometry, the reverse E–t curve and transition time (τ_b), reverse i–t curve shape and amplitude, and reverse Q–t curve shape and amplitude, respectively, are determined by (a) the quantity of species Red generated during the forward electrolysis time and (b) any factors tending to change the recoverable part of this amount of Red, such as diffusion of Red away from the electrode surface; any chemical instability of Red; and any adsorption of Red on the electrode surface. A primary use of the reverse step experiments is studying chemical decay pathways of Red.

1.2   Elements of Chronomethods Theory

Three of the important kinetic rates in electrochemical processes include (a) diffusion rates of electrode reactants and products (Ox and Red), (b) *the rate of the heterogeneous electron-transfer process itself* (1), and (c) *the rates of any chemical reactions coupled to the electron transfer through their generation or consumption of Ox or Red near the electrode surface.* In the simplest electrode processes and experiments, only the diffusional event need be considered; such instances yield analytically tractable relations between the measured chronomethod response and the bulk reactant concentration (C_{Ox}^b). The main virtues of the chronomethods, however, include their transient experimental capabilities for study of the physical phenomena of heterogeneous electron transfer and coupled chemical reactions, and for application to the adsorption of Ox or Red species on the working electrode surface and electrochemical reactions of films on electrode surfaces that contain Ox and Red species.

Let us consider some basic theoretical relations useful for description of the above-mentioned kinetic events (a–c) and of adsorption. Theoretical characterization of the diffusion rate dependency of a chronomethod requires specification of the physical geometry of the working electrode and its surrounding body of solution. Diffusion conditions are either *semi-infinite* or *finite* (bounded). In the former, the layer of solution within which the electrode reaction alters the concentrations of Ox or Red (the *diffusion layer*) is thin compared with the electrode-to-cell wall distance. In finite diffusion, the thickness of the solution volume that contains Ox or Red is much smaller, and the diffusion layer impinges more or less severely on the outer boundary of this volume. Cases of finite diffusion fall within the realms of "thin-layer electrochemical" experiments and of thin films of electroactive polymers on electrodes.

Chronomethods experiments have been most often performed under semi-infinite diffusion conditions and are emphasized in this chapter. A recent text [1] may be consulted for supplementary descriptions of semi-infinite aspects of chronomethods. Finite diffusion has also been reviewed elsewhere in the context of thin-layer electrochemistry [2].

Semi-infinite conditions are subdivided into cases of *planar*, *spherical*, and *cylindrical* working electrode geometries (Figure 6.2). Planar, or linear, diffusion is mathematically the simplest and is generally striven for in practical experiments. Fick's Second Law is the basic equation of linear diffusional motion:

$$\frac{\partial C_{(x,t)}}{\partial t} = D \frac{\partial^2 C_{(x,t)}}{\partial x^2} \tag{2}$$

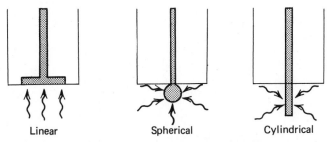

Figure 6.2   Illustration of linear, spherical, and cylindrical diffusions. Arrows represent average diffusional paths.

where D (cm^2/s) is the diffusion coefficient. Theoretically describing any electrochemical experiment involving linear mass transport of Ox or Red requires solving this differential equation under initial and boundary conditions descriptive of the experiment. For an experiment controlled by Ox diffusion, the usual *initial condition* is that of a homogeneous solution at $t = 0$:

$$C_{Ox(x,0)} = C_{Ox}^b \tag{3}$$

The boundary condition statement of *semi-infinite diffusion* is

$$C_{Ox(\infty,t)} = C_{Ox}^b \tag{4}$$

The remaining boundary condition specifies the electrochemical excitation (current or potential step) used in the particular chronomethod. Solving Fick's law for these conditions yields an equation for $C_{Ox(x,t)}$ that describes the time variation of concentration-distance-profiles of Ox during the electrochemical experiment. This equation is then manipulated to produce a relation for the desired response term in the experiment.

Many interesting electrode reactions include a chemical process coupled to reaction (1). As an example, consider an initial equilibrium of reactant Ox with an electroinactive species Z:

$$Z \underset{k_b}{\overset{k_f}{\rightleftharpoons}} Ox; \qquad Ox + ne \rightarrow Red \tag{5}$$

It is evident that for Z$\rightarrow$Ox conversion rates very fast or slow compared with the electrolytic consumption rate of Ox, the effective electrode reactant concentrations become the equilibrium solution concentrations ($C_z^b + C_{Ox}^b$) and C_{Ox}^b, respectively. Chemical Z$\rightarrow$Ox rates comparable to the electrolytic consumption rate, however, cause control of the supply of Ox by k_f and k_b and by the diffusion rate of Z. Study and evaluation of the reaction kinetics then become possible.

Theoretical description of reaction (5) (called a *CE* case, for *chemical–electrochemical reaction sequence*) requires simultaneous solution of Fick's law equations for Ox and Z, each modified to include the chemical reaction rate term:

$$\frac{\partial C_{Ox(x,t)}}{\partial t} = D_{Ox} \frac{\partial^2 C_{Ox(x,t)}}{\partial x^2} + k_f C_{Z(x,t)} - k_b C_{Ox(x,t)} \tag{6}$$

$$\frac{\partial C_{Z(x,t)}}{\partial t} = D_Z \frac{\partial^2 C_{Z(x,t)}}{\partial x^2} - k_f C_{Z(x,t)} + k_b C_{Ox(x,t)} \tag{7}$$

Numerous other chemical reactions coupled to electron transfer are known; important examples include:

$$EC_{irrev} \qquad \begin{aligned} Ox + ne &\rightarrow Red \\ Red &\rightarrow Y \end{aligned} \tag{8}$$

$$EC_{rev} \qquad \begin{aligned} Ox + ne &\rightarrow Red \\ Red &\rightleftharpoons Y \end{aligned} \tag{9}$$

$$ECE \qquad \begin{aligned} Ox + n_1 e &\rightarrow Red \\ Red &\rightarrow Y \\ Y + n_2 e &\rightarrow X \end{aligned} \tag{10}$$

$$Catalytic \qquad \begin{aligned} Ox + ne &\rightarrow Red \\ Red + Z &\rightarrow Ox + X \end{aligned} \tag{11}$$

$$Dimerization \qquad \begin{aligned} Ox + ne &\rightarrow Red \\ 2Red &\rightarrow Red_2 \end{aligned} \tag{12}$$

$$Dimerization \qquad \begin{aligned} Ox + n_1 e &\rightarrow Red \\ Ox + Red &\rightarrow OxRed \\ OxRed + n_2 e &\rightarrow Red_2 \end{aligned} \tag{13}$$

These forms are often studied with *reverse excitation chronomethods*. When the chemical reaction step does not *precede* any electron-transfer step, then a Fick's law solution for diffusion of Ox [as in (6) and (7)] is not responsive to the chemical reaction kinetics. An example is the EC reaction (8), where application of reverse excitation is necessary to follow the fate of Red.

Investigating chemical processes by observing their influence on an electrochemical response constitutes an important branch of electrochemical research because rather labile and often unique solution chemistry can be examined and quantitatively characterized. Reviews of organic electrochemical reactions that illustrate this point have been prepared by Peover [3], Adams [4], Evans [5], and Baizer and Lund [6].

The third important kinetic rate in electrode processes is that of the electron-transfer reaction (1) itself. The rate constant for charge transfer is denoted k_s when the potential scale is referenced to the standard potential E^0 of (1). The net current flow (amperes per square centimeter) at the electrode is the sum of a cathodic (positive) and anodic (negative sign) current component, or

$$i = i_c + i_a = nFk_s \left\{ C_{\text{Ox}(0,t)} \exp\left[\frac{-\alpha n_a F}{RT}(E - E^0) \right] \right.$$
$$\left. - C_{\text{Red}(0,t)} \exp\left[\frac{(1 - \alpha)n_a F}{RT}(E - E^0) \right] \right\} \tag{14}$$

where F is the Faraday constant, α the transfer coefficient (values of 0 to 1, a measure of the electron-transfer energy barrier's symmetry), and n_a the electrons per mole for the rate-determining step in the overall electrode reaction {recognizing that microscopically a sequence of chemical events and charge transfers may occur during (1) at the electrode surface, n_a may be less than n [7]}, R the gas constant (joules per mole-degree), and T the absolute temperature. An alternative statement of (14) is

$$i = i_0 \left\{ \exp\left[\frac{-\alpha n_a F}{RT}(E - E_{\text{eq}}) \right] - \exp\left[\frac{(1 - \alpha)n_a F}{RT}(E - E_{\text{eq}}) \right] \right\} \tag{15}$$

where i_0 is the *exchange current* and $E - E_{\text{eq}}$ is often called η, the *activation overpotential*. The exchange current is the current that flows equally in the cathodic (i_c) and anodic (i_a) directions when the net current is zero ($i_c = -i_a$, at $E = E_{\text{eq}}$) and is a reflection of the dynamic nature of a heterogeneous charge-transfer process (analogous to a chemical electron self-exchange reaction at equilibrium in a solution). Exchange current is concentration dependent and is related to the more fundamental rate parameter k_s by

$$i_0 = nFk_s[C_{\text{Ox}(0,t)}]^{1-\alpha}[C_{\text{Red}(0,t)}]^\alpha \tag{16}$$

Under the condition $i_{\text{net}} = 0$, (14) and (15) reduce to the familiar Nernst equation:

$$E = E^0 - 2.303 \frac{RT}{nF} \log\left(\frac{C_{\text{Red}(0,t)}}{C_{\text{Ox}(0,t)}} \right) \tag{17}$$

Note that the concentrations in the above equations are at the electrode surface ($x = 0$). The term $2.303RT/nF$ is $0.0591/n$ V at 25°C.

If during an actual electrolysis, the $i_{\text{net}} \ll i_0$, then (14) again reduces to the Nernst equation (17) with reasonable accuracy. That is, the electron-transfer rate is sufficiently rapid that displacement of reaction (1) from an equilibrium condition is small and the surface concentrations $C_{\text{Ox}(0,t)}$ and $C_{\text{Red}(0,t)}$ extant during electrolysis are described by the equilibrium equation (17). Thus, in this case the electrode process is said to be reversible. Criteria for reversibility of an electrode reaction depend on comparisons of experiment to theory based on (17).

A value of i_{net} significant in comparison to i_0, however, produces a displacement of Ox and Red concentrations from those given by (17). The electrode process is then said to be *quasi-reversible*, or *irreversible*. The connotation of reversible versus irreversible is fast versus slow, not directional. When $i_c \gg -i_a$

also, the electrode process is "totally irreversible." In the electrochemical context, the words reversible and irreversible are entirely qualitative. Reversibility can depend on the conditions of electrolysis; an electrode reaction that is reversible at low current flow may exhibit substantial irreversibility under experimental conditions demanding much higher currents. Such high-current experimental conditions are sought in the study of electron-transfer rates, which is important in studying detailed charge-transfer processes. The chronomethods are often applicable to such kinetics.

Another important electrochemical event is the occasional tendency of electrode reactant Ox and/or product Red to adsorb on the working electrode surface. The fundamental factors governing such adsorption are of interest; adsorption can play a role in the detailed charge-transfer process and in charge-transfer-coupled chemical reactions (coupled chemical reactions may occur heterogeneously on the electrode surface as well as homogeneously in the solution). The presence of adsorbed reactant or product becomes reflected in transient electrochemical experiments, including the chronomethods, typically through alteration of the electrochemical response from that anticipated for purely diffusion-controlled mass transport. For instance, if the reactant Ox is adsorbed and both solution and adsorbed reactant are electroactive (the usual case)

$$\tag{18}$$

and the adsorption equilibrium is established prior to $t = 0$ [so that $C_{\text{Ox}(x,\,0)\text{soln}} = C_{\text{Ox}}^{b}$], then the charge (coulombs per square centimeter) passed to reduce both forms of reactant in a chronocoulometric experiment exceeds that expected for reaction of diffusing reactant only (Q_t) by

$$Q_{\text{ads}} = nF\Gamma_0 \tag{19}$$

where Γ_0 is the surface excess (moles per square centimeter) of adsorbed reactant. Measurement of Γ_0 depends on dissection of the total measured charge into its Q_{ads} and time-dependent Q_t parts.

2   CHRONOAMPEROMETRY AND CHRONOCOULOMETRY

As noted earlier, the electrochemical excitation in both the chronoamperometric and chronocoulometric experiments is a potential step applied suddenly to an electrode. The response of the ensuing electrode reaction is a current decaying with time; in chronoamperometry this current–time curve is the measured response. In chronocoulometry one measures the integral of the current–time response, as a charge–time curve. Because of the basic similarity of the two methods, they are considered together.

The chronoamperometric experiment (also known as the potentiostatic or potential step method) has been known for many years. Its most basic equation, the Cottrell equation [8], was subjected to experimental scrutiny over four decades ago [9, 10]. The potential step method has experienced steadily increasing use, especially after the application of operational amplifiers revolutionized electrochemical instrumentation (producing, among other things, increasingly excellent potentiostatic instruments). Chronoamperometry and its relatively youthful chronocoulometric [11] offspring are today regarded by many as methods of choice for transient electrochemical experiments probing the kinetics of mass-transport charge-transfer and coupled chemical reaction rates and adsorption.

2.1   Mass-Transport Control by Diffusion

2.1.1   Single Potential Step Experiment

Consider applying to a working electrode a potential step sufficiently negative that the concentration of reactant species Ox is forced immediately to an essentially zero value:

$$C_{Ox(0,t)} = 0 \qquad (20)$$

Applying this and the boundary conditions (3) and (4) to Fick's law for Ox [see (2)] yields an expression for the concentration of Ox as a function of time and distance from the electrode surface

$$C_{Ox(x,t)} = C_{Ox}^b \, \text{erf}\left[\frac{x}{2D_{Ox}^{1/2} t^{1/2}}\right] \qquad (21)$$

where erf denotes the *error function*. Some concentration profiles are plotted in Figure 6.3. Note that the depth of the depleted region around the electrode surface, the *diffusion layer*, is quite small (only 10^{-3} cm after 10 ms) but grows with the square root of time. Diffusion layer thickness is crudely approximated by the term $(Dt)^{1/2}$ cm. The current flow is proportional, through Fick's First Law, to the instantaneous concentration gradient of Ox at the electrode surface:

$$i = nFD_{Ox} \frac{\partial C_{Ox(0,t)}}{\partial x} \qquad (22)$$

which, when applied by differentiation to (21) gives the *Cottrell equation*:

$$i = nFC_{Ox}^b \left(\frac{D_{Ox}}{\pi t}\right)^{1/2} \qquad (23)$$

The Cottrell equation shows that a diffusion-controlled electrolysis of reactant Ox at a planar, stationary electrode gives a current decaying as a $t^{-1/2}$ function that is proportional at any given time to the bulk concentration of electrode reactant. A current versus $t^{-1/2}$ plot is called a Cottrell plot, and its linearity

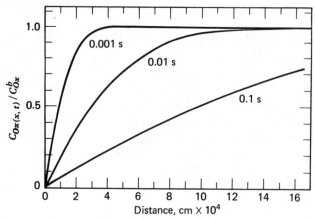

Figure 6.3   Potential step concentration–distance profiles at various times for reactant Ox; $D_{Ox} = 1.0 \times 10^{-5}$ cm^2/s.

(with zero intercept) constitutes a criterion for simple diffusion control of the electrolysis rate.

Consider next applying a potential step sufficiently negative to cause electrolysis of Ox but insufficiently negative to yield the condition of (20); that is, yields $C_{Ox}^b > C_{Ox(0,t)} > 0$. Now, the boundary condition (20) must be replaced with one describing the exact nature of the E_{final}–$C_{Ox(0,t)}$ dependency. Two situations arise: (a) The Ox $+ ne \rightarrow$ Red electron-transfer rate is very fast (reversible reaction). E_{final} is related in that case to $C_{Ox(0,t)}$, and $C_{Red(0,t)}$, through the Nernst equation [see (17)]. (b) The reaction involves a low k_s (irreversible), and the statement of (14) is applied. The latter, irreversible case is considered later in this section. For reversible charge transfer, simultaneous solution of Fick's law for Ox and Red and use of (17) and other appropriate boundary conditions [12] give

$$i = nFC_{Ox}^b \left(\frac{D_{Ox}}{\pi t}\right)^{1/2} \left\{ 1 + \left(\frac{D_{Ox}}{D_{Red}}\right)^{1/2} \exp\left[\frac{nF}{RT}(E - E^0)\right] \right\}^{-1} \quad (24)$$

In the limit of sufficiently negative potential, this expression reduces to (23). The effect of E_{final} in a reversible reaction is seen to simply *scale the current–time response curve*.

An exemplary series of chronoamperometric i–t curves is plotted in Figure 6.4. Note that if currents at any given time t_x (1 s in the inset) are plotted against the value of applied potential E_{final}, a current–potential "wave" for the reversible reaction results that has as its $E_{1/2}$ (value of potential halfway up the wave)

$$E_{1/2} = E^0 - \frac{RT}{nF} \ln \left(\frac{D_{Ox}}{D_{Red}}\right)^{1/2} \quad (25)$$

The inset of Figure 6.4 illustrates the relation of the chronoamperometric experiment to *polarography*. In the polarographic experiment, instead of apply-

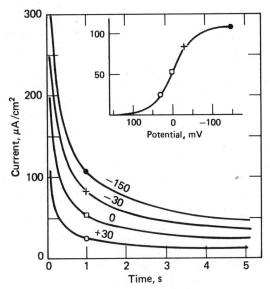

Figure 6.4 Chronoamperometric current–time curves calculated from (24) for various $E-E^0$ (numbers on curves are $E-E^0$ in mV) and $n=1$; $D_{Ox}=D_{Red}=4.0\times10^{-6}$ cm^2/s; $C^b=1.0\times10^{-6}$ mol/cm^3; $T=25°$C. Inset is current at 1 s.

ing a potential step to an existing electrode, a potential is applied and the electrode (a growing mercury drop) is grown. The resulting $t^{1/6}$ variation of current with time is a composite of the $t^{2/3}$ area dependency of the growing mercury drop and the $t^{-1/2}$ dependency of (23) and (24). (A correction for spherical diffusion effects is applied in exact polarographic theory.) A fuller description of polarography and its pulsed-potential variants is given in Chapter 4.

The chronoamperometric experiment historically found its most frequent analytical application in the form of the polarographic experiment. Although chronoamperometric current–time curves at stationary electrodes also provide concentration-proportional currents suitable for analysis, few advantages accrue as compared to polarography; thus, it is rarely applied. Stationary electrode chronoamperometry is very useful however for studying charge-transfer kinetics (k_s) and for analyzing coupled chemical reaction rates. Experimental data are typically compared to (23) and (24) during these applications.

Chronoamperometry is also useful for determining both diffusion coefficients of electroactive compounds [13] and, for compounds where D is known, the active area of an electrode (such as carbon paste [14]). In the latter measurement, the topology and/or spacing between active electrode regions must be dimensionally large compared to the depth of the diffusion layer developed in the experiment [15]. Using the relation $l=(Dt)^{1/2}$, and taking $D=10^{-5}$ cm^2/s and $t=10^{-2}$ s (as the plausibly shortest long-time limit of an experiment), the topology and/or spacing must be $\geqslant 3\times10^{-4}$ cm.

The chronocoulometric, or charge–time, response of the potential step experiment is elicited from (23) and (24) simply by integrating the current–time equation. For the limiting case of sufficiently negative E_{final}, this gives

$$Q = 2nFC_{Ox}^b \left(\frac{D_{Ox}t}{\pi}\right)^{1/2} \tag{26}$$

A charge that is controlled by the diffusion of Ox to the electrode surface exhibits a Q–$t^{1/2}$ proportionality. Actually, when experimental Q data are plotted against $t^{1/2}$, the resulting straight-line plot for a diffusion-controlled system exhibits a positive zero-time intercept. This charge intercept term reflects the Q_{dl} required to charge the double layer on the working electrode from E_{init} to E_{final}. The double-layer charge can also be measured in a blank experiment conducted without sample Ox. A double-layer charging current also appears in the chronoamperometric current–time response, but as a very short-time current "spike" that is often not clearly recorded.

2.1.2   Double Potential Step Experiment

This experiment, of recent vintage, involves application of a positive potential step, causing reduction of Ox and the condition $C_{Red(0,t)} = 0$, followed at some time τ by a negative potential step causing reoxidation of the product Red (generated during $t < \tau$) and the condition $C_{Red(0,t)} = 0$. The shapes of the chronoamperometric and chronocoulometric response curves are illustrated in Figure 6.5. At times $t < \tau$, the current and charge–time curves are described previously by (23) and (26). At times $t > \tau$, the oxidation of Red occurs from the diffusion

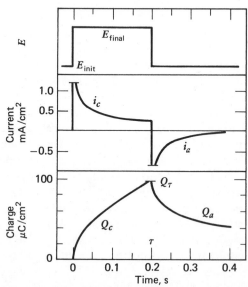

Figure 6.5 Chronoamperometric and chronocoulometric double potential step responses calculated [see (23), (26), (29), and (30)] for $n=1$; $D=4.0 \times 10^{-6}$ cm/s; $C_{Ox}^b = 1.0 \times 10^{-6}$ mol/cm³; $\tau = 0.200$ s.

layer of Red created during the negative potential step. The diffusion profile of Red at time τ is thus in effect the initial condition of the reverse potential step. The Red diffusion profile is connected to that of Ox by the expression stating that the flux of Red leaving the electrode surface equals that of Ox arriving:

$$D_{Ox} \frac{\partial C_{Ox(0,t)}}{\partial x} = -D_{Red} \frac{\partial C_{Red(0,t)}}{\partial x} \tag{27}$$

Applying these conditions to a Fick's law solution yields for the diffusion profile of Red for $t < \tau$

$$C_{Red(x,t)} = C_{Ox}^b \left(\frac{D_{Ox}}{D_{Red}} \right)^{1/2} \operatorname{erfc} \left(\frac{x}{2D_{Red}^{1/2} t^{1/2}} \right) \tag{28}$$

where erfc is the complementary error function $[1 - \operatorname{erf}(Z)]$. (In the event that $D_{Ox} = D_{Red}$, the diffusion profile of Red is an exact mirror image of that of Ox, reflected at the electrode surface.)

With (28) as initial conditions for $t > \tau$, the response to the reverse, positive potential step (consuming Red) was obtained for the chronoamperometric case by Schwarz and Shain [16], and for the chronocoulometric case by Christie [17]:

$$i_a = \frac{nFD_{Ox}^{1/2}C_{Ox}^b}{\pi^{1/2}} \left[\frac{1}{(t-\tau)^{1/2}} - \frac{1}{t^{1/2}} \right] \tag{29}$$

$$Q_a = \frac{2nFD_{Ox}^{1/2}C_{Ox}^b}{\pi^{1/2}} [t^{1/2} - (t-\tau)^{1/2}] \tag{30}$$

where i_a and Q_a represent anodic currents and charges, respectively, at time $t > \tau$.

Comparing experimental data to these expressions is accomplished by the obvious $i_a - f(t)$ or $Q_a - f(t)$ plots, or more conveniently, by examination of i_a/i_c and Q_a/Q_c ratios. For the ratios, one typically takes pairs of forward (reduction, cathodic) and backward (oxidation, anodic) currents and charges measured at equal times following their respective potential steps (measured at times $t = t_f$ on the forward step and $t_r - \tau$ on the reverse step, so that $t_f = t_r - \tau$ always). Also one takes the reverse charge as the quantity $Q_\tau - Q_a$. Taking ratios of (23) and (29) and (26) and (30) gives these current and charge ratios:

$$\left| \frac{i_a}{i_c} \right| = 1 - \left(\frac{\theta}{1+\theta} \right)^{1/2} \tag{31}$$

$$\frac{Q_\tau - Q_a}{Q_c} = 1 - \left(\frac{1+\theta}{\theta} \right)^{1/2} + \left(\frac{1}{\theta} \right)^{1/2} \tag{32}$$

where $\theta = (t - \tau)/\tau$ and the absolute value recalls the opposite signing of i_a and i_c. Since θ expresses the time scale of the forward and backward measurement relative to that of the reverse excitation τ, it is an intuitively significant quantity. If θ is small, the time scale for electrolysis of Red in the reverse step is small

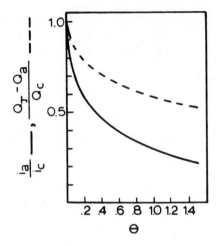

Figure 6.6 Variation of double potential step current and charge ratios with θ [see (31) and (32)].

compared to the time scale τ over which it is generated, so that the diffusion profile for consumption of Red is thin compared to the total $t < \tau$ generated Red profile. This resembles, at very small θ, simply oxidation of a solution of Red, so that i_a/i_c and $(Q_\tau - Q_a)/Q_c$ both approach unity. A second point to note is that, at $\theta = 1$, we measure current and charge for the forward and reverse steps at equal times after the steps [at τ and at $t = 2\tau$, and then $i_a/i_c = 0.293$ and $(Q_\tau - Q_a)/Q_c = 0.586$]. Observing the latter ratios is a quick way to ascertain diffusion-only control of Red.

Figure 6.6 shows further how the ratios depend on θ. Adherence of experimental data to the predictions of (31) and (32) requires that both reactant Ox and product Red react in a diffusion-controlled manner. The discussion following will show that ratios of Figure 6.6 can be quite sensitive to adsorption of reactant Ox and to chemical reactions generating Ox or consuming Red.

2.2   Mass-Transport Control by Coupled Chemical Reactions

Chemical reactions generating Ox or consuming Red that occur with reaction half-lives comparable to the observation time of a chronoamperometric or chronocoulometric response curve can alter the responses from the diffusion-controlled shapes discussed above. Chronoamperometric and chronocoulometric theory have been examined for the reaction types (5), (8)–(13), and many others (see the useful table in [18]). We examine here four important examples.

2.2.1   Preceding Chemical Reactions (CE)

Consider the case of reactant Ox lying at initial equilibrium with an electroinactive substance Z

$$Z \underset{k_b}{\overset{k_f}{\rightleftharpoons}} Ox; \qquad Ox + ne \rightarrow Red \tag{5}$$

with the rates and initial concentrations C_{Ox}^b and C_Z^b connected by the equilibrium constant,

$$K = \frac{k_f}{k_b} = \frac{C_{Ox}^b}{C_Z^b} \qquad (33)$$

For very small k_f, the current–time response to potential step is given by the Cottrell equation (23) where the concentration of Ox is simply the initial equilibrium concentration of Ox. Conversely, if k_f is very large, so that the Z→Ox conversion rate is very fast on the chronoamperometric time scale, then the current–time response is again given by the Cottrell equation, but now the total $C_{Ox}^b + C_Z^b$ concentration determines the current

$$i_\infty = nF(C_{Ox}^b + C_Z^b)(D/\pi t)^{1/2} \qquad (34)$$

Intermediate values of k_f cause significant Z→Ox conversion on the experimental time scale, and the Fickian solution of diffusion equations for Z and Ox [see (6) and (7)] must be obtained, using the boundary conditions of (3) and (4) (written for both Ox and Z), (20), (23), and

$$\frac{\partial C_{Z(0,t)}}{\partial x} = 0 \qquad (35)$$

which is a statement that Z is *electroinactive* at the potential for Ox reduction. There results [12], under the simplifying assumption that $D_{Ox} = D_Z$,

$$i = nF(C_{Ox}^b + C_Z^b)(Dk_f K)^{1/2} \exp[k_f Kt] \, \text{erfc}[k_f Kt]^{1/2} \qquad (36)$$

The limiting case of very large k_f in (36) is (34). It is illustrative to compare the kinetically controlled current of (36) to the rapid Z→Ox conversion current of (34); their ratio is

$$\frac{i}{i_\infty} = (\pi k_f Kt)^{1/2} \exp[k_f Kt] \, \text{erfc}[k_f Kt]^{1/2} \qquad (37)$$

The dependency of the i/i_∞ current ratio on the kinetic parameter $(k_f Kt)^{1/2}$ {the argument of the exp[ ] erfc[ ] term in (37)} is illustrated in Figure 6.7.

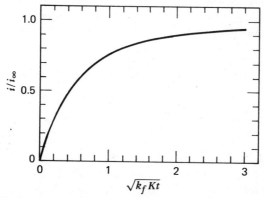

Figure 6.7   Working plot of (37) for chronoamperometric evaluation of prekinetic rate.

The plot confirms intuition: at sufficiently large k_f, K, or time, the observed current approaches the diffusion-controlled value i_∞. Figure 6.7 can be regarded as a *working plot* for comparison of experimental data with (37). Results read from Figure 6.7 for i/i_∞ at various times should yield a constant $k_f K$ to confirm obedience of the experimental situation to the CE reaction case (5). A separate measurement of the equilibrium constant by any appropriate method is required to complete evaluation of the rate constants k_f and k_b.

Chronocoulometry employs an integrated form [19] of (36), and working curves of Q/Q_∞ similar to those of Figure 6.7. It has been applied to study the kinetics of the monomer–dimer equilibrium between Fe(III)–EDTA complexes [20] where it assumes a somewhat different theoretical form because of the second-order character of this reaction.

2.2.2   Following Chemical Reactions (EC, ECE, Catalytic)

Much interesting, transient chemistry becomes accessible by means of electrochemical generation of an unstable species Red. Investigations in this area have been an important facet of electrochemical research. Three important general reaction types following the Ox $+ne\rightarrow$Red electrode reaction are the EC case [examples (8) and (9)], the catalytic case (11), and the ECE case, where the following reaction yields an additional electroactive species [examples (10), (12), (13)].

There are many subcases of these. The chemical reaction steps in the EC and ECE cases can be chemically reversible or irreversible (EC_{rev} and $EC_{rev}E$ versus EC_{irrev} and $EC_{irrev}E$). (The chemical reaction step in the catalytic case is typically irreversible, or else the cell solution decays irreversibly.) Furthermore, the chemical reaction steps may be either first- or second-order processes; this difference has an enormous effect on the mathematical treatment of the electrochemical experiment. The first-order reaction situation is the simplest and is often sought in second-order cases by choosing pseudo-first-order solution conditions (large excess or buffering of other components). Finally, the electrochemical (E) steps can be either reversible (fast-Nernstian) or irreversible (that is, $E_{rev}C_{rev}$ versus $E_{irrev}C_{rev}$, and so forth). The complication of dealing with this extra kinetic equation is often avoided by using large potential steps so that $C_{Ox(0,t)}=0$, and so on, when electron transfer is driven at a fast rate.

EC CASE

Consider the EC reaction where the reaction of Red is first order:

$$Ox + ne \rightarrow Red$$

$$Red \underset{k_b}{\overset{k_f}{\rightleftharpoons}} Y$$

(38)

A single potential step reducing Ox and producing the condition $C_{Ox(0,t)}=0$ would then yield only the Ox-diffusion-controlled current–time responses of (23) and (26). The reverse excitation, double potential step experiment must be applied for detection and kinetic characterization of the reaction of Red.

Obviously, in the EC case, the reverse, anodic currents (or charges) for Red oxidation will be less than the diffusion-only values of (29) and (30), depending on the value of k_f and k_b in (38).

When the chemical reaction is irreversible (EC_{irrev}; $k_b=0$), simultaneous solution of the diffusion equation for Ox (2) and Red

$$\frac{\partial C_{Red(x,t)}}{\partial t} = D_{Red}\frac{\partial^2 C_{Red(x,t)}}{\partial x^2} - k_f C_{Red(x,t)} \tag{39}$$

under appropriate boundary conditions was accomplished by Schwarz and Shain [16]. The chronoamperometric i_c–t behavior prior to reverse potential stepping at time τ is given by (23); that at $t>\tau$ is

$$i_a = nFC_{Ox}^b \left(\frac{D_{Ox}}{\pi}\right)^{1/2} \left[\frac{\phi}{(t-\tau)^{1/2}} - \frac{1}{t^{1/2}}\right] \tag{40}$$

where ϕ is a complex but numerically evaluable function of k_f, τ, and t. This expression reduces to the diffusion-only equation (29) when $k_f=0$; that is, $\phi=1.00$ in the absence of any Red decomposition.

As in diffusion-only controlled reactions with double potential step chrono-amperometry, it is convenient to examine the ratio i_a/i_c where i_a and i_c are, respectively, measured on the current–time response curve (see Figure 6.5) at times $t=t_f$ on the forward step and at $t_r-\tau$ on the reverse step, so that $t_f=t_r-\tau$ always. The ratio of (23) and (40) written at these times is

$$\left|\frac{i_a}{i_c}\right| = \phi - \left(\frac{\theta}{1+\theta}\right)^{1/2} \tag{41}$$

where θ is again $(t-\tau)/\tau$. Figure 6.8 shows working curves prepared [16] for dependency of the current ratio on $k_f\tau$ at various selected measurement times θ. Values of k_f result directly from using these curves and experimental i_a/i_c

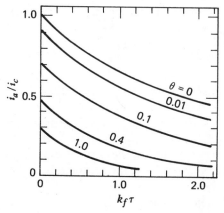

Figure 6.8 Working plot of (41), for selected θ, for double potential step chronoamperometric evaluation of EC postkinetics.

measured at appropriate θ. At short τ, the ratios tend to their diffusion-only values (recall Figure 6.6); at long τ, toward zero as Red becomes more or less completely consumed before reverse potential stepping. It is evident that optimum sensitivity of i_a/i_c data to the reaction of Red is obtained when $\tau = 0.5$ to 1.5 k; that is, the reverse potential step time τ should be of a magnitude comparable to the reaction half-life of the Red→Y reaction.

Marcoux and O'Brien [21] and Cheng and McCreery [22] point out that kinetic data on EC reactions can be obtained using a single potential step if at E_{final}, $C_{Ox(0,t)} > 0$ and the electrode reaction is assumed to be Nernstian. Given availability of the double step method, this seems an unnecessary assumption to endure, unless background current or onset of other waves prevents a sufficiently large potential step from being made.

A better application of single potential step chronoamperometry to EC reactions is simply using the chronoamperometric experiment for generating the unstable reaction product (Red) and using some other method for following the decay of Red or appearance of Y. This can be accomplished spectrophotometrically using optically transparent electrodes. An early example, with application to solvolysis of diphenylanthracene radical anion, was presented by Grant and Kuwana [23]. The theory for this "chronoabsorptometry" experiment, where the light beam is passed through the electrode and the adjacent solution, is closely related to chronocoulometric theory, since one "integrates" the total amount of electrode reaction product by measuring the optical absorption of the entire solution layer next to the electrode that contains it. Theory and applications for this branch of spectroelectrochemistry are found in a series of publications [15, 24–27] and reviews [28–30]. It is also possible to pass the light beam laterally through the diffusion layer [31] but this experiment is less well developed.

The double potential step chronocoulometric response to an EC reaction was reported by Christie [17] with a later correction of the equations by Ridgway and co-workers [32]. The charge–time curve before the reverse potential step at τ is unaffected by the chemical step and is given by the integrated Cottrell equation (26). The Q–t curve at $t > \tau$ is given by

$$Q_a = 2nFC_{Ox}^b \left(\frac{D_{Ox}}{\pi}\right)^{1/2} \left[t^{1/2} - (t-\tau)^{1/2} \int \phi \, dt\right] \tag{42}$$

where $\int \phi \, dt$ is again a complex but numerically obtainable quantity that reduces to 1.00 at $k_f = 0$ [see (30)]. As in the diffusion-only case, we examine the $(Q_\tau - Q_a)/Q_c$ ratio, measuring Q_c and Q_a at equal times after the reducing and oxidizing potential steps, respectively. Figure 6.9 shows how this ratio varies with the kinetic parameter $(k_f \tau)^{1/2}$ for the particular case of $\theta = 1.00$ (Q_c measured at $t = \tau$, Q_a at $t - 2\tau$). For k_f or $\tau \to 0$, the ratio tends to the diffusional value of 0.586.

Double potential step chronoamperometry was applied to an example of (38) by Schwarz and Shain [16], who studied the first-order rearrangement

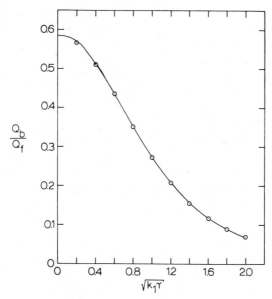

Figure 6.9 Working plot for double potential step chronocoulometric evaluation of EC post-kinetics; $\theta = 1.00$.

reaction of hydrazobenzene (to benzidine), which follows reduction of azo-benzene in aqueous acid. Rate constants measured ranged from 0.6 to 86 s^{-1}, depending on pH. This particular chemical reaction has been frequently employed since as an exemplary EC case. However, the reaction is complicated by the fact that the product hydrazobenzene tends to adsorb on the mercury electrode. Van Duyne and associates [32, 33] pointed out that double potential step chronoamperometry is relatively little affected by reactant or product absorption, since the adsorption current transients decay quickly. The double potential step chronocoulometric method is more sensitive to adsorption, and it must be accounted for to determine the chemical rates. Van Duyne and co-workers showed by application to the azobenzene reduction that the strong hydrazobenzene adsorption on mercury could be accounted for by their theory to obtain an excellent agreement with the chronoamperometric [16] result.

The capability of double potential methods to EC reaction rates was esti-mated to encompass $0.04 < k_f < 10^3$ s^{-1} by Koopman [34, 35] in the course of a study of reaction following electroreduction of nitrophenol. Ascorbic acid undergoes a very rapid hydration reaction following two-electron oxidation, and Perone and Kretlow [36] successfully employed reverse potential stepping times $\tau = 0.5$ to 1.0 ms to evaluate a $k_f = 1.3 \times 10^3$ s^{-1}. The upper limit of acces-sible k_f estimated by Koopman, which itself labels the double potential step chronoamperometric method as a powerful fast kinetics technique, may in fact be too low. An intermetallic compound formation between Zn and

$Co_2(Hg)_x$ occurs upon reduction of $Zn(II)$ into a $Co_2(Hg)_x$ amalgam electrode; Hovespian and Shain [37] employed the double potential step method for its kinetic characterization.

Finally, the above discussion is cast entirely in terms of the chemical reaction $Red \rightarrow Z$, occurring homogeneously in the solution next to the electrode. In some circumstances, however, it is likely that the electrode surface might act as a heterogeneous catalyst to the chemical reaction, which then becomes heterogeneous rather than homogeneous. Distinguishing this case is not straightforward and the possibility is too often ignored in experimental studies. The best test is probably to compare results using different electrode materials, but theoretical analysis can also be done [38, 39]. Guidelli [39] has examined chronoamperometric theory for electrode reactions coupled with preceding (CE), following (EC), catalytic, and parallel first-order, heterogeneous chemical steps.

CATALYTIC CASE

$$
\begin{aligned}
Ox + ne &\longrightarrow Red \\
Red + Z &\xrightarrow{\ k'\ } Ox + X
\end{aligned}
\tag{11}
$$

The special circumstance in which a following chemical reaction regenerates the original reactant Ox is called a catalytic reaction. The chemical reaction increases the cathodic current (or charge) for reduction of Ox in a single potential step experiment and decreases the anodic current (or charge) for reoxidation of Red in a double potential step experiment, as compared to the diffusion-only predictions of (23), (26), (29), and (30).

To solve the single potential step catalytic case for a first-order reaction, the pertinent Fickian equations are for Red, (39) with k_f set equal to $k'C_Z^b$, and for Ox:

$$
\frac{\partial C_{Ox(x,t)}}{\partial t} = D_{Ox} \frac{\partial^2 C_{Ox(x,t)}}{\partial x^2} + k'C_{Red(x,t)}C_Z^b
\tag{43}
$$

where $C_Z^b \gg C_{Ox}^b$ so that pseudo-first-order conditions are obtained and C_Z^b is a constant. Solution of these equations, under the potential step condition $C_{Ox(0,t)} = 0$ and other appropriate conditions, as described by Delahay and Stiehl [40], Miller [41], and Pospisil [42], gives for the *catalytic current*:

$$
i_{cat} = nFC_{Ox}^b D^{1/2} \left[k^{1/2} \, \mathrm{erf}(kt)^{1/2} + \frac{\exp(-kt)}{(\pi t)^{1/2}} \right]
\tag{44}
$$

The catalytic current predicted by (44) is compared to the diffusion-only current (23) for a single potential step reduction of Ox in the working plot, Figure 6.10, as a ratio $i_{cat}/i_{k=0}$ plotted versus the kinetic parameter $(kt)^{1/2}$. The equation for this ratio is

$$
\frac{i_{cat}}{i_{k=0}} = (kt)^{1/2} \left[\pi^{1/2} \, \mathrm{erf}(kt)^{1/2} + \frac{\exp(-kt)}{(kt)^{1/2}} \right]
\tag{45}
$$

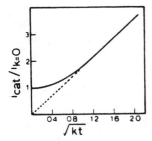

Figure 6.10 Working plot of (45) for chronoamperometric evaluation of catalytic kinetics.

In practice, experimental catalytic currents can be ratioed to the currents for Ox observed in the absence of Z (i.e., $i_{cat}/i_{k=0}$), and values of $(kt)^{1/2}$ for $i_{cat}/i_{k=0}$ at various times read from Figure 6.10.

Note from this figure and from (45) that at large $(kt)^{1/2}$ (>2), the ratio varies linearly with $(kt)^{1/2}$, with slope $\pi^{1/2}$. It is further apparent that at longer times the magnifying effect of the catalytic process on the current increases and more Z is consumed. Thus, in application at longer times (or fast k) the theory holds only for the condition that Z is not seriously depleted. This restriction is readily shown [12] to be

$$\frac{C_Z^b}{C_{Ox}^b} \gg (\pi kt)^{1/2} \tag{46}$$

In classical papers, Delahay and Stiehl [40] and Miller [41] applied the above theory to the catalytic reaction between Fe(II) and hydrogen peroxide following Fe(III) reduction, but in the polarographic chronoamperometric context. Guidelli and Cozzi [43] have developed the theory further. A more recent illustration of chronoamperometry applied to a catalytical reaction was given by Ryan and co-workers [44], who studied the reduction rates of several cytochrome-c redox proteins (Z) by electrogenerated ferrous–EDTA complexes (Red). The cytochromes-c are only slowly reduced directly by the electrode, but react readily with electron-transfer mediators like the iron–EDTA complex, which leads to a catalytic current in the ferric/ferrous–EDTA reaction. This is an important application because of the widespread interest in redox protein electron-transfer chemistry. The chronoabsorptometric version of this catalytic reaction has also been studied, for the mediated reduction of cytochrome-c by electrogenerated methyl viologen, by Mackey and Kuwana [27]. Electrochemistry applied to redox proteins has been reviewed by Feinberg and Ryan [45].

The single potential step chronocoulometric response for the catalytic case is the integral of (44) and, according to Christie [17], is

$$Q_{cat} = nFC_{Ox}^b D^{1/2} \left\{ \left[k^{1/2}t + \frac{1}{2k^{1/2}} \right] \mathrm{erf}(kt)^{1/2} + \left(\frac{t}{\pi} \right)^{1/2} \exp(-kt) \right\} \tag{47}$$

This somewhat more complex expression can be examined and compared to experimental data in the same basic manner as (45) (a plot of $Q_{cat}/Q_{k=0}$ against kt), except that a family of curves representing different values of k is required.

In a double potential step experiment with a first-order catalytic reaction case (11), the current and charge–time responses for times before the reverse step ($t \leqslant \tau$) are given by (44) and (47), respectively; those at $t > \tau$ are [17], respectively,

$$
i_{a(cat)} = nFC_{Ox}^b (kD)^{1/2} \left\{ \mathrm{erf}(kt)^{1/2} + \frac{\exp(-kt)}{(\pi kt)^{1/2}} \right.
$$

$$
\left. - \mathrm{erf}[k(t-\tau)]^{1/2} - \frac{\exp[-k(t-\tau)]}{[k\pi(t-\tau)]^{1/2}} \right\}
$$

(48)

$$
Q_{a(cat)} = nFC_{Ox}^b \left(\frac{D}{k}\right)^{1/2} \left\{ (kt+\tfrac{1}{2})\,\mathrm{erf}(kt)^{1/2} + \left(\frac{kt}{\pi}\right)^{1/2} \exp(-kt) \right.
$$

$$
\left. - [k(t-\tau)+\tfrac{1}{2}]\,\mathrm{erf}[k(t-\tau)]^{1/2} - \left[\frac{k(t-\tau)}{\pi}\right]^{1/2} \exp[-k(t-\tau)] \right\}
$$

(49)

The chronocoulometric response can be evaluated [17] in the same way as the EC equation (42): a plot of $[Q_{\tau(cat)} - Q_{a(cat)}]/Q_{c(cat)}$ against $(k\tau)^{1/2}$ for $\theta = 1.00$. The form of the plot is qualitatively similar to that of Figure 6.9. The charge ratio tends to the diffusion-only value 0.586 at very short times after potential step reversal and to zero at long τ or large k. Like the single step experiment, the effect of the catalytic process is magnified at longer times. Lingane and Christie [46] applied the double potential step chronocoulometric method to the reduction of Ti(IV) to Ti(III) with catalytic regeneration of Ti(IV) produced by reaction of Ti(III) with hydroxylamine; a $k = 43.4\,\mathrm{L/mol \cdot s}$ results. It might be noted, however, that the advantages of the integral response cited by the authors could perhaps have been more simply procured by using the single potential step method (44).

ECE CASE

A postkinetic chemical process (10) that leads to a product Y, which is more readily reducible than the original electrode reactant Ox, is an especially interesting circumstance in electrochemistry and occurs in a wide variety of organic and other electrode reactions. A classically cited example of the ECE case (which has actually proved to be more complex than simple ECE [47–49]) is the one-electron reduction of an aromatic hydrocarbon followed by protonation of the radical anion to yield a further and more readily, one-electron-reducible neutral radical [3, 6]. In a single potential step experiment, the second reduction enhances the observed cathodic current over the diffusion-only value for Ox [see 23]. The maximum current enhancement is $(n+n')/n$, when the chemical rate is rapid; and intermediate, kinetically controlled currents are seen at lesser rates.

There are several categories of ECE processes (see below). For the simplest case of a first-order chemical step in (10), the single potential step chrono-amperometric ECE theory was first given by Koutecky [50] and by Tachi and Senda [51], and later by Alberts and Shain [52] in a more intelligible form. Under the potential step condition that $C_{Ox(0,t)} = C_{Y(0,t)} = 0$, and for the irreversible chemical reaction case ($k_b = 0$), the chronoamperometric result is

$$i = FC_{Ox}^b \left(\frac{D}{\pi t}\right)^{1/2} \{n + n'(1 - \exp[-k_f t])\} \tag{50}$$

The effect of the chemical step in the ECE process on the current–time response is illustrated in Figure 6.11 as a current–$t^{1/2}$ plot of (50). At short times after the potential step, current is mainly controlled by electrolysis of Ox and the slope of the i–$t^{-1/2}$ plot is proportional to n. At sufficiently long times, the chemical process will "keep pace" with the electrochemistry, the i–t response is governed by joint Ox + Y electrolysis, and the i–$t^{1/2}$ slope is proportional to $n + n'$. At intermediate times the rate k_f is also a current-controlling and thus obtainable factor.

Experimental ECE data can be compared to (50) in two ways. First, a series of working curves similar to Figure 6.11 is prepared for varying k_f and compared to experimental data to find the working curve of "best fit." Or, a working plot is prepared of $\log(k_f t)$ plotted against the ratio of (50) to the expression for $k_f = \infty$ [i.e., (23) written for $n = n + n'$], and experimental values of i/i_∞ observed at various times are used to read values of $k_f t$. Various approaches may be used to obtain data for i_∞ for the system under study [52].

The reversible ECE case (for $k_b = 0$) was also treated by Alberts and Shain [52]; but, for reasons they discuss, nonzero values of k_b have little effect on the working curves. They applied the chronoamperometric theory to the reduction of p-nitrosophenol to p-hydroxylaminophenol, which dehydrates with $k_f = 1.4\,s^{-1}$ (the chemical step, pH = 4.8) to further reducible p-benzoquinoneimine. In later examples, Mastragostino and Savéant [53] applied chronoamperometry to an ECE reaction from uranyl ion, and Nelson and Adams [54] applied chronoamperometry in an elegant study of the oxidation of triphenylamines, which undergo dimerization to further oxidizable benzidines in acetonitrile solvent. Booman and Pence [55, 56] considered the theory for a dispropor-

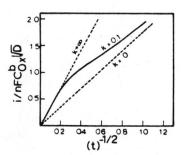

Figure 6.11 Current–time responses for chrono-amperometry with ECE kinetics for $k_f = \infty$, 0.1, and 1; and $n = n' = 1$.

tionation reaction of Red ($2\text{Red} \rightarrow \text{Ox} + \text{Y}$, a second-order process) and applied the chronoamperometric experiment to the case of uranium(V) disproportionation.

2.2.3   Coupled Chemical Reactions That Are Second Order

An important general problem in theoretical treatments of electrode-coupled chemical reactions concerns reactions that are second order. The Fickian equations most often become explicitly insoluble. In the ECE reaction, for instance, the fact that the species Y is reducible at less negative potentials than species Ox means that the reaction

$$\text{Red} + \text{Y} \underset{k_2}{\overset{k_1}{\rightleftharpoons}} \text{Ox} + \text{X} \qquad (51)$$

is thermodynamically favorable and may occur as a supplementary pathway to product X and larger currents. This possibility was overlooked in treatment of the early chronoamperometric ECE theory [52], but was later pointed out by Hawley and Feldberg [57], who called the reaction the *nuance reaction*. The combination of (10) and (51), in which the two chemical reactions compete, is termed the *ECE nuance reaction* [18]. The complication in the solution of the ECE nuance chronoamperometric theory is a mathematical consequence of the second-order nature of [51]. Thus, the Fickian expression that must be solved for Ox becomes in the presence of (51)

$$\frac{\partial C_{\text{Ox}(x,t)}}{\partial t} = D_{\text{Ox}} \frac{\partial^2 C_{\text{Ox}(x,t)}}{\partial x^2} - k_1 C_{\text{Ox}(x,t)} C_{\text{X}(x,t)} + k_2 C_{\text{Red}(x,t)} C_{\text{Y}(x,t)} \qquad (52)$$

Analogous equations are written for $\partial C_{\text{X}(x,t)}/\partial t$, $\partial C_{\text{Red}(x,t)}/\partial t$, and $\partial C_{\text{Y}(x,t)}/\partial t$, each containing *products of concentration terms* (because of the second-order rate law) like those in (52). These concentration term products prove difficult to separate into explicit solutions for C_{Ox}, C_{Red}, C_{X}, and C_{Y}.

For many years, the mathematical barrier to handling cases like the above impeded applications of electrochemical methods (in general) to second-order coupled reaction situations. One approach to the problem was to avoid it, by seeking pseudo-first-order reaction conditions. This obviously is not feasible in (51). Another approach was application of an approximate concept known as the *reaction layer* [12]. This quite effectively simplifies the problem toward a tractable mathematical solution, but the model in both nature and approximate character loses the ability to distinguish the input of the nuance pathway in the ECE nuance case. Some limited progress [19, 58] was made by application of numerical methods like those introduced by Nicholson and Shain [59] in their famous invention and popularization of cyclic voltammetry.

It remained however for introduction of a new, computer-based mathematical approach by Feldberg and Auerbach [60] for this problem to yield to widespread solution of different second-order electrochemical coupled reaction mechanisms. In this approach, the diffusion to/from the electrode is treated by dividing the diffusion profile into thin, parallel volume elements. Species Ox, Red, X, Y, and so on are transported from one volume element to the next and a time-dependent diffusion profile (concentration–distance diagram) is calculated (*digitally simulated*). The thereafter-computed time dependent $\partial C/\partial X$ of this diagram at $x=0$ gives the current–time result. Individual current–time responses are computed at varying concentrations, k values, and so forth, and the results are presented either tabularly or (more useful) graphically. The preferred graphical approach compares the kinetically determined current (or charge) to that for $k=0$ or $k=\infty$. Feldberg, who has reviewed the calculation in some detail [61], has popularized presentation of results as an apparent n value (n_{app}), which effectively is a kinetic current-scaling factor, versus a kinetic parameter involving some product of k and t. Various refinements of the calculation procedure have appeared [18, 62–65], and numerous coupled reactions treated. A very useful general, comparative treatment of different cases, using the digital simulation approach, was recently prepared by Hanafey, Scott, Ridgway, and Reilley [18]. This paper is also useful by tabularizing twenty-four cases with literature citations as to prior theoretical treatments for single potential and double step chronoamperometry and chronocoulometry.

Many of the second-order coupled reaction cases are closely related to one another and occur as mechanistic alternatives in the interpretation of electrochemical behavior of real chemical systems. An early case was deciding, for the ECE nuance reaction, whether the rate of (51), in which Y is homogeneously reduced, is sufficient to compete with (or entirely supplant) the heterogeneous consumption of Y in the simple ECE (10). Hawley and Feldberg [57], applying digital simulation in an early example of this approach, showed that neglect of the nuance reaction (51) causes results [in k_f of (10)] to be biased high or low depending on the value of k_1/k_2. Theoretical $i/i_\infty - k_f t$ working curves for various k_1/k_2 ratios were calculated for the assumption that (51) maintained an equilibrium position throughout the electrolysis. These were applied by Adams and associates [66] to an experimental case, the 1,4-addition of amines or hydrochloric acid to *o*-benzoquinones electrogenerated by catechol reductions, to demonstrate the importance of the nuance reaction.

An even more striking case of ECE versus ECE nuance is found in the work of Amatore and Savéant [47, 48], who in studies of aromatic hydrocarbon reductions in the presence of proton sources show that the nuance cross-reaction, not the ECE pathway, probably dominates in the electrochemistry of these species. Their work also emphasized the utility of double as opposed to single potential step chronoamperometry for distinguishing closely related mechanisms [47, 49].

Distinguishing closely related coupled reaction cases can sometimes require very accurate experimental chronoamperometric or chronocoulometric results, since the theoretically predicted responses do not differ significantly. An example is the reduction of an organic species that leads to dimerization. The first reaction product is a radical anion (R). Subsequently, to make dimer, do two radicals couple (R + R→radical–radical dimer) or does a radical couple with a parent (0 + R→radical–parent dimer) followed by further electrode reduction? This was an active question in studies of electrohydrodimerization [67], a crucial step in a commercially important electrosynthesis [6]. Feldberg [68] and others [55, 56, 69, 70] devised theories for these (and other related) cases. Feldberg compared the theory and concluded that, with chronoamperometry, exceedingly accurate data would be required to distinguish the pathway. Subsequently, the virtues of the double potential step methods were shown [18] to provide somewhat better resolution.

The problem of distinguishing related second-order coupled chemical reactions often remains tenuous, however, when relying on a single electro-chemical method, even when complete, digitally simulated, theoretical responses are available. Therefore, various authors have attacked the problem with further discerning pathways. One approach has been to bring multiple tech-niques to bear on the problem [18, 67], which can be effective, albeit involved. A different approach, which should be adopted more often than it is, is using *low-temperature electrochemistry*. Collecting current–time or charge–time data at a series of temperatures amounts to systematically moving the time frame of the experiment through the time span of the coupled chemical reaction (by varying the rate of the later) and should be very effective in the mechanistic sorting-out process. Van Duyne and Reilley [71–73] described the experimental and theoretical aspects of low-temperature single and double potential step chronoamperometric and chronocoulometric experiments with coupled reac-tions in a classical series of papers. Grypa and Maloy [74] have applied low-temperature experiments with the double potential step chronocoulometric method to study the mechanism of ethyl cinnamate and diethyl fumarate electrohydrodimerization.

An unusual ECE reaction scheme is that in which the second electrochemical step is an oxidation whereas the first was a reduction (or conversely). Feldberg and Jeftic [75] have discussed the theory for this case, in which the "nuance" cross-reaction plays an important role, and point out that the chronoamper-ometric current can change sign during current decay. The reduction of $Cr(CN)_6^{3-}$ in alkaline solution exhibits such a current sign change and is a possible example of this type of ECE reaction.

2.3   Adsorption of Electrode Reactant

2.3.1   Single Potential Step Chronocoulometry

Consider the case in which the reactant Ox lies at $t=0$ in an equilibrium between adsorbed and solution forms

$$Ox_{ads} \quad\diagdown\; ne \atop \diagup\; ne \quad Red \qquad (18)$$

$$Ox_{soln}$$

Both Ox_{ads} and Ox_{soln} are presumed to be electroactive. Application of a potential step sufficiently negative to produce $C_{Ox(soln)(0,t)}=C_{Ox(ads)(0,t)}=0$

yields the normal diffusional current–time response [see (23)] for electrolysis of Ox_{soln}; but the adsorbed species Ox_{ads}, not requiring a mass-transport step prior to reduction and present in a limited amount determined by the surface excess of Ox_{ads} (Γ_0 at $t=0$), reacts in a sharp current spike at very small time. Although the presence of the current spike for reaction of Ox_{ads} can sometimes be discerned, it is difficult to measure this quantitatively. If, however, we record the chronocoulometric ($Q-t$) response of this experiment, the charge attributable to reaction of Ox_{ads}

$$Q_{ads} = nF\Gamma_0 \quad (C/cm^2) \tag{19}$$

is "remembered" at longer times and in fact simply adds at any time to the time-dependent charge accumulated by reaction of Ox_{soln} (26).

The expression for the chronocoulometric single potential step response in the presence of reactant adsorption is then

$$Q = 2nFC_{Ox}^b \left(\frac{D_{Ox}t}{\pi}\right)^{1/2} + Q_{ads} + Q_{dl} \tag{53}$$

where C_{Ox}^b is the bulk concentration of Ox_{soln}, Q_{dl} is the difference in the charge on the electrical double layer between E_{init} and E_{final} of the potential step, and Q_{ads} represents $nF\Gamma_0$ at E_{init}. The experimental $Q-t$ data are analyzed through a $Q-t^{1/2}$ plot, of which Figure 6.12 is an example. The zero-time intercept provides $Q_{ads} + Q_{dl}$, from which, after correction for Q_{dl}, the surface excess Γ_0 can be obtained with (19).

The correction Q_{dl} is obtained from a blank potential step from E_{init} to E_{final} in supporting electrolyte solution alone. Since the electrode double layer is in effect a capacitor, this is a time-independent charge, Q_{dl}^b. It is necessary, however, to account for any change in the double-layer charge at E_{init} caused by the presence of Ox_{ads}. This can be evaluated, for dropping mercury electrodes, by integrating the charge accumulated on a growing mercury drop, at E_{init}, in the presence and absence of Ox_{ads} [76]. The difference in these charges, ΔQ_{dl},

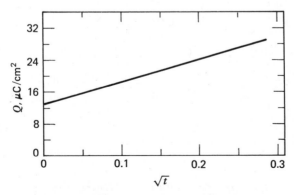

Figure 6.12 Analysis of chronocoulometric data for reactant adsorption by (53): $n=1$; $D = 4.0 \times 10^{-6} \text{ cm}^2/\text{s}$; $C_{Ox}^b = 0.5 \text{ mol/cm}^3$; $\Gamma_0 = 1.0 \times 10^{-10} \text{ mol/cm}^2$; $Q_{dl} = 8 \ \mu C/cm^2$.

when subtracted from Q_{dl}^b, yields the appropriate Q_{dl} for application to (53) and evaluation of Γ_0.

The chronocoulometric method of adsorption measurement was introduced by Christie and co-workers [77]. Because of its clear superiority to the shortcomings of the chronopotentiometric method, then used for adsorption measurements (see below), chronocoulometry quickly supplanted this method. It, in turn, has been supplanted by its double potential step partner (again, below) as the present "faradaic method of choice" for accurate and sensitive surface excess studies.

The single potential step method has been applied to adsorption of metal complexes on mercury electrodes, including studies of adsorption of Co(III)–EDTA complexes by Anson [78]; of Pb(II) iodide, bromide, and thiocyanate by Murray and Gross [76]; of Hg(II) thiocyanate by Anson and Payne [79]; of Cd(II) iodide and In(III) thiocyanate by O'Dom and Murray [80, 81]; and of Zn(II) thiocyanate by Osteryoung and Christie [82].

The above experiments refer to potential steps of such magnitude that the concentrations $C_{Ox(soln)(0,t)}$ and $C_{Ox(ads)(0,t)}$ are driven to zero. An interesting, and for a Nernstian reaction, tractable case is stepping to a lower potential, so that significant concentrations of Ox_{soln} and Ox_{ads} remain at the electrode surface. Reinmuth and Balasubramanian [83] have treated this case, for a Langmurian adsorption isotherm, to show that the chronocoulometric charge contains both an instantaneous and a time-dependent adsorptive component as well as the normal diffusive charge. The presence of an Red_{ads} state was also allowed. The analysis reveals the presence of prewave and postwave adsorptive charges for product and reactant adsorption, respectively, as expected.

2.3.2 Double Potential Step Chronocoulometry

Inasmuch as Γ_0, and consequently Q_{ads}, can often be an exceedingly small term (Γ_0 values measured have ranged from 3×10^{-12} to 8×10^{-10} mol/cm^2), Q_{dl} frequently exceeds Q_{ads}. The necessity for accurate evaluation of Q_{dl} can then become extreme, and the accumulation of errors in the total of four experiments required in the single potential step method can limit the sensitivity of detection and quantitative evaluation of Q_{ads}. The double potential step chronocoulometric experiment, introduced by Anson [84] and theoretically characterized by Christie and associates [85], eliminates much of the experimental error incurred in the Q_{dl} correction by acquiring $Q_{ads} + Q_{dl}$ and Q_{dl} data in a single experiment.

In the double potential step chronocoulometric experiment with reactant adsorption, the charge during $t \leqslant \tau$ is again given by (53) and the charge during $t > \tau$ by [85]

$$Q_a = 2nFC_{Ox}^b \left(\frac{D_{Ox}}{\pi}\right)^{1/2} [t^{1/2} - (t-\tau)^{1/2}] + Q_{ads}\left[\frac{2}{\pi}\sin^{-1}\left(\frac{\tau}{t}\right)^{1/2}\right] \quad (54)$$

This expression, describing charge for oxidation of Red, differs from the diffusion-only equation (30) only in the last term, which accounts for the extra quantity of Red (in the Red diffusion profile) arising from previous reduction of Ox_{ads}. This term is small and well approximated by a simpler expression; making such an approximation and now measuring charge as a difference between Q_a and the cathodic charge at τ [Q_c, (53) written for $t=\tau$] gives

$$Q_\tau - Q_a = 2nFC_{Ox}^b \left(\frac{D_{Ox}}{\pi}\right)^{1/2} \left[1 + \frac{a_1 \Gamma_0 \pi^{1/2}}{2C_{Ox}^b D_{Ox}^{1/2} \tau^{1/2}}\right]$$
$$\times [(t-\tau)^{1/2} + \tau^{1/2} - t^{1/2}] + a_0 Q_{ads} + Q_{dl} \tag{55}$$

The terms a_0 and a_1 are constants resulting from the above approximation over a specified range of t and τ; typical values [85] are $a_0 = -0.0688$ and $a_1 = 0.970$.

Experimental data are treated according to (53) and (55) by plots of Q_c against $t^{1/2}$ and $Q_\tau - Q_a$ against $(t-\tau)^{1/2} + \tau^{1/2} - t^{1/2}$. An example is shown in Figure 6.13. Comparison of (53) and (55) shows that the intercepts of these plots are $Q_{ads} + Q_{dl}$ and $a_0 Q_{ads} + Q_{dl}$; the difference between these intercepts eliminates Q_{dl} and is $Q_{ads}(1 - a_0)$, allowing calculation of Γ.

This theory assumes, in expressing the Q_{dl} for E_{init} at $t > \tau$, instantaneous reestablishment of the Ox_{ads} layer after the reverse potential step. This is not necessarily the case, as pointed out by Elliott and Murray [86], but the difficulty is not a common one.

Evaluation of reactant surface excesses by the double potential step method yields experimental uncertainties in Γ_0 from two to five times smaller than those typical ($\pm 1 \times 10^{-11}$ mol/cm²) [80] for the single potential step experiment. A further, extremely significant advance was the application [87] of on-line computerized data acquisition methods to the double potential step experiment. The considerably greater accuracy of data recording thus possible lowered the uncertainty in adsorption measurements further to about $\pm 1 \times 10^{-12}$ mol/cm² [88].

Anson and co-workers have applied this experiment to adsorptions, on mercury working electrodes, of Cd(II) from thiocyanate [89], thiosulfate [90], chloride [87], iodide and bromide [88] solutions, and of anthraquinone monosulfonate ion [91].

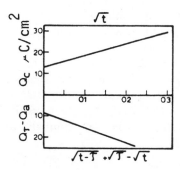

Figure 6.13 Analysis of double potential step chronocoulometric data for reactant adsorption. Parameters same as in Figure 6.12; $\tau = 0.100$ s.

2.4   Charge-Transfer Kinetics

As shown earlier by (14) and (15), the rate of a charge-transfer reaction $Ox + ne \rightarrow Red$ depends on both its rate constant k_s and the value of the applied potential. The influence of k_s on current is removed at sufficiently negative applied potential, as then, the charge-transfer rate process is made fast enough that mass transport of reactant becomes the current-limiting factor. Such was the case in the chronoamperometric and chronocoulometric experiments discussed above; the relations derived there had no dependence on either the value of k_s or the reversibility of the charge-transfer reaction. Consider now use of much smaller potential steps and the resulting charge-transfer rate connotations in chronoamperometric and chronocoulometric experiments.

The chronoamperometric, or *potentiostatic* (as it is commonly called), experiment was developed for charge-transfer kinetic studies by Gerischer and Vielstich [92, 93]. Consider first the experiment as conducted in a solution containing both Ox and Red; E_{init} is thus the equilibrium ($i=0$) potential E_{eq} of this mixture and E_{final} is a potential somewhat more negative than E_{eq}. Solution of the Fickian equations for Ox and Red using (15) as a boundary condition yields

$$i = i_0 \left\{ \exp\left[\frac{-\alpha nF}{RT} \eta \right] - \exp\left[\frac{(1-\alpha)nF}{RT} \eta \right] \right\} \exp(\lambda^2 t)\, \mathrm{erfc}(\lambda t^{1/2}) \qquad (56)$$

where

$$\lambda = \frac{i_0}{nF} \left(\frac{\exp\left[\dfrac{-\alpha nF}{RT} \eta \right]}{C_{Ox}^b D_{Ox}^{1/2}} + \frac{\exp\left[\dfrac{(1-\alpha)nF}{RT} \eta \right]}{C_{Red}^b D_{Red}^{1/2}} \right) \qquad (57)$$

and $\eta = E_{final} - E_{eq}$, the *charge-transfer overvoltage*. The first term of (56) is recognized as identical to (15). The second, time-dependent term represents the *mass-transfer overvoltage*, describing the decay of current resulting from electrolytic depletion of Ox, and enhancement of Red, in a diffusion layer around the working electrode. The time dependency of the mass-transfer term is illustrated in Figure 6.14. At a sufficiently long time ($\lambda t^{1/2} > 5$), the $\mathrm{erfc}(\lambda t^{1/2})$ term is $\sim \exp[-\lambda^2 t]/\eta^{1/2}\lambda t^{1/2}$, λ and i_0 vanish from (56), and the current becomes dominated by a $i - t^{-1/2}$ mass transfer similar to that of the Cottrell equation (23). The most pronounced dependency on the charge-transfer kinetic term i_0 occurs at short electrolysis times, and current–time data aiming at measurement of i_0 are taken with this in mind.

Two approaches are useful for dealing with (56) at short electrolysis times. In one, measurement at times sufficiently small that $1 \gg \lambda t^{1/2}$ allows restatement of the mass-transfer term as $1 - 2\lambda t^{1/2}/\eta^{1/2}$. Thus, a plot of current against $t^{1/2}$ is linear with the zero-time intercept current equal to (15). If, also, the potential step is sufficiently small ($|\eta| < \alpha nF/RT$, a few millivolts), then (15) simplifies to

$$i_{\eta \to 0} = -\frac{nFi_0\eta}{RT} \qquad (58)$$

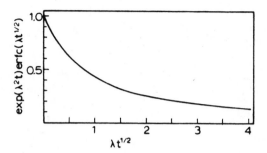

Figure 6.14 Variation of mass-transfer overvoltage term of (56) with its argument $\lambda t^{1/2}$.

from which the exchange current i_0 is calculable. Application of (16) then yields the charge-transfer rate k_s. In the second approach [94, 95], current data ratios at times t and $4t$ over a span of time when $0.14 < \lambda t^{1/2} < 1.0$ are employed with an alternative mathematical expression for mass transfer, to again obtain the zero-time current of (15). Further details and other diffusion geometries can be found in subsequent papers [96–98] and in a review of methods for charge-transfer study by Delahay [99]. Additional variants of the methods for data analysis have also been described [100, 101]. It is estimated that the potentiostatic method is applicable to measuring charge-transfer rates $k_s < 1$ cm/s.

The potentiostatic experiment can also be conducted on solutions containing only reactant $Ox(C_{Red}^b = 0)$, a convenience when solutions of Red are difficult to maintain or prepare. In this case the Fickian solution for mass transport of Ox and Red employs (14) as a boundary condition, and there results [12]

$$i = nFk_s C_{Ox}^b \left\{ \exp\left[\frac{-\alpha nF(E-E^0)}{RT} \right] \right\} \exp(\gamma^2 t)\, \text{erfc}(\gamma t^{1/2}) \tag{59}$$

where

$$\gamma = k_s \left\{ \frac{\exp\left[\dfrac{-\alpha nF(E-E^0)}{RT} \right]}{D_{Ox}^{1/2}} + \frac{\exp\left[\dfrac{(1-\alpha)nF(E-E^0)}{RT} \right]}{D_{Red}^{1/2}} \right\} \tag{60}$$

and E is E_{final} in the potential step. Again, at long times (large $\lambda t^{1/2}$), (59) reduces to a largely mass-transport controlled condition. At shorter times (59) is handled in the same manner as (56); a plot of current against $t^{1/2}$ yields

$$i_{t=0} = nFk_s C_{Ox}^b \left\{ \exp\left[\frac{-\alpha nF(E-E^0)}{RT} \right] \right\} \tag{61}$$

from a logarithmic form of which, with a series of experiments at different E, data for k_s result. See a recent determination [102] of k_s for $[Ru(bpy)_3]^{3+/2+}$ in aqueous acid for an exemplary application of (59), and the text by Albery [7] for further discussions of electrode kinetics.

Chronocoulometric theory for the case of $C^b = 0$ has been given by Christie and associates [103]; the integrated form of (59) is

$$Q = \frac{nFk_s C_{Ox}^b}{\gamma^2} \left\{ \exp\left[\frac{-\alpha nF(E-E^0)}{RT} \right] \right\} \left\{ \exp(\gamma^2 t)\, \mathrm{erfc}(\gamma t^{1/2}) + \frac{2\gamma t^{1/2}}{\pi^{1/2}} - 1 \right\} \quad (62)$$

At sufficiently long times $(\gamma t^{1/2} > 5)$, the term $\exp(\gamma^2 t)\, \mathrm{erfc}(\gamma t^{1/2})$ becomes negligible; thus a plot of Q against $t^{1/2}$ becomes linear with the extrapolated intercept on the charge axis of

$$t_{int}^{1/2} = \frac{\pi^{1/2}}{2\gamma} \quad (63)$$

An example of such a plot is shown in Figure 6.15. Values of γ obtained for a series of potential steps at different E pass through a minimum at $E = E^0$, as is apparent from (60). For $\alpha = 0.5$ and $D_{Ox} = D_{Red}$, this occurs at $E = E^0$, and $\gamma_{min} = 2k_s/D^{1/2}$ (see [103] for more general equations for γ_{min}).

It is interesting that in the chronocoulometric method, data taken at long times are useful for extraction of k_s; just the opposite is true in chronoamperometry. This is a direct consequence of the integral (charge) mode of data recording (effecting a "memorization" of preceding kinetic influences on the flow of current) and is a practical convenience of this method. Difficulties [103] can arise, however, for slower charge-transfer rates, inasmuch as the electrolysis times necessary to achieve linearity in the Q–$t^{1/2}$ plot (to obtain $\gamma t^{1/2} > 5$) may be sufficiently long to be susceptible to convective and other deleterious effects on the diffusion mass-transport process. Lingane and Christie [104], by combining chronoamperometric and chronocoulometric data, seek to eliminate short- and long-time measurement problems, respectively, of these methods.

An analogy to chronocoulometry is chronoabsorptometry, measuring the optical absorbance of Red by passing a light beam through the diffusion layer. The chronoabsorptometry experiment based on (62), expressed in absorbance rather than charge, has been shown and experimentally illustrated by Bancroft and co-workers [105].

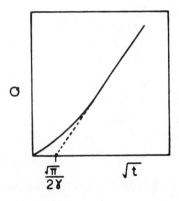

Figure 6.15 Analysis of charge-transfer-controlled chronocoulometric data.

2.5   Nonplanar Diffusion and Disk Microelectrodes

The above theory and experiment applies to the classic context of planar or linear diffusion such as that to a disk electrode in Figure 6.2. Planar diffusion conditions are satisfied when the disk radius (r) is large compared to the diffusion layer dimensions [i.e., $r \gg (Dt)^{1/2}$]. When this condition is not satisfied, a significant increase in the chronoamperometric current occurs because of diffusion to the electrode from the solution volume around the disk edge, the so-called *edge effect*. And when the edge effect is extreme (disk diameters of about 10^{-3} cm), a dramatic change in the experimental behavior occurs. Above very short times (about 0.1 s), the chronoamperometric current becomes *constant*. This occurs because the rate of the effectively hemispherical diffusion to the "pinpoint" electrode is so fast that it completely balances out the consumption of the reactant at the electrode. Another way to regard the effect is that the volume of solution from which diffusion occurs to the electrode is now a hemisphere, which is much larger than the cylindrical volume in simple planar diffusion. Under these circumstances, by slowly scanning the electrode potential, *steady-state current–potential voltammograms* are seen that resemble those observed at rotated disk electrodes.

Chronoamperometry and slowly scanned voltammograms at disk microelectrodes are of considerable recent interest because of a variety of practical factors. Observing steady-state currents is one obvious practical advantage [albeit counterbalanced by the very small (nanoampere) currents one must often measure]. Another is the minimizing of uncompensated resistance effects in low-conductivity media caused by the small currents that flow [22]. Freedom from cross-reaction catalysis in analysis of mixtures, caused by the effective loss of electrolysis product diffusing into the hemispherical void, is another [106, 107]. The original and still considerable impetus to the use of microelectrodes has been their application to *in vivo* electrochemistry [108].

The theoretical diffusion problem for mass transport to a disk microelectrode of vanishingly small diameter is not simple and remains unsettled in exact detail. For a reaction uncomplicated by coupled chemical reactions, it is clear however that the general equation has the form

$$i = \frac{nFC_{Ox}^b D_{Ox}}{(\pi t)^{1/2}} + brnFC_{Ox}^b D_{Ox} \tag{64}$$

where r is the disk electrode radius. The time-dependent term is the Cottrell response (19), which at a sufficiently long time becomes smaller than the right-hand, radius-dependent, and time-independent term. The right-hand term thus expresses the steady-state voltammetric current for the microelectrode. The numerical value of b, and previous literature thereon, has been discussed by Heinze [109]; see also papers by Oldham [110] and Pons and co-workers [111].

2.6   Finite Diffusion

The preceding situations have used exclusively semi-infinite diffusion conditions [(3) and (4)]. Chronoamperometry and chronocoulometry can also be

performed under finite or bounded diffusion conditions. Finite diffusion mathematics is more complex than semi-infinite, but analytical solutions have been
obtained for the three experiments illustrated in Figure 6.16. In these experiments, the electroactive species either diffuses to the electrode from within a
thin film only, as in Panel A, or to the electrode from within and across the thin
film, as in Panels B and C.

In Panel A, the electroactive species occupies only the thin film of solution
and no reactant exists at distances beyond the film thickness, d. In Panel B,
the electroactive species is present as a solute both in the film and in the unstirred solvent medium beyond distance d. Its concentration may be different
in the film and medium, if there is an equilibrium caused by partitioning between
the film and solvent medium ($P = C_{film}/C^b$). The diffusion coefficients of the
electroactive species may also be different in the film and solvent medium (that
is, $D_{film} \neq D_{soln}$). In Panel C, electroactive species are present in both film and
contacting solvent medium, and a partitioning effect may again be present.
Conditions of the experiment in Panel C, in contrast to Panel B, are such that
concentration polarization in the medium beyond the film ($x > d$) is not important (i.e., transport in that medium is very much faster than that in the film).

Labeling the electroactive component as a reducible Ox, for a potential
step producing $C_{Ox(0,t)} = 0$, the equations for the experiments of Panel A [112,
113], Panel B [114–116], and Panel C [117, 118] are, respectively,

$$i = \frac{nFAD_{Ox,film}^{1/2}PC_{Ox}^b}{\pi^{1/2}t^{1/2}}\left[1 + 2\sum_{j=1}^{\infty}\left(\frac{1-\alpha}{2-\alpha}\right)^j \exp\left(\frac{-j^2d^2}{D_{Ox,film}t}\right)\right] \quad (65)$$

where $D_{Ox,film}$ is the diffusion coefficient of Ox in the film of thickness d in
Panel A,

$$i = \frac{nFAD_{Ox}^{1/2}C}{\pi^{1/2}t^{1/2}}\left[\sum_{k=0}^{\infty}(-1)^k\left\{\exp\left(\frac{-k^2d^2}{D_{Ox}t}\right) - \exp\left[\frac{-(k+1)^2d^2}{D_{Ox}t}\right]\right\}\right] \quad (66)$$

where $\alpha = P(D_{Ox,film}/D_{Ox})^{1/2}$, and $D_{Ox,film}$ and D_{Ox} are the diffusion coefficients

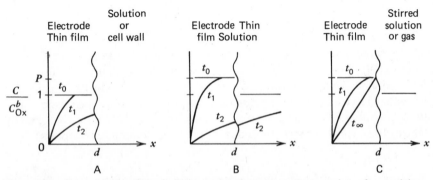

Figure 6.16 Concentration–distance diagrams for times $t = t_0 = 0$, t_1, and t_2 after applying a
potential step causing $C_{Ox(0,t)} = 0$, for different finite diffusion cells (see text).

of Ox at $0 < x < d$ and at $x > d$, respectively, in Panel B, and

$$i = \frac{nFAD_{Ox,film}^{1/2}PC_{Ox}^b}{\pi^{1/2}t^{1/2}}$$

(67)

$$i = \frac{nFAD_{Ox,film}PC_{Ox}^b}{d}$$

(68)

where (67) applies to the short-time response and (68) to the steady-state current observed at long time in Panel C.

Note that at a sufficiently short time, where the diffusion layer does not yet impinge on the boundary d, the three experiments in panels A–C all give the same Cottrellian response. Namely, at short times, (65) reduces to (23), and (66) reduces to (67), which is the same as (23) except for the concentration of Ox being $C_{Ox,film} = PC_{Ox}^b$. These short-time equations can also of course be expressed as charge–time, as in (26), if the data recording is chronocoulometric.

Now let us place the experiments of Panels A–C in Figure 6.16 into real context. Panel A corresponds to the experimental situation of *thin-layer electrochemistry* in which the cell wall opposite the working electrode is placed at distance d. See [119] for a review of thin-layer electrodes and theory. The current–time and charge–time theories of potential steps in thin layers of solutions are seldom invoked, however, because the uncompensated resistance problems of thin-layer cells are usually considerable.

Panel A also corresponds to the experimental situation of thin, polymeric coatings on electrodes that contain electroactive substances. The film contacts an electrolyte solution at $x > d$, which contains no electroactive substances. Such coatings, studied over the past several years (see a review in [120]), are typically from 5 to 10,000 nm thick, and their electroactivity may come either from the redox groups affixed to the polymer (redox polymer) or from redox counterions of an ion-exchange polymer film. The diffusion coefficient $D_{Ox,film}$, measured from chronoamperometry [112] or chronocoulometry [121] of the electroactivity of these films, in part (mainly for redox polymers) reflects electron-hopping chemistry within the films; that is, $D_{Ox,film}$ is an electron diffusion coefficient (D_{ct} or D_e). Measurements of D_e have provided a way to study this special kind of polymer electron-transfer chemistry, and this subject has been intensely scrutinized.

The experiment of Panel B also refers to a polymer film (in this case a non-electroactive film) on an electrode surface in contact with a solution in which an electroactive constituent is dissolved. The electroactive constituent partitions into the film. The Panel B experiment is useful for measuring the permeability of the polymer film to the test molecule Ox, whose charge and/or size are varied as part of a study of permeability [114, 122]. Chronoamperometric theory on pinholes [123, 124] is relevant to such studies.

Panel C corresponds to so-called *voltammetric membrane electrodes*, developed long ago by Bowers and co-workers [118, 125]. Concentration polarization was avoided (approximately) by stirring the solution. The experiment and theory were resurrected by McCallum and Pletcher [117] for the situation where the medium beyond d is a gas stream (which is nonpolarizable), the electrode (porous) is placed at $x = 0$, and the electrolyte solution at (minus) $-x$. This experiment is then an analytical gas-sensor electrode.

As electrochemists continue to probe spatial microstructuring [120] of the electrode surface, finite diffusion experiments will enjoy increasing significance. A coupled chemical reaction theory should appear in time.

3   CHRONOPOTENTIOMETRY

The elementary theory for current step, or constant-current, chronopotentiometry was proposed in the early twentieth century [126–129]. The term "transition time" was coined in later work [130, 131). Although some (albeit inaccurate) experiments were described in the early reports, it was not until the work of Gierst and Juliard [132], followed closely by a timely chapter on the subject [12] and coincident papers by Delahay and Mamantov [133] and Reilley and co-workers [134], that a more searching interest in chronopotentiometry was aroused. Research into this subject soon circumscribed the significant aspects of chronopotentiometric theory and experiment [135], the areas of applicability, the advantages, and the limitations. Today it is understood that chronopotentiometry offers advantages in some circumstances, but that for many situations the difficulties in accurate transition time measurements cause one to seek alternative methods (see Section 4).

3.1   Mass-Transport Control by Diffusion

3.1.1   Constant-Current Experiment

The boundary condition describing application of a current step causing reduction of electrode reactant Ox is expressed by Fick's First Law:

$$\frac{i}{nFD_{Ox}} = \frac{\partial C_{Ox(0,t)}}{\partial x} \tag{22}$$

Solution of the linear diffusion equation (2) under this and the conditions of (3) and (4) yields the concentration–time–distance profile of species Ox as

$$C_{Ox(x,t)} = C_{Ox}^b - \frac{2it^{1/2}}{nFD_{Ox}^{1/2}\pi^{1/2}}\exp\left(-\frac{x^2}{4D_{Ox}t}\right) + \frac{ix}{nFD_{Ox}}\operatorname{erfc}\left(\frac{x}{2D_{Ox}^{1/2}t^{1/2}}\right) \tag{69}$$

The analogous Fickian solution for product Red, but with initial condition $C_{Red(x,0)} = 0$, yields for the profile of Red

$$C_{Red(x,t)} = \frac{2it^{1/2}}{nFD_{Red}^{1/2}\pi^{1/2}}\exp\left(-\frac{x^2}{4D_{Red}t}\right) - \frac{ix}{nFD_{Red}}\operatorname{erfc}\left(\frac{x}{2D_{Red}^{1/2}t^{1/2}}\right) \tag{70}$$

Concentration–distance profiles calculated from (69) are shown in Figure 6.17. The response for this chronopotentiometric electrolysis is called the *transition time* (τ), and it occurs when $C_{Ox(0,t)} = 0$. In the figure this is 1.9 s, and

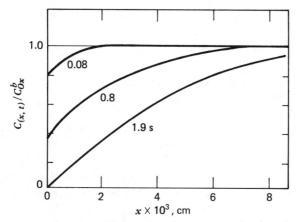

Figure 6.17 Chronopotentiometric concentration–distance curves for electrolysis with transition time 1.9 s; and $n=1$; $D=1.0\times10^{-5}$ cm^2/s; $C_{Ox}^b=1.0\times10^{-6}$ mol/cm^3; and $i=2.0\times10^{-3}$ A/cm^2.

a break in the potential–time curve occurs at this point since the electrolysis of Ox can no longer support the impressed current and the potential rises to that of an additional electroreduction reaction. Setting $x=0$ and $t=\tau$ in (69) yields the *Sand equation*, the basic expression for many chronopotentiometric applications:

$$\frac{i\tau^{1/2}}{C_{Ox}^b}=\frac{nFD_{Ox}^{1/2}\pi^{1/2}}{2} \tag{71}$$

Several features of the Sand equation require comment. First, the $\tau^{1/2}$–C_{Ox}^b relationship can be the basis for quantitative chronopotentiometric assay of concentration. Analytical applications of chronopotentiometry have been reviewed by Davis [136]. Secondly, for known n and C_{Ox}^b, evaluation of the diffusion coefficient D_{Ox} becomes possible. Next, although transition time τ decreases at increased applied current i, the product $i\tau^{1/2}$ should remain constant. Experimental constancy of $i\tau^{1/2}$ is the chronopotentiometric criterion that mass transport of reactant Ox is diffusion controlled over the range of times inspected. Finally, the factor $i\tau^{1/2}/C_{Ox}^b$, the *chronopotentiometric constant*, is characteristic of the species Ox. Equation (71) has been verified by numerous workers; see [134] for an early example.

Equation (71) is derived from mass-transfer considerations alone, and therefore the chronopotentiometric transition time τ is not altered by charge-transfer reversibility considerations. The charge-transfer rate constant does influence, however, how potential varies with time as $C_{Ox(0,t)}$ changes (i.e., the shape of the chronopotentiometric E–t curve). Reversibility, thus, indirectly affects the accuracy of τ determination since the potential–time break is typically less distinct with an irreversible reaction.

For a reversible electrode reaction, insertion of (69) and (70) (at $x=0$) into the Nernst equation yields, after rearrangement and use of (71),

$$E = E^0 + \frac{0.0591}{n} \log \frac{D_{Red}^{1/2}}{D_{Ox}^{1/2}} + \frac{0.0591}{n} \log \frac{\tau^{1/2} - t^{1/2}}{t^{1/2}} \quad (72)$$

The shape of a reversible chronopotentiogram is illustrated in Figure 6.17. The chronopotentiometric criterion for reversibility is a linear plot of E versus $\log[(\tau^{1/2} - t^{1/2})/t^{1/2}]$ with slope $0.0591/n$. An irreversible reaction gives in this plot a slope larger than $0.0591/n$, or a curved plot. The value of $t = \tau/4$ is a characteristic potential, the *quarter-wave potential* $E_{1/4}$, which can be identified with the polarographic half-wave potential $E_{1/2}$.

The potential–time relation for a totally irreversible reduction of Ox is found by insertion of (69) (at $x = 0$) into (14), retaining only the i_c term, which yields [137]

$$E = E^0 + \frac{0.0591}{\alpha n_a} \log \frac{2k_s}{\pi^{1/2} D_{Ox}^{1/2}} + \frac{0.0591}{\alpha n_a} \log(\tau^{1/2} - t^{1/2}) \quad (73)$$

A totally irreversible chronopotentiometric wave is displaced negatively on the potential axis from its reversible counterpart (Figure 6.18) according to both the values of k_s and of the applied i (through τ). The irreversible chronopotentiogram also has a different shape, and usually exhibits a less distinct transition time break. It is evident that (73), and the more complex equation that results from retention of both terms of (14) (quasi-reversible case), provide a route to chronopotentiometric evaluation of the charge transfer rate k_s.

Another current step experiment, the *galvanostatic* experiment, should be mentioned. This experiment is useful for measurement of faster charge-transfer

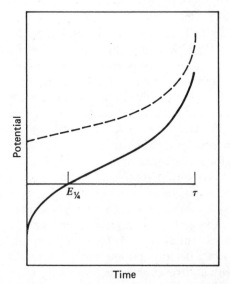

Figure 6.18   Reversible (——) and irreversible (- - -) chronopotentiograms.

rates, and in effect obtains the kinetic information from the very-short-time portion of the chronopotentiometric potential–time curve. In a solution containing both Ox and Red, a current step is applied to produce only a small overpotential change (η, recall Section 2); η should be small to allow linearization of (15). Over a very short time, an η–$t^{1/2}$ response is obtained that is extrapolated to zero time to yield an η related to the applied current by (15). Beyerlein and Nicholson [138] have discussed the usual [99] i–t linearization of (15) in galvanostatic quasi-reversible theory and have included effects of double-layer corrections in the determination of τ [139]. More details on the galvanostatic experiment and a related one (double current pulse) can be found in the review by Delahay [99]. An important critique of the double-pulse technique has been given by Kooijman and Sluyters [140] and Van Leeuwen and Sluyters [141], who see little advantage as compared to single current steps.

In chronopotentiometry of solutions containing more than one electroactive component,

$$\text{Ox}_1 + n_1 e \rightarrow \text{Red}_1; \qquad \text{Ox}_2 + n_2 e \rightarrow \text{Red}_2 \qquad (74)$$

multiple waves are observed when the reactions are separated by >0.1 V. While the transition time for the more easily reduced component Ox_1 is again described by (71), that of the second species is elongated by the continued influx and reduction of Ox_1. Correction of this "residual diffusion" effect gives, for the transition time of Ox_2 [134, 142]

$$(\tau_1 + \tau_2)^{1/2} - (\tau_1)^{1/2} = \frac{n_2 F D_2^{1/2} \pi^{1/2} C_2^b}{2i} \qquad (75)$$

where τ_2 is measured from τ_1. For the general case of m components, the expression for the mth wave is [134]

$$\left(\sum_{j=1,2}^{m} \tau_j \right)^{1/2} - \left(\sum_{j=1,2}^{m-1} \tau_j \right)^{1/2} = \frac{n_m F D_m^{1/2} \pi^{1/2} C_m^b}{2i} \qquad (76)$$

These equations permit translation of transition time data into concentration for analysis of multicomponent systems.

Multiple chronopotentiometric waves also result when a single electroactive species is reduced in several stages:

$$\text{Ox} + n_1 e \rightarrow \text{Red}_1; \qquad \text{Red}_1 + n_2 e \rightarrow \text{Red}_2 \qquad (77)$$

Examples of such behavior are found in oxygen, vanadium (V→III→II), and copper (II→I→0) electrochemistry. A concentration-independent relation [134, 142] exists between τ_1 and τ_2 in such cases:

$$\tau_1 \left(\frac{n_1 + n_2}{n_1} \right)^2 - \tau_1 = \tau_2 \qquad (78)$$

For $n_1/n_2 = 1, 0.5$, and 2, $\tau_2/\tau_1 = 3.0, 8.0$, and 1.25, respectively. Actual values of τ_1 and $(\tau_1 + \tau_2)$ are also given by the Sand equation written for $n = n_1$ and

$n = (n_1 + n_2)$, the latter expression is applicable whether or not one can accurately discern the individual value of τ_1 in an experimental case.

3.1.2   Current Reversal Experiment

The current reversal chronopotentiometric experiment involves replacement of the initially applied current step excitation, at a time $t_r < \tau$, with reverse current step excitation. For a case where Ox is initially reduced, the reverse current step causes the product Red obtained during the forward step to be reoxidized. The resulting reverse chronopotentiometric wave terminates in a transition time τ_b when the supply of Red at the electrode surface is exhausted $[C_{\text{Red}(0,t)} = 0]$. The overall forward and reverse chronopotentiometric curve is illustrated in Figure 6.19 for reversible (—) and irreversible (---) electrode reactions. Description of the reverse transition time τ_b was first given by Berzins and Delahay [142]. For the case in which equal forward and reverse current steps are employed ($i_c = -i_a$), one obtains the remarkably simple and C_{Ox}^b, D_{Ox}, D_{Red}, and i_c independent result:

$$\frac{t_r}{\tau_b} = 3.00 \tag{79}$$

In the more general case, for any relative value of i_c and i_a,

$$\frac{i_a^2 - 2i_a i_c}{i_c^2} = \frac{t_r}{\tau_b} \tag{80}$$

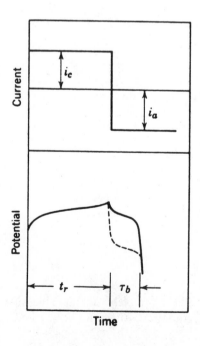

Figure 6.19   Reversible (——) and irreversible (---) current reversal chronopotentiograms.

Equation (79) provides an experimental criterion for the diffusion-controlled behavior of the electrode product species Red. As long as Red diffuses freely in the solution (or in the electrode, in the case of a metal amalgam produced at a mercury electrode), exactly one-third of the Red generated during the forward current step is recovered in the reverse step. The remainder of Red is "lost" by diffusion into the solution. If Red is insoluble and thus is quantitatively retained at the electrode surface, $\tau_b = t_r$. An example is the anodic (forward) generation of silver chloride on a silver working electrode in a chloride-containing solution. Values of t_r/τ_b intermediate between 1.00 and 3.00 are found in instances of specific adsorption of Red on the electrode surface, as then, depending on the strength of the adsorption, a mixture of diffusing Red (approximately one-third recoverable during τ_b) and adsorbed Red (quantitatively recoverable, usually during τ_b) is generated during t_r. For a chemical decomposition of Red into a less readily reoxidized substance, less Red becomes recoverable during τ_b, and t_r/τ_b then exceeds 3.00. This case is further considered later in this section.

Although the mass-transport equations (79) and (80) for transition times are independent of charge-transfer kinetics, the relative placement of the forward and reverse chronopotentiometric waves on the potential scale (see Figure 6.19) and their shapes are functions of k_s. If k_s is sufficiently large (the reversible case), the potential–time behavior of the reverse wave is given by [142]

$$E = E^0 + \frac{0.0591}{n} \log \frac{D_{Red}^{1/2}}{D_{Ox}^{1/2}} + \frac{0.0591}{n} \log \frac{t_r^{1/2} - t^{1/2} + 2(t - t_r)^{1/2}}{t^{1/2} - 2(t - t_r)^{1/2}} \tag{81}$$

The potential analogous to $E_{1/4}$ during the forward wave (that at which the extreme right-hand term vanishes) occurs at $t - t_r = 0.215\,\tau_b$. The equality $(\text{forward})E_{1/4} = (\text{reverse})E_{0.215}$, constitutes an additional reversibility criterion, but it is very susceptible to uncompensated resistance effects and is thus not very useful.

Smaller values of k_s in current reversal experiments result in eventual irreversibility, and $E_{1/4}$ becomes more negative than $E_{0.215}$ (for a cathodic–anodic reversal sequence). This splitting is greater at smaller k_s or greater applied current and can be used to measure k_s [12, 143, 144]. It would seem that, except for highly irreversible reactions, current reversal should be superior to the forward current experiment [see (73) and quasi-reversible modification] in terms of experimental sensitivity to k_s, but some of the advantage is removed by charging current limitations [144].

3.2   Mass-Transport Control by Coupled Chemical Reactions

As with chronoamperometry and chronocoulometry, chemical reactions generating or consuming Ox or Red alter constant-current and current reversal chronopotentiometric responses from the diffusion-only values. Because of the limitations of transition time measurements, chronopotentiometric theory for such cases is less highly developed. Four of the important reaction coupled types that have been theoretically characterized are discussed next.

3.2.1   Preceding Chemical Reactions (CE)

The boundary value problem for reactant Ox lying in initial equilibrium with an electroinactive species Z [see (5)] is solved, for chronopotentiometry, by application of conditions (3), (4), (22), (33), and (35) to (6) and (7). The result, by Delahay and Berzins [145], is

$$i\tau_k^{1/2} = \frac{nFD^{1/2}\pi^{1/2}[C_{Ox}^b + C_Z^b]}{2} - \frac{i\pi^{1/2}}{2K(k_f + k_b)^{1/2}} \, \text{erf}[(k_f + k_b)^{1/2}\tau_k^{1/2}] \quad (82)$$

In the second term the chronopotentiometric $i\tau^{1/2}$ product depends on the value of applied current. Figure 6.20 illustrates this for several degrees of kinetic control.

For very small chemical rates or short transition time, conversion of Z to Ox is insignificant during the chronopotentiogram; and (82) reduces, through expansion of the error function at small argument to $\text{erf}(\beta) = \beta/\pi^{1/2}$, to the Sand equation (71) written for Ox, or

$$i\tau^{1/2} = \frac{nFD^{1/2}\pi^{1/2}[C_{Ox}^b + C_Z^b]}{2 + (2/K)} \quad (83)$$

Attaining this limiting condition is determined of course by experimentally measuring τ sufficiently short compared to the chemical reaction half-life; it also depends on whether K is large or small since both k_f and k_b appear in (82).

At the other extreme (large chemical rate, large τ, small current), $\text{erf}(\beta) = 1$ for $\beta > 2$, so that (82) becomes

$$i\tau_k^{1/2} = \frac{nFD^{1/2}\pi^{1/2}[C_{Ox}^b + C_Z^b]}{2} - \frac{i\pi^{1/2}}{2K(k_f + k_b)^{1/2}} \quad (84)$$

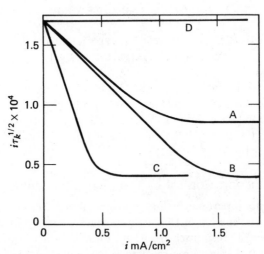

Figure 6.20 Effect of prekinetics on chronopotentiometric response: $n=1$; $D = 4.0 \times 10^{-6}$ cm^2/s; $C_{Ox}^b + C_Z^b = 1.0 \times 10^{-6}$ mol/cm^3; Curve A: $K = 1.0$, $k_f + k_b = 100$; Curve B: $K = 0.3$, $k_f + k_b = 900$; Curve C: $K = 0.3$, $k_f + k_b = 100$; Curve D: very large k_f or K.

and a linear relation between $i\tau_k^{1/2}$ and i prevails. Extrapolation of the relation to zero current yields the $i\tau^{1/2}$ limit for limiting fast CE kinetics, and $K(k_f + k_b)^{1/2}$ can be obtained from the slope. Assuming that a variation of 10% in $i\tau^{1/2}$ is detectable over the range of usable chronopotentiometric transition times (see below), one can estimate [12] that values of $K(k_f + k_b)^{1/2} < 500\,s^{-1}$ are measurable. Reinmuth [146] has discussed further analysis of chronopotentiometric CE data.

A few chronopotentiometric applications to CE reactions have been reported, mainly where the CE reaction involves either deprotonating a weak acid [147–150] or losing a ligand from a metal complex to yield a more readily reducible coordination state [132, 151]. Also, for metal complex reductions, absence of a chronopotentiometric CE effect can provide a criterion for whether the metal complex is reduced directly without ligand predissociation [12, 137]. Theory for chronopotentiometric CE reactions of the form $pZ \rightarrow Ox$ and $Z \rightarrow pOx$ has also been presented [152].

3.2.2   Following Chemical Reactions

EC CASE

According to (38) a cathodic chronopotentiogram leading to a decomposing species Red yields a transition time no different from that observed in the absence of the chemical reaction, inasmuch as τ is controlled by mass transport of reactant Ox not product Red. On the other hand, the shape and $E_{1/4}$ of the chronopotentiometric wave are changed by the EC chemical process, the general effect being a positive shift of $E_{1/4}$ of magnitude depending on values of k_f/k_b, k_f, and the applied current. The theory for the potential shift effect has been derived and applied to a chemical example [153–155], but is limited by its requirement for a reversible charge-transfer process. Transition times in current reversal chronopotentiometry however do not suffer from this reversibility requirement.

The qualitative effect of (38) on the chronopotentiometric current reversal ratio t_r/τ_b is to lower the quantity of recoverable charge-transfer product Red and to increase the ratio above 3.00. The theory describing this effect, for an irreversible first-order EC chemical reaction ($k_b = 0$), is [156–158]

$$2\,\text{erf}(k_f\tau_b)^{1/2} = \text{erf}[k_f(t_r + \tau_b)^{1/2}] \tag{85}$$

The analytic behavior of the reversal ratio, τ_b/t_r, can be described, for evaluating k_f, as a working curve as in Figure 6.21. The τ_b/t_r ratio approaches the diffusional value of 0.33 at short t_r or small k_f and dwindles to zero at sufficiently long t_r. Figure 6.21 clearly shows that values of $t_r k_f = 1$ to 3 provide the optimum experimental situation for measurement of k_f. Dracka [156] has estimated that first-order rates of about $10^2\,s^{-1}$ can be evaluated by current reversal chronopotentiometry. The method has been applied to the hydrolysis reaction of electrogenerated benzoquinoneimine by Testa and Reinmuth [158]. Ehman and Sawyer [159] applied both forward and reverse current chronopotentiometry to study the electrode kinetics for reduction and hydrolytic reactions of iron(II) and (III) bipyridine and oxo-bridged bipyridine complexes. In one of the few

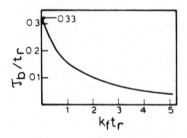

Figure 6.21 Working plot of (85) for current reversal chronopotentiometric evaluation of EC postkinetics.

continuing theoretical studies of chronopotentiometry, Dracka [160–165] has extended the EC theory to include cases where Red decomposes by two parallel reaction paths, where Red and Y are both electroactive during the reverse transitions, and where there are (second-order) dimerization and disproportionation reactions of Red.

CATALYTIC CASE

Regeneration of the original reactant Ox by a reaction of Red [see (11)] produces kinetic control of both Ox and Red mass transport. The cathodic transition time in forward current chronopotentiometry becomes elongated by the supply of regenerated Ox, and the reverse transition time τ_b in current reversal chronopotentiometry is by the same token abbreviated. The solution for the catalytically controlled cathodic transition time for Ox was derived and experimentally tested by Delahay and co-workers [153] and is

$$i\tau_{cat}^{1/2} = \frac{nFD^{1/2}C_{Ox}^b(k\tau_{cat})^{1/2}}{\mathrm{erf}(k\tau_{cat})^{1/2}} \tag{86}$$

The behavior of (86) is illustrated in Figure 6.22. At sufficiently slow kinetics or small transition time (large current), the error function reduces to $2(k\tau_{cat})^{1/2}/\pi$, and the system displays Sand equation behavior [current-independent

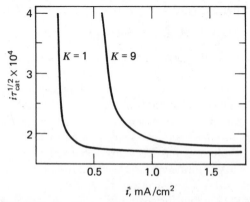

Figure 6.22 Effect of catalytic kinetics on chronopotentiometric response: $n=1$; $D=4.0\times10^{-6}$ cm^2/s; $C_{Ox}^b = 1.0\times10^{-6}$ mol/cm^3.

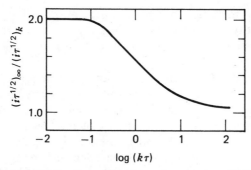

Figure 6.23 Working plot for chronopotentiometric evaluation of ECE kinetics; $n = n'$.

$(i\tau_{cat})^{1/2}]$. At faster kinetics or larger transition time, the value of $(i\tau_{cat})^{1/2}$ rises sharply.

The ratio t_r/t_b in current reversal for the catalytic reaction is again described [157] by (85), and a kinetic study can be conducted in the same manner as for the simple EC postkinetic case. Applications of chronopotentiometry to catalytic kinetics have emphasized the chemical regeneration of iron(III) or titanium(IV) following their electroreduction [153, 157, 166, 167].

ECE CASE

The chronopotentiometric forward current theory for an ECE chemical process, which leads to a product more readily reducible than by the original electrode reactant [see (10)], was presented by Testa and Reinmuth [168] for $k_b = 0$. The solution for $i\tau_k^{1/2}$ is complex and best examined and compared with experimental data by using a computed working curve [169, 170] such as that shown in Figure 6.23. Analogous to similar curves in chronoamperometry, the vertical axis is presented in a normalized fashion as the ratio of $i\tau_\infty^{1/2}$ (for infinitely fast kinetics $k_f \to \infty$) to the kinetically controlled $i\tau_k^{1/2}$. The former fast kinetics $i_\infty \tau^{1/2}$ product is given by the Sand equation (71) written for $n + n'$ electrons. The figure shows that the fast kinetics limit is 1.00 and is approached at large transition times (low current), and the slow kinetics limit is $(n + n')/n$ and is produced by transition times short compared to the reaction half-life of the kinetic process so that only the $Ox + ne \to$ reaction is important. At intermediate times, experimental $i\tau_\infty^{1/2}/i\tau_k^{1/2}$ ratios can be used with the working curve to read values of k.

Chronopotentiometric studies of ECE cases include the reductions and subsequent reactions of o-nitrophenol [169] and p-nitrosophenol [170]. Chronopotentiometric theory has not been extended in the manner of chronoamperometry (see above) to include the ECE nuance other schemes, or second-order processes, although the mathematical tools to do so are now available.

3.3 Adsorption of Electrode Reactant

An adsorbed, electroactive reactant or product on a working electrode surface alters the electrochemical response from the diffusion-controlled value, since the adsorbed material requires no mass-transport time and is available in restricted quantity. In chronopotentiometry, an adsorbed reactant increases the chronopotentiometric product $i\tau^{1/2}$. The effect is magnified by increasing the applied current. A qualitative rationale for the $i\tau^{1/2}$ increase, recognized and applied by Lorenz [171] and Lorenz and Muhlberg [172, 173], is that a diffusion-controlled transition time τ_s (by solution reactant Ox, or "SR") decreases according to i^{-2}, whereas transition time τ_a for an adsorbed reactant Ox, or "AR," decreases with i^{-1}. Since the products $i\tau_s^{1/2}$ and $i\tau_a$ are constants, a measured $i(\tau_s + \tau_a)^{1/2}$ increases at short times or high applied currents.

How much $i\tau^{1/2}$ increases at large currents depends on the relative values of Γ_0 (surface excess of Ox) and the solution concentration C_{Ox}^b. The precise shape of the $i\tau^{1/2}$ versus i curve is, however, also a function of the *order* in which adsorbed reactant (hereafter denoted as AR) and solution reactant (denoted SR) become reduced during the chronopotentiogram, an *unfortunate* functionality in that assessment of this order of reaction becomes an implicit part of a Γ_0 measurement. Figure 6.24 illustrates four situations that can logically arise. The labels SR,AR and AR,SR refer to attainment of transition time for SR prior to reduction of any AR, and the reverse of this, respectively. As shown, this leads to split waves. The SAR case corresponds to reaction of SR and AR at similar potentials and gives no clear wave splitting. These three models assume that any re-equilibration between SR and AR is slow on the experimental time scale. The EQUIL case permits an equilibrating interconversion of AR and SR throughout the chronopotentiogram so that they have an intrinsically common transition time.

Differentiating among these cases depends on establishing their theory and comparing experimental data thereto to observe one, unique fit. Only then can Γ_0 be reliably established.

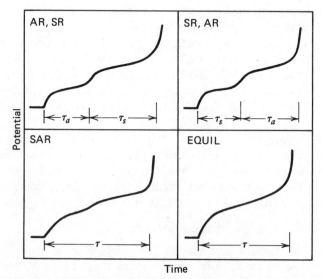

Figure 6.24 Chronopotentiograms typical of the various adsorption models.

3.3.1   SR, AR

The Sand equation (71) gives the first transition time, for SR, and the second transition time τ_a is given by an equation first described (approximately) by Lorenz [171] and thereafter rigorously described by Reinmuth [174] and Anson [175]:

$$\frac{nF\pi\Gamma_0}{i} = (\tau_s + \tau_a)\arccos\left[\frac{\tau_s - \tau_a}{\tau_s + \tau_a}\right] - 2(\tau_s\tau_a)^{1/2} \tag{87}$$

For comparison to experimental results, individual values of τ_a may be inserted to seek a current-independent value of Γ_0. When τ_s is indistinct, a computer solution for Γ_0 can be sought from (87) using values of $\tau_s + \tau_a$ at several currents [176]; alternatively an operational test using current sweep chronopotentiometry may be employed [177].

The SR,AR case is the one "thermodynamically expected" for strong adsorption of reactant with charge-transfer reversible reaction of both SR and AR. According to Brdicka [178], the adsorbed reactant AR, being in a lower free energy state and having a more negative standard potential (E^0) than solution reactant SR, yields a wave negatively displaced from that of SR. Clean-cut cases of SR,AR chronopotentiometry are rare, an example being the adsorption of lead iodide complex from iodide supporting electrolyte [76].

3.3.2   AR, SR

Here, the transition time for AR is governed by $i\tau_a = nf\Gamma_0$, and the second by the Sand equation (71) for SR. Thus, if the wave splitting is distinct, constant $i\tau_a$ and $i\tau_s^{1/2}$ products must result. Indistinct splitting is treated by the combined expression given by Lorenz [171]:

$$i\tau = nF\Gamma_0 + \frac{D\pi(nFC_{Ox}^b)^2}{4i} \tag{88}$$

where $\tau = \tau_s + \tau_a$. Data comparison is accomplished by plotting i versus $1/i$. This case can occur with irreversible charge transfers where the charge-transfer rate for SR is lower than that for AR; an example (approximated) is the adsorption of phenylmercuric ion [179] on mercury electrodes.

3.3.3   SAR

This case is an approximation for the situation in which equilibrium between SR and AR does not proceed rapidly and the reduction potentials for AR and SR are very similar. A fixed division between the current reducing SR and that reducing AR is arbitrarily presumed, the division being such as to lead to a common transition time. The theory [171] is

$$i\tau = nF\Gamma_0 + \frac{nFC_{Ox}^b(\pi D\tau)^{1/2}}{2} \tag{89}$$

and experimental data are examined by a plot of $i\tau$ versus $\tau^{1/2}$.

3.3.4   EQUIL

This case results from a combination of weak to moderately strong adsorption and a labile adsorption equilibrium between SR and AR. For a linear adsorption isotherm, Lorenz [171] gives the following theoretical description:

$$\tau^{1/2} + \frac{\pi^{1/2}\Gamma_0}{2D^{1/2}C_{Ox}^b}\,\phi = \frac{nFC_{Ox}^b(D\pi)^{1/2}}{2i} + \frac{\pi^{1/2}\Gamma_0}{2D^{1/2}C_{Ox}^b} \tag{90}$$

where $\phi = \exp(C_{Ox}^{b2}D\tau/\Gamma_0^2)\,\mathrm{erfc}(C_{Ox}^b D^{1/2}\tau^{1/2}/\Gamma_0)$. Experimental data are compared to this expression by a computer fitting or through an iterative graphical approach [76]. The adsorptions of several mercury(II) complexes on mercury electrodes [76] were reported to adhere to (90).

Following the initial work by Lorenz [171–173], a variety of chronopotentiometric adsorption measurements were conducted: that of riboflavin on mercury by Tatwawadi and Bard [176] and Herman and Blount [180], of Alizarin Red S and cobalt chloride–ethylenediamine on mercury by Laitinen and Chambers [181], of carbon monoxide on platinum by Munson [182, 183], of iodine on Pt by Herman and Blount [180], and of the others mentioned above. This considerable activity in adsorption measurements, however, was short-lived, because of important experimental shortcomings of chronopotentiometry and the emergence of a superior alternative: chronocoulometry [77]. The necessary "single-fit" comparison of i–τ data to the various theoretical expressions [(87)–(90)] is often not clear-cut, because the apparently mathematically diverse equations in reality give similar i–τ dependencies. Lingane [184] ventures to say it is difficult, if not impossible, to obtain a clear-cut, single-fit identification of the proper model for assessment of precise surface excess values. Factors optimizing possibilities for correct identification have been discussed [76], and Herman and Blount [180] have noted that current reversal data are also valuable. The fact that a series of experiments (20 or more) is required to describe a system's i–τ behavior adequately for model comparison makes the experimental labor per value of Γ_0 considerable. Finally, and probably most importantly, experimental bias in τ can occur through the positive contribution of double-layer charging [185] to τ at short times and a possible potential dependency of the adsorption [186].

3.4   Other Chronopotentiometric Techniques

Constant current and current reversal are the most commonly employed chronopotentiometric experiments. Theory for several other experiments has been developed, for special purposes. These experiments depend on other current excitation shapes, such as programmed currents that increase with powers of time [187, 188] (applied $i = \beta t^r$) or exponentially with time [188]. Current increasing as the square root of time ($r = \frac{1}{2}$) is interesting since it yields a chronopotentiometric transition time τ directly proportional to concentration and inversely proportional to current ($\beta\tau/C^b = $ constant). This offers certain advantages in analytical applications compared to the constant-current experiment.

Under the label "nonlinear relaxation methods," Rangarajan [189] has extended the generality of the programmed current chronopotentiometric theory; and as "nonlinear perturbations," Kontturi and colleagues have reinvented the power of time and exponential current excitations [190] and extended them to coupled EC, ECE, and catalytic mechanisms [190, 191]. Despite the mathematical advantages of programmed currents, the topic has no significant applications to date except for the cyclical variant.

Cyclical (current) chronopotentiometry is useful when studying coupled chemical reactions and is illustrated in Figure 6.25. The experiment consists of a series of current steps with current reversals occurring at each successive cathodic and anodic transition time. The theory for cyclic chronopotentiometry was developed by Herman and Bard [192–195] for diffusion control of Ox and Red and also for various coupled reaction causes, such as CE, EC, catalytic, and ECE kinetics. For the diffusion-controlled case, the ratio of the first two transition times (τ_1 and τ_2, $n = 1, 2$, respectively) has the familiar current reversal value of 3.00 [see (79)]. Inasmuch as reactant Ox consumed during the first current step is not completely recovered during current reversal, the second cathodic transition time τ_3 is less than τ_1 and also $\tau_3/\tau_4 < 3.00$. At large n, $\tau_n/\tau_{n+1} \rightarrow 1.00$. The following table gives computed values for transition time ratios for EC and catalytic cases, for a selected decomposition rate of Red: $k = 1.00$. Such theoretical data (for various k) can be compared with experimental τ_n values through dependency of either τ_n/τ_{n+1} or τ_n/τ_{n+2} ratios on n; the functionality of these ratios with n is different for the different reaction schemes and is also sensitive to k.

Relative Cyclic Chronopotentiometric Transition Times [192, 193]

n	Diffusion Only	EC Case $k = 1.00$	Catalytic Case $k = 1.00$
1	1.000	1.000	1.000
2	0.333	0.167	0.167
3	0.588	0.384	0.673
4	0.355	0.138	0.169
5	0.546	0.292	0.658
6	0.366	0.120	0.169
7	0.525	0.243	0.655
8	0.373	0.108	0.169
9	0.513	0.211	0.654
10	0.378	0.099	0.169

In principle, cyclic chronopotentiometry yields no extra kinetic information beyond that available solely from $i\tau^{1/2}$ or τ_1/τ_2. For EC, in fact, τ_n/τ_{n+1} is less and less affected by the chemical process (as compared to the diffusion-only value) as n increases. The value of the cyclic experiment lies in its collection, in one experiment, of a series of data points, all of which contain some kinetic

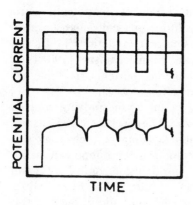

Figure 6.25 Cyclic chronopotentiometry: excitation and response. Current reversed at preselected potential at transition time.

information; random errors are to some extent averaged by relying on the larger data set to evaluate k. Cukman and Pravdic [196] point out that the successive transition time ratios are useful in the difficult mechanistic ECE versus disproportionation distinction, for example. Also, the series of data points, collectively, is more sensitive to deviations from ideal adherence of a chemical system to a supposed coupled chemical reaction scheme since such deviations have a cumulatively larger effect on transition times at larger n.

The ideas of cyclic chronopotentiometry have been extended to cyclical chronocoulometry by Vukovic [197, 198], who gives digitally simulated theory, for the diffusion-controlled, CE, EC, and catalytic coupled reaction cases. Vukovic points out that advantages of cyclic chronopotentiometry also apply to the charge–time method.

Another form of chronopotentiometry, which uses a constant current step, is derivative chronopotentiometry, where one electronically measures and records dE/dt rather than potential E against time in the chronopotentiogram. As discussed by Peters and Burden [199], and as is evident from the reversible chronopotentiogram of Figure 6.18, dE/dt passes through a minimum at $t < \tau$. For a reversible reaction, $t_{min} = 4\tau/9$. Detecting τ_{min} permits a more reliable measurement of τ, but unfortunately suffers from requiring reversibility of the electrode reaction (for a completely irreversible reaction $\tau_{min} = \tau/4$). Sturrock and co-workers [200, 201] have extended the reversible theoretical relations for dE/dt to multicomponent systems; to low-concentration analysis; and, most recently, to CE reactions, as in dissociations of Cd(II) cyanide and nitrilotriacetate complexes [202].

Sturrock [203] had actually, but without commensurate theory, earlier introduced the derivative chronopotentiometric readout in reference to cyclic chronopotentiometry, recording in this case dE/dt against E. This experiment is a quantitation of the older oscillographic polarographic method [204] formerly popular among European electrochemists.

Constant current chronopotentiometry has recently [205] been applied to a rather novel transport problem, that of charge transport across the interface between two immiscible liquids, such as nitrobenzene and water. The charge transport (passage of ions) is driven by a current step applied to a pair of electrodes, one in each liquid. The interfacially transferred ion may be solubilized in one of the phases by an ionophore. The theory for this experiment involves accounting for both concentration gradient (as with Fick's law) and potential gradient in the transport problem.

Finally, ac chronopotentiometry employs an ac current superimposed upon a dc forward or reverse current. In this variant, the ac component of the potential time response is recorded. Intuitively, one can see that the ac response will resemble a dE/dt derivative, and instrumentally this mode of derivative-taking is easier to manage without signal/noise complications. Theory was introduced some time ago [206–208] but no significant applications have appeared.

4   THE PRACTICE OF CHRONOAMPEROMETRY, CHRONOCOULOMETRY, AND CHRONOPOTENTIOMETRY

4.1   Response Quality

Chronomethods theory for diffusion-controlled, coupled chemical reactions, charge-transfer kinetics, and adsorption was detailed above. Use of the theory of course requires an ability to conduct experiments that faithfully reflect the boundary conditions and other assumptions implicit in the theory. This can be regarded as the response quality of chronoamperometric current–time, chronocoulometric charge–time, and chronopotentiometric transition time data. The general sources of deviation of experiment from theory, or imperfection in response quality, are:

1. Each theoretical chronomethods relation is grounded on a certain geometry of mass transport, which is usually linear diffusion [Equation (2)]. Response quality is lowered by any unrecognized nonlinearity of the profiles of electrode reactant and product.

2. Chronomethods theory is also predicated on application to the working electrode of current or potential steps of "infinitely fast" rise time and of precisely prescribed magnitude. Imperfections can arise by means of electrode double-layer capacitance, uncompensated ohmic resistance of the cell solution, and inadequate quality of the electrochemical instrument employed.

3. The accuracy of the data recording system is also important; at fast electrolysis times in chronoamperometry, for instance, the compressed data display of an oscilloscope screen is considerably less precise than that obtainable by storage of transient data in a digital (microcomputer) memory with later, slower output on an expanded paper copy.

4. With chronopotentiometry, a limitation on response quality occurs through the uncertainty of how best to extract a numerical value for transition time from the recorded chronopotentiogram.

5. Finally, when coupled chemical reactions are involved, whether the presumed and theoretically treated reaction scheme is actually that occurring in the real chemistry must always be considered.

The relative importance of these various sources of response quality imperfection depends also on the demands of a useful time scale placed by the chronomethod employed and the properties of the electrochemical reaction. Each chronomethod in practice with a particular electrode, cell, solvent, instrument, and so on provides a usable experimental time window, and response quality beyond an upper and lower time value steadily diminishes. The long-time limit is most often imposed by effects associated with maintenance of ideal diffusion conditions. The short-time limit is usually a function of the felicity with which the desired current or potential excitation signal can be forced on the working electrode.

Further comments on the details of chronomethods experiments and their relation to response quality follow.

4.2   Working Electrodes and Nonlinear Diffusion

Three working electrode geometries—disk, spherical, and cylindrical—usable for chronomethods experiments are illustrated in Figure 6.2. These electrodes may be fabricated from a great variety of materials, including mercury, platinum, gold, SnO_2 and In_2O_3 films, and many forms of carbon (glassy, pyrolytic, paste). A spherical electrode can be a pendant or hanging mercury drop (HMDE) or a fused bead of Pt or Au. A disk or cylindrical mercury electrode is made by amalgamating the surface of a disk or wire. Such thin-film mercury electrodes are useful in stripping analysis. Mercury may also be used as a pool electrode. Disk electrodes may also be fashioned from SnO_2 and In_2O_3, n-type semiconductors, which in highly doped form have metal-like characteristics, or from carbon. Carbon can be used in its glassy (vitreous, usually gives lowest background currents), pyrolytic (basal plane or edge plane), and paste (carbon powder plus a mulling agent like Nujol) forms.

Strictly speaking, linear diffusion theory is applicable only to a planar working electrode surface. In practice, however, any nonplanar working electrode can yield experimental results conforming to linear diffusion theory if the results are obtained at electrolysis times *sufficiently short that the diffusion layer developed in the experiment* [dimension about $(Dt)^{1/2}$ cm] *remains thin in comparison to the radius of curvature of the nonplanar electrode surface.* Thus the spherical and cylindrical geometries (Figure 6.1) can be used in short-time chronomethod experiments. The spherical HMDE is a popular form of mercury electrode because of this. The HMDE electrode was developed by DeMars and Shain [209] and Kemula and Kublik [210], who employ different procedures for preparing the pendant mercury drop. An HMDE of area $0.05 \, cm^2$ can provide chronoamperometric data adhering to within 1 or 2% of the linear diffusion Cottrell equation (23) at times less than about 1 s. For HMDE chronomethod data at longer electrolysis times, the appropriate spherical diffusion boundary value problem must be solved to allow interpretation of the result; an alternative preferable procedure for long electrolysis times is to select a more truly planar working electrode surface like the disk electrode.

A typical preamble to chronomethod experimentation with an unfamiliar electrode is a careful assessment, over the electrolysis time interval of interest, of the diffusion conditions prevailing at that electrode. This is done using a model, diffusion-controlled electrode reaction; and, for linear diffusion geometry, is based on the time independence of the chronoamperometric $it^{1/2}$, chronocoulometric $Q/t^{1/2}$, and chronopotentiometric $i\tau^{1/2}$ parameters of (23), (26), and (71), respectively. Model reactions might be Cd(II) or benzoquinone reduction on mercury, ferrocene oxidation on a solid electrode in nonaqueous solvent, or ferrocyanide oxidation on a solid electrode in aqueous medium. Onset of an appreciable spherical diffusion component is signaled in this test by an increase in the measured parameter with increasing electrolysis time.

The model reaction test of a working electrode can also reveal time variation of the $it^{1/2}$, $Q/t^{1/2}$ diffusion parameters caused by other diffusional nonidealities. At sufficiently long electrolysis times, with any working electrode, convective disturbances eventually introduce an additional mass-transport mode, causing the diffusion parameters to increase. Convection arises from *vibrational* motions and from solution motion induced by a *density gradient* within the diffusion layer. Vibrations can be minimized by proper cell mounting and damping and by density gradients partially controlled by proper spatial orientation of the disk electrode surface and shielding of its perimeter (with a skirt), as shown by Bard [211] in a chronopotentiometric study. In general, however, freedom from convective disturbances at electrolysis times exceeding 100 s is rarely attained and they often occur at much shorter (a few seconds) times. Another convective disturbance unique to mercury electrodes can occur, in the short-time domain, from surface tension changes (pool or drop shape changes) accompanying abrupt alterations in the electrode potential. The data of Shain and Martin [212] provide a relatively severe example of this. A third type of solution motion at mercury electrodes, convective streaming (analogous to that causing "maxima" in polarography), can occur unpredictably. Streaming can be absent in the model electrode reaction but may appear in a chemical system subsequently examined using the same working electrode. Attempts to eliminate this effect with "maximum suppressors" (added surfactants) require considerable caution, as interference with the kinetic or adsorption process under study can result.

Shielding can occur around the "neck" of an HMDE (see Figure 6.2) when the distance between the mercury neck and the capillary from which it is hung becomes comparable to the diffusion-layer depth. This is most likely at long electrolysis times and causes a diminution of the chronomethod diffusion parameter [212, 213]. Ideally, the capillary is tapered at its end to minimize shielding. Shielding is less noticeable in double potential step experiments, since shielding effects during forward and reverse electrolysis times tend to cancel in the current or charge ratios.

And, if a disk electrode radius becomes comparable to (or smaller than) the diffusion layer, diffusion to the edges of the disk from the solution volume outside the cylindrical projection of the disk becomes important. This *edge effect* increases the chronomethods $it^{1/2}$, $Q/t^{1/2}$, or $i\tau^{1/2}$ parameters. Edge diffusion can be treated analytically [110, 111, 214] or by digital simulation [109, 215]. If the edge effect is extreme, as it will be when the electrode disk diameter is say 10^{-3} cm, the entire character of the chronomethods experiment changes, as is discussed in Section 2.5 about *microelectrodes*. Microelectrodes are one of the more exciting advances in recent electroanalytical techniques.

Lastly, other factors besides diffusional properties can enter into selecting a choice working electrode. Mercury HMDE electrodes are prized for their provision of a clear and exceedingly reproducible surface, valuable in any experiment where heterogeneous events dominate (adsorption, electron-transfer kinetics). Mercury, on the other hand, has a limited positive potential range; and in recent years, as oxidation reactions in nonaqueous media have been explored, many solid electrode studies have appeared. Pt disk electrodes provide an excellent positive range in, for example, acetonitrile solvent. However, careful attention must then be given to the resurfacing, polishing, and general cleanliness of the Pt electrode. This can range from a minor problem (when the events of interest occur homogeneously in the diffusion layer) to a debilitating difficulty when heterogeneous events are the objects of interest. An excellent illustration of solid electrode pretreatment dependencies in electron-transfer kinetics has been given by Daum and Enke [216].

4.3   Control and Measurement—Chronopotentiometry

Historically, the introduction of operational amplifiers by DeFord and co-workers and Booman and co-workers in the 1950s for control and measurement in electrochemical experiments had an enormous effect. These electronic devices gave versatility to the electroanalytical chemists' choice of experiment and were used by most researchers to design and construct their own equipment. The use of "homemade" operational amplifier-based electrochemical equipment is still very widespread, but is now usually augmented by microcomputers that control the timing, and so on of control and data reading. Commercial equipment is also now common, and its advent has facilitated electrochemical research by chemists lacking much prior experience.

Figure 6.26 shows a simple operational amplifier chronopotentiometric circuit for a three-electrode cell. The experiment is initiated by sudden application of potential V, which, dropping across resistance R to the virtual ground of the operational amplifier CON summing point (*), determines the current flowing between the working (—O) and auxiliary (—|) electrodes in the amplifier feedback loop. The three-electrode circuit avoids much of the ohmic potential drop through the solution by monitoring the potential of the working electrode versus a reference electrode (→) with a high input impedance voltage follower. The ohmic potential drop remaining in the potential output depends on the working electrode geometry and is minimized by working electrode–reference electrode proximity. For current reversal chronopotentiometry, the sign of the

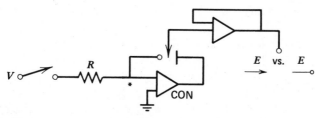

Figure 6.26   Elementary operational amplifier apparatus for chronopotentiometry; $i_{cell} = -V/R$.

constant potential V (also typically generated by an operational amplifier device) is simply altered at the desired time t_r. The equipment used to record the chronopotentiometric potential–time output depends on the particular application. For charge-transfer kinetics, expansion and accuracy of the recorded potential scale is crucial, as is minimization of residual ohmic potential drop. For other work a relatively coarse potential scale expansion is usually sufficient.

This brings us to the measurement of transition times; chronopotentiometric applications to analysis, adsorption, and coupled chemical reaction kinetics are founded upon τ values. Unfortunately, transition time measurements are often rendered uncertain by contributions from processes other than electrolysis of the cell reactant. The main villain is double-layer charging, another is reaction of surface oxides or extraneous adsorbed impurities. The (portion of) applied current required to charge the double layer is related to the rate of potential change and the double-layer capacitance C_{dl} by

$$i_{dl} = C_{dl} \frac{dE}{dt} \tag{91}$$

where C_{dl} also depends (mildly) on the potential E. Thus the current reducing Ox during the chronopotentiogram is less than that applied, according to C_{dl} and the instantaneous value of dE/dt. The latter term varies widely and throughout the chronopotentiogram, being largest at early times and around τ. A double-layer charging current is of course necessary in any electrochemical experiment where the working electrode potential is altered (except coulostatics [1]). The point in interpretation of experimental results is whether correction for double-layer charging is easily and accurately accomplished. In chronopotentiometry, the complex variation of i_{dl} and its persistence throughout the experimental curve makes such correction quite difficult. The relative value of i_{dl} is largest (a) when τ is short (large dE/dt), when C_{Ox}^b is low (small applied current), and (b) near τ itself (E changes there). The qualitative illustration of Figure 6.27 shows that the potential breaks are smeared and the time elapsing between them elongated.

Several approaches have been suggested to deal with double-layer charging problems in obtaining reliable τ data. One is instrumental and attempts to electronically supply and add the current of (91) to the desired applied current. A method for doing this employs two identical cells, one containing sample and

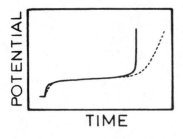

Figure 6.27 Illustration of effect of double-layer charging on chronopotentiograms. Ideal reversible curve (——); with double-layer effect (---).

the other only supporting electrolyte [217]. A chronopotentiometric circuit applies the desired constant current to the sample cell; its potential output simultaneously controls the working electrode potential in the blank cell. The blank cell current is thus i_{dl} (if the cells are perfectly matched); this is added to the constant sample cell current. Another method could use a single cell [218] in which the output of the reference electrode is electronically differentiated and returned to the cell as a current through a second input resistance to the CON amplifier. A constant value of C_{dl} is assumed. Both methods rely on positive electronic feedback and thus experience instrument instability when τ is short.

Another approach to double-layer charging correction relies on various graphical procedures empirically judged to compensate τ to some extent for the distortion. Several such procedures have been used [133, 150, 219–222], but none has gained uniform acceptance. Given transition time data measured by some graphical method, in a further treatment proposed by Bard [223] and Lingane [224], the term i_{dl} is assumed to be a constant fraction of applied current throughout the chronopotentiogram. A plot of $i\tau$ against $\tau^{1/2}$ yields an intercept term representative of the double layer, which is then applied as a correction to each individual transition time data point. Evans [225] has illustrated the efficacy of this method. Of course scrutiny of any graphical procedure for dealing with the double-layer problem is greatly facilitated by a knowledge of actual theoretical curves for the distortion. These have been obtained by numerical computation methods by DeVries [226] and Rodgers and Meites [227]; Olmstead and Nicholson [139] have given explicit theory. All involve assumption of a potential-independent C_{dl}, but nonetheless are very revealing of the seriousness of the double-layer influence on τ. DeVries [228] has compared his theory to the graphical transition time measurement schemes, as have Olmstead and Nicholson [139], who conclude that the method of Laity and McIntyre [220] is the best empirical approach.

In summary about τ measurement, accurate work can be accomplished under circumstances (long τ and higher concentration levels) in which double-layer distortion is minimal. For work at short τ (about 10 ms, for example), however, measurement of τ entails so many uncertainties that chronopotentiometry becomes poorly competitive, for quantitative studies, with the other chrono-methods.

4.4   Control and Measurement—Chronoamperometry and Chronocoulometry

Chronoamperometry and chronocoulometry approach in practice the ideal boundary value form of a (potential) step excitation more closely than is typical in chronopotentiometry. This is a consequence of localizing the change in electrode potential (concurrently localizing the double-layer charging) to the beginning, short-time portion of the experiment. Thus, the lower time limit for good response quality in a potential step experiment depends strongly on the quality of the potentiostat instrument and of the cell design. Potentiostat design has been continually improved and with careful cell design (working–reference electrode proximity, high electrolyte concentration, small working electrode [214, 215]) measurements at a few microseconds can be achieved. Using small electrodes reduces the size of the current flowing and thus the solution resistance consequences, but too small an electrode also yields appreciable edge diffusion effects.

An elementary three-electrode operational amplifier potentiostat for chrono-amperometry is shown in Figure 6.28. The experiment is initiated by sudden addition of the desired step potential E_{step} to the initial potential E_{init}. Potential control of the working ($-\bigcirc$) relative to reference ($\rightarrow$) electrode potentials is accomplished by amplifier CON, which senses any imbalance between the reference electrode potential input and E_{init} (or $E_{init}+E_{step}$) and provides a voltage output and resultant current flow at the auxiliary electrode to maintain equality of the reference–working and E_{init} (or $E_{init}+E_{step}$) potentials. The current flow at the working electrode is monitored by the current follower amplifier CUR, whose output E_1 is related to the cell current as shown. The circuit for chronocoulometry is a simple extension; voltage E_1 is integrated by

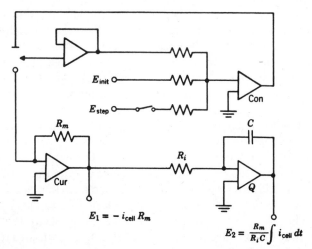

Figure 6.28   Elementary operational amplifier potentiostat for chronoamperometry and chronocoulometry.

amplifier Q to yield a voltage E_2 proportional to the total charge passed in the cell. A double potential step experiment is accomplished by, following the above, removal of E_{step} at the desired time.

Short-time chronoamperometric currents are quite large as the double layer charges to its new potential and because the Cottrell equation has a current–$t^{-1/2}$ dependency (23). The output current limits of the operational amplifiers are of obvious importance. A more useful measure of potentiostat performance, however, is its rise time, or the time required for the relative working and reference electrode potentials to achieve precisely the desired value of $E_{init} + E_{step}$. Potentiostat rise time is a function both of the amplifiers employed and of the electrical properties of the cell, particularly with respect to the value of $R_{uncomp}C_{dl}$, where R_{uncomp} is the residual solution ohmic resistance and C_{dl} the electrode capacitance. More precisely, performance must be stated with reference to the potentiostat circuit in combination with the array of resistance (compensated and uncompensated solution and faradaic resistance) and capacitance (double layer) that constitutes the particular electrochemical cell used. In fact, the electrochemical cell can be represented by a physical analog resistance–capacitance network; such "dummy cells" are convenient for testing the rise time of a given potentiostat under various potential step conditions. Evaluation of the potentiostat rise time is important not only from the standpoint of circuit optimization for minimal rise time, but also to define the lower time limit on which current or charge–time data of acceptable quality can be expected. In general, deviation from theory exceeding several percent is encountered at times less than about 10X (rise time).

A fundamental approach can be taken to optimize potentiostat performance by writing the transfer function [229–232] of the cell, which is a mathematical representation of the frequency response of the cell–potentiostat to a specified potential excitation. This gives insight into adjusting potentiostat parameters (and cell parameters where possible) to achieve minimal rise time. In a classical set of papers, Perone and co-workers [233–235] have considered the relationship of the rise time of the cell–potentiostat combination to charging current. They both experimentally and theoretically evaluate the degree of distortion of very-short-time chronoamperometric currents and suggest corrections of these.

R_{uncomp} is a very significant cell parameter in terms of influence on the rise time of the potentiostat–cell system. R_{uncomp} is the portion of the cell ohmic drop sensed by the reference electrode; the potential operative at the working electrode is less than that applied, by the product of R_{uncomp} and instantaneous cell current. At short electrolysis times, when cell current is large, this loss from the control potential can be large and can provide a major fraction of the overall system rise time [236]. In good cell design, therefore, the reference electrode junction with the cell solution, usually in the form of a capillary salt bridge (a Luggin capillary), is positioned as close to the working electrode as is practical. Theory relating the interelectrode distance to the uncompensated term has been given by Nemec [237]; this theory shows that the spacing must be quite close to achieve a major reduction. For instance, only a 50% reduction results for a spacing of one electrode radius from a spherical working electrode [236]. Further relief from the uncompensated resistance effect on rise time can be obtained through feedback and addition of a voltage proportional to the output of amplifier CUR to the control potential, which cancels a portion of the uncompensated potential drop [236, 238]. This involves a positive feedback loop, so full compensation is not consistent with maintaining potentiostat stability, but nevertheless valuable lowering of rise can be attained. This *uncompensated resistance feedback* has become very widely, but not always properly, used.

Other important features of cell and experiment design that influence rise time include the working electrode area and sample concentration, to which cell current and thus iR_{uncomp} are proportional. One thus avoids using sample concentration significantly larger than necessary for accurate correction of background current (obtained in a blank experiment). The working electrode area is kept as small as is consistent with obtaining conformity to the desired diffusion conditions [214, 215]. Large supporting electrolyte concentrations lower R_{uncomp}; studies in nonaqueous media generally experience poorer rise times because of the higher solution resistances typical of such media. Last, there can be significant differences in the required potentiostat rise times for chronoamperometric or chronocoulometric experiments directed at charge-transfer measurements and those elewhere directed. In the former, charge-transfer experiment much smaller potential steps are used, and small losses in potential control at short electrolysis times produce a relatively greater uncertainty in the transient data.

Measurement of current and charge–time transients in the potential step experiment were in early days accomplished by oscilloscope recording when data at short times were required, and by strip chart recording for data taken at approximately >1 s. Now, however, digital data acquisition systems [87, 239] (minicomputers and microcomputers) are rather common. These have had more than simply a convenience/versatility impact on electrochemical experimentation. The digital memory can act as a super-recorder, giving far better accuracy of current or charge measurement, at short or long times, than possible with oscilloscopic or paper recorder devices. This in effect allows more sensitive detection of interesting phenomena. Adsorption measurements with double potential step chronocoulometry are an excellent example. Additionally, the data, once stored in digital memory, can be compared to ready-programmed theoretical expressions in the computer. The data analysis can be initiated immediately following termination of the experiment; this can reduce the total time required for a transient experiment, including the entire sequence of computer data collection, computer data interpretation, and output of "answers," to 1 to 2 min in favorable cases. Such immediate data interpretation produces efficiency in allowing a series of potential step experiments on a given sample solution to be tailored and adjusted as the series progresses to yield a more optimum experimental characterization.

References

1. A. J. Bard and L. R. Faulkner, *Electroanalytical Methods*, Wiley, New York, 1980.
2. C. M. Reilley, *Pure Appl. Chem.*, **1**, 137 (1968).
3. M. E. Peover, "Electrochemistry of Aromatic Hydrocarbons and Related Substances," in A. J. Bard, Ed., *Electroanalytical Chemistry*, Vol. 2, Dekker, New York, 1967.
4. R. N. Adams, *Acc. Chem. Res.*, **2**, 175 (1969).
5. D. H. Evans, *Acc. Chem. Res.*, **10**, 313 (1977).
6. M. M. Baizer and H. Lund, *Organic Electrochemistry*, 2nd ed., Dekker, New York, 1983.
7. W. J. Albery, *Electrode Kinetics*, Clarendon, Oxford, 1975.
8. K. F. Herzfeld, *Phys. Z.*, **14**, 29 (1913).
9. H. A. Laitinen and I. M. Kolthoff, *J. Am. Chem. Soc.*, **61**, 3344 (1939).
10. H. A. Laitinen, *Trans. Electrochem. Soc.*, **82**, 289 (1942).
11. J. H. Christie, G. Lauer, R. A. Osteryoung, and F. C. Anson, *Anal. Chem.*, **35**, 1979 (1983).
12. P. Delahay, *New Instrumental Methods in Electrochemistry*, Interscience, New York, 1954.
13. H. Ikeuchi, Y. Shiwa, H. Tsujimoto, M. Kakihama, S. Takekawa, and G. P. Sato, *J. Electroanal. Chem.*, **111**, 287 (1980).
14. R. Neeb and W. Ranly, *Chem. Anal. (Warsaw)*, **17**, 969 (1972).
15. M. Petek, T. E. Neal, R. L. McNeely, and R. W. Murray, *Anal. Chem.*, **45**, 32 (1973).
16. W. M. Schwarz and I. Shain, *J. Phys. Chem.*, **69**, 30 (1965).
17. J. H. Christie, *J. Electroanal. Chem.*, **13**, 79 (1967).
18. M. K. Hanafey, R. L. Scott, T. H. Ridgway, and C. N. Reilley, *Anal. Chem.*, **50**, 116 (1978).
19. J. R. Delmastro and G. L. Booman, *Anal. Chem.*, **41**, 1409 (1969).
20. H. J. Schugar, A. T. Hubbard, F. C. Anson, and H. B. Gray, *J. Am. Chem. Soc.*, **91**, 71 (1969).
21. L. Marcoux and T. J. P. O'Brien, *J. Phys. Chem.*, **76**, 1666 (1972); L. Marcoux, *J. Phys. Chem.*, **76**, 3254 (1972).
22. H. Y. Cheng and R. L. McCreery, *Anal. Chem.*, **50**, 645 (1978).
23. G. C. Grant and T. Kuwana, *J. Electroanal. Chem.*, **24**, 11 (1970).
24. H. N. Blount, *J. Electroanal. Chem.*, **42**, 271 (1973).
25. D. Meyerstein, F. M. Hawkridge, and T. Kuwana, *J. Electroanal. Chem.*, **40**, 377 (1972).
26. R. P. Van Duyne, T. H. Ridgway, and C. N. Reilley, *J. Electroanal. Chem.*, **69**, 165 (1976).
27. L. N. Mackey and T. Kuwana, *Bioelectrochem. Bioenerg.*, **3**, 596 (1976).
28. T. Kuwana and W. R. Heineman, *Acc. Chem. Res.*, **9**, 241 (1976).
29. N. Winograd and T. Kuwana, "Spectroelectrochemistry at Optically Transparent Electrodes," in A. J. Bard, Ed., *Electroanalytical Chemistry*, Vol. 7, Dekker, New York, 1974.
30. W. R. Heineman, *Anal. Chem.*, **50**, 390A (1978).
31. R. Pruiksma and R. L. McCreery, *Anal. Chem.*, **51**, 2253 (1979).
32. T. H. Ridgway, R. P. Van Duyne, and C. N. Reilley, *J. Electronal. Chem.*, **34**, 267 (1972).
33. R. P. Van Duyne, T. H. Ridgway, and C. N. Reilley, *J. Electroanal. Chem.*, **34**, 283 (1972).

34. Von R. Koopman, *Ber. Bunsenges. Phys. Chem.*, **70**, 121 (1967).
35. Von R. Koopman and H. Gerischer, *Ber. Bunsenges. Phys. Chem.*, **70**, 127 (1967).
36. S. P. Perone and W. J. Kretlow, *Anal. Chem.*, **38**, 1760 (1966).
37. B. K. Hovespian and I. Shain, *J. Electroanal. Chem.*, **14**, 1 (1967).
38. J. D. E. McIntyre, *J. Phys. Chem.*, **71**, 1196 (1967).
39. R. Guidelli, *J. Phys. Chem.*, **72**, 3535 (1969).
40. P. Delahay and G. L. Stiehl, *J. Am. Chem. Soc.*, **74**, 3500 (1952).
41. S. L. Miller, *J. Am. Chem. Soc.*, **74**, 4130 (1952).
42. A. Pospisil, *Collect. Czech. Chem. Commun.*, **18**, 337 (1953).
43. R. Guidelli and D. Cozzi, *J. Electroanal. Chem.*, **14**, 245 (1967).
44. M. D. Ryan, J. F. Wei, B. A. Feinberg, and Y. K. Lau, *Anal. Biochem.*, **96**, 326 (1979).
45. B. A. Feinberg and M. D. Ryan, *Top. Bioelectrochem. Bioenerg.*, **4**, 225 (1981).
46. P. J. Lingane and J. H. Christie, *J. Electroanal. Chem.*, **13**, 227 (1967).
47. C. Amatore and J. M. Savéant, *J. Electroanal. Chem.*, **107**, 353 (1980).
48. C. Amatore and J. M. Savéant, *J. Electroanal. Chem.*, **86**, 227 (1978).
49. C. Amatore and J. M. Savéant, *J. Electroanal. Chem.*, **102**, 21 (1979).
50. J. Koutecky, *Collect. Czech. Chem. Commun.*, **18**, 183 (1953).
51. I. Tachi and M. Senda, "Polarographic Current of Stepwise Electrode Process Involving Chemical Reaction," in I. Longmuir, Ed., *Advances in Polarography*, Vol. 2, Pergamon, New York, 1960, p. 454.
52. G. S. Alberts and I. Shain, *Anal. Chem.*, **35**, 1859 (1963).
53. M. Mastragostino and J. M. Savéant, *Electrochim. Acta*, **13**, 751 (1968).
54. R. F. Nelson and R. N. Adams, *J. Am. Chem. Soc.*, **90**, 3925 (1968).
55. G. L. Booman and D. T. Pence, *Anal. Chem.*, **37**, 1366 (1965).
56. D. T. Pence and G. L. Booman, *Anal. Chem.* **38**, 1112 (1966).
57. M. D. Hawley and S. W. Feldberg, *J. Phys. Chem.*, **70**, 3459 (1966).
58. J. R. Delmastro, *Anal. Chem.*, **41**, 747 (1969).
59. R. S. Nicholson and I. Shain, *Anal. Chem.*, **36**, 706 (1964).
60. S. W. Feldberg and C. Auerbach, *Anal. Chem.*, **36**, 505 (1964).
61. S. W. Feldberg, "Digital Simulation: A General Method for Solving Electrochemical Diffusion-Kinetic Problems," in A. J. Bard, Ed., *Electroanalytical Chemistry*, Vol. 3, Dekker, New York, 1969.
62. J. R. Sandifer and R. P. Buck, *J. Electroanal. Chem.*, **49**, 161 (1974).
63. T. Joslin and D. Pletcher, *J. Electroanal. Chem.*, **49**, 171 (1974).
64. I. Ruzic and S. Feldberg, *J. Electroanal. Chem.*, **50**, 153 (1974).
65. N. Winograd, *J. Electroanal. Chem.*, **43**, 1 (1973).
66. R. N. Adams, M. D. Hawley, and S. W. Feldberg, *J. Phys. Chem.*, **71**, 851 (1967).
67. W. V. Childs, J. T. Maloy, C. P. Keszthelyi, and A. J. Bard, *J. Electrochem. Soc.*, **118**, 874 (1971).
68. S. Feldberg, *J. Phys. Chem.*, **73**, 1238 (1969).
69. E. T. Seo, R. F. Nelson, J. M. Fritsch, L. S. Marcoux, D. W. Leedy, and R. N. Adams, *J. Am. Chem. Soc.*, **88**, 3498 (1966).
70. M. Mastragostino, L. Nadjo, and J. M. Savéant, *Electrochim. Acta*, **13**, 721 (1968).
71. R. P. Van Duyne and C. N. Reilley, *Anal. Chem.*, **44**, 142 (1972).
72. R. P. Van Duyne and C. N. Reilley, *Anal. Chem.*, **44**, 153 (1972).
73. R. P. Van Duyne and C. N. Reilley, *Anal. Chem.*, **158** (1972).
74. R. D. Grypa and J. T. Maloy, *J. Electrochem. Soc.*, **122**, 377 (1978).
75. S. W. Feldberg and L. Jeftic, *J. Phys. Chem.*, **76**, 2439 (1972).
76. R. W. Murray and D. J. Gross, *Anal. Chem.*, **38**, 393 (1966).

77. J. H. Christie, G. Lauer, R. A. Osteryoung, and F. C. Anson, *Anal. Chem.*, **35**, 1979 (1963).

78. F. C. Anson, *Anal. Chem.*, **36**, 932 (1964).

79. F. C. Anson and D. A. Payne, *J. Electroanal. Chem.*, **13**, 35 (1967).

80. G. W. O'Dom and R. W. Murray, *Anal. Chem.*, **39**, 51 (1967).

81. G. W. O'Dom and R. W. Murray, *J. Electroanal. Chem.*, **16**, 327 (1968).

82. R. A. Osteryoung and J. H. Christie, *J. Phys. Chem.*, **71**, 1348 (1967).

83. W. H. Reinmuth and K. Balasubramanian, *J. Electroanal. Chem.*, **38**, 79 (1972).

84. F. C. Anson, *Anal. Chem.*, **38**, 55 (1966).

85. J. H. Christie, R. A. Osteryoung, and F. C. Anson, *J. Electroanal. Chem.*, **13**, 236 (1967).

86. C. M. Elliott and R. W. Murray, *Anal. Chem.*, **47**, 908 (1975).

87. G. Lauer, R. Abel, and F. C. Anson, *Anal. Chem.*, **39**, 765 (1967).

88. F. C. Anson and D. J. Barclay, *Anal. Chem.*, **40**, 1791 (1968).

89. F. C. Anson, J. H. Christie, and R. A. Osteryoung, *J. Electroanal. Chem.*, **13**, 343 (1967).

90. D. J. Barclay and F. C. Anson, *J. Electrochem. Soc.*, **116**, 438 (1969).

91. F. C. Anson and B. Epstein, *J. Electrochem. Soc.*, **115**, 1155 (1968).

92. H. Gerischer and W. Vielstich, *Z. Phys. Chem.* (*Frankfurt am Main*), **3**, 16 (1955).

93. W. Vielstich and H. Gerischer, *Z. Phys. Chem.* (*Frankfurt am Main*), **4**, 10 (1955).

94. C. A. Johnson and S. Barnartt, *J. Electrochem. Soc.*, **114**, 1256 (1967).

95. C. A. Johnson and S. Barnartt, *J. Phys. Chem.*, **71**, 1637 (1967).

96. S. Barnartt, *Electrochim. Acta*, **15**, 1313 (1970).

97. S. Barnartt and C. A. Johnson, *J. Electroanal. Chem.*, **24**, 226 (1970).

98. S. Barnartt and C. A. Johnson, *Anal. Chem.*, **43**, 2 (1971).

99. P. Delahay, "Study of Fast Electrode Processes," in P. Delahay, Ed., *Advances in Electrochemistry and Electrochemical Engineering*, Vol. 1, Interscience, New York, 1961.

100. A. Yamada and N. Tanaka, *Anal. Chem.*, **45**, 167 (1973).

101. N. Tanaka, A. Kitani, A. Yamada, and K. Sasaki, *Electrochim. Acta*, **18**, 675 (1973).

102. C. R. Martin, I. Rubinstein, and A. J. Bard, *J. Electroanal. Chem.*, **151**, 267 (1983).

103. J. H. Christie, G. Lauer, and R. A. Osteryoung, *J. Electroanal. Chem.*, **7**, 60 (1964).

104. P. J. Lingane and J. H. Christie, *J. Electroanal. Chem.*, **10**, 284 (1965).

105. E. E. Bancroft, H. M. Blount, and F. M. Hawkridge, *Anal. Chem.*, **53**, 1862 (1981).

106. M. A. Dayton, A. J. Ewing, and R. M. Wightman, *Anal. Chem.*, **52**, 2392 (1980).

107. A. J. Ewing, M. A. Dayton, and R. M. Wightman, *Anal. Chem.*, **53**, 1842 (1981); M. A. Dayton, J. C. Brown, K. J. Stutts, and R. M. Wightman, *Anal. Chem.*, **50**, 946 (1980).

108. R. N. Adams, *Anal. Chem.*, **48**, 1126A (1976); R. M. Wightman, E. Strope, P. M. Plotsky, and R. N. Adams, *Nature* (*London*), **262**, 145 (1970).

109. J. Heinze, *J. Electroanal. Chem.*, **124**, 73 (1981).

110. K. Oldham, *J. Electroanal. Chem.*, **122**, 1 (1981).

111. B. Speiser and S. Pons, *Can. J. Chem.*, **60**, 1352, 2463 (1982).

112. P. Daum, J. R. Lenhard, D. R. Rolison, and R. W. Murray, *J. Am. Chem. Soc.*, **102**, 4649 (1980).

113. D. M. Oglesby, S. H. Omang, and C. N. Reilley, *Anal. Chem.*, **37** (1965) 1312.

114. J. Leddy and A. Bard, *J. Electroanal. Chem.*, **153**, 223 (1983).

115. P. J. Peerce and A. J. Bard, *J. Electroanal. Chem.*, **112**, 97 (1980).

116. M. Lovric, *J. Electroanal. Chem.*, **123**, 373 (1981).

117. C. McCallum and D. Pletcher, *Electrochim. Acta*, **20**, 811 (1978).

118. R. C. Bowers, G. Ward, C. M. Wilson, and D. DeFord, *J. Phys. Chem.*, **65** (1961) 672.
119. A. T. Hubbard and F. C. Anson, in A. J. Bard, Ed., *Electroanalytical Chemistry*, Vol. 4, Dekker, New York, 1970; C. N. Reilley, *Rev. Pure Appl. Chem.*, **18**, 137 (1968).
120. R. W. Murray, *Annu. Rev. Mater. Sci.*, **14**, 145 (1984).
121. N. Oyama and F. C. Anson, *J. Electrochem. Soc.*, **127**, 640 (1980).
122. T. Ikeda, R. Schmehl, P. Denisevich, K. Willman, and R. W. Murray, *J. Am. Chem. Soc.*, **104**, 2683 (1982).
123. T. Gueshi, K. Tokuda, and H. Matsuda, *J. Electroanal. Chem.*, **89**, 247 (1978).
124. T. Gueshi, K. Tokuda, and H. Matsuda, *J. Electroanal. Chem.*, **101**, 29 (1979).
125. R. C. Bowers and R. W. Murray, *Anal. Chem.*, **38** (1966) 461.
126. H. F. Weber, *Wied. Ann.*, **7**, 536 (1879).
127. H. J. S. Sand, *Philos. Mag.*, **1**, 45 (1901).
128. T. R. Rosebrugh and W. L. Miller, *J. Phys. Chem.*, **14**, 816 (1910).
129. Z. Karoglanoff, *Z. Elektrochem.*, **12**, 5 (1906).
130. J. A. Butler and G. Armstrong, *Proc. R. Soc. London*, **406**, A139 (1933).
131. J. A. V. Butler and G. Armstrong, *Trans. Faraday Soc.*, **30**, 1173 (1934).
132. L. Gierst and A. Juliard, *J. Phys. Chem.*, **57**, 701 (1953).
133. P. Delahay and G. Mamantov, *Anal. Chem.*, **27**, 478 (1955).
134. C. N. Reilley, G. W. Everett, and R. H. Johns, *Anal. Chem.*, **27**, 483 (1955).
135. B. B. Damaskin, *The Principles of Current Methods for the Study of Electrochemical Reactions*, McGraw-Hill, New York, 1967.
136. D. G. Davis, "Applications of Chronopotentiometry to Problems in Analytical Chemistry," in A. J. Bard, Ed., *Electroanalytical Chemistry*, Vol. 1, Dekker, New York, 1966.
137. P. Delahay and T. Berzins, *J. Am. Chem. Soc.*, **75**, 2486 (1953).
138. F. H. Beyerlein and R. S. Nicholson, *Anal. Chem.*, **44**, 1917 (1972).
139. M. L. Olmstead and R. S. Nicholson, *J. Phys. Chem.*, **72**, 1650 (1968).
140. D. J. Kooijman and J. H. Sluyters, *J. Electroanal. Chem.*, **13**, 152 (1967).
141. H. P. Van Leeuwen and J. H. Sluyters, *J. Electroanal. Chem.*, **39**, 25 (1972).
142. T. Berzins and P. Delahay, *J. Am. Chem. Soc.*, **75**, 4205 (1953).
143. L. B. Anderson and D. J. Macero, *Anal. Chem.*, **37**, 322 (1965).
144. F. H. Beyerlein and R. S. Nicholson, *Anal. Chem.*, **40**, 286 (1968).
145. P. Delahay and T. Berzins, *J. Am. Chem. Soc.*, **75**, 2486 (1953).
146. W. H. Reinmuth, *J. Electroanal. Chem.*, **38**, 95 (1972).
147. N. Tanaka and T. Murayama, *Z. Phys. Chem. (Frankfurt am Main)*, **21**, 146 (1959).
148. R. P. Buck and L. R. Griffith, *J. Electrochem. Soc.*, **109**, 1005 (1962).
149. H. B. Mark and F. C. Anson, *Anal. Chem.*, **35**, 722 (1963).
150. R. P. Buck, *Anal. Chem.*, **35**, 1853 (1963).
151. V. I. Kravstov, *Elektrokhimiya*, **4**, 486 (1968).
152. W. H. Reinmuth, *Anal. Chem.*, **33**, 322 (1961).
153. P. Delahay, C. C. Mattax, and T. Berzins, *J. Am. Chem. Soc.*, **76**, 5319 (1954).
154. W. K. Snead and A. E. Remick, *J. Am. Chem. Soc.*, **79**, 6121 (1957).
155. A. C. Testa and W. H. Reinmuth, *Anal. Chem.*, **32**, 1518 (1960).
156. O. Dracka, *Collect. Czech. Chem. Commun.*, **25**, 338 (1960).
157. C. Furlani and G. Morpurgo, *J. Electroanal. Chem.*, **1**, 351 (1960).
158. A. C. Testa and W. H. Reinmuth, *Anal. Chem.*, **32**, 1512 (1960).
159. D. L. Ehman and D. T. Sawyer, *Inorg. Chem.*, **8**, 900 (1969).
160. O. Dracka, *Collect. Czech. Chem. Commun.*, **32**, 3987 (1967).
161. O. Dracka, *Collect. Czech. Chem. Commun.*, **36**, 1889 (1971).

162. O. Dracka, *Collect. Czech. Chem. Commun.*, **36**, 1876 (1971).
163. O. Dracka, *Collect. Czech. Chem. Commun.*, **41**, 48 (1976).
164. O. Dracka, *Collect. Czech. Chem. Commun.*, **41**, 953 (1976).
165. O. Dracka, *Collect. Czech. Chem. Commun.*, **42**, 1093 (1977).
166. H. B. Herman and A. J. Bard, *Anal. Chem.*, **36**, 510 (1964).
167. O. Fischer, O. Dracka, and E. Fischerova, *Collect. Czech. Chem. Commun.*, **25**, 323 (1960).
168. A. C. Testa and W. H. Reinmuth, *Anal. Chem.*, **33**, 1320 (1961).
169. A. C. Testa and W. H. Reinmuth, *J. Am. Chem. Soc.*, **83**, 784 (1961).
170. G. S. Alberts and I. Shain, *Anal. Chem.*, **35**, 1859 (1963).
171. W. Lorenz, *Z. Elektrochem.*, **59**, 730 (1955).
172. W. Lorenz and H. Muhlberg, *Z. Elektrochem.*, **59**, 736 (1955).
173. W. Lorenz and H. Muhlberg, *Z. Phys. Chem. (Frankfurt am Main)*, **17**, 129 (1958).
174. W. H. Reinmuth, *Anal. Chem.*, **33**, 322 (1961).
175. F. C. Anson, *Anal. Chem.*, **33**, 1123 (1961).
176. S. V. Tatwawadi and A. J. Bard, *Anal. Chem.*, **36**, 2 (1964).
177. R. W. Murray, *J. Electroanal. Chem.*, **7**, 242 (1964).
178. R. Brdicka, *Z. Elektrochem.*, **48**, 278 (1942).
179. R. F. Broman and R. W. Murray, *Anal. Chem.*, **37**, 1408 (1965).
180. H. B. Herman and H. N. Blount, *J. Electroanal. Chem.*, **25**, 165 (1970).
181. H. A. Laitinen and L. M. Chambers, *Anal. Chem.*, **36**, 5 (1964).
182. R. A. Munson, *J. Electroanal. Chem.*, **5**, 292 (1963).
183. R. A. Munson, *J. Phys. Chem.*, **66**, 727 (1962).
184. P. J. Lingane, *Anal. Chem.*, **39**, 485 (1967).
185. A. J. Bard, *Anal. Chem.*, **35**, 340 (1963).
186. J. H. Christie and R. A. Osteryoung, *Anal. Chem.*, **38**, 1620 (1966).
187. R. W. Murray and C. N. Reilley, *J. Electroanal. Chem.*, **3**, 64 (1962).
188. R. W. Murray, *Anal. Chem.*, **35**, 1784 (1963).
189. S. K. Rangarajan, *J. Electroanal. Chem.*, **62**, 31 (1975).
190. K. Kontturi, M. Lindstrom, and G. Sundholm, *J. Electroanal. Chem.*, **63**, 263 (1975).
191. K. Kontturi, M. Lindstrom, and G. Sundholm, *J. Electroanal. Chem.*, **71**, 21 (1976).
192. H. B. Herman and A. J. Bard, *Anal. Chem.*, **35**, 1121 (1963).
193. H. B. Herman and A. J. Bard, *Anal. Chem.*, **36**, 510, 971 (1964).
194. H. B. Herman and A. J. Bard, *J. Phys. Chem.*, **70**, 396 (1966).
195. H. B. Herman and A. J. Bard, *J. Electrochem. Soc.*, **115**, 1028 (1968).
196. D. Cukman and V. Pravdic, *J. Electroanal. Chem.*, **49**, 415 (1974).
197. M. Vukovic, *J. Electroanal. Chem.*, **133**, 25 (1982).
198. M. Vukovic, *J. Electroanal. Chem.*, **152**, 15 (1983).
199. D. G. Peters and S. L. Burden, *Anal. Chem.*, **38**, 530 (1966).
200. P. E. Sturrock, W. D. Anstine, and R. H. Gibson, *Anal. Chem.*, **40**, 505 (1968).
201. P. E. Sturrock, G. Privett, and A. R. Tarpley, *J. Electroanal. Chem.*, **14**, 303 (1967).
202. R. H. Gibson and P. E. Sturrock, *J. Electrochem. Soc.*, **123**, 1170 (1976).
203. P. E. Sturrock, *J. Electroanal. Chem.*, **8**, 425 (1964).
204. R. Kalvoda, *Techniques of Oscillographic Polarography*, Elsevier, New York, 1965.
205. D. Homolka, L. Q. Hung, A. Hofmanova, M. W. Khalil, J. Koryta, V. Marecek, Z. Samec, S. K. Sen, P. Vanysek, J. Weber, and M. Brezina, *Anal. Chem.*, **52**, 1606 (1980).
206. Y. Takemori, T. Kambara, M. Senda, and I. Tachi, *J. Phys. Chem.*, **61**, 968 (1957).
207. N. P. Bansal and H. L. Jindal, *J. Indian Chem. Soc.*, **49**, 957 (1972).

208. N. P. Bansal, *Indian J. Chem.*, **11**, 1199 (1973).
209. R. D. DeMars and I. Shain, *Anal. Chem.*, **29**, 1825 (1957).
210. W. Kemula and Z. Kublik, *Rocz. Chem.*, **33**, 1431 (1959).
211. A. J. Bard, *Anal. Chem.*, **33**, 11 (1961).
212. I. Shain and K. J. Martin, *J. Phys. Chem.*, **65**, 254 (1961).
213. R. W. Murray and D. J. Gross, *J. Electroanal. Chem.*, **13**, 132 (1967).
214. Z. G. Soos and P. J. Lingane, *J. Phys. Chem.*, **68**, 3821 (1964).
215. J. B. Flanagan and L. Marcoux, *J. Phys. Chem.*, **77**, 1051 (1973).
216. P. H. Daum and C. G. Enke, *Anal. Chem.*, **41**, 653 (1969).
217. W. D. Shults, F. E. Haga, T. R. Mueller, and H. C. Jones, *Anal. Chem.*, **37**, 1415 (1965).
218. C. G. Enke, Princeton University, private communication, 1966; W. H. Reinmuth, *Anal. Chem.*, **38**, 270R (1966).
219. P. Delahay and C. C. Mattax, *J. Am. Chem. Soc.*, **76**, 874 (1954).
220. R. W. Laity and J. D. E. McIntyre, *J. Am. Chem. Soc.*, **87**, 3806 (1965).
221. W. H. Reinmuth, *Anal. Chem.*, **33**, 485 (1961).
222. L. Nadjo and J. M. Savéant, *J. Electroanal. Chem.*, **75**, 181 (1977).
223. A. J. Bard, *Anal. Chem.*, **35**, 340 (1963).
224. J. J. Lingane, *J. Electroanal. Chem.*, **1**, 379 (1960).
225. D. H. Evans, *Anal. Chem.*, **36**, 2027 (1964).
226. W. T. DeVries, *J. Electroanal. Chem. Interfacial Electrochem.*, **17**, 31 (1968).
227. R. S. Rodgers and L. Meites, *J. Electroanal. Chem.*, **16**, 1 (1968).
228. W. T. DeVries, *J. Electroanal. Chem.*, **18**, 469 (1981).
229. G. L. Booman and W. B. Holbrook, *Anal. Chem.*, **35**, 1793 (1963).
230. G. L. Booman and W. B. Holbrook, *Anal. Chem.*, **37**, 795 (1965).
231. I. Shain, J. E. Harrar, and G. L. Booman, *Anal. Chem.*, **37**, 1768 (1968).
232. F. Magno and G. Bontempelli, *Anal. Chem.*, **53**, 599 (1981).
233. L. L. Miaw and S. P. Perone, *Anal. Chem.*, **51**, 1645 (1979).
234. S. S. Fratoni and S. P. Perone, *Anal. Chem.*, **48**, 287 (1976).
235. K. F. Dahnke, S. S. Fratoni, and S. P. Perone, *Anal. Chem.*, **48**, 296 (1976).
236. G. Lauer and R. A. Osteryoung, *Anal. Chem.*, **38**, 1106 (1966).
237. L. Nemec, *J. Electroanal. Chem.*, **8**, 166 (1964).
238. E. R. Brown, T. G. McCord, D. E. Smith, and D. D. Deford, *Anal. Chem.*, **38**, 1119 (1966).
239. G. Lauer and R. A. Osteryoung, *Anal. Chem.*, **40 (10)**, 40A (1968).

Chapter **7**

SPECTROELECTROCHEMISTRY

Richard L. McCreery

1   INTRODUCTION AND SCOPE

There are many techniques involving combinations of optical and electro-chemical phenomena, and new ones continue to be announced. In a majority of cases the spectroscopic technique is used to examine an electrochemical process, although in some cases light excites rather than probes the electrode environment. To define the scope of this chapter, hybrids of optical and electro-chemical methods are classified into three broad categories. The first, referred to as spectroelectrochemistry and the principal subject of this chapter, comprises *in situ* methods in which the spectroscopic probe is intimate to the electrochemical process and monitors the course or dynamics of the electrode reaction. The second classification includes off-line spectroscopic methods in which a conventional spectroscopic technique such as infrared absorption or nuclear magnetic resonance (NMR) characterizes the products of an electro-chemical reaction after the fact. Since these methods are conventional, they will not be discussed. The third group, photoelectrochemistry, consists of various approaches in which an electrode monitors a photochemical process, often at a semiconductor electrode. Though an important area in electro-chemistry, it will not be discussed here.

Therefore "spectroelectrochemistry" refers to techniques in which a spectro-scopic probe is used to probe electrochemical events *in situ*, with intimate coupling of the electrochemical and spectroscopic processes. The optical and electrochemical measurements are often made simultaneously, and the results include qualitative, quantitative, and dynamic information about species involved in the electrochemical reaction. All the methods discussed here fit this definition, although the techniques themselves differ greatly in wavelength region, configuration, and information content.

Within this definition of spectroelectrochemistry, a further classification is useful depending on whether the technique provides information about the electrode surface itself or about the associated solution in the diffusion layer. Solution spectroelectrochemistry probes species within the diffusion layer or nearby solution and provides information about the identities or concentrations of molecules generated or consumed by an electrode process as they diffuse toward or away from the electrode surface. Thus the electrode generates reac-tants or establishes a potential and is not directly important to the optical technique. A common example is ultraviolet–visible (UV–vis) absorption spec-troscopy of a reactive species generated at an optically transparent electrode. On the other hand, interfacial, or surface spectroelectrochemistry, probes surface processes by their effect on the optical probe. For example, light reflected by a polished platinum electrode is affected by a layer of adsorbed material and the effect may be used to examine the adsorption process. Solution and inter-facial spectroelectrochemistry are discussed separately, since their methodolo-gies are usually distinct.

The various solution spectroelectrochemistry types discussed in this chapter are conveniently classified by spectral technique. The most common method involves UV–vis absorption by the materials of interest, with the optical beam passing through the electrode or internally or externally reflected from the electrode. Methods with similar geometry include infrared absorption and UV–vis fluorescence, although these methods are more restricted in terms of electrode material, solvent, or chemical systems. Electron spin resonance (ESR) and NMR have also been developed with ESR being much more common because of its high sensitivity for radical species. Raman spectra have been obtained from electrogenerated species, although a greater development has occurred for applying Raman to surface-adsorbed species. Finally, several optical methods for examining the solution near an electrode are based on interference or diffraction of the probe beam.

Spectroelectrochemical techniques applied to the electrode–solution interface differ considerably from solution techniques mainly because of the high sensitivity required when examining the few atomic layers comprising the interface region. Specular reflectance methods involve electrode reflectivity measurements as a function of electrochemical processes, while ellipsometry measures phase shifts of a light beam upon reflection. Surface Raman spectroscopy provides vibrational spectra of adsorbed species for a limited combination of electrodes and adsorbates. Although not as well developed, infrared reflectance and photothermal spectroscopy may also identify and monitor adsorbed species. These optical methods often complement high-vacuum surface analysis techniques, since the spectroelectrochemical probes provide surface information directly in the solution of interest.

These spectroelectrochemical techniques and their strengths and weaknesses are discussed in greater detail in subsequent sections. At this point the utility of methods combining spectroscopic and electrochemical processes should be considered. What new information, unavailable from an electrochemical experiment, is gained by adding an often complex, *in situ* spectroscopic probe? First, spectroscopy provides qualitative information about the intermediates or products of an electrochemical process. A conventional electrochemical experiment provides very limited structural information about solution or adsorbed species; and identification of solution molecules from their electrochemical redox potentials would be frustrated by electrochemistry's low resolution and by the many variables affecting observed potentials. Spectroscopy, on the other hand, is more specific for the species present and often provides structural information. Second, spectroscopy provides quantitative information about solution species with greater accuracy and selectivity than the simple electrochemical technique. In many cases the electrochemical current may contain contributions from capacitance or background processes resulting in erroneous determination of concentration. An optical probe is free of such errors and also provides selectivity on the basis of the absorption spectrum. Third, spectroelectrochemistry is well suited to studying dynamic processes associated with an electrochemical charge transfer. Using a suitable wavelength, a particular intermediate may be selected and its time course determined. Often the resulting transient yields more information than an analogous current measurement.

Finally, spectroelectrochemical techniques can provide some information about the orientation of adsorbed molecules on the electrode surface. While spectroscopic methods for obtaining such detail are embryonic, no analogous methods exist. Both the promise and the realization of obtaining new information about electrochemical processes have given impetus to the development of spectroelectrochemical methods, and the unique benefits of spectroelectrochemistry will be stressed as individual methods are discussed.

2 SPECTROELECTROCHEMISTRY OF COMPONENTS IN SOLUTION

2.1   UV–Visible Absorption Methods

UV–visible spectroscopic methods, probably the most commonly encountered of the spectroelectrochemical techniques, are very well suited to electrochemical applications, and have been reviewed often during the last decade [1–5]. Their popularity is partially attributable to the optical transmission characteristics of the solvents and electrodes commonly used in electrochemistry.

2.1.1   Transmission Through an Optically Transparent Electrode

The earliest UV–vis transmission method was developed by Kuwana in 1964 [6] and was based on the optically transparent electrode (OTE). The experimental details will be discussed later, but the basic configuration is shown in Figure 7.1. The electrode is a semitransparent metal film or grid through which the optical beam of a conventional or custom UV–vis spectrophotometer is passed. As materials are generated or consumed at the electrode, their spectra may be obtained, or the time course of any reactions of UV–vis absorbers may be monitored.

The simplest example of an OTE experiment involves an electrochemical process with stable components, and the spectral measurements taken under essentially equilibrium conditions. Consider the general reaction of (1) where

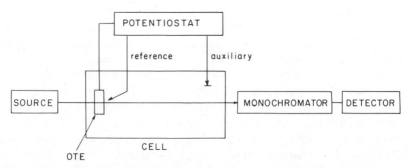

Figure 7.1   Basic arrangement for spectroelectrochemistry for a transmission experiment through an optically transparent electrode (OTE).

both Ox and Red are stable and a charge-transfer reaction involving n electrons is carried out at the OTE.

$$Ox + ne^- \rightleftharpoons Red \tag{1}$$

Further suppose that a constant potential is applied to the OTE and the solution is stirred until equilibrium is reached. At equilibrium, the concentrations of Ox and Red will be dictated by the Nernst equation, as shown in (2), where $E^{0\prime}$ is the formal potential, E_{app} is the applied potential, T is absolute temperature, and F is the Faraday constant. $C_{Ox}^{E\,app}$ and $C_{Red}^{E\,app}$ are the concentrations of Ox and Red at a given E_{app}, assuming Nernstian equilibrium.

$$E_{app} = E^{0\prime} + \frac{RT}{nF} \ln \frac{C_{Ox}^{E\,app}}{C_{Red}^{E\,app}} \tag{2}$$

At applied potentials significantly negative of the $E^{0\prime}$ for the couple, nearly all the material will be in the form of Red, while at potentials positive of $E^{0\prime}$, Ox will predominate. At intermediate potentials, both Ox and Red will have significant concentrations as determined by (2). A spectrum obtained with the solution equilibrated with the electrode at some potential will reflect the composition of the solution, just as it would if Ox and Red were mixed in amounts appropriate to satisfy (2). The Ox/Red ratio is controlled by the electrode rather than by some external chemical reagent, as in the case of a redox titration.

Spectral characteristics of Ox and Red are determined independently by setting the potential significantly positive or negative of $E^{0\prime}$. Thus the absorption spectra of both Ox and Red are conveniently determined by using the electrode reaction rather than a chemical reagent to convert one to the other. This aspect is particularly advantageous when one form is air sensitive or difficult to synthesize by another means. If optical path length is accurately known, the molar absorptivity of a component may be determined at any accessible wavelength.

Given the spectra of pure solutions of Ox and Red, it is possible to determine the concentrations of Ox and Red at any time, assuming their spectra are not severely overlapped. An important application of this approach is the determination of accurate $E^{0\prime}$ and n values for a variety of redox systems. A plot of applied potential versus log(Ox/Red) determined spectrally should be linear with a slope of $2.303(RT/nF)$ and an intercept of $E^{0\prime}$. The advantage of this spectroelectrochemical method over conventional electrochemistry is the lack of a requirement for a fast charge-transfer rate. Since the current is not measured, the time elapsing before the solution and the electrode equilibrate is immaterial. Consequently, Ox and Red may be large biological molecules or other electrochemically recalcitrant materials that are not amenable to the usual electrochemical methods because of very low charge-transfer rates.

A good example of equilibrium spectroelectrochemistry is the determination of n and $E^{0\prime}$ for cytochrome-c [7]. The cell, a fine gold grid sandwiched between two glass plates, created an optically transparent thin-layer electrochemical cell (OTTLE) with a cell thickness of less than 0.2 mm. The thin cell assured fairly rapid mass transport and therefore rapid access of solution species to the electrode. The electrode was about 50% transparent, and the entire cell could be placed in a conventional spectrophotometer modified to permit leads to the electrodes. Solution was drawn into the thin-layer cell by capillary action, with the reference and auxiliary electrodes placed in the solution below (Figure 7.2). Electrode potential was controlled by a conventional potentiostat, similar to those available commercially [8].

When 2,6-dichlorophenol indophenol was placed in the solution, in addition to the usual supporting electrolyte, the series of spectra shown in Figure 7.3 resulted. At low potentials the spectra show only baseline absorbance, since the reduced form of the compound is transparent in this wavelength region. As the potential is increased, the broad absorption centered at 600 nm indicates the presence of oxidized indophenol. When only one redox component absorbs, an expression for Ox/Red as a function of absorbance at various potentials is simply derived:

$$A_{600}^{E\,\text{app}} = \varepsilon_{\text{Ox}}^{600} b C_{\text{Ox}}^{E\,\text{app}} \tag{3}$$

$$A_{600}^{\max} = \varepsilon_{\text{Ox}}^{600} b C_{\text{Ox}}^{\max} \tag{4}$$

where $A_{600}^{E\,\text{app}}$ = absorbance at 600 nm at some applied potential;
   $\varepsilon_{\text{Ox}}^{600}$ = molar absorptivity of Ox at 600 nm;
   $C_{\text{Ox}}^{E\,\text{app}}$ = concentration of Ox at E_{app};
   $A_{600}^{\max}$ = maximum absorbance for the case where $E_{\text{app}} \gg E^{0'}$;
   b = cell thickness, which equals the optical path length.

Since

$$C_{\text{Red}}^{E\,\text{app}} = C_{\text{Ox}}^{\max} - C_{\text{Ox}}^{E\,\text{app}} \tag{5}$$

$$\frac{C_{\text{Ox}}^{E\,\text{app}}}{C_{\text{Red}}^{E\,\text{app}}} = \frac{A_{600}^{E\,\text{app}}}{A_{600}^{\max} - A_{600}^{E\,\text{app}}} \tag{6}$$

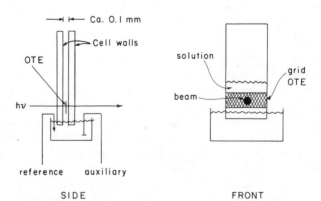

Figure 7.2 Optically transparent thin-layer electrochemical cell employing gold grid electrode. Cell is constructed from glass plate (e.g., microscope slides) and solution is drawn into cell by capillary action from a small reservoir.

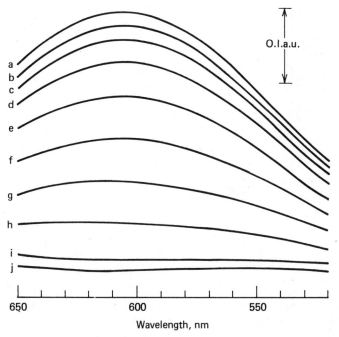

Figure 7.3   Spectra of 1.0 mM 2,6-dichlorophenolindophenol taken with an OTTLE cell at various applied potentials. Solution was allowed to equilibrate with electrode before spectra were run. E_{app} versus SCE were as follows: (a) +0.200 (V), (b) +0.025, (c) +0.015, (d) +0.0045, (e) −0.005, (f) −0.015, (g) −0.0255, (h) −0.0355, (i) −0.066, (j) −0.200. Reprinted with permission, from [7].

Equation (6) expresses the Ox/Red ratio at any potential as a function of easily measured absorbance values.

A plot of the applied potential versus log(Ox/Red) determined in this fashion is shown in Figure 7.4, Curve A. Recall that applied potential is that value imposed on the cell, and the spectra used to calculate the Ox/Red ratio were determined after the solution equilibrated with this potential. Equilibration time depends on the chemical system being examined and is typically a few minutes for each potential. The slope of the line in Figure 7.4, Curve B, is 31.3 mV, which favorably compares with the value predicted from (2) of 29.5 mV ($2.303RT/2F$). The y intercept was 226 mV versus NHE (−18 mV versus SCE), again in good agreement with an average of literature values (224 mV) determined by conventional methods. The standard deviation of $E^{0\prime}$ for a series of runs ($N=4$) was 0.5 mV, indicating the method's excellent precision.

In principle, the same method may be applied to the much larger cytochrome c protein; however, the cytochrome will not interact with the electrode without denaturation. Therefore indophenol was used as a mediator to equilibrate the protein with the electrode. Electrochemical charge transfer actually occurred between the electrode and the indophenol, but the indophenol then homogeneously exchanged electrons with the protein. Both redox systems will equi-

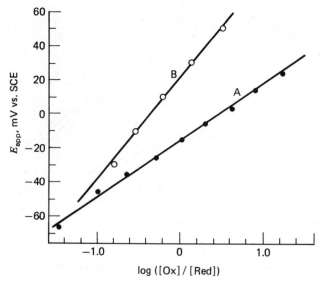

Figure 7.4   Plots of E_{app} versus log([Ox]/[Red]) for (A) 1.0 mM 2,6-dichlorophenolindophenol, and (B) 0.5 mM cytochrome c with 0.1 mM dichlorophenolindophenol. Concentration ratio was determined spectrophotometrically while E_{app} was imposed on the OTE. Reprinted, with permission, from [7].

librate with the applied potential, over a period of time slightly longer than that required for indophenol to equilibrate itself. Spectra for a solution containing both components are shown in Figure 7.5. Strong absorption of the cytochrome 550 nm band is apparent, as well as the underlying indophenol spectrum. A plot of E versus log(Ox/Red) for the cytochrome is also shown in Figure 7.4, Curve B. From this plot, $n = 1.00$ and $E^{0'}$ is 262 mV versus NHE (18 mV versus SCE), again in excellent agreement with values from other methods. The same approach was applied to cobalt complexes related to vitamin B_{12} with impressive results [9].

While the thin-layer approach is quite useful, it suffers from the inherent short optical path length of a thin cell. An alternative technique, predating the thin-layer method, involves a longer path length and somewhat more complex apparatus [10–12]. The electrochemical cell is a 1 cm long cylinder along whose axis the optical beam passes. The electrode is a thin film of platinum or tin oxide coated on glass and is approximately 20% transparent in the visible wavelength range. This cell design (Figure 7.6) is quite versatile and particularly useful for studying redox enzymes because of effective exclusion of oxygen.

This longer cell in principle may be used in the same manner as the thin-layer cell; however, its developers applied a different equilibrium technique based on coulometric titration. The cell receives a known quantity of charge causing a known conversion of Ox to Red, or vice versa. The spectrum is determined after each charge increment and the rate of absorbance change with charge is calculated. Equation (7) simply determines n from such data, and

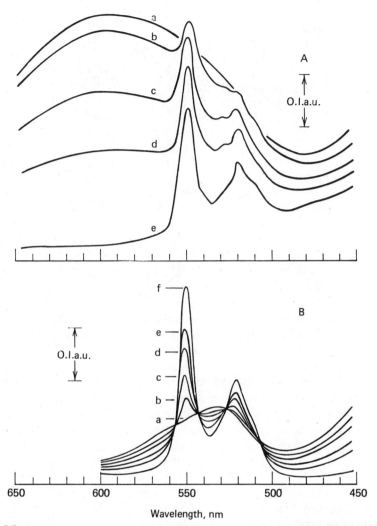

Figure 7.5   Spectra of cytochrome c and 2,6-dichlorophenolindophenol for different values of E_{app}, mV versus SCE. OTTLE cell, thickness = 0.022 cm. (A) 0.4 mM cytochrome c/1.0 mM 2,6-dichlorophenolindophenol. (a) +250.0, (b) +30.0, (c) +1.0, (d) −30.0, (e) −250.0. (B) 0.5 mM cytochrome c/0.01 mM 2,6-dichlorophenolindophenol. (a) 250.0, (b) +50.0, (c) +30.0, (d) +10.0, (e) −10.0, (f) −250.0. Reprinted, with permission, from [7].

the method was successfully applied to cytochrome c.

$$n = \frac{\Delta Q}{\Delta A} \varepsilon \qquad (7)$$

where ΔQ = charge injected per millilitre of solution;
   ΔA = change in absorbance per centimetre of path length;
   ε = molar absorptivity of cytochrome.

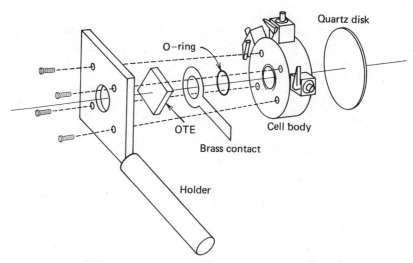

Figure 7.6 Typical cell employing a metal film OTE and having a cell volume of approximately 1 mL. Reprinted, with permission, from [12].

Additional analysis of absorbance versus charge curves successfully determines the $E^{0\prime}$ value. This technique was used in elegant investigation of the multiple redox centers in cytochrome c oxidase [13].

Both the potentiostatic thin-layer method and spectroelectrochemical coulometric titration provide information about n values and formal redox potentials. Their advantages are excellent control over the redox ratio of Ox to Red, good exclusion of oxygen, and highly precise results. They are more complex than some older techniques, but the nature of the chemical systems involved often demands such complexity. Both methods require stable reactant and product, as would any technique that relies on attainment of an equilibrium.

When one member of a redox pair is unstable, spectroelectrochemistry provides a powerful means to monitor the course of any reactions. Consider a case where the oxidized form of a couple is stable, but the reduced form is not. An OTE may generate the reactive Red from a stable solution of Ox and the subsequent reaction may be monitored spectroscopically. This approach is particularly useful because a specific component may be monitored selectively using the spectral probe, in contrast to the usual electrochemical measurement, which monitors the total current. Since the current may result from a variety of processes, faradaic and otherwise, kinetic information is often more easily extracted from a spectroscopic measurement than from a current signal. A spectroelectrochemical experiment's absorbance versus time curve has no errors from charging current or background processes and thus provides more reliable kinetic results. Several examples of the spectroelectrochemical approach applied to a variety of reactions with lifetimes ranging from many minutes to a few tens of microseconds are discussed here.

Consider the thin-layer cell discussed above, containing azobenzene in an ethanol–water solution with 0.05 M hydrochloric acid [14]. With no potential applied the spectrum is that of azobenzene (Figure 7.7). At -0.60 V versus SCE, the azobenzene is reduced to hydrazobenzene, as shown in the curves of Figure 7.7 marked 5–45 s. The curve at 45 s, nearly pure hydrazobenzene, begins to rearrange to benzidine. After a long period of time (750 s), the spectrum no longer changes and corresponds to pure benzidine. A first-order plot of the absorbance at 293 nm for the spectra taken between 25 s and 350 s is shown in Figure 7.8. The plot linearity verifies the pseudo-first-order nature of the reaction. The rate constant, determined from the slope, was 2.7×10^{-3} s^{-1}, in good agreement with other techniques.

This approach is simply an electrochemical analog of a stopped-flow experiment. The thin-layer cell allows reactant generation in under 1 min and yields a homogeneous solution after that time. Unlike stopped-flow or other chemical means, however, reactant generation in the spectrochemical experiment is very flexible in the potential used to generate the reactant. The applied potential is continuously variable over a wide range, corresponding to a continuously tunable free energy change for the reaction generating the reactive species. Furthermore, the amount of "reagent" used to initiate the reaction is carefully controlled in electrochemistry. With chemical generation, a new

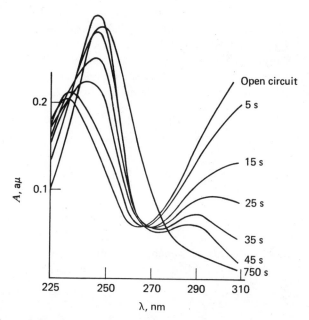

Figure 7.7 Spectra of 1 mM azobenzene in an OTTLE cell before and after a potential step from 0 to -0.6 V versus SCE. Spectra were obtained during a reaction of reduced azobenzene to form benzidine. Spectra were signal averaged 100 times using a rapid scanning spectrometer. Reprinted, with permission, from [14].

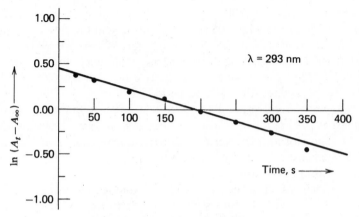

Figure 7.8 Kinetic plot of data from Figure 7.7, taken at 293 nm. Reprinted, with permission, from [14].

reagent is required for each different potential, and an excess or deficit of re-agent may occur if sufficient care is not employed.

Another example exploits the potential control allowed by electrochemistry to alter the observed decay reaction rate [15]. Rather than converting all of the stable form of a redox couple to the reactive form, only part of the conversion takes place, according to the Nernst equation. For example, if Ox is the reactive form, the applied potential in the thin-layer cell is equal to $E^{0'}$, and after the redox reaction has equilibrated only half of the available unreacted material will be in the form of Ox. The observed reaction rate will be slower by a factor of 2 for a first-order reaction. The observed rate constant is given by (8)–(10), where k_1 is the usual first-order rate constant for the decay of pure Ox.

$$\frac{dC_{Ox}^{E_{app}}}{dt} = -\left(\frac{R}{R+1}\right)k_1 C_{Ox}^{E_{app}} \tag{8}$$

$$R = \frac{C_{Ox}^{E_{app}}}{C_{Red}^{E_{app}}} = \exp\left[\frac{nF}{RT}(E_{app} - E^{0'})\right] \tag{9}$$

$$k_{obs} = \left(\frac{R}{R+1}\right)k_1 \tag{10}$$

Experimental data for this approach are shown in Figure 7.9, for the hy-drolysis of quinoneimine formed from paraaminophenol (11) and (12).

$$+2e^- + 2H^+ \tag{11}$$

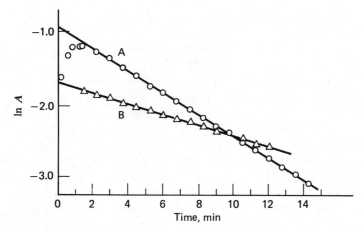

Figure 7.9   Kinetic plots for the hydrolysis of electrogenerated paraquinoneimine monitored at 270 nm. The reactive quinoneimine was generated at a gold grid electrode in an OTTLE cell in a 1.81 M H$_2$SO$_4$ solution. (A) $E_{app} = +0.75$ V versus SCE; (B) $+0.555$ V. Reprinted, with permission, from [15].

$$H_2O + \underset{O}{\overset{NH}{\bigcirc}} \xrightarrow{k_1} \underset{O}{\overset{O}{\bigcirc}} + NH_3 \tag{12}$$

In Curve A the applied potential was sufficiently positive of $E^{0\prime}$ for the conversion of all paraaminophenol to quinoneimine, and the absorbance of quinoneimine had the usual first-order decay with slope equal to the rate constant. When the applied potential was equal to $E^{0\prime}$ (Curve B), the decay was still first order, but the slope was equal to one-half the limiting value. As shown in Figure 7.10,

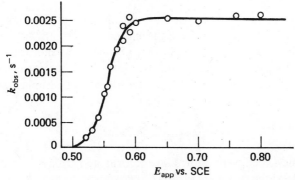

Figure 7.10   Observed rate constant determined from the slope of plots similar to those in Figure 7.9, as a function of applied potential. Solid line is theoretical for a first-order rate constant of 2.56×10^{-3} s^{-1}; points are experimental.

the observed rate constant depended on the applied potential as predicted by the Nernst equation. Both the rate constant and the $E^{0'}$ for a system of this type may be determined from the potential-dependent kinetic data.

Thin-layer spectroelectrochemical techniques are often quite useful for slow reactions, but are severely limited during rapid processes. Sufficient time, at least 20 s for practical optically transparent thin-layer electrochemical cells, must elapse for diffusional equilibration of the solution and the electrode. Therefore, these methods generally are not successful when applied to reactions lasting less than a few minutes. The methods were presented first because of their simplicity; however, the more commonly used methods are older and are based on transient measurements lasting from a few seconds down to a few micro-seconds. Since transient measurements do not occur with diffusional equi-librium, the mathematical description of the processes is more complex; but, correspondingly, the accessible time scale and information content are far greater.

Consider a planar OTE immersed in a solution containing the reduced form of a redox pair along with the usual supporting electrolyte and reference and auxiliary electrodes. An optical beam passes through the solution and elec-trode, perpendicular to its plane, and has a wavelength that Ox absorbs but Red does not. Finally, suppose a potential is applied to the OTE that is suf-ficiently positive to completely oxidize any Red at the electrode surface, resulting in diffusion-limited production of Ox. The resulting concentration versus distance profiles are time dependent with the diffusion layer extending deeper into the solution as the oxidation progresses. These profiles are determined from Fick's laws of diffusion and are valid provided Ox is stable and the time is sufficiently short to avoid convection (less than about 30 s, depending on cell geometry and other variables).

Since absorbance adds in series, the absorbance obtained with the optical beam is given by (13).

$$A(t) = \log \frac{I_0}{I} = \int_0^\infty \varepsilon_{Ox} C(x, t) dx \tag{13}$$

I_0 and I are the beam intensities before and during the application of the poten-tial; ε is the molar absorptivity of Ox at the chosen wavelength; and $C(x, t)$ is the concentration distribution of the absorber, Ox, with x being the coordinate perpendicular to the electrode plane, and t the time from initiation of electrolysis. Given a knowledge of Fick's laws, (13) can be solved using Laplace transforms [1] to yield (14).

$$A(t) = \frac{2}{\pi^{1/2}} \varepsilon_{Ox} C_{Red}^{bulk} (D_{Red} t)^{1/2} \tag{14}$$

D_{Red} is the diffusion coefficient of the *reduced* form, and C_{Red}^{bulk} is its bulk concen-tration before electrolysis began.

The absorbance is dependent on the integral of the concentration of the absorbing species, so it depends on D_{Red}, which determines the generation rate, rather than D_{Ox}, which determines the rate at which Ox diffuses away from the electrode. Consequently, absorbance is dictated not by diffusion of the electrode reaction products, but by the rate of reactant mass transport to the electrode. By analogy to Beer's law the effective optical path length is given by $(2/\pi^{1/2})(D_{Red}t)^{1/2}$. Since diffusion coefficients in liquids are small (ca. 1×10^{-5} cm^2/s) and t is rarely more than 10 s, the effective path length is short, typically less than 0.01 cm. Thus the absorbance from an OTE experiment will be 0.01 or less as large as the analogous experiment in a typical 1 cm spectrophotometer cell.

An absorbance versus time transient for generation of a stable absorber is shown in Figure 7.11. As indicated in (14), a plot of absorbance versus $\sqrt{t}$ should be linear (Figure 7.12). Nonlinearity of an A versus $\sqrt{t}$ plot implies that the charge-transfer process is not progressing at a diffusion-limited rate or that the absorbing product is undergoing a chemical reaction.

Absorbance measurement with an OTE during the diffusion-controlled generation of an absorbing species parallels the electrochemical technique of chronocoulometry. In this method, the current is integrated over the time from

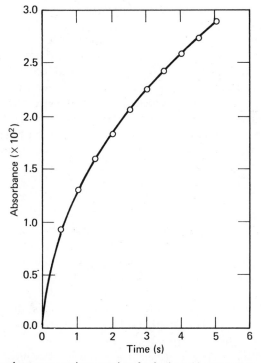

Figure 7.11   Absorbance versus time transient for ferricyanide generation at a SnO$_2$ OTE (transmission geometry) from a solution containing 5.3 mM K$_4$Fe(CN)$_6$. The line is the experimental result for diffusion-limited production of Fe(CN)$_6^{3-}$ (monitored at 420 nm). Circles are theoretical points from (14). Adapted from [16].

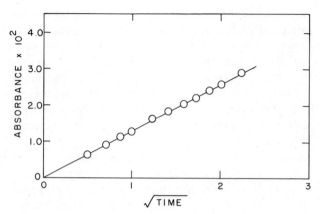

Figure 7.12 Absorbance versus $t^{1/2}$ plot for the data of Figure 7.11. Slope is equal to $(2/\pi^{1/2})\varepsilon_{Ox}D^{1/2}C_{Red}^{bulk}$. Calculated from data in [16].

the beginning of electrolysis, and the resulting charge is

$$Q = \frac{2}{\pi^{1/2}} nFAC_{Red}^{bulk}(D_{Red}t)^{1/2} \qquad (15)$$

Both absorbance and charge have the same dependence on $(Dt)^{1/2}$, since both represent integrals of the total amount of electrolysis. As pointed out by Kuwana and Winograd [1], the two are not, however, identical, because absorbance is proportional to the integrated concentration profile, while charge is related to the integrated flux of electroactive species reaching the electrode.

After an absorbing species is generated at the electrode, it may be removed by returning the electrode potential to its initial value. Following potential reversal, the absorbing species is converted back to its colorless form, and absorbance will decrease according to

$$A_{reverse} = \frac{2}{\pi^{1/2}} \varepsilon_{Ox}C_{Red}^{bulk}D_{Red}^{1/2}[t^{1/2} - (t-\tau)^{1/2}] \qquad (16)$$

At a time equal to 2τ, where τ is the forward electrolysis time, absorbance is 0.414 times its maximum value for a simple case of a stable absorber. This value is quite sensitive to complications introduced by chemical reactions of the absorber and is therefore a useful diagnostic of such processes. Note that all of the absorber is not returned to its colorless state in the double step experiment, however. Since diffusion occurs both toward and away from the electrode after the second potential step, some absorber diffuses out into the solution and does not undergo another charge-transfer reaction.

A double potential step spectroelectrochemical experiment is shown in Figure 7.13 for a stable electrogenerated absorber. In Figure 7.14 the same data are plotted in a more diagnostically useful form. If such a plot is linear with slope equal to that of the forward A versus $\sqrt{t}$ plot, the absorber is stable and its

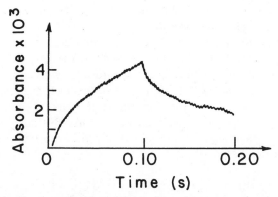

Figure 7.13 Absorbance versus time transient for a double step experiment to the diffusion-limited potential for generation of ferricyanide (5 mM), using a SnO$_2$ OTE in transmission mode. at $t=0$, potential was stepped from 0.0 to 1.5 V versus SCE, at $t=0.1$ s, it was returned to 0.0 V. Reprinted, with permission, from [26].

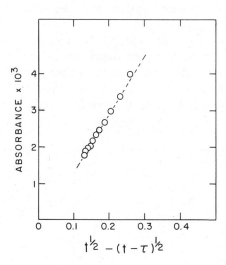

Figure 7.14 Plot of absorbance versus $t^{1/2} - (t-\tau)^{1/2}$ for the reverse step of the experiment depicted in Figure 7.12. The slope of the plot is equal to that for the line in Figure 7.12. Calculated from data in [26].

generation is carried out at a diffusion-controlled rate. Deviation from this ideal behavior indicates either a chemical reaction or some other complication.

The transient spectroelectrochemistry techniques discussed so far involve diffusion-controlled charge-transfer reactions, in which the oxidation or reduction reaction at the electrode occurs instantaneously with its rate controlled only by mass transport to the surface. However, the approach is not limited to this case, and spectroelectrochemistry may be used to determine heterogeneous charge-transfer rates [16, 17]. If an OTE is immersed in a solution of a reduced species, such as ferrocyanide ion, and its potential set at a value significantly positive (>400 mV) of the $E^{0'}$ for ferro/ferricyanide, absorbance caused by ferricyanide will have the usual $\sqrt{t}$ dependence of (14). The ferrocyanide is instantly converted to ferricyanide at the electrode surface, and the absorbance has the shape expected for diffusion-controlled generation of a stable absorber. Alternatively, suppose that the potential applied at the OTE is equal to $E^{0\prime}$, such that the Ox/Red ratio at the surface should ideally be equal to unity, as predicted from the Nernst equation. If the heterogeneous charge-transfer rate is fast relative to the time scale of the experiment, the Nernst equation applies, and absorbance is exactly half that predicted for the diffusion-controlled case [17]. Furthermore, assuming rapid charge transfer, the absorbance has the form of (17) for any applied potential.

$$A(t) = \frac{2}{\pi^{1/2}} E_{Ox} C_{Red}^{bulk} (D_{Red}t)^{1/2} \left(\frac{R}{R+1} \right) \tag{17}$$

$$R = \exp\left[\frac{nF}{RT} (E_{app} - E^{0\prime}) \right] \tag{18}$$

Absorbance always equals a constant times the diffusion-limited value, the constant equaling the fraction of total electroactive species in the absorbing form.

If heterogeneous charge transfer is slow on the experimental time scale, however, (17) does not apply, and the "constant" is time dependent. At long times, absorbance eventually reaches the value of (17); at short times absorbance lags the value for rapid charge transfer. This process has been described in detail for both irreversible [16] and quasi-reversible [17] charge-transfer reactions,

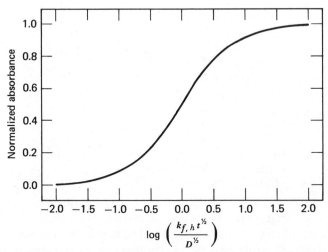

Figure 7.15 Theoretical working curve for the determination of heterogeneous charge-transfer rate constants by transmission spectroelectrochemistry at an OTE. Normalized absorbance is the ratio of the observed absorbance to that for the diffusion-limited case. Reprinted, with permission, from [16].

along with procedures for determining charge-transfer rate constants from absorbance data. Figure 7.15 presents kinetic information for the irreversible reaction in which the reverse of a charge-transfer reaction is assumed not to occur. Normalized absorbance A_n is the observed absorbance at some experimental time divided by the absorbance at the same time expected for the diffusion-limited case. The diffusion-limited value is easily determined from an experiment where the applied potential is sufficiently far from $E^{0\prime}$ to assure diffusion control. The abscissa of Figure 7.15 is a dimensionless rate parameter that includes the rate constant and time. Once an experimental A_n is known the rate parameter $k_{f,h}t^{1/2}/D^{1/2}$ may be determined from the curve. With a known time and diffusion coefficient, the heterogeneous rate constant is determined easily. Figure 7.16 demonstrates an excellent fit between theoretical and experi-

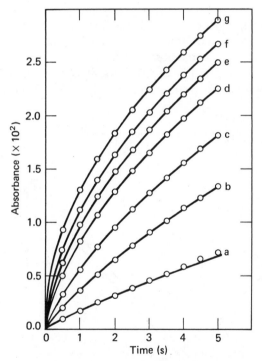

Figure 7.16 Absorbance–time behavior for potential step electrooxidation of 5.34 mM ferrocyanide at tin oxide OTE at pH 7. Solid lines are experimental transients; open circles are theoretical responses. The overpotential is η, defined as $E_{app} - E^{0\prime}$, and transients were obtained for a range of η values. In this figure, they were: Curve a: $\eta = 20$ mV, theoretical response calculated for $k_{f,h} = 4.02 \times 10^{-4}$ cm/s; Curve b: $\eta = 70$ mV, theoretical response calculated for $k_{f,h} = 1.05 \times 10^{-3}$ cm/s; Curve c: $\eta = 120$ mV, theoretical response calculated for $k_{f,h} = 2.00 \times 10^{-3}$ cm/s; Curve d: $\eta = 170$ mV, theoretical response calculated for $k_{f,h} = 4.57 \times 10^{-3}$ cm/s; Curve e: $\eta = 220$ mV, theoretical response calculated for $k_{f,h} = 7.13 \times 10^{-3}$ cm/s; Curve f: $\eta = 270$ mV, theoretical response calculated for $k_{f,h} = 1.41 \times 10^{-2}$ cm/s; Curve g: $\eta = 820$ mV, and is the diffusion-limited absorbance transient. Reprinted, with permission, from [16].

mental absorbance transients, and Figure 7.17 shows the expected potential dependence of rate constants determined in this manner. Other comparisons demonstrate compatibility with rate constants obtained from different techniques.

Although theoretically more complex, a similar approach may be used to examine quasi-reversible charge-transfer reactions, that is, those in which the reverse charge-transfer reaction must be considered. Equally good results were obtained, thus confirming theory and experimental technique. The spectroelectrochemical approach offers freedom from complications caused by other faradaic or nonfaradaic processes. Since the optical beam monitors only the absorbing species, absorbance reflects only the charge transfer associated with this species. Conventional electrochemical techniques monitor current and may erroneously determine a rate constant from the current because of interference from extraneous processes. As with the spectroelectrochemistry applications discussed earlier, the selectivity of the optical probe provides the major advantage over conventional methods. Additional advantages include freedom from errors caused by background faradaic processes and the comparatively small effects of charging current on the observed response.

In addition to potential step experiments, it is often useful to scan the potential while monitoring absorbance with the OTE. This method allows

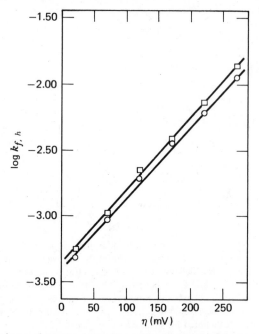

Figure 7.17   Dependence of spectroelectrochemically determined heterogeneous rate constants (□) and chronocoulometric (○) rate constants on overpotential. Reprinted, with permission, from [16].

assignment of spectral features to particular electrochemical processes (Figure 7.18). At pH 2, orthotolidine is oxidized by a two-electron process to a quinoid form with $\lambda_{max} = 437$ nm. In Curve 4, Figure 7.18, absorbance at 437 nm increases after the onset of oxidation current, indicating that the absorber results from the current-generating process. At pH 4 orthotolidine undergoes two sequential one-electron oxidations, with the intermediate absorbing at 630 nm. By comparing Curves 2 and 3 it is apparent that the 630 nm absorber is generated first at the first peak in Curve 2. The 437 nm quinoid absorber is formed next, as indicated in Curve 3. The value of simultaneous scanning and optical monitoring in this case is mainly qualitative, in that the connection between a voltammetric peak and the generation of an absorber is established. Therefore the spectral probe may provide additional information about the origin of the electrochemical response.

A more quantitative relationship for absorbance during a potential scan exists, based on the first derivative of an absorbance versus time curve [18]. Absorbance at an OTE has the same time dependence as faradaic charge, since both reflect the integrals of the electrolysis current. Current is the time derivative of charge, so the first derivative of a chronocoulogram with respect to time should yield a current versus time transient. Similarly, the time derivative of absorbance should be analogous to the current, and should have the same shape as a current versus time curve. In the case of a potential sweep experiment, the time derivative of absorbance should have the shape of a voltammogram and

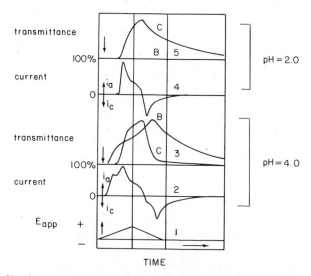

Figure 7.18 Simultaneous monitoring of transmittance and current at an OTE during oxidation of orthotolidine. Curve 1, potential sweep, 20 mV/s; Curve 2, current observed (anodic up) at pH 4; Curve 3, transmittance, $B = 630$ nm, $C = 437$ nm; pH 4.0; Curve 4, current, pH 2.0; Curve 5, transmittance pH 2.0, $B = 630$ nm, $C = 437$ nm. Note absence of the 630 nm chromophore at pH = 2. Adapted from [1], Fig. 7.14.

should contain comparable information about $E^{0'}$ values, coupled chemical reactions, and so on.

This argument is validated in Figure 7.19, for the generation and monitoring of trianisylamine cation radical. Curve B is the absorbance measured during the potential scan, with a scan direction reversal at 740 mV. Curve A is a conventional voltammogram for the same system, and Curve C is the derivative of Curve B. The first absorbance derivative has the shape of a normal voltammogram and may be used in the same way to determine $E^{0'}$, concentration, and so on. The optical signal has a different scan rate dependence, however, with the peak absorbance signal being given by (19), for the generation of colored Ox from colorless Red.

$$\left(\frac{dA}{dE}\right)_{peak} = -0.0881 n^{1/2} \varepsilon_{Ox} D_{Red}^{1/2} C_{Red}^{bulk} v^{-1/2} \qquad (19)$$

where C_{Red}^{bulk} is in mol/L and v is scan rate in mV/s. Unlike the voltammetric

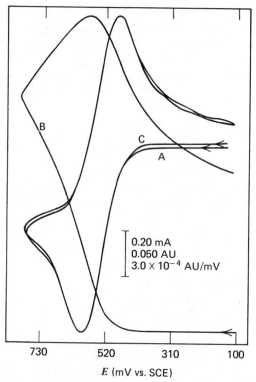

730   520   310   100

E (mV vs. SCE)

Figure 7.19 Simultaneously acquired current–potential (A) and absorbance–potential (B) responses for the cyclic linear sweep oxidation of 3.16 mM trianisylamine at a platinum OTE in acetonitrile/0.100 M TEAP at $v = 551$ mV/s. Trace C is the derivative of Trace B with respect to scan potential. AU = absorbance unit; arrows indicate initial scan direction. Reprinted, with permission, from [18].

response, which increases linearly with $v^{1/2}$, the optical signal decreases with increasing scan rate.

As with all spectroelectrochemical methods, the advantage of this technique is freedom from extraneous current response caused by charging current or undesired faradaic processes. Furthermore, response depends only on those species that absorb at the chosen wavelength, whereas the current response may have contributions from several species. Since the already small optical signal becomes smaller as the potential is swept more rapidly, the optical technique is usually unsuitable at high scan rates. Currently the optical technique appears quite complementary to conventional voltammetry and will be particularly useful when its selectivity is exploited.

The methods described so far represent the most common OTE techniques involving UV–vis transmission and remain highly useful in modern spectroelectrochemistry. Procedures and instrumentation required for OTE experiments in the UV–vis transmission mode will be discussed later.

2.1.2   Spectroelectrochemical Monitoring of Reaction Kinetics Using the OTE

In addition to the applications involving identification and quantification of electrogenerated species, spectroelectrochemistry is extensively used to examine reactions of species associated with charge transfer. Most commonly observed is the decay of a reactive species generated at an OTE usually under diffusion-controlled conditions. For example, during generation and decay of 9,10-diphenylanthracene cation radical ($DPA^{+\cdot}$) in wet acetonitrile [19], $DPA^{+\cdot}$ may be generated electrochemically at an OTE and monitored with an optical beam at 653 nm. In the absence of water DPA^{+} is stable and, as expected, absorbance increases linearly with $t^{1/2}$. In the presence of water DPA^{+} undergoes a pseudo-first-order reaction to form products that do not absorb at 653 nm. Clearly the absorbance will be lower than the usual diffusion-controlled value when the reaction proceeds, and the spectroelectrochemical response may be analyzed to determine the rate constant by comparing the absorbance in the presence of a reaction to that in its absence. This approach is extensively used for various reactions and mechanisms, ranging from first-order reactions of small molecules [19] to charge-transfer processes of large biological components such as cytochrome c [20]. The many literature examples have been reviewed often [1, 2, 4, 5] and are not listed here; however, several features of the approach are discussed.

Compared to conventional kinetic techniques such as stopped-flow or ordinary spectrophotometry, spectroelectrochemistry has three major advantages. First, if a reactive species is a member of a redox pair, it may be conveniently generated at the electrode. With other methods, it is often necessary to start with stable reactants that will exist long enough in solution to start the kinetic run after mixing. Since spectroelectrochemical methods involve *in situ* generation and subsequent monitoring, such stability is unimportant. Second, the time scale of spectroelectrochemistry is very wide, ranging from microseconds to minutes or hours. There are few methods that operate reliably over such a wide range. Third, electrochemistry provides good control over reactant generation, since the process usually proceeds under controlled-potential conditions. Compared to pulse radiolysis or flash photolysis, there is less uncertainty about what is generated by the potential pulse initiating the reaction.

On the other hand, these advantages carry with them a price in terms of complexity and generality. A reaction studied by spectroelectrochemistry occurs in the diffusion layer, except for the slow equilibrium methods. This layer is inherently inhomogeneous, with large variations in the reactant concentrations with distance from the electrode. Such concentration variations complicate the kinetics, and the behavior of the absorbance must be predicted with computer simulations. Determining the rate constant from absorbance versus time transients is not always simple, and reasonable care must be exercised to avoid pitfalls. A second deficiency of spectroelectrochemistry is its generality. While many different reactions have been successfully studied with this method, it is limited to redox reactions involving a fairly strong chromophore. Therefore the number of reactions that are amenable to spectroelectrochemical monitoring is not infinite, but the method is not alone in this respect. Spectroelectrochemistry is particularly useful for examining fast reactions such as charge transfer, or electrochemically generated unstable reactants.

In UV–vis transmission at an OTE, three experiments are commonly used for examining reaction rates and mechanisms: single potential step, double potential step, and open circuit relaxation. In all three cases a theoretical absorbance versus time transient is calculated from analytical solutions or from finite difference simulation techniques [21]. The theoretical calculations take into account the effects of the inhomogeneous concentration distribution of the observed response. By comparing the experimental and theoretical transients, the rate constant may be determined.

Consider a solution containing 9,10-diphenylanthracene and water in acetonitrile, placed in a spectroelectrochemical cell with an OTE placed in a spectrophotometer set at 653 nm. With no potential applied, the system is stable, and no absorbance is observed. When a potential is applied that generates DPA^+ at a diffusion-controlled rate, the absorbance transient shown as points in Figure 7.20 is observed. Initially, the curve has the parabolic shape predicted for a stable species, indicated by (14). As time progresses, the pseudo-first-order reaction with water occurs and the absorbance is lower than the $t^{1/2}$ dependence for a stable species. Figure 7.20 also shows the computer simulation for a rate constant of 0.096 s^{-1}. A similar examination of the pyridination of diphenylanthracene with single potential step spectroelectrochemistry was carried out by Blount [22].

An alternative to the single potential experiment is the open circuit relaxation method, in which the electrode is disconnected after some reactant is generated by a potential step to the diffusion limit. Absorbance is monitored both during and after the generation period, and a rate constant is determined from the shape of the curve. It has been shown that for a simple first-order decay reaction, the

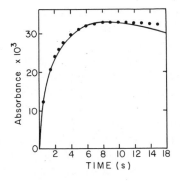

Figure 7.20 Single potential step absorbance versus time transient for the generation of diphenylanthracene cation radical in wet acetonitrite ($[H_2O] = 1.8\ M$). Solid curve is theoretical for a pseudo-first-order reaction with $k = 0.096\ s^{-1}$. Points are experimental for monitoring of DPA^+ at 653 nm. Deviation of theory and experiment at times greater than 10 s was attributed to convection. Reprinted, with permission, from [19].

absorbance following cessation of electrolysis decays exponentially, exactly as it would for a homogeneous solution [19]. The decay process has a half-life of $\ln(2)/k$, where k is the first-order rate constant. This simple result occurs only in first-order processes, and the decay curve for a second- or higher-order reaction is not simply related to the rate constant [23].

The third common method for examining electrochemically initiated reactions is double potential step spectroelectrochemistry. The potential is first stepped to the diffusion limit for absorber generation, then returned to a value where the absorber is converted back to starting material. If the absorber reacts to form a nonabsorbing product, the shape of the absorbance versus time transient is altered and kinetic information can be derived from the curve. Experimental examples include studies of the reactions of several heterocyclic radical ions [24], and reduction of electrogenerated ferricyanide by ascorbate [25, 26].

Single and double potential step and open circuit relaxation techniques are commonly used with transmission mode spectroelectrochemistry, and their counterparts for other geometries are discussed in sections on other optical arrangements. Before we leave the subject, several points about spectroelectrochemical examination of kinetic processes should be noted. First, the rate data derived from these methods are often unobtainable from other methods. The speed of spectroelectrochemistry, combined with its ability to generate and monitor unstable reactants, often makes it the method of choice. Second, different mechanisms often have similar responses when examined by spectroelectrochemical methods. The technique sometimes makes subtle mechanistic distinctions difficult; and if possible, mechanistic conclusions should be supported by independent experimental results. Third, once the mechanism is established, spectroelectrochemistry provides accurate rate constants over a wide variety of conditions and reactants. It is very useful for determining the effects of pH, reagent structure, or reaction conditions on the reaction of interest.

2.1.3   External Reflection Spectroelectrochemistry

In addition to direct transmission of the optical beam through an OTE there are several alternative arrangements for UV–vis absorption measurements at an electrode. Three techniques are discussed, each with a quite distinct optical geometry (Figure 7.21). The first is external reflection, often called specular reflectance, in which the optical beam reflects directly off the electrode. The intensity of the reflected beam indicates the presence of absorbers generated by an electrochemical process. A second approach is internal reflection at an OTE, where total internal reflection occurs at the OTE–solution interface. A portion of the beam enters the solution and interacts with the diffusion layer, permitting spectroscopic measurement. Finally, the beam may pass parallel to a planar electrode and is then analyzed optically to obtain spectral information about electrogenerated materials. In all cases, the end result is UV–vis spectral information about identities, concentrations, and lifetimes of solution species involved in electrochemical processes. Despite their common objectives, the approaches vary greatly in sensitivity, ease of use, and time resolution.

External reflection geometry has several similarities to the OTE transmission experiment, because in both cases the optical beam passes through the entire diffusion layer (Figure 7.21). At normal incidence, the absorbance for an electrogenerated absorber monitored with external reflection is exactly twice that for OTE transmission, assuming a constant electrode reflectivity. While reflectivity indeed changes with potential [27], the effect is quite small and usually negligible. Therefore the OTE equations apply to external reflection at normal incidence, with doubled absorbance.

The advantage of external reflection is that it does not require an optically transparent electrode. Bulk metal or carbon [28] electrodes are acceptable if they are polished sufficiently well to reflect the beam. In addition, bulk metal electrodes are usually more robust than the metal film or grid OTEs and therefore much longer lived. The requirement for high and constant reflectivity may

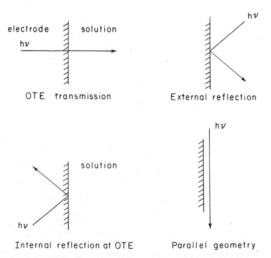

Figure 7.21 Alternative geometries for spectroelectrochemistry. Hashed lines represent electrode–solution interface.

be a problem with external reflection spectroelectrochemistry, since careful electrode fabrication is required. Also, external reflection experiments are slightly more complex than those with an OTE, since most spectrometers are designed for transmission rather than reflection geometry.

Enhancements of external reflection geometry include the use of a thin-layer cell [29], glancing incidence [28, 30], multiple reflection [31, 32], modulated potential techniques [33, 34], and the use of microelectrodes [35, 36]. The glancing incidence approach exploits simple geometry to improve sensitivity (Figure 7.22). The absorbance for generation of absorbing Ox from colorless Red is given by (20), which indicates enhanced sensitivity by a factor of $1/\sin \alpha$ over the normal incidence case, or $2/\sin \alpha$ compared to an OTE transmission case.

$$A = \left(\frac{2}{\sin \alpha}\right) \frac{2}{\pi^{1/2}} \varepsilon_{Ox} C_{Ox}^{bulk} (D_{Red})^{1/2} \tag{20}$$

The equation is valid for an enhancement of about 100 over OTE [28, 30] and is therefore useful for weak absorbers or low concentrations. Since the enhancement is a simple geometric effect, all the equations for the OTE case are used, except absorbance will be up to 100 times larger, all else being equal.

An enhancement of similar magnitude is obtained with multiple reflections of the beam off two parallel electrodes [31]. Figure 7.23 shows an arrangement

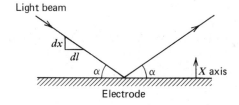

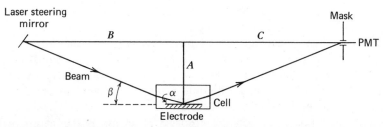

Figure 7.22   Experimental arrangement for external reflection spectroelectrochemistry. Beam is from either a laser or a continuum source–monochromator combination. PMT is a photodetector connected to an appropriate data monitor such as a computer or oscilloscope. B and C are 10–70 cm, A is 0.3–3.0 cm, electrode length along optical axis is 0.1 to 3 cm (not to scale). Reprinted, with permission, from [30].

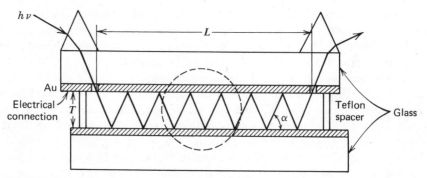

Figure 7.23 Electrode and cell arrangement for multiple external reflection spectroelectro-chemistry. Hashed areas represent gold electrodes, and beam enters cell through prisms. Drawing is not to scale. Reprinted, with permission, from [31].

for multiple reflections of a He–Ne laser beam from two gold film electrodes. The number of reflections within the thin-layer cell is calculated from the geometry of the system, and enhancements of 100–1000 over the OTE case are observed. In addition to possessing high sensitivity, the method may be applied to solutions that are initially colored. The thin-layer cell and its many reflections assure that the beam does not pass through a long length of "inactive solution"; thus a major fraction of the solution sampled by the beam is electrolyzed. This is not the case for the usual OTE experiment in which a majority of the solution sampled by the beam is unaffected by the electrode. A difficulty with the multiple reflection experiment is its requirement for a laser source. Less collimated sources lead to beam divergence in the cell and poor quantitative accuracy. Additionally a multiple reflection method is highly sensitive to electrode reflectance changes with potential, an effect amplified by the number of reflections.

Modulated specular reflectance is useful for examining adsorbed species, and also for examining short-lived solution components generated at an electrode [33, 34]. Typically the potential is a square wave oscillating between two potentials on either side of $E^{0\prime}$ for the redox system of interest. The magnitude of the potential is usually sufficient to drive the oxidation or reduction reactions to their diffusion limits. Such an applied potential generates a steady-state concentration of an electrogenerated absorber, with the magnitude of the concentration being determined by the lifetime of the species. Short-lived absorbers have low but finite steady-state concentrations, while the concentration of stable species slowly increases with time. For any reactive absorber, the decay rate is at some point equal to the generation rate and a constant absorbance is observed. Under this condition the absorber spectrum is scanned, slowly if desired, since the concentration of the reactive species is at steady state.

A spectrum of radical anion obtained from CO_2 reduction is shown in Figure 7.24. Note the spectrum agrees well with that obtained from a non-electrochemical method. While this modulation technique is suitable for

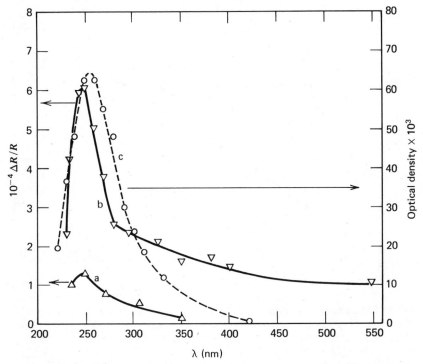

Figure 7.24 Modulated specular reflectance spectra of saturated CO_2 in water at a lead cathode. Curve a, modulation from -1.0 to -1.6 V; Curve b, -1.0 to -1.8 V at 30 Hz. Curve c is a spectrum of CO_2 obtained by another technique. Reprinted, with permission, from [28].

obtaining spectra, it is difficult to use for obtaining kinetic information since it involves a steady state. The spectrum is used to confirm the identities of intermediates; transient experiments are then conducted at fixed wavelength to observe decay kinetics. Since the change in electrode reflectance varies with both potential and wavelength, it is important to obtain a background spectrum separate from the species of interest.

A very small electrode at which a single external reflection occurs offers significant advantages [35, 36]. Consider an optical beam reflected from a 25 μm diameter platinum wire. The experiment is performed as usual, and absorbance is measured as a function of time following the potential step generating an absorber. The resulting absorbance versus $t^{1/2}$ curve is shown in Figure 7.25. The curve is nearly linear, as expected for the planar case, with the curvature caused by cylindrical diffusion. The microelectrode technique offers two major advantages when compared to the usual large planar electrodes. First, electrode capacitance is approximately 1000 times less than for a large electrode; therefore, the time constant for cell response is very short—on the order of 0.5 μs. The resulting transient response is excellent, with theoretically expected absorbance obtained 4 μs after the initiation of electrolysis. Second, the small

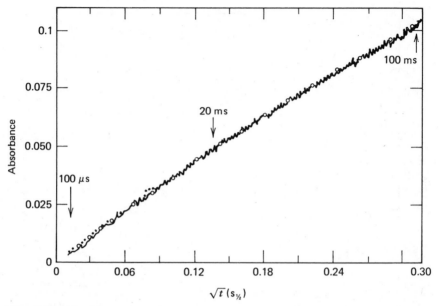

Figure 7.25   Absorbance versus time plots for external reflection spectroelectrochemistry at a 12.5 μm radius Pt wire microelectrode. Open circles are theory, with slight curvature caused by cylindrical diffusion. Solid line is an experimental transient for a three-electrode potentiostat, points are for a two-electrode system. Solution contained 3.5 mM orthodianisidine in 1 M H$_2$SO$_4$, which is oxidized to a quinoid form absorbing at 515 nm. Reprinted, with permission, from [36].

electrode area results in a very small electrolysis current, eliminating the need for a three-electrode potentiostat. Typical electrolysis of currents less than 1 μA result in negligible ohmic potential error when using a two-electrode cell. Figure 7.25 includes A versus $t^{1/2}$ transients recorded with both three- and two-electrode arrangements, which are indistinguishable from theory. Since there is no electronic feedback required, two-electrode arrangements are not only simpler, but also more stable. Although the microelectrode experiment is conceptually identical to any single external reflection experiment, it has special features that should be of value when studying fast reactions, including transient response and increased duty cycle.

Examples of reactive systems studied by external reflection spectroelectro-chemistry include the electrophilic [37] and charge-transfer [38] reactions of chlorpromazine cation radical and the reduction of carbon dioxide [39]. As with OTE experiments, experimental response and simulated curves are compared to obtain rate constants. Figure 7.26 illustrates the compatibility of theory and experiment for a charge-transfer reaction. Normalized absorbance is defined as the ratio of absorbance in the presence of reducing agent to that in its absence. In Figure 7.26 the reducing agent was dopamine, which reacted with electrogenerated chlorpromazine radical, monitored at 525 nm. The wide sensitivity range of external reflection techniques is an advantage for kinetic

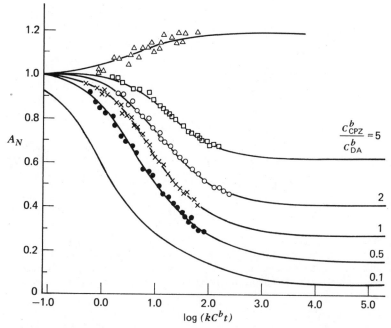

Figure 7.26 Theoretical and experimental plots of normalized absorbance versus a rate parameter for the oxidation of dopamine by electrogenerated chlorpromazine cation radical. In all but the top curve, the cation radical was monitored at 525 nm, while in the upper curve oxidized dopamine was monitored at 400 nm. Reprinted, with permission, from [38].

studies, allowing examination of a wide concentration range of reactants and slowing second-order reactions by operating at low concentrations.

2.1.4   Internal Reflection Spectroelectrochemistry (IRS)

Conceptually and experimentally internal reflection geometry is more complex than external reflection geometry, but its special features offer an attractive alternative for many applications [1, 40–44]. Consider a metal film OTE on a glass substrate, with the optical beam approaching the OTE from within the glass (Figure 7.27). At the critical angle, the beam will be totally internally reflected and will not enter the bulk of the solution. However, the electric field of the light beam does penetrate the solution to some degree and forms an evanescent wave. The penetration depth is on the order of 10^{-5} cm, a value highly dependent on the refractive indices of glass, metal film, and solution, and on the wavelength [44]. The evanescent wave may be absorbed by electrogenerated materials, and absorbance may be measured using the internally reflected light.

The electric field of the light decreases to $1/e$ of its incident value at a distance δ from the electrode surface (21) [1]:

$$\delta = \frac{\lambda_0}{4\pi I_m(n_c \cos \phi_c)_{\text{solution}}} \tag{21}$$

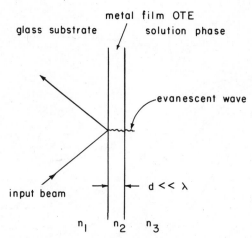

Figure 7.27 Optical arrangement for internal reflection spectroelectrochemistry at a metal film OTE. The refractive indices of substrate, metal film, and solution are n_1, n_2, and n_3. If the film thickness is much smaller than the wavelength, the metal refractive index does not markedly affect penetration depth. Spectroscopic information about the solution is derived from absorption of the evanescent wave.

where λ_0 = wavelength in vacuum;

n_c = complex refractive index in solution;

ϕ_c = complex angle in solution.

At distances greater than about 2000 Å the electric field is nearly zero, and at these distances the presence of absorber no longer affects the absorbance. Therefore the IRS method samples only the region within less than one wavelength from the electrode. Absorbance of internally reflected light for a stable absorber (Ox in this case) is given by (22) [1]. An effective path length, b_{eff}, is analogous to Beer's law and is proportional to the penetration depth. Once the diffusion layer extends beyond the penetration depth, absorbance is constant since the absorber concentration in the region sampled by the internally reflected light is no longer changing. Absorbance at such times is given by (23).

$$A_{Ox}(t) = \varepsilon_{Ox} b_{eff} C_{Red}^{bulk} \left(\frac{D_{Red}}{D_{Ox}} \right)^{1/2} [1 - \exp(a^2 t)\mathrm{erfc}(a t^{1/2})] \qquad (22)$$

where $a = D_{Ox}^{1/2}/\delta$;

$b_{eff} = \delta N_{eff}$ with N_{eff} a geometric factor;

erfc = the error function complement.

$$A_{Ox}(t > 1 \text{ ms}) = \varepsilon_{Ox} b_{eff} C_{Red}^{bulk} \left(\frac{D_{Red}}{D_{Ox}} \right)^{1/2} \qquad (23)$$

Steady-state absorbance is small because of the short penetration depth and is reached quickly, typically within 1 ms. Therefore IRS is a fast but low sensitivity method that samples only the interfacial region. Figure 7.28, Curve a, shows absorbance versus time curves for IRS experiments for generation of a stable chromophore. The addition of charge injection electronics and a high power potentiostat achieved time resolution of 4 μs [41]—the best transient response reported for potential step methods, with or without optical monitoring. Figure 7.28 also shows a transient obtained for the same absorber in the presence of a second-order reaction. Analysis of the reduction in absorbance under conditions where a reaction occurs yields a rate constant. The fact that the sampled region can be filled or dumped quickly is of great value when time averaging, since initial conditions are easily restored between time-averaged runs. The price one pays for these advantages is significant mathematical complexity because of the evanescent wave, and a rather sophisticated experimental apparatus. In addition, it is difficult to obtain spectra of electrogenerated materials, because the penetration depth and other optical parameters are highly wavelength dependent, leading to significant distortion of any spectra obtained. At fixed wavelength, however, IRS has yielded impressive results.

2.1.5   Spectroelectrochemistry With the Beam Parallel to the Electrode Surface

A more recent and therefore less common alternative to external and internal reflection involves an optical beam passing parallel to the electrode surface [45–48] (Figure 7.29). Thus optical path length is determined by the electrode

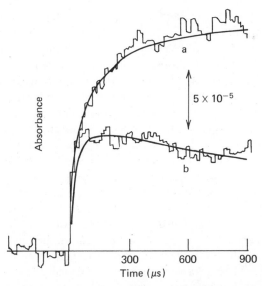

Figure 7.28   Absorbance–time transients for IRS of methyl viologen dication reduction, monitored at 605 nm. Curve a is a diffusion-limited reduction of MV^{2+} to its cation radical; Curve b is for reduction to MV^0. Solid lines are theoretical transients; noisy curves are experimental. Reprinted, with permission, from [41].

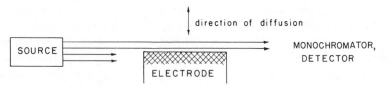

Figure 7.29   General arrangement for parallel geometry with beam passing parallel to a planar electrode surface. Species diffuse across the beam, and various optical elements may precede the monochromator to select portions of the beam.

dimension along the optical axis, not by diffusion-layer thickness. This dimension may be quite long (up to 1 cm) compared to the diffusion-layer thickness (0.01 cm or less) leading to high sensitivity. Such sensitivity requires more complex experimental apparatus, since the beam must be analyzed optically after leaving the electrode region.

In the earliest examples of the parallel geometry the variable to be measured is refractive index rather than absorptivity [48]. In an interferometric experiment, for example, a monochromatic collimated beam passes parallel to a planar electrode, and portions of the beam are phase shifted because of the local refractive index gradient in the diffusion layer. The concentration gradient within the diffusion layer affects different portions of the beam to different degrees. An interference pattern constructed from the beam changes during the diffusion process, and the change is used to learn about diffusion or other types of mass transport. Although the process is mathematically complex, the theory has been worked out and the interference pattern is predictable if the concentration profile is known.

Interferometry is attractive because of the spatial information contained in the interferogram, and unlike almost all other spectroelectrochemical methods interferometry depends on the distribution of electrogenerated material as a function of a distance away from the electrode surface rather than on its integral. Therefore, in principle, interferometry may determine concentration profiles for complex mass-transport conditions such as those in large-scale electrolytic processes. The applications of interferometry or any refractive index technique to most problems of interest to chemists have been limited by several factors. First, refractive index changes are small, approximately 0.001 M^{-1}, and difficult to detect for the millimolar concentrations usually employed in electrochemistry. Second, the method is nonselective, since any dissolved species may change the solution refractive index. Third, spectral information is difficult to obtain from a refractive index as a function of wavelength, so identification of electrogenerated species is difficult. Fourth, examining a single component of a mixture is almost impossible because of the complex dependence of refractive index on all components present. A reaction involving an electrogenerated species produces a hopelessly complex refractive index change and interferogram. In conclusion, refractive index methods are actively pursued for examining mass transport, but their applications to chemical problems involving reactions or analysis are nonexistent.

Several methods with parallel geometry are available that are based on absorptivity rather than refractive index. The simplest involves a planar electrode positioned parallel to the beam from a commercial spectrophotometer [46]. A small fraction of the beam passes by the electrode, sampling the region within about 1 mm of the electrode surface. As absorber is generated at the electrode, it diffuses into the beam and is detected by the spectrophotometer in the usual manner. A typical absorbance versus time curve for such an experiment is shown in Figure 7.30. Although it was demonstrated empirically that absorbance increases linearly with the square root of time, such a conclusion is not theoretically obvious. High sensitivity was demonstrated for the method, with 10^{-8} M levels detected for strong chromophores. Thus the high sensitivity expected for the long (1 cm) path length is realized, but countered by the indefinite width of the region sampled by the beam. The electrogenerated species diffuses across a beam of uncertain width and absorbance cannot be predicted quantitatively. Furthermore, the beam is always much wider than the diffusion layer, so absorbance grows slowly and several minutes are required to reach a constant value.

A more recent approach employed a slit positioned at the trailing edge of the electrode that allowed only a well-defined portion of the beam to pass through to the detector [45, 49]. As shown in Figure 7.31, the slit samples a particular region of the diffusion layer, and the optical path length equals the electrode length along the optical axis. Although experimentally more complex, the method has two significant advantages. First, the distance away from the electrode surface that is monitored is accurately defined by the position of the slit, which may be as close as 25 μm. Figure 7.32 shows theoretical and experimental absorbance curves for particular distances, demonstrating the rise and fall of absorber during a double potential step experiment. Absorbance at a particular distance is easily calculated from the diffusion equations and reaches

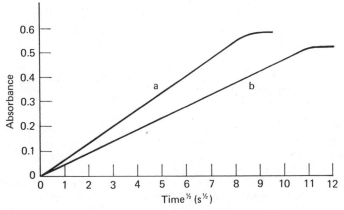

Figure 7.30 Absorbance–time curves for parallel geometry with a spectrophotometer beam passing parallel to a planar electrode, with no slit to define sampling distance. Curve a, nitroferrin; Curve b, o-tolidine. Reprinted, with permission, from [46].

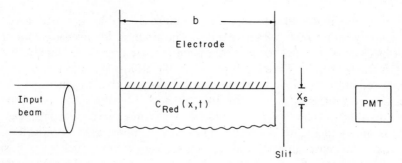

Figure 7.31 Parallel geometry employing a slit to select a particular region of the diffusion layer. C_{Rep} is the electrogenerated absorber; x_s is the distance away from the electrode selected by the slit. The input beam in this case is a laser, and therefore monochromatic. Wavy line represents the approximate boundary of the diffusion layer.

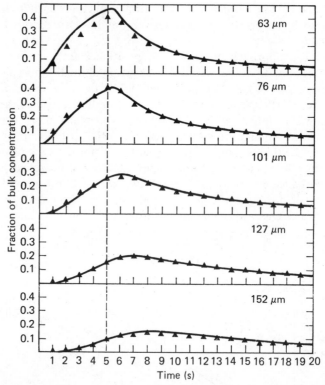

Figure 7.32 Observed and theoretical absorbance transients for an experiment using the approach of Figure 7.31. At $t = 0$, the blue tetramethylparaphenylene diamine cation radical was generated at a diffusion-controlled rate, and at $t = 5$ s the potential was returned to a value that reduced the radical back to its colorless precursor. The radical was monitored with a He–Ne laser, and the results are expressed as a fraction of bulk concentration, calculated from Beer's law. Dimensions near the plots are the distances between the electrode and the slit center. Points are experimental, curves are theoretical, calculated from Fick's laws. The length of the electrode along the optical axis (0.45 cm) led to high absorbance values in the range of 0.05–2.0 units. Reprinted, with permission, from [45].

626

a constant value at long times when the reactant at that distance is completely converted to absorber. This limiting value is governed by Beer's law, using the path length of the electrode along the optical axis. Also, the time required to reach this limiting value may be quite short (ca. 300 ms), since the sampled region may be close to the electrode. The second advantage is the new information about concentration versus distance profiles. The profile shape for a given species is a function of mass transport and any chemical reactions and provides insight into processes unavailable from techniques supplying only the integrated absorber concentration. Unfortunately the slit technique has fairly low resolution, approximately 25 μm; therefore, distances closer than this value cannot be monitored effectively. The method is best applied to experiments having time scales of greater than 0.5 s to allow the electrogenerated species time to cross the region sampled by the slit.

A simpler and more convenient arrangement using parallel geometry [50] retains the advantage of long path length, but is used with a conventional spectrophotometer. A thin-layer cell with one wall composed of the working electrode was interfaced to a spectrometer with fiber optics. The fibers directed the beam down the interior of the thin-layer cell, with light passing parallel to the electrode. The optical path length was 1 cm, yet the thin-layer arrangement permitted a small cell volume of only 20 μL. Furthermore, electrolysis was completed within the cell in about 30 s. The fiber optic cell is a simple alternative to the slit arrangement, since it requires no spectrometer modification and is easily used with a continuum source. Response time required for electrolysis of species in the sampled region is slower than for the slit arrangement (30 s versus 0.3 s). Major applications of the fiber optic thin-layer cell will be for obtaining spectra of stable species at potentially low concentration; however, time response limits transient applications to experiments lasting several tens of seconds.

A fourth example of the parallel geometry is based on optical diffraction and is the best method for sampling the region close to the electrode [47, 51]. A laser beam passing by a planar electrode is diffracted and falls on a target some distance past the electrode (Figure 7.33). The diffraction pattern is the Fourier transform of the beam cross section at the electrode, in this case a Gaussian beam truncated by the opaque electrode. The light at higher diffraction angles corresponds to higher Fourier frequencies of the beam cross section. These high frequencies define the sharpness of the electrode edge, and therefore are most sensitive to electrochemical processes. If an absorber is generated at the electrode, the electrode edge appears gradual rather than sharp, and the absorber's effect on the diffraction pattern is more pronounced at higher diffraction angles. As shown both theoretically and experimentally high diffraction angles are attenuated rapidly in the presence of electrogenerated absorber (Figures 7.34 and 7.35). Furthermore, the limiting attenuation of the high-frequency components is calculated by Beer's law, with a path length equal to the electrode length along the optical axis.

Functionally, absorbance measurements made with diffracted light can be

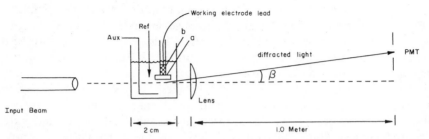

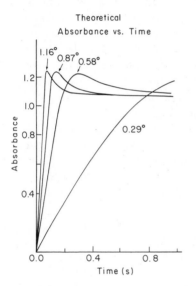

Figure 7.33 Experimental arrangement for spectroelectrochemistry using diffracted light. The photomultiplier tube (PMT) is placed in the geometric shadow of the electrode, and the diffracted intensity it collects is used for absorbance measurements. Absorber generated at the working electrode (a) diffuses across the beam and attenuates the diffracted light. Input beam is a laser; β ranges from 0 to about 3°.

Figure 7.34 Calculated absorbance versus time transients for diffracted light measured at various diffraction angles for diffusion-limited generation of a chromophore from a colorless precursor. Parameters were: electrode length, 0.027 cm; bulk concentration of precursor, 2.6 mM; molar absorptivity of chromophore, 15,100 M^{-1} cm^{-1}; diffusion coefficient, 1.6×10^{-5} cm^2/s; wavelength = 6328 Å. Transients were calculated using the Fourier transform approach described in [47, 51].

made to sample only the region very close to the electrode surface. After 200 ms from electrolysis initiation, the diffracted light behaves as if it passed through a narrow "cell" placed against the electrode (Figure 7.36). The "cell" length is equal to that of the electrode along the optical axis with a width of about 5–10 μm. The absorbance follows Beer's law for path lengths ranging from 0.014 to 1.1 cm and concentrations of 12 μM to 6 mM. Thus the region sampled by the diffracted light is the region of greatest interest to the electrochemist, since it is immediately adjacent to the electrode. A further feature of the diffractive technique is the possibility of spatially resolving the diffusion layer. Since the diffion pattern is the Fourier transform of the beam cross section, it contains information about chromophore distribution within the diffusion layer.

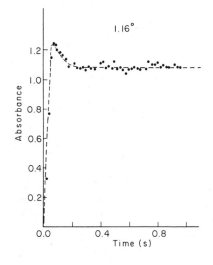

Figure 7.35 Observed and calculated transients for diffractive spectroelectrochemistry. Points were observed for the generation of trisbromophenylamine cation radical from a solution containing 2.6 mM of its reduced form. Dashed line was calculated from parameters measured independently.

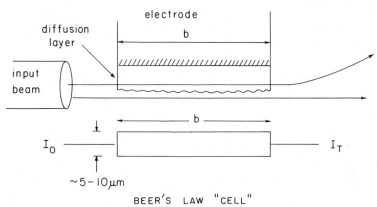

Figure 7.36 Approximate behavior of diffracted light in diffractive spectroelectrochemistry. The upper beam effectively samples the diffusion layer with ca. 10 μm of the electrode surface. It reaches a limiting absorbance dictated by Beer's law, as if an imaginary cell were placed against the electrode and filled by the electrochemical reaction. The length of the "cell" is equal to the electrode dimension along the optical axis.

Therefore, it should be possible to invert the diffraction pattern to generate a concentration versus distance profile.

2.1.6   Summary of UV–Vis Absorption Methods

The various UV–vis spectroelectrochemical techniques vary widely in sensitivity, time resolution, and ease of application. Table 7.1 summarizes the comparative features of the various geometries with some strengths and weaknesses. Generally the simplest methods are based on transmission through an OTE and offer compatibility with conventional spectrophotometers. The more

Table 7.1 Comparison of Approaches for UV–Vis Spectroelectrochemistry

	Effective Optical Path Length[a] (cm)	Time Response[b]	Electrolysis Time[c]	Continuum Source Permissible?[d]	Amenable to Commercial Spectrometer?	Relative Difficulty[e] (1 = Simple, 10 = Complex)
Transmission through metal film OTE	$1.13\sqrt{Dt}$	50 μs	NA	Yes	Yes	3
Transmission through OTTLE	$(\sim 10^{-2})$	30 s	30 s	Yes	Yes	1
Transmission through grid OTE	$2.26\sqrt{Dt}$	50 ms[f]	NA	Yes	Yes	2
External reflection	$2.26\sqrt{Dt}$	4 μs	NA	Yes	With difficulty	4
Internal reflection	$<10^{-4}$	4 μs	1 ms	No	No	9
Parallel geometry with slit	0.1–1.0	0.5 s	1–10 s	No[g]	No	7
Parallel geometry, fiber optic cell	1.0	~1 s	30 s	Yes	Yes	5
Parallel geometry, diffraction	0.2–1.0	10 ms	0.02 s	No[g]	No	8

[a]By analogy to Beer's law.
[b]Time required for absorbance response for a potential step to reach theoretically predicted value.
[c]Time required to reach constant, limiting absorbance.
[d]Is the method amenable to wavelength scanning to produce spectra?
[e]Relative ease of experiments on a subjective scale.
[f]Depends on grid size, can be much slower.
[g]Requires laser source, wavelength range limited to that of laser.

complex methods of IRS and parallel geometries yield impressive time response and sensitivity, but also increased experimental complexity. Obviously the technique selected depends on the demands of the application, however, UV–vis techniques clearly offer a wide range of capabilities with respect to time scale, sensitivity, and information content.

2.2   Vibrational Spectroscopy of Solution Species

While UV–vis spectroelectrochemistry is the most developed and widely used technique, it suffers from fairly low resolution. The spectral bands are broad and usually featureless and often make accurate observation of a mixture of components difficult. Furthermore, the only structural information provided is whether particular chromophores are present within the molecule. Vibrational spectroscopy is therefore an attractive alternative to UV–vis methods, particularly when structural information is required. Both infrared absorption and Raman spectroscopy are combined with electrochemistry to examine both surface and solution species. This section deals with spectral examination of solution species generated or consumed electrochemically and surface methods are discussed later.

2.2.1   Infrared Absorption Spectroelectrochemistry

Because of difficulty with electrode and solvent transmission, infrared absorption spectroelectrochemistry developed slowly. Metal film transparent electrodes of common use in UV–vis spectroelectrochemistry absorb or reflect IR radiation efficiently and are unsuitable in the IR region. In addition, solvent IR absorption is often strong, and necessitates short optical path lengths in a carefully designed electrochemical cell. A thin-layer cell with a grid electrode circumvents these constraints [52]. The IR beam passes through NaCl windows surrounding a cell containing three gold grids as working electrodes. The optical path length was on the order of 100 μm and the solvent was acetonitrile. This approach was used in conjunction with a commercial IR spectrometer to study the reduction of nihydrin, with the spectra shown in Figure 7.37 permitting conclusions about the products that were made. Another approach used a conventional absorption spectrometer and employed a flow system with an electrochemical reactor upstream from the spectrometer [53]. The cell was much simpler than the optically transparent versions, but time resolution is degraded since time must elapse between the electrochemical process and spectroscopic measurement.

Internal reflection spectroelectrochemistry has been used in the infrared wavelength region for studying both solution and surface species [54–56]. Germanium has sufficient IR transmission to be used as an IRS substrate and is a viable electrode within certain potential limits. Germanium is simply used directly as the electrode, with a germanium plate forming one wall of the electrochemical cell. While IR absorption spectra may be obtained in this manner, the narrow potential window of germanium makes it unattractive for many experiments. Germanium may be coated with a thin layer of carbon, which

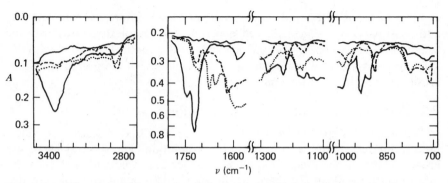

Figure 7.37   Infrared absorption spectra obtained with an OTE during the electrochemical reduction of ninhydrin in acetonitrite. Faint solid line is solvent alone; heavy solid line is unelectrolyzed ninhydrin; dotted line is a spectrum taken after 5 min of electrolysis at −1.1 V, dashed line after 5 min at −1.9 V versus SCE. Initial ninhydrin concentration was 30.5 mM; path length was 400 μm; three 100 wire/in. gold grids served as the electrode. Reprinted, with permission, from [52].

maintains optical transparency but improves the electrochemical characteristics [57, 58]. An alternative IRS arrangement uses a gold grid electrode wrapped around a germanium rod IRS element. The electrochemical characteristics are those of gold, while the ability to carry out IRS experiments in the infrared region is maintained [59]. Such an electrode was used to monitor the generation of benzoquinone radical anion in acetonitrile.

2.2.2   Raman Spectroelectrochemistry

The principal difficulty with infrared absorption spectroelectrochemistry is the incompatibility of solvents and electrode materials with light in the infrared region. Since Raman spectroscopy uses visible light, these difficulties are alleviated, and simple glass cells and electrodes may be employed. Furthermore, because water is a poor Raman scatterer, species in aqueous solution are studied with minimal solvent interference. Thus Raman scattering is used as an alternative to IR absorption for obtaining vibrational information about solution and surface species in electrochemical processes.

Raman spectroscopy is based on the scattering of light by objects that are much smaller than the wavelength. When light strikes a molecule, about one photon in 10^4 is scattered at the same wavelength as the incoming light. This process is called Rayleigh scattering and has an intensity proportional to the fourth power of the incident light frequency. Less than one in 10^6 photons striking the molecule will be inelastically scattered, gaining or losing energy during the interaction with the molecule. This process is Raman scattering, and the energy loss or gain is derived from vibrational molecular transitions. Raman scattering appears at discrete energy shifts away from the incident energy for a monochromatic input. The energy shifts observed equal those of the vibrational modes in the molecules, therefore Raman and infrared absorption provide similar information. Although the selection rules for the two processes are different, both provide vibrational information that is more structurally specific than electronic transitions available from UV–vis absorption spectroscopy.

Unfortunately, Raman scattering is a very weak process, and the technique is usually too insensitive for examining the millimolar levels of species often encountered in electrochemical experiments. However, Raman scattering is greatly intensified by factors as high as 10^6 when the exciting radiation is within an absorption band of the molecule of interest. This technique is called resonance Raman spectroscopy and provides sufficient sensitivity to observe electrochemically generated species. Raman spectroelectrochemistry first involved the resonance Raman technique applied to electrogenerated tetracyanoethylene (TCNE) anion radical [60]. Figure 7.38 shows the input laser beam reflected off the electrode and its wavelength close to the absorption maximum of the radical (430 nm). The radical was generated either by controlled potential electrolysis or by repetitive potential steps. Figure 7.39 shows Raman spectra of both TCNE and its radical anion. Note the sharp features of the spectrum compared to those of a typical UV–vis absorption spectrum, allowing structural features of the absorbing species to be deduced.

The resonance Raman spectroelectrochemical technique used in the millisecond time scale has been reported [61] and demonstrates the monitoring of specific vibrational bands as the molecule undergoes charge transfer. An excellent example of the value of vibrational information for diagnosing mechanistic

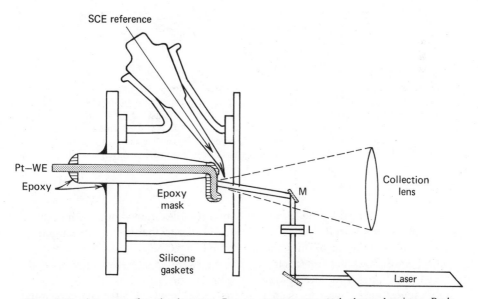

Figure 7.38 Apparatus for simultaneous Raman spectrometry and electrochemistry. Backscattered Raman light was collected by a large lens after being excited by a laser. Except for the window, the electrochemical cell was conventional. Reprinted, with permission, from [60].

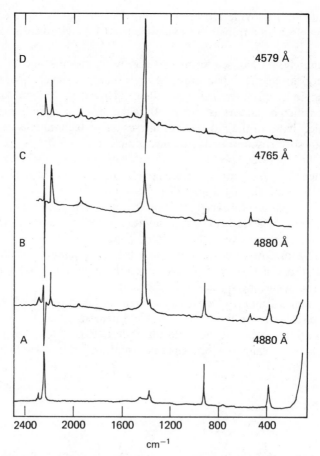

Figure 7.39 Resonance Raman spectra of tetracyanoethylene anion radical (TCNE⁻) electro-generated by controlled potential coulometry at -0.20 V versus SCE in 0.1 M TBAP/CH$_3$CN: (A) background scan [TCNE°] = 2.1 mM, cell at open circuit, laser power = 106 mW, band-pass = 2.0 cm^{-1}; (B) [TCNE⁻] = 2.1 mM, laser power = 106 mW, band-pass = 2.1 cm^{-1}; (C) (TCNE⁻) = 3.3 mM, laser power = 80 mW, band-pass = 2.1 cm^{-1}; (D) [TCNE⁻] = 3.3 mM, laser power = 30 mW, band-pass = 2.3 cm^{-1}. All spectra were scanned at 50 cm^{-1} using a 1 s counting interval per unit wavenumber increment. Plasma lines have been manually removed from the spectra excited at 4765 and 4579 Å. Reprinted, with permission, from [60].

problems is provided by the Raman examination of tetracyanoquinodimethane (TCNQ) reduction to its radical anion [62]. Prior to the Raman investigation, an absorption band at 477 nm was attributed to TCNQ⁻. However, when the radical anion was generated electrochemically, no band at 477 nm was observed. By comparing the resonance Raman spectra with those of known stable materials, it was deduced that the reaction of TCNQ⁻ with residual oxygen caused the 477 nm band. Exclusion of oxygen in the electrochemical cell prevented formation of the 477 nm chromophore, while introduction of oxygen produced it. The Raman spectrum of the product of the reaction of $TCNQ^{\overline{\cdot}}$ with oxygen was identical to that of an authentic sample of a TCNQ derivative prepared by other means. The high resolution of Raman spectroscopy provided definitive structure proof, whereas UV–vis spectroscopy misled the original investigators. Application to large molecules using a circulating cell further illustrates the utility of resonance Raman spectroscopy coupled to electrochemistry [63].

Despite their potential power, neither IR absorption nor Raman are extensively used for examining solution species. IR has the difficulties already mentioned and Raman is fairly expensive and insensitive. When resonance enhancement exists, Raman is an excellent tool, but this requirement does limit its applicability. Conversely, application of Raman to electrode surfaces is widespread and will be discussed later.

2.3   Magnetic Resonance Spectroelectrochemistry

In principle, both electron spin resonance (ESR) and nuclear magnetic resonance are attractive techniques to combine with electrochemistry, since they are both high-resolution techniques that provide structural information. In practice, however, ESR is substantially more amenable to studying electrochemical processes and has been developed much more extensively.

2.3.1   Electron Spin Resonance

The significant synergism between electrochemistry and ESR provides some of the driving force for the development of methods combining the two techniques. ESR, with detection limits in the $10^{-8}\ M$ range for radical species, can be very sensitive and can easily monitor electrochemically generated materials. Many electrochemical reactions transfer one electron, producing a species detectable by ESR. ESR provides information highly important to the electrochemist about the unshared electron density on the molecule. Furthermore, ESR is a high-resolution technique and distinguishes radicals with very slightly different structures. Finally, electrochemistry is an excellent method for generating radicals for ESR study, since electrons may be transferred to or from a solution species in a very well-controlled manner. Both the driving force (electrode potential) and quantity (charge) of the electrochemical process may be controlled easily, allowing more convenient radical generation than could be obtained chemically or by other means.

Details of the ESR experiment are available in numerous sources and are not repeated here. The application of ESR to electrochemistry and background discussions on the ESR technique appeared in two reviews over the last five years [64, 65]. However, several points about ESR spectrometry have special significance to the electrochemical application. First, ESR is only sensitive to molecules containing an unpaired electron, typically organic radicals or paramagnetic inorganic complexes. While a large number of electrochemical processes do involve transfer of a single electron, they are probably in the minority of all organic electrochemical processes. Second, ESR transition involves interaction of the magnetic vector of microwave radiation with the molecule to produce microwave absorption. However, the electric vector of the radiation causes absorption and heating upon interaction with the solvent. Losses of radiation to the solvent must be avoided in an ESR experiment. Electrochemical experiments require high dielectric constant solvents to solvate ions, and these are the same solvents that most strongly absorb microwaves. Thus the problem of solvent losses is particularly important for combined ESR–electrochemistry experiments but is alleviated with proper cell design. Third, the insertion of metal electrodes in an ESR cavity does not significantly disturb the ESR signal, so it is possible to generate radicals with simultaneous monitoring of the ESR spectrum.

An ESR spectrum may be obtained on an electrochemical product generated either outside or inside the ESR cavity. External generation has the advantage of experimental simplicity, since an electrode need not reside in the cavity. Any disturbance to the ESR measurement caused by cells and electrodes in the cavity are eliminated. However, external generation has the disadvantage of a time lag between generation and observation that may allow short-lived radicals to decay significantly before reaching the cavity. Internal generation is therefore more desirable for observing unstable radical species, and this feature is often worth the price of greater experimental complexity. Figure 7.40 shows ESR spectra obtained from electrolysis of tetrabutylboride anion in the presence of a spin-trapping agent. Identical spectra were obtained for both internal and external generation. Of significant concern with internal generation is the usually high ohmic potential error caused by the thin-layer cell geometry necessary for the ESR experiment. Resistance errors lead to slow product generation and severe errors can result in time-resolved experiments. One solution to the problem is the generation of materials at constant current rather than constant potential, allowing simultaneous electrochemical and ESR experiments with good time resolution [66]. A different approach is a radical change in cavity design [67] to yield a coaxial cavity and cell. Such an arrangement has minimal error from uncompensated resistance and produces excellent ESR spectra.

Information available from ESR monitoring of electrochemical processes is qualitative, quantitative, and dynamic. Qualitative information about structure is available from the appearance of the ESR spectrum, its coupling constants, and so on. This feature is the most widely exploited of the three and has been used to verify the formation of radicals by electrochemical charge-transfer reactions. Since ESR is an inaccurate quantitative method, quantitative analysis is difficult with ESR. However, relative strengths of signals from different radicals may be used to infer relative abundance. The dynamics of radical population may be used to deduce decay rates and mechanisms. An example [66] is the electrochemical generation of diethylfumarate anion radical by constant current, using ESR to monitor radical concentration. The rate of dimerization of the radical was deduced from the shape of the ESR signal after an electrochemical current pulse. The time resolution of the method ranged

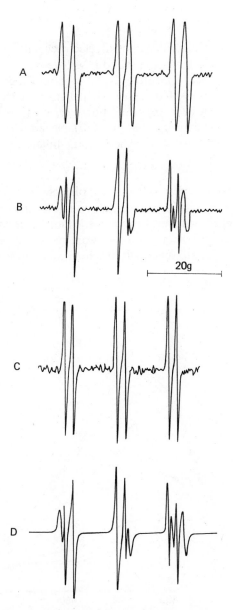

Figure 7.40 ESR spectra of α-phenyl *N-tert*-butylnitrone adducts with products of the electro-oxidation of tetrabutylboride ion under various conditions: (A) rigorous exclusion of oxygen during electrolysis; (B) no exclusion of oxygen during electrolysis; (C) saturation with oxygen during electrolysis; (D) computer simulation of a mixture of 68% A and 32% C (cf. spectrum B). Reprinted, with permission, from [69].

from a few hundred milliseconds up. A more recent technique involving ESR combined with electrochemistry is spin trapping of electrogenerated radicals [68, 69].

2.3.2   Nuclear Magnetic Resonance

While NMR has been combined with electrochemistry for determining product structures, the electrochemical and NMR steps are very rarely directly coupled because of technical difficulties. NMR requires a high concentration of spinning sample, two features that make design of an *in situ* electrochemical cell difficult. Such a cell has been reported, however, with good NMR spectra obtained during electrochemical generation [70]. Nevertheless, the method has not grown because of its substantial experimental difficulties.

2.4   Fluorescence Spectroelectrochemistry

Like ESR, fluorescence spectroscopy is applicable to a limited number of materials but is extremely sensitive and quite suitable for trace concentrations of species involved in electrochemical processes. Fluorescence spectroscopy has been combined with electrochemistry to study aromatic hydrocarbons [71] and peptides [72], and a long path spectroelectrochemical fluorescence cell has been

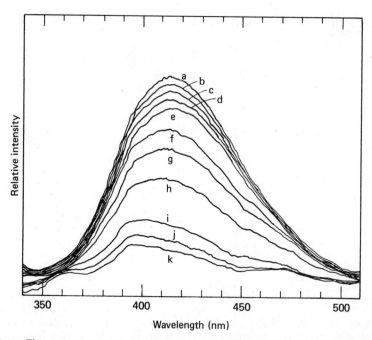

Figure 7.41   Fluorescence spectra of *o*-tolidine at various applied potentials, obtained with a solution of 5.0 μM *o*-tolidine in 1.0 M HClO$_4$ and 0.5 M CH$_3$COOH. (a) 0.40 V; (b) +0.580 V; (c) +590; (d) +0.596; (e) +0600; (f) +0605; (g) +0611; (h) +0.620; (i) +0.659 V, (j) +0.800. λ_{ex} was 270 nm (all potentials are in V versus SCE). Reprinted, with permission, from [72].

designed and tested [73]. As with absorption spectroelectrochemistry, fluorescence is monitored as a function of electrode potential and the effect of redox state on the fluorophore is deduced. Figure 7.41 shows the fluorescence spectra of *o*-tolidine as a function of potential with an excitation wavelength of 270 nm. The appropriate Nernst plot for these spectra has a slope of 29 mV, in agreement with the two-electron nature of the process. The formal potential calculated from the plot is in excellent agreement with that from other techniques. Though first reported in 1968 [71], fluorescence is not widely used in spectroelectrochemistry, perhaps because of its limited generality. Nevertheless, it is a very sensitive and specific technique when applied and is capable of yielding useful information.

2.5   Summary of Solution Spectroelectrochemical Techniques

The several spectroelectrochemical methods discussed may be compared according to their capability to provide three types of information about chemical systems: qualitative, quantitative, and dynamic. Qualitative information identifies species by comparisons of spectra or by structural inference from spectra. Quantitative information includes concentration, and the performance of a technique is indicated by accuracy and detection limit. Dynamic information is derived from the time course of a spectroelectrochemical variable and is used to deduce kinetic and mechanistic parameters for reactions associated with charge transfer.

Table 7.2 compares several techniques in terms of their relative ability to determine structure, concentration, and dynamic properties of solution species undergoing an electrochemical process. A few generalizations from the table

Table 7.2   Comparison of Spectroelectrochemical Methods

Spectroelectrochemical Method	Structural Specificity	Useful Time Scale	Concentration Range	Quantitative Accuracy
UV–vis	Poor	$4\ \mu s \rightarrow \infty^a$	$10^{-8} \rightarrow 10^{-2} M$	Excellent
IR	Excellent	$1\ s \rightarrow \infty^a$	—	Fair
Raman	Excellent	$5\ ms \rightarrow \infty$	$10^{-3} \rightarrow 10^{-1} M$	Fair
ESR	Excellent for radical species	$1\ s \rightarrow \infty$	$10^{-6} \rightarrow 10^{-3} M$	Poor
NMR	Excellent	—	$10^{-2} \rightarrow 1 M$	Poor
Fluorescence	Poor	$\sim ms \rightarrow \infty$	$10^{-8} \rightarrow 10^{-3} M$	Good
Interferometry	None	$100\ ms \rightarrow min$	$10^{-3} \rightarrow 10^{-1} M$	Poor

[a] An "infinite" time scale corresponds to diffusional equilibrium. In any transient experiment convection will contribute to mass transport after a few tens of seconds depending on the cell design. Therefore any theory assuming mass transport is purely diffusional will fail after convection interferes.

deserve special note. UV–vis absorption techniques have excellent dynamic and quantitative characteristics but provide relatively little structural information because of the broad nature of electronic absorption spectra. A particular strength of UV–vis absorption is a wide available time scale, allowing numerous kinetic experiments to be carried out. Raman and infrared absorption techniques are highly structurally specific but do impose constraints on the systems to be examined and are semiquantitative in presently used forms. ESR is sensitive and may provide structural information for paramagnetic species but is of limited quantitative value. In most cases, the time resolution of ESR is insufficient for kinetic experiments. Obviously the method of choice from the variety available will depend on the particular demands of the problem.

3   SPECTROELECTROCHEMICAL TECHNIQUES FOR EXAMINING ELECTRODE SURFACES

In the methods described in the previous section the region of interest was the solution near the electrode surface, with the electrode controlling the charge-transfer reaction. The experiments were designed to minimize any effects of changes of the electrode surface on the spectral technique involved. For example, electrode reflectivity changes were minimized in an external reflection experiment when solution components were examined. The techniques discussed in this section are intended to study surface changes, so the methods are designed to maximize the sensitivity of the method to processes occurring at the electrode-solution interface. There is a high driving force for the development of surface probes, since events occurring at the interface are of fundamental importance to electrochemistry. If the details of the interaction between solution species and the solid or liquid electrode are available, the fundamental charge-transfer process can be understood.

Unfortunately, formidable problems are encountered when developing optical probes of the electrode–solution interface. Thus the breadth of applications is not as wide as that for solution spectroelectrochemical methods. The amount of material present at a surface is exceedingly small, typically 10^{-10} mol/cm^2 for monolayer coverage. Additionally, the electric field of light striking a conducting electrode surface is attenuated to the point where it is near zero at the surface for good conductors. Therefore the sensitivity of optical probes involving absorption is reduced for surface bound molecules, making their detection and characterization difficult. Despite these hurdles, several optical methods have been developed that provide information about the electrode-solution interface.

The majority of the recently developed and powerful surface analysis techniques, such as X-ray photoelectron spectroscopy, secondary ion mass spectrometry, and low-energy electron diffraction, require a vacuum and are not amenable to examining electrodes in solution. Such methods have been applied extensively to studying electrodes after their removal from the solution, but our discussion is limited to those spectroscopic methods that can be employed simultaneously with the electrochemical process. These *in situ* methods include UV–vis and infrared reflectance, ellipsometry, Raman spectroscopy, and photo-thermal spectroscopy.

The *in situ* spectroelectrochemical methods for examining surfaces are logically categorized according to how the light interacts with the electrode surface. Reflectance methods involve measurements of changes in reflectivity caused by events at the electrode surface such as film formation or adsorption of a species from solution. The primary measurement is the intensity of light undergoing specular reflection. Ellipsometry is based on measurements of the reflected light phase, particularly the phase shifts of different polarizations of incident light induced during reflection. Both specular reflectance and ellipso-metry provide information about the optical constants of films at the electrode–solution interface and in some cases provide spectral information about ad-sorbed species. Raman scattering and photothermal spectroscopy do not rely on reflection from the surface, but rather depend on secondary effects of the inci-dent light, namely, scattering and thermal effects. Raman provides vibrational information and is amenable to single layers of adsorbed material in certain cases. Photothermal spectroscopy is based on temperature changes at the electrode caused by absorption and provides an indirect probe of the absorption spectrum of the surface. All four methods provide information about the electrode–solution interface but differ widely in application.

3.1   Reflectance Techniques

When an optical beam is reflected off a smooth electrode in solution, there are four effects of interest to electrochemists that may change the intensity of reflected light [74]. First, the optical constants of the electrode material, usually a metal, determine the inherent reflectivity of the surface. These optical properties are determined by the free electron density near the interface, which in turn depends on the electronic environment adjacent to the electrode. Second, the refractive index of the solution within the double layer affects the reflectivity, and this process is dependent on the concentration and arrangement of ions within the double and diffusion layers. Third, solution absorptivity determines how much light is absorbed before and after electrode reflection. Fourth, the presence of surface-adsorbed molecules changes the optical characteristics of the electrode and affects reflectivity. Of these four effects, the methods described in the section on solution spectroelectrochemistry are designed to exploit only one, the absorptivity of the solution. The experiments were constructed to minimize the other three effects, since they contributed errors to the measure-ments. In contrast, reflectance experiments for studying surfaces are designed to maximize interfacial effects on reflectivity so the nature of the interface may be probed.

3.1.1   UV–Vis Reflectance

McIntyre [27] presented a thorough discussion of theory and practice of reflectance techniques employing light in the UV–vis spectral region. He points out variables that are particularly important for using reflectance techniques to determine optical characteristics of the solution–electrode interface, several of which will be noted here. First, the optical constants of electrode, solution, and adsorbed layer are of obvious importance, and these quantities may vary with potential. Second, the incident angle and polarization of the light have large effects on reflectance, in part because of electric field changes at the surface. Incident light may be polarized either parallel to the incident plane (perpendicular to the surface) or perpendicular to the incident plane (parallel to the surface, Figure 7.42). At normal incidence, the two polarizations interact identically with the surface; at any other angle, their interactions differ substantially.

Reflectance techniques of greatest interest to electrochemists contain information about surface films, including monolayers; thus, this discussion is limited to such experiments. A beam polarized parallel to the incident plane with a 45° incident angle is least sensitive to refractive index changes in the double layer [27, 74], so this case is often used for surface studies. Figure 7.43 shows a typical experimental arrangement. If a small modulation is applied to the electrode potential, phase-sensitive detection may be employed to measure small reflectivity changes. Figure 7.44 is a reflectance spectrum for a platinum electrode in aqueous solution, showing the effects of oxide formation on the reflectance.

3.1.2   Ellipsometry

Ellipsometry [75] was adapted from surface science methods for electrochemistry and provides information of a nature similar to that from UV–vis reflectance methods. When light is reflected from a surface, the amplitude and phase of the reflected beam depend on the nature of the surface and the polarization of the incident beam. When linearly polarized light is reflected, the two components are phase shifted to different extents and the resulting light is elliptically polarized. Analysis of the reflected light reveals optical constants of the surface and possible surface films, with the data analysis consisting of fitting experimental results with computer-simulated theoretical plots. In addition to optical constants, ellipsometry provides information about film thickness.

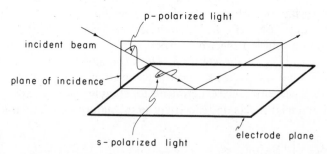

Figure 7.42   Polarization components of externally reflected light. p-Polarized light has electric vector parallel to the incident plane; s-polarized light has electric vector perpendicular to incident light.

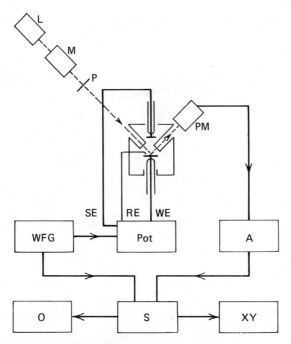

Figure 7.43 Block diagram of modulated specular reflectance apparatus. L, Xe/Hg arc lamp; M, monochromator; P, polarizer; PM, photomultiplier; A, amplifier; Pot, potentiostat; WFG, waveform generator; S, signal averager or phase-sensitive detector; O, oscilloscope; XY, X-Y recorder. Reprinted, with permission, from [29].

UV–vis reflectance and ellipsometry are useful for examining the optical properties of the electrode–solution interface and any associated films. The technique can be used to deduce the electronic characteristics of the interface, and inferences may be made about double-layer properties. However, they are not well suited to identifying the components of surface films and cannot supply very much information about complex films involving more than one component. Recent developments in infrared reflectance and surface Raman scattering should allow more molecular specificity for surface spectroscopic probes.

3.1.3   Infrared Reflectance

As with other types of absorption spectroscopy, conventional infrared reflectance is too insensitive for direct application to monolayer films on electrode surfaces. However, with modulation and Fourier transform techniques IR becomes useful on surfaces, and the information obtained is well worth the added complexity. IR techniques, though fairly recent additions to surface spectroelectrochemical methodology, are expected to reveal identities of absorbed species and the nature of their interactions with the surface.

In electrochemically modulated infrared reflectance spectroscopy [76], the electrode potential is modulated by an approximately 10 mV sinusoidal wave

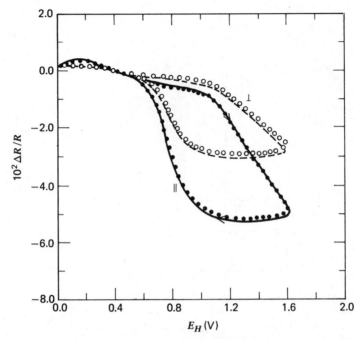

Figure 7.44 External reflectance changes for a platinum electrode in 1 M HClO$_4$. Symbols indicate perpendicular and parallel orientations relative to the incident plane. Curves are for argon-saturated solutions; points are for O$_2$. A small modulation was superimposed upon a potential scan to allow determinators of reflectance as a function of potential. Reprinted, with permission, from [73].

during wavelength scanning. With a phase-sensitive detector, only that portion of the reflectance modulated by the potential is detected; absorption by the solvent, electrode, or optical components is not detected. This results in high sensitivity and the capability of observing monolayer films. The technique is completely analogous to UV–vis electroreflectance experiments, except the spectral features are derived from vibrational rather than electronic transitions.

When the magnitude of the potential modulation is increased, it is possible to observe adsorbed intermediates [77]. The modulation process effectively produces a difference spectrum of the electrode at two widely separated potentials. For example, Figure 7.45 shows a reflectance spectrum for a platinum electrode in a methanol solution with the applied potential modulated at 8 Hz between −0.2 and +0.2 V versus SCE. The 2051 cm^{-1} band was attributed to adsorbed carbon monoxide bonded to the Pt surface at one site, while the 1850 band was assigned to bridging CO bonded at two sites. It was also pointed out that no adsorbed —COH species was observed spectroscopically.

Fourier transform IR reflectance (FTIR) has also been successfully applied to electrodes [78]. FTIR has the advantage of speed, resolution, and extensive data analysis capabilities, but is not as sensitive as the potential modulation techniques. Figure 7.46 shows an FTIR spectrum of a platinum electrode in

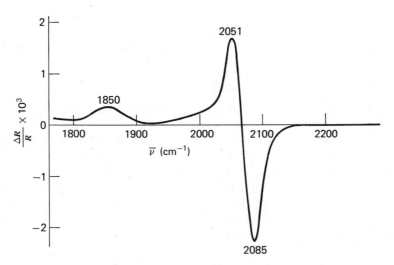

Figure 7.45 Infrared reflectance spectrum from a platinum electrode in 0.5 M methanol in 1 M H_2SO_4. Potential was modulated from -0.2 to $+0.2$ V versus SCE at 8.5 Hz during a wavelength scan of 0.0127 μm/s. Features at 1850 and 2060 cm^{-2} were attributed to adsorbed carbon monoxide. Reprinted, with permission, from [77].

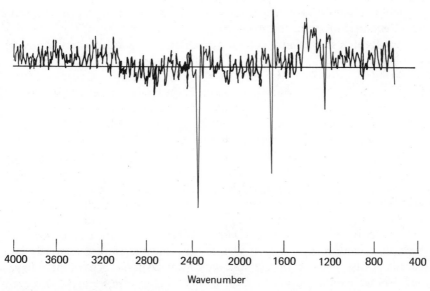

Figure 7.46 Fourier transform infrared reflectance spectrum of a platinum electrode in CD_3CN containing 0.1 M LiClO$_4$. Spectrum is a difference between spectra obtained at 0.0 and 2.5 V. Reprinted, with permission, from [78].

CD_3CN. Spectral features were assigned to adsorbed CD_3CN, and vibrational frequencies of the bands depended on potential, confirming the surface specificity of the method.

3.2   Surface Raman Spectroelectrochemistry

Raman spectrometry as a probe of events at electrode surfaces has received a great deal of attention [79, 80] during the last nine years for primarily two reasons. As mentioned earlier, it can provide vibrational information about surface molecules, unlike UV–vis absorption or high-vacuum techniques. Vibrational spectra not only provide structural information about molecules present at the surface, but may also provide insight into surface orientation. In addition to being a more structurally specific method, Raman spectrometry may be conducted *in situ* during an electrochemical process. The technique's high information content coupled with its ability to directly monitor electrode processes in real time provides great incentive for developing Raman surface probes.

Unfortunately, the low sensitivity of Raman spectrometry is a major barrier to its application to surfaces. For example, carbon monoxide adsorbed on a planar platinum surface has a maximum predicted Raman intensity of about two photons per second, with a signal-to-noise ratio (S/N) of about 0.6 [81]. Thus even with good scattering conditions, the signal is barely detectable, and one would be very hard pressed to derive any useful information from such an experiment. If Raman is to provide spectra surface species, there must be a significant enhancement of the scattered light intensity. Two enhancements of sufficient magnitude have been observed: resonance Raman spectrometry and surface enhanced Raman spectrometry.

When Raman scattering is excited by radiation close to a molecular absorption band, a large enhancement of up to 10^6 is observed in the Raman intensity for vibrations associated with the absorbing chromophore. Calculated values for S/N for a resonance enhanced system, I_2 on Pt, are about 1000, indicating more than sufficient sensitivity. Inherent to resonance Raman enhancement is the potential danger of sample photolysis, and care should be taken to assure the spectrum does not degrade rapidly with time.

There are few examples of surface resonance Raman mainly because interest in the topic is overshadowed by the other enhancement mechanism, surface enhanced Raman spectrometry (SERS). In 1974 [82, 83], a Raman spectrum of pyridine was obtained when it was adsorbed onto an electrochemically roughened silver surface. The silver was cycled in a KCl solution and a layer of AgCl was repeatedly formed and reduced back to silver. The result is a microscopically rough silver surface with a high surface area and therefore a relatively large quantity of adsorbed pyridine per unit area. Raman spectra of pyridine with very good S/N were obtained with such a sample as shown in Figure 7.47. Several pieces of evidence indicated the spectra were derived from surface-adsorbed molecules and not from solution. First, the peaks were shifted slightly in frequency from those of pyridine dissolved in aqueous solution. Second, peak

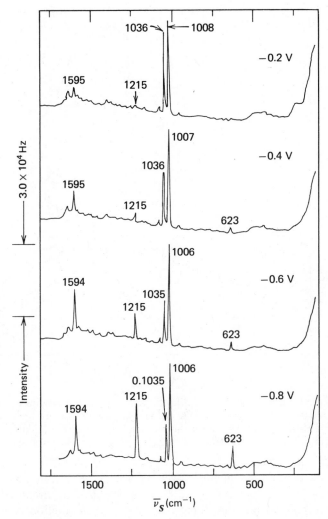

Figure 7.47 Raman spectra of pyridine adsorbed on an electrochemically roughened silver electrode. Electrode potential was held at the indicated value while the spectroscopic scan was made. Reprinted, with permission, from [84].

intensities varied with the electrode potential, with some peaks varying more than others (Figure 7.48). There is no conceivable way for electrode potential to affect molecules in bulk solution, since the double layer is only a few tens of angstroms thick. Third, the spectra were very weak on an unroughened electrode, an observation totally inconsistent with the notion that solution molecules are being observed.

The original observers of high-intensity Raman scattering from surfaces, Fleischmann and Hendra, attributed the enhancement to the high surface area

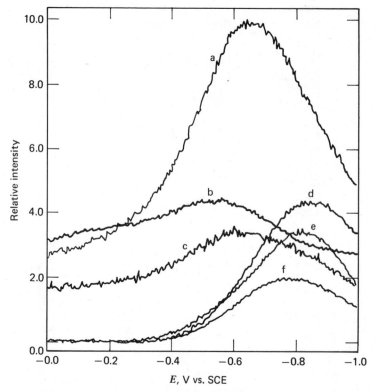

Figure 7.48   Potential dependence of Raman intensity for six bands of pyridine adsorbed on a silver electrode. Laser excitation was at 5145 Å, and the potential was scanned while the spectrometer monitored a single wavelength. Raman shifts were as follows: (a) 1006 cm^{-1}; (b) 1035 cm^{-1}; (c) 3056 cm^{-1}; (d) 1215 cm^{-1}; (e) 1594 cm^{-1}; (f) 623 cm^{-1}. Reprinted, with permission, from [84].

of the electrochemically roughened silver surface [82, 83]. Later, Van Duyne recognized the effect was too large to be caused merely by enhanced surface area [84]. He pointed out that there must be some interaction between the surface and adsorbate which contributes to the enhancement, and from this conclusion came the name surface enhanced Raman spectrometry (SERS).

Elucidation of the SERS mechanism has been intensely studied in order to explain the origin of the approximately four to six orders of magnitude of observed enhancement over the intensity calculated for a monolayer film. While the complete picture remains unresolved, it is generally agreed that three effects contribute to the higher Raman intensity observed for adsorbates on certain metal surfaces, particularly silver. First, the electrochemically roughened surface has a higher microscopic surface area than does a planar surface, and therefore has more scattering molecules in the beam. Second, there are electric field effects that enhance the local field in the region of the adsorbate. This field varies with the electrode material and its topography but will be most pronounced for small particles such as those on an electrochemically roughened surface. Third, there are polarizability effects related to interactions of the electrons in the metal with molecular orbitals in the adsorbate. Raman intensity is determined by the polarizability of the scattering center, that is, on how easily the electron distribution is perturbed by the electric field of the incoming light. If electrons from the electrode are coupled to the adsorbate orbitals, adsorbate polarizability is enhanced. Therefore the SERS effect is most prominent for metals with high conductivity at optical frequencies, such as silver in the entire visible wavelength region.

Regardless of the origin of SERS, it is useful to consider the information available from the technique. First, a surface enhanced spectrum indicates the presence of adsorbed molecules, so the method may be used to study surface coverage as a function of potential. Since not all surface molecules can be detected by SERS, the absence of a signal does not necessarily mean that adsorption has not occurred; but if a spectrum is present, it can be used to examine potential dependence. Second, the qualitative appearance of the spectra can indicate which molecules are present, a point of particular interest with reactive systems. It might be possible to identify adsorbed intermediates during a reaction sequence if the Raman scattering is sufficiently strong. Third, the SERS of a given molecule is dependent on orientation, particularly on which groups interact with the surface. Therefore details of surface interactions may be studied, and their changes with potential revealed.

While the potential of SERS for providing new surface information is enormous, there are some important limitations. The enhancement magnitude is directly related to the optical properties of the solid, and large enhancements are observed for only a few materials in certain wavelength regions. Silver works well in the visible while copper and gold provide good enhancement at wavelengths longer than about 550 nm. Although platinum is commonly used as an electrode, it has insufficiently high optical conductivity to provide high enhancement, and as yet strong SERS signals have not been observed on platinum. SERS on mercury has been reported [85, 86], but this observation is presently in dispute. In addition, the technique requires a strong Raman scatterer for good S/N; therefore, the best adsorbates are those with fairly high polarizability. While many molecules may be studied, the technique is not presently considered general. Finally, the roughened surface increases surface area and scattered intensity, thereby somewhat restricting the substrate choice to those materials that can be roughened effectively.

3.3   Photothermal and Photoacoustic Spectroelectrochemistry

These two methods are quite recent additions to the list of spectroelectrochemical techniques and are derived from the same phenomenon. When an electrode absorbs incident light, it is heated slightly as the absorbed light is thermalized. With a fairly intense source, the temperature change can be measured with a thermistor placed close to the electrode. The resulting technique is called photothermal spectroscopy and has been applied to absorbing films on platinum electrodes [87] and on semiconductor electrodes [88].

In photoacoustic spectroscopy the thermal effects of light absorption are manifested differently, as an acoustic wave in the solvent. The light beam is chopped at a convenient rate and sample absorption generates a series of pressure waves in the solvent [89]. These waves are detected by a microphone or a piezoelectric transducer, and the signal is related to the amount of light absorbed. If an intense monochromatic source that can be scanned is used, an absorption spectrum may be obtained by monitoring the photoacoustic signal as a function of wavelength. Figure 7.49 shows a typical example of a photoacoustic experiment and demonstrates the formation of an absorbing film of heptylviologen bromide. During reduction of the soluble, colorless precursor to the viologen film, the photoacoustic signal increases, followed by film removal by oxidation and coincident decrease in the photoacoustic response.

Photothermal and photoacoustic methods provide absorption spectra of

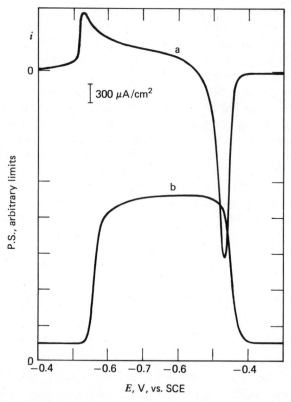

Figure 7.49 Photoacoustic response of a platinum electrode during a cyclic potential scan in a solution of heptylviologen bromide. Curve a is the current, showing formation of an absorbing film at about -0.5 V, then removal of the film again on the return scan. Curve b is the response of a piezoelectric transducer, which detects absorption of light ($\lambda = 550$ nm) while the film is present. Reprinted, with permission, from [89].

nontransparent or rough electrodes that are not amenable to conventional OTE or reflectance techniques. It is too soon to tell, but they may be of particular value in studying semiconductors and photoelectrochemical processes. The instrumental requirements are fairly demanding, with a high-intensity, tunable light source required for either method. Although their use is not widespread, new developments are actively being pursued.

3.4   Summary of *In Situ* Surface Spectroelectrochemical Methods

Table 7.3 compares several spectroelectrochemical techniques. UV–vis reflectance and ellipsometry are fairly limited in information content from a chemist's point of view, with limited applicability to identification of surface-bound species. For example, it would be difficult to determine the adsorbed intermediates in an organic oxidation by one of these methods, since the effects of adsorbed material on visible reflectance would be subtle. Raman and infrared reflectance provide structural information about adsorbed species and are highly promising for elucidating details of surface processes. The requirement for the enhancement mechanism restricts the use of Raman to certain substrates and adsorbates, but it can be very powerful when applicable. IR reflectance may become more useful than Raman for examining surface processes because of its broader applicability. Photothermal techniques are currently too new to assess their future impact.

Because high vacuum techniques cannot be applied *in situ* they were excluded from this discussion. However, they continue to be used extensively for examining electrode surfaces and much has been learned about electrochemical interfaces despite the high vacuum requirement.

Table 7.3   Subjective Comparisons of Surface Spectroelectrochemical Methods

Method	Physical Basis	Principal Information Supplied	Approximate Resolution	Generality
UV–vis reflectance	Reflectivity, absorption	Optical properties of interface	<1 ms possible	Any reflective surface
IR reflectance	Absorption	Vibrational spectrum of adsorbates	~ 1 s	Absorption must be modulated by potential
Ellipsometry	Phase shifts upon reflection	Optical constants, film thickness	>1 s	Any surface film on reflective surface
Surface Raman	Raman scattering	Vibrational spectra of adsorbates	$\sim$ ms	Limited to materials exhibiting enhance enhancement
Photothermal, photoacoustic	Absorption	Absorption spectra	~ 10 s	Any material with sufficient absorption

4   INSTRUMENTATION FOR SPECTROELECTROCHEMISTRY

4.1   General

Since the spectrometer, cells, and potentiostats required for spectroelectrochemistry are highly varied, only a few examples are presented to illustrate general features of the techniques. The design of a spectroelectrochemical experiment obviously is dictated by the optical geometry and the wavelength region to be employed. The transmission, reflection, and parallel geometries are discussed separately, but all share certain general features inherent in the optical and electrochemical aspects of the experiment. A block diagram of a typical spectroelectrochemical apparatus was shown in Figure 7.1. Often, all components except the electrochemical cell are included in a conventional spectrometer modified to accept the cell and electrode leads. The convenience of a commercial instrument is an important factor in experiment design, and many cell arrangements were designed with such spectrometers in mind. Generally, however, commercial spectrometers lack the time response required for transient experiments and are employed for equilibrium spectroelectrochemical experiments or for experiments with time scales greater than a few seconds.

For transient experiments more specialized instruments are generally employed, either custom-built or commercial rapid scanning or fixed wavelength spectrometers. A vast majority of transient experiments occur at fixed wavelength, with instrumental time response determining the useful experimental time scale. A common configuration is a single beam spectrophotometer consisting of source, monochromator, cell, and photodetector. Such a device may be constructed for less than $5000, and may be very versatile. Fixed wavelength monitoring of absorber generation with an OTE in transmission mode uses such an instrument. With the electrode potential at a value where no generation occurs, the intensity reaching the photodetector is monitored on a transient recorder or storage oscilloscope. When the potential is stepped to a value where electrolysis occurs, the intensity decrease at the photodetector is monitored as a function of time. The resulting intensity versus time transient is then converted to absorbance versus time, either by hand or by an on-line data system. An alternative is a commercial rapid-scanning instrument manufactured by Harrick, which includes the required optics for dual-beam operation with two photodetectors. The instrument is not only suitable for fixed wavelength, transient spectroelectrochemical experiments, but also provides rapid wavelength scanning in the 100 nm/ms time scale.

A commercial potentiostat often handles the electrochemical portion of the spectroelectrochemical experiment. If sufficient current is available to control the large electrode areas, the demand on the potentiostat is fairly minor for equilibrium experiments. Demands are significantly greater for transient experiments, since accurate potential control and time response are essential. Resistance compensation is usually employed on thin-film electrodes because of their substantial internal resistance. Despite this difficulty, outstanding time response is achieved by injecting a large charge pulse at the beginning of a potential step. For an IRS experiment employing a metal film OTE and a powerful (400 W) potentiostat, spectroelectrochemical results reached expected values within 4 μs after the charge injection pulse [43]. Microelectrode experiments often dispense with a potentiostat completely, and the smaller electrode area greatly decreases power demands. The currents are small enough so ohmic resistance errors are negligible, and a two-electrode arrangement is possible without sacrificing accuracy [36].

At the heart of spectroelectrochemical apparatus are the cell and the electrode, and cell design varies greatly with application. The three common geometries used in UV–vis and infrared absorption include transmission of the optical beam through the electrode, reflection from its surface, or passage parallel to the electrode surface. Since these configurations impose very different constraints on cell and electrode design, they will be discussed separately.

4.2   Transmission Through an OTE

The most common geometry for UV–vis spectroelectrochemistry is transmission through an optically transparent electrode. The electrode is either a thin film on a transparent substrate or a partially transparent grid. Thin-film electrodes [1, 4] have been made from gold, platinum, tin oxide, and carbon, usually vapor-deposited on glass. A mercury-coated platinum OTE allows use in the negative potential region [90]. A new OTE consisting of a metal film on a plastic substrate may prove to be useful, since the metal-coated plastic is commercially available and easy to use [91]. Film transmission depends on its thickness, but is typically 20%. An additional important parameter is film resistance, which can be fairly high. There is a trade-off between optical transmission and film resistance, with thicker films yielding low transmission and low resistance. High film resistance yields poor potential control, since electrode potential will change across the optical beam as electrochemical current flows through the film. A reasonable compromise between transmission and resistance occurs in the region of a few hundred angstrom thickness for platinum.

Thin-film electrodes offer planarity of the active surface, a boundary condition for many solutions to the differential equations required when describing the spectroelectrochemical response. Planarity is unimportant for equilibrium experiments; however, for transient experiments it is a significant advantage over the grid electrodes discussed below. Film resistance is a disadvantage of thin-film electrodes, but is often alleviated by carefully choosing experimental conditions. If the electrolysis current is kept low or if resistance compensation is employed, film resistance effects are reduced, but not completely eliminated. Intolerance to abuse is a second drawback of thin-film electrodes, and they are fairly easily damaged by current or potential extremes. With care, however, thin-film electrodes are very useful and have provided a large amount of spectroelectrochemical data.

The first use of grid electrodes involved gold "minigrids" consisting of fine (100–2000 wires/in.) gold mesh made by photolithographic techniques [92]. They transmit 75% (100 wires/in.) to 22% (2000 wires/in.) of the light and, of course, have the electrochemical properties of gold. They can be coated with mercury to provide a more extended range in the negative potential region [93]. The principal advantage of minigrids is the ease of fabrication of electrodes, and the comparative sturdiness of the finished product. The disadvantage of minigrids as compared to thin-film electrodes is the inherent nonplanar nature of the diffusion process. At short times the diffusion layer lies within the grid holes and the diffusion field shape is complex. The usual equations for planar diffusion obviously do not apply and predicting absorbance is exceedingly difficult. Fortunately, as the diffusion layer grows with time, it fills in the holes and the electrode behaves like a plane with regard to diffusion. For a 2000 wire/in. grid, this planar behavior is reached in 10–20 ms [94]; thus the electrode is assumed to be planar at times longer than this induction period. For coarser grids planar behavior takes substantially longer to develop, requiring 100 ms for a 1000 wire/in. grid.

The disadvantages of minigrids are all derived from their shape. To assure planar behavior, the grid must be very fine; however, fine grids (1000–2000 wires/in.) are mechanically weak. The 100–200 wire/in. grids are robust but reach planar behavior slowly, requiring several seconds. Processes in the 10–20 ms or less time scale cannot be quantitatively compared with theory and extraction of meaningful kinetic data is difficult at best. The most fruitful area for minigrid application is in equilibrium measurements where time response is not a consideration, and ease of use is important.

A large number of cell designs for OTE transmission experiments have been developed, but two types are used most commonly. Figure 7.6 shows an exploded view of a cell used for thin-film OTEs in both equilibrium and transient experiments. The cell volume is about 1 mL, and the entire apparatus may be mounted in the beam of a commercial spectrometer. Figure 7.2 depicts an OTTLE cell employing a gold minigrid as the working electrode. The cell is a "sandwich" consisting of microscope slides and Teflon spacers and may be constructed in less than an hour. A significant concern with OTTLE cells is the ohmic resistance caused by the small cross section of the cell. One of the major advantages of OTTLE cells is the rapid diffusional equilibrium occurring with cells of 100 μm or less thickness. However, the large ohmic resistance in thin cells leads to potential inhomogeneity across the face of the minigrid. The resistance problem is particularly severe in nonaqueous electrolytes which have typically ten times the resistance of their aqueous equivalents.

4.3   Reflection Geometry

For external reflection experiments the only requirements of the electrode are smoothness and sufficiently high reflectivity. Common examples are polished gold or platinum surfaces, but glassy carbon and electrodeposited mercury may also be employed. The electrodes are usually prepared from bulk metals by the same polishing procedures used for conventional electrochemistry, with the last step commonly being submicron alumina on glass. With external reflection from microelectrodes, the light beam is reflected off the side of a cylindrical wire and polishing is neither necessary nor possible, since the wire is about 10 μm in diameter.

Apparatus for external reflection from a large planar electrode is shown in Figure 7.43. The choice of incident angle depends on the experiment but is typically 45°, with the light polarized parallel to the incident plane to minimize reflectivity changes caused by perturbations in the solution refractive index near the electrode. A chopper is sometimes employed to improve the accuracy of the reflectivity measurement, since the observed reflectivity changes are often in the region of 1 part in 10^5. In addition, modulation of the potential is accomplished with a function generator and a lock-in amplifier detects the reflectivity change caused by the potential modulation.

Glancing incidence external reflection experiments are carried out with the apparatus in Figure 7.50, which is usually constructed on an optical table or bench. The source is a laser or collimated continuum source whose sufficiently small angular divergence avoids a wide spread of incident angles. The electrode length along the optical axis is typically 1–2 cm and the incident angle of the beam is 0.7–2° [31].

For external reflection from microelectrodes, a collimated beam is focused onto the electrode with a lens, and a fiber optic probe is used to collect reflected light in order to minimize stray light pickup [36]. An alternative to the cylinder is a microdisk embedded in glass or epoxy. The disk has smaller area and longer life, reflected light is more easily collected, and improved cell time constant permits submicrosecond time response [95].

Internal reflection experiments differ greatly from external reflection in both the cell and the electrodes employed. A typical arrangement is shown in Figure 7.51 for visible light and a thin-film electrode. The light must be well collimated

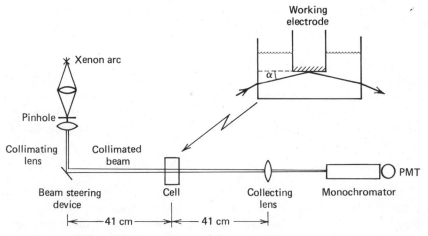

Figure 7.50 Apparatus for glancing incidence external reflection spectroelectrochemistry with a xenon arc continuum source. Entire apparatus is mounted on an optical table; PMT indicates photomultiplier tube (not to scale). Reprinted, with permission, from [28].

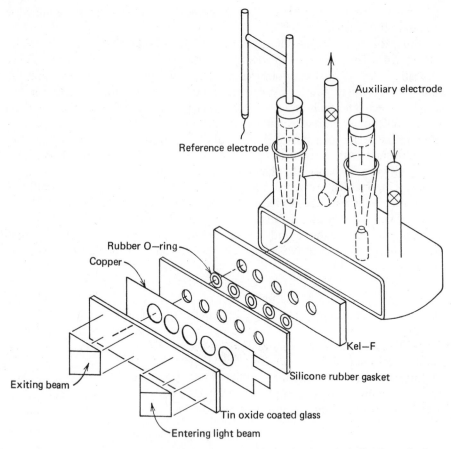

Figure 7.51   Cell for multiple internal reflection spectroelectrochemistry, including five reflections for this case. Reprinted, with permission, from [96].

and the angle of incidence on the electrode plane is slightly larger than the critical angle. Since the penetration depth of the evanescent wave depends on many factors, including wavelength, refractive index, and incident angle, the sensitivity of the apparatus is calibrated with some known system before use. Of course the optical apparatus differs for IRS in the infrared region, but the various arrangements are not discussed here.

4.4   Parallel Geometry

Experiments employing the parallel geometry are not as well developed as previously discussed approaches, and their instrumental arrangements vary substantially. Figure 7.31 shows an experiment where a slit defines the region sampled by an optical beam passing by a planar electrode. The slit may be moved relative to the electrode plane to sample different distances from the surface, all with the high sensitivity inherent in the long optical path length. However, because of the low beam intensity that passes through the slit, this technique is not amenable to commercial spectrometers and laser sources have been employed. A fiber optic cell that may be used in a conventional spectrometer has been developed with the beam passing parallel to an electrode in a thin-layer cell [50]. The path length is 1 cm, the cell has small volume, and the electrochemical behavior is satisfactory. While the cell is not well suited to fast experiments because of internal resistance, it has good sensitivity for equilibrium measurements.

An apparatus for diffractive spectroelectrochemistry is shown in Figure 7.33. This method currently requires a laser beam and a stable optical bench or table. The cell consists of an electrode with a short dimension along the optical axis sandwiched between optical windows. The direction of diffusion is toward and away from the electrode edge, with linear diffusion assured by the cell walls. The electrode surface must be well polished and aligned so that the beam travels parallel to a flat electrode. In addition resistance compensation is usually employed because of the fairly high resistance inherent in the thin layer of solution near the electrode. Longer electrodes avoid this problem and lead to higher sensitivity with some degradation of the time response and agreement with theoretical absorbance transients [97].

4.5   Computer Techniques in Spectroelectrochemistry

A large fraction of all spectroelectrochemical experiments involves a computer in some way, so their utility in such experiments is discussed briefly here. There are three applications of computer techniques to spectroelectrochemistry that are particularly common: on-line data acquisition and control, analysis of results, and simulation of experimental responses.

The desirability of on-line computer control of spectroelectrochemical experiments stems from the many experimental variables of potential, time, absorbance, and so on, coupled to the often large amount of data. For example, a rapid scanning absorbance spectrometer may repetitively acquire spectra during a potential scan, resulting in many different absorption spectra. In order to analyze these spectra later, several thousands of data points must be stored. Furthermore, the entire experiment may take less than 1 min, requiring fast acquisition. Clearly a computer is well suited to any such experiment generating large amounts of experimental data. An additional area where an on-line computer is of high value is the study of fast reaction kinetics. Since fast spectroelectrochemical experiments require signal averaging to observe small absorbances, a computer or similar signal averager is essential.

Once a spectroelectrochemical experiment is complete, significant data analysis is usually necessary before interpretations may be made. Data treatment may range from a simple plot of absorbances versus $t^{1/2}$ to a complex treatment of ellipsometric data, and usually such analysis is assisted by a computer. Often the same computer that controlled the experiment carries out the analysis directly, or in some cases the data are transferred to a larger computer. The tabulation and plotting features of most data acquisition systems are similar to those required in spectroelectrochemistry, with the additional requirement of extensive numerical analysis capability.

The third function of computers in spectroelectrochemistry is the simulation of experimental response, mentioned earlier in Section 2.1.2. If an absorbing species generated at an electrode undergoes a chemical reaction to a non-absorber, for example, the observed absorbance will depend on the rate constant for the reaction, diffusion to and away from the electrode, and the rate of charge transfer. Each of these processes is fairly complex, but can be described by rigorous mathematical expressions. Computer simulation of the absorbance response must take all such effects into account and generate some theoretical response. By comparing theory with experiment, a rate constant or reaction mechanism may be deduced. Finite difference simulation methods have been successfully applied to this problem, and many reactions have been elucidated by examining spectroelectrochemical results.

4.6   Summary

Table 7.1 briefly listed the comparative complexity of several spectroelectrochemical methods, but some mention of relative instrumental difficulty is useful here. Of the optical absorption techniques an OTTLE cell in a conventional spectrophotometer is the most easily employed, and meaningful experiments may be conducted after a few hours invested in cell construction. Since both potentiostat and spectrometer are commercial and the cell is easily fabricated, there are no significant hurdles to first-time use of the technique. The drawbacks of the OTTLE configuration are slow response and potential errors caused by solution resistance, and OTTLE cells are not useful for fast experiments. Techniques based on metal film electrodes of the type shown in Figure 7.6 also use a commercial spectrometer and potentiostat, have good transient response, but are more difficult to fabricate than the OTTLE configuration. A third method amenable to commercial spectrometers is the fiber optic thin-layer cell, which has high sensitivity, modest time response, but is fairly difficult to build.

When a technique requires a custom spectrometer, the experimental complexity obviously increases significantly. Microelectrodes, diffractive methods, IRS, and most external reflection methods require a substantial investment in building an instrument and cell. As always, the special advantages of these methods must be assessed against required development effort. The trade-off between simplicity and performance is as important with spectroelectrochemistry as with other instrumental methods.

References

1. T. Kuwana and N. Winograd, "Spectroelectrochemistry at Optically Transparent Electrodes, I," in A. J. Bard, Ed., *Electroanalytical Chemistry*, Vol. 7, Dekker, New York, 1974.
2. W. R. Heineman, *Anal. Chem.*, **50**, 390A (1978).
3. A. J. Bard and L. R. Faulkner, "Spectrometric and Photochemical Experiments," in *Electrochemical Methods*, Wiley, New York, 1980.

4.   T. Kuwana and W. R. Heineman, *Acc. Chem. Res.*, **9**, 241 (1976).
5.   W. R. Heineman, F. M. Hawkridge, and H. N. Blount, "Spectroelectrochemistry at Optically Transparent Electrodes, II," in A. J. Bard, Ed., *Electroanalytical Chemistry*, Vol. 13, Dekker, New York, 1983.
6.   T. Kuwana, R. K. Darlington, and D. W. Leedy, *Anal. Chem.*, **36**, 2023 (1964).
7.   W. R. Heineman, B. J. Norris, and J. F. Goeltz, *Anal. Chem.*, **47**, 79 (1975).
8.   EG&G Instruments, Princeton Applied Research Model 173.
9.   W. R. Heineman, T. M. Kenyherz, T. P. Angelis, B. J. Norris, and H. B. Mark, *J. Am. Chem. Soc.*, **98**, 2469 (1976).
10. W. R. Heineman, T. Kuwana, and C. Hartzell, *Biochem. Biophys. Res. Commun.*, **49**, 1 (1972).
11. W. R. Heineman, T. Kuwana, and C. Hartzell, *Biochem. Biophys. Res. Commun.*, **50**, 892 (1973).
12. F. M. Hawkridge and T. Kuwana, *Anal. Chem.*, **45**, 1021 (1973).
13. J. L. Anderson, T. Kuwana, and C. Hartzell, *Biochemistry*, **15**, 3847 (1976).
14. F. Blubaugh, A. Yacynych, and W. R. Heineman, *Anal. Chem.*, **51**, 561 (1979).
15. R. McCreery, *Anal. Chem.*, **49**, 206 (1977).
16. D. E. Albertson, H. N. Blount, and F. M. Hawkridge, *Anal. Chem.*, **51**, 556 (1979).
17. E. Bancroft, H. N. Blount, and F. M. Hawkridge, *Anal. Chem.*, **53**, 1862 (1981).
18. E. Bancroft, J. Sidwell, and H. N. Blount, *Anal. Chem.*, **53**, 1390 (1981).
19. G. Grant and T. Kuwana, *J. Electroanal. Chem.*, **24**, 11 (1970).
20. L. Mackey and T. Kuwana, *Bioelect. Bioenerg.*, **3**, 596 (1976).
21. A. J. Bard and L. R. Faulkner, *Electrochemical Methods*, Wiley, New York, 1980, Appendix A.
22. H. N. Blount, *J. Electroanal. Chem.*, **42**, 271 (1973).
23. J. F. Evans and H. N. Blount, *J. Electroanal. Chem.*, **102**, 289 (1979).
24. G. Gruver and T. Kuwana, *J. Electroanal. Chem.*, **36**, 85 (1972).
25. N. Winograd, H. N. Blount, and T. Kuwana, *J. Phys. Chem.*, **73**, 3456 (1979).
26. H. N. Blount, N. Winograd, and T. Kuwana, *J. Phys. Chem.*, **74**, 3231 (1970).
27. J. D. E. McIntyre, "Specular Reflection Spectroscopy of the Electrode–Solution Interphase," in P. Delahay and C. W. Tobias, Eds., *Advanced Electrochemistry and Electrochemical Engineering*, Vol. 9, Wiley-Interscience, New York, 1973.
28. J. P. Skully and R. L. McCreery, *Anal. Chem.*, **52**, 1885 (1980).
29. P. T. Kissinger and C. N. Reilley, *Anal. Chem.*, **42**, 12 (1970).
30. R. L. McCreery, R. Pruiksma, and R. Fagan, *Anal. Chem.*, **51**, 749 (1979).
31. C. E. Baumgartner, G. T. Marks, P. A. Aikens, and H. H. Richtol, *Anal. Chem.*, **52**, 267 (1980).
32. D. C. Walker, *Anal. Chem.*, **39**, 896 (1967).
33. A. Bewick and A. M. Tuxford, *Symp. Faraday Soc.*, **4**, 116 (1970).
34. A. W. Aylmer-Kelly, A. Bewick, P. Cantrill, and A. M. Tuxford, *Faraday Discuss. Chem. Soc.*, **56**, 96 (1973).
35. R. L. Robinson and R. L. McCreery, *Anal. Chem.*, **53**, 997 (1981).
36. R. L. Robinson, C. W. McCurdy, and R. L. McCreery, *Anal. Chem.*, **54**, 2356 (1982).
37. J. S. Mayausky, H. Y. Cheng, P. M. Sackett, and R. L. McCreery, "Spectroelectrochemical Examination of the Reactions of Chlorpromazine Cation Radical with Physiological Nucleophiles," in K. Kadish, Ed., *ACS Adv. Chem. Ser.*, **201**, 442 (1982).
38. J. S. Mayausky and R. L. McCreery, *Anal. Chem.*, **55**, 308 (1983).
39. A. Bewick, J. Mellor, and B. Pons, *Electrochim. Acta*, **25**, 931 (1980).
40. N. Winograd and T. Kuwana, *J. Am. Chem. Soc.*, **93**, 4343 (1971).

41. N. Winograd and T. Kuwana, *J. Am. Chem. Soc.*, **92**, 224 (1970).
42. N. Winograd and T. Kuwana, *Anal. Chem.*, **43**, 252 (1971).
43. J. E. Davis and N. Winograd, *Anal. Chem.*, **44**, 2152 (1972).
44. W. N. Hanson, "Internal Reflection Spectroscopy in Electrochemistry," in P. Delahay and C. W. Tobias, Eds., *Advanced Electrochemistry and Electrochemical Enginering*, Vol. 9, Wiley-Interscience, New York, 1973.
45. R. Pruiksma and R. L. McCreery, *Anal. Chem.*, **51**, 2253 (1979).
46. J. F. Tyson and T. S. West, *Talanta*, **27**, 335 (1980).
47. P. Rossi, C. W. McCurdy, and R. L. McCreery, *J. Am. Chem. Soc.*, **103**, 2524 (1981).
48. R. H. Muller, "Double Beam Interferometry for Electrochemical Studies," in P. Delahay and C. W. Tobias, Eds., *Advanced Electrochemistry and Electrochemical Engineering*, Vol. 9, Wiley-Interscience, New York, 1973.
49. R. Pruiksma and R. L. McCreery, *Anal. Chem.*, **53**, 202 (1981).
50. J. D. Brewster and J. L. Anderson, *Anal. Chem.*, **54**, 2560 (1982).
51. P. Rossi and R. L. McCreery, "Optical Diffraction by Electrodes," in A. G. Marshall, Ed., *Fourier, Hadamard, and Hilbert Transforms in Chemistry*, Plenum, New York, 1982.
52. W. R. Heineman, J. N. Burnett, and R. W. Murray, *Anal. Chem.*, **40**, 1974 (1968).
53. B. Clark and D. H. Evans, *J. Electroanal. Chem.*, **69**, 181 (1976).
54. P. R. Tallant and D. M. Evans, *Anal. Chem.*, **41**, 835 (1969).
55. J. S. Mattson and C. A. Smith, *Science*, **181**, 1055 (1973).
56. D. Laser and M. Ariel, *Anal. Chem.*, **45**, 2141 (1969).
57. J. S. Mattson and C. A. Smith, *Anal. Chem.*, **47**, 1122 (1975).
58. J. S. Mattson and T. Jones, *Anal. Chem.*, **48**, 2164 (1976).
59. D. Laser and M. Ariel, *J. Electroanal. Chem.*, **41**, 381 (1973).
60. D. L. Jeanmaire, M. R. Suchanski, and R. P. Van Duyne, *J. Am. Chem. Soc.*, **97**, 1699 (1975).
61. D. L. Jeanmaire and R. P. Van Duyne, *J. Electroanal. Chem.*, **66**, 235 (1975).
62. M. R. Suchanski and R. P. Van Duyne, *J. Am. Chem. Soc.*, **98**, 250 (1976).
63. J. L. Anderson and J. R. Kincaid, *Appl. Spectrosc.*, **32**, 356 (1978).
64. T. M. McKinney, "Electron Spin Resonance and Electrochemistry," in A. J. Bard, Ed., *Electroanalytical Chemistry*, Vol. 10, Dekker, New York, 1977.
65. A. J. Bard and L. R. Faulkner, *Electrochemical Methods*, Wiley, New York, 1980, p. 614.
66. A. J. Bard and L. R. Faulkner, *Electrochemical Methods*, Wiley, New York, 1980, p. 620.
67. R. D. Allendoerfer, G. A. Martinchek, and S. Bruckenstein, *Anal. Chem.*, **47**, 890 (1975).
68. E. E. Bancroft, H. N. Blount, and E. J. Janzen, *J. Am. Chem. Soc.*, **101**, 3692 (1979).
69. G. L. McIntyre, H. N. Blount, H. J. Stronks, R. V. Shetty, and E. G. Janzen, *J. Phys. Chem.*, **84**, 916 (1980).
70. J. A. Richard and D. M. Evans, *Anal. Chem.*, **47**, 964 (1975).
71. A. Yildiz, P. T. Kissinger, and C. N. Reilley, *Anal. Chem.*, **40**, 1018 (1968).
72. M. J. Simone, W. R. Heineman, and G. P. Kreishman, *J. Colloid Interface Sci.*, **86**, 295 (1982).
73. M. J. Simone, W. R. Heineman, and G. P. Kreishman, *Anal. Chem.*, **54**, 2382 (1982).
74. J. D. E. McIntyre and D. M. Kolb, *Symp. Faraday Soc.*, **4**, 50 (1970).
75. R. H. Muller, "Principles of Interferometry," in P. Delahay and C. W. Tobias, Eds., *Advanced Electrochemistry and Electrochemical Engineering*, Vol. 9, Wiley-Interscience, New York, 1973.

76. A. Bewick, K. Kunimatsu, J. Robinson, and J. W. Russell, *J. Electroanal. Chem.*, **119**, 175 (1981).

77. B. Beden, C. Lamy, A. Bewick, and K. Kunimatsu, *J. Electroanal. Chem.*, **121**, 343 (1981).

78. T. Davidson, B. S. Pons, A. Bewick, and P. Schmidt, *J. Electroanal. Chem.*, **125**, 235 (1981).

79. R. K. Chang and T. F. Furtak, *Surface Enhanced Raman Spectrometry*, Plenum, New York, 1982.

80. R. F. Candane, J. M. Gilles, and A. A. Lucas, *Vibrations at Surfaces*, Plenum, New York, 1982.

81. R. P. Van Duyne, "Laser Excitation of Raman Scattering from Adsorbed Molecules on Electrode Surfaces," in C. B. Moore, Ed., *Chemical and Biochemical Applications of Lasers*, Vol. IV, Academic, New York, 1979.

82. M. Fleischmann, P. Hendra, and A. McQuillan, *J. Electroanal. Chem.*, **65**, 163 (1974).

83. A. J. McQuillan, P. Hendra, and M. Fleischmann, *J. Electroanal. Chem.*, **65**, 933 (1975).

84. D. L. Jeanmaire and R. P. Van Duyne, *J. Electroanal. Chem.*, **84**, 1 (1977).

85. R. Naaman, S. Buelow, O. Cheshnovsky, and D. Herschbach, *J. Phys. Chem.*, **84**, 2692 (1980).

86. L. Sanchez, R. Birke, and J. Lombardi, *Chem. Phys. Lett.*, **79**, 219 (1981).

87. G. H. Brilmyer and A. J. Bard, *Anal. Chem.*, **52**, 685 (1980).

88. A. Fujishima, Y. Maeda, K. Honda, G. H. Brilmyer, and A. J. Bard, *J. Electroanal. Soc.*, **127**, 840 (1980).

89. R. E. Malpas and A. J. Bard, *Anal. Chem.*, **52**, 109 (1980).

90. W. R. Heineman and T. Kuwana, *Anal. Chem.*, **43**, 1075 (1971).

91. R. Cieslinski and N. R. Armstrong, *Anal. Chem.*, **51**, 565 (1979).

92. R. W. Murray, W. R. Heineman, and G. W. O'Dom, *Anal. Chem.*, **39**, 1666 (1967).

93. W. R. Heineman, T. P. DeAngelis, and J. F. Goelz, *Anal. Chem.*, **47**, 1364 (1975).

94. M. Petek, T. Neal, and R. Murray, *Anal. Chem.*, **43**, 1069 (1971).

95. R. S. Robinson and R. L. McCreery, *J. Electroanal. Chem.*, **182**, 61 (1985).

96. W. Winograd and T. Kuwana, *J. Electroanal. Chem.*, **23**, 333 (1969).

97. P. Rossi and R. L. McCreery, *J. Electroanal. Chem.*, **151**, 47 (1983).

Chapter **8**

CONDUCTANCE AND TRANSFERENCE DETERMINATIONS

Michael Spiro

1   INTRODUCTION AND DEFINITIONS

Conductance and transference are complementary properties. The electrical conductance of an electrolyte solution is a measure of the extent to which *all* the ions present move when an electric field is applied and so carry the resulting electric current. The transference number of one of the ionic constituents, on the other hand, records the relative extent to which that constituent transports the electric charge. Detailed definitions are given in Section 1.2. Thus conductance essentially depends on the sum of the ionic conductances and the transference number on their ratio. The product of these two quantities therefore yields the conductance (or the mobility) of individual ions. This is the only property of an individual ion that can be accurately determined without any arbitrary assumptions having to be made.

The limiting values of the molar conductance, transference number, and ionic conductance at infinitesimal ionic strength [1] are functions only of ion–solvent interactions, described in Sections 2.2 and 2.3. By contrast, their variations with concentration depend essentially on ion–ion interaction (Section 2.1). Much information on the structure of electrolyte solutions and on ionic compositions and equilibria can therefore be derived from these properties. A great variety of systems is currently being studied by conductance and transference measurements: they include electrolytes in a wide range of organic and inorganic solvents, ionic solutions at high temperatures and pressures, and very concentrated aqueous solutions. Many of these are of considerable industrial importance.

The numerous applications of conductance and transference experiments are listed and discussed in Section 3. Conductance continues to be a major analytical tool, whether in a traditional field like conductometric titrations or in such modern techniques as ion chromatography. Conductance forms the basis of an important method of determining equilibrium constants such as solubility products and acid dissociation constants, and is frequently used in kinetic work to study the rates of both homogeneous and heterogeneous reactions involving ions. Very fast reactions can now be followed by employing conductometric detection in stopped-flow and pressure-jump methods. Moreover, conductance sensing increasingly serves to monitor and hence control a variety of industrial processes.

Conductances are relatively easy to measure, particularly because self-balancing conductance bridges and some conductivity cells are now commercially available. However, the determination of transference numbers requires apparatus that, even if not complicated, has to be fabricated specially. Sections 4 and 5 of this chapter are devoted to the experimental aspects of conductance and transference measurements, respectively. The emphasis throughout will be on precision measurement.

1.1   Units

In this chapter the various properties are defined in SI units, which are now standard in many countries [2]. The SI (for Système International) was given official status in a resolution of the 11th General Conference on Weights and Measures in Paris in 1960 [3]. It is essentially a rationalized mks (meter-kilogram-second) system although it is of course permissible by SI rules to express measured numerical quantities in accepted subunits such as centimeter. It is worth emphasizing that the Debye–Hückel–Onsager interionic attraction theory has been derived and is still often used in the unrationalized cgs (centimeter-gram-second) form. Care must therefore be taken in applying equations from the literature.

The SI system uses the mole instead of the equivalent as the basic unit of chemical quantity. This procedure is followed here but, because of their widespread use in the conductance field, the main equivalent properties will also be defined [4] and will be designated by the subscript eq. In the current literature the problem of conversion from equivalent to molar standards is made more difficult by some workers redefining their "mole" as $1/v_+z_+$ [see (3) below] of a formula weight, which allows them to retain the original "equivalent" equations with "mole" designations. In this chapter, mole means formula weight. Concentrations will often be expressed in mol/dm^3 (frequently but unofficially called molar or M) both because it is customary and because 1 mol/dm^3 is an accepted standard state in solution. The new IUPAC recommended standards for conductance cell calibrations in terms of absolute ohms are given in Section 4.1.3.

1.2   Macroscopic Definitions and Their Microscopic Significance

1.2.1   Conductance

The conductance G (siemens or ohm^{-1}, or in old texts the jocular mho) of a conductor is the inverse of its resistance R (ohm). It is the resistance between two electrodes immersed in the solution under investigation that is normally measured in the laboratory. If the electrodes possess an effective cross-sectional area A (m^2) and are a distance l (m) apart,

$$R = \frac{\rho l}{A} = \frac{l}{\kappa A} = \frac{1}{G} \tag{1}$$

where ρ (ohm·m) is the specific resistivity and κ (ohm^{-1} m^{-1}) the specific or electrolytic conductivity. It should be noted that κ and ρ are intrinsic properties of the solution at the given temperature and pressure while G and R depend also on the dimensions of the measuring system. In practice l/A, called the cell constant k (m^{-1}), is not determined with a ruler but by measuring the resistance when the cell is filled with a solution of known specific conductivity.

It is obvious that κ will depend on the characteristics and concentration c (mol/m^3) of the electrolyte present in the solution. As chemists we are interested in properties per mole (or in the past, per equiv) and it is therefore usual to express conductance results in the form of the molar conductance Λ ($m^2 \cdot S/mol$ or $m^2/ohm \cdot mol$) defined by

$$\Lambda = \kappa/c \tag{2}$$

Let us consider a single electrolyte $C_{v_+}A_{v_-}$, which is present to the extent of c mol/m^3. If it dissociates according to

$$C_{v_+}A_{v_-} \rightleftharpoons v_+C^{z+} + v_-A^{z-} \tag{3}$$

the equivalent concentration (equiv/m^3) is given by

$$c_{eq} = v_+z_+c = v_-|z_-|c \tag{4}$$

The z_i symbols refer to the algebraic charge number of the subscripted species i while $|z_i|$ is its arithmetic charge number. For the H^+ ion both z_+ and $|z_+|$ are $+1$; for the SO_4^{2-} ion $z_i = -2$ whereas $|z_i| = +2$. Thus the equivalent conductance Λ_{eq} (m^2/ohm·equiv) is defined by

$$\Lambda_{eq} = \frac{\kappa}{c_{eq}} = \frac{\Lambda}{v_+ z_+} = \frac{\Lambda}{v_- |z_-|} \tag{5}$$

For example, with CaCl$_2$ $\Lambda = 2\Lambda_{eq}$ since $z_+ = 2$ and $v_- = 2$. Equation (4) can be generalized for individual ions of type i by writing

$$c_{i\,eq} = |z_i| c_i \tag{6}$$

Thus if the above electrolyte is completely dissociated in solution,

$$\begin{aligned} c_+ &= v_+ c = c_{+\,eq}/z_+ \\ c_- &= v_- c = c_{-\,eq}/|z_-| \end{aligned} \tag{7}$$

The significance of the various conductance parameters can be understood in microscopic terms by looking at the movements of the actual ionic charge carriers in the solution. If an electric field X (V/m) is applied between two electrodes in a uniform solution, any ion i moves in the direction of the field with a steady-state velocity v_i (m/s) that is proportional to the potential gradient; that is,

$$v_i = u_i X \tag{8}$$

where u_i is the mobility of the ion (m^2/s·V). Now consider an imaginary reference plane (Figure 8.1) of area A in the solution and perpendicular to the applied field. In time t seconds all the ions of type i contained in the volume $v_i t$ (m) $\times A$ (m^2), or $v_i t A c_i$ mol of i, pass across the plane. These carry an electric charge of $|z_i| F v_i t A c_i$ coulomb (C), where F is the Faraday constant (96,487 C/mol [2] for singly charged ions). The total flux of charge in coulombs transferred across the reference plane in both directions by all the anions and cations is therefore

$$\sum_i |z_i| F v_i t A c_i$$

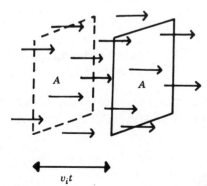

$v_i t$

Figure 8.1 Movement of ions of type i across a plane.

The current I (A) is obtained on dividing by t. Now according to Ohm's law

$$IR = Xl \tag{9}$$

Thus, by using (1) and (8), the specific conductivity κ is given by

$$\kappa = \frac{I}{AX} = \sum_i |z_i| Fu_i c_i \tag{10}$$

It is now convenient to introduce the molar conductance of i, λ_i (m²/ohm·mol), which is directly proportional to the mobility u_i according to the equation

$$\lambda_i = |z_i| Fu_i = |z_i| \lambda_{i\,eq} \tag{11}$$

Here λ_{ieq} is the equivalent ionic conductance (m²/ohm·equiv). Thus with (6),

$$\kappa = \sum_i \lambda_i c_i = \sum_i \lambda_{i\,eq} c_{i\,eq} \tag{12}$$

For example, in an aqueous solution of oxalic acid (H_2Ox) containing the ionic species H^+, HOx^-, and Ox^{2-}

$$\kappa = \lambda_{H^+} c_{H^+} + \lambda_{HOx^-} c_{HOx^-} + \lambda_{Ox^{2-}} c_{Ox^{2-}} \tag{13}$$

If only one electrolyte is present in the solution and this dissociates completely into only two kinds of ions, then by (2), (5), and (7)

$$\kappa = \Lambda c = \sum_i \lambda_i c_i = \lambda_+ v_+ c + \lambda_- v_- c \tag{14}$$

The equation $\Lambda = \kappa/c$ defines the molar conductance Λ of the electrolyte. Similarly, $\Lambda_{eq} = \kappa/c_{eq}$ defines the equivalent conductance whence

$$\kappa = \Lambda_{eq} c_{eq} = \sum_i \lambda_{i\,eq} c_{i\,eq} = \lambda_{+eq} v_+ z_+ c + \lambda_{-eq} v_- |z_-| c \tag{15}$$

It follows from (4) that

$$\Lambda = v_+ \lambda_+ + v_- \lambda_- \tag{16}$$

$$\Lambda_{eq} = \lambda_{+eq} + \lambda_{-eq} \tag{17}$$

In the case of $CaCl_2$, for instance, $\Lambda = \lambda_{Ca^{2+}} + 2\lambda_{Cl^-}$. The relative simplicity of (17) compared with (16) is one reason why many conductance workers still use equivalent rather than molar conductances.

Let us also briefly look at a binary symmetrical electrolyte that is incompletely dissociated in solution, for example CH_3COOH or $MgSO_4$. If the degree of dissociation is α,

$$c_+ = c_- = \alpha c \tag{18}$$

$$\Lambda = \frac{\kappa}{c} = \sum_i \lambda_i \left(\frac{c_i}{c} \right) = \alpha(\lambda_+ + \lambda_-) \tag{19}$$

Similarly

$$\Lambda_{eq} = \frac{\kappa}{c_{eq}} = \alpha(\lambda_{+eq} + \lambda_{-eq}) \tag{20}$$

1.2.2   Transference Number

As shown above, the charge carried across the reference plane in Figure 8.1 by ions of type i is $|z_i|Fv_itAc_i$ coulomb or $|z_i|v_itAc_i$ faraday. Now the number of faradays carried by ions of type i across the reference plane (which is taken as fixed relative to the solvent) when 1 faraday of electricity passes across that plane is the ionic or electric transport number t_i. Thus from (8), (11), and (12)

$$t_i = \frac{|z_i|v_itAc_i}{\sum_i |z_i|v_itAc_i} = \frac{|z_i|u_ic_i}{\sum_i |z_i|u_ic_i} = \frac{\lambda_ic_i}{\sum_i \lambda_ic_i} = \frac{\lambda_ic_i}{\kappa} \tag{21}$$

If the solution contains only one electrolyte, then by (2)

$$t_i = \lambda_ic_i/\Lambda c \tag{22}$$

It is clear that t_i is also the fraction of the total current carried by the ionic species i. This is actually the definition given in many elementary textbooks but it is of little use in practice since we possess no devices for measuring current fractions.

It follows that t_i is dimensionless, either positive or zero, and that

$$\sum_i t_i = 1 \tag{23}$$

For example, in an aqueous solution of oxalic acid there are present as solutes the (hydrated) species H^+, HOx^-, and Ox^{2-} and the molecular species H_2Ox. Thus from (21), the transport number of the H^+ ion is given by

$$t_{H^+} = \frac{\lambda_{H^+}c_{H^+}}{\lambda_{H^+}c_{H^+} + \lambda_{HOx^-}c_{HOx^-} + \lambda_{Ox^{2-}}c_{Ox^{2-}}} \tag{24}$$

The basic principle of measurement is evident from the definition. The number of moles of i passing some reference plane (such as an etch mark on a tube) per faraday is determined and, with the charge number of species i known, the number of faradays carried by these ions follows immediately. This is sufficient when the electrolyte dissociates into only two ions, as in the case of aqueous HCl. However, a serious difficulty arises with a solution containing complex ions; aqueous oxalic acid is a typical example. Here, because of the rapid dynamic equilibrium between the ionic and molecular species in the solution, it is impossible to measure experimentally the number of moles of H^+, HOx^-, or Ox^{2-} *ions* that pass a reference plane. Hydrogen is carried toward the cathode as H^+ ions and toward the anode as part of HOx^- ions, and the difference between the two flows, the *net* number of moles of hydrogen transferred in one direction, is the only quantity we can determine experimentally (by titration with alkali, for example). Similarly, we can obtain the net transfer of oxalate in the form of both HOx^- and Ox^{2-} ions but not the transfer of each ionic species separately. Clearly, to define the quantity actually measured in the laboratory, another kind of transference number—the ion constituent transference number—must be introduced.

It is convenient to begin with the concept of "ion constituent" or radical. Thus in oxalic acid solution the ion constituent H^+ exists in the forms of the ions H^+, the ions HOx^-, and the molecules H_2Ox, while the ion constituent Ox^{2-} exists in the form of the species HOx^-, Ox^{2-}, and H_2Ox. It is the concentration of H^+ ion constituent, not that of H^+ ions, that is determined by chemical analysis such as titration with alkali; and likewise it is the transfer of an ion constituent, and not that of an individual ion, that can be measured when an electric current flows. This brings us to the formal definition: "The transference number, T_R, of a cation or anion constituent R is the net number of faradays carried by that constituent in the direction of the cathode or anode, respectively, across a reference plane fixed with respect to the solvent, when 1 faraday of electricity passes across the plane." Clearly, T_R is dimensionless and

$$\sum_R T_R = 1 \tag{25}$$

Numerically, T_R equals the net number of equivalents of R traversing the plane per faraday, or the net number of moles of R passing for every $|z_R|$ faradays. T_R may be positive, zero, or negative. A well-known example is a concentrated aqueous solution of cadmium iodide in which the presence of many mobile CdI_3^- and CdI_4^{2-} ions results in a *net* transfer of cadmium *ion constituent* toward the anode. The ionic or electric transport number of cadmium *ions* is of course still positive though, as explained above, it cannot be measured.

A quantitative relation between T_R and the properties of the ions in solution is easily derived by supposing that each mole of ion i contains $N_{R/i}$ mol of ion constituent R. For example, in 1 mol of CdI_3^- ions there are 3 mol of iodide ion constituent and 1 mol of cadmium ion constituent. The electric charge on any ion is then the algebraic sum of the charges of its component ion constituents; that is,

$$\sum_R z_R N_{R/i} = z_i \tag{26}$$

It was shown earlier that $v_i t A c_i$ mol of ion i cross the reference plane in Figure 8.1 in time t. The number of moles of ion constituent R crossing the plane as part of the ionic species i is therefore $N_{R/i} v_i t A c_i$. Before summing over all species to find the total transfer of R we must take two directional effects into account. First, cations and anions carry a given ion constituent in opposite directions and so there is need of a factor $z_i/|z_i|$ or $|z_i|/z_i$ that is $+1$ for any cation and -1 for any anion. Second, a factor $z_R/|z_R|$ is required as well because the definition of transference number distinguishes between cation and anion constituents.

Thus the net number of moles of R crossing the reference plane is

$$\sum_i \frac{z_R}{|z_R|} \cdot \frac{|z_i|}{z_i} N_{R/i} v_i t A c_i$$

To obtain T_R we must multiply by $|z_R|$ to obtain the net number of faradays transferred by R and divide by the total flux of charge, in faradays, transferred across the plane in both directions. Canceling common factors and utilizing (8) and then (11) we finally arrive at

$$T_R = \frac{\sum_i z_R(|z_i|/z_i)N_{R/i}u_ic_i}{\sum_i |z_i|u_ic_i}$$

$$= \frac{\sum_i (z_R/z_i)N_{R/i}\lambda_ic_i}{\sum_i \lambda_ic_i} \tag{27}$$

By (6) and (11), the products λ_ic_i may be replaced by the equivalent properties $\lambda_{ieq}c_{ieq}$. Furthermore, the relation between T_R and t_i is

$$T_R = \sum_i \left(\frac{z_R}{z_i}\right) N_{R/i} t_i \tag{28}$$

The two transference numbers are equal only for an electrolyte dissociating into not more than two ionic species, for then ion constituent R and ion i become synonymous.

A quick check proves that (27) and (28) do indeed fulfill condition (25):

$$\sum_R T_R = \sum_R \sum_i t_i N_{R/i} \frac{z_R}{z_i} = \sum_i \sum_R t_i N_{R/i} \frac{z_R}{z_i}$$

$$= \sum_i t_i \sum_R N_{R/i} \frac{z_R}{z_i} = \sum_i t_i \quad [\text{by (26)}]$$

$$= 1 \quad [\text{by (23)}]$$

Let us now apply these equations to find the transference numbers of the hydrogen and oxalate ion constituents in an aqueous oxalic acid solution:

$$T_{H^+} = \frac{\lambda_{H^+}c_{H^+} - \lambda_{HOx^-}c_{HOx^-}}{\lambda_{H^+}c_{H^+} + \lambda_{HOx^-}c_{HOx^-} + \lambda_{Ox^{2-}}c_{Ox^{2-}}} \tag{29}$$

$$T_{Ox^{2-}} = \frac{2\lambda_{HOx^-}c_{HOx^-} + \lambda_{Ox^{2-}}c_{Ox^{2-}}}{\lambda_{H^+}c_{H^+} + \lambda_{HOx^-}c_{HOx^-} + \lambda_{Ox^{2-}}c_{Ox^{2-}}} \tag{30}$$

A further small term should be added or subtracted to allow for any net migration of H_2Ox molecules carried along with the ions [5] but this is normally omitted.

If the concentration of the oxalic acid is so high that the relative concentration of Ox^{2-} is negligible, it would be possible to choose as ion constituents either H^+ and Ox^{2-} as above or else H^+ and HOx^-. In the latter case:

$$T_{H^+} = 1 - T_{HOx^-} = \frac{\lambda_{H^+} c_{H^+}}{\lambda_{H^+} c_{H^+} + \lambda_{HOx^-} c_{HOx^-}} = \frac{\lambda_{H^+}}{\lambda_{H^+} + \lambda_{HOx^-}} \qquad (31)$$

This is a typical equation for a simple 1:1 electrolyte. The only restriction in the choice of ions as constituents is that they must not dissociate into smaller entities under the given experimental conditions: no ambiguity arises as long as a given choice is clearly stated and thereafter consistently adhered to. Where the solute reacts appreciably with the solvent, solvonium and solvate ion constituents (e.g., H^+ and OH^- in water) must be included. Solutions of amino acids are examples of this [6].

Further applications of (27) to other inorganic and organic electrolytes may be found in [6–8]; of particular interest is its use in a review of transport properties of polyelectrolyte solutions [9]. It must be emphasized once again that what is experimentally measured is always the transference number of an ion constituent, and only for solutions containing no more than two ionic species (e.g., strong binary electrolyte solutions) is the transference number also numerically the same as the ionic or electric transport number. Most elementary textbooks mention only the latter, and it is for this reason that in the present chapter the ion constituent transference number has been treated in some detail.

The frame of reference used in the definition of transference numbers in solution is an imaginary plane fixed with respect to the solvent as a whole. This is an operational concept suitable for macroscopic experiments, yet it leaves something to be desired from the microscopic viewpoint. Solvent molecules may be crudely classified into those that are free and those that are solvating and traveling with the ions; could one but set up a plane stationary with respect to the free part of the solvent that does not move on the passage of current, one would be able to determine "true" or "absolute" transference numbers [10]. Unfortunately, no one has yet been able to do so. Much effort has been expended on this subject, but both of the ingenious techniques employed have been shown to be invalid. In the first a small amount of a so-called inert reference substance such as sucrose was added and the movement of solvent measured relative to it, until it was found that the results depended on the substance chosen and on its concentration [11]. The reason was simple— the reference substance, too, was polar, and so to some extent it also solvated and traveled with the ions. In the second approach the electrolyte solution was divided into two (or more) sections by a diaphragm or membrane stretched across the cell, and the amount of solvent carried through during electrolysis was measured volumetrically or gravimetrically. Apart from the problem of electroosmosis, the main obstacle to success was that the type of diaphragm influenced the results, and indeed by using ion-exchange membranes one can obtain at will transference numbers ranging from 0 to 1. Similar problems have been encountered in the search for suitable inert porous frits for diaphragm cell diffusion experiments [12, 13]. Thus the solvent as a whole must be used to fix the position of the reference plane. In moving boundary experiments one measures the transference number relative to the cell and then an appropriate volume correction must be applied; this is discussed further in Section 5.3.1.

Molten salts, for which the solvent is also the electrolyte, form a special case and are discussed elsewhere [14, 15].

1.2.3   Ion Constituent Conductance and Mobility

The molar conductance $\bar{\lambda}_R$ of an ion constituent R is given in terms of its concentration c_R by

$$\bar{\lambda}_R = \frac{T_R \kappa}{c_R} = \sum_i \left(\frac{z_R}{z_i}\right) N_{R/i} \lambda_i \left(\frac{c_i}{c_R}\right) \tag{32}$$

and its mobility $\bar{u}_R$, by analogy with (11), by

$$\bar{u}_R = \frac{\lambda_R}{F|z_R|} = \sum_i \left(\frac{z_R}{|z_R|}\right)\left(\frac{|z_i|}{z_i}\right) N_{R/i} u_i \left(\frac{c_i}{c_R}\right) \tag{33}$$

This equation is simpler in terms of equivalent concentrations:

$$\bar{u}_R = \sum_i \left(\frac{z_R}{z_i}\right) N_{R/i} u_i \left(\frac{c_{i\,eq}}{c_{R\,eq}}\right) \tag{34}$$

Equations (32)–(34) apply both to solutions of a single electrolyte and to solutions of mixed salts, as in the case of seawater [16]. For a simple strong electrolyte solution like aqueous $CaCl_2$, $\bar{\lambda}_{Cl^-} = \lambda_{Cl^-}$ and $\bar{u}_{Cl^-} = u_{Cl^-}$. On the other hand, in a solution of a weak electrolyte like aqueous acetic acid the concentration of the hydrogen *ion constituent* (c_R) equals the stoichiometric concentration of the solute, c, while the concentration of the hydrogen *ions* (c_i) is αc, where α is the degree of dissociation of the acid. Accordingly,

$$\bar{u}_{H^+} = \alpha u_{H^+} \quad \text{and} \quad \bar{u}_{OAc^-} = \alpha u_{OAc^-} \tag{35}$$

These conclusions are applied later to determine the stability or otherwise of possible moving boundary systems.

2   VARIATION WITH PHYSICAL FACTORS

2.1   Variation with Concentration

2.1.1   Conductances

Molar conductances fall rapidly with rising concentration for electrolytes that are strongly associated. However, for electrolytes that are slightly associated or completely dissociated in solution, Λ values decrease quite gently with rising concentration. Clearly, the molar conductance would not change at all with concentration if the ions did not interact with each other. But they do. Coulombic interaction tends to produce order, with a given cation (say) repelling other cations and attracting anions. Counteracting this phenomenon is thermal motion, which tends to randomize the positions of the various species in the liquid. The result is a compromise: the Debye–Hückel "ionic atmosphere."

Here the cation in question has on average more negative and fewer positive ions around it than if the ionic distribution were quite random. This atmosphere acts as if the net ions of opposite charge were spherically distributed around the central ion at a distance from its surface called the Debye length, $1/\kappa$, where

$$\kappa = B\sqrt{I} \tag{36}$$

$$I = \frac{1}{2}\sum_i c_i z_i^2 = cz^2 \quad \text{(for a } z:z \text{ valent electrolyte)} \tag{37}$$

$$B(\text{dm}^{3/2}/\text{mol}^{1/2} \cdot \text{m}) = 50.29 \times 10^{10}/(\varepsilon_r T)^{1/2} \tag{38}$$

Here I is the ionic strength (mol/dm^3), ε_r the dielectric constant or relative permittivity of the solvent, and T the absolute temperature. In water at 298.15 K and 1 bar, $B = 0.329 \times 10^{10}$ [17]. It follows that the distance between the center of the ion in question and the effective spherical shell of opposite charge is $1/\kappa$ for a point ion but $(1/\kappa) + a$ for a real ion, where a is the "distance of closest approach" of cation and anion. The various physical and mathematical assumptions of the theory have been critically examined by Robinson and Stokes [17].

When an electric field is applied, the central ion and its ionic atmosphere move in opposite directions toward the electrodes. This produces two effects that slow down the movement of the central ion. First, the ions of the atmosphere collide with the solvent molecules and transfer to them part of their momentum so that the central ion moves against a counterflow of solvent [18]. This is called the electrophoretic effect, expressed symbolically by reducing the molar conductance of the central ion i by an amount λ_i^e. Second, an asymmetry effect arises because the opposite motion of the ion and those in its atmosphere disturb the latter's spherical distribution. When the central ion and the electrical center of the atmosphere no longer coincide, they are dragged back by a restoring force. This dies away as the atmosphere rearranges itself, a process that requires a short but finite time of relaxation. The average restoring force experienced by the ion is therefore called the asymmetry or relaxation effect [17]. It is commonly symbolized as a relaxation field ΔX acting in the opposite direction to the applied field X. Thus, we may write for the molar conductance of the electrolyte

$$\Lambda = (\Lambda^0 - \Lambda^e)\left(1 - \frac{\Delta X}{X}\right) \tag{39}$$

where Λ^0 is the limiting molar conductance at infinitesimal ionic strength where the ions are infinitely far apart. The same equation applies to every ion since even the ions in the atmosphere of the chosen ion are themselves centers of other time-averaged atmospheres. For $z:z$ valent electrolytes the Onsager theory gives [17]

$$\Lambda^e = \frac{B_2 z^2 \sqrt{I}}{1 + Ba\sqrt{I}} \tag{40}$$

where [17]

$$B_2(\text{dm}^{3/2} \cdot \text{m}^2/\text{mol}^{3/2} \cdot \text{ohm}) = \frac{8.249 \times 10^{-4}}{\eta(\varepsilon_r T)^{1/2}} \tag{41}$$

The viscosity of the solvent, η, must here be given in kg/m $\cdot$ s. In aqueous solutions at 298.15 K and 1 bar, $B_2 = 60.64 \times 10^{-4}$.

Equation (39) forms the basis of the Debye–Hückel–Onsager conductance theory. It is derived from the so-called primitive model [19] of rigid charged unpolarizable spheres in a dielectric continuum whose properties are those of the bulk solvent—a model sometimes jocularly referred to as "brass balls in the bathtub" or BBB. The original theory, developed and extended in recent years by several theoreticians, now accounts satisfactorily for the concentration variations observed in dilute solutions up to at least $\kappa a = 0.2$, that is, approximately 0.03 mol/dm^3 for 1:1 electrolytes in water at 298.15 K [20]. Most of the new equations can be cast into the form

$$\Lambda = \Lambda^0 - S\sqrt{c} + Ec \ln c + J_1 c - J_2 c^{3/2} \tag{42}$$

The coefficients S and E depend only on the properties of the solvent and the charges of the ions, whereas J_1 and J_2 depend also on the distance of closest approach, a. In extremely dilute solutions the dominant concentration term is that in $\sqrt{c}$. For $z:z$ valent electrolytes, the limiting Onsager slope S is given by

$$S = z^3[B_1\Lambda^0 + B_2] \tag{43}$$

where

$$B_1(\text{dm}^{3/2}/\text{mol}^{1/2}) = \frac{82.04}{(\varepsilon_r T)^{3/2}} \tag{44}$$

In water at 298.15 K and 1 bar, $B_1 = 0.2300$. The other parameters E, J_1, and J_2 are discussed below, as well as the methods for evaluating Λ^0.

Many electrolytes in water, and virtually all electrolytes in solvents of low permittivity, are only partly dissociated in solution. If the degree of dissociation is α at a concentration c, (42) must be rewritten as

$$\Lambda/\alpha = \Lambda^0 - S\sqrt{\alpha c} + E\alpha c \ln(\alpha c) + J_1\alpha c - J_2(\alpha c)^{3/2} \tag{45}$$

However, investigators do not want to find just α but rather the thermodynamic association constant K_A, where

$$K_A = \frac{(1-\alpha)f_A}{\alpha^2 c f_\pm^2} \tag{46}$$

The activity coefficient f_A of the associated species (ion pair or molecule) is usually taken as unity, and the mean ionic activity coefficient $f_\pm$ expressed by the Debye–Hückel equation [17, 20]

$$\ln(f_\pm) = -\frac{q\kappa}{1+\kappa a} \tag{47}$$

where the Bjerrum distance q equals

$$q = \frac{|z_+ z_-| e^2}{8\pi\varepsilon_0 \varepsilon_r kT} \qquad (48)$$

Here z_+ is the algebraic charge number of the cation and z_- that of the anion, e the charge on the electron (in coulombs), ε_0 the permittivity of vacuum, and k the Boltzmann constant [2]. In water at 298.15 K and 1 bar, $q = 3.57|z_+ z_-| \times 10^{-10}$ m. Alternatively, the Debye–Hückel equation is often written

$$\log_{10} f_\pm = -\frac{A|z_+ z_-|\sqrt{I}}{1 + Ba\sqrt{I}} \qquad (49)$$

where [17]

$$A(\text{dm}^{3/2}/\text{mol}^{1/2}) = 1.8246 \times 10^6/(\varepsilon_r T)^{3/2} \qquad (50)$$

which equals 0.5115 in water at 298.15 K and 1 bar. For a partially dissociated $z:z$ valent electrolyte, $I = \alpha c z^2$. It should be noted here that the mean stoichiometric activity coefficient as determined by thermodynamic methods like vapor pressure and emf is not $f_\pm$ but $\alpha f_\pm$ [21].

In practice it is often convenient to combine (45) and (46) and rearrange in the form

$$\Lambda = \Lambda^0 - S\sqrt{\alpha c} + E\alpha c \ln \alpha c + J_1 \alpha c - J_2(\alpha c)^{3/2} - \Lambda \alpha c K_A f_\pm^2 \qquad (51)$$

This is the way almost all accurate conductance data in dilute solutions are processed nowadays. The object is to fit a given set of experimental Λ, c points to (51) to obtain values of the three unknown parameters Λ^0, K_A, and a. A computer-operated least squares program is generally employed, with the standard deviation σ quoted to show the goodness of fit [20]. In a major variant of this method, forcefully advocated by Justice [22], the distance a is replaced by the Bjerrum distance q. In the Bjerrum model of ion association [17], q is the closest distance of approach of the "free" or unpaired ions whose movement contributes to conductance, whereas ions separated by distances between q and a are ion paired or associated. If the cation and anion are of equal numerical charge, the ion pair is neutral and no longer contributes to the carrying of electric charge. The distance a is the closest distance of approach of the "bound" ions in the ion pair, and it depends on the radii of the ions and the degree of their solvation. Thus the value of a is not known a priori whereas that of q is given by (48). In the Justice approach, adopted by several workers, the number of unknown variables is reduced to two: Λ^0 and K_A.

At least three major versions of (51) have been and are in common use. Their theoretical derivations differ mainly in the use of slightly different boundary conditions and in personal decisions as to the cut-off points in mathematical series. In brief, we have:

1. The theory of Fuoss and Onsager [23], later amended [24]. It was applied, without a J_2 term, to a large series of conductance data by Kay [25] who explained the least squares and weighting procedure employed.

2. The theory of Fuoss and Hsia [26] cast into the mold of (51) by Fernández-Prini [27]. This version incorporates an improved treatment for the relaxation field and has now superseded the earlier Theory 1. The necessary parameters E, J_1, and J_2 are given in Table 8.1. They are based on the Justice approach of taking $a = q$, and the appropriate numerical values were taken from a review by Fernández-Prini [18] and converted to SI units. This review should also be consulted if conductances are to be analyzed with variable values of a, in which case J_1 and J_2 are expressed in terms of the parameter $b = 2q/a$. Parameters for unsymmetrical electrolytes have been listed by Evans and Matesich [28] and more recently by Quint and Viallard [29] who have also extended the theory to mixed electrolyte solutions. The basis of computer-fitting programs has been described by several authors [18, 25, 30–32].
3. The theory of Pitts and his coworkers [33, 34], transformed into the standard expression (51) by Fernández-Prini and Prue [30].

The corresponding parameters are also listed in Table 8.1.

Recent comparisons carried out with several electrolyte systems [18, 35–37] indicate that the degrees of fit by Theories 2 and 3 are very similar, both usually better than 0.1%. There is therefore little to choose between them in practice. An excellent summary of these theories has been provided by Justice [20].

During the last few years Fuoss [38, 39] and especially Lee and Wheaton [40–42] have published new conductance equations based on a more realistic physical picture than the primitive BBB model. Lee and Wheaton [40] envisage three regions around the ion itself, which is treated as a rigid conducting sphere.

Table 8.1 Coefficients E, J_1, and J_2 of (51) for Conductance Theories 2 and 3 (See Text) Applied to Symmetrical $z:z$ Valent Electrolytes Taking the Distance of Closest Approach as q

For both theories

$$E = E'\Lambda^0 - E'', \quad E'(\text{dm}^3/\text{mol}) = 2.943 \times 10^{12} z^6/(\varepsilon_r \text{T})^3$$
$$E''(\text{dm}^3 \cdot \text{m}^2/\text{ohm} \cdot \text{mol}^2) = 433.2 z^6/(\varepsilon_r \text{T})^2 \eta$$
$$J_1 = J_1'\Lambda^0 + J_1''$$
$$J_2 = J_2'\Lambda^0 + J_2''$$

Theory 2[a]	*Theory 3[a]*
$J_1' = 2E'(\ln \phi + 2.2824)$	$J_1' = 2E'(\ln \phi + 2.7718)$
$J_1'' = 2E''(3.6809 - \ln \phi)$	$J_1'' = 2E''(4.0139 - \ln \phi)$
$J_2' = 3.8037 E' \phi$	$J_2' = 8.8784 E' \phi$
$J_2'' = 5.2872 E'' \phi - 1.5527 B_2 z^3/\Lambda^0$	$J_2'' = 10.815 E'' \phi$

[a] $\phi(\text{dm}^{3/2}/\text{mol}^{1/2}) = 4.202 \times 10^6 z^3/(\varepsilon_r \text{T})^{3/2}$.

In region 1 all the solvent molecules are aligned by the ion's field and there is dielectric saturation, in region 2 (the Gurney co-sphere of outer radius R_i) the solvent structure is still modified by the ion's field though to a lesser extent, and in region 3, beyond R_i, the solvent retains its bulk properties. The resulting conductance equation is complicated in the general case of a mixture of ions but simplifies considerably [41] for a single associated symmetrical electrolyte:

$$\Lambda = \alpha\Lambda^0(1 + C_1(\kappa q) + C_2(\kappa q)^2 + C_3(\kappa q)^3) - \left(\frac{\rho\kappa}{1+\kappa R}\right)\left(1 + C_4(\kappa q) + C_5(\kappa q)^2 + \frac{\kappa R}{12}\right)$$

(52)

where $\rho = z^2 Fe/3\pi\eta$ for molar conductances. The distance R is the maximum separation between a cation and an anion that still allows them to behave as an ion pair. The C_n symbols represent complex functions of κq tabulated by Lee and Wheaton [41] and simplified in the form of power series in κq by Pethybridge and Taba [43]. Computer programs are now available [42, 44] for fitting experimental data to the theory to derive values of the three unknown parameters Λ^0, K_A, and R. Tests conducted so far [42, 44] show the theory to be promising, particularly since it is also applicable to unsymmetrical and mixed electrolytes.

Getting the small concentration dependence exactly right is important in determining association constants K_A of slightly associated electrolytes [45]. For strongly associated electrolytes the ionic strength becomes small and the Debye–Hückel–Onsager interionic effects make a relatively minor contribution to the much greater fall of Λ with increasing concentration. The conductance behavior is then well described by simpler equations. The best of these is that of Shedlovsky [46, 47] which combines (46) with a truncated form of (45) that includes only the $S\sqrt{\alpha c}$ term and part of the term linear in αc. Suitable rearrangement yields

$$\frac{1}{\Lambda S(Z)} = \frac{1}{\Lambda^0} + \frac{c\Lambda f_{\pm}^2 K_A S(Z)}{(\Lambda^0)^2}$$

(53)

where $Z = S\sqrt{\Lambda c}/(\Lambda^0)^{3/2}$, $\alpha = \Lambda S(Z)/\Lambda^0$, and the function $S(Z)$ is given by

$$S(Z) = \left[\frac{Z}{2} + \sqrt{1 + \left(\frac{Z}{2}\right)^2}\right]^2 = 1 + Z + \frac{Z^2}{2} + \frac{Z^3}{8} + \cdots$$

(54)

Values of $S(Z)$ have been tabulated by Daggett [48]. A plot of $1/\Lambda S(Z)$ against $c\Lambda f_{\pm}^2 S(Z)$ is then a straight line from whose intercept and slope one may derive values of K_A and Λ^0. Since one needs to guesstimate a value of Λ^0 in the first place in order to evaluate the various Z values, a second approximation is often needed but rarely a third. A modified version of (53) is useful when the solvent correction is appreciable [49]. With extremely weak electrolytes the slope of the Shedlovsky plot is so steep that the value of the intercept $1/\Lambda^0$ becomes uncertain. It may then be possible to obtain Λ^0 by combining the limiting conductances of other, stronger, electrolytes. The traditional example is acetic acid in water for which, by the Kohlrausch law of independent ionic migration [cf. (64)],

$$\Lambda^0_{\text{HOAc}} = \Lambda^0_{\text{HCl}} + \Lambda^0_{\text{NaOAc}} - \Lambda^0_{\text{NaCl}} \tag{55}$$

The electrolytes HCl, NaOAc, and NaCl are completely dissociated in dilute aqueous solutions so that their Λ^0 values are easily obtained by extrapolation.

In solvents of very low permittivity ($\varepsilon_r < 12$), multiple ion association may occur. Ion pairs first solvate single ions to form triple ions of type C_2A^+ and CA_2^- with further association to quadruplets C_2A_2 and higher species. Appropriate analysis of such solutions can yield (approximate) values of association constants [50–52]. However, by a recent theory [53] it was possible to fit conductances in low-permittivity solvent without invoking triple ions. Certain electrolytes form triple ions even in solvents of quite high permittivity, examples being acids in water ($A^- H^+ A^-$) [54, 55] and sodium in liquid ammonia and amine solvents ($e^- Na^+ e^-$) [56], and this is reflected in the conductances. Quite marked changes occur in the conductance when still higher complexes form, as happens around the critical micelle concentration in polyelectrolyte solutions (Figure 8.2) [57]. Conductance is therefore a good indicator of structural change.

2.1.2   Transference Numbers

Transference numbers are essentially ratios of ionic conductances and so usually vary less with concentration than do conductances themselves. As shown in Figure 8.3, the concentration dependence of the transference numbers of most completely dissociated symmetrical electrolytes follows a standard pattern in dilute solutions. As the concentration increases, the transference numbers increase when they exceed 0.5, decrease when they are less than 0.5, and change hardly at all when they are close to 0.5. Moreover, the rate of change of transference number with concentration is greater the larger the value of $(T-0.5)$. The explanation is given by the Debye–Hückel–Onsager interionic attraction theory. The ionic analogue of (39) is

$$\lambda_i = (\lambda^0_i - \lambda^e_i)\left(1 - \frac{\Delta X}{X}\right) \tag{56}$$

For a $z:z$ valent electrolyte λ^e_i and $\Delta X / X$ are the same for both cation and anion, so that [58, 59] by (31)

$$T_{\pm} = \frac{\lambda_{\pm}}{\lambda_+ + \lambda_-} = \frac{\lambda^0_{\pm} - \frac{1}{2}\Lambda^e}{\Lambda^0 - \Lambda^e} \tag{57}$$

Note that the relaxation field term has completely canceled out so that the concentration dependence of these transference numbers depends exclusively on the electrophoretic effect. Rearrangement yields

$$T_{\pm} = T^0_{\pm} + \frac{(T_{\pm} - \frac{1}{2})\Lambda^e}{\Lambda^0} \tag{58}$$

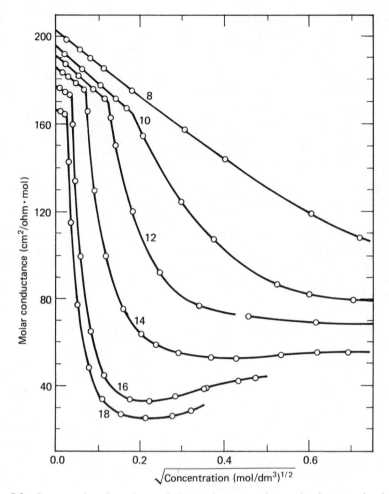

Figure 8.2 Concentration dependence of the conductances of several primary amine hydro-chlorides in water at 60°C [57]. The number on each curve designates the number of carbon atoms in the electrolyte.

Insertion of the Onsager value of Λ^e from (40) and of I from (37) leads to

$$T_\pm = T_\pm^0 + \frac{(T_\pm - \frac{1}{2})B_2 z^3 \sqrt{c}}{\Lambda^0 (1 + Baz\sqrt{c})} \tag{59}$$

In extremely dilute solutions $Ba\sqrt{I} \ll 1$ and (59) tends to the limiting form

$$T_\pm = T_\pm^0 + \frac{(T_\pm^0 - \frac{1}{2})B_2 z^3 \sqrt{c}}{\Lambda^0} \tag{60}$$

This is the equation of the dotted lines in Figure 8.3, and it can be seen that these serve well as limiting tangents to most of the experimental $T_\pm$ versus $\sqrt{c}$ plots, AgNO$_3$ being one of the few exceptions. The presence of the $(T_\pm - \frac{1}{2})$ factor in

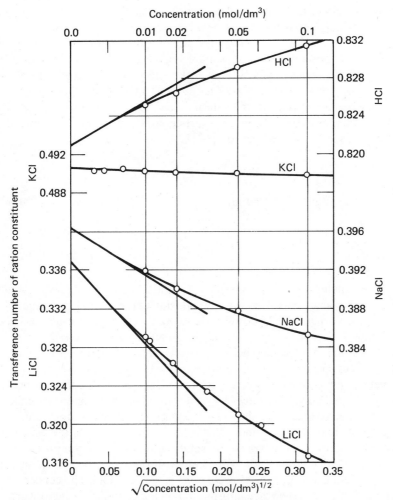

Figure 8.3 Variation with concentration of the transference numbers of some 1:1 electrolytes in water at 25°C [60]. The straight lines are the theoretical limiting slopes.

the $\sqrt{c}$ term explains the signs of the concentration dependence mentioned at the beginning of this subsection. A semiempirical extension of these equations was devised by Longsworth [60] and has frequently been employed in the literature. It involves calculating $T_{\pm}^{0}$ values (called $T_{\pm}^{0\prime}$) from a theoretical equation like (59) or (60) for a set of experimental $T_{\pm}$, c points, and then plotting $T_{\pm}^{0\prime}$ versus c. This procedure usually gives a good straight line up to about 0.1 mol/dm³ in water and extrapolation to $c = 0$ allows easy determination of the limiting transference number $T_{\pm}^{0}$. This usually agrees well with $T_{\pm}^{0}$ values obtained by more sophisticated extrapolation methods [61, 62].

Despite the success of the limiting equation (60), Sidebottom and Spiro [61] showed that neither the Onsager electrophoretic term nor that provided by the Pitts theory adequately describes the concentration dependence of $T_\pm$ at moderate concentrations. In many cases the experimental data could be fitted only by the use of physically unreasonable values of a. The trouble was traced by Justice, Perie, and Perie [62] to an "echo effect" in the theory. The electrophoretic effect for a given ion was found [63] to consist of two parts: λ^{e1}, which arises exclusively from the external field, and λ^{e2}, caused by the space-dependent part of the interionic two-particle force. With the sign convention used here, this leads to the equation [62]

$$T_\pm - \tfrac{1}{2} = (T_\pm^0 - \tfrac{1}{2}) \left(\frac{1 + 2\lambda^{e2}/\Lambda^0}{1 - 2\lambda^{e1}/\Lambda^0} \right) \tag{61}$$

which in turn can be developed into

$$\frac{(T_\pm - \tfrac{1}{2})}{(T_\pm^0 - \tfrac{1}{2})} = 1 + \frac{B_2 \sqrt{c}}{\Lambda^0} + \frac{E''}{\Lambda^0} c \ln c + \left(\frac{B_2}{\Lambda^0} \right)^2 c - \frac{2E''c}{\Lambda^0} [Q_e(b) - \ln(2\phi)] + \cdots \tag{62}$$

for $1:1$ electrolytes, where B_2 is given by (41) and E'' and ϕ by Table 8.1. A novel feature of this transference theory is the inclusion of the $c \ln c$ term. Values of the function $Q_e(b)$, where $b = 2q/a$, were found by fitting a wide range of experimental data. For most systems $Q_e(b)$ lay in a relatively narrow band between 0.2 and 4, indicating that transference numbers are fairly insensitive to the value of a and hence to specific short-range interactions [20, 62].

When symmetrical electrolytes are incompletely dissociated, the concentration c in (59)–(62) must be replaced by αc. It follows that transference numbers of associated electrolytes vary less with concentration than do transference numbers of fully dissociated ones. This is the exact opposite of the behavior of the corresponding conductances. In solutions containing complex ions the transference numbers must reflect the compositions, concentrations, and conductances of all the ionic species present, as in (29) and (30). The concentration dependence of $T_\pm$ can then give information on the complex species. Hydrogen-bonded triple ions in aqueous acid solutions [8, 54] have been detected in this way, and evidence has been obtained about triple ions in solvents of very low permittivity [64]. Classic examples are provided by concentrated aqueous zinc and cadmium halide solutions in which the cation constituent transference number is negative, indicating net transfer of metal from cathode to anode, because a large fraction of the metal ion constituent is present in the form of electrically mobile and negatively charged complexes such as $ZnCl_3^-$ and $ZnCl_4^{2-}$. In polyelectrolytes the formation of micelles surrounded by counterions leads to a sudden drop in the transference number of the counter ion constituent and a corresponding rise in that of the polymeric ion constituent (Figure 8.4) [57].

2.1.3   Concentrated Solutions

All the Debye–Hückel–Onsager based equations break down at high concentrations [65]. Above about 1 mol/dm^3 in water the curves of Λ and of T_R

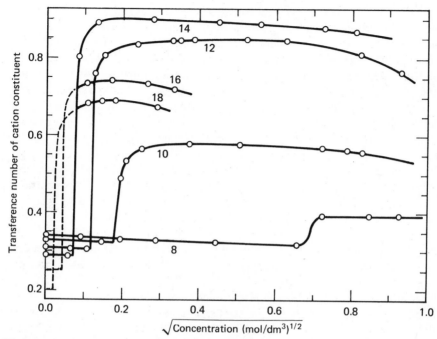

Figure 8.4 Concentration dependence of the transference numbers of several primary amine hydrochlorides in water at 60°C [57]. The number on each curve designates the number of carbon atoms in the electrolyte.

against concentration become purely empirical plots from which relatively little information on solution structure can be gleaned. A more fruitful approach is to combine all the transport properties for a given system—conductance, transference, diffusion—and to calculate from them certain other parameters that better reflect the ion–ion and ion–solvent interactions. Two major sets of representation exist. The first is provided by the thermodynamics of irreversible processes in which the flows of the ion constituents are related to electrochemical potential gradients by linear "Onsager" phenomenological or transport coefficients l_{ij}. Subscript 1 conventionally refers to the cation constituent and subscript 2 to the anion constituent. According to the Onsager Reciprocal Relations, which have several times been confirmed within experimental error [65], $l_{12} = l_{21}$. Numerical values of the three independent coefficients l_{11}, l_{22}, and l_{12} at any specified concentration c of an electrolyte $C_{v_1}^{z_1} A_{v_2}^{z_2}$ can then be calculated from the equation [66]

$$\frac{l_{ij}}{c} = \frac{T_i T_j \Lambda}{z_i z_j F^2} + \frac{v_i v_j D_v}{v R T (1 + d \ln \gamma / d \ln m)} \tag{63}$$

where Λ is the molar conductance, D_v the diffusion coefficient on its usual volume frame of reference [12], γ the mean molal activity coefficient at the molality m (mol solute per kg solvent), F the Faraday constant, R the gas constant, T the Kelvin temperature, and $v = v_i + v_j$.

Miller [66, 67] has reviewed much of the data available in concentrated aqueous solutions. Briefly, the values of l_{11}/c, which essentially represent interactions of the cations with themselves and with the solvent, generally decrease slowly with rising concentration. However, the curve for H^+ in HCl drops sharply at higher concentrations where its proton-jumping ability is impaired [66, 68]. Similarly, the values of l_{22}/c decrease gradually but in a specific way for each electrolyte. Thus the l_{22}/c values for the chloride ion constituent of different chlorides diverge as the concentration increases, and their behavior depends upon whether the salt is completely dissociated (e.g., NaCl) or not (e.g., CdCl$_2$). The l_{12}/c parameter, however, is primarily a measure of the degree of coupling between the motions of cation and anion. Hence l_{12}/c is zero at infinitesimal concentration, and the values rise with increasing concentration in a manner specific for each electrolyte. For salts of a given charge type, l_{12}/c will tend to be larger when there is significant ion pair formation (e.g., CdCl$_2$ compared with BaCl$_2$). These few examples illustrate the greater insight that can be gained by recasting the traditional transport properties into the phenomenological coefficient mold. The theory has recently been extended to transport in multicomponent solutions [69].

A more recent and alternative approach is based on linear response theory and involves time correlation functions. This method of representation was first applied to electrolyte solutions by Hertz [70, 71] and yields velocity correlation coefficients (vcc), f_{ij}. These calculations require not only conductances, transference numbers, and mutual diffusion coefficients but also self (tracer)-diffusion coefficients of the cation and anion constituent and of the solvent (D_1, D_2, and D_0, respectively). The Hertz formulation therefore leads to six independent f_{ij} coefficients that reflect cation–cation, anion–anion, cation–anion, salt–salt, salt–solvent, and solvent–solvent interactions but not ion–solvent interactions [72]. A later derivation by Woolf and Harris [73], who converted all the experimental properties to a solvent-fixed reference frame, and subsequently by Miller [72], led to velocity correlation coefficients that reflected cation–cation, anion–anion, cation–anion, cation–solvent, anion–solvent, and solvent–solvent interactions but no salt–salt or salt–solvent ones. Application to the data for various halides in water showed [71, 73] that the f_{11} and f_{22} coefficients for strong electrolytes all declined from zero at infinitesimal concentration to negative values as the concentration increased, followed in some cases by a rise. This indicates that it takes a comparatively long time for an ion to diffuse into the disturbed surroundings recently vacated by another ion of the same kind [71]. However, for electrolytes that exhibit strong complexation like CdCl$_2$ [74] and CdI$_2$ [75], the f_{22} values rise from zero to positive values and pass through a maximum. The anion–anion velocity correlations are positive here because of the presence of species such as CdI$_3^-$ and CdI$_4^{2-}$ in which the anions must move together. The values of f_{12} are also positive at low concentrations, particularly for ions that readily form ion pairs or complexes. The theory and interpretation of the vcc approach are still being actively developed.

2.2   Variation with Solute and Solvent

The general interionic attraction theory, discussed above, allows us to separate the values of conductances and transference numbers into limiting values (designated by a zero superscript) and terms due to association. The latter lead to association constants. Their values range over many powers of ten and depend on whether the association arises from covalent bonding (as in the CH_3COOH molecule) or from purely coulombic interactions (e.g., $Cu^{2+}SO_4^{2-}$ ion pairs in water). Theories for the latter effect [17] explain why association increases markedly as the charge on the ions increases and as the permittivity of the solvent decreases. However, since association constants may also be determined by other techniques [76, 77] such as spectroscopy, colligative properties, distribution methods, and emfs of concentration cells, their structural variations and specific ion–ion interactions are not further discussed in this chapter.

The limiting values of Λ^0 and T_R^0 apply at infinitesimal ionic strength (not at "infinite dilution" since even the pure solvent contains ions [1, 78]) where the electrolyte is completely dissociated and the ions are an infinite distance apart. Here the limiting conductances and transference numbers are known functions of the limiting ionic conductances λ_i^0. For the electrolyte $C_{v_+}A_{v_-}$, by (2), (7), (11), (16), (27), and (32):

$$\Lambda^0 = (\kappa/c)_{c\to 0} = v_+\lambda_+^0 + v_-\lambda_-^0 \tag{64}$$

$$T_+^0 = \frac{v_+\lambda_+^0}{v_+\lambda_+^0 + v_-\lambda_-^0} \tag{65}$$

$$\lambda_+^0 = \frac{T_+^0\Lambda^0}{v_+} = |z_+|Fu_+^0 \tag{66}$$

where the subscript + refers to the cation constituent C. Therefore only the behavior of the limiting conductances of individual ions needs to be considered in discussing the variations of the limiting transport properties with solute and solvent.

Inspection of tables of ionic conductances at 25°C in various solvents allows certain broad generalizations to be made [79]:

1. For any given ion (except small inorganic ones), the more viscous the solvent the slower the ion.
2. In almost all solvents, the conductances of the alkali metal cations increase as the crystal radii increase, that is, $\lambda_{Li+}^0 < \lambda_{Na+}^0 < \lambda_{K+}^0 < \lambda_{Rb+}^0 < \lambda_{Cs+}^0$.
3. Iodide is the slowest halide ion in most aprotic solvents and the fastest halide ion in all protic solvents.
4. In many dipolar aprotic solvents, the fastest anions possess appreciably greater conductances than the fastest cations. This tends not to happen in protic solvents.
5. The conductances of R_4N^+ cations tend to decrease as the size of R increases [in contradistinction to 2].

6. Organic anions, too, tend to move more slowly the larger they are.
7. The H^+ ion is outstandingly mobile in hydroxylic solvents such as the alcohols, and in a few solvents the solvate anion is also exceptionally mobile (e.g., water [17], hydrogen fluoride [80], and sulfuric acid [78]).
8. In certain amine and ether solvents, the solvated electron exhibits exceptional mobility [81].
9. The molar conductances of ions of similar size rise as the charge number of the ion increases.

The theory of ionic conductances may be conveniently treated in terms of the friction coefficient ζ (kg/s). Let an ion i move with a steady-state velocity v_i^0 (at infinitesimal ionic strength) when subject to a force $\mathscr{F}$. Then

$$v_i^0 = \frac{\mathscr{F}}{\zeta_i} \tag{67}$$

If the force arises from the application of a potential gradient X,

$$\mathscr{F} = |z_i|eX \tag{68}$$

Hence, by (8) and (11),

$$u_i^0 = \frac{v_i^0}{X} = \frac{|z_i|e}{\zeta_i} \tag{69}$$

$$\lambda_i^0 = \frac{z_i^2 eF}{\zeta_i} = \frac{z_i^2 F^2}{L\zeta_i} \tag{70}$$

where $-e$ is the charge on the electron and L is the Avogadro number. For spherical ions of radius r_i moving through a continuous incompressible fluid of viscosity η, a hydrodynamic equation derived by Sir George Stokes in 1850, later applied by Sutherland, and more recently discussed by Tyrrell [82], leads to

$$\zeta_i = n\pi\eta r_i \tag{71}$$

Here n is a numerical factor that ranges from 4 for conditions of "slip" (i.e., if the sphere is completely slippery and does not drag along any liquid with it) to 6 for conditions of "stick" (where the solvent immediately adjacent to the ion "wets" the sphere and so moves along with it) [79]. Combination of (70) and (71) with insertion of numerical values [2] gives, in SI units,

$$\lambda_i^0(m^2 \cdot S/mol)\eta(kg/m \cdot s)r_i(m) = 4.920 \times 10^{-15}z_i^2/n \tag{72}$$

The theory shows that this frictional coefficient, unlike that in gases, depends upon the size and shape (see below) of the solute species but not upon its mass [83]. Experiment has confirmed that the differences in the ζ values for isotopic ions in water are slight, around 0.1–0.3% [84–86]. However, the Stokes equation holds quantitatively only for ions much larger than the solvent molecules or agglomerates because only then can the medium be justifiably regarded as a structureless continuum. For ions at least as large as Bu_4N^+, such as Am_4N^+ or BPh_4^-, the corollary

$$\lambda_i^0 \eta = \text{constant} \qquad (73)$$

applies well in most solvents, with water being slightly anomalous [79, 87]. This is known as Walden's rule. It can be extremely useful for making reasonable predictions of the individual ionic conductances, to the order of ± 1–2%, in solvents and at temperatures for which no transference numbers are available to split the molar conductances into their ionic parts [87]. Once the conductance of one particular ion is known, of course, those of all the other ions can be calculated solely from limiting electrolyte conductances through the additivity relation (64). An alternative and improved method was originally suggested by Fowler and Kraus [88]. This is based on the use of a reference electrolyte such as $(i\text{-}Am)_4N^+B(i\text{-}Am)_4^-$ [89] whose two ions are massive, effectively spherical, of equal radius, and present a similar chemical "face" to the solvent: they are then presumed to possess equal mobilities in any given medium. The limiting molar conductance of this salt can therefore be divided by 2 to yield the limiting conductances of each of the component ions. This method offers a better chance of success than Walden's rule because of a certain parallelism in the variations of the ionic Walden products in going from solvent to solvent: such variations then partially cancel out in the ratio λ_+^0/λ_-^0 of the reference electrolyte [79]. Clearly this "equitransferent" electrolyte can be a most helpful device for those solvents (not hydroxylic ones) in which it is sufficiently soluble and stable. However, it must be pointed out that ionic conductances so derived are only known to an accuracy of about 1%, whereas a proper transference study can improve on this by a factor of 10 [79].

Even for smaller spherical ions, the Stokes equation (72) explains at least qualitatively why ionic conductances decrease as the solvent viscosity rises and why, in a given solvent, the conductances of organic ions are smaller the larger the ion. That the mobilities of alkali metal cations increase with increasing crystallographic radius is traditionally interpreted in terms of the strong solvation of the smaller ions like Li^+ where the migrating entity is the solvated and not the bare ion. This is sometimes described as the "solventberg" picture. The degree of solvation is also greater the larger the charge on the ion, and for this reason molar conductances do not rise as fast as the z_i^2 factor suggests.

Before dealing with recent refinements of the Stokes equation, we should examine how ζ is affected if the ion is not spherical. Much of our knowledge here comes from diffusion coefficients D_i since, by an argument due to Einstein [82],

$$D_i^0 = k\text{T}/\zeta_i \qquad (74)$$

where k is the Boltzmann constant. The diffusion coefficient route allows one to apply the theoretical equations below to molecular as well as ionic species and so obtain more information on shape factors. Let us consider first the friction coefficients of rigid solute species starting with ellipsoids whose semiaxes are of length a, b, and c. The appropriate equations were given by Gans and by Perrin [90]. For simplicity we will quote only the cases of ellipsoids of revolution around the a axis so that $b=c$. Then the translational friction coefficients in the various directions are given by

$$\zeta_a = \frac{16\pi\eta(a^2-b^2)}{(2a^2-b^2)S-2a} \tag{75}$$

$$\zeta_b = \zeta_c = \frac{32\pi\eta(a^2-b^2)}{(2a^2-3b^2)S+2a} \tag{76}$$

For prolate ellipsoids where $a>b$ the function S equals

$$S = \frac{2}{\sqrt{a^2-b^2}} \ln\left(\frac{a+\sqrt{a^2-b^2}}{b}\right) \tag{77}$$

while for oblate ellipsoids where $a<b$,

$$S = \frac{2}{\sqrt{b^2-a^2}} \tan^{-1}\left(\frac{\sqrt{b^2-a^2}}{a}\right) \tag{78}$$

If the solute moves with random orientation, the mean translational friction coefficient is given by [91, 92]

$$\frac{1}{\zeta_e} = \frac{1}{3}\left(\frac{1}{\zeta_a}+\frac{1}{\zeta_b}+\frac{1}{\zeta_c}\right) \tag{79}$$

whence

$$\zeta_e = 12\pi\eta/S \tag{80}$$

These equations have been applied with fair success to small rigid ions such as benzene-1,4-disulfonate [92, 93].

Two special cases of the ellipsoidal model have been quoted by Barr [94]. The first is that of an oblate ellipsoid in which $a \ll b=c$ so that it assumes the shape of a thin circular disk of radius $r=b$. Hence for the disk moving broadside on through the solvent (in "stick" or "nonslip" fashion)

$$\zeta_a = 16\eta r \tag{81}$$

whereas edgewise movement leads to

$$\zeta_b = \zeta_c = 32\eta r/3 \tag{82}$$

Thus a species shaped like a flat disk can maximize its velocity by migrating edgewise, and there is evidence that it does so [82, 95]. The other special shape is that of a rigid rod which can to a first approximation be regarded as a prolate ellipsoid in which $a \gg b=c$. Its length is therefore $l=2a$ and its maximum radius of cross section $r=b$. If the rod moves lengthwise,

$$\zeta_a = \frac{4\pi\eta l}{2\ln(l/r)-1} \tag{83}$$

whereas broadside motion gives

$$\zeta_b = \zeta_c = \frac{8\pi\eta l}{2\ln(l/r)+1} \tag{84}$$

Similar relationships have been derived by Broersma [96] for the motions of closed cylinders. If the rod moves with random orientation, then from (79)

$$\zeta_e = \frac{3\pi\eta l}{\ln(l/r)} \tag{85}$$

The same equation has been given by Rice [97].

Kirkwood and Riseman [98, 99] and Peterlin [100] later developed formulas for the friction coefficients of chainlike solutes. The usual model is of N monomeric units, often regarded as spherical beads of radius a_0 and with individual friction coefficients ζ_0 [$=6\pi\eta a_0$, cf. (71)], set a distance b_0 apart. For a rigid rodlike solute [99, 97, 101]

$$\zeta = \frac{6\pi\eta b_0 N}{\ln N - (1 - 6\pi\eta b_0/2\zeta_0)} \tag{86}$$

This equation has been applied to p,p'-polyphenyldisulfonates [101] and other bolaform ions or bolions. Their name derives from the bola [102], a missile weapon consisting of balls of stone or iron attached to the ends of a cord or thong. Thrown properly it can wrap itself around the legs of an animal and bring it down. The two charges of like sign at the end of a bolaform ion often repel each other so much that the ion is completely stretched out [101]. At the other extreme lie randomly coiled macromolecules. If the chain is sufficiently long the friction coefficient is given by [97, 98, 101]

$$\zeta = \frac{N\zeta_0}{1 + [8N^{1/2}\zeta_0/3\pi(6\pi)^{1/2}\eta b]} \tag{87}$$

Here b is the effective bond length which, in the case of free rotation, is related to the actual bond length b_0 by a function of the fixed skeletal bond angle θ [98, 100]

$$b = b_0 \left(\frac{1 - \cos\theta}{1 + \cos\theta}\right) \tag{88}$$

Different forms of (87) have appeared in other publications [97, 101]: they can be interconverted by algebraic manipulation and by using the relation $Nb^2 = \langle h^2 \rangle$, the mean square separation of the chain ends. More complicated equations arise for hindered rotation around each bond and for short chains where Peterlin's more elaborate formulas must be applied [97, 100, 101].

Let us return to spherical solutes. Many workers [17, 79, 103–105] have shown that the Stokes equations (71) and (72) do not hold quantitatively unless the ionic radius exceeds 5.5 Å in water and 10 Å in a solvent such as sulfolane. The reason lies in the hydrodynamic model which has treated the solvent as a structureless continuum. In real life, therefore, the Stokes equation cannot be expected to apply well unless the ion is considerably larger than a molecule of the solvent, which is rarely the case for electrolyte systems of common interest. Many attempts have therefore been made to develop empirical and theoretical modifications of the Stokes treatment. Among the better empirical methods is that of Robinson and Stokes [17] who plotted a correction factor curve, based on crystal radii and molar volumes, for ions in water. Similar plots are now available for certain other solvents [79]. On the theoretical side it has been recognized that the ion experiences not only the hydrodynamic friction considered in the Stokes model but also dielectric friction. The moving ion orients the solvent dipoles around it, and these can relax again into a random distribution only after the ion has passed. Thus an electrostatic drag force, proportional to the dielectric relaxation time τ of the solvent, is exerted on the ion. The theory is difficult, and earlier treatments by Zwanzig [106] were found greatly to overestimate the effect [79, 105, 107]. A later analysis by Hubbard and Onsager [108] has led to the following equations

$$\zeta_i = 6\pi\eta r_i + \left(\frac{17}{280}\right)\frac{\tau e^2}{r_i^3}\left(\frac{\varepsilon_s - \varepsilon_\infty}{\varepsilon_s^2}\right) \quad \text{(for "stick")} \tag{89}$$

$$\zeta_i = 4\pi\eta r_i + \left(\frac{1}{15}\right)\frac{\tau e^2}{r_i^3}\left(\frac{\varepsilon_s - \varepsilon_\infty}{\varepsilon_s^2}\right) \quad \text{(for "slip")} \tag{90}$$

where $-e$ is the charge on an electron and ε_s and ε_∞ are the low-frequency (static) and high-frequency permittivities, respectively, of the solvent. Dielectric friction thus effectively acts as an increase in the local viscosity around an ion [109]. Equations (89) and (90) were recently tested against literature data for a wide range of aprotic and hydrogen-bonded solvents [109]. The results are disappointing. Particularly for the smaller ions, where the dielectric friction effect is greatest, the discrepancies are wide with the predicted conductances generally much larger than the experimental ones. The introduction of dielectric saturation does little to bridge this gap [110]. Moreover, because the theory is based on a continuum model, it does not explain why certain anions in dipolar aprotic solvents possess considerably higher molar conductances than the fastest cations [79].

The ideal theory of ionic conductance would explain ionic motion in terms of the ion–solvent interaction, the structure of the solvent surrounding the ion, and the nature of motion of the solvent molecules [109]. Wolynes has recently begun to develop such a molecular approach [111–113]. He expressed the friction coefficient microscopically in terms of the correlation function of random forces, and then separated the forces acting on the ion into a rapidly varying part caused by hard collisions with solvent molecules and a more slowly varying part arising from the softer attractive ion–solvent forces. In the limit of strong but short-range ion–solvent interactions the theory led to a rigid solvation or "solventberg" picture while at the other limit, for very weak but long-range potentials, the theory reduced to the early continuum dielectric friction formula of Zwanzig [106]. Numerical calculations [112] using only hard sphere radial distribution functions and ion–dipole potentials led to predicted ionic conductances that (as for the continuum theory) were too high for small ions. However, the molecular theory is clearly capable of further refinement and can, in principle, allow one to distinguish between ions of different sign.

In many hydroxylic solvents the hydrogen (solvonium) cation and/or the solvate anion possess unusually large conductances [79]. In water at room temperature, for example, the H_3O^+ ion moves almost five times as rapidly as the next fastest singly charged cation while the limiting mobility of OH^- is about two and one-half times as large as that of the next fastest anion [17]. Size cannot be the explanation since both experimental evidence and theoretical calculations show that the proton in solution is attached to at least one molecule of solvent whose radius is comparable with that of ions such as K^+ [114]. Clearly hydrodynamic or bodily transport of these solvent ions is supplemented by an additional conductance mechanism. This feature is particularly pronounced in a solvent like pure sulfuric acid where the high viscosity greatly decreases the rate of bodily transport of "normal" ions [cf. (72)]: here the limiting conductances of $H_3SO_4^+$ and HSO_4^- are more than 100 times greater than those of other ions [78]. It is generally accepted that the additional mechanism involves proton-jumping between favorably oriented solvent molecules. According to Conway, Bockris, and Linton [115], the rate-determining step in water is the rotation of an H_2O molecule in the field of an H_3O^+ ion to a position that allows a fast proton-jump between them, probably by quantum-mechanical tunneling. Isotopic and temperature studies appear to support this Grotthuss-type model. Prototropic conductance relies on the presence of bulk aggregates of the hydroxylic solvent and does not operate when a diluent solvent (like dioxane) is added or in concentrated electrolyte solutions where most of the solvent molecules become bound in the solvation sheaths of the ions [116]. Unfortunately the theory does not allow an easy *a priori* prediction as to which solvent ions will show such "excess" conductance; it is surprising, for instance, that $MeOH_2^+$ in methanol conducts abnormally quickly but OMe^- does not. Several other theories of prototropic conductance have been examined critically by Erdey-Grúz and Lengyel [116]. Recently the Grotthuss principle was questioned by Hertz [117] in one of a series of controversial papers [118, 119] on the concentration and conductance of H^+ in water.

Another solute species, the electron, displays exceptionally high conductances in polar solvents like liquid ammonia, amines, ethers, and alcohols [81] as well as in water where it exists only transiently [120]. The electron lies inside a solvent "cavity," the whole entity being loosely described as a solvated electron. Its high electrical conductance cannot be explained by bodily motion of the whole cavity and it is believed that the electron jumps or leaks away by quantum-mechanical tunneling. However, acceptable estimates of its excess conductance have yet to be made with this model.

2.3 Variation with Temperature, Pressure, and Field

Limiting ionic conductances rise with increasing temperature. This rise largely parallels the corresponding fall in the viscosity of the solvent, as expected from the Stokes equation (72). Thus the limiting molar conductances of most salts in water at 25°C increase by about 2% per 1°C while the limiting transference numbers, which are essentially ratios of ionic conductances, only change by amounts of the order of 0.1% per 1°C. Usually, however, the faster the ion the smaller is the temperature coefficient of its conductance so that the transference numbers of many strong symmetrical electrolytes approach 0.5 as the temperature rises. Solvonium and solvate ions that exhibit protonic conductance possess smaller temperature coefficients of conductance as the bulk solvent structure is progressively broken up in hotter solutions [116]. The value of $\lambda_{H^+}^0$ in water reaches a maximum of approximately 950 cm²/ohm·mol at around 300°C where it is now only twice as large as $\lambda_{K^+}^0$, compared with a factor of 5 at 25°C [121]. However, relatively little work over a wide temperature range has been done in nonaqueous solvents and there is an even greater dearth of transference data. Conductances in molten salts were recently discussed by Smedley [122].

A quantitative relation between λ_i^0 and the absolute temperature T is provided by the transition state theory. If the ion is pictured as migrating through the liquid by a series of jumps of length L from one equilibrium position to another, and if each jump requires a characteristic standard partial molar free energy of activation $\Delta G^{0\ddagger}$, then it can be shown that [123, 124, 79]

$$\lambda_i^0 = \frac{z_i^2 e F L^2}{6h} \exp\left(\frac{-\Delta G^{0\ddagger}}{RT}\right) \tag{91}$$

where h is the Planck constant and R the gas constant. The pre-exponential factor for a univalent ion equals 0.350 m²/ohm·mol if $L = 3\,\text{Å}$. It follows that

$$\left(\frac{\partial \ln \lambda_i^0}{\partial T}\right)_P = \frac{\Delta H^{0\ddagger}}{RT^2} + \frac{2\alpha}{3} = \frac{E_P}{RT^2} \tag{92}$$

$$\left(\frac{\partial \ln \lambda_i^0}{\partial T}\right)_V = \frac{\Delta U^{0\ddagger}}{RT^2} = \frac{E_V}{RT^2} \tag{93}$$

$$\left(\frac{\partial \ln \lambda_i^0}{\partial P}\right)_T = -\frac{\Delta V^{0\ddagger}}{RT} - \frac{2\beta}{3} \tag{94}$$

where $\Delta H^{0\ddagger}$, $\Delta U^{0\ddagger}$, and $\Delta V^{0\ddagger}$ are the standard partial molar enthalpy, internal energy, and volume of activation, respectively, and E_P and E_V are the activation energies at constant pressure and constant volume. In most solutions E_V is much smaller than E_P and varies less with temperature. The parameters α and β are the coefficients of cubical expansion and of compressibility, respectively, of the solvent, and arise from the variation of L with temperature and pressure

$$\left(\frac{\partial \ln L}{\partial T}\right)_P = \frac{1}{3}\left(\frac{\partial \ln V}{\partial T}\right)_P = \frac{\alpha}{3} \tag{95}$$

$$\left(\frac{\partial \ln L}{\partial P}\right)_T = \frac{1}{3}\left(\frac{\partial \ln V}{\partial P}\right)_T = -\frac{\beta}{3} \qquad (96)$$

A refined analysis [124, 125] shows that the differentiation in (93) should properly be carried out at constant volume of the activated state rather than at constant total volume. The usefulness of (92)–(94) lies in the derived thermodynamic properties of activation whose experimental values provide insight into the phenomenon of ionic transport.

The limiting conductances of normal ions in water change very little with increasing pressure [126]. At 25°C they initially increase by a few percent to 1 kbar and then decrease almost linearly above 2 kbar until they reach about 70% of their original value at 10 kbar where water freezes. In the high-pressure range this parallels the change in the fluidity of water, $1/\eta$, as might be expected from Stokes' law. However, the conductance of the H^+ ion rises by 8% up to 2 kbar and then slowly decreases. In organic solvents, where the viscosities increase appreciably with rising pressure, the conductances fall accordingly.

At finite concentrations in water the conductances of weak electrolyte solutions increase markedly with pressure because the dissociation constants become larger. A striking example is water itself; its autoprotolysis constant K_w has risen fourfold at 2 kbar and has leapt from $10^{-14}\,mol^2/kg^2$ at 25°C and 1 bar to about 1 mol^2/kg^2 at 1000°C and 150 kbar [127]. This results in large solvent corrections for aqueous conductance measurements at high pressures.

Another physical parameter that can be varied at will is the strength of the electric field, X. When X becomes sufficiently large (1–10 MV/m), the ions move so fast that the ionic atmosphere does not have time to build up completely. The electrophoretic and relaxation effects are correspondingly diminished and Λ then approaches Λ^0 [17]. This is the so-called first Wien effect whose theory was developed by Onsager's group [128]. With weak electrolytes a second Wien effect comes into play—an increase in the dissociation constant of the electrolyte. This dissociation field effect typically increases the conductance by a few units percent at 10 MV/m. Experimental measurements by Bailey and Patterson [129] confirmed the theoretical predictions of Onsager [128]. Further developments of the theory have appeared recently [130]. The Wien effects are unlikely to have any significant influence in moving boundary determinations of transference numbers since the potential gradient along the tube rarely exceeds 20 kV/m.

Conductance measurements are customarily carried out with alternating current to reduce polarization disturbances at the electrodes. When the usual frequency of 10^3–10^4 Hz is raised to several MHz, the molar conductance is found to increase and to approach a limiting value of less than Λ^0. The explanation again lies in the interionic effects: when the frequency is so large that the period of oscillation becomes comparable to the time of relaxation of the ionic atmosphere, the latter departs less from the spherical symmetry of the unperturbed state [128]. The asymmetry or relaxation effect therefore diminishes at high frequencies while the electrophoretic effect is retained. A quantitative theory has been derived by Debye and Falkenhagen and the results are in good agreement with experiment [128].

3   APPLICATIONS

3.1   Fundamental Information

Solvation and other aspects of ion–solvent interaction are understood most easily from properties of individual ions rather than from those of complete electrolytes. In fact the only property of single ions at infinitesimal ionic strength that we can determine unambiguously is the limiting molar ionic conductance λ_i^0. As shown by (66), it is obtained by multiplying the limiting molar conductance by the limiting transference number. Some structural interpretations based on ionic conductances were described in Sections 2.2 and 2.3. Once λ_i^0 values have been tabulated for a given solvent, the limiting conductances and transference numbers of many other electrolytes can be predicted from (64) and (65). Furthermore, from such a table one may calculate the limiting trace diffusion coefficient of an ion i by combining (70) and (74):

$$D_i^0 = \frac{RT\lambda_i^0}{z_i^2 F^2} \tag{97}$$

By another well-known equation due to Nernst [17], one can also calculate the limiting diffusion coefficient of an electrolyte $C_{v_+}A_{v_-}$,

$$D^0 = \frac{RT}{F^2} \left(\frac{z_+ + |z_-|}{z_+|z_-|} \right) \left(\frac{\lambda_+^0 \lambda_-^0}{|z_-|\lambda_+^0 + z_+\lambda_-^0} \right) \tag{98}$$

Ion–ion interaction is studied by the variation of electrolyte properties with concentration. Conductances provide information on both the electrophoretic and relaxation effects [(39)], whereas transference numbers depend upon the electrophoretic effect alone [cf. (57)]. This allows the two effects to be investigated independently of each other [59]. Another, but less common, way of separating the two effects is by measuring conductances at extremely high frequencies where the relaxation effect becomes very small.

Of cardinal importance in the field of irreversible thermodynamics is the validity of the Onsager Reciprocal Relations. A useful test of these, for the case of isothermal electrical and matter transport, is the numerical equality at high concentrations of the transference numbers determined by the Hittorf method and from the emfs of cells with transference. In recent years this reciprocal relation has been tested and confirmed by several such experiments [131–133].

3.2   Analysis

3.2.1   Direct Measurement of Conductance

The specific conductivity κ of a solution is a sensitive function of its ionic content. For a strong electrolyte, (12) shows that κ rises almost proportionately to concentration c provided the latter is not so high that interionic forces significantly decrease the molar conductance. If the electrolyte is very weak, it follows by a combination of (12) and (46) that κ is approximately proportional to $\sqrt{c}$. In either case κ is a good measure of concentration. Conductance is therefore a preeminently suitable method for determining the concentration of a given electrolyte provided a calibration curve has first been established so that the Λ–c relationship is known [cf. (42)]. By attention to good technique, including close control of temperature, an accuracy of 0.1% or better in c is fairly readily attainable. The amount of electrolyte present in mixed solutions is also frequently estimated by conductance, an example being the salt content of brackish waters. The salinity (the mass of mineral salts per kg) of different samples of seawater has been routinely determined by conductivity on the assumption that the relative composition of seawater is constant [16]. Conductance is increasingly employed to monitor industrial processes and has been used, for instance, to control the alkali concentration in a large-scale method of peeling tomatoes for canning [134].

A specific instance of conductance sensing in analysis is provided by the novel technique of ion chromatography [135]. This is particularly useful for determining the type and concentration of anions present in a dilute mixture. The method consists of passing the solution through a separator column and then through a suppressor column filled with cation-exchange resin in the H^+ form. The anions thus emerge consecutively in their highly conducting acid form. The eluent flows through a conductance cell, and an attached conductivity meter and recorder then show up a series of peaks that correspond to the various anions. After calibration, the position of each peak signals the type of anion present (Cl^-, NO_3^-, SO_4^{2-}, etc.) and its height or area is a measure of the anion's concentration in the original mixture.

3.2.2   Conductometric Titrations

Conductance is one of the main electrochemical methods for following the course of certain titrations. It is a feasible method whenever the rate of change of conductance (with amount of reagent added) occurring *before* the equivalence point differs significantly from the rate of change occurring *after* it. The graph of conductance versus volume of reagent added will then show a change of slope around the equivalence point [136, 137]. The conductometric method is therefore often suitable for acid–base titrations since in water and several other solvents the solvonium (H^+) and solvate (e.g., OH^-) ions possess much higher ionic conductances than other "normal" ions.

In practice the conductance cell may serve as the reaction vessel (Figure 8.5a) or else a conductance electrode holder can be dipped into a beaker (Figure 8.5b). The solution must be well mixed with a magnetic or mechanical stirrer after each successive addition of reagent from the burette. Determining the cell constant is unnecessary; it is quite sufficient to plot the overall conductance G or the reciprocal of the resistance, $1/R$. The reagent solution is usually 10–50 times more concentrated than the solution being titrated and the extent of dilution is then small enough to be ignored in the plot. If dilution is significant,

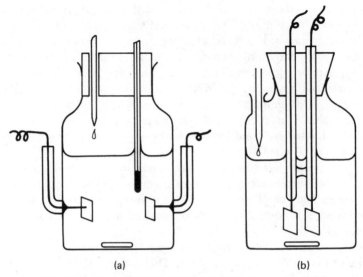

Figure 8.5   Apparatus for conductometric titrations, after [136].

however, $G(V_0 + V)/V_0$ should be plotted against V, where V is the volume of reagent added and V_0 the initial volume in the titration vessel. Conductance titrations can be performed with dilute or highly colored solutions where ordinary dye indicators fail, but the solutions must not contain an excess of some inert supporting electrolyte because this greatly reduces the sensitivity of endpoint detection. In a variation of the normal technique [136, 137], high-frequency (>1 MHz) or oscillometric titrations can be carried out with no electrodes inside the reaction vessel (Section 4.2).

Figure 8.6 illustrates some typical conductometric acid–base titrations. In Figure 8.6a the addition of NaOH results in the reaction

$$H^+ + Cl^- \xrightarrow{\text{Na}^+ + \text{OH}^-} Na^+ + Cl^- + H_2O$$

so that the highly conducting H^+ ions are progressively replaced by much slower Na^+ ions. The conductance therefore falls before the endpoint and, after it, the addition of excess NaOH produces a rise. The endpoint itself is obtained by extrapolating the two linear portions. In Figure 8.6b the conductance increases when alkali is added because the slightly dissociating acetic acid is replaced by the completely dissociated salt Na^+OAc^-. (The small decrease at the beginning is caused by the initial suppression of the dissociation of the acid by the common ion effect of the extra acetate ions formed.) The curvature around the endpoint arises from hydrolysis of the sodium acetate and the final increase in conductance is, as before, caused by the addition of excess Na^+ and OH^- ions. Conductance–volume plots often display some curvature around the equivalence point because of hydrolysis, dissociation, or the finite solubility of an insoluble product, and so the points in its vicinity should be disregarded when extrapola-

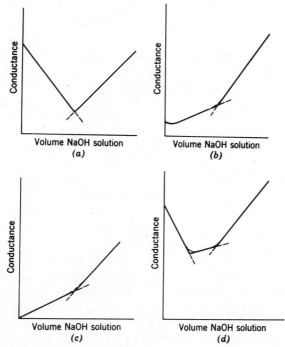

Figure 8.6   Typical conductometric titration plots in aqueous media. (*a*) Strong acid (e.g., HCl) + strong base. (*b*) Weak acid (e.g., CH_3COOH) + strong base. (*c*) Very weak acid (e.g., H_3BO_3) + strong base. (*d*) Mixture of strong and weak acid (e.g., HCl + CH_3COOH) + strong base.

ting the linear sections of the plots. Precipitation reactions may also be followed conductometrically. That of $AgNO_3$ by NaCl produces a nearly horizontal plot before the endpoint as $Ag^+ + NO_3^-$ ions are replaced by the almost equally conducting $Na^+ + NO_3^-$ ions, but after the equivalence point the conductance increases as excess NaCl is added. Such plots also tend to be curved around the endpoint if the solubility product is insufficiently small, and decreasing the solubility by adding methanol or ethanol is then beneficial. Conductometric titrations can also be carried out in nonaqueous solvents. Further details may be found elsewhere [136–138].

3.3   Structural Investigations

The limiting molar conductances of single electrolytes offer clear pointers to the way in which they dissociate. Take ML_2Cl_2 in water as an example, where M is a metal atom and L a nondissociating ligand like NH_3. If this salt dissolves as a molecular species, Λ^0 will be zero, whereas dissociation into $[ML_2Cl]^+ + Cl^-$ or $[ML_2]^{2+} + 2Cl^-$ would give rise to Λ^0 values of approximately 140 and 260 cm^2/ohm · mol, respectively, at 25°C [17, 139]. Measurements at somewhat different temperatures could be corrected to 25°C by

Walden's rule (73). Confirmatory evidence may be gained from the concentration dependence of the molar conductance in very dilute solutions since, in any case, Λ has probably been plotted against $\sqrt{c}$ to obtain Λ^0. Feltham and Hayter [140] have presented graphs of $\Lambda_{eq}^0 - \Lambda_{eq}$ versus $\sqrt{c_{eq}}$ for a variety of solutes of different structural types in dilute solutions in water, methanol, and nitromethane, and have demonstrated that clear demarcations in the slopes exist between salts of types such as 1:1, 2:1 or 1:2, 3:1 or 1:3. The slopes of the various structural types differ even more in molar terms since, by (4) and (5),

$$\frac{d\Lambda}{d\sqrt{c}} = (v_+ z_+)^{3/2} \frac{d\Lambda_{eq}}{d\sqrt{c_{eq}}} \tag{99}$$

For this method to succeed the counterion (such as Cl^-) must be large to decrease the possibility of ion association and there should be no significant solvolysis or other side reaction. One of the main practical difficulties in nonaqueous solvents lies in the fact that compounds that dissociate into highly charged ions are often only slightly soluble.

Transference numbers can often serve as clear structural indicators. As early as 1859, Hittorf electrolyzed aqueous potassium silver cyanide solutions and found that the transference number of the silver ion constituent was negative and that its numerical value was always one-half that of the transference number of the cyanide ion constituent. In other words, silver and cyanide both migrated from cathode to anode in the molar ratio of 1:2 [cf. (27)], clearly pointing to the existence of the complex ion $Ag(CN)_2^-$. The same method has been employed to discover the existence of complexes of the transuranium metals [141] and of complex ions present in aluminum plating solutions [142]. Anionic complexes in concentrated aqueous zinc and cadmium halide solutions have been revealed by the negative transference numbers of the metal ion constituents (cf. Section 2.1.2), and the formation of micelles in polyelectrolyte solutions is dramatically shown by the breaks in the curves in Figures 8.2 and 8.4. Indeed, complex ions should always be suspected when the concentration variations of conductances and transference numbers depart markedly from those predicted by the Debye–Hückel–Onsager theory.

Conductance and transference determinations can be used together in the "mobility" approach to structural questions. An interesting early example was provided by aqueous solutions of HF, which were believed to contain the species HF, H^+, F^-, and HF_2^- [143]. At any given concentration five equations were available: mass balance, electroneutrality, the dissociation constant for HF (46), the specific conductivity (12), and the transference number (27). From these the five unknowns (the concentrations of the four solute species and $\lambda_{HF_2^-}$) could be calculated since the conductances of H^+ and F^- were already known. Although today more sophisticated corrections would be applied for the interionic effects, the results obtained by this simple analysis nicely explained the somewhat curious behavior of aqueous HF solutions [143]. This method of attack has been successfully employed in studying the constitution of ion-exchange resins [144]; of solutions of soaps, detergents, dyes, and other polyelectrolytes [9]; and of solutions of the sodium salt of deoxyribonucleic acid (DNA) [145] where such properties as the polymer charge fraction and the fraction of free gegenions were derived.

3.4   Determination of Equilibrium Constants

Only ions, and not molecules or ion pairs, affect the conductance. This property is therefore ideal for determining dissociation (or association) equilibria between molecules (ion pairs) and ions. Modern procedures based on slightly different versions of (51) or on (52) were described in Section 2.1.1. If the electrolyte is strongly associated and the interionic corrections are small, as for instance with benzoic acid in water, the simpler Shedlovsky equation (53) is perfectly adequate for obtaining both K_A and Λ^0. Finally, with even stronger association it is best to derive the Λ^0 value separately by means of Kohlrausch's law of independent ionic migration as exemplified by (55). Special measures must be adopted if the solvent itself is extensively self-dissociated, as with pure sulfuric acid [146].

The self-dissociation (autoprotolysis) constants of water $(K_w = c_{H^+} c_{OH^-})$ and other pure solvents are of special interest and have been accurately measured conductometrically [147, 148]. Above 50°C the specific conductivity κ of water is at least two orders of magnitude greater than at 25°C, which makes the measurements less sensitive to ionic contamination. From (2) and (42), since the ionic strength is very low,

$$\kappa = \Lambda^0 c_{H^+} = \Lambda^0 K_w^{1/2} \tag{100}$$

The small interionic corrections may easily be applied if higher accuracy is desired. The value of Λ^0 is obtained by the analogue of (55):

$$\Lambda^0_{HOH} = \Lambda^0_{HCl} + \Lambda^0_{KOH} - \Lambda^0_{KCl} \tag{101}$$

The solubility product K_{sol} of a slightly soluble substance may be determined in a similar way. A saturated solution possesses a specific conductivity which, for a 1:1 solute, can to a first approximation be written

$$\kappa = \Lambda^0 c_s = \Lambda^0 K_{sol}^{1/2} \tag{102}$$

if the dissolved electrolyte is completely dissociated in solution and forms no complexes. If, however, the electrolyte is also associated in solution (e.g, benzoic acid in water), then it follows by a combination of (19) and (46) that

$$K_{sol}^{1/2} = c_s = \frac{\kappa}{\Lambda^0} + \frac{\kappa^2 K_A}{(\Lambda^0)^2} \tag{103}$$

The value of K_A can be determined separately from the conductances of the undersaturated solutions. The relatively small interionic corrections for the electrophoretic and relaxation effects on Λ and for the activity coefficients on K_A [see (49)] can be inserted by a short series of successive approximations; in the first step, the ionic strength is simply given by κ/Λ^0 from (103).

Equilibrium constants of reactions involving ions on both sides can also be determined conductometrically provided the molar conductances of the reactant and product ions differ significantly. In practice, this usually means that one side of the reaction involves proton-jumping ions. As examples one may cite hydrolysis reactions [138] like

$$RNH_3^+ + H_2O \, (+Cl^-) \rightleftharpoons RNH_2 + H^+ (+Cl^-)$$

and the ionization of alcohols in alkaline solutions [149, 150]

$$ROH + OH^- (+Na^+) \rightleftharpoons OR^- + H_2O \, (+Na^+)$$

The equilibrium constant of the latter reaction equals K_{ROH}/K_w, which allows the dissociation constant K_{ROH} of the alcohol to be determined. The limiting conductances of common ions like H^+, Na^+, OH^-, and Cl^- can be found from the literature (see Appendix) and that of the remaining ion (RNH_3^+, OR^-) is obtained by suitable data manipulation [149] and equations like (42). An advantage in working with reactions that involve the same number of ions on each side is that their activity coefficients approximately cancel out in the equilibrium expression. Another reaction that involves both ionic reactants and products is the formation of triple ions as in

$$A^- + HA \, (+H^+) \rightleftharpoons AHA^- (+H^+)$$

The equilibrium constants and triple ion conductances can be derived either from conductance or transference experiments [8, 54–56] or from a combination of the two (Section 3.3). The multiple ionic association in micelle formation has also been referred to earlier.

3.5   Determination of Rate Constants

The kinetics of general chemical reactions

$$\nu_A A + \nu_B B + \cdots \rightarrow \nu_Y Y + \nu_Z Z + \cdots$$

are normally studied under pseudo-first-order conditions for one of the reactants. Let this be A. Then, if the reaction is followed by measuring the specific conductivity κ (which, to a first approximation, is a linear function of concentration), it follows that [151]

$$\ln\left(\frac{c_{A,0}}{c_A}\right) = \ln\left(\frac{\kappa_\infty - \kappa_0}{\kappa_\infty - \kappa_t}\right) = k_{obs} t \tag{104}$$

where c is concentration, t time, k_{obs} the pseudo-first-order rate constant, and subscripts 0, t, and ∞ represent conditions initially, at time t, and at equilibrium, respectively. Since only ratios of κ are involved one could equally well use the overall conductance G, and plots of $\ln(\kappa_\infty - \kappa_t)$ or $\ln(G_\infty - G_t)$ versus t will yield straight lines of slope $-k_{obs}$. Modifications of (104) have been proposed to correct for the small variations of molar conductances with ionic strength [152]. Analogous equations can be derived for second-order and other kinetic situations [151]. Thus any chemical reaction that produces or consumes ions, or whose product and reactant ions possess considerably different molar conductances, can in principle be followed conductometrically. Typical examples are the solvolysis of alkyl halides [152]

$$RX + H_2O \rightarrow ROH + H^+ + X^-$$

and the saponification of an ethyl ester

$$RCOOEt + OH^-(+Na^+) \rightarrow RCOO^- + EtOH(+Na^+)$$

Manually balanced conductance bridges suffice to study slow reactions; however, it is advisable to check that the platinum conductance electrodes are not acting as catalysts [153]. Reactions whose half-times are of the order of 1 min can be carried out in a suitable reaction cell involving rapid solution flow past the electrodes [154] and the readings are most conveniently taken with an autobalance bridge. The rates of many reactions with half-times from several seconds down to a few milliseconds, such as

$$H^+ + HCO_3^- \rightarrow H_2CO_3$$

may be studied in commercial continuous flow or stopped-flow apparatus [155] using conductance monitoring by storage oscilloscope or transient recorder. Even faster reactions can be followed by the pressure-jump technique. For conductance monitoring, a sample and reference cell are both placed inside the pressure autoclave and form two arms of an ac Wien bridge whose off-balance potential is read with an oscilloscope. The bridge must be operated at a frequency greater than the reciprocal of the relaxation time of the reaction, usually 40 kHz or higher [156]. Conductometric detection is less favorable in temperature-jump cells that rely on Joule heating because of the high background electrolyte concentration needed.

Conductance is also a convenient way of studying the kinetics of heterogeneous reactions involving ions. Chief among these are crystallization processes: the rates of dissolution of slightly soluble salts like AgCl or PbSO$_4$ and the rates of their precipitation from supersaturated solutions have often been followed conductometrically [157]. In an interesting recent example, two relaxation methods (pressure-jump and electric field pulse) with conductometric detection were used to study another type of process involving two phases: the kinetics of adsorption–desorption of Pb^{2+} ions on suspended γ-Al$_2$O$_3$ [158].

3.6   Separations and Preparations

The direct moving boundary method for determining transference numbers (Section 5.3) is the logical forerunner of the moving boundary method for investigating electrophoresis. Studies of mixtures of strong electrolytes and of buffer mixtures [159] have provided much insight into the significance of the schlieren patterns recorded in the electrophoresis of proteins and other substances [160]. Just as its electrophoretic counterpart, the moving boundary method has been employed to separate mixtures of ions (e.g., caproate and valerate ions) both in solution [161] and in ion-exchange resins [144]. Since isotopic ions display small conductance differences [84, 85], these too can be separated by appropriate ion migration techniques in either solutions or melts [162]. The moving boundary method may also be used to synthesize electrolytes in solution. As explained in Section 5.3.3, a system like AZ/BY becomes AZ/BZ/BY when current flows, and samples of the newly created electrolyte BZ can then be removed from the cell.

4   MEASUREMENT OF ELECTROLYTIC CONDUCTANCE

4.1   Alternating Current Method

The most common technique for measuring conductance is the audiofrequency ac method pioneered by Kohlrausch over a century ago. This method will therefore be described in some detail followed by only a brief mention of other techniques.

4.1.1   The Wien Bridge

The conductance (or, in practice, the resistance) of a solution is determined by immersing in it two (or more) electrodes, usually of platinum. Several cell designs are illustrated in Section 4.1.2. Such a conductance cell possesses both resistance (R) and capacitance (C) and therefore an impedance (Z). The basic ac network for measuring its impedance (Z_1) is the lozenge-shaped Wien bridge depicted in Figure 8.7. In the literature the bridge is often called after Wheatstone whose resistive network predated its more general ac use by Wien [163]. Here an oscillator drives a sinusoidal ac through the circuit, and the detector shows no signal (or, in reality, a minimum signal) when the alternating potentials

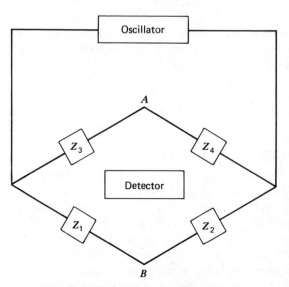

Figure 8.7   Basic circuit of a Wien bridge.

at A and B are of equal amplitude and exactly in phase [17]. This is the condition of balance at which

$$\frac{Z_1}{Z_2} = \frac{Z_3}{Z_4}$$

(105)

In the laboratory the impedances Z_3 and Z_4 are replaced by known pure resistors R_3 and R_4, usually equal and often 1000 ohm each. To balance the cell impedance Z_1 then requires in the remaining arm of the bridge both a variable resistor R_2 and a variable capacitor C_2, either in series or in parallel (see Figure 8.9). These are adjusted until a minimum signal is produced in the detector.

To understand the significance of R_2 and C_2 it is necessary to take a closer look at the electrical components of the conductance cell. The essential features are shown in Figure 8.8, based on a diagram in an excellent review by Braunstein and Robbins [164]. We wish to measure R_1, the resistance of the solution, but to do so requires some knowledge of four other electrical contributions introduced by the electrodes. The first and most obvious one turns out to be relatively unimportant: the capacity C_{el} produced by the electrodes themselves behaving as a condenser with the solution as dielectric. Its impedance of $1/2\pi f C_{el}$ (where f is the ac frequency) is around 2×10^7 ohm at 1000 Hz, and since it acts parallel to R_1 its effect is very small unless $R_1 > 10^4$ ohm [164]. The second contribution—and the most important one—originates in the double layer at each electrode–solution interface. When an electrode is positively charged by the oscillator it will attract a layer of negative ions and when negatively charged, a layer of positive ions. This charge and recharge resembles the action of a condenser of capacitance C_{dl} in series with R_1. For smooth electrodes in either aqueous solutions or molten salts the impedances $1/2\pi f C_{dl}$ so produced is of the order of 10 ohm at a 1 cm^2 electrode at 1000 Hz [164]. Because the double layer makes such a serious contribution it is common practice to try to reduce it by platinizing the electrodes (which increases the area and therefore C_{dl}) and by working at fairly high frequencies.

So far each electrode has been treated as ideally polarized, with no charge actually crossing the interface. The third contribution comes into play when the surge of current actually causes chemical oxidation at the positively charged electrode and reduction at the negatively charged one. Such processes that

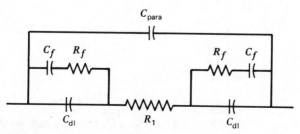

Figure 8.8   Equivalent electrical circuit of a two-electrode conductance cell, following [164].

involve interfacial current flow are termed "faradaic." They can arise when the system in question possesses a high exchange current density i_0 (or electrode rate constant) [165], as with Ag^+ ions at a silver electrode, or when the discharge of the ubiquitous H^+ ions is aided (depolarized) by the presence of dissolved or adsorbed oxygen [166]. In modern work, where sensitive detectors are available, the chance of actual electrolysis is minimized by using low applied potentials of amplitude less than 100 mV which is smaller than the decomposition potential for many systems. In an ac circuit a faradaic process behaves like a faradaic impedance Z_f [167, 168] composed of a resistance R_f plus a series capacitance C_f which, as shown in Figure 8.8, shunt the double-layer capacitance. The theory shows that R_f is not a true resistance but varies with angular frequency ω ($= 2\pi f$, where f is the frequency in Hz) according to

$$R_f = \theta + \frac{\sigma}{\sqrt{\omega}} \tag{106}$$

The resistance θ is inversely proportional to the exchange current density i_0. Neither is C_f a true capacitance for it equals $1/\sigma\sqrt{\omega}$. The faradaic impedance is therefore often represented by the sum of the resistance θ and a "Warburg impedance" composed of a frequency-dependent resistance $(\sigma/\sqrt{\omega})$ and a frequency-dependent capacitance in series.

The last contribution in Figure 8.8 arises from various parasitic effects designated by C_{para}. These result from stray capacitances in leads, etc., and the so-called Parker effect caused by a capacitative shunt between parts of the cell of opposite polarity [169]. The latter is largely removed by proper cell design (Section 4.1.2).

This complex electrical network within the cell can be balanced in the Wien bridge by one of the arrangements shown in Figure 8.9. Matters are simplified by putting $Z_3 = R_3 = R_4 = Z_4$, so that by (105) $Z_1 = Z_2$. The detailed analysis for either of the circuits shown is best handled by the method of complex notation [163, 164]. To make the problem more tractable, we omit consideration of the small capacity C_{el} and of the parasitic effect C_{para}, and treat the two electrodes as one. Then for Figure 8.9a where the variable components in arm 2 are placed in series, R_2 equals the "real" part of Z_1 while $-1/\omega C_2$ can be equated to the "imaginary" part of Z_1. Working through the complex algebra involved leads to

$$R_2 = R_1 + \frac{\theta + \sigma/\sqrt{\omega}}{1 + 2C_{dl}\sigma\sqrt{\omega} + 2C_{dl}^2\sigma^2\omega + 2C_{dl}^2\theta\sigma\omega^{3/2} + C_{dl}^2\theta^2\omega^2} \tag{107}$$

For Figure 8.9b with the parallel arrangement in arm 2, $1/R_2$ equals the "real" part of $1/Z_1$. It follows that [170, 171]

$$R_2 = R_1 + \frac{R_1\theta + \theta^2 + (\sigma/\sqrt{\omega})(R_1 + 2\theta) + 2\sigma^2/\omega}{R_1(1 + C_{dl}^2\theta^2\omega^2) + 2R_1 C_{dl}\sigma(\sqrt{\omega} + C_{dl}\sigma\omega + C_{dl}\theta\omega^{3/2}) + \theta + \sigma/\sqrt{\omega}} \tag{108}$$

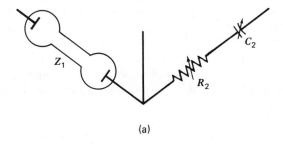

(a)

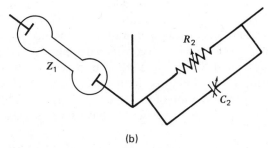

(b)

Figure 8.9 Methods of balancing the cell impedance in arm 1 by series or parallel connections in arm 2 of the Wien bridge. Arms 3 and 4 comprise equal pure resistances.

Although the latter equation contains more terms, the parallel arrangement in arm 2 is usually preferred in the laboratory because it requires much smaller values of C_2 for balance. Smaller capacitances can be obtained with higher accuracy and less frequency dependence than larger ones [164].

Let us now examine a number of special cases that are often encountered [17, 170].

1. C_{dl} is extremely large, which makes its impedance very small compared with that of the faradaic path. Then

$$R_2 = R_1 \quad \text{(Figures 8.9a and 8.9b)} \tag{109}$$

This condition is closely fulfilled when the electrodes are heavily coated with platinum black. In such cells R_2 is found to be virtually independent of the ac frequency.

2. The resistance θ is infinitely large, so that the electrodes are ideally polarized:

$$R_2 = R_1 \quad \text{(Figure 8.9a)} \tag{110a}$$

$$R_2 = R_1 + \frac{1}{R_1 C_{dl}^2 \omega^2} \quad \text{(Figure 8.9b)} \tag{110b}$$

In dilute aqueous solutions the magnitude of the second term in (110b) tends to be vanishingly small. This is not true, however, for highly concentrated solutions or molten salts [164]. Under these circumstances R_2 should be extrapolated against $1/\omega^2$ to infinite frequency, or alternatively it may be more convenient to adopt the series balancing procedure in Figure 8.9a.

3.   The Warburg impedance is very small. Hence $\sigma/\sqrt{\omega} \ll \theta$ and, probably, $1/\sigma\sqrt{\omega} = C_f \gg C_{dl}$. Consequently,

$$R_2 = R_1 + \frac{\theta}{1 + C_{dl}^2 \theta^2 \omega^2} \qquad \text{(Figure 8.9a)} \qquad \text{(111a)}$$

$$R_2 = R_1 + \frac{R_1 \theta + \theta^2}{R_1(1 + C_{dl}^2 \theta^2 \omega^2) + \theta} \qquad \text{(Figure 8.9b)} \qquad \text{(111b)}$$

The latter equation is identical with one derived from the equations of Feates, Ives, and Pryor [166, p. 582]. If in addition $R_1 \gg \theta$, (111b) becomes identical with (111a). This situation corresponds to dilute solutions in a cell with gray-platinized electrodes [166] and is also considered applicable to bright platinum electrodes [17]. Steel and Stokes have found in practice that the value of R_1 obtained by solving (111a) for three frequencies agrees well with the value obtained by a linear extrapolation of R_2 versus $1/\omega$ [17, 170]. This finding helps to explain why linear plots of R_2 versus $1/\omega$ have several times been reported in the literature [172].

4.   The Warburg impedance is very large so that $\sigma/\sqrt{\omega} \gg \theta$ and, probably, $1/\sigma\sqrt{\omega} = C_f \ll C_{dl}$. This leads to

$$R_2 = R_1 + \frac{1}{2C_{dl}^2 \sigma \omega^{3/2}} \qquad \text{(Figure 8.9a)} \qquad \text{(112a)}$$

$$R_2 = R_1 + \frac{\sigma(R_1 + 2\theta) + 2\sigma^2/\sqrt{\omega}}{R_1\sqrt{\omega} + 2R_1 C_{dl}^2 \sigma^2 \omega^{3/2} + \sigma} \qquad \text{(Figure 8.9b)} \qquad \text{(112b)}$$

If, moreover, R_1 is much larger than the Warburg impedance, the latter equation simplifies to

$$R_2 = R_1 + \frac{\sigma}{\sqrt{\omega} + 2C_{dl}^2 \sigma^2 \omega^{3/2}} \qquad \text{(Figure 8.9b)} \qquad \text{(112c)}$$

which in many cases will approximate to

$$R_2 = R_1 + \frac{\sigma}{\sqrt{\omega}} \qquad \text{(Figure 8.9b)} \qquad \text{(112d)}$$

Small values of θ are to be expected whenever i_0 is large, and therefore it is not surprising that Jones and Christian [173] observed linear R_2 versus $1/\sqrt{\omega}$ behavior when studying solutions of $AgNO_3$ between silver electrodes and solutions of $Ni(NO_3)_2$ between nickel electrodes.

These examples of the application of (107) and (108) to particular systems should help in selecting both suitable electrodes for a given solution and the most appropriate extrapolation procedure. It must be emphasized that for accurate work the conductance should be measured at several frequencies around and preferably above 1000 Hz. Unless a frequency-independent range can be found, the conductance (resistance) of each solution must be extrapolated to infinite frequency by the function that best fits the data [17, 164]. Although simple plots of R_2 versus $1/\sqrt{f}$, $1/f$, or $1/f^2$ have often represented data successfully, quadratic functions in $1/\sqrt{f}$ [174, 175] or the empirical formula

$$R_2 = R_1 + af^{-n} \tag{113}$$

have also been applied [176, 177]. The three unknowns here are most easily solved by fitting (113) to sets of three points whose frequencies lie in geometric progression; that is, $f_2^2 = f_1 f_3$. However, in 1970 Hoover [171] criticized all such empirical equations as leading to extrapolated R_1 values that were too low. Instead he recommended fitting by the more complex equations such as (111) which are directly based on circuit models of the faradaic process. If the highest precision is not required, a resistance measurement at 5 or 10 kHz should give a satisfactory answer [171].

Let us now return to the design of the bridge itself, as depicted in Figure 8.10. The basic principles were established by Grinnell Jones and co-workers [178, 179], Shedlovsky [180], and Dike [181], and have been well summarized

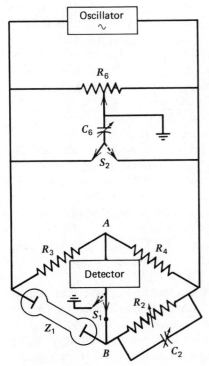

Figure 8.10 Wien bridge circuit incorporating the Wagner earthing device, after [17].

by Robinson and Stokes [17]. The oscillator should produce a pure sine wave and be adjustable in frequency from 500 Hz to several thousand hertz. The amplitude should also be adjustable: the lower the applied potential the smaller the effects of Joule heating and electrode polarization. (For high-field Wien effect experiments, Joule heating is allowed to dissipate by generating the large potentials in microsecond pulses [182].) The generator should be isolated from the bridge by a good quality transformer and be shielded and preferably placed some distance away from the detector. Another transformer isolates the bridge from the detector circuit which incorporates an amplifier to increase the strength of the off-balance signal near the balance position. An operational amplifier has recently been described [183] and lock-in amplifiers have been recommended as null detectors [36, 172]. The historical audiodetection by a telephone headset has now generally been replaced by a visual cathode ray oscilloscope display. The generator is connected to the horizontal detection plates and the amplifier to the vertical ones. If the unknown were a true resistance, the bridge imbalance would be indicated by a sloping oscilloscope trace whose vertical projection is proportional to the out-of-balance voltage while its horizontal projection is proportional to the amplitude of the applied sinusoidal wave [164]. For a conductance cell, however, the trace is usually an ellipse. Proper bridge balance of both R_2 and C_2 is then registered by a horizontal straight line. Resistance variations of about 1 part in a million can readily be detected in this way even with generator potentials as low as 100 mV [36].

The bridge must be constructed of high-quality components. The resistors in particular must be extremely stable otherwise frequent recalibration is necessary [166]. Time constants below 10^{-8} s are required [36]. The carefully matched resistors R_3 and R_4 are usually 1000 ohm each and are specially wound to minimize induction effects [17, 28]. In parallel with the variable resistance R_2 is the variable capacitor C_2 with a maximum capacity of 1000 pF. An important feature is the Wagner earthing device [178] composed of R_6, C_6, and S_1: this eliminates errors due to resistance–capacitance shunts between elements of the bridge and improves the sharpness of the balance point. First, switch S_1 is connected to point B (as shown) and the bridge is balanced by adjusting R_2 and C_2. Then S_1 is turned to the earth (ground) position and R_6 and C_6 are adjusted until the detector registers a minimum signal. This sequence of operations is repeated until the lowest balance point is obtained. The potentials at A and B are then not only equal but are also at earth potential so that the pickup of stray electrical noise by the detector is minimized [17]. Careful shielding and electrostatic screening are advisable [180, 138, 176] to avoid current leaks through stray couplings and shifting capacities introduced through the hands of the operator.

Classical Wien bridges based on the above principles have been widely employed for several decades. A precision Grinnell Jones Conductivity Bridge [178, 181] was for many years manufactured by Leeds and Northrup Ltd. Janz and McIntyre [184] have described a research bridge incorporating an impedance comparator that could be constructed from unitized market components available in 1961. Shortly afterwards Walisch and Barthel [185] published details of a self-balancing bridge in which a multiturn resistance potentiometer was automatically rotated by a servomotor to the balance position. This could be accomplished in less than 1 s for a 10% change of conductance, as in a conductometric titration. With the advent of the silicone chip it is now possible to balance bridges automatically by electronic rather than mechanical means. In 1981 Kiggen and Laumen [186] described a Wien bridge in which, in place of the usual detector, the balancing voltage sends an electrical signal via a 1:1 transformer and a band-pass amplifier to a microprocessor. The latter progressively changes the values of R_2 and C_2 by means of reed relays, and the value of R_2 that corresponds to the minimum balancing voltage is printed out. The long average time of 2 min for one balancing procedure could be decreased by using a faster analog/digital converter. Also required is a more versatile sine-wave generator capable of operating at several frequencies and at an amplitude lower than the 5 V source employed.

Recently a powerful new ratiometric approach based on microprocessors has been devised by Bond and Brown [187, 188]. The aim is no longer to balance a ratio-arm bridge but simply to compare the signals from the unknown impedance Z_u with those from a known reference impedance Z_s. The basic measuring system is shown in Figure 8.11. An ac wave is first sent through Z_s and the voltage across it, E_s, is amplified. The ensuing signal passes in turn through two phase-sensitive detectors (PSD) that allow the magnitudes of the inphase and the 90° out-of-phase (quadrature) components to be recorded. Switching between the two PSDs is carried out by MUX2, a computer-controlled switch. The first multiplexer, MUX1, then switches to the signal E_u from the unknown impedance, and its orthogonal components are similarly determined via MUX2.

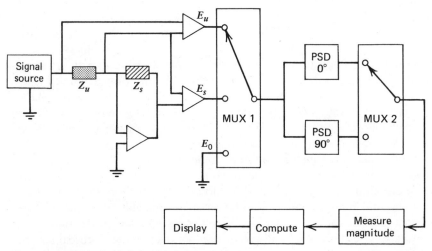

Figure 8.11 Basic circuit for ratiometric method of measuring conductances, following [188].

The third MUX1 connection allows for zero errors, E_0. Thus if Z_s is a pure resistor R_s, the "real" or resistive part of Z_u is given by

$$|Z_u| = R_s \left(\frac{E_u - E_0}{E_s - E_0} \right)$$ (114)

Series resistances in the leads are measured by shorting Z_u and, for high impedances where shunt capacity is the dominant error, a reading is first taken by open-circuiting Z_u. An automatic trim correction is then applied to subsequent readings. For complex impedances the microprocessor applies transformation equations to derive not only R but also C, L, or other desired properties such as G. The result is then displayed. These features and others are incorporated in the B905 Wayne Kerr Automatic Precision Bridge. This applies only 50 mV for resistances below 100 kΩ, operates at several frequencies, and provides output readings to better than 0.1% in a fraction of a second. Good accuracy and flexibility have been achieved here by completely integrating a microcomputer control system into the hardware, and further developments in this direction may be expected in the near future.

4.1.2   Design of Conductance Cells

According to (1), the resistance of a cell is given by

$$R = \frac{l}{\kappa A} = \frac{k}{\kappa}$$ (115)

where k is the cell constant. It is advisable to use cells of dimensions such that R does not fall far below 1000 ohm (to keep the relative errors due to polarization and cell leads within bounds) or much above 30,000 ohm (to reduce errors caused by insulation leakage) [138]. Should it be necessary to measure very high resistances, as in determining the solvent correction, it is common practice to shunt the cell with a known high resistor R' (say 10^4 ohm) and to calculate R from the equation [189]

$$\frac{1}{R_{obs}} = \frac{1}{R} + \frac{1}{R'}$$ (116)

For work with concentrated solutions of high specific conductivity κ, a suitably large cell constant (up to 500 cm^{-1}) is achieved with a long thin cell (Figure 8.12a). Solutions of moderate concentrations require an intermediate shape as in Figure 8.12b, and very dilute solutions can be investigated in cells with large electrodes close together as in Figures 8.13 and 8.14. Their cell constants could be made as small as 0.1 cm^{-1} or even lower.

These diagrams illustrate another important point. If the cell is poorly designed, the cell constant will be found to vary with the specific conductivity of the solution. This so-called Parker effect [192] was thoroughly investigated by Jones and Bollinger [169] who traced it to capacitative coupling between different parts of the cell. As a result the observed resistance is less than the true resistance. The effect is virtually eliminated by placing the electrical leads and the filling tubes well away from each other: a separation of 15 cm between parts

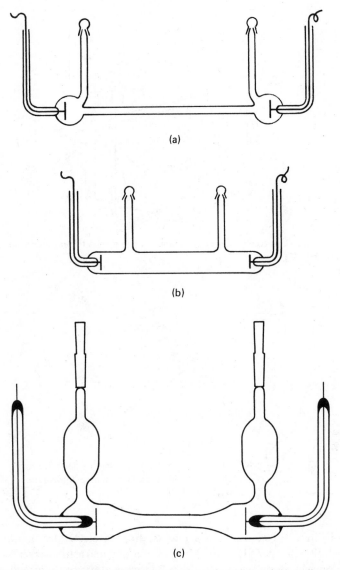

(a)

(b)

(c)

Figure 8.12   Standard cell designs suitable for solutions of (a) high and (b) medium conductivity, after [17]. Diagram (c) shows Stokes' modified cell that incorporates mixing bulbs for removing the Soret effect [190].

of the cell of opposite polarity has been recommended [169]. The filling tube should also be full of solution. In the past the lead tubes contained mercury which is a nuisance when the cell is inverted, and nowadays electrical contact is usually made by welding thick lead wires of silver [17] or gold [28] to the outer platinum ends of the electrodes.

The electrical leads should not be placed close to each other even in the

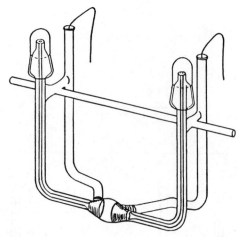

Figure 8.13   Shedlovsky's conductance cell suitable for use with detergent and protein solutions that tend to froth [138].

Figure 8.14   Flask cell with the electrodes in the measuring chamber close together, suitable for solutions of low conductivity [191].

absence of filling tubes, as when a pair of dipping electrodes is immersed in a flask cell. Shedlovsky [189, 138] has shown that parasitic currents can then flow from one electrode lead to the other by a capacitance path directly between the leads, and by a series capacitance and resistance bypath through the solution behind and below the electrodes. The electrode leads should therefore be removed from within the cell. However, simple dipping electrodes are still commonly used in conductometric titrations where experimenters only need to record sizable changes in resistance (Figure 8.5b). The above restriction does not apply to the dipping electrode assembly of Nichol and Fuoss [174] which is constructed of two concentric platinum cylinders. The lead to the outer electrode is a platinum tube that acts as an electrical shield for the lead to the inner electrode, so eliminating stray electrical paths.

Two small disturbing effects should also be considered when designing the cell. The first is the Soret effect investigated by Stokes [193]. When a conductance cell containing solution at one temperature (e.g., ambient) is immersed in a thermostat bath at another temperature, the initial large temperature gradient in the solution produces a corresponding concentration gradient [194], which may take days to disperse completely. Although the resistance of the cell may come to an almost constant value after an hour or so, this will not be the correct value. The latter is obtained by thoroughly mixing the solution, either by means of the large mixing bulbs in Figure 8.12c, by passing the solution in and out of the cell section in a flask cell as in Figure 8.14, or by stirring the solution with a magnetic stirrer bar [195]. The effect is of the order of 0.1%, is larger the greater the initial temperature difference, and depends upon the electrolyte. In aqueous solutions acids, hydroxides, and salts of NR_4^+ and polyvalent ions possess particularly large Soret coefficients [196].

The second troublesome effect was studied by Prue [197]. Conductivities of very dilute solutions often drift upward by $10^{-6}–10^{-5}$ ohm^{-1} m^{-1} per hour, a phenomenon wholly or partially reversed on shaking the solution. This "shaking effect" disappears when the interior walls of the cell are waxed and is slower if the glass surface is siliconed. The phenomenon appears to be caused by concentration of electrolyte at the polar glass surface as a result of ion-exchange processes: this low-resistance layer of solution at the walls is dispersed by shaking. A recent alternative explanation in terms of energy storage in hierarchical levels of the solution [198] does not sound tenable. The effect can be prevented by ensuring that the edges of the electrodes are not too close to the glass walls, and by making the diameter of the electrode chamber considerably larger than the diameter of the electrodes. On the other hand, positioning small electrodes in a large bulk produces lines of current flux behind the electrodes that give rise to a strong frequency dependence [199].

It is sometimes desirable to keep the total cell volume small, and appropriate designs have been published [189, 200, 138]. The cell in Figure 8.13 with truncated cone electrodes is especially suited for detergent and protein solutions that are liable to froth [201]. On the other hand, flask cells of large volume are convenient for preparing solutions *in situ* and for carrying out a series of dilution or concentration runs in which progressive increments of solvent or concentrated solution are added [36]. From what has been said already, it is best (unless special precautions are taken [202]) to divide such a cell into a large preparation chamber and a smaller electrode chamber. Figure 8.14 illustrates a popular design based on an Erlenmeyer flask; another is shown in Figure 8.15. Cells based on similar principles have been built by many other workers [189, 195, 203, 204] and special reference must be made to the ingenious doughnut-shaped cell devised by Mysels [200]. The contents of the electrode chamber must be mixed thoroughly with the solution in the main body of the cell, either by repeated pressure, by suction through one interconnecting tube [189, 203], or by including both an upper and a lower connecting tube. In the latter case, circulation of liquid between the two sections can be achieved by placing a magnetic stirrer bar in a well at the bottom of the reservoir vessel [195, 200, 204, 172] (Figure 8.15). The solution resistance of the cell in Figure 8.14 (and presumably of the other cells also) is independent of the liquid level in the flask provided it is above the upper connecting tube [28].

Several enterprising researchers have invented novel electrode configurations that largely overcome the perennial problem of electrode polarization. These devices may be grouped into guarded electrodes, double cells, and four-electrode cells. Hawes and Kay [202] designed the first guarded electrode. Its solution face is a circular 10 mm diameter platinum disk surrounded by a concentric platinum wire guard ring 16 mm in diameter. The copper lead wire connected to the central disk is insulated and leaves the cell inside a 9.5 mm diameter brass tube that is connected to the guard ring and that is itself insulated by a glass sheath. The nearby facing electrode is simply a 16 mm platinum disk.

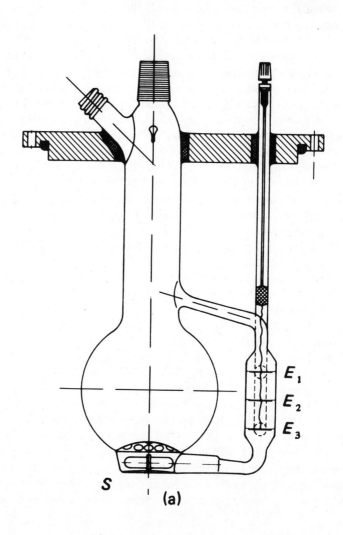

E_1

E_2

E_3

S

(a)

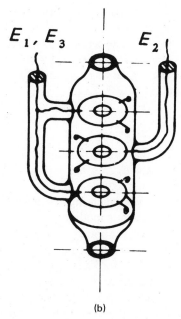

(b)

Figure 8.15   (a) Flask cell with stirrer well used by Barthel, et al. [36, 228]. (b) The guarded three-electrode measuring assembly is shown separately.

The current lines to the center disk of the guarded electrode are now parallel and independent of the environment because they are surrounded by a fringe field which is carried to earth. The electrodes can therefore be sited right inside the flask cell, and the solution can be continuously stirred by a magnetic stirrer bar without any ill effect. However, the cell constant is necessarily of the order of 1 cm^{-1}. A more easily fabricated electrical guarding device is being used by Barthel's group [36, 172]. As Figure 8.15 illustrates, the two outer electrodes E_1 and E_3 of a three-electrode cell are simply connected together to restrict the electric field to the intervening space.

Two other groups of workers have sought to obviate all polarization problems at the electrodes by recording the *differences* in resistance between two cells similar in all respects except for the distance between the electrodes. In Ives and Pryor's double cell [205], the four electrodes were platinum hemispherical cups center-drilled to allow free flow of solution through both cells. The frequency dependence of the resistance difference was, as expected, extremely small. More recently, Saulnier [206] has invented a double differential cell. Each of the two cells was made of Pyrex tubing of uniform cross-sectional area A (1 cm^2) with flat circular electrodes of the same area. The cells were placed vertically in a common container that held the solution, and the bottom electrode of each cell was fixed in position. The tubes could be moved vertically to allow solution to enter the cells. The upper electrodes could also be moved vertically within the tubes by means of attached pistons. Cell 1 in series with a variable resistor R_1 formed arm 1 of a Wien bridge (Figure 8.7), and cell 2 plus a variable resistor R_2 formed arm 2. The spacings between the two pairs of electrodes were first made approximately equal and the bridge was balanced by adjusting R_1. The spacing between the electrodes in cell 2 was then increased by a known amount Δl, and the bridge was rebalanced by increasing the resistance of R_1 by ΔR_1. Thus from (1),

$$\Delta R_1 = \frac{\Delta l}{\kappa A} \tag{117}$$

This arrangement therefore permits the *absolute* determination of the specific conductivity of the solution, and at 25°C the agreement with accepted standards has been shown to be better than 0.08% [207].

A quite different way of circumventing the effects of polarization and of platinization is by the four-electrode method [208–210]. Although well known in solid-state physics, it has been applied only rarely to ac measurements in electrolyte solutions. The plan is shown in Figure 8.16. The two outer electrodes C_1 and C_2 are used to pass a known sinusoidal current $\tilde{I}$ through the solution. The two bright platinum probe electrodes P_1 and P_2 positioned between C_1 and C_2 develop a potential difference ΔE over a stretch of the solution in the middle of the cell well removed from the current-carrying electrodes. Then [210]

$$\Delta E = \Delta \tilde{E} + \Delta E_{\text{mix}} \tag{118}$$

$$\Delta E = \tilde{I} Z \tag{119}$$

where ΔE_{mix} arises from any asymmetry in the mixture potentials [211] of the probe electrodes, and Z is the impedance of the solution between P_1 and P_2. The solution resistance between P_1 and P_2 can be obtained by measuring the inphase component of $\Delta \tilde{E}$ with respect to the current $\tilde{I}$ by a phase-sensitive detector. In an alternative potentiostatic arrangement, the ac potential difference $\Delta \tilde{E}$ is controlled in amplitude and frequency and the resulting inphase alternating current is measured. The four-electrode technique is the normal approach in dc conductance measurements (Section 4.3).

Once the type of cell has been decided upon, choices have to be made as to the cell material and how physically to introduce the electrodes. The number of recommended ways of sealing platinum into glass is legion because no ideal method exists. The nub of the problem is that the coefficient of linear thermal expansion α of platinum $(9 \times 10^{-6} \, \text{K}^{-1})$ is almost the same as that of soda

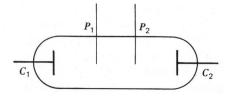

Figure 8.16 Diagram illustrating the principle of the four-electrode method.

glass—undesirable because conducting ions are easily leached from it—and quite different from that of normal Pyrex glass ($3.5 \times 10^{-6} \, K^{-1}$) with which most cells are built. Solutions include the use of tungsten ($\alpha = 4.5 \times 10^{-6} \, K^{-1}$) as an intermediate substance [203], soda-to-Pyrex glass graded seals [212], Jena 16 III glass ($\alpha = 8.1 \times 10^{-6} \, K^{-1}$ [213]), Corning 707 [214] or 7070 glass [202] which readily wet platinum, uranium glass [182, 267], and epoxy resin like Araldite [215, 216]. The advantage of the resin lies in its physical flexibility which is useful for high-pressure measurements; its disadvantage is its lack of chemical inertness. It can even be slowly dissolved by a solvent like DMSO [216]. The other organic polymer frequently tried is Teflon (PTFE): it is certainly inert enough and sufficiently plastic to allow it to deform around a platinum seal without breaking up to at least 12 kbar [126]. Both Teflon plugs [217, 218] and gaskets [219] have been employed to fit platinum electrodes. However, the α value of the Teflon polymer is some ten times that of platinum and one must remember that air bubbles cling to it tenaciously [203].

Since most seals are of the platinum–glass type, their construction requires great care. Pyrex glass pinch-seals with thin platinum wire or foil have several times been advocated [166, 205, 220] and are said to be vacuum tight. The resulting lead resistance can be allowed for in several ways (Section 4.1.4). The details of making a durable seal between platinum and Corning 707 glass (which can be sealed onto Pyrex glass by using Corning 3321 glass as an intermediate) have been described in words by Hnizda and Kraus [214] and in words and pictures by Evans and Matesich [28]. The latter have also repeated the detailed instructions given by Marsh and Stokes [203] for a Pyrex–tungsten–platinum seal. Briefly, one end of a 1 mm diameter tungsten rod is fused to the stem of the electrode disk and the other end (via a piece of platinum) to the gold or silver lead wire. The tungsten itself is carefully sealed into the Pyrex tube in such a way that only the platinum electrode will be exposed to the solution. The original descriptions should be consulted before embarking on any of these procedures.

The cells are usually constructed of Pyrex glass although quartz has been used [216] to exclude ion-exchange processes at the surface. The reservoir flask for preparing the solutions need not be made of the same material as the electrode capsule: in fact, stainless steel or polythene flasks have been preferred for alkaline solutions [203]. At high temperatures and pressures water is a powerful solvent [219] and Teflon cells have therefore proved popular here despite the fact that the Teflon material undergoes several phase transitions both with increase in temperature [175] and with increase in pressure [126]. Teflon and polyethylene cells were employed for measurements with aqueous HF solutions [218] and, for HF as solvent, sapphire cells with Kel-F fittings have been recommended [221].

4.1.3   Cell Calibration

Apart from Saulnier's absolute method [206], all conductance measurements are relative. To convert the measured cell resistance R or conductance G to absolute values of specific conductivity κ requires a proportionality factor, the cell constant k. Although this is formally given by l/A (Section 1.2.1), the effective area A is closer to the internal cross-sectional area of the glass tubing connecting the electrodes than to the actual electrode area in cells like those in Figure 8.12; the length l will include the contribution made by the lines of current flux passing behind the electrodes. The determination of k with a ruler is therefore impossible (except in Saulnier's apparatus) so that each cell must be calibrated with one or more standard solutions of known specific conductivity. Such standards were provided by Jones and Bradshaw in 1933 [213]. In a classic piece of work, they compared the resistances of 0.01, 0.1, and 1 D (demal: see Table 8.2) aqueous KCl solutions with that of pure mercury by using a 3 M solution of sulfuric acid as a substance of intermediate resistance. Their research required a whole series of conductance cells whose constants spanned a factor of 10^3. Since the international ohm had been defined in terms of the resistance of a fixed length and mass of mercury at $0°C$, they were accordingly able to determine the specific conductivities of these KCl solutions in the units (int ohm)$^{-1}$ cm^{-1}. The demal concentration unit was chosen in order to specify the compositions of the KCl solutions in terms of the masses of KCl per kg of aqueous solution, both corrected to vacuum, so that the concentrations would be independent of any later changes in atomic weights or volume standards: a 0.01 D KCl solution is almost 0.01 M. The Jones and Bradshaw tables have been faithfully followed for half a century [222], either as they stand or incorporated into derived fitting equations of the form $\kappa = f(c_{KCl})$ [223–228] to increase the flexibility of calibration and to extend it to cells of very small cell constant.

The time has now come for a change. Under the SI system [2], resistance must be expressed in absolute ohm whose reciprocal is the siemens. In symbols

$$1 \text{ S} = 1 \text{ (abs ohm)}^{-1} = 1.00049 \text{ (int ohm)}^{-1}$$

The current International Practical Temperature Scale, agreed in 1968, is also slightly different from the 1927 scale in force during the Jones and Bradshaw measurements. As a result, new calibration standards were recommended by IUPAC in 1976 and have been published in 1981 by Juhasz and Marsh [229]. The new procedure specifies the use of potassium chloride of purity not less than 99.99 mass %: analytical grade KCl recrystallized twice from conductivity water and dried at 770 K for 24 h is usually suitable. The solvent should be water that has been distilled or passed through an ion-exchange resin to reach a conductivity of less than 1.2×10^{-4} S/m at 25°C. The specific conductivity of the solution is then given by

$$\kappa_{soln} = \kappa_{ref} + \kappa_{water} \tag{120}$$

The appropriate κ_{ref} values at 0, 18, and 25°C are listed in Table 8.2, which also reproduces the recommended fitting equations. All conductance work in the future should be based on these internationally agreed calibrations. It would be helpful if from now on the symbol S were always employed for (abs ohm)$^{-1}$ to avoid ambiguity as to what sort of ohm is meant.

For conductance work at temperatures other than those specified in Table

Table 8.2 IUPAC Recommended Reference
Conductivities κ at Various Temperatures[a]

Solution[b] $T_{1968}(K)$	1 D κ (S/m)	0.1 D 10κ (S/m)	0.01 D $10^2\kappa$ (S/m)
273.15	6.514	7.134	7.733
291.15	9.781	11.163	12.201
298.15	11.131	12.852	14.083

[a]Reference [229].
[b]1 D solution: 71.1352 g potassium chloride in 1 kg of aqueous
solution. 0.1 D solution: 7.41913 g potassium chloride in
1 kg of aqueous solution. 0.01 D solution: 0.745263 g potassium chloride in 1 kg of aqueous solution. All values given
above refer to true mass in vacuum. Alternatively, conductance cells can be calibrated with a potassium chloride solution
of known concentration together with an equation representing the molar conductance over the appropriate concentration range. The recommended equations below are given in
terms of the molar conductance Λ where $\Lambda = \kappa/c$, and reproduce the conductivities of the appropriate reference solutions
to within 0.015%.

1. In the concentration range between $c = 10^{-4}$ and 0.04
   mol/dm^3 at 298.15 K the Justice equation [225]

   $$\Lambda(\text{mS} \cdot \text{m}^2/\text{mol}) = 14.984 - 9.484c^{1/2}$$
   $$+ 5.861c \log c + 22.89c - 26.42c^{3/2} \quad \text{(a)}$$

2. In the concentration range between $c = 0.01$ and 0.10
   mol/dm^3 at 298.15 K the Chiu–Fuoss equation [226]

   $$\Lambda(\text{mS} \cdot \text{m}^2/\text{mol}) = 14.988 - 9.485c^{1/2}$$
   $$+ 2.547c \ln c + 22.0c - 22.9c^{3/2} \quad \text{(b)}$$

3. In the concentration range between 0.05 and 1.0 mol/dm^3
   at 298.15 K the Rostock equation [227]

   $$\Lambda(\text{mS} \cdot \text{m}^2/\text{mol}) = 14.995 - 9.925c^{1/2} + 13.575c$$
   $$- 12.075c^{3/2} + 5.787c^2 - 1.172c^{5/2} \quad \text{(c)}$$

4. In the concentration range between 10^{-4} and 0.05 mol/
   dm^3 at 291.15, 283.15, and 273.15 K the Barthel, et al.
   equations [228]
   291.15 K
   $$\Lambda(\text{mS} \cdot \text{m}^2/\text{mol}) = 12.945 - 8.035c + 3.286c \log c + 15.43c$$
   $$- 14.30c^{3/2} \quad \text{(d)}$$

   283.15 K
   $$\Lambda(\text{mS} \cdot \text{m}^2/\text{mol}) = 10.7314 - 6.495c + 2.706c \log c + 12.54c$$
   $$- 11.03c^{3/2} \quad \text{(e)}$$

   273.15 K
   $$\Lambda(\text{mS} \cdot \text{m}^2/\text{mol}) = 8.1659 - 4.778c + 2.059c \log c + 9.38c$$
   $$- 7.93c^{3/2} \quad \text{(f)}$$

8.2 one should apply a temperature correction to the cell constant k. This arises from the thermal expansion of platinum and glass and so depends on the cell geometry [17, 228]. The simplest case is a long narrow cell as in Figure 8.12a with large electrodes at each end. Here virtually all the resistance resides within the narrow tube, and it follows from (115) that

$$\frac{1}{k}\left(\frac{dk}{dT}\right) = \frac{1}{l}\left(\frac{dl}{dT}\right) - \frac{1}{A}\left(\frac{dA}{dT}\right)$$

$$= \alpha_{glass} - 2\alpha_{glass} = -\alpha_{glass} = -3.5 \times 10^{-6}\, K^{-1}\quad \text{(Pyrex glass)}$$

The analysis is more complex for two large electrodes relatively close together and sealed into glass (Figure 8.14). If the stems holding the circular electrodes are short, then to a first approximation [17]

$$\frac{1}{k}\left(\frac{dk}{dT}\right) \approx \alpha_{glass} - 2\alpha_{Pt} = -15 \times 10^{-6}\, K^{-1}$$

A similar result is obtained for the short ring electrode cell in Figure 8.15 [228]. However, the correction becomes much larger if the electrodes in Figure 8.14 are held to the glass by platinum stems whose lengths are comparable to the electrode separation [17].

The cell constants of Teflon cells employed for high-pressure research show small hysteresis effects. This is partly due to some irreversibility in the properties of the polymer brought about by the phase changes that it undergoes [126] and in part arises when the cell is dismantled after a run and then refilled [219]. In precise work it is therefore necessary to take an initial conductance reading when the cell is at 25°C and atmospheric pressure. Since the conductance under these conditions will either be known or can readily be measured in normal conductance cells, this procedure amounts to a redetermination of the cell constant at the start of each run [219]. An 0.01 D KCl solution has been suggested [230] as a conductance reference up to 800°C and 12 kbar with a probable uncertainty of ca. 0.5% at 100°C and 1 kbar and ca. 5% at the highest temperatures and pressures.

4.1.4   Run Technique

Most workers platinize their electrodes. This increases the surface area by a factor of several hundred [231] and with it the double-layer capacitance. It also decreases the current density by the same factor and so greatly reduces the degree of electrode polarization. These two beneficial effects ensure almost complete frequency independence as (109) shows. On the other hand, platinized electrodes can adsorb significant quantities of electrolyte and are thus unsuitable for work with very dilute solutions or for measuring solvent conductances. (The walls of the cell may themselves adsorb electrolyte and in working at extreme dilutions it is as well to test this by adding more glass to the cell [232].) The experimenter should also be alert to the danger of inadvertent catalysis by the electrodes, especially if they are platinized. Redox reactions are particularly vulnerable to platinum catalysis [153]. The outcome might be decomposition of the solute (e.g., reduction of $Co(NH_3)_5Br^{2+}$ to Co^{2+} [233], interaction between Ag^+ and mannitol [234], inversion of sucrose at 50°C [170]), solute–impurity interaction (e.g., oxidation of formate ions by oxygen adsorbed on the platinized electrodes [235]), solute–solvent interaction (e.g., decomposition of sodium dissolved in liquid ammonia [236]), or decomposition of the solvent itself. Hydrogen peroxide is a prime example—the problem was solved by using tin electrodes instead [237]—and even as stable a substance as ethanol can be decomposed by platinum at around 245°C [238]. Nevertheless, in the majority of systems studied, platinized electrodes engender no chemical instability and can be used with advantage.

Before platinization the electrode surface must be cleaned. Brief immersion in aqua regia (comprising ca. 5 M HCl and 2 M HNO_3) is effective and also strips off any old platinum black [220]. This must be followed by treatment with concentrated nitric acid and plentiful washing with water. It is helpful, just prior to the platinization itself, to reduce the surface by cathodic electrolysis in dilute sulfuric acid for 10 min [220]. There are many different recipes for the platinizing solution. That of Jones and Bollinger [239] is a mixture of 0.3% chloroplatinic acid and 0.025% lead acetate in 0.025 M HCl. The current density is 1.5 [239] or 10 [17] mA/cm^2 with polarity reversal every 10 s until approximately 0.4 C/cm^2 has been passed for light platinization and approximately 10 C/cm^2 for a heavy coating. However, later work has shown that the optimum effects of lead (see below) are achieved with a concentration five times lower, and Feltham and Spiro [231] recommend 3.5% chloroplatinic acid with only 0.005% lead acetate. Addition of hydrochloric acid reduces the electrical resistance. A current of 30 mA/cm^2 for 5 min cathodic treatment is likely to produce optimum results (9 C/cm^2). If possible the solution should be stirred throughout and no gas evolution should occur at the cathode. A drawback of the traditional current reversal method is that each electrode in turn also acts as an anode and so adsorbs oxygen and chlorine. This can be avoided in certain cell designs by connecting the two platinum electrodes together as parallel cathodes and placing the common anode symmetrically between them. One end of a salt bridge leading to a well-removed anode would serve, or else a large silver foil (previously lightly chloridized to prevent oxidation by chloroplatinate ions) could be inserted. The solution must then contain added chloride ions, for example, up to 2 M HCl. Whatever the chosen arrangement, the deposit should possess a uniformly smooth black appearance and adhere firmly to the electrodes. These are the qualities promoted by the presence of lead. Research has shown [240] that the lead is firmly incorporated within the deposit but if its presence is undesirable, all surface lead capable of being removed can be dissolved out by soaking the electrodes for 24 h in well-aerated 1 M $HClO_4$. Satisfactory platinization without any lead additive has been achieved with a current density of 10–20 mA/cm^2 and a 2% chloroplatinic acid solution 2 M in HCl; this produced a light gray or golden film with a lifetime of several weeks [220]. The platinized electrodes must be thoroughly washed, especially if the plating solution contained acid, and always kept immersed in distilled water.

Conductance cells are commonly cleaned with fuming or hot nitric acid and then repeatedly washed with distilled water [191, 202]. Some workers also steam the cell [28], others [166] regard this practice as deleterious. It is not advisable to soak the cells with a chromic–sulfuric acid mixture because the chromium penetrates the glass and requires several changes of boiling water for its removal [241]. It also etches the glass. So does an alkaline surfactant, 2% Decon 75 solution, which has recently been suggested [242] for removing ions adsorbed on the glass walls. Its prolonged use is likely to change the cell constant of the cells depicted in Figure 8.12.

Rarely does one wish to obtain the conductance of just one solution of a given electrolyte. For almost every purpose—be it the determination of the limiting molar conductance, the association constant of the electrolyte, or a study of interionic effects—one needs to measure the conductivity over a range of concentrations. There are three main ways of proceeding [28, 36].

1. Individual runs. Each solution is made up separately by weight from the solvent and either the salt itself or a concentrated stock solution. Although this is the only feasible method for small conductance cells without an attached flask compartment, it is time consuming when investigating a series of solutions and can be wasteful of material.

2. Dilution runs (the Ostwald–Arrhenius technique [243]). A known amount of a concentrated solution is placed in the cell and progressively diluted with increments of solvent. This procedure clearly requires a flask cell with a magnetic stirrer or some other means of mixing. The solvent can be added by means of a weight burette [243, 28] and both it and the solution protected from atmospheric contamination by a stream of solvent-saturated nitrogen or argon. An extra experiment must be carried out at the end to obtain the solvent conductivity.

3. Concentration runs. In this, probably the most popular method today, a given quantity of solvent is first placed in the flask-type cell and its conductivity measured. Successive additions are then made of a concentrated stock solution (preferably by using a well-designed weight burette [243, 28]) or of the pure salt. The solid is most conveniently introduced inside a small weighing bottle or cup which is either released by a magnetic switch [36] or dropped through a hole in the special cup-dispensing device invented by Hawes and Kay [202]. This allows eight salt-containing cups to be dropped consecutively into the liquid without exposing it to the atmosphere. Details of its construction [244] and use [245] are given in the references cited and have been well summarized by Evans and Matesich [28]. In some cases the desired electrolyte is best generated within the conductance cell; alkali metal alcoholates, for instance, were formed in situ by adding known amounts of the metal amalgam to the alcohol [246].

In every conductance measurement care must be taken to avoid errors introduced by the Soret and shaking effects described in Section 4.1.2. Agitation with a magnetic stirrer is a good remedy for both. Once thermal equilibrium has been reached, the resistance readings should be constant and reproducible. Except for the solvent conductivity, an accuracy of 1 part in 1000 is easily attained and 1 part in 10^4 or even 10^5 is quite common in much conductance research. As was emphasized in Section 4.1.1, the frequency dependence should always be checked: a large variation points to a deficiency in the cell design or the bridge circuit, and a small one ($<0.1\%$) should be removed by appropriate extrapolation to infinite frequency. This extrapolated resistance should also be independent of the applied bridge voltage. However, with very high resistances such as that of the solvent, the resistance sometimes *increases* with increasing frequency. A leakage path to ground has then developed along a resistance (that of the solvent in the cell) in series with a capacitance (between the bottom of the flask and the magnetic stirrer mechanism). Mysels and co-workers [247] have analyzed this situation and shown that here the resistance should be extrapolated to *zero* frequency against f^2 at low frequencies ($<7\,\text{kHz}$). The significance of the solvent correction is discussed at the end of this section.

A correction must always be applied for the resistance of the electrode leads. A simple way of doing this is to fill the cell with mercury, preferably before platinizing the electrodes. The contribution of the mercury itself is negligible except in a capillary cell and so the measured resistance is just that of the leads [190]. In the sophisticated four-leads method used in platinum resistance thermometry and described by Feates, Ives, and Pryor [166], the resistances of the cell leads cancel out. However, incorporation of four leads does entail greater structural complexity in the cell.

For accurate work it is essential to fill the thermostat bath with a good transformer oil [178] or a suitable silicone oil [172]. Most conductance work to date has been restricted to one temperature (18°C in the last century, 25°C in this, and 30°C or higher in hot countries like India), but this situation is changing. A well-designed thermostat assembly has been described by Wachter and Barthel [172] that is capable of ranging between -70 and $+50$°C with a reproducibility of 0.001°C. Barthel and his group in particular [36] are now routinely determining conductances over a wide concentration and temperature range in order to construct Λ–c–T diagrams. The concentration run technique (see 3) is employed but the temperature is changed in regular steps before the next solute increment is added. This procedure is termed the method of isologous sections. To cover eight temperatures in 10°C steps takes 6 h for a given concentration so that some 10 days are needed to measure the conductances of 8–10 concentrations over a 70°C spread. The stability of the solutions in the cell over this long period is safeguarded by an inert gas atmosphere and by the exclusion of light [172]. One small practical point to conclude with: thermostat motors frequently interfere electrically with the measuring circuit. Should extra shielding prove inadequate, the motors must be switched off during the actual resistance readings [178].

The solvent itself possesses a finite conductivity. Its ionic content stems partly from its own self-dissociation (which in the case of water contributes 5.5×10^{-6} ohm^{-1} m^{-1} at 25°C and much more at higher temperatures and pressures

[127]) and partly from dissolved impurities. Thus the usual specific conductivity of approximately 1×10^{-4} ohm^{-1} m^{-1} for distilled water at 25°C is mainly caused by the dissolved carbon dioxide and the H^+ and HCO_3^- ions in equilibrium with it. If the solvent and solute ions simply carry their own quota of the current independently of each other, it follows from (12) that

$$\kappa_{solute} = \kappa_{solution} - \kappa_{solvent} \tag{121}$$

An example is provided by aqueous KCl solutions in (120). However, $\kappa_{solvent}$ is not exactly the same as the measured value because introduction of solute increases the ionic strength. The mobilities of the solvent ions therefore decrease slightly and, more important, the activity coefficients and hence the degrees of dissociation of water and of dissolved H_2CO_3 are affected [248]. Moreover, changes in solvent conductivity can occur through ion-exchange processes with the cell walls—the shaking effect [197]. In the method of isologous sections, allowance must be made for the variation of the solvent conductivity with time and with temperature [249].

Quite a different situation arises if the solvent and solute ions interact. In aqueous solutions of acids or bicarbonates the dissociation of H_2CO_3 is repressed by the common ion effect and the value of $\kappa_{solvent}$ in (121) must be decreased accordingly. In alkaline solutions the OH^- ions react with H_2CO_3 to form the more poorly conducting HCO_3^- ions so that a negative solvent correction must be applied! Its value must of course be obtained by calculation. Obviously it is good policy in accurate work to remove the dissolved carbon dioxide and to carry out all the operations in an inert atmosphere and, in general, to reduce the levels of all such interacting impurities to the lowest values possible [36].

A further correction is sometimes applied for nonionic impurities in the solvent. The prime example is the presence of small residual quantities of water in nonaqueous solvents, a problem usually tackled by determining the effect on the solution conductivity of adding further small amounts of water. A linear variation with water content then allows extrapolation to anhydrous conditions [250]. No such extrapolation is justified when the variation is far from linear as with solutions of acids in alcohols [251].

4.2   Electrodeless Methods

The vast majority of conductance measurements are performed with cells containing two (or more) electrodes inside them. Yet their presence poses many problems, not the least of which is the constructional difficulty of sealing platinum into glass (Section 4.1.2). On a more fundamental level, the passage of current across the metal–solution interface gives rise to polarization with its undesirable frequency dependence, and relief through platinization introduces the risks of catalysis and of significant adsorption in dilute solutions (Section

4.1.4). The liquid under study may also contaminate and poison the electrodes, as happens when industrial fluids containing oil, tar, or slime are conductometrically monitored [208]. Certain solutions, especially highly acid ones, can lead to electrode corrosion. All these difficulties are overcome—albeit at the price of some simplicity—by various electrodeless or noncontacting arrangements. Many different circuits and devices have been proposed over the years [252]. This section briefly describes two main types: the audiofrequency transformer bridge and radiofrequency measurements (oscillometry).

The transformer bridge method was independently developed into a precision tool by Gupta and Hills [253] and by Calvert, Griffiths, and co-workers [254, 255]. It is best suited to work with solutions of moderate or high conductivity. The principle can be explained with the help of Figure 8.17 and by assuming that both transformers behave ideally. The primary of the input or voltage transformer T_1 is connected to an audiofrequency source. Its two secondary windings of N_u and N_s turns, respectively, are wound in phase opposition. For ideal behavior the ratio of these turns equals the ratio of the potentials produced in the upper and lower central circuits; that is,

$$\frac{E_u}{E_s} = \frac{N_u}{N_s} \tag{122}$$

The current flowing through the unknown impedance Z_u is then $I_u = E_u/Z_u$ and that flowing through the standard impedance Z_s is $I_s = E_s/Z_s$. These currents pass through the n_u and n_s primary windings, respectively, of the output or current transformer T_2. Its secondary winding is connected to a detector which could be a tuned amplifier and an oscilloscope. If Z_s is appropriately adjusted, the detector will give a null indication. Under this condition, zero core flux is produced in the current transformer. The algebraic sum of the ampere-turns must therefore be zero [254, 255] so that

$$I_u n_u = I_s n_s \tag{123}$$

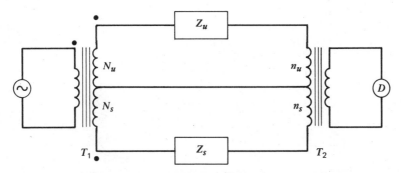

Figure 8.17   Basic circuit diagram of a transformer bridge.

Combining these equations leads to

$$\frac{Z_u}{Z_s} = \left(\frac{E_u}{E_s}\right)\left(\frac{I_s}{I_u}\right) = \left(\frac{N_u}{N_s}\right)\left(\frac{n_u}{n_s}\right) \tag{124}$$

Both the unknown and the standard impedances can be divided into parallel resistive and reactive components. The usual "standard" circuit between the transformers includes various tapping arrangements along N_s to provide a wide range of standard resistances for balance and a variable capacitor [253, 254]. We may now rewrite (124) in purely conductance terms

$$G_u = \frac{G_s N_s n_s}{N_u n_u} \tag{125}$$

Such a transformer bridge can clearly be used to measure the conductivities of solutions in the normal type of conductance cell containing two internal platinum electrodes. It also permits the determination of conductance without any contacting electrodes. Consider the case where N_u and n_u are made single turns, which converts the balance equation to

$$G_u = G_s N_s n_s \tag{126}$$

The physical nature of this single winding is unimportant as long as it is adequately conducting, and indeed a closed loop of an electrolyte solution in a glass or plastic tube serves perfectly well. Figure 8.18 shows how this can be done by winding the two transformers on toroidally shaped cores. In effect, then, the two transformers are coupled with the annulus of solution, and the degree of coupling is proportional to the conductivity of the solution. This explains the need for the solution to be reasonably well conducting since otherwise leakage inductance effects occur [257]. The sizes of the test probe and the toroidal cores are less at higher frequencies and 20 or 50 kHz are therefore preferable to the 1 kHz frequency employed in some of the earlier work. Several other loop designs are feasible including flow-through cells. In other devices the transformer unit can be directly immersed in a large vessel filled with the test solution [254, 258]. This electrodeless technique deserves wider recognition than it has received so far.

When accuracies of only 1–2% are required, one can dispense with the "standard" circuit and link the toroidal transformers solely with the solution loop (Figure 8.18). The voltage output from the second transformer is then

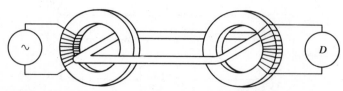

Figure 8.18 Illustration of the electrodeless method in which two toroidal transformers are coupled by an annulus of solution, after [256]. The "standard" circuit is not shown.

measured. A theoretical analysis of this situation [256] has shown that in an intermediate-frequency region, the voltage transfer by the loop is directly proportional to the sample conductance over a range of several powers of ten. Commercial instruments are available for this type of measurement [257, 258].

Electrodeless conductivity measurements with radio frequencies (1–300 MHz) have been employed for several decades but are unsuitable for precision work. Burkhalter [137] presents a good review of this technique. In essence, RF electromagnetic energy is transmitted through the walls of the container and stored inside the vessel. The magnitude of the energy absorption, which depends upon the solution composition, is detected by an appropriate parameter of the field generator such as the variation of load or of frequency. Several types of circuit have been described [137, 252] and are incorporated in various commercial instruments. There are two main cell designs: a cylindrical glass container that sits snugly inside a coil in the oscillator circuit, and a glass cylinder with two separated metal bands firmly attached to its outside. Leads from these metal bands couple the cell into the oscillator circuit. The main disadvantage of the RF method lies in the fact that the response (whether of voltage or frequency) is not a linear function of solution conductance or resistance [137, 252]. At any given frequency, the conductive component of the cell admittance (the reciprocal of its impedance) first rises with increasing conductivity of the solution and then falls again at higher values of κ. The curve itself changes with frequency. The capacitative component increases with κ and eventually reaches a plateau region; the shape of this curve, too, is a function of frequency. Furthermore, the sensitivity of response for a given system depends on both the frequency range employed and the cell dimensions. An increase in sensitivity can often be achieved only at the expense of limiting the concentration range that can be investigated, and vice versa. It is hardly surprising, therefore, that the RF method has found application mainly in industrial monitoring and in conductometric titrations [136, 137]. Among the useful analyses that can be carried out this way are titrations of various phenols in benzene + methanol solvents by sodium methoxide, and of amines in glacial acetic acid with perchloric acid. However, because of the nonlinear response, the shapes of these titration plots are not necessarily the same as those described in Section 3.2.2 for the corresponding audiofrequency titrations. The original literature should therefore be consulted for the most suitable experimental conditions for any particular analysis.

4.3   Direct Current Method

Ohm's law

$$G = \frac{I}{\Delta E} \tag{127}$$

is the basis of the direct current method. Known since the time of Kohlrausch, it was not until 1942 that Gunning and Gordon [259] developed it into a technique whose precision rivals that of the standard ac method. A four-electrode cell is required (Figure 8.16). The products of the electrode reactions at the current carriers C_1 and C_2 should contaminate the solution as little as possible and it is essential that no composition change reaches the central part of the cell near the potential probe electrodes P_1 and P_2. The latter should be reversible with small bias potentials. In aqueous and alcoholic halide solutions these conditions are best fulfilled by using four silver/silver halide electrodes. Gunning and Gordon carefully specified the geometry of the cell and their design is reproduced in Figure 8.19. The current carriers were placed in deep wells at the ends with narrow connecting tubes leading to the central body of the cell to minimize contamination by the electrode products. The potential probe electrodes were made of vertical platinum disks covered in fused glass except for a narrow vertical slit in each which was silver plated and halogenized. These electrodes were inserted into sidearms off the central body, and their positions in the sidearms were accurately and reproducibly fixed by thin etched lines on the male and female joints. Elias and Schiff [260] subsequently devised modified probe compartments that can hold a solution different from that which fills the main cell. Thus even with aqueous $NaNO_3$ solutions for which no reversible electrode existed, or in work with nitromethane solutions that dissolve silver halides [250], it proved possible to retain reversible Ag–AgX electrodes in

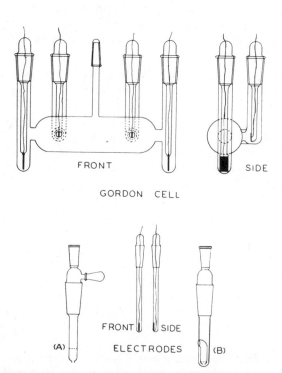

Figure 8.19 Gordon's direct-current conductivity cell together with the modified probe electrodes designed by Elias and Schiff [260].

aqueous halide solutions in the probe chambers. The two liquid junction potentials largely cancel each other out in the potential difference ΔE between the two probes. Any remaining liquid junction or bias potential difference between P_1 and P_2 is removed by taking the average value of ΔE when the direction of the current is reversed [259]. (This trick is not available for the corresponding four-electrode ac method described in Section 4.1.2.) The value of G calculated from (127) should be independent of the magnitude of the current, and this also provides a check on whether electrode products have diffused out into the central region.

The Elias and Schiff modifications make it easy to determine the conductivity of the solvent. An alternative procedure [261] is to measure the conductivity of an extremely dilute electrolyte solution, say 5×10^{-6} mol/dm³ KCl. A knowledge of the contribution of the electrolyte to κ_{solution}, calculated from experiments in more concentrated solutions and the requisite theoretical $\Lambda-c$ relationship, allows κ_{solvent} to be evaluated. It is curious that the dc method is so rarely used despite its precision and simplicity: a constant-current device and accurate digital multimeters are the only electrical instruments required.

5   MEASUREMENT OF TRANSFERENCE NUMBERS

5.1   Comparison of Methods

Transference numbers can be determined by several different methods. The decision as to which one should be used depends upon a number of factors. Chief among these are the accuracy desired, the concentration range of interest, and the properties of the solution in question. Comparative information regarding the first two points is given in Table 8.3, in which mb stands for moving boundary method. The principal requirements for any given method, outlined in Table 8.4, must be compatible with the properties of the solution under investigation. For aqueous solutions, an accuracy of about 2–5% can usually be readily obtained with almost any method. It must be emphasized that many precautions and corrections enjoined in the text below are necessary only when the required precision warrants their adoption. The Hittorf method is probably the simplest one. The mb methods, or the Hittorf method for concentrated solutions, are the best choices if high accuracy is desired. The emf method is most useful for a survey over a wide temperature and concentration range.

Two other methods, previously described in some detail [262], were not included in this edition because they have been seldom used in the past decade. The first is the analytical boundary method. Here two solutions are separated by a mechanical plane such as a porous glass frit, and the number of ion constituents transported through the frit per faraday is determined by analysis. This technique becomes particularly useful when the migrating ion constituent is radioactively labeled [262, 263], as in the recent determination of the transference numbers of the ionic constituents of seawater [16]. In the case of a

Table 8.3 Concentration Ranges of Aqueous Solutions[a] for Which the Main Transference Methods Are Best Suited, and the Usual Maximum Accuracy Obtainable

		Transference Method		
Approximate Optimum	*Hittorf*	*Direct mb*	*Indirect mb*	*Emf*
Upper concentration	No limit	$1\ M^b$	No limit[c]	No limit[d]
Lower concentration[e]	$0.01\ M$	$0.001\ M^b$	$0.001\ M$	$0.001\ M^d$
Error limits	$\pm 0.1\%$	$\pm 0.03\%$	$\pm 0.03\%$	$\pm 0.2\%$

[a]In solvents of lower permittivity, difficulties are introduced by the higher electrical resistance of the solutions and the resulting Joule heating and convection, the relative scarcity of suitable electrodes, the decreased solubility of many electrolytes, and certain other specific points mentioned in the footnotes.

[b]In nonaqueous solvents lower concentrations must be used. The solvent correction restricts the accuracy at low concentrations, the volume correction and the Soret effect at high concentrations.

[c]The method is restricted mainly by the availability of data for the leading solution.

[d]Provided the electrodes remain reversible, reproducible, and insoluble. The stability and/or reproducibility of electrodes tend to decrease in nonaqueous solutions. Some electrodes become quite soluble at high concentrations.

[e]Lower concentrations can often be investigated at the expense of decreased accuracy.

micellar electrolyte, tagging is possible by an insoluble dye. Modifications of the method include the use of two frits with the intervening solution stirred throughout the run.

The second method omitted this time is that of emf cells in force fields. When a cell containing two identical electrodes bathed by a common solution is spun in an ultracentrifuge, an emf of the order of 1 mV develops. This is a function of the transference number of the electrolyte. The centrifugal emf method requires complex equipment and has so far been applied only to iodide solutions.

Transference measurements in molten salts have been reviewed elsewhere [264] and are not discussed in this chapter.

5.2   Hittorf Method

5.2.1   Introduction and Theory

In the method named after Hittorf, who introduced it in 1853, a known quantity of electricity is passed through a cell filled with the solution whose transference numbers are to be determined. The solutions in various sections of the cell are then directly or indirectly separated, weighed, and quantitatively analyzed.

Figure 8.20 illustrates a general Hittorf cell, with E_A and E_C representing the anodic and cathodic electrode sections or compartments and M_A, M_M, and M_C three middle sections (anode middle, middle middle, and cathode middle).

Table 8.4 Basic Requirements for the Application of the Main Transference Methods

| | | Transference Method | | |
| | | Direct | Indirect | |
Necessity for Suitable	Hittorf	mb	mb	Emf
Analytical method[a]	Yes	—	Yes	—
Leading solution[b]	—	—	Yes	—
Following solution[c]	—	Yes	—	—
Method of following mb	—	Yes	—	—
Electrode(s)[d]	Yes	Yes	Yes	Yes
Electrical measurements[e]	Yes	Yes	—	Yes
Activity coefficients	—	—	—	Yes

[a]In the indirect mb method the transference number is directly proportional to the one analysis made in each run. In the Hittorf method more than one analysis is performed per run, and the transference number is proportional to the relatively small difference between two such analyses.

[b]Containing a faster noncommon ion constituent. The transference number in the leading solution must be known.

[c]Containing a slower noncommon ion constituent. For autogenic boundary formation this requirement merges with the conditions for a suitable electrode.

[d]In the emf method the electrode reaction must be reversible on the passage of very small and momentary currents; in all other methods the electrodes need not act reversibly but they must stand up to the passage of appreciable quantities of electricity.

[e]Of potential in the emf method, of the quantity of electricity (current and time) in the other methods.

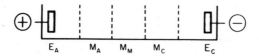

Figure 8.20 Schematic diagram of a Hittorf cell.

The dotted lines stand for the section boundaries, which, as discussed in Section 1.2, are considered fixed with respect to the solvent throughout the experiment. The basic principle follows directly from the definition of transference number. For every faraday of electricity passed through the cell, $T_R/|z_R|$ mol (T_R equivalents) of a cation constituent R, carrying T_R faradays of charge, migrate across every section boundary in the direction of the cathode, that is, out of E_A and into M_A, out of M_A and into M_M, out of M_M and into M_C, out of M_C and into E_C. Clearly the compositions of the middle sections remain the same and only those of the electrode sections change. Analysis of either of these for the constituent R, therefore, yields T_R. A similar description can be given for an anion constituent migrating in the opposite direction. The electrode reactions must also be considered, since they may involve the constituent of interest; an example demonstrates the reasoning employed.

Suppose the anode is a thick silver wire, the cathode a silver wire thickly coated with AgCl, and the cell is filled with a solution of $BaCl_2$. The following chemical changes then take place when 1 faraday is allowed to flow through the cell. In the anode compartment, $\frac{1}{2}T_{Ba}$ mol (T_{Ba} equivalent) of barium ion constituent (which carries T_{Ba} faradays of charge) migrate out and T_{Cl} mol of chloride ion constituent migrate in. Moreover, 1 mol of chloride ion constituent disappears from the solution through the electrode reaction

$$Cl^- + Ag \rightarrow AgCl + e^-$$

This produces a net loss in the anode solution of $1 - T_{Cl}$, or T_{Ba} [see (25)] mol of chloride. The solution remains electrically neutral with a net loss of $\frac{1}{2}T_{Ba}$ mol of $BaCl_2$. The converse occurs in the cathode section: $\frac{1}{2}T_{Ba}$ mol of barium migrate in, T_{Cl} mol of chloride migrate out, and 1 mol of chloride enters the solution by virtue of the electrode process

$$e^- + AgCl \rightarrow Ag + Cl^-$$

There is thus a net gain of $\frac{1}{2}T_{Ba}$ mol of $BaCl_2$. It must be emphasized, however, that the chemical changes in the anode and cathode compartment solutions are not equal and opposite in every Hittorf experiment. Quite a different result would have emerged if, for example, a cadmium anode had been chosen for the present example instead of a silver one. In the middle sections as much chloride or barium ion constituent enters by migration on one side as leaves by migration on the other, and there is therefore no net change in solution composition. The purpose of these middle sections is to ensure that all concentration changes have taken place entirely within the two electrode compartments and that no intermixing has occurred. One middle compartment alone is not sufficient since equal diffusion from the more concentrated cathodic side and toward the more dilute anodic one would leave no net change in concentration and so give the experimenter the false impression that all is well.

An illustration taken from the literature will clarify the procedure used. Jones and Dole [265], in their sixteenth run, electrolyzed an aqueous 0.24745 M (4.949 wt%) solution of $BaCl_2$ (molecular weight 208.27) with a silver anode and a AgCl cathode in a cell similar to that in Figure 8.21. They stopped the experiment when 0.024644 faradays of electricity had passed and then isolated the electrode compartments by closing the stopcocks S and S'. The stoppers were removed and the solutions in the sections M_C, M_M, and M_A carefully pipetted out and gravimetrically analyzed. They were found to contain 4.946, 4.947, and 4.946 wt% $BaCl_2$, respectively, so that no significant mixing had occurred and the initial concentration can be taken as 4.947 wt%. The solutions in the electrode compartments were then weighed and similarly analyzed. The anode solution weighed 121.58 g and contained 4.148 wt% $BaCl_2$ or, to put it another way, it was made up of 116.54 g water and 5.043 g (or $5.043/208.27 = 0.02421_3$ mol) $BaCl_2$. In order to find the decrease in the number of moles in the

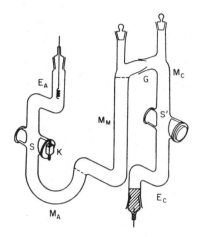

Figure 8.21 Hittorf cell used by MacInnes and Dole [270]. Reprinted, with permission, from [278].

anode compartment, we must now calculate the number of moles of $BaCl_2$ that were present initially in the *same mass of water*—transference numbers are defined in terms of migration across a plane fixed with respect to the solvent. Since 4.947 g $BaCl_2$ were present originally in $(100 - 4.947)$ or 95.053 g water, it is clear that 116.54 g water must at the start have contained

$$(116.54 \times 4.947)/(95.053 \times 208.27)$$

or 0.02912_1 mol $BaCl_2$. Thus on the passage of 0.024644 faradays, the anode compartment lost $(0.02912_1 - 0.02421_3)$ or 0.00490_8 mol $BaCl_2$. The theoretical discussion above shows that 1 faraday would have led to the loss of $\frac{1}{2}T_{Ba}$ mol and so the cation constituent transference number is

$$2 \times 0.00490_8/0.024644 = 0.398$$

An analogous treatment of the cathode department figures gives $T_{Ba} = 0.399$. The agreement is very good and illustrates how analysis of both electrode sections provides an excellent check on the internal consistency of a run. The calculation also highlights the principal weakness of the Hittorf method—the fact that the final result depends upon the relatively small difference between two quantities.

It was tacitly assumed in the above treatment that all the current is carried by the solute ions. In fact the solvent itself contains some ions, partly from self-dissociation and partly from dissolved impurities, and if their concentrations and conductances are not altered by introducing the solute, the corrected trans-ference number is given by [60]

$$T_R(\text{cor}) = T_R(\text{obs})\left(1 + \frac{\kappa_{\text{solvent}}}{\kappa_{\text{solute}}}\right) \tag{128}$$

where κ is the specific conductivity in $\text{ohm}^{-1} \text{ m}^{-1}$. Such a *solvent correction* is appreciable only in fairly dilute solutions. It must be modified if there is any chemical interaction between the solute and the solvent ions, as with aqueous solutions of acids, bases, and carbonates because water contains the ions H^+, OH^-, and HCO_3^-. An acid solute, for example, suppresses the dissociation of dissolved carbon dioxide, and the specific conductivity of the solvent must be changed accordingly (Section 4.1.4). However [266], the solvent correction may not be simply that given by (128) because fast impurity ions such as H^+ and OH^- migrate in different proportions to those of the solute, and the safest procedure for dilute solutions is to reduce the solvent conductance to as low a level as possible.

The problem of the solvent correction disappears if the sole purpose of the research is to obtain the molar conductance of the ion constituent R [78, 267]. This follows by combining (32) and (128):

$$\bar{\lambda}_R c_R = T_R(\text{cor})\kappa(\text{cor}) = T_R(\text{obs}) \left(\frac{\kappa_{\text{solution}}}{\kappa_{\text{solute}}} \right) (\kappa_{\text{solute}}) = T_R(\text{obs})\kappa(\text{obs}) \quad (129)$$

This result is particularly useful when the solvent corrections are large for both the transference and conductance measurements.

5.2.2   Analysis

The transference number depends upon a relatively small difference between two concentrations, and thus the accuracy of the final result depends mainly upon the accuracy of the analyses. Four- to five-figure analyses yield three-figure transference numbers. Any suitable analytical procedure may be used— gravimetric, volumetric, electrometric, spectrophotometric, counting methods for radioactive isotopes, and so on. If the solute consists of a mixture of substances, an analysis should be performed for each one.

After a run, samples from the various middle sections are analyzed first, and if the concentrations in these differ appreciably from each other or from the initial concentration, the run should be rejected. If, however, the concentrations are the same, then, depending upon the system, the number of moles of one or more ion constituents or of the electrolyte as a whole, in one or both electrode compartments, is determined by one of three main methods:

1.   If the cell is made in several pieces attached during the run, then the solutions in the middle sections can be drained off and the electrode section separated from the remainder of the cell. The section is dried on the outside and weighed. Either the whole solution is washed out for analysis, or the solution in the section is carefully homogenized by tilting and shaking without spilling any solution and a sample is taken, weighed, and analyzed. In either case the electrode section is then washed out completely, dried, and weighed, unless the weight of the empty section has been determined before the run. Thus the total weight of the electrode compartment solution is known as well as the concentration of a weighed aliquot, so that both the mass of solvent and the number of moles of ion constituent can be calculated [268].

2.   The entire solution in the electrode section is transferred to a weighed flask. The section is carefully rinsed out, not with solvent, but with some of the original electrolyte solution with which the cell had been filled, and the wash liquid is also added to the flask [269]. This is equivalent to including some of the unchanged solution in the middle compartments with the electrode solution, so that in effect the boundary line of the electrode compartment is taken somewhat further away from the electrode; this has no effect at all on the results. The flask is reweighed, and either the whole or a weighed aliquot of the solution is analyzed. Alternatively, if the density of the solution is known, volume rather than weight can be measured by transferring the solution and the wash liquid to a volumetric flask and filling to the mark with the original solution. In accurate work or in work with volatile solvents, it is necessary to minimize solvent evaporation during these operations.

3.   The compartments are isolated from each other after the run by closing stopcocks or by some other procedure. The solutions are carefully homogenized in each section and are then analyzed *within* the cell. This can be done, for example, by measuring the conductivity of the solutions with platinum electrodes that are already sealed into each compartment, provided the conductivity–concentration relationship is known [266]. The total mass of solution in at least one electrode compartment must also be found, either by weighing or, if the density of the solution is known, from a previous calibration of the volume of the electrode section itself.

If possible, both anode and cathode compartment solutions should be weighed and analyzed, as this provides an excellent check on the internal consistency of the run.

5.2.3   Electrodes

The choice of suitable electrodes is very important, and the following criteria should be applied in making a selection:

1.   The electrode reaction should be known, so that the transference number can be derived by an examination of the cell processes (see Section 5.2.1). Secondary reactions can often be eliminated by using electrodes of large surface area and sometimes by excluding oxygen from the solution.

2.   An ideal electrode reaction introduces no foreign ions into the solution [268]. If any foreign ions are introduced, they must not interfere with the analysis of the solution nor must they reach the middle compartments. This restricts the quantity of electricity that can be passed, and therefore the concentration changes that may occur in the electrode section are smaller. Electrodes producing highly mobile ions, such as H^+ and OH^- in aqueous solution, should therefore be avoided.

3.   If possible, the electrode reaction should not be associated with any physical action tending to stir and mix the solution, such as gas evolution or the formation of any precipitate that does not adhere to the electrode [268].

If these disturbances are unavoidable, their effect can be reduced by suitable cell design.

Anodes in common use are ones composed of the metal whose cation constituent is present in the solute (e.g., Cd in $CdCl_2$ solution), or of a metal whose cation reacts with an anion in solution to form an insoluble salt that adheres to the anode (e.g., Ag in NaCl solution), or of a metal whose cation constituent, although foreign to the solution, moves slowly and does not complicate the analysis (e.g., Zn or Cd). In difficult cases (e.g., solutions of soaps or detergents), electrodes with suitable guard solutions around them can be employed, but the electrode compartment should be designed to minimize mixing with the main solution. If a gassing electrode must be used, an appropriate guard substance can prevent the formation of very mobile ions (e.g., in aqueous solution, Li_2CO_3 surrounding a Pt anode substitutes slow Li^+ ions for mobile H^+ ions). Ion-exchange beads around the electrode, with perhaps a membrane as well, prevent the introduction of undesirable ions into the solution.

Cathodes in common use are those that liberate an anion constituent already present in the solution (e.g., silver halide electrodes in halide solutions), or ones composed of a metal upon which a metallic ion in the solution can plate out (e.g., Pt in $AgNO_3$ solution). A guard solution in a well-designed electrode compartment may be necessary [e.g., Pt surrounded by concentrated ferric acetate solution, Pt in a KI solution with dissolved iodine, or Hg in strong $Zn(NO_3)_2$ solution]. If a gassing cathode must be employed in aqueous solution, a suitable substance (e.g., benzoic acid crystals, or anion-exchange resin beads in the nitrate form) around the electrode can prevent the escape of the fast OH^- ions.

At least one electrode reaction in the cell should be free from objections so that the analysis of one electrode compartment is reliable. However, from the point of view of handling the cell after the run, as described in Section 5.2.2, the simpler the electrode compartments the better, so that guard tubes should not be used unless necessary. Aqueous chloride solutions have often been electrolyzed with Ag anodes and AgCl cathodes, and there is much useful advice in the literature [265, 266, 268, 270, 271] on the construction of such electrodes.

5.2.4   Disturbing Effects

The various effects that cause mixing between compartments can be decreased as follows.

Diffusion can be reduced by increasing the cell length and decreasing the electrolysis time.

Electrical Joule heating sets up a radial temperature distribution with a maximum temperature at the axis of the tube [272, 273], and the resulting density differences cause convective stirring. Concentration changes also develop because of the Soret effect [271]. The mean temperature T_{in} inside the tube exceeds the external temperature T_{ext} by an amount [274, 275]

$$T_{in} - T_{ext} = \frac{I^2}{2\pi^2 R_{in}^2 \kappa_{soln} k_{soln}} \left\{ \frac{1}{4} + \frac{k_{soln}}{k_{glass}} \ln \left(\frac{R_{ext}}{R_{in}} \right) \right\} \qquad (130)$$

where R_{in} is the inner radius of the tube, R_{ext} the outer radius, κ the specific conductivity, and k the thermal conductivity of the subscripted material. For water and Pyrex glass k is 0.60 and 1.1 J/m·s·K, respectively [275]. Thus the effect will be decreased if the current is small and if tubing of uniform and moderately large bore is used throughout the cell [268]. A series of right-angled bends in the tubing has been recommended for breaking up convective flow [270]. The thickness of the glass wall does not alter the radial temperature gradient but does influence the mean temperature inside the tube [272]. Thermostating reduces temperature fluctuations. If the experimental temperature is not important, advantage can be taken of the almost complete absence of convective mixing in aqueous solutions near 4°C, where the density of water is a maximum [160]. Convective stirring becomes pronounced in nonaqueous solutions of low conductance, and a preliminary investigation using colored indicators may reveal the extent of the effect [276].

Electrical migration of ions from the electrode reactions into the middle compartments is reduced by increasing the volumes of the electrode compartments and by decreasing the quantity of electricity passed through the cell. Other undesirable electrode effects are discussed above.

Vibrational disturbances are considerably reduced by hanging either the cell [170] or the thermostat [270] from a special suspension or by mounting the thermostat motor on a stand separate from the thermostat. Then vibrations from the motor are not transmitted directly to the apparatus.

The changed solubility of air in the solution at the experimental temperature may lead to the formation of air bubbles. This is overcome by partially degassing the solution, just prior to filling, by shaking it for a few moments at a reduced pressure. Sometimes dissolved oxygen must be removed to avoid side reactions at the electrodes.

5.2.5   Cell Design

In a large cell many of the above disturbing effects are small. However, the molarity changes in the electrode compartments per faraday are then also small, especially in concentrated solutions [277], and the analyses must be very precise. A compromise cell size is therefore selected in practice. The cell should be made of Pyrex rather than soda glass so as not to increase the solvent correction.

Good cell design is very important, and it is unfortunate that much of the early Hittorf work at the end of the nineteenth century was carried out with cells so poorly designed that the resulting data are now of little value. It is even more unfortunate that diagrams of such early apparatus are still faithfully reproduced in general textbooks and even in laboratory manuals. Some examples of well-designed cells are given in Figures 8.21–8.25 in which most of the lettering follows that in Figure 8.20. The cell in Figure 8.23 is easily modified to include two middle compartments. As far as possible, the bore of the tubing is fairly large and uniform throughout each cell and includes several bends to reduce the disturbances caused by convection and Joule heating.

Cells differ mainly in the methods devised for separating the compartments and removing the solutions. If a stopcock is used, the bore should be the same as in the remainder of the cell and the barrel should be hollow and open to allow the thermostat liquid to circulate freely through it and prevent local heating. The handle should also be hollow with openings in it, or else both ends of the barrel can be open and a removable metal key (K, Figure 8.21) used for turning. Choosing a suitable lubricant may prove difficult with nonaqueous solvents or emulsifying solutes [282] and it is worth trying silicone grease [170] or a fluorocarbon grease [78]. Diaphragms cause electroosmosis unless one electrode compartment is closed, relatively low currents must be used to avoid local heating and convection, and surface conductance effects may appear at concentrations below about 0.01 M. Most membranes other than sintered glass frits lead to selective transmission of ions [283] and will produce a zone of altered concentration nearby [284]. Teflon membranes have recently been advocated [285]. With the pressure head device in Figure 8.23, the solutions are

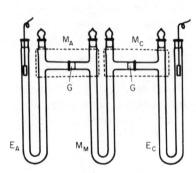

Figure 8.22   Hittorf U-tube cell used by Laing [279].

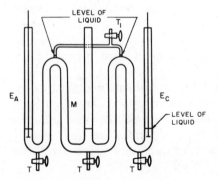

Figure 8.23   Hittorf cell used by Wall, et al. [280].

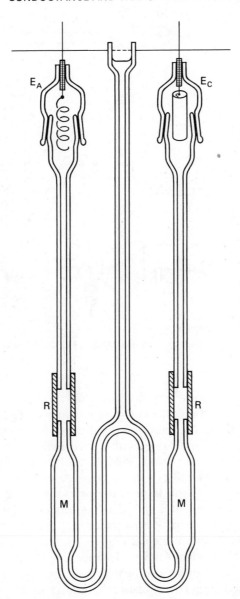

Figure 8.24   Small Hittorf cell employed by Stokes, Phang, and Mills [281] for studying concentrated solutions.

joined initially by suction at T_1 and are separated again after the run by opening T_1. In the Hittorf apparatus in Figure 8.24 the silicone rubber tubes R are clamped after a run with weighed screw clips and then cut off from the middle compartment M for weighing. The solutions in the middle compartments in Figures 8.21 and 8.22 can be removed by pipeting off the requisite amount of solution, and in the latter cell this automatically isolates the electrode solutions.

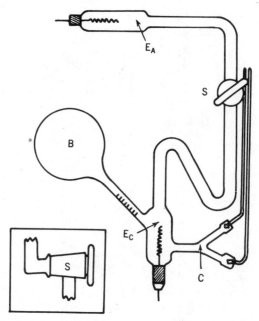

Figure 8.25 Hittorf cell designed by Steel and Stokes [170, 266] for internal conductometric analysis.

Alternatively, the solutions may be drawn off through taps at the bottom (Figure 8.23). If lubricant problems arise, taps with inert Teflon (PTFE) barrels should be employed. Ground-glass joints (G) allow the electrode sections to be separated for weighing or rinsing as described in Section 5.2.2. To prevent seizing of the joints a smear of some suitable lubricant (see above) is helpful or else the latter may be replaced by a thin Teflon sleeve that fits over the standard taper joint. It may be advisable to press the joints together during the run with the usual glass hooks and steel springs arrangement or by means of special clamps. Electrode guard or baffle systems for special purposes often consist of upright and inverted tubes inside each other, but regions of varying electrical resistance should be avoided. A given electrode must be introduced at the top if the products of electrolysis are lighter than the main solution, and vice versa, to eliminate mixing effects caused by gravitational instability (see also Figure 8.40 and 8.41).

Figure 8.25 shows a design allowing internal analysis by means of the conductivity cell C. The bulb B is a mixing chamber sufficient to hold the entire contents of the rest of the apparatus, and its interior and that of its calibrated stem are coated with a silicone layer to ensure complete drainage. With B full of air, the volume of E_C and C and the N-shaped tube up to the large stopcock S is first measured by weighing the amount of water needed to fill these parts with the meniscus standing at a known position on the calibrated stem. For a run the cell is filled with enough solution to bring the level of liquid within the range of the calibrated stem (with B containing air). The entire solution is then tipped several times into the mixing bulb and returned to its original position until adsorption changes are complete and the conductance is constant. Then S is opened, a known current passed for a known time, and S is closed again. The contents of the lower compartment are once more thoroughly mixed via the bulb, B, and the conductance remeasured. The volume should not have changed. As a check, S can be opened and the conductance of the entire cell solution determined: if the chemical changes in the anode and cathode compartments have been equal and opposite, the reading should be the same as the original one. From the known, previously determined, conductance–concentration relationship, the change in concentration in E_C can now be calculated and, bringing in the measured volume, so can the change in the number of moles and hence the transference number. The main defects of this otherwise elegant apparatus are that it allows one to determine the concentration change in only one electrode compartment (in Figure 8.25, of the one whose contents increase in density during electrolysis, and this is either E_A or E_C depending on the circumstances) and there are no middle compartments. Introducing middle compartments, and the appropriate conductance attachments, might complicate the cell unduly and there is therefore no direct proof that convection and diffusion have not contributed to the transport of material in the region of the stopcock. One must instead rely on the indirect evidence provided by constancy of the transference number when the current and time of electrolysis are varied.

5.2.6   Quantity of Electricity

A larger current and a longer time of electrolysis produce, on the one hand, larger changes in concentration in the electrode compartments and therefore a more precise transference number and, on the other hand, increases in many of the disturbing effects. A compromise must be reached on the basis of the magnitudes of the factors involved; normally currents ranging from about 1 to 100 mA are passed for several hours to give concentration changes in the electrode solutions of around 5–50%. For the cell in Figure 8.21 with tubing of diameter 1 cm or less, Pikal and Miller [131] suggest that the current in amperes be less than $0.01\sqrt{\kappa}$ where κ is the specific conductivity in ohm^{-1} m^{-1}. A 50% variation in current or time should not affect the transference number.

Current leakages in the cell obviously vitiate the measurements, and by far the safest procedure is to employ an oil-filled thermostat. As a check, the quantity of electricity is often measured at both ends of the cell.

Three different methods of measurement are available:

1. *Chemical coulometers.* These are based on Faraday's laws of electrolysis: the quantity of electricity passed is calculated from the extent of the chemical reaction at one or both electrodes in the coulometer. The silver and the water coulometers have become classic, and these and others suitable for the micro range have been referred to previously [262, 286]. Nowadays most workers choose instead one of the following electronic devices.

2. *Mechanical and electronic coulometers or current–time integrators.*
Milner and Phillips [287] have given the circuits and designs of a variety of these
instruments, and a few are commercially available. Some years ago Lingane
[288] published a fairly inexpensive mechanical integrator with a compensating
circuit capable of measuring to better than $\pm 0.1\%$ the quantity of electricity
over a wide current range. Several electronic constant-current coulometers
have been described [289, 290] and others could now be built with operational
amplifiers [291] and standard integrated circuit components.

3. *Separate current and time measurements.* The concentration and hence
the resistance changes in a Hittorf experiment are relatively small, and any
simple electronic device can maintain a constant current. Numerous galvanostats have been reported in the literature [287, 289, 290] and several such
instruments can be purchased commercially. The current is determined by
reading a high-impedance digital voltmeter connected across a calibrated resistor in series with the cell. The time of electrolysis may be measured by electric
or electronic means or, more simply, with a watch or stopwatch. See also
Section 5.3.6.

5.3 Direct Moving Boundary Method

5.3.1 Introduction and Theory

In the direct mb method, an electric current maintains and moves a fairly
sharp boundary between the solution under investigation (the leading solution)
and a solution of a suitable indicator electrolyte (the following solution). The
volume traversed by the boundary is measured when a known quantity of
electricity has passed through both solutions.

The theory is explained with reference to Figure 8.26 in which AZ is the
leading and BZ the following electrolyte, so that B is the noncommon indicator
constituent. The mb moves in the direction of the arrow, and a rising (Figure
8.26a) or a falling (Figure 8.26b) boundary is employed depending upon whether
the indicator solution behind the boundary is heavier or lighter than the leading
solution, respectively. The two-salt boundary AZ/BZ is termed a cation boundary if A and B are cation constituents (E_1 in Figure 8.26 is then the cathode and

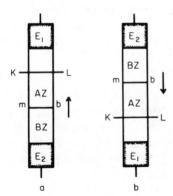

Figure 8.26 Schematic diagram of a general moving boundary cell.

E_2 the anode), and an anion boundary if A and B are anion constituents (E_1 is then the anode and E_2 the cathode). The following literature examples refer to aqueous solutions and should clarify these classifications of the AZ/BZ boundary:

Rising cation boundaries	$KCl/CdCl_2$; $K_3Fe(CN)_6$/methylene blue ferricyanide
Falling cation boundaries	$KCl/LiCl$; $K_2C_2O_4$/cetylpyridinium oxalate
Rising anion boundaries	KCl/KIO_3; H_3PO_4/citric acid
Falling anion boundaries	$KCl/KOOCCH_3$; sodium salt of benzopurpurine 4B dye/sodium benzoate

It is a rather remarkable phenomenon that these boundaries exist at all, for one might expect diffusion to work toward their destruction. The explanation lies in the so-called electrical restoring effect [292]. This comes into operation automatically whenever, in both solutions, the leading noncommon ion constituent A has a higher mobility than the following noncommon ion constituent B, that is,

$$\bar{u}_A > \bar{u}_B \tag{131}$$

For suppose that some A ion constituent drafts back into the BZ solution: at once it finds itself in an environment (usually of higher potential gradient) in which it moves more swiftly than its fellow B ion constituent and so it shoots ahead and overtakes the boundary. Again, a trace of B ion constituent diffusing into the leading AZ solution is slowed down by the conditions there (usually a lower potential gradient, or possibly a change in pH) and is eventually overtaken by the boundary [293]. In practice a steady state is soon established by the tug-of-war between the electrical restoring effect and diffusion, and boundaries therefore possess a small but finite "thickness" δ of the order of 0.1 mm. MacInnes and Longsworth [292] have derived an equation for this thickness:

$$\delta = \frac{4RT}{Fv} \frac{\bar{u}_A \bar{u}_B}{(\bar{u}_A - \bar{u}_B)} \tag{132}$$

where v is the velocity of the boundary. It is clear that the boundary is the "sharper" the more A and B differ in their mobilities, and also the faster the boundary. High currents therefore sharpen the boundary.

The relation between the boundary velocity and the transference number of the leading noncommon ion constituent A can be simply derived with reference to Figure 8.26. Let 1 faraday of electricity pass. The boundary then moves along the tube through a region of volume V_F and in so doing "sweeps forward" $c_A^{AZ} V_F$ moles of ion constituent A carrying $|z_A| c_A^{AZ} V_F$ faradays of charge. All this must in turn have migrated across the imaginary horizontal plane KL ahead of the boundary and is therefore, by definition, equal to T_A^{AZ}. Thus,

$$T_A^{AZ} = |z_A| c_A^{AZ} V_F = \frac{|z_A| c_A^{AZ} V F}{It} \tag{133}$$

As regards the last term, in most experiments less than 1 faraday passes and the volume V traversed by the boundary is then proportionately smaller than V_F by the ratio of the number of coulombs passed (current $I \times$ time t) to the number in 1 faraday (96,487). No assumptions about the degree of dissociation in either solution were made in deriving (133) because both the transference number and the concentration are properties, not of the ion, but of the ion constituent.

Moving boundary theory has been generalized and extended to systems of mixed strong electrolytes [294] and of buffer mixtures [159]. The basic equation common to all mb systems is, for a given ion constituent R,

$$T_R^\alpha - T_R^\beta = |z_R| V_F (c_R^\alpha - c_R^\beta) \tag{134}$$

where α and β designate the leading and following solutions, respectively. Equation (133) is clearly a special case for an ion constituent present on one side of the boundary only.

These equations are subject to two corrections. The *solvent correction* is applied as in (128), bearing in mind that the solvent conductivity can be appreciably greater inside the cell [295]. This is due in part to contamination during filling and to leaching of electrolytes from the cell walls and partly to preferential migration of impurities. The solvent correction increases with decrease in concentration, and the uncertainty inherent in it sets a lower limit on the applicability of the mb method. However, as shown in (129), no solvent correction need be applied to either the transference number or the conductance if only the ion constituent conductance is required. The *volume correction* is necessary because, from the definition of transference number, the boundary movement should be determined relative to a plane in the AZ solution fixed with respect to the solvent instead of relative to plane KL in Figure 8.26 which is fixed with respect to the cell. Since solvent is displaced relative to the cell by the volume changes at the electrodes, it follows that

$$T_A^{AZ}(cor) = T_A^{AZ}(obs) \pm |z_A| c_A^{AZ} \Delta V \tag{135}$$

where ΔV is the volume increase, in dm^3 per faraday of electricity, between a point in the AZ solution that the boundary does not pass and the outer or air face of whichever electrode is closed [296]. In mb experiments it is therefore important to keep one electrode closed and the other open to allow for volume changes during the run. The plus or minus sign in (135) refers to the boundary moving toward or away from the closed electrode, respectively. The correction clearly becomes more important as the concentration increases, and the uncertainty in it limits the accuracy of the mb method at higher concentrations. Where both volume and solvent corrections are large, as in mb work in sulfuric acid as solvent [78], the cross terms between the corrections must not be neglected. The full equation is then

$$T_A^{AZ}(cor) = (T_A^{AZ}(obs) \pm |z_A| c_A^{AZ} \Delta V)\left(1 + \frac{\kappa_{solvent}^{AZ}}{\kappa_{AZ}^{AZ}}\right) \tag{136}$$

where κ_{AZ} is the specific conductivity of the solute in the leading solution [295].

As an example of the volume correction, consider the effect of 1 faraday of electricity passing through the cell in Figure 8.27 in which XY is a hypothetical plane fixed relative to the solvent. Between XY and the Ag/AgCl cathode there is a gain (in moles) of $+1$ of Ag(s), -1 of AgCl(s), and $+1$ of Cl$^-$ attributable to the electrode reaction, and of $+T_K^{KCl}$ of K$^+$ and $-T_{Cl}^{KCl}$ of Cl$^-$ as a result of migration across XY. Thus if the cathode is closed and the anode open to the air,

$$\Delta V = + V_{Ag(s)} - V_{AgCl(s)} + T_K^{KCl} \phi_{KCl}^{KCl}$$

since

$$T_K^{KCl} + T_{Cl}^{KCl} = 1$$

V and ϕ are the molar and apparent molar volumes, respectively. Between XY and the Cd anode, there is a gain (in moles) of $-\frac{1}{2}$ of Cd(s) and $+\frac{1}{2}$ of Cd^{2+} because of the electrode reaction, and of $-T_K^{KCl}$ of K$^+$ and $+T_{Cl}^{KCl}$ of Cl$^-$ because of migration across XY. There is also a transfer, relative to the mb, of 1 of Cl$^-$ from KCl to CdCl$_2$. Hence if the anode is closed and the cathode open,

$$\Delta V = -\tfrac{1}{2}V_{Cd(s)} + \tfrac{1}{2}\phi_{Cd}^{CdCl_2} - T_K^{KCl}\phi_K^{KCl} + T_{Cl}^{KCl}\phi_{Cl}^{KCl} + \phi_{Cl}^{CdCl_2} - \phi_{Cl}^{KCl}$$
$$= -\tfrac{1}{2}V_{Cd(s)} + \tfrac{1}{2}\phi_{CdCl_2}^{CdCl_2} - T_K^{KCl}\phi_{KCl}^{KCl}$$

Until a few years ago the partial molar volume $\bar{V}$ was used in calculating volume corrections. This procedure was then shown to be erroneous [297] because it wrongly implied constancy in the partial molar volume of the solvent. Use of the apparent molar volume ϕ of the solute overcomes this difficulty [296]. The value of ϕ for an electrolyte like AZ at a given concentration c_{AZ} can easily be calculated from the density ρ of the appropriate solution by the equation

$$\phi_{AZ} = \frac{M_{AZ}}{\rho_0} - \left(\frac{\rho - \rho_0}{\rho_0 c_{AZ}}\right) \tag{137}$$

where ρ_0 is the density of the pure solvent at that temperature and M_{AZ} is the molecular weight of AZ. At infinitesimal ionic strength ϕ^0 is identical with $\bar{V}^0$. Values of the latter for many common electrolyte solutions have been tabulated in the literature [298] and may also be found in mb papers.

Because ϕ varies with concentration, a more precise calculation of ΔV requires knowledge about the spatial distribution of the electrolyte, especially

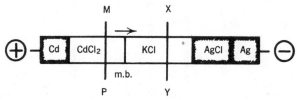

Figure 8.27   Sketch of a specific mb experiment to illustrate the calculation of the volume correction.

around the electrodes [271]. This is particularly important in work with concentrated solutions since, as (135) shows, the volume correction is proportional to the concentration of the leading electrolyte. Three steps that help to overcome this problem [271] are (a) to choose for the closed side of the cell an electrolyte whose apparent molar volume changes little with concentration, (b) to stir the closed electrode chamber to avoid the build-up of regions of different concentrations, and (c) if necessary, to presaturate the solution in this chamber with the insoluble salt involved in the electrode reaction, such as AgCl.

5.3.2   Observation of the Boundary

Three main techniques are known, a fourth is under development, and a fifth has been suggested.

OPTICAL METHODS

The boundary readily stands out if one of the solutions is brightly colored (e.g., potassium permanganate or tetraiodofluorescinate). If the solutions differ in pH and if high accuracy is not required, the same effect may be achieved by adding a little acid–base indicator to both solutions although this does decrease the transference number slightly [299]. However, a simple, more general, and better procedure is to utilize the difference in refractive index between the solutions. The essential equipment is shown in Figure 8.28. The telescope T, which can be slid up and down on the stand D, is focused on the calibrated mb tube C in the thermostat B, which is fitted with parallel glass windows W and W'. At temperatures below that of the room, double windows are advisable [300]. The person peering through the telescope can raise or lower the light source A either by means of a counterweight and string passing over pulleys on the ceiling or by a more complicated thread-and-screw arrangement. L is a lamp bulb (25 W is usually ample), and N is a cylinder (an empty can, blackened inside) which prevents interference from surrounding lights. S is a ground-glass screen which can be made by grinding the faces of a microscope slide and which diffuses the light so that the rectangular opening R is uniformly illuminated. R should be about 4 mm high and at least 6 cm long. A should be set up so that R and C are exactly at right angles to each other as in Figure 8.29 where R appears as a rectangular bar of light. G and G' are the

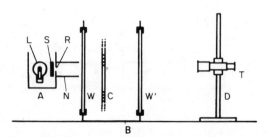

Figure 8.28   Simple assembly for optical observation of the boundary (not to scale).

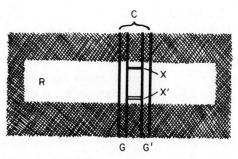

Figure 8.29   View of the boundary through the telescope T of Figure 8.28.

glass sides of tube C. The inside walls of the tube often appear thick and black, and this can be overcome either by employing a bath liquid of similar refractive index to the solution [8] or, more easily, by a suitable lens system. A vertical 12 cm wide cylindrical glass tube filled with water, between the light and the thermostat, is most effective [8]. The boundary becomes visible only when it is close to one of the edges of bar R; if it is near one edge, it appears as a dark line (X), and if it is near the other, as a bright line (X'). The former is the easier to use experimentally and it can often be seen directly with the naked eye. These images are caused by the refraction of the light beam in the optically inhomogeneous boundary region, and for a given electrolyte system they are sharper the greater the current. Since not all boundaries give visible images by this method (e.g., in water KCl/LiCl and NaCl/LiCl give visible images, KCl/NaCl does not), it is worth trying out several different following electrolytes to find one with optimum optical characteristics.

The rate of boundary movement can be measured by selecting a glass tube of completely uniform inner cross section (0.05–0.1 cm^2) provided the telescope in Figure 8.28 is mounted on a calibrated cathetometer stand. However, it is better to use a mb tube arrayed with a series of fixed marks, which permits periodic recalibration of the volume. For work of relatively low precision, the tube can be a graduated commercial pipet of uniform cross section and of 1 or 2 mL volume [301] (the etched marks are generally too broad for very accurate definition of the boundary). For accurate work, fine parallel lines are etched completely around a well-aged uniform piece of Pyrex tubing 15–25 cm long and 2–5 mm ID (a volume of about 1 cm^3 is convenient). In Le Roy's [302] procedure, the tube is drawn to a fine point at one end, and the part to be graduated is covered with a thin film of wax (1 part hard paraffin and 1 part beeswax). The tube is then placed horizontally on two V-shaped metal cradles and the pointed end is held tightly against a flat piece of metal. While the tube is rotated slowly, the end of a small needle held in a metal arm is brought down on the tube to trace a fine line through the wax. Finally, the tube is rotated for 1 min in contact with a few drops of a 1 : 1 HF + HCl solution. Some plain tubing must be left at each end for sealing to the rest of the cell. Enough lines should be etched on to permit several independent measurements of the boundary velocity per run. For example, with 12 lines (called 1, 2, 3, . . . , 12) spaced at roughly equal intervals, six independent determinations are possible (the differences $7-1$, $8-2$, . . . , $12-6$), and any change in the boundary velocity is easily detected. If desired, a larger interval can be left in the middle of the tube (with 12 lines, between lines 6 and 7); this increases the relative precision of each determination but makes it harder to detect any trend in the boundary velocity. The thin, etched lines appear horizontally in G and G' in Figure 8.29. Their positions are more readily spotted during the run if, after the cell is filled, a black grease-pencil mark is made on the tube a few millimeters above or below each line.

The optical method of following boundaries, despite its inherent simplicity, suffers from certain limitations [267]. It requires optically transparent windows in the thermostat bath, which makes it difficult though not impossible to apply at very low [300] or very high temperatures, and at high pressures [303] where the cell must be kept inside a sealed container. Moreover, in solutions more dilute than about 0.005 M, the difference in refractive index becomes very small and the boundary is then too faint for visual observation. Yet for work in non-aqueous solvents of lower dielectric constant, only the very dilute solutions can be investigated at all, boundaries at higher concentrations being disrupted by Joule heating [304]. Another, though relatively minor, restriction is that some solutions that form easily visible boundaries are somewhat light-sensitive; certain picrate solutions are an example [8]. However, it may be possible to track boundaries with polarized light if one of the solutions is optically active [305]. Should the simple optical method fail, recourse can be had to one of the more elaborate optical procedures such as the Longsworth schlieren scanning method [306–308]. This is described elsewhere [160, 309]. Most types of boundary can be followed in this way (see Figure 8.43), but exacting experimental requirements are involved for the optical system.

ELECTRICAL METHODS

The indicator solution always possesses a higher electrical resistance than the leading solution [cf. (131)]. This fact enables one to follow the passage of a boundary by suitable electrical signals at various points along the length of the tube. Until relatively recently this involved sealing platinum microelectrodes into the cell walls. In one recipe [310] a thin platinum ribbon 1 mm wide is stretched across the upper end of a 2.5 mm ID glass tube, another glass tube is pressed onto it, and the pieces are fused together. The piece of ribbon inside the tube is removed by cutting it with a sharp steel rod and rubbing with Carborundum powder. These sections are then sealed together to form a mb tube containing about five pairs of horizontally opposite probes. Rather more satisfactory probes can be produced [267] by sealing 0.2 mm diameter platinum wire into uranium (not Pyrex) glass. The degassed but unbeaded piece of wire is inserted into a tiny hole, and heat is applied to one edge until the glass melts onto the wire. It is advantageous if the wire hangs downward. That part of the wire inside the tube is sheared off with a sharp metal rod, and the probe is reheated so that the glass melts and fills any crevices. The whole tube is finally annealed. Great care must be taken to insulate the leads to these probes [309]. Another group [311] have manufactured their microelectrodes from 0.15 mm diameter iridium wire sealed into prebored 0.2 mm holes in Jena D 50 glass tubing which possesses a similar coefficient of thermal expansion [312].

The passage of the boundary past each probe or probe pair may be detected by measuring the electrical resistance. An ac bridge or ratiometric measurement can be employed (Section 4.1.1) at, say, 10 kHz [311] or 20 kHz [310], and a suitable block diagram was recently published [311]. Large capacitances of 0.02 μF [310] or 0.001 μF [312] on each side of the probe protect the ac circuit from the dc flowing through the transference cell. Figure 8.30a illustrates the type of signal obtained as the boundary passes a pair of horizontally opposite probes: the moment of passage can be taken as the point of inflection of the curve. The precision of timing is much improved by increasing the speed of the chart recorder as the boundary passes the probe region [311]. Figure 8.30b, in contrast, shows the change in resistance produced between two probes mounted several centimeters apart along the tube [313]. The initial and final horizontal lines represent the resistances of the leading and following solutions, respectively, and the points of intersection of these lines with the rising line in between mark the times at which the boundary passes each probe.

Other authors have preferred to record changes in potential. In one device [314], pairs of thin platinum foil electrodes, 0.3 mm apart, are connected through a 1 μF condenser to a recording galvanometer. As long as the probes face only one kind of solution they are subject to a constant potential difference. As the boundary moves past, this potential difference changes from the value characteristic of the leading solution to that in the following solution. The resulting readjustment of electrical charge on the condenser plates gives rise to a current peak (Figure 8.30c). However, the magnitude of this peak is ten times too high for the cell and circuit characteristics given, a curious discrepancy [309]. The main potentiometric method is undoubtedly that developed by Kay's group [267, 309]. Here the potentials between probes spaced vertically along the mb

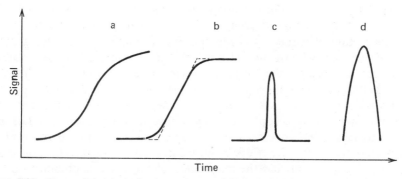

Figure 8.30 Types of signal obtained on passage of the boundary past several kinds of probes.

tube are monitored with a high-impedance digital voltmeter connected to a recorder. The circuit includes a switching network to permit any probe combination to be connected, and a balancing network to reduce the sometimes large potential differences to more easily manageable values. The boundary event is taken as the intersection of the base line and the sloping line, as indicated by the dotted sections in Figure 8.30b. Slight dc current leakage across the cell has the same effect on the transference number as an increase in the conductivity of the solvent [315, 316] and is corrected for in that way. As shown in (129), the solvent correction cancels out if only the ion constituent conductance or mobility is required. Here the mobility of the leading noncommon ion constituent A is by definition given by

$$\bar{u}_A = \frac{l^2}{tE} \tag{138}$$

when the boundary traverses a distance l in time t and where E is the potential drop across this same distance [267]. However, a volume correction must be applied so that $\bar{u}_A$ refers to movement relative to the solvent. The usefulness of this approach was illustrated by measurements in aqueous urea solutions where slow hydrolysis produced large uncertainties in the solvent correction [317].

All the above electrical detection methods are based on platinum microelectrodes sealed into the mb tube and therefore suffer from two inherent weaknesses. The first is structural weakness: the probes are easily broken off in handling and the seals are apt to cause surface cracking, especially when exposed to changes in temperature [309] or pressure [318]. The second difficulty arises from electrode polarization which affects the quantitative interpretation of the ensuing electrical signals. Both these problems were cleverly overcome by Pribadi [318] by using radio frequencies to detect the boundary *outside* the tube walls (cf. Section 4.2). Each "probe" now consists of a set of five parallel metallic rings attached to the outside of the mb tube, as shown in Figure 8.31. In the original work these rings were platinum films 1 mm wide and with 1 mm separations. To prepare them the glass was first etched with an approximately 10% aqueous ammonium hydrogen fluoride solution for 3 h, and a thin layer of commercial platinizing solution was then applied and heated to remove the solvent [319]. Further layers could be applied on top. More recently [320], the construction of a ring was simplified by winding one turn of a slightly flattened nichrome wire tightly onto the glass and fixing it with epoxy cement. In each set of five rings the outermost ones must be earthed (grounded) to act as guards that isolate each probe set from the others. Rings 2 and 3 and rings 4 and 3 essentially form two conductance cells connected back to back, and constitute two arms in a capacitance–resistance transformer bridge. The imbalance signal is followed at the center ring by a phase-sensitive detector via a lock-in amplifier [320]. The bridge balance is disturbed as soon as the boundary enters the probe region, and the maximum voltage signal is obtained on the chart recorder as the boundary passes the center ring (Figure 8.30d). The

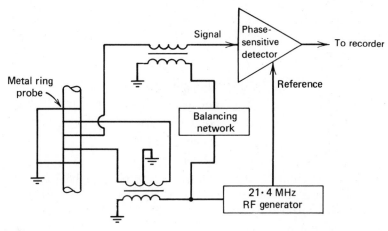

Figure 8.31 Simplified diagram of a five-ring probe outside the moving boundary tube and the attached radiofrequency detector circuit [320].

balancing network in Figure 8.31 is a later development and employs a double-balanced mixer. The amount of inphase and out-of-phase signal can then be regulated by changing the dc input current to the mixer, and this greatly extends the balancing range of the bridge [320]. The later circuit uses the 21.4 MHz generator shown as well as a second RF generator set at 21.405 MHz. Mixing this with the operating signal produces a beat frequency of 5 kHz for amplification and phase detection. This relatively low frequency ensures detection stability and sensitivity while the high RF operating frequency reduces the impedance of the cell walls and leads to a high signal-to-noise ratio. For transference measurements the mb tube holds several probe sets spaced at least 1 cm apart to allow several independent determinations per run, and equivalent rings in each set can be connected in parallel [318]. A boundary passing a probe set can be timed within 1 s. Although the position of the first change of potential gradient as the mb approaches a probe is quite reproducible, the same does not necessarily apply to the position of the maximum in the potential gradient [316].

Electrical methods of sensing boundary passage are well suited for the remote control operations necessary in determinations at high temperature [321], low temperature [311], and high pressure [313, 322]. Current leakage in the leads connecting the probes to the measuring circuit can be avoided by using driven-guard circuits [322]. These employ multilayer cables whose inner copper lead is electrically shielded by a concentric metal sheath kept at the same potential as the probe by operational amplifiers with a gain of 1. This effectively eliminates leakage to ground from the inner conductor.

RADIOACTIVE TRACER METHOD

This technique has been used for ion-exchange resins in which mb systems such as AZ/‡BZ or ‡AZ/BZ (where ‡ indicates radioactivity) were followed by

Geiger–Müller counters [144]. Marx and his co-workers [323–329, 331–334] have successfully applied this idea to solutions, and have followed the movement of boundaries such as:

$HCl \leftarrow ^{42}KCl$ (hard β-emitting following ion [323])

$(^{14}C_2H_5)_4NI \leftarrow CoI_2$ (weak β-emitting leading ion [324, 325])

$^{24}NaCl \leftarrow CdCl_2$ (γ-emitting leading ion [326, 327])

$CoCl_2 \leftarrow ^{60}CoCl_2$ (γ-emitting following ion [328]).

As mentioned in Section 2.2, the mobilities of isotopic ions differ only slightly from each other and not at all in ions such as $H_2{}^{32}PO_4^-$ where the central isotopic atom is well shielded by other atoms [329].

The $CoCl_2$ example above illustrates the "method of identical solutions" [84], identical except that one contains a small fraction of radioactively labeled ions. This method possesses the advantage that no new indicator substance need be sought, and is particularly valuable for solutions in which hydrolysis rules out the use of conventional two-salt boundaries. The price that must be paid for this apparently ideal arrangement is the complete absence of an electrical restoring effect. The boundary therefore becomes more diffuse as it travels along [84] and can be easily and irreversibly upset by small mechanical or convective disturbances. Identical solution experiments should always be carried out with the radioactive solution following. If the tracer leads, its slow desorption from the glass walls produces asymmetric counts-versus-time curves whose midpoints are difficult to locate [328].

Suitable techniques are now available for working with the three main types of radioactive isotope [330]. Consider first α and weak β emitters like ^3H and ^{14}C which allow organic ions to be labeled. For these emitters an apparatus has been constructed that serves simultaneously as transference cell and scintillator [324]. Figure 8.32 shows how it is made from an appropriately shaped slab of plastic scintillator with a 3 mm vertical bore through which the boundary moves, from an autogenic metal anode at the bottom to a cathode chamber at the top. At 8 cm from each end a photomultiplier is attached to detect and amplify the emissions. The counting rate changes when a boundary between a radioactive and a nonradioactive solution passes an observation point. The next group contains tracers like ^{32}P giving off hard β rays or β emitters with a weak ($<20\%$) γ background such as ^{42}K. Their maximum β energy must exceed 1.5 MeV for most of the radiation to penetrate the glass walls of the mb tube. Next to the tube is placed a large screen of blackened Perspex plastic (or an Al/Pb/Al sandwich if the γ content is appreciable) of a thickness 10% less than that needed to absorb all the radiation [331]. The screen contains at least two horizontal slits several centimeters long and about 2.8 mm high in front of which Geiger–Müller end-window counters are located. The counting rate changes as the boundary passes a given slit. The third and largest group of possible nuclides comprises γ emitters, examples being ^{24}Na and ^{60}Co [326]. Instead of the radia-

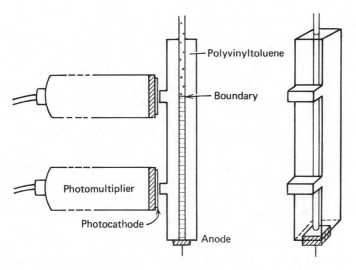

Figure 8.32 Apparatus for detecting boundaries when the leading solution contains an α or weak β emitter [325].

tion being collimated by slits in lead shielding, it was found much more effective here to use as a scintillator a 25 mm wide, 4 mm high, and 2 mm thick NaI crystal mounted at right angles to the mb tube (Figure 8.33). The large width of the crystal then captures the γ rays emitted from a thin horizontal layer of solution. The γ scintillation counter could be moved 20 cm vertically alongside the mb tube [326].

The counts obtained by these various detectors are registered and appropri-

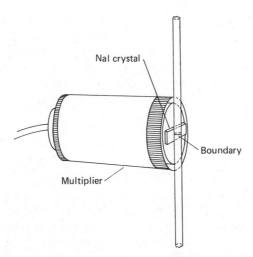

Figure 8.33 Arrangement for detecting the boundary when one solution contains a γ-emitting isotope [333].

ately recorded or displayed. Their shapes resemble the curves in Figures 8.30a [327] or 8.30b [326], and the inflection or half-wave points mark the moments of passage. The recorder sensitivity must be increased between readings if there is a noticeable decrease in the isotope activity during the run. This was found to be necessary with $^{24}Na^+$ whose $t_{1/2} = 15.4$ h. The half-life of the chosen isotope must be at least 1 h and preferably much longer.

For β and γ emitters, activities in the range $0.1-5$ $\mu Ci/cm^3$ are recommended [327] so that the relative tracer content is larger the more dilute the solution [328]. For safety reasons manipulations may need to be carried out behind a thick lead screen. Lead shielding is also advisable for the electrode chambers and for other sections of the cell holding sizable quantities of active solution. However, the total amount of radioactivity can be reduced by confining the active solution to the relevant parts of the cell such as the mb tube [326]. The cells employed so far have mainly been based on either the autogenic design in Figure 8.40 or the sheared versions in Figures 8.36 and 8.37. Extra stopcocks have sometimes been added as an aid to filling a particular section with active solution. Both electrode compartments are best placed on the side away from the detectors to allow the latter unhindered access to the mb tube. Most tracer experiments have been operated with only two counting sites on the tube so that progression of boundary movement could not be detected, and the precision has been less than with optical or electrical detection which involve multiple independent determinations during a run. This disadvantage could be overcome with more detectors. The other handicap of these experiments has been the need to use an air rather than an oil thermostat. However, it has often been the practice—particularly in aqueous or nonaqueous systems involving bright blue $CoCl_2$ solutions—to measure boundary movements optically to as low a concentration as possible and then, as a check, to follow one particular concentration both optically and isotopically before studying still lower concentrations by the tracer method only [332, 333]. The latter technique allows solutions as dilute as 10^{-4} M to be investigated under favorable circumstances. Since radioactive tracers are isotopes of a specific ion constituent, they offer greater selectivity than either optical or electrical sensing which depend upon properties of the whole solution. Tracers can therefore be used for mixed electrolytes such as dilute uranium(VI) in excess nitric acid to suppress hydrolysis [334].

NMR DETECTION

The interesting idea of following boundaries by NMR has recently been explored by Holz and co-workers [335, 336]. The signals, which are not seriously affected by small direct currents, are received by winding two equivalent coils around the mb tube (Figure 8.34). In "direct NMR detection" the frequency of a pulsed NMR spectrometer is adjusted to the resonance frequency of a nucleus in the noncommon leading or following ion constituent, for example, to that of ^{23}Na in the system $NaCl \leftarrow CdCl_2$. Initially the mb tube is full of NaCl solution and each receiver coil yields a maximum ^{23}Na NMR signal. This is measured by

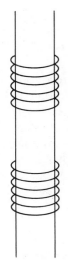

Figure 8.34   Two-coil system for detecting boundary movement by direct NMR [336].

applying a regular 90° RF pulse, and the amplitude of the following free induction decay (fid) is recorded or printed out. It is shown in Figure 8.35a as the small curve to the right of each sharp rectangular pulse. The separation time between pulses is $t_1 > 5T_1$ where T_1 is the longitudinal relaxation time of the nucleus under observation: by making t_1 so much longer than T_1 the nuclei can relax fully after each observation pulse. Usually T_1 is less than 10 s and so t_1 can be 1 min. Because their amplitude is proportional to the number of nuclei (of ^{23}Na) within a coil, the fid curves become smaller and smaller as the leading NaCl solution passes through that coil. The characteristic decrease in the ^{23}Na signals is shown in the figure. The second coil further along the tube is

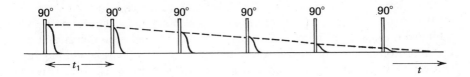

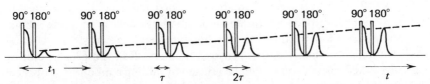

Figure 8.35   (a) $-90°-t_1-90°-t_1-$ pulse sequence showing the behavior of the fid amplitude when the nuclei under observation (e.g., ^{23}Na) are flowing out. (b) $-90°-\tau-180°-t_1-90°-\tau-180°-t_1-$ pulse sequence showing the increase of the echo amplitude when paramagnetic ions are flowing out [336].

then switched into the probe circuit instead, and in due course the phenomenon is repeated. Each moment of passage can be taken as the midpoint of the flowing-out traces, and the difference between these is the time required for the boundary to traverse the volume between the planes in the centers of the two coils.

The second NMR method focuses on the solvent protons. Their relaxation times are markedly reduced by the presence of even very low concentrations of paramagnetic ions, a fact that forms the basis of "indirect NMR detection by relaxation discrimination." Provided either the leading or the following non-common ion is paramagnetic and the other ions diamagnetic, as in the aqueous system $NiCl_2 \leftarrow CdCl_2$, passage of the boundary through a coil will change the ^1H relaxation time. One can observe either T_1 or, as Holz and Radwan suggested [336], the relaxation time T_2 of transversal magnetization. The latter requires the pulsing sequence

$$90°-\tau-180°-t_1-90°-\tau-180°-t_1-$$

where $\tau \leqslant T_2$ and $t_1 > 5T_1$. A spin-echo in the shape of a peak then appears at a time 2τ after each $90°$ pulse, as shown in Figure 8.35b. The echo amplitude is given by the equation

$$A(2\tau) = A(0)\exp\left(\frac{-2\tau}{T_2}\right) \tag{139}$$

Thus the ratio of the echo amplitude of the solvent protons in a diamagnetic solution to that in a paramagnetic solution is equal to

$$\frac{A^{dia}(2\tau)}{A^{para}(2\tau)} = \exp\left(-\frac{2\tau}{T_2^{dia}} + \frac{2\tau}{T_2^{para}}\right) \approx \exp\left(\frac{2\tau}{T_2^{para}}\right) \tag{140}$$

since $T_2^{para} \ll T_2^{dia}$. If one chooses $\tau = T_2^{para}$, the ratio becomes $e^2 = 7.4$. Replacement of a paramagnetic solution by one that is diamagnetic will therefore increase the spin-echo amplitude at 2τ by a factor of 7.4. This is illustrated in Figure 8.35b. With the system $NiCl_2 \leftarrow CdCl_2$ the values chosen for τ were 20 or 40 ms and for t_1 120 s, and the resulting plots of $A(2\tau)$ versus time for each coil exhibited the shape of Figure 8.30b [336]. The mb cell had been thermostatted with a proton-containing liquid (water) but the section of the tube holding the receiver coils was enclosed in a separate jacket filled with CCl_4. The transference numbers for $NiCl_2$ at 25°C so obtained agreed well with those determined by the Hittorf method [281].

The NMR method is clearly capable of further refinement. For a start, the precision can be improved by winding more than two coils around the mb tube to permit several independent velocity measurements in every run. Although NMR equipment is bulky and expensive, Holz and Radwan [336] point out that older NMR instruments could be utilized by adding a modified probehead. The ^1H solvent resonance measurements could be performed with a lower cost pulsed NMR minispectrometer. It is worth noting that the NMR technique possesses features in common with two of the methods discussed earlier: it relies on radiofrequency signals from metal rings outside the tube as does the latest electrical detection method, and like the radioactive tracer method its signals are specific to a given nucleus.

Quite recently, Holz and Müller [337] have suggested an ingenious extension of NMR technology for determining the ion constituent mobilities themselves. The method depends on the production of a phase shift in the spin-echo signal when a magnetic field gradient (G) is applied along the direction in which the ions move in an electric field. The echo amplitude $A(2\tau)$ measured in phase-sensitive detection then changes by a factor $\cos(\gamma G \bar{v} \tau^2)$, where γ is the gyromagnetic ratio of the monitored nuclei and $\bar{v}$ the ion constituent velocity. This basic situation was made more amenable to measurement by various modifications: the replacement of the steady gradient G by two pulsed magnetic field gradients, a counterflow arrangement whereby the volume of the NMR coil was divided into two halves inside which the ions flow in opposite directions, the use of pulses of electric field to reduce Joule heating, the stabilization of the electrolyte solution in agar gel to prevent convective flow, and the use of D_2O as solvent to ensure that the ^1H signals emanated only from the $(C_2H_5)_4N^+$ solute. By means of these devices very reasonable mobilities were obtained for $(C_2H_5)_4N^+$ ions in D_2O set with 0.6 wt% agar. The method is limited to ions with nuclei that possess a nuclear magnetic moment and a fairly long transversal magnetic relaxation time, such as species containing ^1H, ^7Li, ^{19}F, ^{27}Al, ^{31}P, and ^{133}Cs. High-field cryomagnets would render many other nuclei accessible to such experiments.

THERMAL METHOD

Joule heating developed by the electric current causes the temperature of both solutions to rise, especially that of the more poorly conducting following solution. The effect is particularly pronounced with weak electrolyte solutions in which the temperature rise can amount to several degrees centigrade with currents of the order of 1 mA [275]. Normally conditions are arranged to minimize this temperature rise, but two Russian workers [338] have pointed to the possibility of utilizing it for detecting the passage of the boundary. The response of a microthermistor in the tube should give a plot similar to that in Figure 8.30a.

5.3.3 Indicator and Initial Indicator Concentration

A suitable indicator must meet the following requirements:

1. It must not react chemically with the leading solution. However, the term chemical reaction is not intended to exclude a rapid reversible shift in an association–dissociation equilibrium such as $H^+ + OAc \rightleftharpoons HOAc$ which occurs in the NaOAc/HOAc system.

2. The mobility of the following noncommon ion constituent B must be less than that of the leading noncommon ion constituent A in both solutions and in any possible mixture of the solutions that can exist in the boundary [see (131)]. Take the case of a completely dissociated electrolyte, such as KCl in water, for which a cation indicator is required. LiCl is very suitable since the mobility of Li^+ is much less than that of K^+, but HCl is not since the H^+ ion

is faster than K^+. As a second example, consider an incompletely dissociated electrolyte such as aqueous H_3PO_4 for which an anion indicator is needed. Picric acid is not a suitable indicator even though the mobility of the picrate *ion* is less than that of the $H_2PO_4^-$ ion because picric is a stronger acid than phosphoric and the mobility of the picrate *ion constituent* is greater than that of the phosphate *ion constituent* [see (35)]. However, formic acid could be selected as a following electrolyte: it is a weaker acid than phosphoric and, despite the fact that the formate *ion* is faster than the $H_2PO_4^-$ ion, the mobilities of the respective *ion constituents* obey condition (131). A third and quite famous example [339] is the stable aqueous boundary system NaOAc←HOAc in which the intrinsically slow but free Na^+ ion leads hydrogen, a very fast ion considerably hampered by strong association [293].

3.   The leading and indicator solutions should differ appreciably in some property (conductance, refractive index increment, radioactivity) that enables the boundary to be followed. An indicator giving an easily visible boundary with the assembly in Figure 8.28 requires the simplest apparatus.

4.   The indicator solution (or, for that matter, the leading solution) should not be appreciably hydrolyzed [340]. The hydrolysis can sometimes be repressed, for example, in KOAc solutions by adding a little HOAc [341]. Alternatively, the method of identical solutions (with a tracer ion in one of them) can be employed.

Certain other points should also be considered in choosing an indicator substance, such as the interaction between it and the selected electrode in the indicator solution (see below). If the closed electrode is in the following solution, it is advisable to choose as the indicator an electrolyte whose apparent molar volume changes relatively little with concentration [271]. It should be possible to purify the indicator so as to remove from it any fast ion constituents that might overtake the boundary and so affect its velocity. Boundaries are sharper if there is a greater mobility difference between the leading and following ion constituents, as shown by (132).

It is an important property of moving boundaries that the concentration of the indicator BZ behind the boundary adjusts itself automatically to the value given by the Kohlrausch equation

$$\frac{|z_B|*m_B^{BZ}}{|z_A|m_A^{AZ}} = \frac{*T_B^{BZ}}{T_A^{AZ}} \tag{141}$$

The asterisk characterizes the properties of this adjusted or Kohlrausch solution. The theory is treated in more detail in Section 5.4.1 where it is pointed out that the use of molalities m (mol per kg solvent) in place of molarities c (mol per dm^3 solution) obviates the need for a volume correction in (141). If the Kohlrausch solution of BZ is lighter than the leading AZ solution, a falling boundary must be used; if it is heavier, a rising boundary. The boundary system is the more stable the greater the density difference between the leading and the adjusted indicator solutions.

Provided the cell is carefully designed, indicator solutions of widely differing initial concentrations adjust to the Kohlrausch concentration behind the mb [342]. However, every system must be tested on its own merits and much trouble may be saved by adopting the following rules:

1.   The initial concentrations should not differ from the Kohlrausch values by more than about 25% in aqueous solutions and by less in other solvents. A rough estimate of the Kohlrausch concentration can normally be derived from (27) and (141) and from whatever transference and conductance data are available.

2.   For falling boundaries it is safer to use initial concentrations equal to or less than the Kohlrausch concentration so that the system in Figure 8.26b— BZ(init)/BZ(Kohlr)/AZ—is gravitationally stable. This system is usually thermally stable also, because the solution of highest resistance, which is warmed most by Joule heating, will lie at the top of the tube.

3.   For rising boundaries it is wise to use initial concentrations equal to or greater than the Kohlrausch concentration, so that the system in Figure 8.26a— AZ/BZ(Kohlr)/BZ(init)—is gravitationally stable. This system is generally not thermally stable since cooler solutions lie above warmer ones, and convective disturbances are somewhat more likely.

4.   At least two indicator concentrations, differing by a few percent, should be used; if the transference numbers do not agree, further work is necessary to find a range of indicator concentrations over which the transference number is constant and independent of current. Should no such range be discovered, another following electrolyte is called for.

It is not essential that the indicator and leading electrolytes possess an ion constituent in common, for the ion constituent in the leading solution that moves in the opposite direction to the boundary automatically becomes the common ion constituent. Thus AZ/BY becomes AZ/BZ/BY on the passage of current. As an example [343], we wished to study the aqueous anion boundary $NaCl \leftarrow NaIO_3$ but found that KIO_3 was the only iodate salt available in a pure form. The $NaIO_3$ was therefore formed *in situ* by passing current through the system $\oplus NaCl/KIO_3 \ominus$ to produce

$$\oplus NaCl \leftarrow NaIO_3 \rightarrow KIO_3 \ominus$$

The appropriate initial concentration of KIO_3 was then calculated by combining two Kohlrausch relationships

$$m(KIO_3) = \frac{m(NaIO_3)T_+(KIO_3)}{T_+(NaIO_3)} = \frac{m(NaCl)T_-(NaIO_3)T_+(KIO_3)}{T_-(NaCl)T_+(NaIO_3)} \tag{142}$$

The main restriction in these cases is the avoidance of gravitationally unstable density differences within the three electrolyte solutions [344]. The procedure can be used to advantage if discharge of the Z ion at the electrode would cause some disturbance such as gassing, or generate an ion that would overtake the boundary [345]. The idea may also be applied to the synthesis of a solution of the intermediate electrolyte BZ, as suggested in Section 3.6.

5.3.4   Electrodes

One of the electrode compartments must be open to allow for the expansion or contraction of the whole system, and the other should be closed to avoid net electroosmosis and to permit calculation of the volume correction. Good temperature control must be maintained in the thermostat as volume fluctuations in the closed side of the cell seriously affect the boundary velocity.

The electrode in the closed compartment must meet certain requirements:

1.   No gas evolution must occur.

2.   No foreign ions generated by the electrode reaction must reach the boundary, as this progressively decreases the boundary velocity. Since the time taken by such an ion to migrate to the boundary is proportional to the volume of solution it has to pass through, it is sometimes helpful to use a large or extended electrode compartment. Improved temperature regulation is then advisable.

3.   The electrode reaction and the physical state of the products should be known so that the volume correction can be calculated. The extent of any appreciable side reaction can often be found by weighing the electrode before and after a run.

For the open electrode, only condition 2 is applicable. The mixing effect of gas evolution can be reduced by appropriate cell design.

Silver halide electrodes, made electrolytically or by dipping a platinum wire at the end of a glass tube repeatedly into the molten salt, are often used as cathodes. Platinum electrodes immersed in an aged slurry of silver succinate, or some other insoluble silver or lead salt, are convenient cathodes for work with aqueous acid solutions [8, 54]. Pure silver or cadmium normally make excellent anodes, and mercury [311] and cobalt [324] have also been used. Various other electrodes described in Section 5.2.3 are equally suitable here and, since the currents and times of electrolysis are usually smaller than in Hittorf work, the electrode areas need not be as large. If a simple nongassing electrode cannot be found for either the leading or the following solution, an intermediate guard solution may have to be introduced; for example, a KCl solution can serve around a cadmium anode or an AgCl cathode. However, care must be taken during filling to prevent mixing between the guard solution and the main solution, density instabilities must be avoided, foreign ions from the guard solution must not catch up with the boundary, and the guard solution must be taken into account when the volume correction is calculated.

Methods of introducing the electrodes into the cell are discussed below.

5.3.5   Cell Designs and Filling Techniques

The tube volumes between etch marks or probes must be periodically recalibrated because of aging of the glass and the inadvertent distortions caused by glass-blowing. Calibration is achieved either by mercury weighing [292] or by carrying out runs with a substance whose transference numbers are accurately known, for example, KCl in water. In the latter case, the literature values should be "uncorrected" again for the solvent and volume corrections to give the directly observed value that is appropriate for equation (133).

FORMATION OF THE BOUNDARY

Most boundaries of the so-called "sheared" type are formed initially by suddenly joining the leading and indicator solutions by some mechanical device (see below). The advantage of sheared boundaries is that the initial indicator concentration can be varied at will—the exact Kohlrausch concentration not usually being known in advance—until a range of concentrations is found over which the velocity of the boundary is constant. For "autogenic" boundaries, however, the initial junction is between the leading solution and an electrode, and the indicator is formed by the electrode reaction. Thus KCl over a Cd anode forms an indicator solution of $CdCl_2$, and the boundary KCl$\leftarrow$ $CdCl_2$, on passage of current. Autogenic boundary formation is particularly favored for experiments at high temperatures [321] and high pressures [313, 322] in which a minimum of outside manipulation is desirable. The autogenic technique has so far been applied only to rising cation boundaries. Experiments in a number of instances have shown [346] that the solution following behind an autogenic boundary automatically adjusts itself to the Kohlrausch concentration, and this is now always taken for granted. Independence of current is one indirect way of checking that convective mixing has not occurred behind the boundary with the more concentrated solution in the vicinity of the electrode. The concentration stratification that normally exists above an autogenic electrode introduces some uncertainty into the value of the volume correction [60, 271]. More serious difficulties with autogenic runs are not unknown [347, 348] in which case one has to change the anode metal or use sheared boundaries.

Several devices have been invented for the creation of sheared boundaries. The first, described in detail by MacInnes and Longsworth [292], consists of two (glass) disks or plates to which tubes filled with the two solutions are attached; to start a run the plates are slid over each other until their openings coincide. This technique is rarely used nowadays. Neither is the "air bubble" method [344, 349] whose name describes the agent by which the solutions are initially separated. To start a run the bubble is forced up a side tube and the boundary is produced without appreciable mixing. The most popular shearing device is a stopcock. To reduce local Joule heating, the barrel should be hollow and open to permit circulation of the thermostat liquid, and the inner bore should have the same cross-sectional area as the mb tube. The stopcock shape depends upon the type of cell (Figures 8.36 and 8.37). The choice and purity of the lubricant in both these shearing methods are important. Many simple lubricants are essentially hydrocarbons: where these fail, recourse may be had to silicones (which can be removed with triethylamine) [8], fluorocarbons [350], special formulations such as lithium stearate–oil for methanolic solutions

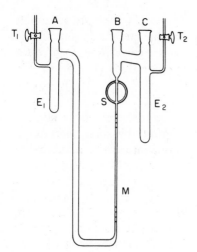

Figure 8.36 Sheared falling boundary cell. The handle of stopcock S has been omitted from the drawing, and should be on the side facing the reader. The cell can be modified to incorporate a four-way stopcock instead, as in Figure 8.37.

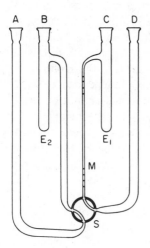

Figure 8.37 Sheared rising boundary cell, following [353]. The handle of stopcock S has been omitted from the drawing and should be on the side facing the reader.

[304], or special techniques for dealing with solutions of detergents [351]. The suitability of any given grease should be tested by seeing if there is any effect on the boundary velocity when more grease has been deliberately smeared inside the uncalibrated portion of the mb tube. The problem can be overcome with a four-way glass barrel fitted with a specially designed solid PTFE key [78] that requires no lubricant. As shown in Figure 8.38, two grooves, each covering 90° of arc, are cut into the circumference of the key. The solution inside the grooves, which is subject to Joule heating, is cooled on the glass side by the thermostat oil. How the key is used to form a boundary is described below in connection with Figure 8.37. A lubricant is also not required in the "flowing junction" method. One solution streams up the tube, the other down it, and at the point where the boundary is to form the mixed solution leaves through an

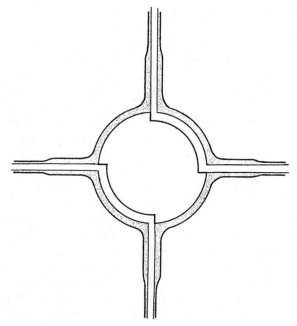

Figure 8.38 Grooved four-way stopcock employing a solid PTFE key [78]. The handle has not been shown, and should be on the side facing the reader.

auxiliary tube. At the appropriate moment the latter is removed—by simply withdrawing it [267] or by sliding a plate [352]—and the boundary remains.

SPECIFIC CELLS

The cells in the diagrams can be used for any method of boundary detection. The tiny marks on the mb tubes M can be imagined as etch marks for optical observation or as platinum probes for electrical detection. The designs in Figures 8.36 and 8.37 are based on a careful study [342, 354] of boundary stability.

In Figure 8.36, tap T_1, tap T_2, and stopcock S are opened, and leading solution is poured into A until the liquid level is just above S. Air bubbles clinging to the lubricant around S can be removed by successively squeezing and releasing a piece of clean flexible tubing inserted just inside A. Jiggling a platinum wire through B is also effective. S is then closed, and the E_1 compartment filled up with leading solution. Solution above S is removed by suction with a drawn-out piece of glass tubing, and the indicator compartment above and to the right of S is repeatedly rinsed, first with solvent, then with indicator solution, the liquid being sucked out each time. The compartment is finally filled with indicator solution, the electrodes are introduced through A and C, B is closed with a ground-glass plug, and the cell is allowed to come to temperature equilibrium in the thermostat. The sidearm tap of the closed electrode section is then shut, S is opened so that bore and tube are well lined up, and the current is switched on immediately.

In Figure 8.37, with stopcock S aligned as shown, indicator solution is added through A and leading solution through D until the cell is full. Air bubbles may be removed from inside S as described above. The electrodes are inserted into B and C, and the cell is allowed to reach thermal equilibrium. Then, to connect E_1 with E_2, S is turned through $90°$, either clockwise for "top shearing" or counterclockwise for "side shearing." In the latter case the boundary must round a corner, but stable indicator adjustment appears to occur more easily [342]. The current is turned on immediately. The cells in Figures 8.36 and 8.37 can be combined to form a cell suitable for both falling and rising boundaries, but it is often found more satisfactory to use two separate cells.

Figure 8.39 depicts a "flowing junction" cell, which can be used for either rising or falling boundaries. With the joints at F and tap T_2 closed, solution I (following solution for a rising boundary, leading solution for a falling one) is added to E_1 and stops after filling about 1 cm of the narrow tube M. The openings on the left-hand side are then closed and the other solution, II, is poured into E_2 until it just overflows into M. With aqueous solutions a bubble usually forms in M, for other solvents the solution normally flows slowly down the tube and mixes with the first solution. In either case the open tube U is inserted, and a thin-walled capillary pipet, rigidly held on a traveling catheto-meter and connected to a siphoning system, is inserted through U into M.

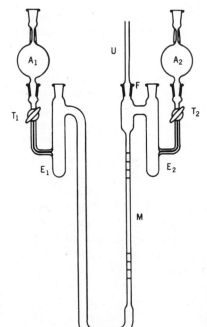

Figure 8.39 Moving boundary cell designed by Kay, et al. [267] for either rising or falling boundaries. The boundary is formed in the tube M by means of a flowing junction.

If a bubble has formed, the pipet is slowly lowered to allow solution II to flow down its sides until a point about 1 cm above solution I is reached. At this stage the siphoning system is turned on and the gas bubble completely removed, bringing the two solutions into contact. The right-hand side of the cell is then completely filled, and the 75 cm^3 bulb A_2 is added with solution II filled well up into the narrow capillary at the top. The level of solution in U is brought to this same height, tap T_2 opened, and the solution is slowly siphoned out through the pipet which is now placed at the point where the boundary is to be formed—near the bottom of tube M for a rising boundary, near the top for a falling one. In the meantime, the 75 cm^3 bulb A_1 has been filled with solution I and placed in position. When the solution level in U and A_2 reaches that in the capillary neck in A_1, tap T_1 is opened and both solutions are permitted to flow into the cell and, at the boundary, out into the siphoning system. When, after about 15 min, bulbs A_1 and A_2 are almost empty, the taps T_1 and T_2 and the siphon on the pipet are closed. The pipet is very slowly retreated about 5 mm, T_2 again opened, and solution II in A_2 permitted to flow again through the pipet very slowly as the latter is removed from the cell.

If a bubble is not initially formed in M when solutions I and II have been poured in, some solution II is allowed to overflow from E_2 into M and the pipet is lowered to a point well below that at which the boundary is to be formed. Most of this added solution is removed and the process repeated but with the pipet about 5 mm higher after each addition. If this procedure is repeated at least ten times, the top solution is entirely free of the bottom solution. The pipet is then placed at the point where the boundary is to be formed and the solutions allowed to flow as before.

Figure 8.40 gives two basic designs for an autogenic cell. The closed electrode is often formed [301, 313, 341, 355] by a tapered plug of metal (e.g., Cd, Ag) ground to fit tightly into the lower end of the mb tube M and sealed in with a suitable cement that must not interact with the solution. Araldite, for instance,

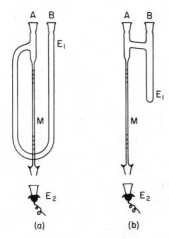

(a)   (b)   **Figure 8.40**   Autogenic rising boundary cells.

is slightly soluble in methanol [327]. Alternatively, the metal anode can be pressed firmly onto a glass flange at the end of the calibrated tube by a Lucite clamp, a leak-proof seal being effected with a small rubber O-ring [322]. In the type [304, 354] shown in Figure 8.40 a long slightly tapered male joint is sealed onto the bottom of M. A tungsten wire is sealed through a small well at the base of the corresponding female joint E_2, and the well is filled with metal shavings, which are fused in a hydrogen atmosphere. The metal surface is scraped clean with a steel wire between runs. Since cavities can form between metal and glass, any air should be removed before runs by filling E_2 with solution and evacuating. To fill the cell, E_2 is slightly greased and pressed home, secured with steel springs on the glass hooks shown, and leading solution is poured in. Any air trapped in the narrow tube M can be allowed to escape through a fine capillary inserted through A. The other electrode is placed in position, A plugged up, and the current started. An even simpler autogenic design is possible if mercury or an amalgam can be used as the anode. The bottom of the mb tube is then enlarged and sealed, and a short tungsten tip sealed through the bottom (Figure 8.41b). Enough mercury or amalgam is poured in to cover the tungsten completely [311, 312].

GENERAL POINTS

Cells should be made of Pyrex glass to prevent appreciably increasing the solvent correction. The cell can be strengthened by a bar of glass tubing sealed horizontally across its back, and a stout vertical glass tube can be sealed onto

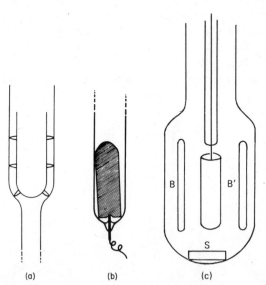

(a)   (b)   (c)

Figure 8.41 Electrode section designs. (a) Device for accommodating a dense electrode solution near the top of the cell. (b) A simple device [304] for introducing a metal anode. (c) Electrode chamber modified to allow stirring during the run to keep the electrode solution of uniform concentration [271].

this as a handle. Vibrational disturbances can be counteracted as described in Section 5.2.4. Careful cleaning of the cells is important. Any lubricant employed must be removed by a suitable solvent so that it does not contaminate the mb tube. Great care must be taken with any abrasive cleaning mixture such as $Na_2Cr_2O_7 + H_2SO_4$, which should be applied sparingly and for only short periods to avoid chromium penetration [241] and volume changes in the tube, and its use must be followed by multiple frequent washings with distilled water. After being drained, the cells should be dried by a stream of clean nitrogen and not by heating.

Gravitational stability must be maintained in the electrode compartments; an electrode should be at the bottom if the products of electrolysis are heavier than the main solution, and vice versa. Figures 8.40a and 8.40b illustrate one way of catering, respectively, for lighter and heavier electrode solutions in E_1. A dense electrode solution can be accommodated near the top of a cell by placing the electrode at the bottom of an inner tube as in Figure 8.41a. A small electrode section similar to this has been used [356] for a tiny autogenic mb cell needing only 10–15 cm^3 of solution. This type of design also reduces mixing effects around gassing electrodes. A satisfactory method [304, 354] of introducing a metal anode into sheared cells is to slide a short rod of the metal gently down on top of a thick tungsten wire sealed through the bottom of the compartment, as in Figure 8.41b. The electrical contact is usually sufficiently good to prevent gassing because electrolysis occurs mainly at the top of the rod. But in acid solutions corrosion occurs: even without passage of current, a cadmium rod will dissolve to form Cd^{2+} while simultaneously H^+ ions are reduced to H_2 on the low overpotential tungsten surface [357]. As mentioned earlier, Figure 8.41b is suitable for mercury or amalgam anodes as long as the liquid metal completely covers the tungsten. Figure 8.41c shows a closed electrode compartment designed for work in concentrated solutions in which stirring is necessary to avoid major spatial concentration inhomogeneities around the electrode and consequent large uncertainties in the volume correction [271]. S is a stirrer bar rotated by an immersible magnetic stirrer motor and B and B' are two of the vertical baffles in the walls introduced to aid turbulence.

Solutions should be partially degassed to avoid air bubbles appearing in the closed compartment or in the mb tube. The cell should always be carefully examined for tiny gas bubbles before and after each run. Adsorption of electrolyte on the cell walls may lead to small errors in work with very dilute solutions.

Leakage of solution in the closed part of the cell must be prevented. The cell openings should be closed by well-ground standard taper joints, either in the form of plugs or electrodes, lightly greased and held down with springs and glass hooks or with special clamps. Leakage can be tested for by filling the cell with solvent and adding mercury so as to maintain a positive or negative pressure head of a few centimeters. At constant temperature, a variation in the position of the mercury surface in the mb tube indicates a leak. Irreproducibility is sometimes traceable to cracks in the glass or insecurely stoppered openings.

The cell should be placed in a liquid rather than an air thermostat so that the

Joule heat is dissipated rapidly enough to prevent destruction of the mb by convection [301]. To prevent current leakage, the bath liquid should be a good electrical insulator, for example, Shell Diala BX transformer oil. Even bare electrode leads can then trail safely through the oil. Although the temperature coefficients of transference numbers are relatively small, very good constancy of the thermostat temperature is essential in mb work because a small expansion or contraction of the solution in the relatively large closed electrode compartment causes an appreciable shift in the position of the boundary in its narrow tube. The analogy with an ordinary mercury thermometer is obvious.

For work at higher temperatures, a layer of mineral oil on top of the electrolyte solution in the open compartment reduces evaporation losses [321]. Measurements can be carried out below room temperature by immersing the whole mb apparatus in a large hermetically sealed Dewar flask [311, 312]. In high-pressure experiments, a flexible Teflon extension of a sidearm allows pressure to be transmitted to the system without contaminating the solution with the pressurizing oil [322].

5.3.6   Current and Time

A by-product of the passage of electricity is Joule heating. A theoretical analysis of the resulting temperature rise in a mb tube [272, 275] shows that in solutions of low conductivity (low concentration, ion association) temperatures may rise by several degrees. The transference number must then be measured at several currents and extrapolated to zero current. A twofold current variation should be investigated for every system; in most cases the transference number remains constant. Too high a current produces a curved boundary or even more drastic changes. The large currents required for work with concentrated solutions can give rise to Soret effects at the boundary which alter its velocity [271]. Too low a current gives a very diffuse boundary [cf. (132) and (133)]. Currents may need to be only a few microamperes and seldom exceed 5 mA except for very concentrated solutions. As a rough guide for aqueous solutions, if the transference number is around 0.5, the current should be such that the boundary takes 1–3 h to traverse 1 cm^3 in a tube of 2–3 mm ID.

One of the classic experiments [292] was to interrupt a mb run and then to demonstrate, on restarting the current, that the boundary formed again and moved forward at its old rate. Closer investigation indicates [358], however, that current interruption does disturb the boundary afterward and should therefore be avoided.

Current leakage in the cell is most simply and effectively avoided by employing an oil thermostat. For measurements at high pressure, leakage between the current leads and ground is circumvented by a specially designed guard circuit [309].

Coulometers are not favored in mb work, for one must be thrown into and another out of circuit every time the boundary passes a calibration mark [341]. Separate current and time measurements are used instead. Since the replacement of one electrolyte in the mb tube by another leads to large increases in cell resistance, an electronic current regulator is essential. There are as many circuit designs as there are groups of workers in the field, in part because different researchers have different requirements for current range, cell resistance, and desired current stability. Several suitable high voltage galvanostats are described in [359, 360, 311] and others are now available commercially.

A simple timing method for optical observation requires only an accurate watch and a metronome clicking at a convenient rate, such as 2 beats/s; when the boundary is just coincident with a calibration mark, the experimenter begins counting aloud and continues to do so until an easily remembered time is noted from the watch. For example, if 23 half-second beats are counted until the watch registers 11/14/30, then the boundary passed the mark at $11/14/18\frac{1}{2}$. No appreciable error is introduced since coincidence of boundary and tube mark can seldom be judged to better than $\pm\frac{1}{2}$ s. Another simple device [355] is a "split-second" stopwatch with two second hands, one of which can be stopped and, after the time has been read, resynchronized with the other hand. In electrical methods of following the boundary, time marks can be placed on the recorder chart by an event marker. The exact time of boundary passage is then calculated from these marks and the recorder chart speed [359]. Other timing techniques are described in the moving boundary literature. For electric timers the fluctuations in ac supply frequency may have to be watched.

5.3.7   Reliability of the Results

Transference numbers obtained by the direct mb method can be very precise and are correspondingly sensitive to faulty technique. If accurate data are desired, it is essential to have evidence of reproducibility, absence of progression,* independence of current and, for sheared boundaries, independence of initial indicator concentration (cf. Section 5.3.3). The transference number should also be independent of the indicator if different indicators are tried. If the transference numbers of both the cation and the anion constituent are measured, they should add up to unity [cf. (25)] within experimental error. Lest it seem that too much caution is advocated, it should be noted that in a number of cases apparently reliable work was shown to be in error because independence of current and initial indicator concentration had been taken for granted [353, 347, 275]. Reproducibility and absence of progression do not by themselves guarantee the reliability of the measurements.

5.3.8   The Differential Moving Boundary Method

A slow diffuse "concentration boundary" moves between two solutions of the same salt but of different concentrations. Theory shows [292] that for the boundary AZ(soln′)/AZ(soln″):

$$T'_A - T''_A = V_F|z_A|(c'_A - c''_A) = \frac{VF|z_A|(c'_A - c''_A)}{It} \tag{143}$$

---

*A boundary is said to show progression up (or down) when the velocity of the boundary, in cm^3/C, increases (or decreases) as the boundary moves along the tube.

where the symbols are the same as for (133) and where V is positive if the boundary moves in the same direction as the ion constituent A. The heavier solution, which is usually the more concentrated, must be the lower one to avoid gravitational instability.

Such a system is useful if only the concentration variation of the transference number is wanted or if the transference number is already known at one concentration. Its main advantage is that no indicator is necessary, and the method has therefore been used for polyelectrolyte solutions for which indicators are hard to find [361] (or else use the indirect mb or analytical boundary method, or the isotopic method of identical solutions). There are two disadvantages. First, the boundary is diffuse because the restoring effect is weak, and the boundary cannot be followed by the simple optical method. Second, the method appears to be unsatisfactory below 0.02 M [362], and the volume corrections are of the same order of magnitude as the boundary displacements. Nevertheless, accurate results can be obtained, particularly in aqueous solutions at lower temperatures [362].

5.4   Indirect Moving Boundary Method

5.4.1   Introduction and Theory

This method is complementary to the direct mb method in that the solution following the boundary is the solution under investigation. The transference number in this solution is determined by measuring the adjusted or Kohlrausch concentration behind the boundary. Indeed, the method has sometimes been called the adjusted indicator technique.

The theory is based on the fact that the velocity of the boundary in Figure 8.26 depends only on the properties of the leading solution AZ and is usually quite independent of the type of the following electrolyte BZ (the indicator) and, within fairly wide limits, of its initial concentration. In fact, the concentration of the following solution behind the boundary adjusts itself automatically to a new value. A slow-moving diffuse "concentration boundary" then forms between the original and adjusted indicator solutions (see Section 5.3.8). Most of the tubing between the mb and the place where it began is therefore filled with the adjusted or Kohlrausch solution. Let its properties be characterized by an asterisk. It can then be proved that [363, 296]

$$\frac{|z_B|^* m_B^{BZ}}{|z_A| m_A^{AZ}} = \frac{{}^* T_B^{BZ} P}{T_A^{AZ}} \tag{144}$$

where m is the concentration of the subscripted ion constituent in mol per kg of solvent (molality). Equation (144) is the amended Kohlrausch equation. No *volume correction* whatever is required provided the concentrations are given in molalities. However, as with the Hittorf and direct mb methods there is need for a small *solvent correction* factor which is given by the symbol P. This is equal to [295]

$$P = \frac{1 + \kappa_{solvent}^{AZ}/\kappa_{AZ}^{AZ}}{1 + \kappa_{solvent}^{*BZ}/\kappa_{BZ}^{*BZ}} \tag{145}$$

where κ_{MZ} is the specific conductivity due to the solute MZ in its superscripted solution. The two solvent conductivities are not necessarily equal. For example, in an aqueous system with $HClO_4$ leading and $NaClO_4$ following, $\kappa_{solvent}$ is much smaller in the $HClO_4$ solution where dissociation of dissolved carbon dioxide is repressed [275]. However, there may be preferential migration of dissolved impurities [295] and the only safe solvent correction is one in which both $\kappa_{solvent}/\kappa_{solute}$ terms are very close to zero. Attention should also be drawn to the likelihood of contamination in the cell itself and in the sampling pipet employed in methods of external analysis [295].

It follows that the transference number of BZ in the adjusted solution can be determined by analyzing the solution behind the boundary formed between BZ and a leading solution AZ of known transference number. For a stable boundary the conductance of the leading noncommon ion constituent A must be greater than that of the following noncommon ion constituent B. The indirect method is therefore particularly useful for determining the transference numbers of slow ion constituents such as Cd^{2+} in cadmium halide solutions [364] and of large organic ion constituents in polyelectrolyte solutions [365]. A phenomenon that is relatively rare, but which has been observed [365] in some polyelectrolyte solutions, is that the ratio $*T_B^{BZ}/*m_B^{BZ}$ may have the same value at two different concentrations. Under these circumstances two different values of $*m_B^{BZ}$, hence of $*T_B^{BZ}$, may be obtained with the same leading AZ solution but with different initial concentrations of BZ.

In the direct mb method one of the main requirements for the following electrolyte is that it contain a slower noncommon ion constituent. In the indirect method the corresponding requirement for the leading solution is that it contain a faster noncommon ion constituent. The two methods therefore complement each other so that mb work in general is virtually free from a restriction of this sort. The indirect mb method is more suitable than the direct method for the accurate determination of transference numbers at high concentrations. The removal of doubts about the volume correction is one reason, another is its greater tolerance of diffuse boundaries. Smaller currents may therefore be employed, and convective and Soret effect disturbances are reduced [366, 68].

5.4.2 General Conditions

Many experimental requirements for the indirect method are simpler than for the direct method. No precise volume, time, or current measurements are required although it is essential to demonstrate that the Kohlrausch concentration is independent of current [309]. The position of the boundary need be known only approximately either by visual, electrical, or counting observations, or by calculation from (133). The choice of electrodes is not critical either— even gassing ones may be used—provided the electrolysis products do not reach the boundary or the BZ solution being analyzed. Various electrode suggestions appear in Sections 5.2.3 and 5.3.4. Overall cell designs can be based on those described in Section 5.3.5 and modified according to the chosen method of analysis (see below).

A suitable leading electrolyte (AZ) must meet the following conditions:

1. It must have an ion constituent (Z) in common with the electrolyte under investigation.
2. It must not react chemically with the solution under investigation although rapid reversible changes in association–dissociation equilibria such as $H^+ + OAc^- \rightleftharpoons HOAc$ may take place in the boundary.
3. The mobility of the leading noncommon ion constituent (A) must be greater than that of the following noncommon ion constituent (B) in both solutions and in any possible mixture of the solutions that can exist in the boundary.
4. The transference number of the leading noncommon ion constituent (T_A^{AZ}) should be known over an appropriate concentration range (see Appendix for data sources).
5. The leading solution (and the following solution) should not be appreciably hydrolyzed, although hydrolysis can sometimes be repressed.
6. The choice of the leading electrolyte depends also on certain other considerations, such as purification, interaction with electrodes, and gravitational stability; these were discussed in Section 5.3.3.

The initial concentration of the following solution should be slightly lower than the Kohlrausch concentration given by (144) for falling boundaries, and slightly greater for rising ones, in order to reduce mixing effects between the initial and the Kohlrausch solutions. In either case it is essential to check that the adjusted BZ concentration obtained is independent of a few percent variation in the initial BZ concentration. Although the Kohlrausch concentration is not known to begin with, a good approximation to it can usually be obtained from a preliminary run [367] using a relatively dilute solution of BZ for a falling boundary or a relatively concentrated solution for a rising one.

The main experimental techniques used for finding $*m_B^{BZ}$ are summarized below. It is necessary in all cases to check whether undisturbed adjustment to the Kohlrausch concentration has occurred by repeating the experiment with another current and again with a somewhat different initial concentration of BZ.

5.4.3   External Analysis

Removing a small sample of solution for external analysis must be accomplished without disturbing the mb system. This difficulty can be minimized in work with aqueous solutions near 4°C [368] where substantially convection-free electrolysis can be carried out in a wide Tiselius cell [160]. If certain precautions are used, a thin-walled capillary attached to a syringe can be lowered into the cell while the current is flowing, and a sample of solution can be removed without appreciably disturbing the boundary or mixing the solutions [368].

A more general method of surmounting the problem was given by Muir,

Graham, and Gordon [367]. These workers used a falling boundary cell incorporating the long narrow tube depicted in Figure 8.42. The boundary is formed above S_1, as described for Figure 8.36, the current is started, and the boundary moves down the tube. When it is certain that the boundary has passed through S_2, the current is shut off and both stopcocks are closed, thus isolating a column of the following solution between S_1 and S_2. The solution above S_1 is then removed under suction with a fine pipet, S_1 is opened, and the solution in the tube down to a level 2 cm below S_1 is also sucked out. A second long capillary pipet (or a needle at the end of a syringe [366]) is now inserted well into the tube, and the solution is removed and analyzed. Conductometric analysis using tiny conductance cells has usually been favored, but any accurate analytical procedure suitable for small samples is satisfactory. An analogous experimental procedure can be devised for rising boundaries.

Results agreeing with direct mb data to better than $\pm 0.03\%$ have been obtained for the KCl/NaCl system when the operations were carefully standardized and when certain small handling errors were corrected for [367]. The method has since been successfully applied to dilute aqueous solutions of sodium lauryl sulfate [369] and to concentrated solutions of several alkali halides [366, 370]. In one paper [366] the method was reversed, and the transference numbers of the leading solution were determined by the Kohlrausch equation from a knowledge of the transference numbers of the following solution.

In another version giving results accurate to about $\pm 0.2\%$ [342], a removable section is mounted as part of the mb tube (in Figure 8.36 or 8.37) by means of standard taper ground-glass joints. When the boundary has passed completely through this part of the tube, the cell is taken apart at the ground-glass joints and platinum electrodes on complementary joints are inserted into the ends of the removable section. This turns it into a small conductance cell filled with the Kohlrausch solution.

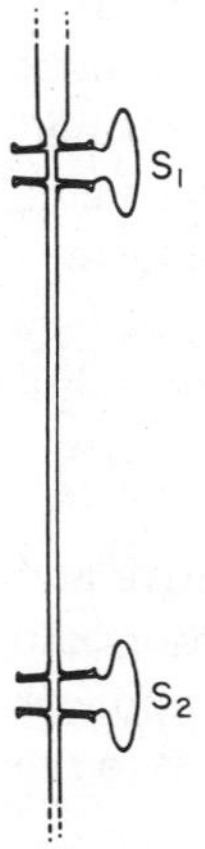

Figure 8.42   Moving boundary tube used for the external analysis of the Kohlrausch solution.

5.4.4   Internal Electrometric Analysis

This idea was developed by Hartley and his co-workers [371] with a "balanced boundary cell" in which a glass piston, moved by a synchronous motor, produces a slow counterflow of solution in the moving boundary tube. This causes the boundary to remain almost stationary while the following Kohlrausch solution is gradually pushed into a sidearm containing platinum wire electrodes; its ac conductance is then measured. The resulting transference numbers are accurate to about $\pm 1\%$.

Hartley's cell, though slightly cumbersome, is advantageously designed to enable the conductance measurements to be carried out in a sidearm through which the main cell current does not flow. It is now possible to determine the concentration of the following solution to 0.1% by using platinum microprobes sealed into the mb tube itself (Section 5.3.2). Determination of the resistance between adjacent probes, which would seem the obvious method, was found to be in error by about 10% [310], probably because of electrode polarization [309]. Even determining the ratio of the resistances of the leading and following solutions with a given pair of probes could give rise to an error of a few percent. Internal conductometric analysis has, nevertheless, been used by the Soviet school [372]. On the other hand, it has proved feasible to use the potentiometric method provided the potential drop between the probes is kept large enough (of the order of 100 V) to render the relative contribution of electrode polarization insignificant [373]. With a constant current flowing, the potential drop is measured between one (or more) probes near the top of the mb tube and one (or more) near the bottom, both before the boundary has reached the first probe (E_1) and after it has passed the last one (E_2). These potentials correspond to the horizontal portions in Figure 8.30b. Then, by Ohm's law,

$$\frac{E_2}{E_1} = \frac{R_{*BZ}}{R_{AZ}} = \frac{\kappa_{AZ}^{AZ} + \kappa_{solvent}^{AZ}}{\kappa_{BZ}^{*BZ} + \kappa_{solvent}^{*BZ}} \tag{146}$$

from which the specific conductivity of the adjusted BZ solution may be derived and hence its concentration. Since the resistances vary linearly with the amount of Joule heating, the ratio E_2/E_1 should be measured at several currents and extrapolated to zero current. With concentrated solutions where higher currents must be employed, it is important to check that the result has not been influenced by convective mixing between the initial and adjusted indicator solutions. This is done by carrying out runs with several initial indicator concentrations until the correct value is bracketed [373].

In the RF method based on the use of ring electrodes outside the mb tube, the response of the detector contains contributions from the capacitative and inductive components of the circuit [309]. It is to be hoped that appropriate electronic modification of the circuit will one day allow the resistive component alone to be captured so that this ingenious technique can also be applied to the determination of the transference numbers of the following solution.

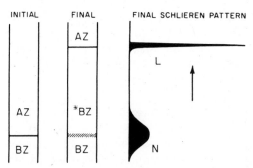

Figure 8.43 Typical optical schlieren patterns for the moving boundary system AZ←*BZ←BZ [374].

5.4.5   Internal Refractometric Analysis

Figure 8.43 shows the type of photograph obtained [307, 374] when the mb system between AZ and BZ is observed by the schlieren scanning optical method [160]. Peak L records the sharp AZ/*BZ two-salt mb. The broad peak N is produced by the diffuse concentration boundary (Section 5.3.8) between the Kohlrausch solution of BZ behind the boundary (*BZ) and the initial solution of BZ. The area under peak N is proportional to the difference in the refractive index between the two BZ solutions; if the refractive index increment of BZ has been determined previously, the Kohlrausch concentration can be calculated from a knowledge of the initial concentration of BZ and the area of peak N. The error involved is probably less than $\pm 1\%$ [307].

5.5   Emf Method Using Cells With Transference

5.5.1   Introduction and Theory

In the methods already described, electricity actually flows and the subsequent transport of an ion constituent is measured. However, even in the absence of actual current flow, the tendency for ions to migrate in an electric field affects the Gibbs free energy change of certain cell processes. The transference number can therefore be determined by measuring the electromotive force (emf) of a galvanic concentration cell with transference. Activity coefficients are also required.

A concentration cell with transference is a cell that includes a liquid junction between two solutions that contain the same electrolyte but at different concentrations. In each solution there is immersed a chemically identical electrode reversible to one of the ions in solution. If, as is usually the case, this ion is the same as one of the ion constituents of the solute, we may write the cell:

$$\text{electrode reversible to } B | A_{\nu_A} B_{\nu_B} | A_{\nu_A} B_{\nu_B} | \text{ electrode reversible to } B \qquad (147)$$
$$\qquad\qquad\qquad\qquad m_1 \qquad m_2$$

where m is the molality of the solute—it is a tradition in emf work to express concentrations in moles per kilogram of solvent. The formula of the binary electrolyte indicates that 1 mol consists of v_A gram-formula-weights of A ion constituent and v_B of B ion constituent. If the algebraic charge numbers of A and B are z_A and z_B, respectively, then by the condition of electroneutrality,

$$v_A z_A + v_B z_B = 0 \tag{148}$$

The equation for the emf is derived by the standard thermodynamic procedure of imagining the cell to be discharged infinitely slowly. Since the cell reaction includes the transfer of ion constituents across the liquid junction, the transference number T_A of the ion constituent A is included in the resulting equation:

$$E_t = \left(\frac{z_B - z_A}{z_B z_A}\right) \frac{RT}{F} \int_2^1 T_A d \ln (m\gamma) \tag{149}$$

where E_t is the emf in volts of the cell as written, R the gas constant, T the absolute temperature, F Faraday's constant, and γ the mean stoichiometric activity coefficient [21]. Numerical values of $2.3026RT/F$ at various temperatures have been tabulated [17, 375]; at 25°C this factor is 0.059158 absolute volt. As an example, for the cell

$$\text{Ag} \mid \text{AgCl} \mid \text{BaCl}_2 \mid \text{BaCl}_2 \mid \text{AgCl} \mid \text{Ag}$$
$$m_1 \qquad m_2$$

$$E_t = \frac{3RT}{2F} \int_2^1 T_{Ba} \, d \ln (m\gamma) \tag{150}$$

because here the electrodes are reversible to the Cl ion constituent so that $A = Ba$, $B = Cl$, $z_A = +2$, and $z_B = -1$. Although no explicit mention is made of diffusion in the above classic thermodynamic derivation, the latter is justified by the fact that the thermodynamics of irreversible processes lead to the same equation provided the Onsager Reciprocal Relations hold (Section 2.1.3). Experimental results are overwhelmingly in favor of the validity of these relations [376, 65] so we can accept the classic equation with some confidence. Moreover, it is well known that the emfs of cells with transference do not change with time nor do they depend upon the manner in which the junction is made provided the composition of the transition solution varies continuously from one side to the other [377].

Equation (149) can be employed as it stands if some functional relationship between T_A and m is assumed and if the experimental data are needed solely to fix the numerical values of the parameters involved. It is much more useful, however, to obtain an explicit value for the transference number without any *a priori* postulate as to the form of its concentration variation, and this can be done through the differential form

$$T_A = \left(\frac{z_B z_A}{z_B - z_A}\right) \frac{F}{RT} \frac{dE_t}{d \ln(m\gamma)} \tag{151}$$

The transference number can therefore be obtained from the emfs, provided activity coefficients are procured from another source. A wide choice of methods is available for their determination [17], and there already exist tables, largely of isopiestic origin, for numerous electrolytes in water [17] and other solvents [378]. Whenever electrodes exist that are reversible to both ion constituents of the electrolyte, it is possible to obtain the required activity coefficients by means of a back-to-back combination of cells without liquid junctions:

electrode reversible to B $|A_{v_A}B_{v_B}|$ electrode reversible to A ⎯⎯⎯⎯⎯⎯
$$m_1$$

⎯⎯electrode reversible to A $|A_{v_A}B_{v_B}|$ electrode reversible to B   (152)
$$m_2$$

whose emf E, in volts, is given by

$$E=\left(\frac{z_B-z_A}{z_B z_A}\right)\frac{RT}{F}\int_2^1 d\,\ln(m\gamma)=\left(\frac{z_B-z_A}{z_B z_A}\right)\frac{RT}{F}\ln\left(\frac{m_1\gamma_1}{m_2\gamma_2}\right) \qquad (153)$$

It is usual to measure the emfs of the individual cells and to produce the effect of the back-to-back combination by subtracting the emfs on paper.

The differential form of (153), together with (151), gives

$$T_A=\frac{dE_t}{dE} \qquad (154)$$

This simple form is a popular alternative to (151) because E can always be calculated from (153) when the activity coefficients are derived from some other physicochemical measurement.

For those relatively few cases in which the electrodes in cell (147) are reversible, not to the simple ion B, but to some complex ion such as A_xB_y, the above equations require modification [379]. Transference numbers can be calculated from E_t data alone provided measurements have been carried out with cells of type (147) with electrodes reversible to B as well as with the complementary cells in which the electrodes are reversible to A [380]. This is possible with electrolytes such as $ZnSO_4$ [380] and HCl.

The concentration to which T_A in (151) or (154) refers is clarified by taking one molality (m_2, for example) as a reference concentration and by keeping it constant in all runs. The transference number then becomes a property of the solution of variable concentration (m_1). An equivalent plan is to use a small number of intermediate reference concentrations and to add up the appropriate emfs so that they all refer to only one reference concentration. This procedure is in fact essential whenever a large heat of mixing at the liquid junction (>3 kJ/mol) makes it necessary to place an upper limit on the difference $|m_1-m_2|$ [381].

The main computational method for obtaining transference numbers from (151) or (154) is to relate E_t and either $\log(m\gamma)$ or E to each other by a simple, empirical, polynomial expression (such as $E_t = a + bE + cE^2$). The coefficients are calculated by least squares or some other curve-fitting procedure, and the equation is differentiated to give the transference number. With modern computers it is a relatively straightforward matter to find an equation to fit either the whole data set or at least overlapping portions of it. It should be noted that differentials at the ends of sections are much less reliable. If the data are rather scattered it is best to repeat the experiments more carefully; as a second best, published schemes for smoothing and interpolating can be consulted [128, 381, 382] and there is also one for checking the calculations [383]. There are special methods for obtaining the slopes at the extreme ends of a curve [384].

Those without access to a computer can use the method of Rutledge [385], which has been described previously [262]. This allows one to find, with a known limit of error, the derivative function dy/dx of a differentiable function y from values of the latter observed for equally spaced values of the independent variable x. A fourth-degree polynomial is employed as a differentiating tool and is applied successively to sets of five consecutive points of the data. It is not implied, however, that the data as a whole can be adequately represented by a polynomial or any other elementary type of function. This method, too, cannot properly be applied to scattered data [386].

The differentiation can be appreciably simplified [131] by introducing a dummy transference number $\bar{T}$ defined by $\bar{T} = E_t/E$, and then transforming (154) to

$$T_A = \bar{T} + E \left(\frac{d\bar{T}}{dE} \right) \tag{155}$$

An equivalent relation can be written for (151). The polynomial fitting is now simpler because $\bar{T}$ is a much more slowly varying function of E than is T_A. Furthermore, only the relatively small term $E(d\bar{T}/dE)$ needs to be evaluated by computer fitting, and the major term $\bar{T}$ is calculated directly from the experimental data. It is therefore important, when using (155), to choose a reference concentration that makes E and E_t large so that small experimental uncertainties in the emfs do not produce large percentage errors in $\bar{T}$ and T_A.

The differentiation in (154) inevitably magnifies any slight errors in the observed quantities, and it has been reported that changes of 0.01 mV in E_t or E affect the transference number to the extent of 0.0004 in the case of HNO_3 [382] and 0.001 in the case of HCl [387] solutions. Emf transference numbers are therefore rarely more accurate than $\pm 1\%$, and probably the most useful application of the emf method is for a relatively rapid survey over a wide concentration and temperature range.

The solvent correction has been discussed in the literature [388].

5.5.2   Electrodes

Electrodes are the prima donnas of emf opera; their performance makes or mars the results and they are apt to be temperamental unless they are carefully handled. Suitable electrodes should be not only reversible to one of the ion constituents in solution, but also insoluble, stable, and reproducible. Ives and Janz [375] give much advice on the preparation and care of a great variety of electrodes and information on electrodes in aprotic organic solvents is reviewed by Butler [389], Difficulties are sometimes encountered until the techniques of making and using electrodes have been perfected, and the following hints may help to track down a source of trouble. Electrodes prepared by plating tend to behave erratically if the whole surface is not covered and respond sluggishly if the deposit is too thick. Many electrodes are adversely affected by the presence of oxygen (which can give rise to mixture potentials [211]), by traces of grease, by shock, by stress, and by certain impurities (e.g., very small amounts of Br^-, I^-, or S^{2-} change the potential of a silver/silver chloride electrode because the more insoluble salt is preferentially formed on the surface). Too large an electrode can adsorb a significant percentage of solute from very dilute solutions. The possibility of chemical interaction between electrode and solute must not be overlooked as this can severely affect the potential. Several types of interaction have been reported for silver/silver halide electrodes: dissolution of the AgX in concentrated halide solutions [363], the formation of a new solid phase such as $2AgI \cdot NEt_4I$ on Ag/AgI electrodes immersed in NEt_4I solutions [390, 391], and chemical attack on the silver metal by oxidizing species such as Co(III) complexes [392]. A good example of electrode choice dictated by the chemical nature of the solution is provided by concentrated aqueous $CuCl_2$ solutions. It was impossible to use either Ag/AgCl electrodes (Ag was oxidized and AgCl dissolved) or Cu electrodes (which reacted with $CuCl_2$ to form CuCl) but satisfactory transference numbers were obtained with chlorine gas electrodes [393].

For several types of electrode (e.g., silver/silver halide) a batch of individual electrodes must be made periodically. Suitable electrodes can be selected by placing them all in one solution and comparing their bias potentials. These can sometimes be decreased by preanodizing or precathodizing the electrodes [131]. Two electrodes are chosen whose potential difference is small (0.01–0.05 mV) and this difference is then eliminated either by correcting for it directly [394] or by interchanging the electrodes (or the solutions) in the two compartments of the cell with transference and recording the mean emf [388, 394, 395]. In another method, which is perhaps not quite as reliable, several selected electrodes are inserted into each compartment and the average cross emf is taken as the best value. The emfs of all cells, no matter what the electrodes, should be constant with time once thermal and chemical equilibria have been established, and they should be reproducible and additive [388].

Many commercial ion-selective electrodes are also now available [396], reversible to cations such as H^+ (the traditional glass electrode), Na^+, and Ca^{2+} and anions like F^-, S^{2-}, and ClO_4^-. However, their stability and reproducibility are usually poorer than those of the traditional homemade variety [397]. A common fault of ion-selective electrodes is a drift of potential with time, of the order of 0.01–1 mV/h. This can be overcome by interchanging electrodes and using a time-extrapolation procedure [398]. Let us take Covington and Prue's example:

glass electrode (P) | HCl (m_1) | glass electrode (Q) Cell A

glass electrode (P) | HCl (m_1) | HCl (m_2) | glass electrode (Q) Cell B

The emf of Cell A was read every minute for 5 min, the two electrodes were transferred (with appropriate washing) to Cell B, and readings continued at 1 min intervals for 10 min. The electrodes were then transferred back to Cell A and potential readings taken for a further 5 min. The whole procedure was then repeated with the position of the electrodes in Cell B reversed. In each case, linear plots of potential versus time allowed one to determine the values of the emfs E_A and E_B that Cells A and B would have had at the moment of transfer of the two electrodes. At this point the difference between the asymmetry potentials of the two electrodes will have been the same in the two cells. It follows that $E_B - E_A$ is the emf of the thermodynamically equivalent cell

$$Pt \mid H_2 \mid HCl\ (m_1) \mid HCl\ (m_2) \mid H_2 \mid Pt$$

and indeed the $E_B - E_A$ values were reproducible to 0.02 mV in several trials. Another but less satisfactory way of coping with electrode drift is to add either the solvent or a concentrated solution in a stepwise fashion to one side of the cell with transference, and to measure the drifts with time of the first and last compositions of the series [399].

5.5.3   Cell Design and Measurement

The emf of a cell with transference is independent of the way in which the liquid junction is formed, provided that no large heat effects occur when the two solutions meet and provided further that the mixed solution does not reach an electrode [400]. The first restriction is overcome, where necessary, by using two solutions whose concentrations do not differ too much (see Section 5.5.1). As regards the second restriction, the solutions should be brought into contact in such a way that the rate of intermixing is reduced as much as possible. This can be done with a flowing junction [401], or with a gravitational free-diffusion junction in which the lighter solution rests on top of the heavier one. The latter is simple and is frequently employed; two examples of cells based on this principle are shown in Figures 8.44 and 8.45.

The cell in Figure 8.44 is set up by pouring less dense solution into the E_1 electrode compartment and then fitting the electrode E_1 while the bent side tube is temporarily closed with a cap. The right-hand compartment is similarly filled with the denser solution to the level shown, and the two parts are joined at the ground-glass joint C. The air space in the bulb below C prevents liquid from touching the lubricant on C. Further stopcocks, preferably of PTFE, can be incorporated on each side to help in filling and thermal equilibration, and the electrode compartments can be modified to allow gas electrodes to be used [68]. Some researchers [395, 398] insert a large central compartment into their design.

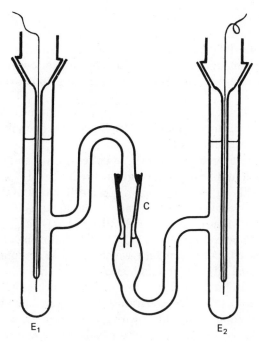

Figure 8.44 Cell with transference based on the designs of Stokes and Levin [402, 386] and Stonehill [166].

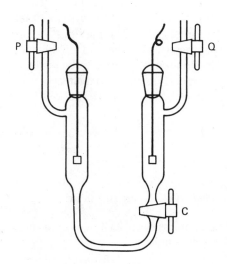

Figure 8.45 Cell with transference used by Phang [403].

Another simple cell is shown in Figure 8.45, where the left-hand side holds the more concentrated solution. After thermal equilibration, taps P and Q are closed and the PTFE stopcock C is opened. Other workers [387, 131] prefer to form the junction inside a three-way stopcock which can be closed except during readings to minimize mixing in long runs over a temperature range.

A helpful visual check on the stability of the liquid junction in a given cell can be obtained by putting a colored test solution into one side of the cell and a colorless one into the other [402].

The emfs are measured to 0.01 mV or better on a high-impedance digital voltmeter which should be regularly standardized. For 0.1% accuracy its input impedance should be at least 1000 times greater than that of the cell. Since the resistance of ion-selective electrodes may be as high as 10^9 ohm, the input impedance of a DVM must here be at least 10^{12} ohm [404]. Much useful practical advice is given by Ives and Janz [375], and the following points may be particularly emphasized. Dirty contacts, which give rise to spurious potentials, must be avoided and all leads should be well insulated as a precaution against electrical leakage. In humid weather, protection against moisture is necessary. Interference by motors or other apparatus can be overcome by enveloping the offending piece of equipment with a wire mesh screen or, more generally, by employing electrically shielded wire for all leads, with the metal shielding connected to ground. A very sensitive test is to observe the emf balance point when a glass rod previously rubbed with silk is brought near one of the wires. There is no risk of picking up stray potentials around the cell when oil is used as the thermostat liquid. The thermostat bath itself must be grounded.

One of the main features of the emf method is the ease and rapidity with which results can often be obtained over a wide temperature range. The cell is either left in the thermostat while the thermoregulator setting is changed or transferred to another thermostat; when the temperature and the cell potential are again constant, their values are once more recorded. The readings should be checked by returning to the original temperature at the end of the run. If the starting temperature is an intermediate one on the range to be covered then two checks are possible, one after the temperature has been decreased and one after it has been increased. Most electrodes behave well from about 0 to about 50°C.

6   APPENDIX

6.1   Useful Data Sources

A1. Landolt-Börnstein, *Zahlenwerte und Funktionen*, Vol. II, Part 7II, 6th ed., Springer-Verlag, Berlin, 1960.
Extensive tables up to 1960 of conductances, transference numbers, and limiting ionic conductances in aqueous and nonaqueous solutions.

A2. H. S. Harned and B. B. Owen, *The Physical Chemistry of Electrolytic Solutions*, 3rd ed., Reinhold, New York, 1958, Appendix A.

Tables of conductances and transference numbers of many common electrolytes in water at 25°C at rounded concentrations. Tables of limiting conductances of various electrolytes in ten organic solvents.

A3.   R. A. Robinson and R. H. Stokes, *Electrolyte Solutions*, 2nd rev. ed., Butterworths, London, 1965, Appendices 6.1–6.3.
Critical compilations of limiting ionic conductances in water at temperatures from 0 to 100°C, and of conductances in concentrated aqueous solutions.

A4.   G. Jander, H. Spandau, and C. C. Addison, Eds., *Chemistry in Nonaqueous Ionizing Solvents*, Vol. I, Part 1, Vieweg + Sohn, Braunschweig, 1966, and subsequent volumes.
Reviews of chemistry in inorganic and organic solvents (including liquid ammonia, hydrogen cyanide, hydrogen fluoride, dinitrogen tetroxide, sulfur dioxide, lower fatty acids and amides) that contain sections on the relevant conductance and transference work.

A5.   J. J. Lagowski, Ed., *The Chemistry of Nonaqueous Solvents*, Vol. II, Academic, New York, 1967, and subsequent volumes.
Chapters on the chemistry in various solvents (including liquid hydrogen halides, sulfuric acid, nitric acid, sulfur dioxide, hydrogen sulfide, antimony chloride, carboxylic acids, amides, cyclic carbonates, pyridine, tetramethylurea, and sulfolane) that contain sections on the relevant conductance and transference work.

A6.   M. Spiro, "Transference Numbers," in A. Weissberger and B. W. Rossiter, Eds., *Physical Methods of Chemistry*, *Part IIA*, 4th ed., Wiley-Interscience, New York, 1971, Chapter IV, Appendix. Critical tables of transference numbers in water and nonaqueous solvents.

A7.   G. J. Janz and R. P. T. Tomkins, *Nonaqueous Electrolytes Handbook*, Vol. I, Academic, New York, 1972, Chapters III, VII, and VIII; Vol. II, 1973, Chapter XI. Extensive tabulations of conductances and transference numbers for several hundred nonaqueous systems. In Chapter III the conductances are listed both by solute and by solvent.

A8.   R. Fernández-Prini, "Conductance," in A. K. Covington and T. Dickinson, Eds., *Physical Chemistry of Organic Solvent Systems*, Plenum, London, 1973, Chapter 5.1, Appendices 5.1.1–5.1.5.
Critical compilation of conductances in many organic solvents.

A9.   M. Spiro, "Transference Numbers" and "Ionic Conductances," in A. K. Covington and T. Dickinson, Eds., *Physical Chemistry of Organic Solvent Systems*, Plenum, London, 1973, Chapters 5.2 and 5.3, Appendices 5.81, 5.12.1–5.12.12.
Selected transference numbers in 20 organic solvents, and tables of limiting ionic conductances in 12 organic solvents.

A10.   V. M. M. Lobo, *Electrolyte Solutions: Literature Data on Thermodynamic and Transport Properties*, Vol. I, Universidade de Coimbra, Portugal, 1975; Vol. II, 1981. A large compilation of literature conductances and transference numbers in aqueous solutions.

A11.   J. Barthel, R. Wachter, and H-J. Gores, "Temperature Dependence of Conductance of Electrolytes in Nonaqueous Solutions," in B. E. Conway and J. O'M. Bockris, Eds., *Modern Aspects of Electrochemistry*, Vol. 13, Plenum, New York, 1979, Table 9.
Survey of publications dealing with nonaqueous conductances at more than one temperature.

A12. M. Spiro and F. King, "Transport Properties in Concentrated Aqueous Electrolyte Solutions," in D. Inman and D. G. Lovering, Eds., *Ionic Liquids*, Plenum, New York, 1981, Chapter 5, Tables 3 and 4.

Lists of conductance and transference measurements in concentrated aqueous solutions up to 1979.

A13. B. S. Krumgalz, *J. Chem. Soc. Faraday Trans. 1*, **79**, 571 (1983).

A table of limiting ionic conductances of 17 common ions in 43 organic solvents at 25°C, as well as a smaller set of values at other temperatures. Some of the λ_i^0 values have been obtained by combining conductances and transference numbers, others by utilizing Walden's rule for large ions.

6.2   Limiting Ionic Conductances

The most common research solvent is undoubtedly water at 25°C and 1 bar. A list of limiting molar conductances λ_i^0 for various ions in this solvent has therefore been compiled in Table 8.5. Half the values quoted were obtained by

Table 8.5   Limiting Molar Conductances of Selected Ions in Water at 25°C in Units of cm^2 S/mol

Cation	λ_+^0	Extrap.[a]	Ref.	Anion	λ_-^0	Extrap.	Ref.
H^+	350.0	P	A14	OH^-	199.2	P	A14
Li^+	38.7$_8$	P	A14	e^-	~185		A21
Na^+	50.1$_0$	P	A14	F^-	55.4$_2$	P	A14
K^+	73.5$_0$	P	A14	Cl^-	76.3$_2$	P	A14
Rb^+	77.3	P	A14	Br^-	78.1$_3$	P	A14
Cs^+	77.0	P	A14	I^-	76.9$_8$	P	A14
Ag^+	62.1	P	A14	NCS^-	66.4		A22
Tl^+	74.8	P	A14	HCO_3^-	45.4	LW	A23
NH_4^+	73.5		A3	NO_3^-	71.4$_1$	P	A14
NMe_4^+	44.3$_7$	P	A14	IO_3^-	40.6	P	A14
NEt_4^+	32.1$_4$	P	A14	$H_2PO_4^-$	32.3		A24
NPr_4^+	23.2$_4$	P	A14	HSO_4^-	45.4		A24
NBu_4^+	19.3$_3$	P	A14	ClO_4^-	67.2	P	A14
NAm_4^+	17.4	P	A14	BPh_4^-	19.8	P	A14
$N(i\text{-}Am)_4^+$	17.9	P	A14	PF_6^-	59.1	P	A14
Mg^{2+}	106.1	LW	A15	$HCOO^-$	54.6		A3
Ca^{2+}	118.0	LW	A15	CH_3COO^-	40.8	P	A14
Sr^{2+}	118.6		A3	$CH_3CH_2COO^-$	35.8		A3
Ba^{2+}	126.9	LW	A15	$CH_3(CH_2)_2COO^-$	32.6		A3
Co^{2+}	107.0		A16	picrate$^-$	30.7$_1$	P	A14
Ni^{2+}	106.1		A17	$PhCOO^-$	32.4		A3
Cu^{2+}	107.2		A3	$PhSO_3^-$	35.1	P	A14
Zn^{2+}	105.6		A3	SO_4^{2-}	159.6	LW	A15
Cd^{2+}	107.0		A18	$S_2O_3^{2-}$	174.3	LW	A15
Al^{3+}	179.0		A19	$S_2O_8^{2-}$	170.5		A25
La^{3+}	209.0		A3	CO_3^{2-}	138.5		A3
$Co(en)_3^{3+}$	222.8		A20	$C_2O_4^{2-}$	148.2		A3

[a]P = Pitts equation (Table 8.1, Theory 3); LW = Lee and Wheaton equation (52). Most other values have been derived by one of a number of earlier theoretical equations such as those of Shedlovsky, Robinson and Stokes, and Fuoss, or from transference data.

Spiers and Pethybridge [A14] who re-extrapolated all seemingly reliable literature conductance data with the full Pitts equation using q as the distance of
closest approach (Theory 3 in Table 8.1). Extrapolations in the literature based
on a variety of earlier theoretical equations yield slightly different values. The
ions whose recalculated values differ by more than $0.2 \ cm^2 \cdot S/mol$ from those in
[A3] are Rb^+, Cs^+, NMe_4^+, NEt_4^+, Co^{2+}, OH^-, HCO_3^-, and picrate.

A14. D. J. Spiers and A. D. Pethybridge, unpublished work.
A15. A. D. Pethybridge and S. S. Taba, *J. Chem. Soc. Faraday Trans. 1*, **78**, 1331 (1982).
A16. G. Marx, W. Riedel, and J. Vehlow, *Ber. Bunsenges. Phys. Chem.*, **73**, 74 (1969).
A17. R. H. Stokes, S. Phang, and R. Mills, *J. Solution Chem.*, **8**, 489 (1979).
A18. R. A. Matheson, *J. Phys. Chem.*, **66**, 439 (1962).
A19. C. R. Frink and M. Peech, *Inorg. Chem.*, **2**, 473 (1963).
A20. H. Kaneko and N. Wada, *J. Solution Chem.*, **7**, 19 (1978).
A21. K. H. Schmidt and S. M. Ander, *J. Phys. Chem.*, **73**, 2846 (1969); G. C. Barker,
      P. Fowles, D. C. Sammon, and B. Stringer, *Trans. Faraday Soc.*, **66**, 1498 (1970).
A22. G. Chittleborough and B. J. Steel, unpublished work.
A23. G. F. Cassford and A. D. Pethybridge, unpublished work.
A24. M. Selvaratnam and M. Spiro, *Trans. Faraday Soc.*, **61**, 360 (1965).
A25. J. Balej and A. Kitzingerova, *Collect. Czech. Chem. Commun.*, **39**, 49 (1974).

References

1. M. Spiro, *Educ. Chem.*, **3**, 139 (1966).
2. M. L. McGlashan (revised by M. A. Paul and D. H. Whiffen), "Manual of Symbols
    and Terminology for Physicochemical Quantities and Units," *Pure Appl. Chem.*, **51**,
    1 (1979); R. I. Holliday, *J. Chem. Educ.*, **53**, 21 (1976).
3. W. J. Hamer and H. J. DeWane, *Natl. Stand. Ref. Data Ser., U.S. Natl. Bur. Stand.*,
    No. 33 (1970).
4. M. Robson and P. G. Wright, *Educ. Chem.*, **2**, 185 (1965).
5. R. Haase, *Z. Phys. Chem. (Frankfurt am Main)*, **39**, 37 (1963); *Angew. Chem. Int.
    Ed.*, **4**, 485 (1965).
6. H. Svensson, *Sci. Tools*, **3**, 30 (1956).
7. M. Spiro, *J. Chem. Educ.*, **33**, 464 (1956).
8. M. Shamim and M. Spiro, *Trans. Faraday Soc.*, **66**, 2863 (1970).
9. T. Kurucsev and B. J. Steel, *Rev. Pure Appl. Chem.*, **17**, 149 (1967).
10. R. Haase, *Z. Phys. Chem. (Frankfurt am Main)*, **14**, 292 (1958); *Z. Elektrochem.*, **62**,
    279 (1958).
11. M. Spiro, *J. Inorg. Nucl. Chem.*, **27**, 902 (1963).
12. R. Mills and L. A. Woolf, *The Diaphragm Cell*, Australian National University,
    Canberra, 1968, pp. 35–36.
13. H. J. V. Tyrrell and K. R. Harris, *Diffusion in Liquids*, Butterworths, London, 1984,
    p. 109.
14. J. Richter and E. Amkreutz, *Z. Naturforsch.*, **27a**, 280 (1972).
15. R. Haase, *Electrochim. Acta*, **23**, 391 (1978).
16. A. Poisson, M. Perie, J. Perie, and M. Chemla, *J. Solution Chem.*, **8**, 377 (1979).
17. R. A. Robinson and R. H. Stokes, *Electrolyte Solutions*, 2nd ed., Butterworths,
    London, 1959.

18. R. Fernández-Prini, "Conductance," in A. K. Covington and T. Dickinson, Eds., *Physical Chemistry of Organic Solvent Systems*, Plenum, London, 1973, Chap. 5.1.

19. R. J. Wheaton, *J. Chem. Soc. Faraday Trans. 2*, **76**, 1093 (1980).

20. J-C. Justice, "Conductance of Electrolyte Solutions," in B. E. Conway, J. O'M. Bockris, and E. Yeager, Eds., *Comprehensive Treatise of Electrochemistry*, Vol. 5, Plenum, New York, 1983, Chap. 3.

21. C. W. Davies, *Electrochemistry*, Newnes, London, 1967.

22. J-C. Justice, *Electrochim. Acta*, **16**, 701 (1971); *J. Phys. Chem.*, **79**, 454 (1975).

23. R. M. Fuoss and L. Onsager, *J. Phys. Chem.*, **61**, 668 (1957).

24. J. E. Lind, Jr., J. J. Zwolenik, and R. M. Fuoss, *J. Am. Chem. Soc.*, **81**, 1557 (1959).

25. R. L. Kay, *J. Am. Chem. Soc.*, **82**, 2099 (1960).

26. R. M. Fuoss and K-L. Hsia, *Proc. Natl. Acad. Sci. USA*, **57**, 1550 (1967); **58**, 1818 (1968).

27. R. Fernández-Prini, *Trans. Faraday Soc.*, **65**, 3311 (1969).

28. D. F. Evans and M. A. Matesich, "The Measurement and Interpretation of Electrolytic Conductance," in E. Yeager and A. J. Salkind, Eds., *Techniques of Electrochemistry*, Vol. 2, Wiley-Interscience, New York, 1973, Chap. I.

29. J. Quint and A. Viallard, *J. Solution Chem.*, **7**, 533 (1978).

30. R. Fernández-Prini and J. E. Prue, *Z. Phys. Chem.*, **228**, 373 (1965).

31. J-C. Justice, R. Bury, and C. Treiner, *J. Chim. Phys. Phys. Chim. Biol.*, **65**, 1708 (1968).

32. M. Linert and P. Rechberger, *Comput. & Chem.*, **6**, 101 (1982).

33. E. Pitts, *Proc. R. Soc. London*, *A*, **217**, 43 (1953).

34. E. Pitts, B. E. Tabor, and J. Daly, *Trans. Faraday Soc.*, **65**, 849 (1969); **66**, 693 (1970).

35. A. D. Pethybridge and S. S. Taba, *Faraday Discuss. Chem. Soc.*, **64**, 274 (1977).

36. J. Barthel, R. Wachter, and H-J. Gores, "Temperature Dependence of Conductance of Electrolytes in Nonaqueous Solutions," in B. E. Conway and J. O'M Bockris, Eds., Vol. 13, *Modern Aspects of Electrochemistry*, Plenum, New York, 1979, Chap. 1.

37. S. D. Klein, Z. Pawlak, R. Fernández-Prini, and R. G. Bates, *J. Solution Chem.*, **10**, 333 (1981).

38. R. M. Fuoss, *Proc. Natl. Acad. Sci. USA*, **75**, 16 (1978).

39. R. M. Fuoss, *J. Phys. Chem.*, **82**, 2427 (1978).

40. W. H. Lee and R. J. Wheaton, *J. Chem. Soc. Faraday Trans. 2*, **74**, 743 (1978).

41. W. H. Lee and R. J. Wheaton, *J. Chem. Soc. Faraday Trans. 2*, **74**, 1456 (1978).

42. W. H. Lee and R. J. Wheaton, *J. Chem. Soc. Faraday Trans. 2*, **75**, 1128 (1979).

43. A. D. Pethybridge and S. S. Taba, *J. Chem. Soc. Faraday Trans. 1*, **76**, 368 (1980).

44. A. D. Pethybridge, *Z. Phys. Chem.* (*Frankfurt am Main*), **133**, 143 (1982).

45. M. Spiro, *Trans. Faraday Soc.*, **55**, 1746 (1959).

46. T. Shedlovsky, *J. Franklin Inst.*, **225**, 739 (1938).

47. R. M. Fuoss and T. Shedlovsky, *J. Am. Chem. Soc.*, **71**, 1496 (1949).

48. H. M. Daggett, Jr., *J. Am. Chem. Soc.*, **73**, 4977 (1951).

49. T. Shedlovsky and R. L. Kay, *J. Phys. Chem.*, **60**, 151 (1956).

50. R. M. Fuoss and F. Accascina, *Electrolytic Conductance*, Interscience, New York, 1959, Chap. XVIII.

51. C. B. Wooster, *J. Am. Chem. Soc.*, **59**, 377 (1937); **60**, 1609 (1938).

52. N. G. Sellers, P. M. P. Eller, and J. A. Caruso, *J. Phys. Chem.*, **76**, 3618 (1972).

53. W. Ebeling and M. Grigo, *J. Solution Chem.*, **11**, 151 (1982); M. Grigo, *J. Solution Chem.*, **11**, 529 (1982).

54. M. Selvaratnam and M. Spiro, *Trans. Faraday Soc.*, **61**, 360 (1965).

55. A. D. Pethybridge and J. E. Prue, *Trans. Faraday Soc.*, **63**, 2019 (1967).

56. E. Arnold and A. Patterson, Jr., *J. Chem. Phys.*, **41**, 3089 (1964); J. L. Dye, *Angew. Chem. Int. Ed. Engl.*, **18**, 587 (1979).
57. A. W. Ralston, *Fatty Acids and Their Derivatives*, Wiley, New York, 1948, pp. 661, 663.
58. R. H. Stokes, *J. Am. Chem. Soc.*, **76**, 1988 (1954).
59. R. L. Kay and J. L. Dye, *Proc. Natl. Acad. Sci. USA*, **49**, 5 (1963).
60. L. G. Longsworth, *J. Am. Chem. Soc.*, **54**, 2741 (1932).
61. D. P. Sidebottom and M. Spiro, *J. Chem. Soc. Faraday Trans. 1*, **69**, 1287 (1973).
62. J-C. Justice, J. Perie, and M. Perie, *J. Solution Chem.*, **9**, 583 (1980).
63. J-C. Justice, *J. Solution Chem.*, **7**, 859 (1978).
64. M. Spiro, "Transference Numbers," in A. K. Covington and T. Dickinson, Eds., *Physical Chemistry of Organic Solvent Systems*, Plenum, London 1973, Chap. 5.2.
65. M. Spiro and F. King, "Transport Properties in Concentrated Aqueous Electrolyte Solutions," in D. Inman and D. G. Lovering, Eds., *Ionic Liquids*, Plenum, New York, 1981, Chap. 5.
66. D. G. Miller, *J. Phys. Chem.*, **70**, 2639 (1966).
67. D. G. Miller, *Faraday Discuss. Chem. Soc.*, **64**, 295 (1977).
68. F. King and M. Spiro, *J. Solution Chem.*, **12**, 65 (1983).
69. J. M'Halla, P. Turq, and M. Chemla, *J. Chem. Soc. Faraday Trans. 1*, **77**, 465 (1981).
70. H. G. Hertz, *Ber. Bunsenges. Phys. Chem.*, **81**, 656 (1977).
71. H. G. Hertz, K. R. Harris, R. Mills, and L. A. Woolf, *Ber. Bunsenges. Phys. Chem.*, **81**, 664 (1977).
72. D. G. Miller, *J. Phys. Chem.*, **85**, 1137 (1981).
73. L. A. Woolf and K. R. Harris, *J. Chem. Soc. Faraday Trans. 1*, **74**, 933 (1978).
74. R. Mills and H. G. Hertz, *J. Chem. Soc. Faraday Trans. 1*, **78**, 3287 (1982).
75. H. G. Hertz, A. V. J. Edge, and R. Mills, *J. Chem. Soc. Faraday Trans. 1*, **79**, 1317 (1983).
76. C. W. Davies, *Ion Association*, Butterworths, London, 1962.
77. J. E. Prue, *Ionic Equilibria*, Pergamon, Oxford, 1966.
78. D. P. Sidebottom and M. Spiro, *J. Phys. Chem.*, **79**, 943 (1975).
79. M. Spiro, "Ionic Conductances," in A. K. Covington and T. Dickinson, Eds., *Physical Chemistry of Organic Solvent Systems*, Plenum, London, 1973, Chap. 5.3.
80. M. Kilpatrick, "Normal and Abnormal Mobilities in Anhydrous Hydrofluoric Acid and Hydrogen Peroxide–Water Systems," in W. J. Hamer, Ed., *The Structure of Electrolytic Solutions*, Wiley, New York, 1959, Chap. 19.
81. U. Schindewolf, *Pure Appl. Chem.*, **53**, 1329 (1981).
82. H. J. V. Tyrrell, *Sci. Prog. (Oxford)*, **67**, 271 (1981).
83. R. H. Stokes, *A.N.Z.A.A.S. Congr. Pap.*, 35 (1957).
84. L. Fischer and K. Hessler, *Ber. Bunsenges. Phys. Chem.*, **68**, 184 (1964).
85. R. Mills and K. R. Harris, *Chem. Soc. Rev.*, **5**, 215 (1976).
86. M. H. O'Leary, *Phytochemistry*, **20**, 553 (1981); private communication (1982).
87. B. Krumgalz, *J. Chem. Soc. Faraday Trans. 1*, **78**, 437 (1982); **79**, 571 (1983).
88. D. L. Fowler and C. A. Kraus, *J. Am. Chem. Soc.*, **62**, 2237 (1940).
89. J. F. Coetzee and G. P. Cunningham, *J. Am. Chem. Soc.*, **86**, 3403 (1964); **87**, 2529 (1965).
90. F. Perrin, *J. Phys. Radium*, **5**, 497 (1934).
91. F. Perrin, *J. Phys. Radium*, **7**, 1 (1936).
92. P. H. Elworthy, *J. Chem. Soc.*, 3718 (1962).
93. M. Yokoi and G. Atkinson, *J. Am. Chem. Soc.*, **83**, 4367 (1961); G. Atkinson and S. Petrucci, *J. Phys. Chem.*, **67**, 1880 (1963).

94. G. Barr, *A Monograph of Viscometry*, Oxford University Press, Oxford, 1931, Chap. VIII.

95. H. J. V. Tyrrell and P. J. Watkiss, *J. Chem. Soc. Faraday Trans. 1*, **75**, 1417 (1979).

96. S. Broersma, *J. Chem. Phys.*, **32**, 1632 (1960).

97. S. A. Rice, *J. Am. Chem. Soc.*, **80**, 3207 (1958).

98. J. G. Kirkwood and J. Riseman, *J. Chem. Phys.*, **16**, 565 (1948); **22**, 1626 (1954).

99. J. Riseman and J. G. Kirkwood, *J. Chem. Phys.*, **18**, 512 (1950).

100. A. Peterlin, *J. Chim. Phys. Phys. Chim. Biol.*, **47**, 669 (1950); **48**, 13 (1951).

101. G. Thomson, S. A. Rice, and M. Nagasawa, *J. Am. Chem. Soc.*, **85**, 2537 (1963).

102. R. M. Fuoss and D. Edelson, *J. Am. Chem. Soc.*, **73**, 269 (1951).

103. J. T. Edward, *J. Chem. Educ.*, **47**, 261 (1970).

104. R. L. Kay, "Ionic Transport in Water and Mixed Aqueous Solvents," in F. Franks, Ed., *Water: A Comprehensive Treatise*, Vol. 3, Plenum, New York, 1973, Chap. 4.

105. D. F. Evans, C. Chan, and B. C. Lamartine, *J. Am. Chem. Soc.*, **99**, 6492 (1977).

106. R. Zwanzig, *J. Chem. Phys.*, **38**, 1603 (1963); **52**, 3625 (1970).

107. R. Fernández-Prini, *J. Phys. Chem.*, **77**, 1314 (1973).

108. J. Hubbard and L. Onsager, *J. Chem. Phys.*, **67**, 4850 (1977); J. B. Hubbard, *J. Chem. Phys.*, **68**, 1649 (1978); B. U. Felderhof, *Mol. Phys.*, **48**, 1003 (1983); **49**, 449 (1983).

109. D. F. Evans, T. Tominaga, J. B. Hubbard, and P. G. Wolynes, *J. Phys. Chem.*, **83**, 2669 (1979).

110. P. J. Stiles, J. B. Hubbard, and R. F. Kayser, *J. Chem. Phys.*, **77**, 6189 (1982).

111. P. G. Wolynes, *J. Chem. Phys.*, **68**, 473 (1978).

112. P. Colonomos and P. G. Wolynes, *J. Chem. Phys.*, **71**, 2644 (1979).

113. P. G. Wolynes, *Annu. Rev. Phys. Chem.*, **31**, 345 (1980).

114. R. P. Bell, *The Proton in Chemistry*, Cornell University, Ithaca, NY, 1959, Chap. III.

115. B. E. Conway, J. O'M. Bockris, and H. Linton, *J. Chem. Phys.*, **24**, 834 (1956).

116. T. Erdey-Grúz and S. Lengyel, "Proton Transfer in Solution," in J. O'M. Bockris and B. E. Conway, Eds., *Modern Aspects of Electrochemistry*, No. 12, Plenum, New York, 1977, Chap. 1.

117. H. G. Hertz, *Ber. Bunsenges. Phys. Chem.*, **84**, 622 (1980).

118. H. G. Hertz, *Ber. Bunsenges. Phys. Chem.*, **84**, 613, 629 (1980); **85**, 456 (1981); **86**, 569 (1982).

119. H. Gerischer and H. Strehlow, *Ber. Bunsenges. Phys. Chem.*, **85**, 455 (1981); G. Anderegg, *Ber. Bunsenges. Phys. Chem.*, **86**, 567 (1982).

120. K. H. Schmidt and S. M. Ander, *J. Phys. Chem.*, **73**, 2846 (1969); G. C. Barker, P. Fowles, D. C. Sammon, and B. Stringer, *Trans. Faraday Soc.*, **66**, 1498 (1970).

121. K. Tödheide, "Water at High Temperatures and Pressures," in F. Franks, Ed., *Water—A Comprehensive Treatise*, Vol. 1, Plenum, New York, 1972, Chap. 13.

122. S. I. Smedley, *The Interpretation of Ionic Conductivity in Liquids*, Plenum, New York, 1980.

123. A. E. Stearn and H. Eyring, *J. Phys. Chem.*, **44**, 955 (see esp. p. 976) (1940).

124. S. B. Brummer and G. J. Hills, *Trans. Faraday Soc.*, **57**, 1816 (1961).

125. F. Barreira and G. J. Hills, *Trans. Faraday Soc.*, **64**, 1359 (1968).

126. S. D. Hamann, "Electrolyte Solutions at High Pressure," in B. E. Conway and J. O'M. Bockris, Eds., *Modern Aspects of Electrochemistry*, No. 9, Plenum, New York, 1974, Chap 2.

127. S. D. Hamann, *Phys. Chem. Earth*, **13–14**, 89 (1981).

128. H. S. Harned and B. B. Owen, *The Physical Chemistry of Electrolytic Solutions*, 3rd ed., Reinhold, New York, 1958, Chaps. 4, 7, 11.

129. F. E. Bailey, Jr. and A. Patterson, Jr., *J. Am. Chem. Soc.*, **74**, 4426, 4428 (1952).
130. D. K. McIlroy and D. P. Mason, *J. Chem. Soc. Faraday Trans. 2*, **72**, 590, 2195 (1976).
131. M. J. Pikal and D. G. Miller, *J. Phys. Chem.*, **74**, 1337 (1970); D. G. Miller and M. J. Pikal, *J. Solution Chem.*, **1**, 111 (1972).
132. A. J. McQuillan, *J. Chem. Soc. Faraday Trans. 1*, **70**, 1558 (1974).
133. A. Agnew and R. Paterson, *J. Chem. Soc. Faraday Trans. 1*, **74**, 2896 (1978).
134. J. T. Stock, *Anal. Chem.*, **56**, 561A (1984).
135. J. E. Girard and J. A. Glatz, *Int. Lab.*, 62 (1981); F. C. Smith and R. C. Chang, *The Practice of Ion Chromatography*, Wiley, New York, 1983.
136. J. W. Loveland, "Conductometry and Oscillometry," in I. M. Kolthoff, P. J. Elving, with E. B. Sandell, Eds., *Treatise on Analytical Chemistry*, Part I, *Theory and Practice*, Vol. 4, Interscience, New York, 1963, Chap. 51.
137. D. G. Davis, "Conductometric Titrations," and T. S. Burkhalter, "High Frequency Conductometric (Impedimetric) Titrations," in C. L. Wilson, D. W. Wilson, with C. R. N. Strouts, Eds., *Comprehensive Analytical Chemistry*, Vol. IIA, *Electrical Methods*, Elsevier, Amsterdam, 1964, pp. 174 and 215, respectively.
138. T. Shedlovsky and L. Shedlovsky, "Conductometry," in A. Weissberger and B. W. Rossiter, Eds., *Physical Methods of Chemistry*, Part IIA, *Electrochemical Methods*, Wiley-Interscience, New York, 1971, Chap. III.
139. Landolt-Börnstein, *Zahlenwerte und Funktionen*, Vol. II, Part 7II, 6th ed., Springer-Verlag, Berlin, 1960.
140. R. D. Feltham and R. G. Hayter, *J. Chem. Soc.*, 4587 (1964).
141. G. K. McLane, J. S. Dixon, and J. C. Hindman, "Complex Ions of Plutonium. Transference Measurements," in G. T. Seaborg, J. J. Katz, and W. M. Manning, Eds., *The Transuranium Elements*, Part 1, McGraw-Hill, New York, 1949.
142. A. Reger, E. Peled, and E. Gileadi, *J. Phys. Chem.*, **83**, 869 (1979).
143. C. W. Davies and L. J. Hudleston, *J. Chem. Soc.*, **125**, 260 (1924); C. W. Davies, *The Conductivity of Solutions*, Chapman and Hall, London, 1930, Chap. XVII.
144. K. S. Spiegler, *J. Electrochem. Soc.*, **100**, 303c (1953).
145. T. Okubo and N. Ise, *Macromolecules*, **2**, 407 (1969).
146. M. Liler, *J. Chem. Soc.*, 4272 (1962); 4300 (1965).
147. H. C. Duecker and W. Haller, *J. Phys. Chem.*, **66**, 225 (1962).
148. W. B. Holzapfel, *J. Chem. Phys.*, **50**, 4424 (1969); G. J. Bignold, A. D. Brewer, and B. Hearn, *Trans. Faraday Soc.*, **67**, 2419 (1971).
149. P. Ballinger and F. A. Long, *J. Am. Chem. Soc.*, **81**, 1050 (1959).
150. R. P. Bell and P. T. McTigue, *J. Chem. Soc.*, 2983 (1960).
151. J. W. Moore and R. G. Pearson, *Kinetics and Mechanism*, 3rd ed., Wiley-Interscience, New York, 1981, Chap. 3.
152. E. A. Moelwyn-Hughes, *The Chemical Statics and Kinetics of Solutions*, Academic, London, 1971, Chap. 12.
153. M. Spiro and A. B. Ravnö, *J. Chem. Soc.*, 78 (1965).
154. W. Walisch and J. Barthel, *Z. Phys. Chem.* (*Frankfurt am Main*), **34**, 38 (1962); J. Barthel and G. Bäder, *Z. Phys. Chem.* (*Frankfurt am Main*), **48**, 109 (1966).
155. E. F. Caldin, *Fast Reactions in Solution*, Blackwell, Oxford, 1964.
156. C. F. Bernasconi, *Relaxation Kinetics*, Academic, New York, 1976, Chap. 12.
157. C. W. Davies and A. L. Jones, *Discuss. Faraday Soc.*, **5**, 103 (1949); *Trans. Faraday Soc.*, **51**, 812 (1955).
158. K. Hachiya, M. Ashida, M. Sasaki, H. Kan, T. Inoue, and T. Yasunaga, *J. Phys. Chem.*, **83**, 1866 (1979).

159. J. C. Nichol, E. B. Dismukes, and R. A. Alberty, *J. Am. Chem. Soc.*, **80**, 2610 (1958), and earlier papers.
160. D. H. Moore, "Electrophoresis," in A. Weissberger, Ed., *Physical Methods of Organic Chemistry*, Part IV, Interscience, New York, 1960, Chap. 47.
161. L. G. Longsworth, *Natl. Bur. Stand. Circ.*, No. 524, 59 (1953); H. Haglund, *Sci. Tools*, **17**, 2 (1970).
162. H. D. Freyer and K. Wagener, *Angew. Chem. Int. Ed.*, **6**, 757 (1967); D. Behne, *Kerntechnik*, **12**, 112 (1970).
163. B. Hague and T. R. Foord, *Alternating Current Bridge Methods*, 6th ed., Pitman, Bath, U.K., 1971.
164. J. Braunstein and G. D. Robbins, *J. Chem. Educ.*, **48**, 52 (1971).
165. R. Tamamushi, *Kinetic Parameters of Electrode Reactions of Metallic Compounds*, IUPAC, Butterworth, London, 1975.
166. F. S. Feates, D. J. G. Ives, and J. H. Pryor, *J. Electrochem. Soc.*, **103**, 580 (1956).
167. J. E. B. Randles, *Discuss. Faraday Soc.*, **1**, 11 (1947).
168. D. C. Grahame, *J. Electrochem. Soc.*, **99**, 370C (1952).
169. G. Jones and G. M. Bollinger, *J. Am. Chem. Soc.*, **53**, 411 (1931).
170. B. J. Steel, "A New Form of Transport Number Apparatus and its Application to the Study of Ionic Mobilities," unpublished doctoral dissertation, University of New England, Armidale, Australia, (1960).
171. T. B. Hoover, *J. Phys. Chem.*, **74**, 2667 (1970).
172. R. Wachter and J. Barthel, *Ber. Bunsenges. Phys. Chem.*, **83**, 634 (1979).
173. G. Jones and S. M. Christian, *J. Am. Chem. Soc.*, **57**, 272 (1935).
174. J. C. Nichol and R. M. Fuoss, *J. Phys. Chem.*, **58**, 696 (1954).
175. O. V. Brody and R. M. Fuoss, *J. Phys. Chem.*, **60**, 177 (1956).
176. H. Gerischer, *Z. Elektrochem.*, **58**, 9 (1964).
177. T. B. Hoover, *J. Phys. Chem.*, **68**, 876 (1964).
178. G. Jones and R. C. Josephs, *J. Am. Chem. Soc.*, **50**, 1049 (1928).
179. G. Jones and G. M. Bollinger, *J. Am. Chem. Soc.*, **51**, 2407 (1929).
180. T. Shedlovsky, *J. Am. Chem. Soc.*, **52**, 1793 (1930).
181. P. H. Dike, *Rev. Sci. Instrum.*, **2**, 379 (1931).
182. J. A. Gledhill and A. Patterson, Jr., *J. Phys. Chem.*, **56**, 999 (1952); G. Atkinson and M. Yokoi, *J. Phys. Chem.*, **66**, 1520 (1962).
183. P. G. N. Moseley and A. White, *Electrochim. Acta*, **22**, 871 (1977).
184. G. J. Janz and J. D. E. McIntyre, *J. Electrochem. Soc.*, **108**, 272 (1961).
185. W. Walisch and J. Barthel, *Z. Phys. Chem. (Frankfurt am Main)*, **39**, 235 (1963).
186. H. J. Kiggen and H. Laumen, *Rev. Sci. Instrum.*, **52**, 1761 (1981).
187. D. F. Bond, *Microelectron. Reliab.*, **18**, 53 (1978).
188. R. F. Brown and D. F. Bond, *Inst. Elec. Eng. Conf. Publ.*, **174**, 125 (1979).
189. T. Shedlovsky, *J. Am. Chem. Soc.*, **54**, 1411 (1932).
190. R. H. Stokes, *J. Phys. Chem.*, **65**, 1242 (1961).
191. H. M. Daggett, Jr., E. J. Bair, and C. A. Kraus, *J. Am. Chem. Soc.*, **73**, 799 (1951).
192. H. C. Parker, *J. Am. Chem. Soc.*, **45**, 1366, 2017 (1923).
193. R. H. Stokes, *J. Phys. Chem.*, **65**, 1277 (1961).
194. J. N. Agar, *Trans. Faraday Soc.*, **56**, 776 (1960); *Adv. Electrochem. Electrochem. Eng.*, **3**, 31 (1963).
195. J. E. Lind, Jr., and R. M. Fuoss, *J. Phys. Chem.*, **65**, 999 (1961).
196. P. N. Snowdon and J. C. R. Turner, *Trans. Faraday Soc.*, **56**, 1409 (1960).
197. J. E. Prue, *J. Phys. Chem.*, **67**, 1152 (1963).

198. G. Resch, V. Gutmann, and H. Schauer, *J. Indian Chem. Soc.*, **59**, 130 (1982).

199. J. E. Prue, private communication (1972).

200. K. J. Mysels, *J. Phys. Chem.*, **65**, 1081 (1961).

201. D. G. Oakenfull and D. E. Fenwick, *J. Phys. Chem.*, **78**, 1759 (1974).

202. J. L. Hawes and R. L. Kay, *J. Phys. Chem.*, **69**, 2420 (1965).

203. K. N. Marsh and R. H. Stokes, *Aust. J. Chem.*, **17**, 740 (1964).

204. H. Corti, R. Crovetto, and R. Fernández-Prini, *J. Solution Chem.*, **8**, 897 (1979).

205. D. J. G. Ives and J. H. Pryor, *J. Chem. Soc.*, 2104 (1955).

206. P. Saulnier, *J. Solution Chem.*, **8**, 835 (1979).

207. K. N. Marsh, *J. Solution Chem.*, **9**, 805 (1980).

208. E. Barendrecht and N. G. L. M. Janssen, *Anal. Chem.*, **33**, 199 (1961).

209. F. P. Anderson, H. C. Brookes, M. C. B. Hotz, and A. H. Spong, *J. Sci. Instrum.*, **2**, 499 (1969); H. C. Brookes, M. C. B. Hotz, and A. H. Spong, *J. Chem. Soc. A*, 2410 (1971).

210. R. Tamamushi and K. Takahashi, *J. Electroanal. Chem.*, **50**, 277 (1974).

211. M. Spiro, "Mixture Potentials in Chemistry," in D. V. Fenby and I. D. Watson, Eds., *The Physical Chemistry of Solutions*, Massey University, Palmerston North, New Zealand, 1984, pp. 1–21.

212. A. B. Gancy and S. B. Brummer, *J. Electrochem. Soc.*, **115**, 804 (1968).

213. G. Jones and B. C. Bradshaw, *J. Am. Chem. Soc.*, **55**, 1780 (1933).

214. V. F. Hnizda and C. A. Kraus, *J. Am. Chem. Soc.*, **71**, 1565 (1949).

215. B. G. Oliver and W. A. Adams, *Rev. Sci. Instrum.*, **43**, 830 (1972).

216. C. Cooke, C. McCallum, A. D. Pethybridge, and J. E. Prue, *Electrochim. Acta*, **20**, 591 (1975).

217. R. P. Clark and J. R. Moser, *J. Electrochem. Soc.*, **118**, 666 (1971).

218. T. Erdey-Grúz, L. Majthényi, and E. Kugler, *Acta Chim. Acad. Sci. Hung.*, **37**, 393 (1963).

219. D. A. Lown and Lord Wynne-Jones, *J. Sci. Instrum.*, **44**, 1037 (1967).

220. D. J. G. Ives and G. J. Janz, *Reference Electrodes: Theory and Practice*, Academic, New York, 1961, pp. 60–61, 106–108.

221. T. A. O'Donnell and P. T. McTigue, private communication (1982).

222. G. J. Janz and R. P. T. Tomkins, *J. Electrochem. Soc.*, **124**, 55C (1977).

223. J. E. Lind, Jr., J. J. Zwolenik, and R. M. Fuoss, *J. Am. Chem. Soc.*, **81**, 1557 (1959).

224. J. J. Zwolenik and R. M. Fuoss, *J. Phys. Chem.*, **68**, 903 (1964).

225. J-C. Justice, *J. Chim. Phys. Phys. Chim. Biol.*, **65**, 353 (1968).

226. Y-C. Chiu and R. M. Fuoss, *J. Phys. Chem.*, **72**, 4123 (1968).

227. R. Sänding, R. Feistel, A. Grosch, and J. Einfeldt, quoted in [222].

228. J. Barthel, F. Feuerlein, R. Neueder, and R. Wachter, *J. Solution Chem.*, **9**, 209 (1980).

229. E. Juhasz and K. N. Marsh, *Pure Appl. Chem.*, **53**, 1841 (1981).

230. A. S. Quist, W. L. Marshall, E. U. Franck, and W. von Osten, *J. Phys. Chem.*, **74**, 2241 (1970).

231. A. M. Feltham and M. Spiro, *Chem. Rev.*, **71**, 177 (1971).

232. N. L. Cox, C. A. Kraus, and R. M. Fuoss, *Trans. Faraday Soc.*, **31**, 749 (1935).

233. R. J. Mureinik, A. M. Feltham, and M. Spiro, *J. Chem. Soc. Dalton Trans.*, 1981 (1972).

234. J. M. Stokes and R. H. Stokes, *J. Phys. Chem.*, **62**, 497 (1958).

235. F. Auerbach and H. Zeglin, *Z. Phys. Chem. Stoechiom. Verwandschaftsl.*, **103**, 178 (1923).

236. R. R. Dewald and J. H. Roberts, *J. Phys. Chem.*, **72**, 4224 (1968).

237. E. S. Shanley, E. M. Roth, G. M. Nichols, and M. Kilpatrick, *J. Am. Chem. Soc.*, **78**, 5190 (1956); D. K. Thomas and O. Maass, *Can. J. Chem.*, **36**, 449 (1958).

238. P. Newton, C. S. Copeland, and S. W. Benson, *J. Chem. Phys.*, **37**, 339 (1962).
239. G. Jones and D. M. Bollinger, *J. Am. Chem. Soc.*, **57**, 280 (1935).
240. A. M. Feltham and M. Spiro, *J. Electroanal. Chem.*, **28**, 151 (1970).
241. E. P. Laug, *Ind. Eng. Chem. Anal. Ed.*, **6**, 111 (1934); H. P. Dibbs, *Can. J. Chem.*, **40**, 565 (1962).
242. A. D. Pethybridge and D. J. Spiers, *J. Electroanal. Chem.*, **66**, 231 (1975).
243. A. M. El-Aggan, D. C. Bradley, and W. Wardlaw, *J. Chem. Soc.*, 2092 (1958).
244. R. L. Kay, B. J. Hales, and G. P. Cunningham, *J. Phys. Chem.*, **71**, 3925 (1967).
245. C. G. Swain and D. F. Evans, *J. Am. Chem. Soc.*, **88**, 383 (1966).
246. J. Barthel and G. Schwitzgebel, *Z. Phys. Chem.* (*Frankfurt am Main*), **54**, 173 (1967); R. Wachter and J. Barthel, *Electrochim. Acta*, **16**, 713 (1971).
247. E. K. Mysels, P. C. Scholten, and K. J. Mysels, *J. Phys. Chem.*, **74**, 1147 (1970).
248. C. W. Davies, *The Conductivity of Solutions*, Chapman and Hall, London, 1930, Chap. IV.
249. G. E. Cassford and A. D. Pethybridge, *J. Electroanal. Chem.*, **147**, 289 (1983).
250. A. K. R. Unni, L. Elias, and H. I. Schiff, *J. Phys. Chem.*, **67**, 1216 (1963).
251. R. De Lisi, M. Goffredi, and V. T. Liveri, *J. Chem. Soc. Faraday Trans. 1*, **75**, 1667 (1979), and earlier papers.
252. K. Cruse, *Z. Chem.*, **5**, 1 (1965).
253. S. R. Gupta and G. J. Hills, *J. Sci. Instrum.*, **33**, 313 (1956); S. R. Gupta, "Some Thermodynamic Properties of Electrolytes," unpublished dissertation, Imperial College, London (1957).
254. R. Calvert, J. A. Cornelius, V. S. Griffiths, and D. I. Stock, *J. Phys. Chem.*, **62**, 47 (1958).
255. V. S. Griffiths, *Anal. Chim. Acta*, **18**, 174 (1958); *Talanta*, **2**, 230 (1959).
256. R. A. Williams, E. M. Gold, and S. Naiditch, *Rev. Sci. Instrum.*, **36**, 1121 (1965).
257. R. A. O. Gross and P. B. Sawyer, *Aust. Process Eng.*, **4**, 29 (1976).
258. I. Fatt, *Rev. Sci. Instrum.*, **33**, 493 (1962).
259. H. E. Gunning and A. R. Gordon, *J. Chem. Phys.*, **10**, 126 (1942).
260. L. Elias and H. I. Schiff, *J. Phys. Chem.*, **60**, 595 (1956).
261. G. C. Benson and A. R. Gordon, *J. Chem. Phys.*, **13**, 470 (1945).
262. M. Spiro, "Transference Numbers," in A. Weissberger and B. W. Rossiter, Eds., *Physical Methods of Chemistry*, Part IIA, *Electrochemical Methods*, Wiley-Interscience, New York, 1971, Chap. IV.
263. M. Perie, J. Perie, and M. Chemla, *Electrochim. Acta*, **19**, 753 (1974).
264. G. J. Janz and R. D. Reeves, "Molten Salt Electrolytes—Transport Properties," in C. W. Tobias, Ed., *Advances in Electrochemistry and Electrochemical Engineering*, Vol. 5, Interscience, New York, 1967.
265. G. Jones and M. Dole, *J. Am. Chem. Soc.*, **51**, 1073 (1929).
266. B. J. Steel and R. H. Stokes, *J. Phys. Chem.*, **62**, 450 (1958).
267. R. L. Kay, G. A. Vidulich, and A. Fratiello, *Chem. Instrum.*, **1**, 361 (1969).
268. E. W. Washburn, *J. Am. Chem. Soc.*, **31**, 322 (1909).
269. A. L. Levy, *J. Chem. Educ.*, **29**, 384 (1952).
270. D. A. MacInnes and M. Dole, *J. Am. Chem. Soc.*, **53**, 1357 (1931).
271. F. King and M. Spiro, *J. Solution Chem.*, **10**, 881 (1981).
272. M. Mooney, "Minimizing Convection Currents in Electrophoresis Measurements," in *Temperature, Its Measurement and Control in Science and Industry*, Reinhold, New York, 1941, p. 428.
273. L. Fischer and K. Hessler, *Ber. Bunsenges. Phys. Chem.*, **68**, 184 (1964).

274. E. A. Kaimakov and V. B. Fiks, *Russ. J. Phys. Chem.*, **35**, 873 (1961).
275. J. R. Gwyther, S. Kumarasinghe, and M. Spiro, *J. Solution Chem.*, **3**, 659 (1974).
276. L. P. Hammett and F. A. Lowenheim, *J. Am. Chem. Soc.*, **56**, 2620 (1934).
277. A. N. Campbell and K. P. Singh, *Can. J. Chem.*, **37**, 1959 (1959).
278. M. Dole, *Principles of Experimental and Theoretical Chemistry*, McGraw-Hill, New York, 1935, p. 134.
279. M. E. Laing, *J. Phys. Chem.*, **28**, 673 (1924).
280. F. T. Wall, G. S. Stent, and J. J. Ondrejcin, *J. Phys. Colloid Chem.*, **54**, 979 (1950).
281. R. H. Stokes, S. Phang, and R. Mills, *J. Solution Chem.*, **8**, 489 (1979).
282. H. W. Hoyer, K. J. Mysels, and D. Stigter, *J. Phys. Chem.*, **58**, 385 (1954).
283. J. W. McBain, *Proc. Wash. Acad. Sci.*, **9**, 1 (1907).
284. M. Spencer, *Electrophoresis*, **4**, 36 (1983).
285. M. Perie, J. Perie, and M. Chemla, *J. Chim. Phys.*, **72**, 148 (1975).
286. M. Spiro, "Transference Numbers," in A. Weissberger, Ed., *Physical Methods of Organic Chemistry*, Part IV, Interscience, New York, 1960, Chap. 46.
287. G. W. C. Milner and G. Phillips, *Coulometry in Analytical Chemistry*, Pergamon, Oxford, 1967, Chaps. 3, 4.
288. J. J. Lingane, *Anal. Chim. Acta*, **44**, 199 (1969).
289. J. C. Quayle and F. A. Cooper, *Analyst*, **91**, 355 (1966).
290. J. A. Pike and G. C. Goode, *Anal. Chim. Acta*, **39**, 1 (1967).
291. G. B. Clayton, *Operational Amplifiers*, 2nd ed., Newnes-Butterworth, London, 1979.
292. D. A. MacInnes and L. G. Longsworth, *Chem. Rev.*, **11**, 171 (1932).
293. M. Spiro, *Trans. Faraday Soc.*, **61**, 350 (1965).
294. V. P. Dole, *J. Am. Chem. Soc.*, **67**, 1119 (1945); H. Svensson, *Ark. Kemi Mineral. Geol.*, **22A** (10) (1946); E. B. Dismukes and E. L. King, *J. Am. Chem. Soc.*, **74**, 4798 (1952); J. C. Nichol and L. J. Gosting, *J. Am. Chem. Soc.*, **80**, 2601 (1958).
295. J. R. Gwyther, M. Spiro, R. L. Kay, and G. Marx, *J. Chem. Soc. Faraday Trans. 1*, **72**, 1419 (1976).
296. J. R. Gwyther and M. Spiro, *J. Chem. Soc. Faraday Trans. 1*, **72**, 1410 (1976).
297. R. J. Bearman and L. A. Woolf, *J. Phys. Chem.*, **73**, 4403 (1969).
298. F. J. Millero, *Chem. Rev.*, **71**, 147 (1971).
299. G. Baca and R. D. Hill, *J. Chem. Educ.*, **47**, 235 (1970).
300. J. L. Dye, R. F. Sankuer, and G. E. Smith, *J. Am. Chem. Soc.*, **82**, 4797 (1960).
301. L. G. Longsworth, *J. Chem. Educ.*, **11**, 420 (1934).
302. D. J. Le Roy, unpublished doctoral dissertation, University of Toronto, Toronto, Canada (1939).
303. S. Claesson, S. Malmrud, and B. Lundgren, *Trans. Faraday Soc.*, **66**, 3048 (1970).
304. J. A. Davies, R. L. Kay, and A. R. Gordon, *J. Chem. Phys.*, **19**, 749 (1951).
305. K. N. Marsh, M. Spiro, and M. Selvaratnam, *J. Phys. Chem.*, **67**, 699 (1963).
306. L. G. Longsworth and D. A. MacInnes, *J. Am. Chem. Soc.*, **62**, 705 (1940).
307. L. G. Longsworth, *J. Am. Chem. Soc.*, **66**, 449 (1944).
308. L. G. Longsworth, *Ind. Eng. Chem. Anal. Ed.*, **18**, 219 (1946).
309. R. L. Kay, "Transference Number Measurements," in E. Yeager and A. J. Salkind, Eds., *Techniques of Electrochemistry*, Vol. 2, Wiley-Interscience, New York, 1973, Chap. II.
310. J. W. Lorimer, J. R. Graham, and A. R. Gordon, *J. Am. Chem. Soc.*, **79**, 2347 (1957).
311. J. Barthel, U. Ströder, L. Iberl, and H. Hammer, *Ber. Bunsenges. Phys. Chem.*, **86**, 636 (1982).

312. H. Hammer, "Überführungsmessungen an Me_4NClO_4 und Einzelionenbeweglich-keiten in Acetonitril im Temperaturbereich $-35°C$ bis $25°C$," Dissertation, Universität des Saarlandes, Saarbrücken (1975).

313. F. T. Wall and S. J. Gill, *J. Phys. Chem.*, **59**, 278 (1955); F. T. Wall and J. Berkowitz, *J. Phys. Chem.*, **62**, 87 (1958).

314. E. Passeron and E. Gonzalez, *J. Electroanal. Chem.*, **14**, 393 (1967).

315. G. A. Vidulich, G. P. Cunningham, and R. L. Kay, *J. Solution Chem.*, **2**, 23 (1973).

316. A. Fratiello and R. L. Kay, *J. Solution Chem.*, **3**, 857 (1974).

317. G. A. Vidulich, F. X. Gleason, J. F. Lynch, W. C. Mattern, and R. McCabe, *J. Solution Chem.*, **1**, 263 (1972).

318. K. S. Pribadi, *J. Solution Chem.*, **1**, 455 (1972).

319. G. A. Vidulich and R. L. Kay, *Rev. Sci. Instrum.*, **37**, 1662 (1966).

320. K. Lee and R. L. Kay, *Aust. J. Chem.*, **33**, 1895 (1980).

321. J. E. Smith, Jr. and E. B. Dismukes, *J. Phys. Chem.*, **67**, 1160 (1963); **68**, 1603 (1964).

322. R. L. Kay, K. S. Pribadi, and B. Watson, *J. Phys. Chem.*, **74**, 2724 (1970).

323. G. Marx, L. Fischer, and W. Schulze, *Radiochim. Acta*, **2**, 9 (1963).

324. J. Vehlow and G. Marx, *Naturwissenschaften*, **58**, 320 (1971).

325. J. Vehlow, Dissertation, Freie Universität Berlin (1972).

326. W. Schulze, M. Hornig, and G. Marx, *Z. Phys. Chem. (Frankfurt am Main)*, **53**, 106 (1967).

327. G. Marx and D. Hentschel, *Talanta*, **16**, 1159 (1969).

328. G. Marx, W. Riedel, and J. Vehlow, *Ber. Bunsenges. Phys. Chem.*, **73**, 74 (1969).

329. G. Marx, L. Fischer, and W. Schulze, *Z. Phys. Chem. (Frankfurt am Main)*, **41**, 315 (1964).

330. C. M. Lederer and V. S. Shirley, Eds., *Table of Isotopes*, 7th ed., Wiley-Interscience, New York, 1978.

331. G. Marx and W. Schulze, *Kerntechnik*, **7**, 13 (1965).

332. G. Marx and M. Mirza, *Z. Naturforsch.*, **32a**, 185 (1977).

333. G. Marx, private communication (1983).

334. G. Marx and H. Bischoff, *J. Radioanal. Chem.*, **30**, 567 (1976).

335. M. Holz and C. Müller, *J. Magn. Reson.*, **40**, 595 (1980).

336. M. Holz and J. Radwan, *Z. Phys. Chem. (Frankfurt am Main)*, **125**, 49 (1981).

337. M. Holz and C. Müller, *Ber. Bunsenges. Phys. Chem.*, **86**, 141 (1982).

338. E. A. Kaimakov and V. I. Sharkov, *Russ. J. Phys. Chem.*, **38**, 893 (1964).

339. W. L. Miller, *Z. Phys. Chem. Stoechiom. Verwandschaftsl.*, **69**, 436 (1909).

340. G. S. Hartley and J. L. Moilliet, *Proc. R. Soc. London*, **A140**, 141 (1933).

341. D. J. Le Roy and A. R. Gordon, *J. Chem. Phys.*, **6**, 398 (1938).

342. A. R. Gordon and R. L. Kay, *J. Chem. Phys.*, **21**, 131 (1953).

343. M. A. Esteso, C-Y. Chan, and M. Spiro, *J. Chem. Soc. Faraday Trans. 1*, **72**, 1425 (1976).

344. G. S. Hartley and G. W. Donaldson, *Trans. Faraday Soc.*, **33**, 457 (1937).

345. J. N. Sahay, *J. Sci. Ind. Res.*, **18B**, 235 (1959).

346. H. P. Cady and L. G. Longsworth, *J. Am. Chem. Soc.*, **51**, 1656 (1929); L. W. Shemilt, J. A. Davies, and A. R. Gordon, *J. Chem. Phys.*, **16**, 340 (1948).

347. A. G. Keenan and A. R. Gordon, *J. Chem. Phys.*, **11**, 172 (1943).

348. T. L. Broadwater and R. L. Kay, *J. Phys. Chem.*, **74**, 3802 (1970).

349. R. H. Davies, *Nature*, **161**, 1021 (1948).

350. A. M. Sukhotin, *Russ. J. Phys. Chem.*, **34**, 29 (1960).

351. H. W. Hoyer, K. J. Mysels, and D. Stigter, *J. Phys. Chem.*, **58**, 385 (1954).

352. C. A. Coulson, J. T. Cox, A. G. Ogston, and J. St. L. Philpot, *Proc. R. Soc. London*, **A192**, 382 (1948); M. B. Smith and S. J. Roe, *Anal. Biochem.*, **17**, 236 (1966).

353. R. W. Allgood, D. J. Le Roy and A. R. Gordon, *J. Chem. Phys.*, **8**, 418 (1940).

354. R. L. Kay, unpublished doctoral dissertation, University of Toronto, Toronto, Canada (1952).

355. A. K. Covington and J. E. Prue, *J. Chem. Soc.*, 1567, 1930 (1957).

356. L. G. Longsworth and D. A. MacInnes, *J. Am. Chem. Soc.*, **59**, 1666 (1937).

357. M. D. Neville and M. Spiro, unpublished work.

358. J. C. Nichol, *J. Phys. Chem.*, **66**, 830 (1962).

359. R. L. Kay, G. A. Vidulich, and A. Fratiello, *Chem. Instrum.*, **1**, 361 (1969).

360. J. R. Gwyther, *J. Phys. E*, **5**, 979 (1972); K. R. Renton and M. Spiro, to be published.

361. H. Lal, *Nature*, **171**, 175 (1953).

362. L. G. Longsworth, *J. Am. Chem. Soc.*, **65**, 1755 (1943).

363. L. J. M. Smits and E. M. Duyvis, *J. Phys. Chem.*, **70**, 2747 (1966); **71**, 1168 (1967).

364. K. Indaratna and A. J. McQuillan, to be published.

365. G. S. Hartley, B. Collie, and C. S. Samis, *Trans. Faraday Soc.*, **32**, 795 (1936).

366. J. Tamás, O. Kaposi, and P. Scheiber, *Acta Chim. Acad. Sci. Hung.*, **48**, 309 (1966).

367. D. R. Muir, J. R. Graham, and A. R. Gordon, *J. Am. Chem. Soc.*, **76**, 2157 (1954).

368. L. G. Longsworth, *J. Am. Chem. Soc.*, **67**, 1109 (1945).

369. P. Mukerjee, *J. Phys. Chem.*, **62**, 1397 (1958).

370. D. J. Currie and A. R. Gordon, *J. Phys. Chem.*, **64**, 1751 (1960). The results have been recalculated in [363].

371. G. S. Hartley, E. Drew, and B. Collie, *Trans. Faraday Soc.*, **30**, 648 (1934); G. S. Hartley, B. Collie, and C. S. Samis, *Trans. Faraday Soc.*, **32**, 795 (1936); C. S. Samis and G. S. Hartley, *Trans. Faraday Soc.*, **34**, 1288 (1938).

372. B. P. Konstantinov, E. A. Kaimakov, and N. L. Varshavskaya, *Russ. J. Phys. Chem.*, **36**, 535, 540 (1962).

373. J. P. Rupert and R. L. Kay, unpublished work, quoted in [309].

374. R. A. Alberty, *J. Chem. Educ.*, **25**, 619 (1948).

375. D. J. G. Ives and G. J. Janz, *Reference Electrodes: Theory and Practice*, Academic, New York, 1961.

376. D. G. Miller, *Chem. Rev.*, **60**, 15 (1960).

377. E. A. Guggenheim, *Thermodynamics*, 4th ed., North-Holland, Amsterdam, 1959, Chap. 9.

378. M. Salomon, "Thermodynamic Measurements," in A. K. Covington and T. Dickinson, Eds., *Physical Chemistry of Organic Solvent Systems*, Plenum, London, 1973, Chap. 2.2.

379. M. Spiro, *Trans. Faraday Soc.*, **55**, 1207 (1959).

380. R. H. Stokes, *J. Am. Chem. Soc.*, **77**, 3219 (1955).

381. W. J. Hamer, *J. Am. Chem. Soc.*, **57**, 662 (1935).

382. H. I. Stonehill, *J. Chem. Soc.*, 647 (1943).

383. R. H. Stokes and B. J. Levien, *J. Am. Chem. Soc.*, **68**, 1852 (1946).

384. C. Lanczos, *Applied Analysis*, Pitman, London, 1957, p. 323; J. C. Amphlett and E. Whittle, *Trans. Faraday Soc.*, **64**, 2130 (1968).

385. G. Rutledge, *Phys. Rev.*, **40**, 262 (1932); H. Margenau and G. M. Murphy, *The Mathematics of Physics and Chemistry*, 2nd ed., Van Nostrand, Princeton, NJ, 1956, p. 473.

386. E. P. Purser and R. H. Stokes, *J. Am. Chem. Soc.*, **73**, 5650 (1951).

387. H. S. Harned and E. C. Dreby, *J. Am. Chem. Soc.*, **61**, 3113 (1939).

388. A. S. Brown and D. A. MacInnes, *J. Am. Chem. Soc.*, **57**, 1356 (1935).

389. J. N. Butler, "Reference Electrodes in Aprotic Organic Solvents," in P. Delahay, Ed., *Advances in Electrochemistry and Electrochemical Engineering*, Vol. 7, Interscience, New York, 1970.

390. R. Fernández-Prini and J. E. Prue, *J. Phys. Chem.*, **69**, 2793 (1965).

391. M. Lucas, *Bull. Soc. Chim. France*, 1792 (1969).

392. M. Archer and M. Spiro, *J. Chem. Soc. A*, 68, 82 (1970).

393. R. N. Ellis, R. H. Stokes, A. C. Wright, and M. Spiro, *Aust. J. Chem.*, **36**, 1913 (1983).

394. T. Shedlovsky and D. A. MacInnes, *J. Am. Chem. Soc.*, **58**, 1970 (1936).

395. W. J. Hornibrook, G. J. Janz, and A. R. Gordon, *J. Am. Chem. Soc.*, **64**, 513 (1942).

396. A. K. Covington, "Introduction: Basic Electrode Types, Classification and Selectivity Considerations," in A. K. Covington, Ed., *Ion-Selective Electrode Methodology*, Vol. I, CRC Press, Boca Raton, FL, 1979, Chap. 1.

397. R. J. Simpson, "Practical Techniques for Ion-Selective Electrodes," in A. K. Covington, Ed., *Ion-Selective Electrode Methodology*, Vol. I, CRC Press, Boca Raton, FL, 1979, Chap. 3.

398. A. K. Covington and J. E. Prue, *J. Chem. Soc.*, 3696, 3701 (1955).

399. G. Biedermann and G. Douheret, *Chem. Scr.*, **16**, 138 (1980).

400. D. A. MacInnes, *Chem. Rev.*, **18**, 347 (1936).

401. A. B. Lamb and A. T. Larson, *J. Am. Chem. Soc.*, **42**, 229 (1920); E. J. Roberts and F. Fenwick, *J. Am. Chem. Soc.*, **49**, 2787 (1927); J. V. Lakhani, *J. Chem. Soc.*, 179 (1932); G. N. Ghosh, *J. Indian Chem. Soc.*, **12**, 15 (1935).

402. R. H. Stokes and B. J. Levien, *J. Am. Chem. Soc.*, **68**, 333 (1946).

403. S. Phang, *Aust. J. Chem.*, **32**, 1149 (1979).

404. K. Cammann, *Working with Ion-Selective Electrodes*, Springer-Verlag, Berlin, 1979, Chap. 4.

Chapter **9**

POLAROGRAPHY

Antonín A. Vlček, Jiří Volke, Lubomír Pospíšil, and Robert Kalvoda

1   INTRODUCTION

Polarography was discovered by J. Heyrovský in 1922. The polarographic method, defined by Heyrovský as electrolysis with a polarizable dropping mercury electrode, gained rapid acceptance both as a method of investigating electrode processes and as an important analytical tool. Polarography's widespread use was made possible by the construction of the polarograph by Heyrovský and Shikata in 1925, a device that automatically records current–potential curves. The polarograph was one of the very first automatic apparatus in physical and analytical chemistry.

The theoretical basis was laid by Ilkovič in 1934 (see the Ilkovič equation for diffusion current, below) and Heyrovský and Ilkovič in 1935 (see the equation for the current–potential curve—polarographic wave—and definition of half-wave potential, below).

In electrochemistry, the polarographic method has provided new and unexpected ways for studying the mechanism of electrode processes and has triggered a wide research of reactions taking place at mercury, and later also on other metallic, electrodes. To gain a deeper insight into the nature of the electrode reaction, polarography has been modified and new methods derived. In most cases the basic principle of polarography, however, is used: the measurement of the response of the electrode to its polarization. Currently, a wide variety of methods exists for studying electrode processes and for analytical purposes. The literature may advocate the dominant role of one method over another. However, to solve the mechanism of the electrode reaction in all its details, including adsorption, double-layer effects, and coupled chemical processes, a complex electrochemical approach is necessary because each method yields only specific information and a complete picture is only reached by combining results obtained by various ways of polarization of the electrode. Moreover, electrochemistry is not a self-consistent method for full elucidation of the chemism of the electrode process, and it has to be combined with other physicochemical, mainly spectroscopic, methods. In analytical application selecting the proper electrochemical method depends on the nature of the problem to be solved; consequently, no preference is generally advocated.

The term polarography is sometimes used rather vaguely, either in a very broad sense or in a very narrow, almost historical manner. The use of the term is, interestingly, geographically dependent and reflects the background of the particular electrochemical school.

Similarly, the expression polarograph is not always used for commercial instruments, even if they perform solely in the polarographic mode of polarization. This may be attributed primarily to licensing rights and, occasionally, even to the influence of modish trends.

Polarography has a distinct and very well-defined place among the various electrochemical techniques. It uses potentiostatic control of the working electrode polarization under conditions of convective diffusion, which measures the current response i of the electrode system in dependence on the polarizing voltage E, whose time rate of change, dE/dt, is negligible with respect to the time constant of the mass transport to the electrode surface; that is, the measurable i does not depend upon the time rate of change of the polarizing voltage.

The IUPAC recommendation prefers to limit polarography to the use of liquid electrodes whose surfaces are periodically or continuously renewed (e.g., dropping or streaming mercury electrode). Since other electrodes, such as the rotating disk electrode, may have their contacting solutions continuously renewed and therefore fulfill generally the same conditions met by the dropping mercury electrode, the present chapter is not limited to "liquid electrodes with renewable surfaces." In addition, reference to and use of other electrochemical methods will be made.

Methods directly derived from polarography primarily increase the analytical sensitivity or detect new information about the electrode process. Various pulse polarographic techniques use discontinuous polarization potential and measure the current response in given time intervals only. The ac polarographic techniques are derived from dc polarography, as defined above, by imposing a small perturbation voltage of various frequencies upon the main dc polarizing voltage.

2   THEORETICAL FUNDAMENTALS

The measurement of electrolytic currents as a function of applied potential imposes certain requirements on experimental conditions that are determined by inherent properties of the electrode–solution interface [1–4]. A metallic electrode in the solution of an electrolyte is surrounded by an electrode double layer having the capacity C, which must be charged when the electrode is placed in contact with the solution. The resultant electric current flowing through the electrode–solution interface is termed *charging, capacity,* or *nonfaradaic* current. To ensure the least possible variation of the double-layer structure in the course of the electrolytic process and also to obtain sufficient electric sample conductivity, a suitable strong electrolyte is added. This strong electrolyte, also called an *indifferent, supporting,* or *base* electrolyte, must not interfere with the investigated system.

The definition for polarographic techniques specifies the type of working electrode whose response may be termed "polarographic." In experiments performed under the conditions of convective diffusion the application of any type of stationary electrode is excluded. In practice, the dropping mercury (or amalgam) electrode (DME) and the rotating disk electrode (RDE) are most widely used. An important parameter of an electrode is its geometrical surface area, which in the case of the DME is given as a mean area over the drop life,

$$\bar{A} = 0.51(mt_1)^{2/3} \tag{1}$$

or as the instantaneous value at time t

$$A = 0.85(mt)^{2/3} \tag{2}$$

where m is the mercury flow rate and t_1 the drop time. The macroscopic surface area for the RDE is only an approximate parameter because of the microscopic roughness of the solid metal surface.

The value of the charging current depends on the time change of the electrode charge density Q

$$i_c = \frac{dQ}{dt} \tag{3}$$

Thus, the time change of the electrode area brings about the charging current, which would decay rapidly if the electrode were stationary,

$$i_c = (E - E_z)C\frac{dA}{dt} = \frac{2}{3} \times 0.85(E - E_z)Cm^{2/3}t^{-1/3} \tag{4}$$

where E is the electrode potential and E_z is the zero charge potential. From (4) it follows that the charging current should be infinitely large at the beginning of the drop and should decay with time. When the double-layer capacity sharply changes with potential, as in the case of the electrode adsorption–desorption phenomenon, the charging current may show breaks or spurious small *waves* at those potentials that should not be misinterpreted as faradaic processes.

When the electrode potential in the presence of electroactive species reaches a value characteristic of the nature of the substance being reduced or oxidized, the current starts to increase exponentially with the potential causing the depletion of electroactive species (depolarizer) in the vicinity of the electrode surface. The current increase decelerates at sufficiently large potentials and finally levels off forming a constant plateau called the limiting current. Under conditions of a limiting current the concentration of electroactive species at the electrode surface reaches zero and only the particles supplied by the mass transport toward the electrode surface are subjected to the electrode reaction.

Currents caused by the electron exchange between a metallic phase of an electrode and a solution species, which result in the change of the overall number of electrons of the reacting species, are called faradaic currents. From Faraday's law we can write a relationship for the faradaic current:

$$i = nF \frac{dN}{dt} \tag{5}$$

where n is the number of electrons exchanged by a single particle in the course of the electrode reaction, F equals 96,500 coulombs, and dN/dt is the number of particles reaching the electrode per unit time. Correct derivation of dN/dt based on the mass transfer is a theoretical subject of various polarographic techniques and will be outlined in the following sections.

2.1   Mass-Transfer Contributions

Generally, mass transport toward the electrode has all three components: diffusion, migration, and convection.

Migration current components are undesirable, but are observed when the indifferent electrolyte is absent and the reducible species are charged. At that time the limiting current is controlled not only by the diffusion, but also by the voltage drop between the electrode and the solution through which the electroactive ions migrate. Depending on the mutual direction of the migration and the diffusion, the resulting limiting current is either enhanced or depressed. Migration currents are effectively suppressed by a 50–100 times surplus of indifferent electrolytes over the concentration of the electroactive ions.

An important convective component arises from the growth of the dropping mercury electrode or from a rotation of a metallic disk and will be described later.

Diffusion currents are the electrolytic currents that are determined by the rate of diffusion of electroactive species from the bulk of the solution and are a consequence of the concentration gradient created by the electrode reaction depletion on the electrode surface. Hence the rate of the electrode reaction must be high enough to obtain currents controlled solely by diffusion. Then, using Fick's First Law, (5) is written

$$i = nFAD \left(\frac{\partial c}{\partial x} \right)_{x=0} \tag{6}$$

where D is the diffusion coefficient. The concentration gradient at the electrode surface may be obtained from Fick's Second Law. A solution based on linear diffusion, that is, diffusion in one direction only, is not fully satisfactory because the motion of the electrode surface for the dropping mercury electrode, together with the concentration gradient, must be considered. Fick's Second Law is then written

$$\frac{\partial c}{\partial t} = D \frac{\partial^2 c}{\partial x^2} + \frac{2}{3} \frac{x}{t} \frac{\partial c}{\partial x} \tag{7}$$

with the initial and boundary conditions

$$c = c^* \quad \text{for } t = 0$$
$$c = c_0 \quad \text{for } t > 0, x = 0 \tag{8}$$

where c^* is the concentration in the bulk and c_0 is the constant concentration at the electrode surface and is determined exclusively by the applied potential. For the limiting diffusion current, $c_0 = 0$. The solution of (7) and (8) gives the concentration c as a function of time and distance from the electrode and hence the concentration gradient needed for the electrolytic current calculation is obtained. Finally,

$$i = nFAD \frac{c - c_0}{\sqrt{\frac{3}{7}\pi Dt}} \tag{9}$$

By substituting (1) for surface area and $c_0 = 0$ for the limiting diffusion current, the Ilkovič equation is obtained [5, 6]:

$$i_d = 0.732nFcD^{1/2}m^{2/3}t^{1/6} \tag{10}$$

or

$$i_d = 706ncD^{1/2}m^{2/3}t^{1/6} \tag{11}$$

if the Faraday constant is included in the numerical value. By using the mean surface area, (2) during drop life, we derive the mean limiting diffusion current

$$\bar{i}_d = 0.627nFcD^{1/2}m^{2/3}t_1^{1/6} \tag{12}$$

The Ilkovič equation was further refined by considering the curvature [7, 8] of the electrode and the screening influence [9, 10] of the capillary tip. This results in a correction factor $(1 \pm 3.97D^{1/2}t^{1/6}m^{-1/3})$ by which i_d is multiplied—the plus or minus sign depends on whether the original species are present in the solution or in the form of an amalgam, respectively.

A similar relationship for the diffusion-controlled current can be derived for the rotating disk electrode. Since the description of convective diffusion toward a rotating disk must properly consider the hydrodynamics of the solution, we will deal with this subject separately. All relationships for the DME given in the following paragraph are easily reformulated for the rotating disk electrode.

2.2   Current–Potential Relationship for a Reversible Electrode Reaction

Equations (10)–(12) describe the potential-independent part of a current–potential curve (the polarographic curve) under the condition that the surface concentration c_0 of reducible or oxidizable species is zero. Along the rising part of a reversible polarographic wave the Nernst equation relating the concentration of reduced and oxidized forms [Red]/[Ox] and the electrode potential is valid:

$$E = E^0 + \frac{RT}{nF} \ln\left(\frac{[\text{Red}]_0}{[\text{Ox}]_0}\right) \tag{13}$$

where E^0 is the standard redox potential, n is the number of exchanged electrons, and the subscript 0 denotes the concentrations at the electrode surface. Cases in which the Nernst equation is not applicable because of slow electrode reaction or a chemical reaction influencing either $[\text{Red}]_0$ or $[\text{Ox}]_0$ will be discussed later.

Three different cases must be distinguished depending on which is present in the bulk of a solution: (a) only the oxidized form which is reduced, (b) only the reduced form which is oxidized (may be present also in a form of amalgam), (c) both the oxidized and the reduced forms. Resulting i–E curves or polarographic waves are then denoted as cathodic, anodic, or cathodic–anodic waves, respectively. At potentials where the surface concentration is nonzero, the product formed by the faradaic process must diffuse away from the electrode because of the resultant concentration gradient. The net current flowing across the electrode interface is given by (12) and is expressed using either the concentration of oxidized species

$$i = \kappa([\text{Ox}] - [\text{Ox}]_0) \tag{14}$$

or the concentration of the reduced form

$$i = \kappa'([\text{Red}]_0 - [\text{Red}]) \tag{15}$$

Here the concentration-independent parameters from (12) are denoted as

$$\kappa = 0.627 nF D_{\text{Ox}}^{1/2} m^{2/3} t^{1/6}$$
$$\kappa' = 0.627 nF D_{\text{Red}}^{1/2} m^{2/3} t^{1/6} \tag{16}$$

and are sometimes called the Ilkovič constants [11]. Simple arithmetic operations on (13)–(15) will yield the polarographic wave [12]:

$$E = E^0 \mp \frac{RT}{nF} \ln\left(\frac{\bar{i}}{\bar{i}_d - \bar{i}} \sqrt{\frac{D^{\text{Ox}}}{D_{\text{Red}}}}\right) \tag{17}$$

where minus and plus signs denote cathodic and anodic waves, respectively, and $\bar{i}_d$ is the limiting diffusion current. Similarly, for a cathodic–anodic wave, we have

$$E = E^0 - \frac{RT}{nF} \ln\left(\frac{\bar{i} - \bar{i}_{da}}{\bar{i}_{dc} - \bar{i}} \sqrt{\frac{D_{\text{Ox}}}{D_{\text{Red}}}}\right) \tag{18}$$

where $\bar{i}_{da}$ and $\bar{i}_{dc}$ are anodic and cathodic limiting currents, respectively. From (17) an important parameter deduction follows: the potential at which the current is $\bar{i} = \bar{i}_d/2$. It is denoted as the half-wave potential $E_{1/2}$ and, assuming $D_{\text{Ox}} = D_{\text{Red}}$ for a reversible system, is equal to the standard redox potential E^0.

The influence of activity coefficients, not considered in the derivation given above, is usually negligible for all but the most accurate measurements.

2.3   Properties of Diffusion-Controlled Currents

From the Ilkovič equation given by (10)–(12) and from equations describing the *i–E* curves, (17) and (18), some important features can be summarized, which are useful either for the diagnostics of this type of current or for analytical applications:

1.   Current at a given constant potential is linearly proportional to the bulk concentration of the redox form undergoing the electrode reaction.

2.   Current is proportional to the square root of the diffusion coefficient, which may be evaluated if the concentration is precisely known (or by using the limiting current of a reference substance, the diffusion coefficient of which is known).

3.   Mean limiting diffusion current is proportional to the square root of the height of the mercury column, $h^{1/2}$, with zero intercept, or in the case of a pressurized electrode assembly, to the hydrostatic pressure under which the drop of mercury is formed. This property follows from the flow rate $m = \text{const} \times h$, whereas the natural drop time $t_1 = \text{const}/h$ and is bound to the mean currents and does not apply to the currents measured by means of sampling at a certain time of the drop life or to the mechanically controlled drop time (tast polarography).

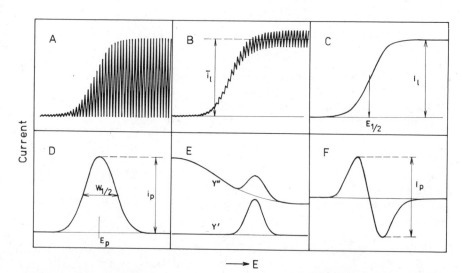

Figure 9.1   The potential-dependent response of various polarographic techniques: A—instantaneous dc current; B—mean dc current (damped); C—sampled dc current, same response for normal pulse polarography or the rotating disk electrode; D—first derivative response (the differential pulse polarography); E—phase-sensitive ac polarography; F—second derivative response (e.g., phase-sensitive second harmonic ac polarography).

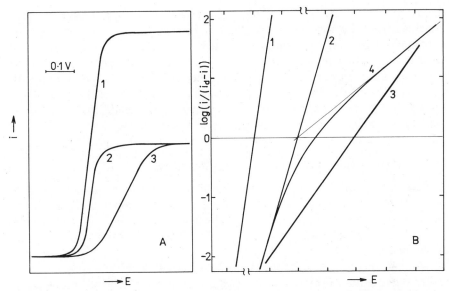

Figure 9.2 A—dc polarographic waves for (1) a two-electron reversible reduction, (2) a one-electron reversible reduction, and (3) a one-electron irreversible reduction. B—log-plot analysis of curves 1–3 from Figure 9.2A. Curve 4—quasireversible reduction showing estimation of reversible $E_{1/2}$ by means of limiting straight lines intersection. Curve 3 is an irreversible wave.

4. The shape of the i–E curve (17) and (18) determines the dependence of $\ln[i/(i_d - i)]$ versus E or by analogy $\ln[(i - i_{da})/(i_{dc} - i)]$ versus E, which should be linear with the slope $\mp RT/nF$, the so-called log-plot analysis of a polarographic wave. This plot is often used for testing the reversibility or estimation of the number of electrons n involved in a given electrode reaction. It fails in the case of ill-defined limiting currents or with overlapping waves.

5. The half-wave potential $E_{1/2}$(E for $i = 0.5i_d$) is independent of the concentration and for $D_{Ox} = D_{Red}$ equals the standard redox potential E^0 of a corresponding redox couple.

Figures 9.1A–9.1C show a typical example of a polarographic response for an instantaneous, mean, and sampled current, respectively. The one- and two-electron reduction waves are shown in Figure 9.2A and their logarithmic plots in Figure 9.2B as Curves 1 and 2.

2.4   Special Diffusion-Controlled Redox Processes

This section discusses a faradaic current possessing all properties described in the previous section, as well as additional characteristic properties. This behavior occurs with the reduction of complexes to metals and is influenced by the complexing agent concentration, the electrode processes of organic compounds involving protonization, the processes involving the mercury ions, and the reversible reduction of quinones. The predominant importance of these systems is that the polarographic data may be used for evaluation of thermodynamic equilibrium constants and related parameters. Theoretical relationships in this section are derived from the Nernst equation by incorporating the chemical equilibria between the species in the solution and those that directly participate in the heterogeneous charge-transfer process at the electrode. Then the equations for the current–potential curves also include thermodynamic equilibrium constants and concentrations of electroinactive ions, such as H^+ ions or ligand concentrations, which participate in those equilibria.

2.5   Reversible Reduction of Complex Ions

Besides the condition of electrochemical reversibility the equations given below also assume (a) the metal being formed yields an amalgam, (b) equilibrium between the metal cation M^{n+} and the ligand X is sufficiently mobile, and (c) the ligand concentration is so high it is considered uniform up to the electrode surface.

When a complex of a formula MX_p is reversibly reduced, the corresponding half-wave potential of this complex $(E_{1/2})_{compl}$ is more negative than the $(E_{1/2})_{free}$ of the free M^{n+} ion because of the additional energy required for the complex decomposition. The half-wave potential shift depends on the ligand concentration according to [13–15]

$$E_{1/2} = (E_{1/2})_{compl} - (E_{1/2})_{free} = \frac{RT}{nF} \ln\left(\frac{D_M}{D_{MX}}\right) - \ln\beta - \ln[X]^p \qquad (19)$$

where $\beta = [MX_p]/[M][X]^p$ is the stability constant of a complex, [X] is the ligand concentration, and D_M and D_{MX} are the diffusion coefficients of the free and complexed species, respectively. Equation (19) determines both β and the number of ligands p in a given complex. When the $E_{1/2}$ versus $\log[X]$ plot is not exactly a straight line, correction to zero ionic strength is necessary [16] because (a) the requirement of the surplus of [X] is unfulfilled or (b) a consecutive complex forms, which is discussed below. The surplus needed depends on n, p, and in most cases is between ten- and fiftyfold.

If a metal ion forms more than one complex with the same ligand, consecutive complex formation proceeds by the following equilibria:

$$M^{n+} \rightleftharpoons MX^{n+} \rightleftharpoons MX_2^{n+} \cdots \rightleftharpoons MX_j^{n+} \rightleftharpoons \cdots \qquad (20)$$

According to Bjerrum [17] the stability constant of a complex containing j ligands is defined as

$$\beta_j = k_1 k_2 \cdots k_j = \frac{[MX_j^{n+}]}{[M^{n+}][X]^j} \qquad (21)$$

where k_1 to k_j are the consecutive stability constants

$$k_i = \frac{[MX_i^{n+}]}{[MX_{i-1}^{n+}][X]} \qquad i = 1, 2, \ldots, j \qquad (22)$$

The half-wave potential shift of a complex reduction is then

$$E_{1/2} = -\frac{RT}{nF} \ln \sum_{i=1}^{j} \beta_i[X]^i \tag{23}$$

De Ford and Hume [18] established a procedure that determines consecutive stability constants by defining the set of recurrent functions based on (23):

$$F_0(X) = \sum_{i=1}^{j} \beta_i[X]^i = \exp\left(-\frac{nF}{RT}\Delta E_{1/2}\right)$$

$$F_1(X) = \frac{F_0(X) - 1}{[X]} \tag{24}$$

$$F_i(X) = \frac{F_{j-1}(X) - \beta_{j-1}}{[X]}$$

The procedure usually detects the limit

$$\lim_{X \to 0} F_i(X) = \beta_i \qquad i = 0, 1, \ldots, j \tag{25}$$

by graphic extrapolation. Starting with the $F_0(X)$ versus $[X]$ plot, evaluation of β_0 enables calculation of $F_1(X)$, and so on. The plot of $F_{j-1}(X)$ versus $[X]$ is linear and $F_j(X)$ is independent of $[X]$, thereby indicating the highest number of ligands.

The reversible electrode reaction of a complex changing both valency and the number of ligands at the same time, provided the ligand is in a surplus, is schematically written

$$MX_p^{n+} + (n-m)e \Leftrightarrow MX_q^{m+} + (p-q)X \tag{26}$$

If β_{Ox} and β_{Red} are the stability constants of the oxidized and reduced forms, respectively, then for equal diffusion coefficients

$$E_{1/2} = -\frac{RT}{(n-m)F} \ln\left(\frac{\beta_{Ox}}{\beta_{Red}}\right) - (p-q)\frac{RT}{(n-m)F} \ln[X] \tag{27}$$

Consequently, the half-wave potential shift yields only the stability constants ratio and the difference in the number of ligands. Under certain conditions $(E_{1/2})_{compl}$ may be independent of $[X]$.

2.6   Reversible Reactions With Proton Participation

Many reversible reactions of organic molecules occur together with protonization equilibria that are usually fast enough to avoid influencing the reaction's overall reversibility. Generally, both the oxidized and reduced forms may bind several protons as well as all intermediate species if more than one electron is transferred. A complicated reaction scheme is usually termed *schema carré* according to the original work of Jacq [19]:

$$
\begin{array}{ccc}
\text{A} \overset{e}{\Longleftrightarrow} \text{A}^- & \overset{e}{\Longleftrightarrow} \text{A}^{2-} & \\
\text{H}^+\!\!\downarrow\!\!\uparrow \qquad \downarrow\!\!\uparrow\text{H}^+ & \downarrow\!\!\uparrow\text{H}^+ \cdots (K_1)_i \\
\text{AH}^+ \overset{e}{\Longleftrightarrow} \text{AH}^{\cdot} & \overset{e}{\Longleftrightarrow} \text{AH}^- & \\
\text{H}^+\!\!\downarrow\!\!\uparrow \qquad \downarrow\!\!\uparrow\text{H}^+ & \downarrow\!\!\uparrow\text{H}^+ \cdots (K_2)_i \\
\text{AH}_2^{2+} \overset{e}{\Longleftrightarrow} \text{AH}_2^{+} & \overset{e}{\Longleftrightarrow} \text{AH}_2 & \\
(E_1^0)_i & (E_2^0)_i &
\end{array}
\tag{28}
$$

Horizontal reactions (denoted by $\Leftrightarrow$) are reversible heterogeneous charge-transfer steps characterized by the corresponding E^0 values for the first and second electron transfer. Vertical pathways are the acid–base equilibria described by respective dissociation constants for the first and second proton dissociation. Assuming the absence of the reduced form, a surplus of H^+ ions in the bulk of the solution, and E_2^0 more positive than E_1^0 and applying the Ilkovič diffusion treatment [20], one can derive the relation

$$
E_{1/2} = E^0 - \frac{RT}{2F} \ln K_1 K_2 + \frac{RT}{2F} \ln([H^+]^2 + K_1[H^+] + K_1 K_2)
\tag{29}
$$

which describes the dependence of the reversible half-wave potential on solution pH. If pH is changed over a broad range, (29) may be simplified to three different relations depending on whether $[H^+]^2$, $K_1[H^+]$, or $K_1 K_2$ is the predominant term in the expression $RT/2F \ln([H^+]^2 + K_1[H^+] + K_1 K_2)$. The $E_{1/2}$ versus pH plot then consists of three linear portions with slopes 58, 29, and 0 mV, respectively. The intersections of these linear $E_{1/2}$–pH plots may be used for the approximate determination of corresponding pK values.

2.7   Insoluble Mercury Salts or Formation of Complexes

Electrode reactions involving mercurous or mercuric ions differ to some extent from other processes because the reduction product, or the starting substance in the case of anodic oxidations, is the material of the electrode itself; hence, no chemical polarization of the electrode occurs. Consequently, current–voltage curve shape is not described by (17). For example, when reducing mercurous ions, the half-wave potential depends on $[Hg_2^{2+}]$:

$$
E_{1/2} = E_{Hg^{2+}}^0 + \frac{RT}{2F} \ln \left(\frac{[Hg_2^{2+}]}{2} \right)
\tag{30}
$$

Mercury complexes HgX_p reduction is also characterized by a concentration-dependent $E_{1/2}$ [21]

$$
E_{1/2} = E_{Hg^{2+}}^0 - \frac{RT}{2F} \ln \left(\frac{\bar{i}}{2} \right)^{p-1} \frac{\kappa_{HgX_p}}{\kappa_X} \beta_p
\tag{31}
$$

where β_p is the stability constant and κ_{HgX_p} and κ_X are the Ilkovič constants based on the diffusion coefficient of the complex and the ligand, respectively.

At sufficiently positive potentials the mercury electrode undergoes anodic oxidation and produces, if no subsequent complex formation occurs, Hg_2^{2+}

and Hg^{2+} ions in the ratio of about 120:1. There is no depletion of the reduced form; thus, limiting diffusion current cannot be reached and the observed current exponentially increases

$$i = -\kappa \exp\left(\frac{(E - E^0_{Hg/Hg_2^{2+}})2F}{RT}\right) \tag{32}$$

If we substitute $E^0_{Hg/Hg_2^{2+}} = +0.556$ V versus SCE and $[Hg_2^{2+}] = 10^{-4}$ mol/L, the current increase is first detectable at about $+0.44$ V versus SCE. This potential is the positive limit for DME application in aqueous media. In the presence of anions forming low solubility salts with Hg_2^{2+} (like halides, N_3^-, OH^-, S^{2-}, SH^-, and others), or soluble complexes with Hg^{2+} (like CN^-, SO_3^{2-}, $S_2O_3^{2-}$, SCN^-), the anodic current increases at less positive potentials than $+0.4$ V, with an observed limiting current caused by anion diffusion (if anion concentration is not too high). Again the $i–E$ relationships are derived from the solubility product or the stability constants [20].

2.8   Reversible Currents for Semiquinones and Dimers Formation

Electrode reactions involving a reversible uptake of two electrons in two consecutive steps are coupled with the intermediate disproportionation equilibrium. The resulting diffusion limiting current corresponds in all cases to a two-electron reduction process; the slope of the $i–E$ curve depends, however, on the magnitude of the disproportionation constant. For $K_d > 16$ two separate one-electron waves are observed; for $K_d = 16$ the curve has a shape corresponding to an uptake of two-thirds electron; and for $K_d = 4$ one two-electron wave whose slope corresponds to $n = 1$ is found. For $K_d = 0$ a single, two-electron wave with $n = 2$ appears [20]. Alternatively, the intermediate, usually a radical, may participate in a dimerization reaction.

2.9   Finite Charge-Transfer Rate in Electrode Processes

The electrode reactions in the preceding sections were characterized by such a high rate of electron exchange between the metallic phase and the reacting species in the solution that the equilibrium between the reduced and oxidized forms was established according to the Nernst equation, and the faradaic current was controlled only by the electroactive species diffusion. If the electron-transfer rate is not high enough, it assumes control, the observed electrode processes are called irreversible, and the Nernst equation is not obeyed. Reversibility or irreversibility is not an absolute property of the redox system but depends on the mutual ratio of the electrochemical method time constant and the charge-transfer reaction rate. In dc polarography the method's time constant is given by drop time, the time available for equilibrium establishment, which may be varied only over a limited interval. Electrode reactions with charge-transfer control,

$$Ox \underset{k_{-e}}{\overset{k_{+e}}{\rightleftharpoons}} Red \tag{33}$$

are described by the rate constants k_{+e} and k_{-e} for reduction and oxidation, respectively. Both constants are potential dependent,

$$k_{+e} = k^0 \exp\left(-\frac{\alpha nF}{RT}(E-E^0)\right)$$

$$k_{-e} = k^0 \exp\left(\frac{(1-\alpha)nF}{RT}(E-E^0)\right)$$

(34)

where k^0 is the standard heterogeneous rate constant corresponding to the rate at E^0 and is expressed in cm/s and α is the transfer coefficient. The derivation of the faradaic current [22–28] for the irreversible case i_{irrev} at potential E is based on the formulation for the diffusion-controlled case, except a different boundary condition takes into account that the difference in the rate of a forward and a backward charge-transfer reaction yields net current. At potentials far from E^0 the charge-transfer rate becomes so high that the diffusion rate assumes control and a limiting diffusion current is observed. The relation between i_{irrev} and the reversible current i_{rev}, measured for an infinitely fast reversible electrode reaction at a given potential E, is given by a tabulated function [27]

$$i_{irrev} = i_{rev} F(\chi)$$

(35)

where

$$\chi = \sqrt{\frac{12}{7}\left(\frac{k_{+e}}{D_{Ox}} + \frac{k_{-e}}{D_{Red}}\right)}\sqrt{t}$$

(36)

Similarly the function $\bar{F}(\chi_1)$ is used for mean currents [28], and χ_1 is related to the rate constants in the same way as χ except t_1 replaces t. Occasionally an approximate formula is used instead of functions $F(\chi)$ or $\bar{F}(\chi_1)$ and derivation of the current–voltage curve shape is possible. For a simple case of unidirectional cathodic reduction $(k_{-e} \ll k_{+e})$

$$\frac{i_{irrev}}{i_{rev} - i_{irrev}} = 0.886 k_{+e}\sqrt{\frac{t_1}{D}} = 0.886 k^0 \sqrt{\frac{t_1}{D}}\exp\left(\frac{-anF}{RT}(E-E^0)\right)$$

(37)

Also, the half-wave potential

$$E_{1/2} = E^0 + \frac{2.3RT}{\alpha nF}\log\left(0.886 k^0\sqrt{\frac{t_1}{D}}\right)$$

(38)

From these equations we easily deduce some important features of irreversible polarographic waves:

1. Half-wave potential differs from E^0, depends on t_1, and anodic and cathodic waves of the same redox couple have different $E_{1/2}$.

2. The polarographic wave is more protracted when compared with a reversible wave (Curve 3 in Figure 9.2A).

3. Log-plot analysis does not yield the appropriate n value because the slope also includes the transfer coefficient (Curves 3 and 4, Figure 9.2B).

4. The upper limit of the heterogeneous charge-transfer rate, distinguishable from the diffusion control, is about $k^0 \leqslant 0.02$ cm/s. Larger values of k^0 must be measured by faster techniques (like ac or pulse polarography).

5. Irreversible electrode reactions of complex ions or organic substances with the participation of hydrogen ions cannot be analyzed according to (23), (27), or (29); more complicated expressions must be used.

6. The $E_{1/2}$ value may also depend on the concentration and on the type of indifferent electrolyte used as will be outlined in the next section.

2.10   Electrode Double Layer and Irreversible Processes

Indifferent electrolyte ion distribution in the space adjacent to a metallic electrode's charged surface differs from that distributed in the bulk of the solution and depends on the charge (or potential) of the electrode. This space, the electrode double layer, is mathematically treated in a way similar to the ionic atmosphere in the strong electrolyte theory. The electrode double layer's characteristic property is the potential drop in the diffuse part of the double layer, usually denoted as ψ potential, or sometimes ϕ_2 potential, which in uncomplicated cases may be calculated for a given potential and electrolyte concentration from the charge of the electrode Q [1–4],

$$Q = -11.72\sqrt{c} \sinh\left(\frac{zF}{2RT}\psi\right) \tag{39}$$

where z is the charge of a symmetrical z-z valent electrolyte and the constant 11.72 holds for 25°C and aqueous solutions. Typical values of ψ potentials range from a few millivolts to several hundred millivolts. The influence of the ψ potential on the electrode reaction was first considered by Frumkin [29] in the case of H^+ reduction. Frumkin postulated that the charged electroactive species distribution at the electrode surface is modified by the influence of ψ potential and only the species in contact with the electrode can participate in the electrode reaction. Furthermore, the electrode reaction is not governed by the value of the electrode potential E, but rather by the so-called effective $E - \psi$. If both effects are included in the derivation of simple i–E relationships, in most cases we obtain

$$\Delta E_{1/2} = \left(1 - \frac{z}{\alpha n}\right)\Delta\psi \tag{40}$$

for $E_{1/2}$ shift or for the heterogeneous rate constant the dependence

$$k^0_{app} = k^0 \exp\left(-\frac{z'F}{RT}\psi\right) \tag{41}$$

where the suffix app denotes the apparent rate constant observed for the electrode reduction or oxidation of species with a charge z'. The corrections according to (40) and (41) are summarily called the *Frumkin correction*. Since these effects reflect the actual charge of reduced or oxidized species they may be occasionally used to identify the rate determining ion charge. The ψ effects are sometimes responsible for rather unusual shapes of polarographic waves or for more or less deep minima on the plateau of the diffusion limiting currents. Such minima are observed when highly charged anions are reduced and the electrostatic repulsion caused by ψ potential becomes substantial at negative potentials. For full interpretation of these phenomena references [20, 30, 31] should be consulted.

2.11   Coupled Chemical Reactions

The simple heterogeneous reaction scheme Red $\Leftrightarrow$ Ox is not applicable in numerous examples because one or both participants are further transformed by a homogeneous chemical reaction. When the electroactive species are formed from inactive predominant species, the reaction is a preceding chemical reaction; when the primary product reacts chemically, it is a follow-up chemical reaction. A third type of chemical reaction involves chemical regeneration of the depolarizer by a reaction parallel to the electrode reaction proper. There are other possible confirmed and theoretically solved reaction schemes.

A typical simple example is the reduction of various organic acids with a reducible undissociated form and with an inactive anion, or one reduced at more negative potentials. Then, according to pH, one may observe a limiting current smaller than the diffusion-controlled values with specific properties by the controlling recombination rate of anion with H^+, which yields the undissociated form. Dissociation of complexes is similar and produces the free metal ions reduced again at more positive potentials. If A is the inactive form and k and k' are the rate constants of the homogeneous chemical reactions in forward and backward direction, respectively, we have the scheme

$$
\begin{array}{c}
A \\
k \downarrow \uparrow k' \\
Ox \xleftrightarrow{\;ne\;} Red
\end{array}
\qquad (42)
$$

By incorporating the chemical equilibria into the mass-transfer equation the kinetically controlled current i_k was calculated by Koutecký [32] in the form of a tabulated function

$$
\frac{i_k}{i_d} = F(\chi) \qquad (43)
$$

where the rate parameter is

$$
\chi = \sqrt{\frac{12}{7} \frac{t_1 k}{K}} \qquad (44)
$$

and the equilibrium constant is defined as

$$
K = \frac{[A]}{[Ox]} \qquad (45)
$$

The function $F(\chi)$ mirrors the slow charge-transfer step (35) and a similar solution is given for mean currents. If the equilibrium constant K is known, polarography offers a simple method for measuring rapid chemical reactions. Function $\bar{F}(\chi_1)$ may be approximated by [32, 33]

$$\bar{F}(\chi_1) = \frac{1}{1.5 + \chi_1} \tag{46}$$

which gives for the mean current

$$\frac{\bar{i}_k}{\bar{i}_d - \bar{i}_k} = 0.886 \sqrt{\frac{kt_1}{K}} \tag{47}$$

Alternatively, one can write

$$\bar{i}_k = \frac{0.886\sqrt{kt_1/K}}{1 + 0.886\sqrt{kt_1/K}} \, \bar{i}_d \tag{48}$$

In a slow chemical reaction the second term in the denominator of (48) may be neglected, and $\bar{i}_k \ll \bar{i}_d$. It follows that $\bar{i}_k \simeq \sqrt{t}\,\bar{i}_d$, or by substituting the Ilkovič constant, $\bar{i}_k \simeq (mt_1)^{2/3}$, which implies for the dropping mercury electrode, the independence of $\bar{i}_k$ from the height of the mercury head. Under dc polarographic conditions this is an important criterion for distinguishing kinetic currents and should not be confused with that used for tast polarography or sampled instantaneous currents where $i_k \simeq t^{2/3}$.

The double-layer structure can influence the observed rate constant by ψ effects, as in irreversible charge-transfer reactions. The theory for these corrections is fully developed [30, 34–37] and is given as a multiplier in the right-hand side of (47).

Catalyzed electrode processes, or the chemical reactions paralleling the heterogeneous electrode process, may be described according to the reaction scheme

$$\mathrm{Ox} \overset{ne}{\Longleftrightarrow} \mathrm{Red}$$
$$\mathrm{Red} + \mathrm{X} \underset{k'}{\overset{k[\mathrm{X}]}{\rightleftharpoons}} \mathrm{Ox} \tag{49}$$

where X is the solution component that regenerates the oxidized form. Usually the chemical reaction is unidirectional, that is, $k' \ll k$. Furthermore, if X is in sufficient surplus, the reaction is monomolecular and solved by Koutecký for the mean currents [38–40]

$$\frac{\bar{i}_l}{\bar{i}_d} = \bar{\psi}(\chi_1) \tag{50}$$

where $\bar{i}_l$ denotes the catalytic current, now higher than it would be for diffusion control, function $\bar{\psi}$ is tabulated, and the rate parameter is defined as

$$\chi_1 = kt_i \tag{51}$$

When $\chi_1 > 10$, which corresponds to a value at least three times higher than $\bar{i}_l$ over $\bar{i}_d$, then the function $\bar{\psi}$ may be approximated by

$$\bar{\psi}(\chi_1) = 0.81\chi_1 \tag{52}$$

When the pseudo-monomolecular character of the reaction (48) is unfulfilled, the depletion of X at the electrode and its diffusion must also be considered. Cases, such as a partial regeneration of electroactive species (the disproportionation reactions), were solved and corresponding rate equations are in the literature [41, 42].

Chemical reactions following the electrode process are usually inactivation type—for a monomolecular inactivation

$$A \underset{}{\overset{ne}{\rightleftharpoons}} B$$
$$B \overset{k}{\longrightarrow} P$$

$$(53)$$

where P is the final product. Inactivation following a reversible electrode process results in a polarographic wave that still has a reversible shape; however, the wave is displaced in oxidations toward more positive and in the case of reductions toward more negative potentials than potentiometry's E^0. Evaluation of k is possible [43] only if the E^0 value for the redox couple A/B is known and is to be calculated from the relationship

$$E_{1/2} = E^0 - \frac{RT}{nF} \ln \left(0.886 \sqrt{\frac{D_B}{D_A}} kt_1 \right) \tag{54}$$

In a bimolecular chemical inactivation, like the dimerization reaction

$$A \underset{}{\overset{ne}{\rightleftharpoons}} B$$
$$2B \overset{k}{\longrightarrow} P$$

$$(55)$$

the expression for the half-wave potential [44]

$$E_{1/2} = E^0 - 0.36 \frac{RT}{nF} + \frac{RT}{3nF} \ln(ckt_1) \tag{56}$$

differs from (53) because $E_{1/2}$ now depends on the c of the depolarizer.

2.12   Adsorption on Mercury and Its Influence on Polarographic Currents

Often a solution component interacts in the form of the faradaic current with the electrode surface by means other than a complete electron exchange. A substance from the solution may be bound to the electrode surface by physical, chemical (the formation of a band), or coulombic forces causing the component's accumulation at the solution–metal interface. The theory of adsorption from the solution phase to the electrode is a complicated field of electrochemistry with many problems still to be solved. This section will be limited to a qualitative description of adsorption effects on polarographic currents, as distinguished by two cases:

1.   The electroactive species are adsorbed (the reduced form, the oxidized form, or both).

2. The component of the solution, which is electroinactive at a given potential range, is adsorbed.

Generally, adsorption effects are expected when the solution contains highly hydrophobic compounds (those with a longer alkyl chain) that covalently interact with Hg (like sulfur compounds), quaternary ions with bulky substituents, and many others. Adsorption properties, also strongly dependent on the solvent dielectric constant, occasionally may be eliminated or suppressed by replacing water with another suitable solvent.

Adsorption of either redox form in a reversible redox process alters the properties of that couple because adsorption energy contributes significantly to the overall energy required for an electron exchange in the faradaic process [45–52]. Thus, adsorption of the oxidized form is responsible for the reduction of adsorbed species to proceed at more negative potentials than E^0, while the reduced form adsorption has an opposite effect. Since full coverage of the electrode surface limits the amount of adsorbed species, adsorbate reduction usually produces a small adsorption wave, but corresponding to the reduction of the species supplied by diffusion. Therefore adsorption of the reduced form creates an adsorption prewave (which is more positive than E^0); and adsorption of only the oxidized form causes an adsorption postwave.

Once adsorption equilibrium is reached, redox behavior may be investigated over a wide concentration range. This investigation will show, at the lowest concentrations, an adsorption wave that increases nonlinearly with bulk concentration, until full coverage is achieved, at which time the diffusion-controlled wave evolves. Until full coverage is reached, the adsorption wave $E_{1/2}$ is strongly dependent on the concentration. The limiting value of the adsorption wave at full coverage is

$$\bar{i}_a = 0.85nFZm^{2/3}t^{-1/3} \tag{57}$$

where Z is the maximum number of molecules adsorbed per square centimeter. This equation distinguishes $\bar{i}_a$ from $\bar{i}_d$: $\bar{i}_a$ is linearly dependent and $\bar{i}_d$ changes with the square root of the height of the mercury head.

The situation is further complicated when an irreversible electrode process is accompanied by the adsorption of electroactive species or products of the electrode reaction. Generally, product adsorption does not facilitate electron transfer but merely causes blocking or autoinhibitory effects, characterized by instantaneous current–time curves with anomalous shapes [20].

When an electroinactive substance is adsorbed forming the inhibitive layer at the electrode in the potential range where the faradaic process proceeds, the formation of this layer influences the shape of an i–E curve [53–60] causing dips or minima on the limiting currents. The polarographic wave may split into two waves as a result of multiple redox couples rather than inhibitor desorption from the surface. Higher inhibitor concentrations may completely suppress the wave. Usually the electrode processes, including multiple electron transfers like Cd^{2+}/Cd^0, among others, are more sensitive to inhibition than a one-electron reduction such as Tl^+/Tl^0; however, many compounds inhibit the latter as well.

2.13   Rotating Disk Electrodes (RDEs)

The invention of rotating electrodes greatly enhanced the utility of solid electrodes. A disk-shaped electrode vertically immersed into the solution while rotating around its center at a constant speed is a reproducible way to generate convective mass transport from the bulk. Comparable to the DME theory is the rigorous theory [61] describing RDE $i-E$ relationships and explaining why RDEs are employed for many electrochemical problems. A simple outline of the complicated hydrodynamic theory, fully explained in the RDE literature [62] is discussed in this and other sections.

Formulation of the convective diffusion calculation of the concentration gradient of electroactive species at the electrode is based on

$$\frac{\partial c}{\partial t} = D\,\Delta c - v\,\mathrm{grad}\,c \tag{58}$$

where v is the velocity vector for the flowing fluid. The Navier–Stokes equation describes the fluid flow. The problem is easily solved under certain simplifying conditions, such as steady-state concentrations and unstirred solutions.

The equations given below discuss current flowing through the RDE and are based on several assumptions, particularly on the laminar flow of fluid toward the disk. The flow mode is determined by the value of the Reynolds number, defined for the rotating disk as

$$\mathrm{Re} = \frac{\omega r^2}{v} \tag{59}$$

where $\omega = 2\pi f$ is the angular rate of rotation, r is the disk diameter, and v is the coefficient of kinematic viscosity. In aqueous solutions the turbulent flow starts at about $\mathrm{Re} \approx 10^5$, which should be noted in experimental work. Furthermore, the axial symmetry of rotation is assumed; hence, the mechanical perfection of the electrode design is necessary to avoid additional stirring or vibrations. By solving the mass-transport equations one can show that the mass flux as well as the thickness of the diffusion layer at the electrode (where there is a concentration gradient) are the same for all points at the RDE surface thus making this type of solid electrode very attractive for quantitative electrochemical research. The limiting current for the diffusion-controlled electrode process is given by

$$i_d = \pm 0.620 nFD^{2/3} v^{-1/6} \omega^{1/2} c \tag{60}$$

A diagnostic test for a RDE diffusion-controlled process based on (60) consists of the linear i versus $\omega^{1/2}$ plot, which should pass through the origin of coordinates. This test's failure may occur either because of the experimental set-up—an inhomogeneous disk surface covered by deposits or bubbles, turbulence, vibrations, or stirring—or a more complicated mechanism. Equation (60) may also be used to determine diffusion coefficients if v is known—for example, from the viscosity measurements.

Total charge-transfer control by a slow electrode process yields the current ($i_{irrev} < i_d$)

$$i_{irrev} = nFkc^\mu \tag{61}$$

which is independent from the rotation speed. In (61) k is the heterogeneous charge-transfer constant and μ is the reaction order of the c concentration. For the mixed control by diffusion and the slow charge-transfer process, the i versus $\sqrt{\omega}$ plot consists of two linear asymptotes, one passing through the origin and the other independent from $\sqrt{\omega}$ and equal to i_{irrev}. Reaction order may be determined by extrapolation of the diffusion line toward two points on the $i-\sqrt{\omega}$ curve from

$$\mu = \frac{\log(i_{irrev}) - \log(i)}{\log(i_d) - \log(i_d - i)} \tag{62}$$

The heterogeneous rate constant is then evaluated according to

$$k = \frac{i}{c^* nF(1 - i/i_d)^\mu} \tag{63}$$

This relationship is simplified for first-order reactions because (63) may be rewritten

$$\frac{1}{i} = \frac{1}{nFkc^*} + \frac{1.6v^{1/6}}{nFD^{2/3}c^*} \frac{1}{\sqrt{\omega}} \tag{64}$$

and hence k is evaluated from a linear plot of $1/i$ versus $1/\sqrt{\omega}$. However, this procedure is applicable only for potentials sufficiently different from the E^0 of the redox couple in order to be able to avoid the backward electrode reaction rate, that is, $E - E^0 \gg RT/\alpha nF$. If both forward k_{+e} and backward k_{-e} rate constants must be considered, use

$$\frac{1}{i} = \frac{b^\alpha}{(1-b)i^0} + \left[\frac{1}{D_{Ox}^{2/3}c_{Ox}^*} + \frac{1}{D_{Red}^{2/3}c_{Red}^*} \right] \frac{1.61v^{1/6}}{(1-b)nF} \frac{1}{\omega^{1/2}} \tag{65}$$

where $b = \exp(nF/RT)(E - E^0)$ and i^0 is the exchange current density related to the standard heterogeneous charge-transfer constant by

$$i^0 = k^0(c_{Ox}^*)^{1-\alpha}(c_{Red}^*)^\alpha \tag{66}$$

Values of i^0 and α are evaluated from the intercept of the straight lines of $1/i$ versus $1/\sqrt{\omega}$ plot at different potentials.

The RDE is also suitable for the study of coupled chemial reactions. The solution for various reaction mechanisms is often similar to that given for the polarographic current. In one example of a preceding chemical reaction according to (42) the rate constant k of a chemical reaction may be determined by

$$\frac{i}{\sqrt{\omega}} = B + C\,\frac{i}{\sqrt{k}} \qquad (67)$$

from the slope of the $1/\sqrt{\omega}$ versus i plot. The constants B and C are defined as

$$B = 0.620nFD^{2/3}v^{-1/6}c$$
$$C = \frac{-K^{1/2}D^{1/6}}{1.61v^{1/6}} \qquad (68)$$

The RDE was significantly modified by further construction with a ring separated from the disk by a small insulator gap, resulting in the rotating ring-disk electrode (RRDE). By applying different potentials to the ring and to the disk and by measuring the ratio of the ring and disk currents, this technique may be used for the identification of reaction intermediates. The RRDE theory was again developed [63].

2.14   Pulse Polarography

While during a classic polarographic experiment the applied voltage is continuously changing and the resulting current is continuously measured, there are numerous methods based on the pulse change of the potential and on measuring the response at suitable time referred to the pulse. This concept was developed by Barker [64, 65] and ranks the three most widely used techniques, the normal pulse polarography (NPP), the differential pulse polarography (DPP), and the square-wave polarography (SWP) [66–72]. These techniques were mainly introduced to increase the faradaic-to-charging current ratio and hence to increase the lowest detection limit for analytical applications.

The NPP mode usually maintains the DME at a constant potential well before the faradaic wave; then, almost at the end of the drop life, a voltage pulse, the amplitude of which gradually increases from drop to drop, is applied. Typical pulse duration is 50 ms. The current is measured shortly before the pulse end (like 40 ms after the pulse edge) and is plotted against the pulse amplitude. The current–potential relationship for a reversible process is

$$i = nFc^{*}A \sqrt{\frac{D}{\pi\tau}\left(\frac{1}{1+P}\right)} \qquad (69)$$

where $P = \exp(nF/RT)(E - E_{1/2})$ and τ is the pulse duration. As the pulse voltage becomes more negative than $E_{1/2}$, the value of P approaches zero, yielding the limiting current given by the Cottrell relation:

$$i_l = nFc^{*}A \sqrt{\frac{D}{\pi\tau}} \qquad (70)$$

In comparison with the Ilkovič equation of normal polarography this mode effects a sevenfold increase in sensitivity. The NPP shape is a wave, similar to that of a normal polarogram. In an irreversible process NPP may determine kinetic parameters k^0 and α in much the same way as normal polarography except the upper limit for k^0 measurement is higher—time τ is now of the order of milliseconds in contrast with the drop time in dc polarography. Also, coupled chemical reactions and adsorption effects for pulse polarography were considered and the necessary theoretical background was published. In normal polarography, maxima on the limiting currents are caused by streaming phenomena—in NPP, such maxima may be caused by adsorption of electroactive species, may enhance sensitivity, and, therefore, are of analytical importance.

Differential pulse polarography DPP differs from NPP when prior to pulse application the potential is no longer constant, but is replaced by a ramp voltage with dc polarographic characteristics. Pulse amplitude is constant and measured current is displayed as the difference between the current sampled closely before pulse application and the current sampled at the pulse end. If the polarographic i–E equation is now differentiated [73] as

$$i = i_d \frac{1}{1+P} \tag{71}$$

the Cottrell expression substituted for i_d, the relationship for the DPP wave shape with a reversible process is

$$i = \frac{n^2 F^2}{RT} Ac^* \Delta E \sqrt{\frac{D}{\pi \tau}} \frac{P}{(1+P)^2} \tag{72}$$

Equation (72) is valid only for small values of pulse amplitude $\Delta E < RT/nF$. The i–E curve shape is the derivative of the polarographic wave. Peak current is

$$i_p = \frac{n^2 F^2}{4 RT} Ac^* \Delta E \sqrt{\frac{D}{\pi \tau}} \tag{73}$$

Peak potential is related to polarographic $E_{1/2}$ by

$$E_p = E_{1/2} - \frac{\Delta E}{2} \tag{74}$$

The solution for i–E curves is also available [74, 75] for a slow electrode reaction or for considering the electrode sphericity.

The SWP polarization potential is a linearly varying voltage ramp over which a small amplitude square-wave signal, usually of 225 Hz and 1–50 mV amplitude, is superimposed. This method differs from NPP or DPP in that perturbation continues during the whole drop life and the response readout is positioned at a time of negligible time change of the electrode surface. The signal, measured as a function of the linearly varying potential, is the difference of current samples obtained at the end of each half-cycle of the square-wave signal (after filtering off the dc signal component) [76]. The resulting i–E curve is a peak described for a reversible process by

$$i_{sw} = \frac{n^2 F^2}{RT} c^* \Delta E \sqrt{\frac{D}{\pi \tau}} \frac{P}{(1+P)^2} L \tag{75}$$

where L is the constant of the instrument defined as

$$L = \sum_{m=0}^{\alpha} \frac{(-1)^m}{(m+t/\tau)^{1/2}} \tag{76}$$

ΔE is the square-wave amplitude, t is the time measured from the half-cycle beginning, and τ is the half-period of the square-wave. The peak half-width is given for small ΔE as

$$W_{1/2} = 3.52 \frac{RT}{nF} = \frac{90.5}{n} \text{ mV at } 25°C \tag{77}$$

The theory of slow charge-transfer reactions, given by Barker [76], reveals that peak height strongly decreases in the range of the rate constants $k^0 = 10^{-1}$–10^{-4} cm/s; below this it is almost insensitive to the k^0 change and yields peaks about twenty times smaller than in the reversible case.

2.15   Kalousek Commutator Polarography

Pulse and square-wave techniques were preceded by a simple method that periodically switches between either the linear polarization voltage or a constant auxiliary voltage (Types I and II) or between two parallel, linearly changing potential levels (Types III and IV) [20]. The polarization regimes III and IV parallel that used in square-wave polarography and regimes I and II resemble normal pulse polarography, except the duty factor of pulses is unity and switching occurs during the whole drop life. Switching frequencies are from approximately 3 to 100 Hz. Compared to the recent techniques Kalousek polarography records mean currents during the half-period of perturbation; therefore, each of two polarization regimes has two modes of current recording:

Type I   During voltage ramp polarization (the most frequently used technique)
Type II   During auxiliary polarization
Type III   During the lower polarization level
Type IV   During the upper polarization level

Because Kalousek and normal polarographic curves are always compared, reversibility of product reoxidation and unstable intermediates of electrode reactions may be investigated. In reductions, for example, Type I curves yield an anodic–cathodic wave for a reversible case, separate anodic and cathodic waves for a partly irreversible case, and only a cathodic wave for a totally irreversible case. The commutated wave is independent of switching frequency for a product back-transformation kinetic current obtained from a chemical reaction.

The detailed theoretical description of commutated curves is in [70].

2.16   Alternating Current Polarography

This method, usually referred to as ac polarography, is based on super-position of a small-amplitude sine-wave voltage $E_\sim$ over the linear voltage ramp [77–79]. $E_\sim$ frequency ranges from 10 Hz to approximately 10 kHz based on experimental demands. The resulting current contains, in addition to the filtered off dc component, the alternating current component $i_\sim$, which is proportional to the total cell admittance

$$Y = Y' + jY'' \tag{78}$$

where Y' and Y'' are the real and imaginary admittance components, respectively. After proper amplification and rectification, the total alternating current magnitude was originally recorded as a function of dc voltage. Because cell impedance in the simplest case is composed of an ohmic solution resistance R_s in series with the parallel combination of the electrode double-layer capacity C and the faradaic impedance component Z_F, the alternating current is a vector with a definite phase relation to the applied ac perturbating voltage. Measurement of only total alternating current offers no particular advantage during a detailed electrochemical investigation. Alternating current polarography became increasingly attractive after introduction of more sophisticated instrumentation based on phase-sensitive detection with automatic solution resistance compensation [80–82]. This procedure consists of measuring the two curves—one current component, also known as the real or the resistive component, which is inphase with $E_\sim$, and the other, the imaginary or the quadrature component, which is phase-shifted 90° with respect to the $E_\sim$ vector. If the amplitude of $E_\sim$ is precisely known, cell admittance $Y(\omega)$ or impedance $Z(\omega)$ at a given frequency ω may be calculated as

$$i_\sim = \frac{E_\sim}{Z(\omega)} = Y(\omega)E_\sim \tag{79}$$

This information may be analogously obtained by finding the absolute value of the alternating current vector and the phase shift of $i_\sim$ in reference to $E_\sim$. Appropriate R_s compensation is crucial in phase-sensitive ac polarography because the iR_s drop causes phase error, which is eliminated only after laborious vector analysis [81, 83]. The inphase component of a phase-sensitive ac polarogram of a blank solution without any electroactive compounds should be a zero line, since

$$Y'(\omega) = \frac{1}{R_s} \to 0 \tag{80}$$

and the double-layer capacitance for the imaginary component

$$Y''(\omega) = \omega C(E) \tag{81}$$

Double-layer structure and adsorption of various surface-active substances may be examined by this method, which replaces cumbersome balancing bridge measurements historically used for such purposes.

The faradaic peak-shaped contributions on both real and imaginary ac curves, denoted as Y_F' and Y_F'', respectively, are found in the potential region of the polarographic wave. A general case with a finite charge-transfer rate relates these faradaic components to the kinetic parameters by [83, 84]

$$Y_F' = nFA \frac{\xi+1}{2\xi^2+2\xi+1} \left[c_{\text{Red}}' \frac{\partial k_{+e}}{\partial E} - c_{\text{Ox}}' \frac{\partial k_{-e}}{\partial E} \right] \tag{82}$$

$$Y_F'' = nFA \frac{\xi}{2\xi^2+2\xi+1} \left[c_{\text{Red}}' \frac{\partial k_{+e}}{\partial E} - c_{\text{Ox}}' \frac{\partial k_{-e}}{\partial E} \right] \tag{83}$$

where c' is surface concentration and

$$\xi = \frac{k_{+e}}{\sqrt{2\omega D_{\text{Ox}}}} + \frac{k_{-e}}{\sqrt{2\omega D_{\text{Red}}}} \tag{84}$$

Both faradaic components are proportional to bulk concentration; and Y_F' is a signal completely without charging components, with a zero line ideally suited to analytical applications. Furthermore, phase-sensitive ac polarography gives instantaneous information about reversibility at a given ω by comparing the ratio of Y_F' and Y_F''. A reversible electrode reaction corresponds to the limit $\xi \to \infty$; hence, $Y_F' = Y_F''$. In contrast, the totally irreversible process is characterized by $\xi \to 0$, and $Y_F'' \to 0$. This ac technique also determines double-layer capacity in the presence of a quasi-reversible electrode reaction by utilizing sufficiently high frequency since $\omega \to +\infty$ is $\xi \to 0$; and hence $Y_F'' \to 0$ and $Y'' \to \omega C$. The kinetic parameters k^0 and α are evaluated by plotting ξ versus E:

$$\xi = \frac{Y_F''}{Y_F' - Y_F''} = \frac{k_{+e}}{\sqrt{2\omega D_{\text{Ox}}}} + \frac{k_{-e}}{\sqrt{2\omega D_{\text{Red}}}} \tag{85}$$

yielding two linear asymptotes at far positive and negative potentials with respect to $E_{1/2}$. The transfer coefficient is calculated from the slope and k^0 and E^0 from the coordinates of the asymptotes' intercept. The literature describes several methods for determining kinetic parameters [80, 81], for example, those based on the faradaic phase angle

$$\tan \varphi = \frac{Y_F''}{Y_F'} \tag{86}$$

which is plotted versus $\sqrt{\omega}$ for k^0 determination; the potential of the maximum on a cotan φ versus E plot yields the α value. Admittance or impedance data displayed as complex plane plots are often used [81, 83, 85].

Electroactive species adsorption is sensitively indicated by this method, which thus determines whether the redox system properties are obscured by a specific type of interaction with the electrode [86–88]. Strongly adsorbable electro-inactive compounds yield peaks on the ac curves located at potentials of the adsorption–desorption process. However, they are much narrower and their frequency dependence markedly differs from the faradaic peaks. These peaks were also used for determination of surface-active compounds.

Other ac methods [80] such as second-harmonic ac polarography and ac polarography in frequency multiplex mode are only seldom utilized.

A dc polarographic response and one obtained by methods giving the derivative or the second derivative response are compared in Figure 9.1.

3   EXPERIMENTAL ARRANGEMENT AND INSTRUMENTATION

3.1   Introduction

The experimental arrangement in polarography is one of the simplest in chemical instrumentation: in principle it consists of the polarizable working electrode, the unpolarizable reference electrode, the voltage source, the current measuring device, and the electrolytical vessel.

A device such as that in Figure 9.3 can perform a polarographic measurement. The polarizable mercury dropping electrode WE is polarized by the voltage source RG of a low-output resistance. A manually operated voltage divider or a ramp generator polarizing the electrode with continuously increasing voltage—ramp voltage—may be used. An unpolarizable electrode with a defined potential (like SCE and so on) serves as reference electrode, RE. This electrode system is immersed into an electrolytical vessel EV containing examined solution. The current flowing through the electrode system is measured by a current-measuring device A and is dependent on the voltage applied to the WE.

3.2   Sensors

3.2.1   Mercury Electrodes

The most frequently used sensor in polarography is the mercury dropping electrode, which consists of mercury droplets flowing out under constant pressure from a mercury reservoir-connected capillary (with an inner diameter around 0.05–0.07 mm and a length of 10–15 cm). When the capillary is immersed in the solution, head height (ca. 50 cm) should be arranged so that the natural drop time is within the range 3–10 s. DME is characterized by its drop time t_1 and flow rate of mercury m, which should be in the range 1–4 mg/s. The drop time inversely depends on the height of the mercury column, the inner diameter of the capillary, and surface tension under the respective experimental conditions. Surface tension depends on electrode potential and solution composition. Thus drop time measurement must occur under reproducible

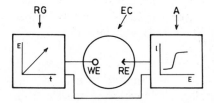

Figure 9.3 Fundamental schema of the polarographic circuit. RG—ramp generator; EC—electrolytic cell; A—device for current measurement; WE—working electrode; RE—reference electrode. *Left:* schematic course of the ramp voltage; *right:* polarographic curve.

conditions—mercury column height, supporting electrolyte, electrode potential—that is, 0 V versus SCE. Drop time may be also externally controlled by a device, which periodically dislodges the mercury drop from the capillary as in sampled polarography and other advanced methods. The mercury flow rate (mg/s) is directly proportional to the mercury column height—when measuring m the capillary tip is dipped into a weighed amount of mercury that is again weighed after a certain time interval, which may be approximately 3 min. Since t_1 and m depend on column height, the mercury drop weight mt_1 and its size A are constant and independent of the column height. The weight changes only with changes of the surface tension of mercury; therefore, when the surface tension is changed, for example, by changing the potential, solution, surfactants, and so on, drop time changes, but the flow rate remains constant.

In newly developed capillary types the end of the capillary has a spindle-shaped inner space; similar drawn-out capillaries were used by Heyrovský in his first polarographic experiments [89]. This spindle-shaped capillary ensures a drop time of high reproducibility. An improvement over the classical DME is the static mercury drop electrode, SMDE, in which a needle [90] or other kind of valve controls mercury flow through the capillary and drop growth. (The current measurement is made when the drop area is no longer changing.) The up and down movement of the needle or other capillary-closing device is electronically controlled from the polarograph. This type of electrode acts either as a hanging mercury drop electrode, used for stripping analysis, or as a periodically renewed hanging electrode which may replace DME with the advantage of a constant surface area after drop formation. When SMDE is used as a "dropping" electrode, a tapper periodically dislodges the old mercury drop from the capillary. (Such a device is produced by Princeton Applied Research Corp., USA, and Laboratorní přístroje, Czechoslovakia.)

Other mercury electrode types are primarily used for stripping analysis [62], such as the hanging mercury drop electrode, which presses a constant amount of mercury from the reservoir through the capillary by a micrometer screw or stationary mercury electrodes which deposit it on a metallic or graphite support.

The DME is sensitive to mishandling. The capillary should not be left in the examined solution when mercury is not flowing; this precaution prevents penetration of the solution into the tip of the capillary. A large positive voltage should not be applied to the electrode because the mercury salts formed by mercury oxidation may damage the capillary orifice. This also applies to high negative voltages at which mercury loses its surface tension. After use the capillary must be washed with distilled water while the mercury is still dropping, then dried, and the mercury reservoir lowered until the mercury flow ceases when the capillary is left on air. In contrast, drawn-out and spindle-shaped capillaries should be kept in distilled water or 0.1 M H_2SO_4 after use.

3.2.2   Microelectrodes

At microelectrodes constructed by sealing, for example, a carbon fiber (diameter 1–10 μm) into a glass tube, depolarizer depletion at the electrode surface at slow scan rates 1–10 mV/s does not occur and a plateau of diffusion-controlled current (a potential-independent current) rather than a peak-shaped curve is obtained. The relative insensitivity of these electrodes to the movement of the solution is a further advantage. Because of small currents, static measurements may also be performed in high-resistance solutions.

3.2.3   Rotating Disk Electrodes

Various disk electrodes are also used. The exact hydrodynamic theory for solution flow at the rotating disk electrode allows mathematical treatment of measured data with DME's precision and accuracy. Electrode material, such as platinum, silver, gold, graphite, glassy carbon, and so on, in rod form with a diameter of 2–5 mm is inserted into Teflon insulation connected to a stainless steel shaft. These types of electrode are located on a special support comprising the motor, the electronic circuit for controlling rotation speed, and other necessary device functions. With electrodes other than mercury, problems arise with the regeneration of the electrode surface. Electrode pretreatment is generally performed mechanically, by polishing, or electrochemically, by periodically switching the electrode to positive and negative voltages. On the electrode surface a mercury layer can be also deposited electrolytically from a mercury(II) nitrate solution, resulting in a surface-coated mercury film electrode. This mercury layer can be formed also *in situ* by means of mercury codeposition from the examined solution to which approximately 10^{-3} M $HgNO_3$ has been added.

3.2.4   Reference Electrodes

As in other electrochemical methods the working electrode potential is referred to a nonpolarizable reference electrode with a defined potential, which must be independent of the current flowing through the cell in polarographic measurements. In most measurements this current is actually very small (less than 10^{-6} A); consequently, no polarization of the reference electrode occurs. Furthermore, conductivity of the measured solution is frequently very high and the resistance of the reference electrode minimal. Therefore the iR drop in the measuring circuit is negligible, and the potential difference between the working E_{WE} and reference electrode E_{RE} practically equals the applied voltage $E_{applied}$. The most widely used nonpolarizable reference electrode is the saturated or normal calomel electrode. The above-mentioned optimum conditions are not always fulfilled—mainly at higher currents, and with limited conductivity of the base solution. In this case the iR drop becomes significant and affects the potential difference $E_{WE} - E_{RE}$ according to

$$E_{applied} = E_{WE} - E_{RE} + iR \tag{87}$$

In such cases a distorted polarographic curve with a half-wave potential shift practically equal to the iR drop results. To prevent this unwanted potential shift, iR compensation must be ensured. A third electrode—the auxiliary electrode AE—is introduced into the circuit in such a mode that the measured current passes through it and not through the reference electrode (Figure 9.4).

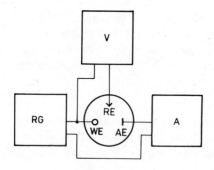

Figure 9.4 Fundamental schema of a three-electrode polarographic circuit for IR drop correction. RG—ramp generator; A—current measurement device; V—voltmeter; WE—working electrode; RE—reference electrode; AE—auxiliary electrode.

The potential difference between WE and RE is then measured and plotted against the current.

When the calomel electrode must not contact the analyzed solutions, for example, in analysis of Cl^- ions in the solution and in nonaqueous solvents, measurements are obtained with a salt bridge between the cell and reference electrode. Other reference electrode types are primarily used with nonaqueous solvents.

3.2.5   Polarographic Cells

Electrolytic cells are always supplied with polarographs. The most popular cells are glass vessels fitted to a cell cover with openings containing O-ring adapters at which the electrodes are fastened so their tips will group near the bottom of the vessel. The solution volume may be 2–50 mL. Since oxygen dissolved in the base solution is polarographically reducible, the cell cover connection contains glass tubing used as inlet and outlet sources for the inert gas of deaeration. The cells are usually arranged in such a way that after deaeration, the gas may be led over the surface of the examined solution. If necessary, thermostated cells may also be used.

3.3   Dc Polarography—Polarizing Unit and Signal Modifier

A simple type of polarograph may be built using operational amplifiers [91] (Figure 9.5). Amplifier 1 is connected as an integrator supplying ramp voltage for the DME monitored on voltmeter V. After a preset value is reached, the charging supply voltage E may be reversed to provide a uniformly decreasing voltage. If the polarizing voltage is to be kept constant, it is sufficient to disconnect the input voltage E: the integrating capacitor C remains in a charged state and the output voltage is thus constant. A relay or switching transistor operated by a comparator circuit may also short-circuit the integrating capacitor. A relay may also change the voltage supply polarity to the integrator input: thus a cyclic curve (cyclic voltammogram) is recorded. An OA wired as a current serves as a signal modifier. The current follower amplifies small currents without introducing an undesirable iR drop into the system. By changing the feedback resistor R_f value, device sensitivity may be adapted to the magnitude of

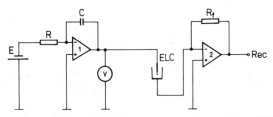

Figure 9.5 Schema of a simple dc polarograph; 1- and 2-FET amplifiers. ELC—electrolytic cell; Rec—recorder.

the current signal, which depends linearly on the depolarizer concentration. If the DME is used, current oscillations caused by dropping of mercury from the capillary tip occur. Damping of these current oscillations is performed by adding a capacitor in parallel to the feedback resistor R_f; the time constant of this RC circuit should equal the drop time of the DME. A better solution is a low-pass filter connected to the output of the current-to-voltage follower. This circuit guarantees proportionality of the signal to the average value of the limiting current.

For amperometric titrations a simple constant voltage source, for example, a battery with a $100\,\Omega$ potentiometer, and a current follower are sufficient. Damping of oscillations is again achieved by adding a capacitor ($0.1\text{--}2\,\mu F$) parallel to R_f. For rapid monitoring of polarographic $i\text{--}t$ curves on individual DME drops a basically identical polarizing circuit is used. However, there is no damping circuit with the amplifier.

3.3.1   The Potentiostat

The potentiostat is a device that compensates for the iR voltage drop in the solution and maintains the WE–RE voltage difference at the desired programmed value despite changes in the current passing through the electrolytic vessel, the concentration of the electroactive substance, the solution resistance, and so on. The potentiostat continuously compares the potential of the WE measured against the RE with the programmed voltage and instantly changes the AE potential to compensate for any voltage difference. Figure 9.6 depicts a polarographic circuit equipped with a potentiostat, where amplifiers 1 and 4 have the same function as in Figure 9.5. The potential difference between the RE and WE is sensed by the RE and fed as a voltage signal to the summing circuit at the amplifier input 2 wired as a potentiostat. The voltage follower 3 protects the RE from overloading; the function of the other parts of the circuit are explained below along with some properties and parameters of the potentiostat.

Rise time, one such important parameter, yields information on how quickly a potentiostat can respond to an error in the WE potential and cancel it. If the potentiostat is used for constant potential electrolysis, rise time is not critical; a value of about 1 ms or longer is sufficient. However, demands increase when

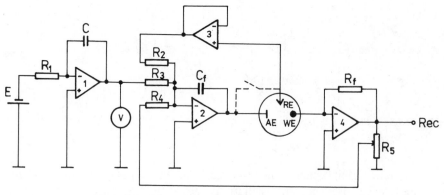

Figure 9.6   Schema of a three-electrode dc polarograph.

the electrode is polarized by a time-varying voltage, particularly in fast sweep methods and methods with ac or square-wave programmed voltage. Rapid charging of the electrode to the required potential depends not only on a fast potentiostat, but also on the capacitance properties of the electrode system: the time constant $\tau = RC$ expresses the time needed for the charging capacitance C through resistance R to reach 63% of the voltage E applied to the potentiostat input. Thus the time required for the electrode voltage to reach the value $0.63\,E$ is equal to CE/i_{max}, where i_{max} is the maximum current that can be supplied by the amplifier. For example, if we have as potentiostat an OA with a slewing rate of 1 V/μs connected to an electrode with a capacity of 1 μF, a time interval corresponding approximately to $4\,CE/i_{max}$ is the time required to charge the electrode from zero potential to 1 V. If the i_{max} of the amplifier is 5 mA, the electrode can be charged to the required potential in only 0.8 ms. Efficient use of a potentiostat with a slewing rate of 1 V/μs would require, in this example, an output current of 4 A, even if the electrolytic current itself was very small.

The potentiostat is capable of compensating the iR drop in the solution, except for the small part of it occurring between the reference (RE) and working electrode (WE). Namely, the current flow through the solution causes a solution voltage gradient that is sensed by the RE. For the voltage measured between the WE and RE to accurately correspond to the working electrode potential, the RE capillary tip, the Luggin capillary, must be in close proximity, of the order of tenths of a millimeter, to the WE electrode. Otherwise, solution resistance between the WE and RE would remain important as an uncompensated part of the working potential value. In general, the RE should be placed outside the electric field of the electrode, that is, in a direction opposite to the WE–AE link. In very accurate measurements, especially in media with poor conductance, the remaining resistance, uncompensated by the potentiostat, between the WE and RE may be eliminated by the positive feedback (Figure 9.6). The positive feedback link from the current-to-voltage convertor 4 is connected to the input of amplifier 2 through resistor R_2. Resistor R_5 is adjusted to the measured or assessed value of the solution resistance between the tip of the Luggin capillary and WE, and the corresponding voltage is brought to the input of amplifier 2 (potentiostat). Great care is necessary in adjusting the amount of positive feedback, since the whole circuit begins to oscillate with over-compensation. The necessary amount of positive feedback is established by an oscilloscope connected to output amplifier 2 (Figure 9.6) in such a way that the potentiometer P_5 is positioned where potentiostat oscillation disappears after minor overcompensation. Because of capacitive load the potentiostat oscillates even without positive feedback. This undesirable phenomenon is prevented by inserting a capacitor C_f into the amplifier 2 feedback (Figure 9.6); if capacitance is as low as possible, on the order of hundreds of pF, the dynamic properties of the potentiostat will be largely unaffected. Resistors R may be 10 kΩ. Before changing the solution in the electrolysis vessel the controlling amplifier output to the voltage follower input must be short-circuited and this link must not be broken until the vessel has been filled, or the output voltage of both amplifiers will reach saturation. The same is valid when using commercial polarographs: one should switch from the two-electrode to the three-electrode configuration only if the solution is prepared for measurement. Various potentiostatic circuits are given in [92].

3.3.2   Advanced Polarographic Methods

The sensitivity handicap of the dc method is in the current needed to charge the polarographic electrode to the desired potential: at a depolarizer concentration of 10^{-5} M, the time-averaged charging current is comparable to the faradaic current. A sensitivity increase of the polarographic method thus requires discrimination against the charging current. Two methods will solve the problem, and in both a sinusoidal voltage or voltage pulse, or train of pulses, of several millivolts amplitude is superimposed on the ramp voltage polarizing the electrode. In sinusoidal ac polarography the alternating current is first separated from the direct current component and then measured after phase-sensitive rectification. The charging current, which is 90° out of phase, is thus eliminated. This method enables determination of depolarizers down to a concentration of 10^{-7} mol/L, but is only applicable for reversible processes; in irreversible processes sensitivity is lower. Unlike the waves of the dc method a peak-shaped curve is obtained; the peak potential approximately equals $E_{1/2}$ and the height is concentration dependent.

The second method, up to now the most effective, is based on the fact that the charging current decays with time more rapidly than the electrolytic current when the electrode is polarized by a voltage pulse. Thus when measuring the current following the voltage pulse, the faradaic component of the signal is predominantly recorded (Figure 9.7). Square-wave polarography [93] and pulse polarography are based [94] on this principle.

To illustrate the charging current that flows through the system at the electrode potential's instantaneous change consider the electrode a condensor C with an ohmic resistance R in series. The polarizing voltage E is constant

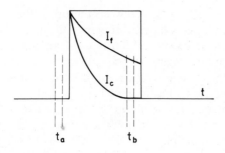

Figure 9.7 A schematic course of the faradaic I_f and charging I_c currents after a voltage pulse; t_a, t_b, sampling periods.

and there is no charging current; if this voltage abruptly changes to $E + \Delta E$, a current pulse arises,

$$i_c = \frac{\Delta E}{R} \exp\left(-\frac{t}{RC}\right) \tag{88}$$

where t denotes the time interval starting at the voltage change. After a time interval of $5\,RC$ the charging current decreases to 0.68% of its original value. The electrolytic current also decreases but at a much slower rate than the charging current, which after a time lapse of $5\,RC$ is practically negligible.

In square-wave polarography a square-wave frequency of about 200 Hz is modulated on the ramp voltage during the whole life of the mercury drop. The amplitude of this square-wave voltage is 1–50 mV. The current is sampled after the delay time following every potential change. The previous current sample is always subtracted from the current sample and the difference, dependent on polarization voltage, is recorded. The measurement is taken only when drop life of the mercury drop ends. Depolarizers characterized by a reversible electrode process down to $10^{-7}\,M$ concentrations, and in some cases even to $10^{-8}\,M$, are determined by this method.

Pulse polarography polarizes the electrode to a potential E_p by constant voltage, usually to a value at which no reduction of the compound occurs, for almost the entire drop life of the mercury electrode. A voltage pulse ΔE_p amplitude and 100 ms duration is applied to the polarized electrode (Figure 9.8) only at the end of drop life. The value of ΔE_p increases with time on every new drop of the DME; thus, ΔE_p represents the ramp voltage in some way. The polarographic current is sampled at the end of the pulse duration and is recorded as a function of the applied polarization voltage $(E_p + \Delta E_p)$. This method produces waves such as those in dc polarography and may be used for determining reversible or irreversible depolarizers down to concentrations of 10^{-6}–10^{-7} mol/L. This technique is now termed normal pulse polarography (NPP) to distinguish it from differential pulse polarography (DPP), currently the most frequently used method in analytical chemistry—the discipline that substantially increased worldwide interest in polarography. In this method the DME is polarized at the end of its lifetime with one voltage pulse—lasting approximately 100 ms and of constant amplitude, 5–100 mV—superimposed on the ramp

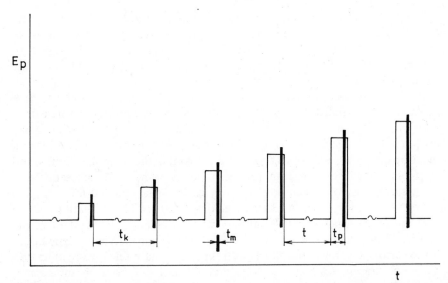

Figure 9.8 A schematic polarization voltage form in normal pulse polarography; t_k drop time of the DME; t_m current sampling period; t_p voltage pulse duration.

voltage. The current is sampled just before pulse application (20 ms) and again at the end of the polarization pulse (Figure 9.9). The difference of both values is then recorded as a function of the applied voltage and a peak-shaped curve is obtained. Differential pulse polarography may be used to determine all types of ions and compounds as in classical dc polarography, but has the advantage of higher sensitivity down to a depolarizer concentration 10^{-7} mol/L. Consequently, DPP is used not only to determine inorganic ions, such as trace metal ions in environmental chemistry, but also to analyze organic compounds.

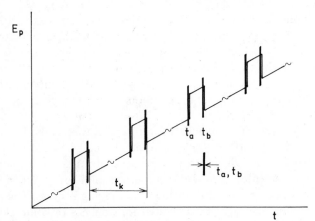

Figure 9.9 A schematic polarization voltage form in DPP: t_k DME drop time; t_a, t_b current sampling periods.

The market offers excellent instrumentation for NPP and DPP, some with incorporated circuits for dc polarography. Nevertheless, published sensitivity increase advances are generally based on an improvement of the signal-to-noise ratio and on elimination of the last traces of charging current and are reviewed in [95].

The "fast sweep" or "rapid" differential pulse voltammetry, a useful routine analysis method, records the voltammogram within the lifetime of a single drop of the DME (thus voltammetry) with a scan rate of 50–100 mV/s [96] and delay between pulses of about 100 ms. This technique needs a convenient DME with a long drop life (or the SMDE), a short time constant of the sampling circuits, and a fast recorder. The high scan rate also causes some increased sensitivity. Direct current as well as ac polarography can also be adapted for a "rapid" method using short controlled drop time down to 50 ms. The oscillations on the curve are much smaller, usually absent, because of the time constant of the recorder. If the electrode process is complicated by adsorption of the depolarizer on the electrode surface, the short drop-time and phase-selective rectification in ac polarography eliminates any resultant difficulties. Short drop times sometimes suppress kinetic currents.

Sampled dc, or tast, polarography further improves the dc polarographic method by sampling the current after the DME drop time; maximum current is recorded, and a curve without oscillations is obtained. Because of a better S/N ratio, SMDE used as a dropping electrode with a constant surface area and sampled current integration [97] affords practically the same sensitivity as DPP.

3.3.3   Stripping Analysis

A substantial increase in sensitivity mainly for the determination of various metal ions in the nanogram range, and below, can be achieved by combination of the above-mentioned polarographic techniques with stripping analysis, where the analyzed metal ions or compounds are first electrolytically accumulated on the electrode and then polarographically analyzed.

Many stripping analysis mode possibilities exist. The metal cation is first reduced to metal at a constant voltage, which forms an amalgam on the hanging mercury electrode or a metal film at solid electrodes. The deposited metal is then anodically dissolved. The compound to be determined may also be accumulated by reduction or oxidation as a sparingly soluble compound formed by reaction of the deposited ion with the components in the supporting electrolyte; for example, Tl^+ is oxidized to Tl^{3+}, whose reaction with OH^- ions forms insoluble thallic oxide. The insoluble compound is then cathodically reduced and dissolved. Some surface-active compounds may be accumulated at the mercury electrode through adsorptive preconcentration, for example, detergents, alkaloids, remedies, crude oil components, and so on. The adsorption–desorption peak is then recorded during the stripping process. The adsorptive accumulation effect can also be exploited for analysis of some electroactive compounds like complexed metal ions (Ni^{2+}, Co^{2+}, UO_2^{2+}) or organic compounds (aromatic nitro compounds, herbicides, remedies, and so on). In this case the reduction peak of the adsorbed compound is recorded.

Stripping analysis may be performed with every type of polarograph by applying a constant voltage to the electrode in a stirred solution within the time required, usually several minutes, for accumulation at the electrode; after stopping the stirring the voltage scan is applied.

Various types of electrodes, electrode systems, and techniques have been developed for stripping analysis [98]. Automation of the whole working process—sequence and duration of all operations—is advisable: such automation not only simplifies the analysis but also contributes to higher precision because all parameters are held constant. Computerization of the system enables signal averaging and correction for background currents. There are microprocessor-controlled polarographs available that automatically perform the entire stripping analysis, completed by a printout of the obtained results.

The combination of the stripping method with DPP represents one of the most sensitive methods for trace metal analysis in the sub-ppb region.

3.3.4   Automation

From the historical point of view the Heyrovský–Shikata polarograph is the first automatic instrument introduced to analytical chemistry that yields a permanent record of the performed analysis. Full automation of polarographs, as that of other physicochemical or physical instruments used in analytical chemistry, is made possible by mini- and microcomputers. The computer controls the entire equipment function and also enables data acquisition. With the polarographic curve stored in the computer's memory many mathematical operations may be performed, such as averaging in repeated accumulation, base line subtraction, resolution of overlapping peaks, and some transformations of the primary data obtained. The instrument frequently prints peak potential and sample concentration onto the record of the polarographic curve. This analog presentation of the polarographic curve remains important because the digital readout (which occurs with many measuring instruments yielding only digital data) sometimes masks deficiencies that are readily apparent on the polarographic curve. However, it should be kept in mind that the microprocessor cannot improve improperly collected data.

3.3.5   Single Purpose Analyzers

Many commercially available single purpose gas detectors and monitors, for SO_2, CO, O_2, NO_x, CN^-, and Cl_2, among others, are based on polarographic or voltammetric principles, usually with electrocatalytic porous membrane, and sometimes amperometric enzyme, electrodes.

An important application of polarography and voltammetry is for measurements in flowing liquid systems, mostly in the amperometric mode; this means that the current is measured at a fixed electrode potential. In the flow-through systems the electrodes in the vessel are the most important parts of the whole device, because the measuring electrode here often works under stringent conditions. This is mostly the case with solid electrodes, which can, and mostly do, lose their sensitivity and long-term stability due to passivation or adsorption. It is therefore often necessary to provide for reactivation or regeneration of the electrode in measurements continuing for long periods, as is usual in industrial applications, environmental analysis, and so on. Sometimes a device is included in the system for a periodical calibration and zeroing, with a feedback to the signal processing circuit. Care must also be devoted to temperature control, sample flow rate, oxygen elimination, and the stability of the reference electrode.

The main application field of flow-through systems is the electrochemical detection of electroactive compounds in HPLC, where the sensitivity of polarography is combined with the excellent separation capabilities of HPLC. On the market there are different types of detectors with either a mercury electrode (for electroreduction) or a graphite paste, glassy carbon, or platinum as electrode materials. These last electrodes are used mostly for electrooxidation processes (drugs, alkaloids, pharmaceutical preparations, vitamins, carcinogenic compounds, phenolic compounds, polycondensed hydrocarbons, and so on). The cells are constructed mostly as thin-layer cells or as wall jet arrangements. The detection limit is in the nanogram, sometimes in the picogram, region. Similar detectors are finding application in the flow-injection method (FIA). It can be expected that HPLC with electrochemical detection will soon be one of the most important analytical methods.

3.3.6   Polarographic Measurements Methodology

As it is necessary to minimize the iR drop in the solution in polarographic measurements, a sufficient amount of indifferent supporting electrolyte is added therefore to the solution. Supporting electrolyte concentration and composition affect the potential range in which the electrode processes may be followed; therefore, 0.1 mol/L solutions of salts or hydroxides of alkaline metals and solutions of acids or buffer solutions are generally used. The mercury electrode may be polarized in these solutions up to a value of 0 V and in sulfates up to $+0.3$ V (SCE); the anions react with mercury ions formed by mercury dissolution and an anodic current is observed. At negative potentials the electrode can be polarized to at least -1.9 V (SCE); in this region the alkali metal cations are reduced. If salts of quaternary amines are used, the polarization range can be extended to almost -3 V (SCE), at low temperatures and nonaqueous solvent even beyond that value (-3.4 V). In acidic solutions the polarographic curve is limited by the current of the hydrogen evolution at about -1.4 V (SCE). Of course the polarization range may be influenced or extended by using either other electrolytes or aprotic solvents. An electronic compensation of the solution resistance is sometimes advisable.

Oxygen is polarographically reducible; therefore, all measurements are performed in its absence. Its reduction wave may coincide with the examined depolarizer wave, and the hydroxyl ions formed in oxygen reduction may alter solution pH at the electrode surface in an unbuffered solution. For this reason oxygen must be removed by purging nitrogen, argon, or another inert gas through the solution for at least 5 min, in particular in nonaqueous solvents. To prevent irregularities at the DME the gas stream has to be stopped during the measurement. Before entering the analyzed solution the inert gas for deaeration has to be passed through a gas-washing bottle filled with the same solvent; its evaporation is thus hindered.

In the depolarization of the mercury electrode, maxima at the i–E curve may be sometimes observed. In this case the current increase corresponding to the rising part of the wave continues after reaching the value of the limiting current, and can several times overreach this current. The current increase discontinuously decreases to the value of the limiting current when applied voltage is further increased. To prevent maxima formation a surface-active compound such as gelatin, in a concentration of 0.005–0.01% or 0.001% Triton X-1000, may be added. In some cases use of electrodes with low values of m might also help to remove the maximum.

Since the electrolyte currents are temperature dependent all measurements are performed at constant temperature.

3.3.7   Height Measurement of Polarographic Waves and Peaks

Wave height measurement is not standardized because of the different wave shapes obtainable. Figure 9.10 presents some measurement possibilities, with the charging current corrected by prolongation of the linear part of Curve b. The sample curve may be subtracted from the supporting electrolyte curve as in computerized polarographic methods; however, the same procedure must always be used when analyzing the waves in comparative measurements. When using DME the current value should be measured halfway through the oscillations. For analytical purposes the measurement even at the maximum of these oscillations may be performed under the condition that all measurements are accomplished in the same manner. At peak height measurements a straight line between the preceding and following linear parts of the peak is analogously constructed, and vertical distance from the peak maximum to this line is measured. However, subtractive measurement using the pure supporting electrolyte curve can be used.

3.3.8 Half-Wave Potential Measurement

$E_{1/2}$ is the potential at which the polarographic current equals 50% of the limiting current; its determination can be based on the wave-height measure-

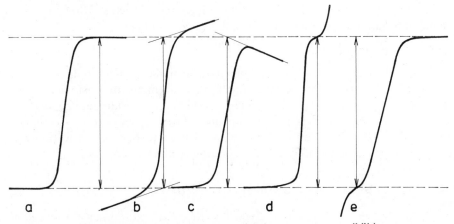

Figure 9.10   Polarographic curves and their measurement possibilities.

ments exemplified in Figure 9.10a, b, d, e. The measurement error increases when waves are not well developed. Measurements are usually taken from the recorded polarographic curves if the potential axis is calibrated and defined with respect to a specific reference electrode. Peak yielding methods measure peak potential E_p and peak current i_p. In some of these methods the half-width of the peak is also measured. This parameter is defined as the width of the peak in millivolts at peak height, where the current is 50% of its maximum height.

3.3.9   Quantitative Analysis

Several quantitative analysis methods are used to compare the analyzed sample and the standard sample curve. A standard sample solution may be used in which the standard yields a wave height approximating the wave height of the examined sample solution. A calibration curve comprised of the curves of several standard solutions in the appropriate concentration range is recommended for routine analysis. When the sample solution contains other, possibly unknown, compounds, the standard addition method is preferred. After the curve of the analyzed sample is recorded, a standard solution is added to the sample and the wave height is again measured; the addition should approximately double the original wave. Linear dependence of the concentration on wave height is necessary. The unknown concentration is calculated from the equation

$$c_{sam} = \frac{c_{st} V_{st} h_{sam}}{h_t V_t - h_{sam} V_{sam}} \tag{89}$$

where c_{sam} = the concentration of the sample solution to be measured,
$\quad$ c_{st} = the concentration of the added standard solution,
$\quad$ V_{st} = the volume of the added standard solution,
$\quad$ V_t = the total volume after addition,
$\quad$ V_{sam} = the volume of the sample solution,
$\quad$ h_t = the wave height after the addition of the standard solution.

In comparative measurements all instrumentally specified parameters along with solution temperature and electrode characteristics, such as DME mercury column height and drop time, are maintained constant.

3.4   Conclusion

Excellent instruments are commercially available for all polarographic techniques discussed. Nevertheless, further improvements in sensitivity, automation, and data acquisition are being studied and the results rapidly applied to new polarograph models. Polarograph manufacturers supply their customers with a detailed description of the instrument use and written application examples. References [99, 100] provide additional information.

4   ANALYSIS OF EXPERIMENTAL READOUT

Polarography is an electroanalytical method for determining the solution composition—primarily, the quantitative concentration determination of one or more solution components; rarely, the quality of unknown substances—or a physicochemical method for elucidating mechanisms of electrode processes, analysis of equilibria in solution, or investigation of chemical reactions of compounds formed by the electrode process.

Any measurable that is sensitively dependent in a known way upon the concentration of the compound in question may be used for electroanalytical purposes. The measurable, usually limiting, current or peak current (in pulse and ac techniques) must be specific for the compound studied and should show as little as possible interference with and influence of other components. When the electroanalytical method is applied to an electrode process, the influence of nature, the concentration of the base electrolyte, the surface-active substances, and the potentially interfering electroactive compounds are to be carefully investigated in the range of experimental conditions that might arise in practical analysis. Consequently, for electroanalytical purposes the elucidation of the electrode process mechanism is a necessary prerequisite that requires deep analysis of electrode behavior using a combination of various techniques such as polarography, ac polarography, fast cyclic voltammetry, commutator techniques coupled with investigation of the influence of various factors—temperature, polarization time scale, concentration of the depolarizer proper, and other solution components—upon electrode behavior. In most cases electrochemistry is not a self-consistent method with regard to mechanism; thus, its combination with other techniques such as ESR, UV–Vis–IR spectra *in situ* at the electrode or in the bulk of the solution, or the method of slight variations of the depolarizer structure is necessary. Only such an analysis produces an adequate answer.

There are many papers in the literature in which the investigation of the electrode process is simply limited to the description of the polarographic wave, limiting current, $E_{1/2}$, and log-plot, or to a single cyclic voltammogram with peak potentials and ratio of peak currents. This kind of limited investigation can lead to erroneous conclusions about the electrode process and does not provide enough information to discover the electrode process characteristics and the nature of its products.

For each electrode reaction investigated it is necessary to determine the following:

1. The nature of the step controlling the rate of the electrode process—diffusion, chemical rate, adsorption.

2. The nature of the electrode reaction proper—fast or slow, that is, electrochemically reversible or irreversible—chemical reversibility of the reaction, the influence of the diffusion layer, and the role of adsorption on the nature of the electrode process.

3. The existence and nature of chemical reactions coupled with the electrode process—preceding, follow-up, or parallel reactions.

4. The nature of primary, intermediate, and final products of the electrode process.

Table 9.1 Qualitative Information from Polarographic i–E Curves

Observation	Interpretation (Intuitive)	Further Approach
1. No marked response	Electroinactive substance in a given range of potentials	Try to enlarge the potential window; check surface activity by ac; increase temperature
2. Single well-developed wave: (a) Anodic (b) Cathodic (c) Anodic–cathodic	Substance undergoes: (a) Oxidation/Hg salt formation (b) Reduction (c) Reversible redox couple present	Verify linearity of i_d vs. concentration and diffusion control; test reversibility, determine n and kinetic parameters if possible; use faster techniques (NPP, DPP, ac, RDE) to get kinetic information; check the influence of solution components (ionic strength, pH, electrolyte type, solvent) and of electrode material; check adsorption by ac
3. Single ill-defined or anomalous wave: (a) Merging with electrolyte decomposition (b) Distorted by maxima or minima (c) Tilted limiting portion	(a) Base electrolyte and/or solvent and/or electrode material not suitable (b) Streaming phenomena present or see (c) (c) Influence of double-layer phenomena	(a) For positive potential change to Pt or Au RDE; for very negative potential use nonaqueous solvent, tetraalkylammonium salts, low T (b) Add surface-active substance to suppress the streaming, decrease temperature (c) Investigate influence of all base electrolyte components at various concentrations, nature of base electrolytes; check inhibition/acceleration by adsorption using ac or i–t curves

838

(d) Nonsymmetrical wave

(e) Discontinuous wave

4. Multiple, well-separated waves without noticeable anomalies

(a) More electroactive species present

(b) Substance with several redox states

(c) Redox process coupled with

5. Anomalous wave followed or preceded by a regular wave

(a) Splitting off a single wave due to inhibition

(b) Coupled chemical reaction influenced by double-layer phenomena

(c) Adsorption of Ox and/or Red form

6. Multiple anomalous ill-defined waves

(a) Apparatus or cell malfunctioning
(b) Nasty system

(d) Overlapping multiple waves, or partially irreversible process, or the same as (c)

(e) Insoluble phase formation or other complications

(a) Check composition of solution

(b) Verify nature of limiting currents and reversibility; determine n and E^0 for each redox step or the kinetic parameters

(c) Confirm the kinetic character of i, check temperature dependence and solution composition influence; use cyclic voltammetry

(a) Remove surfactants from the sample, purify solution with active charcoal; verify inhibition from i–t curves or use ac

(b) Investigate dependence on nature and concentration of base electrolyte

(c) Investigate i_l and $E_{1/2}$ as functions of concentration; use ac, DPP

(d) Try methods with derivative response (like DPP, ac, SWP), check dependence of wave shape on drop-time, concentration, column height, and nature of base electrolyte

(e) Try solvent with better solubility of product

Good luck!

This information may be obtained by investigating the electrode behavior as a function of external parameters such as time constant of the electrochemical technique, composition of the solution, and temperature, and also by analysis of the electrode behavior dependence upon these parameters. Diagnostic criteria were worked out several years ago [101], the use of which leads to the first hypothesis of the electrode process mechanism. A simplified version of most useful diagnostic criteria is given in Tables 9.1–9.6.

The first step of the electrode mechanism investigation poses three questions that need to be answered:

1. What is the actual state of the studied compound in the solution?
2. Is the studied compound at all electroactive?
3. What is the diffusion current of the studied compound corresponding to a one-electron electrode reaction?

The state of the compound in the solution has to be investigated by non-electrochemical methods. It has to be established whether:

The compound does not decompose by dissolution—influence of solvent, influence of supporting electrolyte.

The compound enters equilibria with other components of the solution—complex formation, acid–base equilibria, ion-pair formation, hydration or solvation.

The solution is stable with time, that is, with no slow reactions taking place—interaction with solution components, isomerization, hydrolysis, and so on.

Decomposition or slow reactions under normal experimental conditions might be eliminated by a solvent change, by the supporting electrolyte, or by a temperature decrease. In some cases an electrochemical investigation in a flow-through system can be carried out in such a way that the compound under study is in contact with the decomposing reactant for only a short time before electrochemical measurement. All these possibilities are to be tested because it is imperative to be sure of the actual state of the depolarizer, that is, of the starting point of the whole mechanism. In special cases of coordination compounds, the actual state of the depolarizer and process mechanism may be solved using solely electrochemical methods [101]. However, even in this case quantitative evaluation requires independent data on equilibrium constants.

To answer the second question of our list it is necessary to investigate as broad a range of electrode potentials as possible. For example, the use of supporting electrolytes with alkali metal cations cuts off at least 1 V of available potential in which reduction of many compounds is found. Similarly for electrode material, the use of mercury electrode considerably diminishes the range of potentials available for oxidations. Conversely, a platinum electrode has a short range of negative potentials in many solutions. The choice of solvents is the same; for example, CH_2Cl_2 cuts off reductions more negative than about -1.8 V. An incomplete electrochemical picture is often found in the literature,

Table 9.2 Criteria for Nature of Current Controlling Process

Process Control	Experimental Variable	Method	Criterion
Diffusion	Concentration	For all methods	$\bar{i}_l \sim c$
	Time	DME—dc polarography	$\bar{i}_l \sim \sqrt{h}$
		DME—dc polarography	$\bar{i}_l \sim t_1^{1/6}$
		DME—i–t curves	$i \sim t^{1/6}$
		RDE—dc polarography	$i_l \sim \sqrt{\omega}$
		NPP	$i_l \sim 1/\sqrt{\tau}$
		DPP	$i_p \sim 1/\sqrt{\tau}$
		Ac polarography	$|i_\sim| \sim 1/\sqrt{2\pi f}$
		Ac polarography—phase sensitive	$Y_F' = Y_F''$
Electrode reaction	Concentration	For all methods	$\bar{i}_l \sim c$
	Time	DME—dc polarography	$\bar{i}_l \sim h^0$ (for $\bar{i} \ll \bar{i}_l$)
		DME—dc polarography	$\bar{i}_l \sim t_1^{2/3}$ (for $\bar{i} \ll \bar{i}_l$)
		DME—i–t curves	$i \sim t^{2/3}$ (for $i \ll i_l$)
		Ac polarography—phase sensitive	$Y_F' > Y_F''$
Chemical reaction	Concentration	For all methods	$\bar{i}_l \sim c^y$
			$\bar{i}_l$ function of [X]
	Time	DME—dc polarography	$\bar{i}_l \sim h^0$
		DME—dc polarography	$\bar{i}_l \sim t_1^{2/3}$
		DME—i–t curves	$i \sim t^{2/3}$
		RDE—dc polarography	$i_l \sim \omega^0$
Adsorption of Red and/or Ox	Concentration	For all methods	$\bar{i}_l$–c response like an isotherm; mostly $$\bar{i}_l \sim \frac{c}{(\text{const} + c)}$$
		Dc polarography	$\bar{i}_l \ll \bar{i}_d$
		Ac polarography	Response much more enhanced than in dc polarography
	Time	For all methods	Complicated i–t dependence
		DME—i–t curves	Detailed inspection is recommended
		Ac polarography—phase sensitive	$Y_F' < Y_F''$

Table 9.3 Criteria for Reversible Electrode Processes

Method	Observable	Criterion
DME—dc polarography	Current	$\bar{i}/\bar{i}_d$ independent of h, t_1, ω, τ
	$E_{1/2}$	Independent of c, t_1, ω, τ_1
RDE—dc polarography		$E_{1/2}(Ox) = E_{1/2}(Red)$
NPP	Shape of wave	Log-plot linear, slope RT/nF^a
DPP	E_p	Equals $E_{1/2} \mp \Delta E/2$
	Peak shape	Symmetrical; $W_{1/2} = 90.5/n\ mV^a$
SWP	E_p	Equals $E_{1/2}$
	Peak shape	Symmetrical; $W_{1/2} = 90.5/n\ mV^a$
Ac polarography—phase sensitive	Faradaic phase angle	$45°$ for all frequencies (i.e., $Y_F' = Y_F''$)
	Peak potential	Equals $E_{1/2}$ for both Y_F' and Y_F''
	Width of wave	$W_{1/2} = 90.5/n\ mV$ (for amplitudes $< 8/n\ mV$)
	Shape of wave	Equals first derivative of dc polarographic wave; $E - \log(\sqrt{i_p/i} - \sqrt{(i_p - i)/i})$ plot linear with slope $120/n\ mV$

$^a n$ Identical with that obtained from limiting diffusion current.

Table 9.4 Criteria for Irreversible Electrode Processes

Method	Observable	Criterion
DME—dc polarography	Current	i/i_d a function of h, t_1, ω, τ (for $i \ll i_d$)
RDE—dc polarography	$E_{1/2}$	Function of t_1, ω, τ
		$E_{1/2}(Ox) \neq E_{1/2}(Red) \neq E^0$
NPP	Shape of wave	Log-plot nonlinear or linear with slope $RT/\alpha nF$
DPP	E_p	Function of τ; in some cases shoulders or two peaks
	Peak shape	$W_{1/2} > 90.5/n\ mV^a$ or nonsymmetrical peaks
Ac polarography—phase sensitive	Faradic phase angle	Linear function of $\sqrt{2\pi f}$ with intercept $\cot an\ \varphi = 1$; $Y_F' > Y_F''$; $Y_F'' > 0$ for small k^0 or high ω Not equal to E^0
	E_p	$E_p(Y_F') \neq E_p(Y_F'')$
	Shape of wave	Not equal to first derivative of dc polarographic curve

$^a n$ Identical with that obtained from limiting diffusion current.

Table 9.5 Criteria for Chemical Reactions Coupled to Electrode Processes

Type of Process	i_l/i_d	$E_{1/2}, E_p$	Log-Plot
Preceding slow chemical reaction (electrode process fast)	<1 Function of [X]	Function of t_1, $\omega, \tau, f,$ [X]	$\dfrac{RT}{nF}$
Follow-up slow chemical reaction (electrode process fast)	=1	Function of t_1, $\omega, \tau, f,$ [X]	$\dfrac{RT}{nF}$
Parallel chemical reaction (catalytic regeneration)	>1 Function of [X]	Function of [X]	With increasing rate (k, [X]); decrease of slope from RT/nF to $RT/\alpha nF$
Bimolecular (or higher-order) chemical reaction	All parameters might be function of c with respect to depolarizer		
Chemical reaction coupled to multiple electron transfer	Complicated situation, original literature should be inspected		

Table 9.6 Criteria for Chemical Equilibria Coupled to Electrode Processes

Mechanism	Rate Determining Step	$\dfrac{\partial \log\left(\dfrac{i}{i_d - i}\right)}{\partial \log[X]}$	$\dfrac{\partial E_{1/2}}{\partial \log[X]}$	$\dfrac{\partial i}{\partial t}$	$\dfrac{\partial E_{1/2}}{\partial t_1}$
$\begin{aligned}&M^{n+}\\&\updownarrow\;\overset{ne}{\rightleftharpoons}M\\&MX_p^{n+}\end{aligned}$	Diffusion	p	$-p\dfrac{RT}{nF}$	$\tfrac{1}{6}$	0
$\begin{aligned}&M^{n+}\\&\updownarrow\\&MX^{n+}\overset{ne}{\rightleftharpoons}M\\&\updownarrow\\&MX_r^{n+}\end{aligned}$	Slow electrode reaction	$-r+p$	$-(r-p)\dfrac{RT}{nF}$	$\tfrac{2}{3}^a$	$-\dfrac{RT}{2\alpha nF}$
$\begin{aligned}&M^{k+}\quad\;\; M^{l+}\\&\updownarrow\overset{ne}{\rightleftharpoons}\updownarrow\\&MX_p^{k+}\;\; MX_q^{l+}\end{aligned}$	Diffusion	$+p-q$	$-(p-q)\dfrac{RT}{nF}$	$\tfrac{1}{6}$	0

$M^{k+} \underset{ne}{\rightleftharpoons} \begin{matrix} M^{l+} \\ \updownarrow \\ MX_q^{l+} \end{matrix}$	Diffusion	$-q$	$+q\,\dfrac{RT}{nF}$	$\dfrac{1}{6}$	0
$MX_p^{k+} \underset{ne}{\rightleftharpoons} \left\{\begin{matrix} MX_s^{l+} \\ \updownarrow \\ MX_q^{l+} \end{matrix}\right.$	Slow electrode reaction	0	0	$\dfrac{1}{6}^a$	$\dfrac{RT}{2\alpha nF}$
$\begin{matrix} M^{n+} \\ \updownarrow \\ MX_p^{n+} \underset{k_+ \updownarrow k_-}{\overset{ne}{\rightleftharpoons}} M \\ MX_r^{n+} \end{matrix}$	Slow electrode reaction	$r-p-\dfrac{1}{2}^b$	$-p\,\dfrac{RT}{nF}$	$\dfrac{2}{3}^a$	$\dfrac{RT}{2nF}$
$\begin{matrix} MX_r^{k+} \underset{k_-\ \updownarrow\ k_+}{\overset{ne}{\rightleftharpoons}} MX_r^{l+} \\ MX_p^{l+} \\ \updownarrow \\ M^{l+} \end{matrix}$	Slow follow-up chemical reaction	0	0	$\dfrac{2}{3}^a$	$-\dfrac{RT}{2nF}$

[a] For $i \ll i_d$.

[b] For i equal to the limiting kinetic current only.

because the investigator becomes satisfied with discovering some electro-chemical activity and does not investigate the full accessible range of potentials by varying supporting electrolytes, solvent, or electrode material. Recently, a temperature decrease was shown to increase the accessible range of potentials [102]. In special cases there is no electrochemical activity found under normal conditions. However, a substantial concentration increase of the depolarizer or a temperature increase might result in the appearance of a polarographic wave. A classic example of this behavior is the reduction of sugars at the mercury electrode [20]. This activity is evident in systems for which the dominant form of the depolarizer in the solution is electroinactive with a very low concentration of the electroactive form being slowly formed from the dominant electroinactive form.

One of the most important diagnostic criteria for determining whether or not observed currents are caused by the investigated compound is estimation

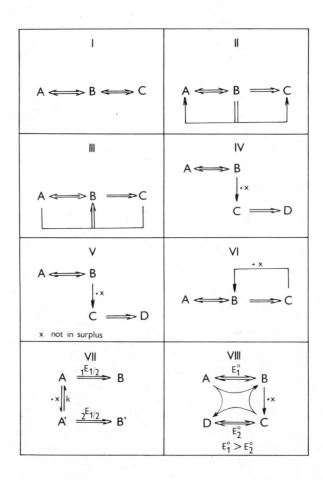

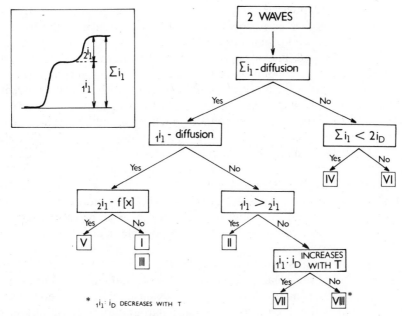

Figure 9.11 Flowcharted analysis of two subsequent waves:

I—Two subsequent one-electron, reversible electrode processes.

II—One-electron, reversible electrode process followed by a one-electron irreversible electrode process, coupled with disproportionation of the intermediate B.

III—Electrode reactions as in II, coupled with reproportionation of A and C yielding B.

IV—One-electron, reversible electrode reaction followed by slow follow-up reaction yielding an electroactive product.

V—Same as IV, reactant causing conversion of B into C not in surplus; that is, diffusion of X is a rate-determining step.

VI—Electrode reactions as in II; final product C reacts under regeneration of B (catalytic process).

VII—Two electroactive forms of depolarizer in sluggish equilibrium undergoing separate electrochemical reduction (i_D defined for the sum of A and A′).

VIII—Electrode-catalyzed substitution reaction; reversible, one-electron electrode reaction, product B is chemically converted to C, which is part of redox couple with more negative redox potential; process is coupled with chemical reaction $A + C \rightarrow B + D$.

In schemes I to VIII species on one vertical line correspond to the same oxidation state; horizontal direction represents a change in the redox state.

⇔   reversible electrode reaction;

⇒   irreversible electrode reaction;

→   a chemical reaction;

i_D one-electron, diffusion-controlled limiting current.

of the diffusion current i_D corresponding to a one-electron, diffusion-controlled electrode reaction of the given compound and its comparison with the actually observed current. A diffusion current much smaller than i_D indicates the presence of an electroactive impurity, the compound investigated being inactive.

The determination of i_D, or generally, the number of electrons exchanged per one particle of the depolarizer, is a complex problem. Criteria such as the log-plot slope or any other approach which analyzes the shape of the current–potential curve are very often misleading (see reduction with formation of "semiquinone"-like intermediates [20]). Coulometric determinations in which the electrode reaction is coupled with special chemical processes, as in disproportionation of intermediates, also might yield incorrect results. Frequently the diffusion coefficient is unknown for i_D calculation. Mostly comparisons with diffusion currents of pilot compounds, which are of similar size and charge type as the compound studied, have to be used. For pilot compounds the value of i_D has to be known unambiguously. It is advisable not to limit the comparison to one pilot compound but to use several such species and thus to frame up the most probable value of i_D. Diffusion coefficients are sensitive to size and charge type and to charge distribution; that is, the influence of localized ion-pairs and specific solvation. Therefore, if higher precision is needed, or values obtained appear doubtful, pilot compounds as near as possible in structure to the depolarizer are to be chosen.

Identification of i_D is the key to utilizing diagnostic criteria. It is always advisable to record the polarographic current–potential curve and to analyze the observed limiting current. Rarely is the mechanism of the electrode process properly solved without knowledge of the polarographic $i–E$ curve and its comparison with the estimated i_D values.

The power of a simple analysis is demonstrated by the flowchart in Figure 9.11 in the frequently encountered situation of a compound yielding two reduction waves.*

The overall limiting current of two waves $\sum i_l$ and the limiting current of the first wave $_1i_l$ are compared with the one-electron diffusion current i_D. When the nature of the rate controlling step for $\sum i_l$ and $_1i_l$ and the dependence of $\sum i_l$ and $_1i_l$ on temperature, and in special cases on composition of solution, are known, comparison with i_D immediately resolves eight basic mechanisms. A more detailed analysis may be performed to confirm the proposed mechanism and to evaluate the kinetic and thermodynamic parameters of the individual steps of the overall electrode process.

Complicated mechanisms may be solved unambiguously only by comparison of experimental dependencies with those predicted from the theoretical

---

*The behavior of the criteria of Figure 9.11 is observed for the whole range of experimental conditions. Therefore, the answers YES for the criterion $_1i_l > _2i_l$ means that under all accessible experimental conditions, such as temperature, concentration of depolarizer, and pH of the solution, the first limiting current is higher than the second one. The answer NO implies that under certain experimental conditions $_1i_l$ might be smaller than $_2i_l$ but does not exclude the possibility that $_1i_l$ might increase with the change of some parameter above $_2i_l$. Such behavior is typical, for example, for mechanism VII, in which $_1i_l$ might vary essentially from zero to $\sum i_l$ with a concentration change of X. Figure 9.11 criteria are not strictly valid under limiting conditions. Thus, for mechanism VII the above situation might arise when $_1i_l \gg _2i_l$. In this case $_1i_l$ would behave as an almost purely diffusion-controlled current and already the second step in our flowchart would be misleading. These examples emphasize the need to study as broad a range of experimental conditions as possible to avoid judgment from limiting cases or incompletely defined criteria.

solution of the corresponding rate law of a model mechanism. The most widely used method for these comparisons is a plot of a measurable, or of a quantity derived from it by simple algebraic transformation, against the variable—either directly against the experimental variable or against a function of it. Thus, function $\chi(t_1)$ is used rather than drop time t_1 for kinetic current. Such plots usually yield an adequate answer; however, in many cases predicted differences in plots for various mechanisms are smaller than the experimental error. Great uncertainty arises, for example, in correcting for double-layer or adsorption effects when quantities describing the interface are calculated from other, independent measurements. For some of the more complicated mechanisms, testing the mechanism by using the plot of the measurable against a properly chosen parameter is impossible because some of the more complicated analytical formulas cannot be obtained. It is then advisable to use more sophisticated methods for the analysis of experimental data. One of the approaches is the direct simulation of the current–potential curves or any other experimentally obtainable dependence of measurables and comparison of curves thus obtained with experimental results. Simulated curves are calculated for a set of parameters such as rate constants to determine optimal agreement between the experimental and simulated dependence (see, e.g., [103]). Thus it is possible to deduce more complicated mechanisms and evaluate corresponding parameters such as rate of equilibrium constants. However, it is still unadvisable to limit the comparison to one curve, and is preferable to investigate the correctness of the fit by comparing a set of experimental and simulated curves dependent on certain variables, for example, temperature, concentration, and so on. Electrode behavior properly solved generally involves the following:

1. Determination of the depolarizer state in the solution, its stability, total electrochemical activity, and estimation of the one-electron diffusion current.

2. Use of simple diagnostic criteria, either qualitatively or quantitatively if analytical expressions are available.

3. Comparison of experimental and simulated curves and best-fit procedure to confirm the mechanism and to evaluate the necessary quantitative parameters.

4. Confirmation of the predicted mechanism by special electrochemical techniques or by independent methods, such as the detection of the nature of intermediates and products by spectroscopic methods *in situ* at the electrode surface or in the course of exhaustive electrolysis.

When deducing the full mechanism of an electrode reaction do not limit the electrochemical method to only one technique, but collect information with a suitable combination of electrochemical methods covering a large time constant rate—that is, from fast, single sweeps up to controlled potential electrolysis.

Following are examples of the situations and interpretations of the individual steps of electrode mechanisms [104].

Considerable information about more complicated polarographic mechanisms, for example, N-alkylpyridinium cations [105], is obtained from controlled-potential coulometric measurements, and the following reaction scheme is probably valid:

$$A \underset{\longleftarrow}{\overset{+n_1 e}{\longrightarrow}} B$$
$$\downarrow \text{slow}$$
$$C \overset{+n_2 e}{\Longrightarrow} D$$

In dc polarography a one-electron wave results in alkaline solutions (pH 11.9, 75% by volume dimethylsulfoxide) at -1.28 V versus SCE with 4×10^{-4} mol/L N-ethylpyridinium iodide. The following homologues were investigated:

N-alkyl	methyl	ethyl	n-propyl	n-butyl	n-pentyl
$E_{1/2}$ (V)	-1.27	-1.28	-1.29_5	-1.30	-1.32

In controlled-potential coulometry with a DME at -1.5 V, the following values of n result:

N-alkyl	methyl	ethyl	n-propyl	n-butyl	n-pentyl
n	1.28	1.21	0.97	0.99	0.81

Controlled-potential (CP) coulometric measurements taken at the potential of the limiting current with a stirred mercury pool electrode yield the following results for n:

N-alkyl	methyl	ethyl	n-propyl	n-butyl	n-pentyl
n	3.19	2.36	1.26	1.37	1.27

Consequently, the generally slow chemical follow-up reaction B→C proceeds more rapidly with the small N-methyl derivative than with the bulky N-n-pentyl. The intermediate radical B is unstable at the applied potential and its existence is only proven by electron spin resonance spectroscopy after a reaction with a spin trap. The reaction B⇒C can be stopped by adding ascorbic acid to the solution before electrolysis; in this case an electroinactive product results from B and the following CP coulometric results are obtained:

N-alkyl	methyl	ethyl	n-propyl	n-butyl	n-pentyl
n	1.60	1.42	0.96	1.03	0.97

The values of n are substantially different from those obtained without blocking, and illustrate that CP coulometry yields additional information if the series of measurements proceeds at varying potentials along the entire i–E curve, as with several p-fluoro-arylalkylketones [106] such as

$$F\!-\!\langle\ \rangle\!-\!CO\cdot R \quad (\equiv A)$$

in buffered solutions at pH 6.5. Here, experimental n_{exp} approaching 1 is obtained at less negative potentials (on the foot of the wave); at more negative potentials, in particular at those corresponding to the limiting current, values of n_{exp} between 1.5 and 1.8, depending on the alkyl group, result (Figure 9.12). Thus, at less negative potentials, a radical ($\equiv B$) is formed ($n_1 = 1$); depending on pH it may be also protonated, and is not further reduced to the corresponding secondary alcohol ($\equiv C$; $n_2 = 1$) since the uptake rate of the second electron is slower at this potential than the dimerization rate leading to the dimer D:

$$A \xrightarrow[n_1]{+e} B \xrightarrow[n_2]{+e} C$$

Reaction with mercury

$$+ B$$

(Mercury compound) D

With increasing negative potentials a competition between the two mechanisms starts and continues until the two-electron uptake $[n_{exp} \to (n_1 + n_2)]$ prevails at

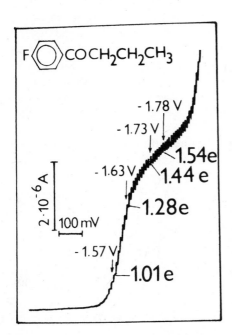

Figure 9.12   The apparent number of electrons n_{app} obtained with constant potential at a stirred mercury pool electrode as a function of the reduction potential. Polarographic i–E curves recorded with a DME: 10^{-3} M p-fluorobutyrophenone in 0.1 M phosphate buffer pH 8.5 (20% ethanol). The figures on the left indicate the potential of the mercury pool electrode versus SCE; those on the right, the value of n_{app} thus obtained.

the limiting current potential with the almost exclusive formation of C:

Acetophenone, propiophenone, butyrophenone, and so on [106] yield the same results. Such reasoning assumes there is no difference between $i-E$ curves recorded with a DME and those recorded with a stirred mercury-pool electrode. Because this is not always true, a repeated recording of the $i-E$ curve with a stationary mercury-drop electrode proved very useful. After several recordings, the change of $E_{1/2}$, and possibly an inactivation of the mercury surface, is sometimes detected, usually when a chemical interaction between the electrode material such as mercury and the electrode process intermediates occurs; for example, in the anodic oxidations of the aldehydic group of aromatic aldehydes at mercury electrodes. Another reason for such a phenomenon is the strong reactant adsorption. Occasionally such interference in coulometric measurements is prevented by rigorous stirring of the mercury-pool electrode, by using a DME, exceptionally by using a solvent in which the substances are more easily dissolved, or by changing the electrode material [107]. If electroactive products are formed in the electrode process, the interpretation becomes easier. The procedures are well known with dc polarography, with cyclic voltammetry, or Kalousek polarography.

The Kalousek commutator not only reveals the electrode process reversibility, as originally derived, but detects electroactive products in general [108]. Thus, electrochemical cleavage of a C—Hal bond is very often proven by the appearance of the halide anodic process commutated wave. The halides are set free at the original reduction potential. This happens, for example, with all reducible halogenopyridines [109] and also with 2-cyano and 4-cyanopyridine [110–112]. With the auxiliary potential of the Kalousek commutator, E_{aux}, maintained at the limiting current potential of the two-electron wave corresponding to the reduction of the C—Hal or C—CN bond, an irreversible anodic wave corresponding to the reaction of Hal$^-$ or CN$^-$ with mercury ions is observed on the commutated curve. In 2,3,5,6-tetrafluoro-1,4-dicyanobenzene [113] this proof is somewhat complicated by the reversible primary formation of a radical anion, very unstable at potentials less negative than that of the limiting current:

$$(90)$$

The more complicated electrode process of 2,2′-pyridil [114] yields 2,2′-pyridoin in a two-electron reduction, if not directly then by a symmetrical primary product, which in a chemical follow-up step rearranges to 2,2′-pyridoin:

2,2′-Pyridil

$2e, 2H^+$
pH 7.5
50% ethanol
$E_{1/2} = -0.59$ V

$-2e, -2H^+$

$E_{1/2} = -0.1$ V

$2e, 2H^+$
$E_{1/2} = -1.2$ V

2,2′-Pyridoin

$2e, 2H^+$
$E_{1/2} = -1.2$ V

With the Kalousek commutator ($E_{aux} = 0.0$ V) 2,2′-pyridoin yields a cathodic reduction wave at -0.59 V corresponding to the product reduction of 2,2′-pyridil. Conversely, 2,2′-pyridil yields an anodic wave at -0.1 V with $E_{aux} = -0.70$ V, thus revealing the formation of 2,2′-pyridoin in the first reduction wave of 2,2′-pyridil. Additionally, the half-wave potentials of the first reduction wave of 2,2′-pyridil and of the only reduction wave of 2,2′-pyridoin coincide regardless of the solution's pH value.

5 POLAROGRAPHIC ACTIVITY: THEORETICAL AND PRACTICAL APPLICATIONS OF POLAROGRAPHY

5.1 Polarographic Activity of Inorganic Compounds

Many inorganic ions (free, but usually coordinated) and inorganic compounds are reducible or oxidizable at the DME and at solid electrodes. Correlations between polarographic activity and chemical structure are involved and almost the entire Periodic Table of Elements could be discussed. However, to illustrate the possibility of studying or determining inorganic compounds, the present knowledge of this field is only alphabetically arranged in Table 9.7,

Table 9.7 Polarographic Activity of Elements

In column A are elements that are in some form (ions, complexes, etc.) polarographically active. In column B are elements that can be determined by stripping methods (anodic, cathodic, adsorptive). In circles are elements frequently determined by polarography or stripping methods.

	A	B		A	B
Actinium					
Aluminum	+	+	Molybdenum	⊕	+
Americium			Neodymium	+	
Antimony	⊕	⊕	Neon		
Argon			Neptunium	+	
Arsenic	⊕	⊕	Nickel	⊕	⊕
Astatine			Niobium	+	
Barium	+	+	Nitrogen	+	
Berkelium			Osmium	+	
Beryllium	+		Oxygen	⊕	
Bismuth	⊕	⊕	Palladium	+	
Boron			Phosphorus	+	
Bromine	⊕	+	Platinum	+	+
Cadmium	⊕	⊕	Plutonium	+	
Calcium	+		Polonium		
Californium	+		Potassium	+	+
Carbon	+		Praseodymium	+	
Cerium	⊕	+	Promethium	+	
Cesium	+		Protactinium	+	
Chlorine	⊕	+	Radium	+	
Chromium	⊕	+	Radon		
Cobalt	⊕	⊕	Rhenium	+	+
Copper	⊕	⊕	Rhodium	+	
Curium	+		Rubidium	+	
Dysprosium	+		Ruthenium	+	+
Erbium	+		Samarium	+	
Europium	+		Scandium	+	
Fluorine			Selenium	⊕	+
Francium			Silicon	+	+
Gadolinium	+		Silver	⊕	⊕
Gallium	+	+	Sodium	+	+
Germanium	⊕	+	Strontium	+	+
Gold	⊕	+	Sulfur	+	+
Hafnium	+	+	Tantalum	+	
Helium			Technetium	+	
Holmium	+		Tellurium	⊕	+
Hydrogen	+		Terbium	+	
Indium	⊕	⊕	Thallium	⊕	⊕
Iodine	⊕	+	Thorium	+	

Table 9.7   *(continued)*

In column A are elements that are in some form (ions, complexes, etc.) polarographically active. In column B are elements that can be determined by stripping methods (anodic, cathodic, adsorptive). In circles are elements frequently determined by polarography or stripping methods.

	A	B		A	B
Iridium	+		Thulium	+	
Iron	⊕	+	Tin	⊕	⊕
Krypton			Titanium	⊕	
Lanthanum	+		Tungsten	⊕	+
Lead	⊕	⊕	Uranium	⊕	⊕
Lithium	+		Vanadium	⊕	+
Lutetium	+		Xenon		
Magnesium	+		Ytterbium	+	
Manganese	⊕	+	Yttrium	+	
Mercury	⊕	⊕	Zinc	⊕	⊕
			Zirconium	+	+

[a]References [20, 115].

which is not exhaustive, but illustrates the possibilities of polarography in inorganic compounds, with special respect to analytical applications.

5.2   Polarographically Reducible and Oxidizable Groups in Organic Substances

Table 9.8 separately exemplifies polarographic reducibility and oxidizability of organic substances on mercury and solid (i.e., platinum, carbon, and gold) working electrodes. The table is not exhaustive, and only demonstrates the possibilities of the method. The activity of a functional group or of the molecule at a given electrode material does not necessarily mean that its polarographic determination or investigation is convenient or recommended. Among the reducible groups most often investigated or determined with the DME are C—N, C—O, C—S, C—Hal, N—N, O—O, S—S, C=C, C=O, C≡N, N=N, N—O, and NO_2 [116–118].

The table does not contain functional groups polarographically active only because of their ability to form insoluble or complex compounds with mercury, such as —SH (e.g., cystein), dithiocarbamates, —NH—CO—NH— (e.g., barbiturates), —NH—CS—NH— (e.g., thiobarbiturates), —NH—NH$_2$ (e.g., phenylhydrazine), R_2N— (ethylenediamine, EDTA). Generally, the systems termed reversible [116, 118]—which even in aqueous solutions behave reversibly, that is, $E_{1/2} = E^0$—exhibit corresponding slopes in the semilogarithmic plot and give a signal with the Kalousek commutator or in cyclic voltammetry;

Table 9.8 Examples of Polarographically Active Bond Groupings in Organic Substances

Bond Grouping	Example and Overall Mechanism

REDUCTIONS

C—N

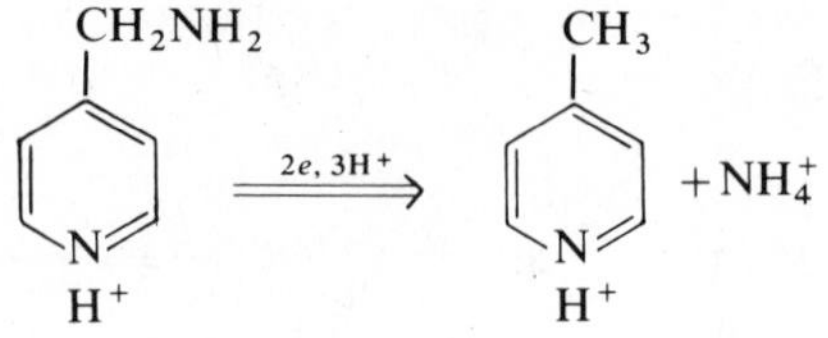

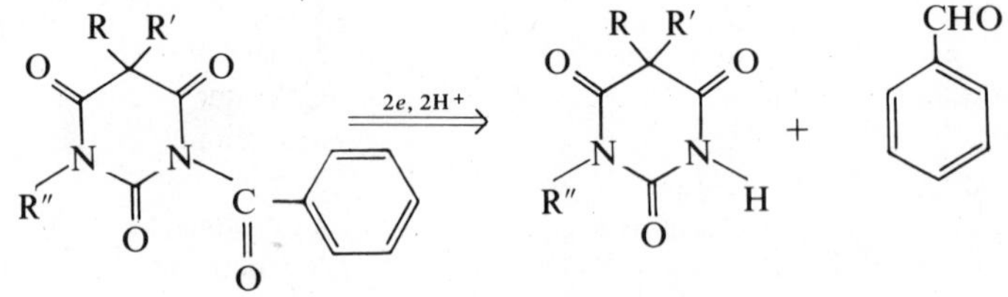

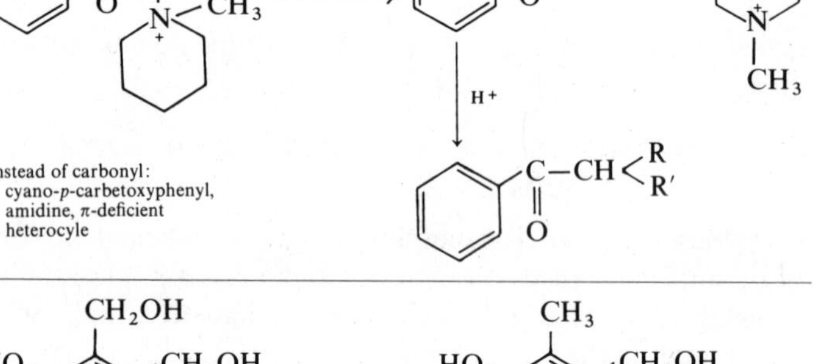

$\xrightarrow{2e,\,3H^+}$ $+ NH_4^+$

$\xrightarrow{2e,\,2H^+}$ $+$ CHO

$\xrightarrow{2e,\,H^+}$ $+$

$\downarrow H^+$

Instead of carbonyl:
cyano-*p*-carbetoxyphenyl,
amidine, π-deficient
heterocyle

C—O

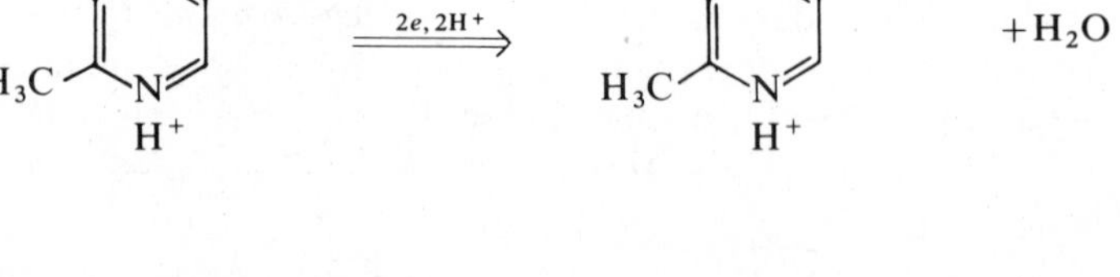

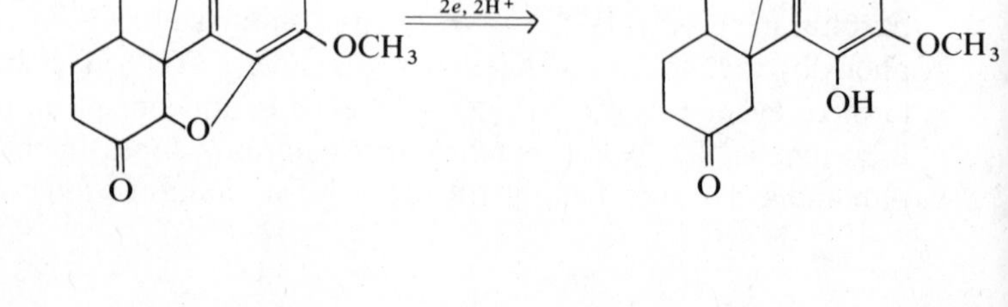

$\xrightarrow{2e,\,2H^+}$ $+ H_2O$

$\xrightarrow{2e,\,2H^+}$

Table 9.8 (*continued*)

Bond Grouping	Example and Overall Mechanism

REDUCTIONS

$$C_6H_5-\underset{O}{C}-\underset{\underset{OH}{|}}{C}\underset{R'}{\overset{R}{<}} \xrightarrow{2e,\,H^+} C_6H_5-\underset{O^-}{C}=C\underset{R'}{\overset{R}{<}} + H_2O$$

$$\downarrow +H^+$$

$$C_6H_5-\underset{O}{C}-CH\underset{R'}{\overset{R}{<}}$$

C—S

$$C_6H_5-\underset{O}{C}-\underset{\underset{SC_6H_5}{|}}{C}\underset{R'}{\overset{R}{<}} \xrightarrow{2e,\,H^+} C_6H_5-\underset{O^-}{C}=C\underset{R'}{\overset{R}{<}} + C_6H_5SH$$

$$\downarrow +H^+$$

$$C_6H_5-\underset{O}{C}-CH\underset{R'}{\overset{R}{<}}$$

C—Hal

$$R-CH_2-Br \xrightarrow{2e,\,H^+} R-CH_3 + Br^-$$

$$C_6H_6Cl_6 \xrightarrow{6e} C_6H_6 + 6Cl$$

(γ-Hexachlorocyclohexane)

N—N

CONHNH$_2$ (pyridine ring) $\xrightarrow{2e,\,2H^+}$ CONH$_2$ (pyridine ring) $+ NH_3$

O—O

(peroxide bridged cyclohexane) $\xrightarrow{2e,\,2H^+}$ HO—(cyclohexene)—OH

S—S

$$HOOC-\underset{\underset{NH_2}{|}}{CH}-CH_2-S-S-CH_2-\underset{\underset{NH_2}{|}}{CH}-COOH \xrightarrow{2e,\,2H^+}$$

$$\xrightarrow{2e,\,2H^+} 2\,HS-CH_2-\underset{\underset{NH_2}{|}}{CH}-COOH$$

857

Table 9.8 (*continued*)

Bond Grouping	Example and Overall Mechanism

REDUCTIONS

$C=C$

$$RO-\!\!\!\bigcirc\!\!\!-CH\!=\!CH-\!\!\!\bigcirc\!\!\!-OR \xrightarrow{2e,\,2H^+}$$

$$\xrightarrow{2e,\,2H^+} RO-\!\!\!\bigcirc\!\!\!-CH_2\!-\!CH_2\!-\!\!\!\bigcirc\!\!\!-OR$$

$C=O$

$$\bigcirc\!\!\!-\underset{O}{\overset{\|}{C}}-Alk \xrightarrow{e,H^+} \bigcirc\!\!\!-\underset{OH}{\overset{\cdot}{C}}-Alk \xrightarrow[\text{Acid}]{e,\,H^+} \bigcirc\!\!\!-\underset{OH}{\overset{|}{C}H}-Alk$$

Alkaline

$$\bigcirc\!\!\!-\underset{\underset{\underset{OH}{|}}{\overset{OH}{|}}{\overset{|}{C}}-Alk$$

$C=N$

$$\underset{N-CH_2-COOH}{CH_3-\underset{\|}{C}-CH_3} \xrightarrow{2e,\,2H^+} \underset{HN-CH_2-COOH}{CH_3-\overset{|}{C}H-CH_3}$$

$N=N$

$$\bigcirc\!\!\!-N\!=\!N-\!\!\!\overset{NH_2}{\underset{}{\bigcirc}}\!\!\!-NH_2 \xleftarrow{2e,\,2H^+} \bigcirc\!\!\!-NH-NH-\!\!\!\overset{NH_2}{\underset{}{\bigcirc}}$$

$N\to O$

$$\underset{H_3C}{\overset{NH-CO-CH_2-N(C_6H_5)_2}{\bigcirc}}\overset{CH_3}{\underset{\overset{\|}{O}}{}} \xrightarrow{2e,\,2H^+}$$

$$\underset{H_3C}{\overset{NH-CO-CH_2-N(C_6H_5)_2}{\bigcirc}}\overset{CH_3}{}$$

$$\xrightarrow{2e,\,2H^+} \qquad +H_2O$$

Table 9.8 (*continued*)

Bond Grouping	Example and Overall Mechanism

REDUCTIONS

N=O

$$\text{CH}_3(\text{CH}_2)_3\overset{\overset{\displaystyle N=O}{|}}{\underset{\overset{|}{H^+}}{N}}\!\!-\!\!\underset{}{\bigcirc}\!\!-\!\!\text{COO}-\text{CH}_2-\text{CH}_2-\text{N}(\text{C}_2\text{H}_5)_2 \xrightarrow{4e,\,4H^+}$$

$$\xrightarrow{4e,\,4H^+}\quad \text{CH}_3(\text{CH}_2)_3\overset{\overset{\displaystyle NH_3^+}{|}}{N}\!\!-\!\!\underset{}{\bigcirc}\!\!-\!\!\text{COO}-\text{CH}_2-\text{CH}_2-\text{N}(\text{C}_2\text{H}_5)_2+\text{H}_2\text{O}$$

$-\text{NO}_2$

$$\text{O}_2\text{N}-\bigcirc \xrightarrow{4e,\,4H^+} \text{HOHN}-\bigcirc +\text{H}_2\text{O}$$

C=N
(Heteroaromatic)

$$\overset{\displaystyle \text{CO}-\text{NH}-\text{CH}_2-\text{CH}_2-\text{N}(\text{C}_2\text{H}_5)_2}{\underset{N}{\bigodot\!\bigcirc}-\text{OC}_4\text{H}_9} \xrightarrow{2e,\,2H^+}$$

$$\xrightarrow{2e,\,2H^+}\quad \overset{\displaystyle \text{CO}-\text{NH}-\text{CH}_2-\text{CH}_2-\text{N}(\text{C}_2\text{H}_5)_2}{\underset{\underset{H}{N}}{\bigodot\!\bigcirc}\overset{H}{-}\text{OC}_4\text{H}_9}$$

OXIDATIONS
(*Mostly at nonmercury electrodes, e.g., platinum, carbon, or gold*)

C—H

$$\text{CH}_3-\text{H} \xrightarrow{-2e,\,-H^+} \text{CH}_3^+$$

$$\downarrow \text{H}_2\text{O} \,\vert\, \text{CH}_3\text{CN}$$

$$\text{CH}_3\text{NHCOCH}_3$$

$$\overset{\displaystyle \text{CH}_2\text{OH}}{\underset{\text{OCH}_3}{\bigcirc}} \xrightarrow{-2e,\,-2H^+} \overset{\displaystyle \text{CHO}}{\underset{\text{OCH}_3}{\bigcirc}}$$

Table 9.8 *(continued)*

Bond Grouping	Example and Overall Mechanism

OXIDATIONS

(Mostly at nonmercury electrodes, e.g., platinum, carbon, or gold)

C—C

$$2\,CH_3COO^- \xrightarrow{\;-2e\;} CH_3CH_3 + 2\,CO_2$$

C—N

Amines

Aliphatic

$$CH_3-CH_2-NH_2 \xrightarrow{\;-e\;} CH_3-CH_2-\overset{(+)}{NH_2} \xrightarrow{\;-e\;}$$

$$\xrightarrow{\quad} CH_3-CH=NH_2$$

Primary aromatic

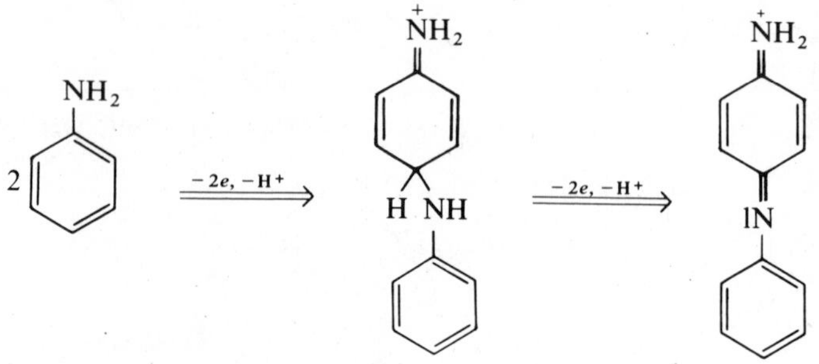

Secondary

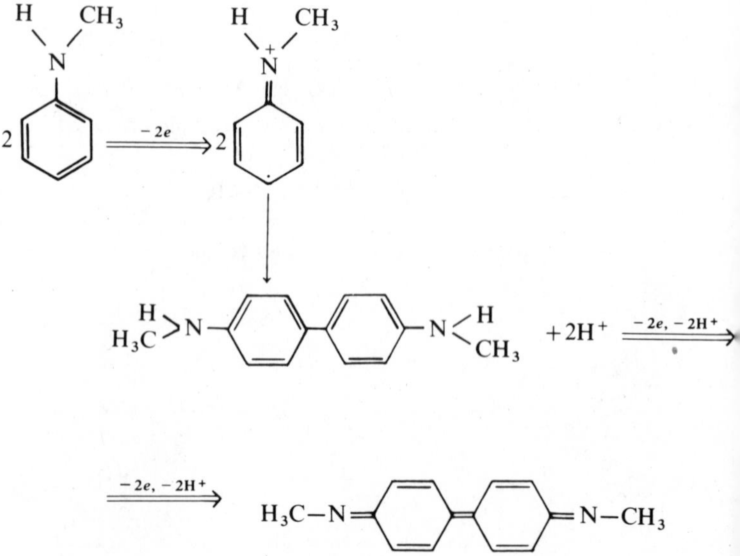

860

Table 9.8 (*continued*)

Bond Grouping	Example and Overall Mechanism

OXIDATIONS

(Mostly at nonmercury electrodes, e.g., platinum, carbon, or gold)

Tertiary

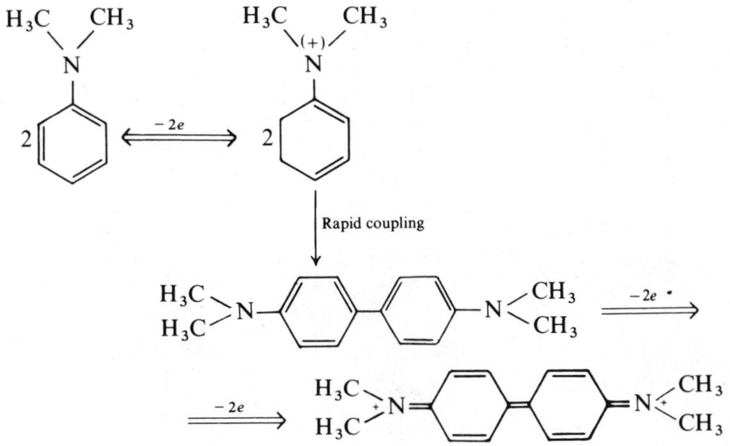

Phenylenediamines

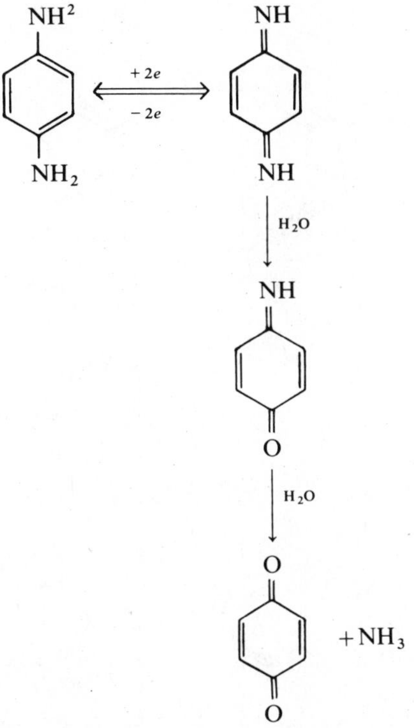

Table 9.8 (continued)

Bond Grouping	Example and Overall Mechanism

OXIDATIONS
(Mostly at nonmercury electrodes, e.g., platinum, carbon, or gold)

Aminophenols

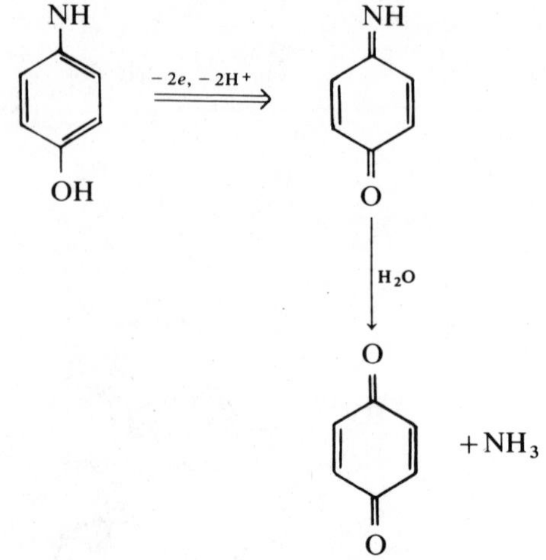

C—O
Phenols

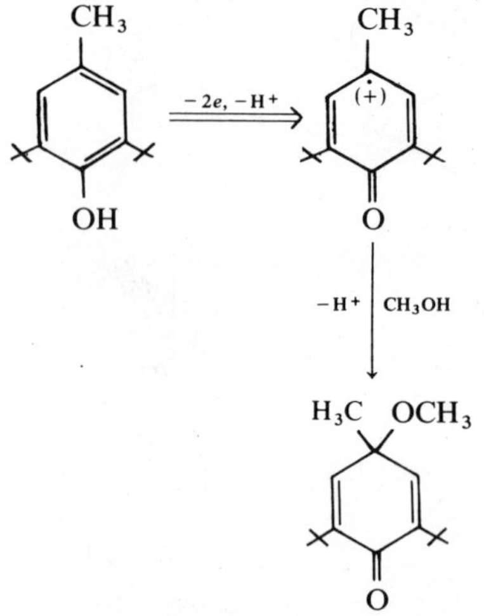

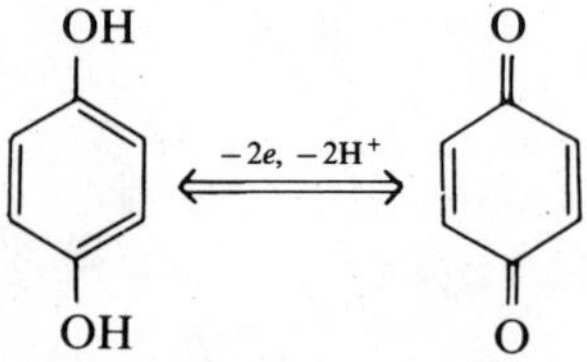

Table 9.8 (continued)

Bond Grouping	Example and Overall Mechanism

OXIDATIONS
(Mostly at nonmercury electrodes, e.g., platinum, carbon, or gold)

Ethers

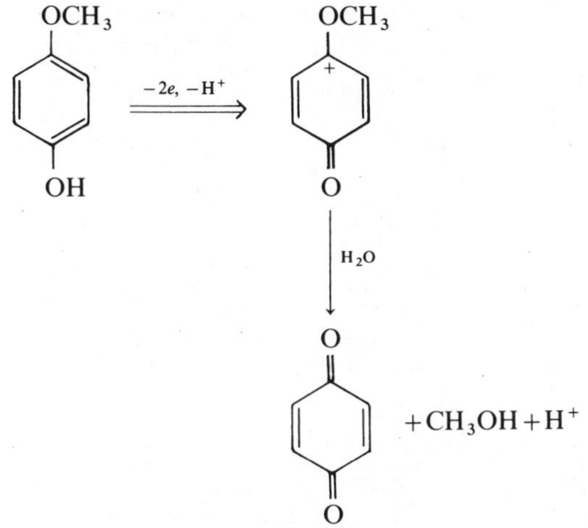

C—S

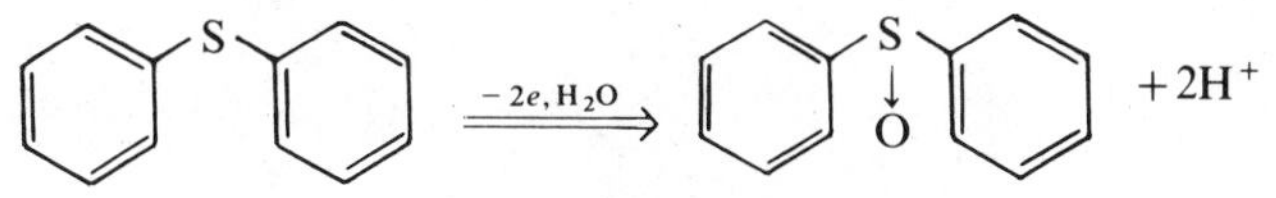

Diphenyl sulfoxide

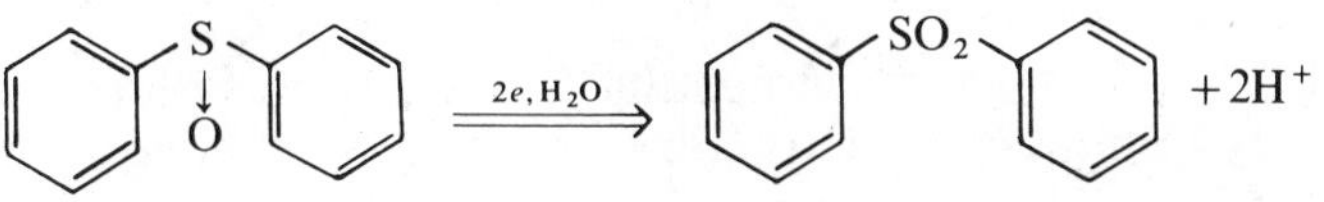

Diphenyl sulfone

they are not emphasized in Table 9.8. These systems are typified by quinone/ hydroquinone, nitrosobenzene/phenylhydroxylamine, some heterocyclic dyes, 4,4'-bipyridines, and 1,2(4-bipyridyl)ethylenes [116, 118]. In the processes identified as partially reversible, the reversible electron uptake or loss is followed by a rapid irreversible chemical follow-up reaction, as with N-alkylpyridinium cations [105] in reductions and with ascorbic acid [119] in oxidations.

Some substances, the polarographic determination (mostly at the DME) of which is important, for some reason (particularly if no other convenient method is available, e.g., in pharmaceutical analysis and/or toxicology) neither reduce nor oxidize at the electrode. In such a case, the so-called functionalization is very often made use of, that is, a transformation of the substance to be determined into an electroactive derivative [117, 120]. The most important cases, characteristic of pharmaceutical applications, are shown in Table 9.9.

Analytical procedures based on functionalization were often underestimated because their accuracy was lower (about 3–5%) than in direct polarography and they were not always specific. However, recent methods are more selective, yield almost 100% in the functionalization chemical step, and are very useful in solving some involved and difficult analytical problems. The determination of the psychotropic metamphetamine [121] proceeds as follows:

$$(92)$$

5.3   Influence of Molecular Structure on Polarographic Behavior of Organic Compounds

Electrochemical behavior of organic substances is necessarily affected by their structure, and this property has at least three different factors: it is the nature of the electroactive group, the nature of molecule substituents, and very often, also the nature of electrode process stereochemistry. The characteristic measurables of the polarographic wave, for example, $E_{1/2}$ or the limiting current i_l, the slope of the wave, the number of waves, and electron consumption are only affected by the whole framework [116, 118] of the molecule to be investigated. Changes in such measurables proceed in a reproducible manner and may even be predicted if the structure and its changes are known (and vice versa).

Hammett, as discussed by Zuman in [122], was the first to derive an equation for the linear dependence of two series of rate and equilibrium constants for aromatic substances substituted in the para- and meta-positions of the reacting

Table 9.9   Functionalization in Organic Polarography

Group Introduced	Method	Substance to be Determined
—NO$_2$	Direct nitration	Phenazone, phenacetine
	Reacting phenols with HNO$_2$	Morphine
	Reacting with picrylfluoride	Methamphetamine
—NO	Nitrosation with HNO$_2$	Secondary amines, local anesthetics
—Br	Substance + H$_2$SO$_4$ + KBr + KMnO$_4$	Citric acid: CBr$_3$COCHBr$_2$ results as the electroactive product
C=N—	Reaction with ammonia, primary amines, hydrazine, hydroxylamine, semicarbazide, Girard D reagent	Aliphatic and alicyclic ketones and aldehydes— irreducible under given conditions: better developed waves with some reducible compounds; simultaneous determination of pyridoxal and pyridoxal-5-phosphate, 17-ketosteroids
Introduction of oxygen into the molecule	Oxidation with Br$_2$, Ce(IV), H$_2$O$_2$, HNO$_3$, NaNO$_3$	Menadiol to the corresponding quinone, lidocaine to its N-oxide, chloropromazine to its sulfoxide

group.

$$\log k - \log k^0 = \rho(\log K - \log K^0) \tag{93}$$

$$\log k - \log k^0 = \rho\sigma \tag{94}$$

$$\log K - \log K^0 = \rho\sigma \tag{95}$$

where k represents the rate constant for a homogeneous chemical reaction, K represents the equilibrium constant (useful in reversible cases), superscript 0 denotes the constant for the reference substance, σ is a constant characterizing the substituent, and ρ is the reaction coefficient, constant for a given reaction. The values of ρ and σ are tabulated for a large number of reactions [122]. Applied in polarography this equation acquires the following form:

$$E_{1/2} - E_{1/2}^0 = \rho\sigma \tag{96}$$

under the assumption that $\alpha n = (\alpha n)^0$. Because the influence of changing the substituents in a process involving the same electroactive group is in the logarithmic relationship between the half-wave potential and the rate constant of an irreversible electrode reaction, the shift of $E_{1/2}$ under the influence of substituents is most often investigated, particularly in quantitative treatments of such problems (Figure 9.13). In a polarographically active compound, a structure change, such as in the position of the same substituent on a benzene or heteroaromatic ring (e.g., pyridine or quinoline), may fundamentally change the electrode process mechanism and yield completely different products. For example, at pH $\simeq 1$, 4-cyanopyridine [112] is transformed to 4-aminomethylpyridine in a four-electron process. However, only one electron is consumed by 3-cyanopyridine; thus, a radical results leading to a dimer.

If only the $E_{1/2}$ shift of the dc polarographic wave of a particular reducible group is followed under the influence of a substituent, certain empirical qualitative rules have been developed [116, 122] that may be easily used as the first orientation. In such a case the tacit assumption is made that the substituent affects the electron density in the reaction center but is sufficiently separated from this area so that the geometry of the transition state remains unchanged. In a nucleophilic reaction (i.e., an electron uptake in polarography) in the benzene series, substituents possessing a $+I$ (CH_3) or $+T$ effect (OCH_3, OH, NH_2) shift $E_{1/2}$ to more negative values, while $E_{1/2}$ of substances with $-I$

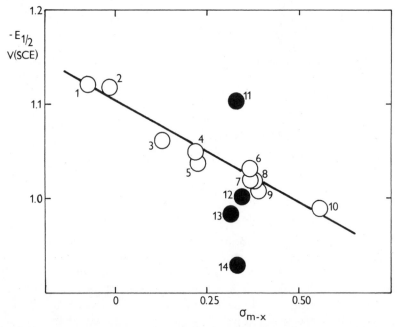

Figure 9.13 Dependence of half-wave potentials of m-substituted phenyl isothiocyanates on the σ constants. Black circles do not fulfill the linear relationship: 1—CH_3; 2—OH; 3—OCH_3; 4—$NHCOCH_3$; 5—C_6H_5; 6—Cl; 7—$COOC_2H_5$; 8—$COCH_3$; 9—$COOC_6H_5$; 10—CN; 11—NCS (2nd w); 12—COC_6H_5; 13—I; 14—NCS (1st w).

substituents (e.g., Cl, Br, I) or a $-T$ substituent (NO_2, COOH) are shifted to more positive values as compared with the unsubstituted compound. The *o*- and *p*-derivatives of a compound with a $-T$ or $-I$ effect are reduced at a less negative potential than the *m*-derivative. The opposite occurs with a $+T$ substituent. The term "I effect" includes both inductive I_s and inductomeric I_d effects, whereas "T effect" includes both mesomeric M and electromeric E effects, all in accord with Ingold's conception. These rules are very useful in polarography of benzenoid compounds and are also applicable in polarography of heterocyles if the changed meaning of the substituent position in the hetero-cyclic nucleus is recognized. The heteroatom effect in the nucleus is specific. In six-membered nuclei the nitrogen atom is looked upon as a strong $-T$ substituent in the benezene nucleus; and, based on this concept, the 2- and 4-substituted pyridines are reduced at less negative potentials than the 3-derivatives [112]. The potential regions in which monosubstituted pyridines with exocyclic electroactive groups are reduced, compared to the corresponding benzene derivatives, are shown in Figure 9.14. Generally, heterocycles with a π-deficient nucleus, such as pyridine and quinoline, are more easily reduced than those with a π-excessive ring, such as pyrrol.

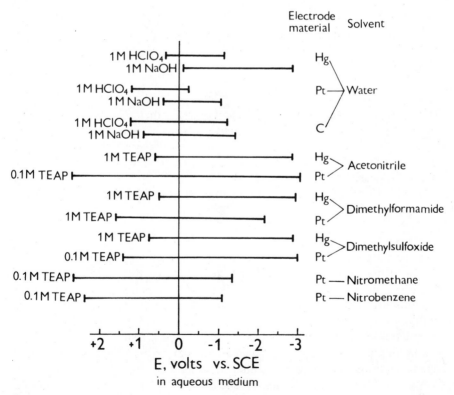

Figure 9.14 Voltage limits for different electrode materials in various solvents with different supporting electrolytes.

Considerable progress has been achieved by using linear free energy relationship equations [122], in particular the Hammett (96) and also the Taft equations. Substitution effects on the free energy of activation change or on the standard free energy change must be resolved into independent contributions of polar, resonance, and, possibly, steric effects. Empirical characteristic constants of the polar, resonance, or steric effects of each substituent have been obtained by comparing the logarithms of rate or equilibrium constants of suitable reactions under the influence of substitution. The differences in standard redox potentials of reversible systems are proportional to the standard free energy change, as are the half-wave potentials of irreversible systems to the change in free energy of activation. According to Taft, as reported by Zuman in [122], the influence of individual contributions to the reaction rate in homogeneous reactions is

$$\log k = P + S + M \tag{97}$$

where P denotes the polar, S the steric, and M the mesomeric contribution. By combining this relationship with the equation for the half-wave potential and the rate constant of an electrode process a new equation results (for irreversible processes only):

$$\Delta E_{1/2} = P + S + M \tag{98}$$

where $\Delta E_{1/2}$ is the difference between the half-wave potential of the substance with the substituent X and the reference substance bearing a hydrogen atom instead of X

$$\Delta E_{1/2} = (E_{1/2})_X - (E_{1/2})_H \tag{99}$$

The values of P, S, and M are independent of each other and cannot be directly determined. The problem may be solved by studying a reaction series where only P is operative and continuing with those compounds in which two effects are expected, such as $P + M$ or $P + S$. In some cases a complete solution of the above equation is possible.

A simple form of the above relation with respect to the polar effect only is the Hammett equation

$$\Delta E_{1/2} = P = \rho_\pi \sigma_X \tag{100}$$

It was first used for benzenoid derivatives bearing an active group and a p- or m-substituent; o-substituents had to be excluded. σ_X is the total polar constant characterizing the nature and position of the aromatic system substituent. These constants are independent of the electroactive group and of polarographic conditions and have been often tabulated, for example, see reference [122]. The reaction constant expressed in volts is $\rho_{\pi,R}$, typical of the polarographically active group attached to a given nucleus, strongly affected by external experimental conditions although independent of other substituents, and considered a measure of the $E_{1/2}$ shift for a given substituent. A positive value points to a nucleophilic mechanism of the electrode process electron uptake as the potential-determining step and a negative value is characteristic of either a dissociation mechanism analogous to $S_N 1$, or a radical mechanism. The latter mechanism has only been observed, for example, with aliphatic halogen derivatives. The application of (100) to heterocyclic systems has also been discussed in detail [122].

A polarographic form of the Taft equation has been proposed for reaction series with prevailing polar effects:

$$E_{1/2} = P = \rho_{\pi,\text{Red}}^* \sigma^* \tag{101}$$

Symbol significance is analogous to that in (100), but the reference substituent X is CH_3. In such a reaction series, the electroactive group is not separated from the substituent, but the steric effects are assumed to be constant in the whole series. The equation is particularly useful with aliphatic compounds.

Equations (100) and (101) enable a decision about the ring or the molecular part bearing the electroactive group and the nature of substituent X, with the latter possibility more often encountered. The following applications are further claimed: prediction of the $E_{1/2}$ of unstudied substances, determination of the reaction mechanism from the sign of the constant ρ, determination of new values of σ, and occasionally the quantitative separation of polar, steric, and resonance substituent effects.

The practical use of these relationships is somewhat restricted by such experimental conditions as the necessity for obtaining all potential values under identical conditions; and for proceeding with an identical electrode mechanism that has the same α and number of electrons and protons. Surprisingly, validity could have been verified despite, for example, the strong adsorption of the electroactive compounds at the electrode, in most cases. Consequently, the half-wave potentials are not exact measures of the factors P, S, and M in (98).

The plots $\Delta E_{1/2}$ versus σ are primary relationships, similar relationships have been found with other physicochemical quantities, such as frequencies in IR and UV absorption spectra, and dissociation constants.

5.4   Isomerism (Steric) Effects in Polarographic Behavior

In addition to electronic effects, polarographic behavior is also influenced by the steric arrangement [123] of the molecule containing an electroactive group (or several groups). However, in this case general rules or generally valid conclusions are not presented, because almost every case requires an individual approach and a solution of the particular interaction. Two substances with both a completely different polarographic effect and a similar steric arrangement occur.

Alkyl groups introduced into aromatic derivatives decrease conjugation and make the substance less easily reducible than its parent substance. This happens, for example, with o-methyl-substituted benzophenones, o-methyl-substituted acetophenones, o-methyl-substituted azobenzenes, and o-methyl-substituted stilbenes. It was attempted to describe such effects by the term $\delta_\pi E_X^s$ in a modified Taft equation.

The I effect of the OH group in isomeric nitrophenols should make the half-wave potentials of *p*-nitrophenol and *o*-nitrophenol more difficult than that of *m*-nitrophenol. However, in acid media $E_{1/2}$ of *o*-nitrophenol is about 120 mV more positive than that of the other two isomers, which means that the formation of an intramolecular hydrogen bond exerts a stronger influence than the I effect. Similar behavior is found with *o*-nitroacetanilide and its isomers. However, the intramolecular bond in *o*-hydroxybenzaldehyde makes the reduction more difficult than that of *p*-hydroxybenzaldehyde. In geometric isomerism, the higher intrinsic energy of the *cis*-isomer should make it more easily reducible than the *trans*-isomer. This actually occurs particularly with reversible systems such as azobenzene in acid solutions, but does not with irreversible processes. One must consider cases individually; thus, maleic acid is more easily reduced than fumaric acid, but the opposite is true with their diethyl esters. Furthermore, the *cis* form is more difficult to reduce with stilbene, 1,4-distyrylbenzene, and 1,2-dibenzoylethylene. Sterically different products can be obtained by changing reduction potentials. This happens, for example, with benzil in neutral or slightly acidic solutions: two products, *cis*- and *trans*-stilbenediol, are formed; the *trans*-isomer partly isomerizes to *cis* under the influence of the electric field, and both isomers are transformed to benzoin at a different rate [124].

Optical antipodes do not differ in their polarographic behavior. The diastereoisomers may differ; that is, erythro $+\alpha, \alpha'$-dibromosuccinic acid reduces more easily than the *threo* form, if they are not dissociated:

$$
\begin{array}{cc}
\mathrm{COOH} & \mathrm{COOH} \\
| & | \\
\mathrm{Br-CH} & \mathrm{Br-CH} \\
| & | \\
\mathrm{Br-CH} & \mathrm{CH-Br} \\
| & | \\
\mathrm{COOH} & \mathrm{COOH} \\
\textit{erythro-} & \textit{threo-}
\end{array}
$$

However, there is no difference in the behavior of their esters.

Questions about the reduction mechanism of organic substances—for example, the reduction of individual isomers and their tendency to be reduced by one rather than another of possible mechanisms—are resolved with quantum chemical calculations; in most cases the LCAO–MO is used (which, strictly speaking, only holds with reversible systems). The willingness of a molecule or particle to accept the first electron in a reduction is adequately described in dc polarography by the energy k_{-1} of the lowest free molecular orbital LFMO [112, 125]. The lower this energy, the more easily the molecule accepts the electron. The treatment proved useful in pyridine-type heterocycles, in sulfur-containing heterocycles, benzenoid hydrocarbons, α,ω-diphenylpolyenes, polyenes, and so on. The energies were calculated by means of HMO approximation. Some of the nitrogen compounds [126] and their $E_{1/2}$ values are shown in Table 9.10.

Table 9.10 Dc Polarographic Data for Pyridinelike Heterocycles

Substance	$E_{1/2}$	$\dfrac{2.3RT}{\alpha nF}$ (mV)	Energy of LFMO $k_{-1}(\beta)$
Acridine	−1.24	188	−0.325
Phenanthridine	−1.64	136	−0.518
Quinoline	−1.69	160	−0.527
Benzo[f]quinoline	−1.70	145	−0.537
Benzo[f]isoquinoline	−1.41	176	−0.603
	−1.78	102	
Isoquinoline	−1.84	175	−0.576
Pyridine	−2.07	191	−0.841

The correlation is expressed by [127]

$$E_{1/2} = 2.127 k_{-1} - 0.555$$

and energy of the lowest vacant molecular orbital is expressed by the value of k_{-1}.

A more complicated example was found for mono- and disubstituted cyano-pyridines [112], in which the sequence of k_{-1} values obtained by HMO does not correspond to the sequence of the half-wave potentials $E_{1/2}$. Therefore k_{-1} values calculated by the self-consistent field (SCF) method were also applied. Except for a single derivative these values give a sequence of $E_{1/2}$ in good agreement with the sequence of k_{-1}. Additionally, the electron densities ^7q for the individual positions of the primary radical with the π-electron septet express the ease of further reduction; that is, the possibility of a nucleophilic attack or the tendency to protonation—an electrophilic attack. The free valence ^7F of the same particle reflects the tendency to dimerization or radical attack, and is greatest in positions 5 of 2-cyanopyridine, and 2 and 3 of 4-cyanopyridine. Both values are clearly smaller than 7F of 3-cyanopyridine in positions 4 and 6, 0.558 and 0.580, respectively. The latter values are the highest found in the three isomers for the particle with the cyclic π-electron septet. This agrees with the one-electron reduction of 3-cyanopyridine over the whole pH region. The tendency toward protonation is expressed by the electron density ^7q in the individual positions of the primary radical. The higher this value the easier protonation. From ^7q for the individual isomers and for the nitrogen in the heterocyclic nucleus and in the nitrile group, respectively, of 2-cyanopyridine (1.362 and 1.398), 3-cyanopyridine (1.250 and 1.404), and 4-cyanopyridine (1.380 and 1.386), one concludes that, generally, protonation of the exocyclic nitrogen atom proceeds more easily. Unless we consider the competitive dimerization, protonation would be easiest with the radical of 3-cyanopyridines. However, protonation does not play a role in the reduction of 3-cyanopyridine.

The high indices ^7F for the radical attack lead to a preference for dimerization. The protonation of the nitrogen atom should be easiest with 4-cyanopyridine because there is only a slight difference between ^7F of both nitrogen atoms. The further reducibility of the radical is shown by the value of ^7q on the nitrile group.

This value does not differ greatly for all three isomers, but the smallest π-electron density was obtained in 4-cyanopyridine $^7q_C + {^7q_N} = 2.370$, followed by 2-cyanopyridine, and finally by 3-cyanopyridine; this agrees with experimental findings. The above-tabulated material and considerations allow an interpretation for the reduction of isomeric cyanopyridines, based on the intermediate formation of a particle with a cyclic π-septet.

5.5   Kinetics of Chemical Reactions

Polarographic measurements may conveniently determine rate constants [20] and investigate the mechanisms of homogeneous chemical reactions in the bulk of the solution and not coupled with an electrode process. These measurements may be carried out when at least one of the participants of the chemical reaction is reducible or oxidizable at the working electrode, in most cases at the DME. Consequently, either the decrease of the reacting component height or the wave-height increase of the resulting component may be followed as a function of time. If the half-wave potentials of the participants are well separated, one may even follow the formation and the disappearance of an active intermediate. Depending on the stability of the participants or on their lifetime the practical performance of the experiment may differ. In the case of slow reactions, $t_{1/2} > 15$ min, whole $i-E$ curves are recorded and the wave heights or their functions plotted versus time. If the half-time of the reaction is between 15 s and 15 min, a continuous record of the decrease or increase in the wave height of a reaction participant with time is employed. This measurement applies a constant potential, corresponding mostly to the limiting current, to the working electrode. The limiting currents thus recorded are usually diffusion controlled. However, kinetic or even catalytic polarographic currents may be used for this purpose, with calibration dependencies plotted and extreme care devoted to maintaining a constant temperature.

With the help of polarography most kinetic measurements are performed directly in the cell in which the reaction takes place. When a reaction is very slow, samples are taken at certain intervals and mixed with the supporting electrolyte solution, and the $i-E$ curve is recorded.

With fast, $t_{1/2} < 15$ s, reactions the reactant's concentration changes even during the life of a single drop; instantaneous currents on a single drop may be used for kinetic measurements [128].

The use of kinetic currents is completely different from the kinetic measurements described in the preceding paragraphs. These currents are controlled by a chemical reaction in the vicinity of the electrode. The combination of the chemical reaction with the electrode process may proceed in three different ways: the chemical reaction may precede the reaction process, run parallel to, or follow the electrode process. Theoretical analysis of the currents resulting from these reactions allows the determination of the rate constant of the chemical reaction. In this way very fast reactions may be followed that are often otherwise very difficult to study by other methods. Calculating rate constants from kinetic currents by both approximate and rigorous methods is also summarized above.

The three types of kinetic currents are exemplified as follows:

1. In preceding reactions the electroactive form of the depolarizer is produced by a chemical reaction from a species that is inactive at the given potential (a CE mechanism). The transformation rate constant of the inactive to the active form can be computed, and has been done, for example, with formaldehyde [20], isomeric pyridine-aldehydes (in both cases the hydrated form is electroinactive), reducible weak acids (only the nondissociated form is reducible), a number of monosaccharides, and dissociating complexes. Half-wave potentials of kinetic currents of complexes enable determination of their stability constants [20].

2. First-order reactions that are parallel to the electrode process and yield kinetic currents are exemplified with the catalytic reduction of hydrogen peroxide in the presence of haemin, the reduction of titanium in the presence of hydroxylamine, by uranium in the presence of nitrates [20], and so on. Second-order reactions are best represented by the reaction of a ferrous EDTA complex with a low concentration of hydrogen peroxide.

3. The reactions subsequent to the electrode process are illustrated by the case where the product of a reversible electrode process is transformed into a polarographically inactive substance by a chemical reaction (an EC mechanism). Such behavior very often occurs with organic systems such as ascorbic acid, which is reversibly oxidized in a two-electron process to an unstable intermediate and is further transformed into the inactive final product. More importantly, however, are the bimolecular follow-up reactions occurring after a reversible single-electron uptake, first illustrated with the one-electron reduction of N-alkylpyridinium cations [105, 129]. The primary resulting radical dimerizes and yields an irreducible dimer. Because of strong adsorption the one-electron reduction wave is distorted; therefore, evaluation of the polarographic curves and rate constant determination of the dimerization are very inaccurate. This often occurs with organic compounds in all the kinetic currents described in the preceding paragraphs. The reduction of aromatic carbonyl compounds such as benzaldehyde or acetophenone is sometimes treated in organic polarography [130, 131]. These reactions occur as follows:

$$2 \begin{array}{c} R \\ \diagdown \\ C=O \\ \diagup \\ R' \end{array} \xrightarrow{2e,\,2H^+} 2 \begin{array}{c} R \\ \diagdown \\ \overset{\displaystyle\cdot}{C}-OH \\ \diagup \\ R \end{array} \Big\downarrow k_2 \tag{102}$$

$$\begin{array}{cc} R\ R' & R\ R' \\ \diagdown\diagup & \diagdown\diagup \\ C & - & C \\ | & | \\ OH & OH \end{array} \tag{103}$$

Equations (102) and (103) only hold for aqueous solutions (in aprotic media radical anions are primarily formed). The half-wave potential of the wave corresponding to the above process was derived [131]:

$$E_{1/2} = E^0 - 0.36 \frac{RT}{nF} + \frac{RT}{3nF} \ln(ck_2 t_1) \tag{104}$$

Unlike a first-order reaction, $E_{1/2}$ depends on c in addition to the drop time t_1 and the rate constant of inactivation k_2.

5.6   Polarographic Analysis

5.6.1   Qualitative Analysis

Although characteristic of the compound or group to be reduced or oxidized at the electrode, the half-wave potential alone or analogous data is seldom used for quality determination of an unknown substance. Solution composition, particularly the solvent and the supporting electrolyte, is an important factor affecting the half-wave potential of the same compound or electroactive group.

The supporting electrolyte is regarded as a complexing agent or an ion-pair forming agent in most inorganic processes and as a buffer solution or a source of hydrogen ions in organic compounds. Buffers are often recommended in aqueous solutions and have two functions: they represent the supporting electrolyte and maintain a constant pH value. To fulfill both tasks, the buffer concentration must be approximately fifty times higher than that of the substance to be determined [132, 133]. If the pH is comparable to pK_a of the buffer acid, a twenty times higher buffer concentration suffices [134].

Originally, most polarographic measurements were carried out with aqueous solutions. Because of the many organic substances to be investigated and the low solubility of some inorganic substances, organic solvents and sometimes their mixtures with water were introduced in polarographic analysis. In addition to their analytical application, pure nonaqueous, particularly aprotic, solvents proved extremely useful in studying and interpreting more complicated electrode mechanisms. In aprotic media the intermediates are more stable in processes where the substance takes up several electrons or where the electron uptake is followed by a chemical reaction of the intermediate, for example, the originally two-electron wave is separated into two one-electron steps that may be more easily interpreted. If radicals result as intermediates, not only electrochemical methods, but also ESR spectroscopy are used [135]. A condition of stabilizing the primary intermediates is the absence of protons and proton donors, as well as that of oxygen, in the solution [135]. Special techniques [136] have been developed for purifying the solvents and for measuring with N,N'-dimethylformamide, acetronitrile, dimethylsulfoxide, dichloromethane, tetrahydrofuran, and so on. Removing the last traces of water is a serious problem. However, the demands for solvent purity and the absence of water are not so large in normal analytical work where the electrode process is already known (Figure 9.14).

It follows from preceding chapters that the half-wave potential of a substance or a species must be jointly given with the conditions under which it was measured. These data are published in the form of tables, with the first volume containing half-wave potentials of inorganic ions and substances published by Vlček [137]. A more detailed series of several volumes with $E_{1/2}$ values for inorganic substances combined with other electrochemical data, for example, limiting currents, number of electrons consumed or set free in the electrode process, mechanism of the electrode process, and so on, is being published by Meites and Zuman [115]. A similar series of as yet five volumes with data about half-wave potentials of organic substances has also been compiled by Meites and Zuman [138].

In aqueous solutions half-wave potentials of organic substances are often a function of the supporting electrolyte pH. These values are only pH independent in pH regions where the prevalently present solution species directly undergoes the electrode process. This is the case, for example, with alkyl halides. Generally, the $E_{1/2} = f(\text{pH})$ plot is expressed by modifying the following equation:

$$E = E^0 + \frac{RT}{\alpha nF} \ln\left(0.886 k_e^0 \sqrt{\frac{t_1}{D}}\right) - \frac{RT}{nF} \ln(K)$$
$$+ \frac{pRT}{nF} \ln([H^+]) - \frac{RT}{\alpha nF} \ln\left(\frac{\bar{i}}{\bar{i}_d - \bar{i}}\right) \tag{105}$$

It follows from this relationship that

$$E_{1/2} = \text{const} - \frac{2.3pRT}{\alpha nF}\, \text{pH} \tag{106}$$

or

$$\frac{\Delta E_{1/2}}{\Delta \text{pH}} = -\frac{2.3pRT}{\alpha nF} \tag{107}$$

According to Mairanovskii [139] it holds further at a given ionic strength (if $p = n$) that

$$E_{1/2} = \text{const} + \frac{RT}{F} \ln([H^+]) \exp\left(-\frac{\psi F}{RT}\right) \tag{108}$$

is the mean value of the electrostatic potential with respect to the bulk of the solution at the distance of an ionic radius from the electrode.

For a change in the ionic strength we can derive (z is the charge of the particle)

$$\Delta E_{1/2} \cong \Delta\psi \left(\frac{\alpha n - z}{\alpha n} + \frac{\Delta E_{1/2}}{\Delta \text{pH}} \frac{F}{2.3RT}\right) \tag{109}$$

With reducible organic cations, $z > 0$: this is why a considerable shift of $E_{1/2}$ to more negative values is observed in irreversible reductions with an increase in ionic strength. The shift may reach 200 mV for a tenfold increase in ionic strength. This value agrees fairly well with that calculated from (109) for $z = 1$,

$\alpha n \doteq 0.3$, and $\Delta E_{1/2}/\Delta \text{pH} \doteq -60$ to -70 mV/pH. Such a value has been observed, for example, in the polarographic reduction of piperidine-N-oxide. Consequently, the half-wave potentials must be measured at a strictly defined composition of the supporting electrolyte. Only then may the experimental data be compared with those tabulated in the literature, where the composition of the solution must also be presented.

When polarographic measurements are applied to qualitative analysis (this is not often the case), the experimental $E_{1/2}$ is compared with that tabulated in the literature for the assumed structure. If the complete structure is unknown, the information is compared with the potential regions in which the individual functional (electroactive) groups on the same framework or nucleus (Figure 9.15) are reduced or oxidized, along with a Hammett or Taft plot for such a compound (Figure 9.13), to determine possible substituents and their positions. Next, $\Delta E_{1/2}/\Delta \text{pH}$ in pH-dependent reactions is determined and followed by calculation or determination of the number of electrons taking part in the electrode process. Combination of these data yields a compound's structure in suitable situation. An important application of polarography in qualitative

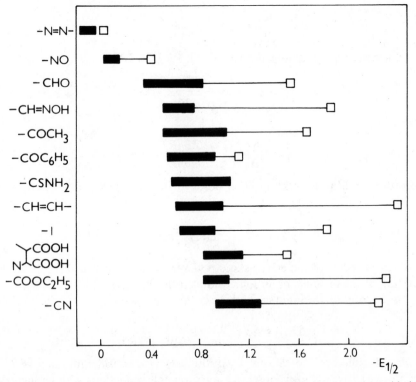

Figure 9.15 Polarographic half-wave potentials of substituted pyridines in comparison with their benzene analogues: pH 4–5; rectangles: pyridine derivatives (sequence of $E_{1/2}$:4>2>3); squares: benzene derivatives.

analysis of organic compounds is the determination of the form in which a substance exists in solution: this is the case, for example, with aldehydes whose hydrated form is nonreducible at the DME [140], with ketones in their free and cyclic form (e.g., sugars) [141], and with enediolic substances such as 2,2'-pyridoin [114]. However, in all situations one must decide whether the two forms are in frozen equilibrium, where the equilibrium constant can be directly determined, or whether the transformation rate is rapid enough to cause kinetic currents.

5.6.2   Quantitative Analysis

The most frequent practical application of polarography is in quantitative analysis. The sensitivity of classic dc polarography is in the 10^{-5} mol/L range as the lower quantitatively determined concentration limit. Thus the method was useful and attractive but not sensitive enough, particularly when analyzing low concentrations of substances in biological matrices, or, generally, in trace analysis. More recent pulse methods increased sensitivity to about 10^{-7}–10^{-8} mol/L as the lower concentration limit in differential pulse polarography. Therefore average organic reducible substance concentrations down to 1–10 ng/mL may be assayed or at least detected. The sensitivity limit may reach even 10^{-9} mol/L with anodic, or cathodic, stripping pulse methods (Figure 9.16).

The evaluation of quantitative measurements is usually based on calibration curves; in complicated samples, such as biological matrices, standard addition is preferred. The absolute method of the Ilkovič equation or of analogous relationships is practically never used. Measuring wave (the limiting current) or peak height is described above.

The reproducibility of quantitative polarographic measurements characterizes the accuracy of polarographic procedures and is a function both of the method and the concentration of the substance to be determined. Thus in dc polarography reproducibility is 1.5–2% for $c > 10^{-4}$ mol/L, 3% for $c < 10^{-4}$ mol/L, and 5% for $c \leqslant 10^{-5}$ mol/L. The minimum absolute quantity determined with this method is about 500 ng. In DP polarography the reproducibility is about 2% at concentrations above 10^{-5} mol/L and 10% in the region approaching 10^{-7} mol/L, and the same for square-wave polarography at very low concentrations. In ac polarography the reproducibility is 2% for concentrations above 10^{-5} mol/L. With sufficiently small volumes (0.05–0.5 mL) DP polarography determines quantities as low as 0.1 ng with substances exhibiting $M = 100$.

When calculating concentrations of polarographically active components in mixture, resolution power of the method is quantitatively and qualitatively important. In dc polarography at equal concentrations (equal wave heights) of both electroactive components the $\Delta E_{1/2}$ difference must be more than 100 mV if quantitative evaluation is performed. However, wave separation generally depends on three independent factors: $\Delta E_{1/2}$, wave slopes (in the rising portion), and limiting current ratios. Meites [142] suggested the following relationship for the least acceptable $\Delta E_{1/2}$ when two substances, A and B, are to be deter-

LOWER SENSITIVITY LIMITS OF ANALYTICAL METHODS USED IN
ORGANIC ANALYSIS

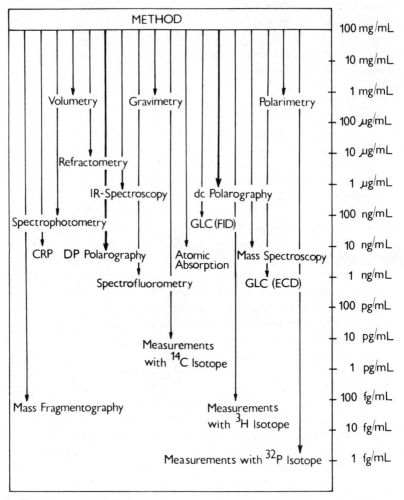

Figure 9.16   Lower sensitivity limits of analytical methods used in organic analysis.

mined simultaneously:

$$|E_{1/2}| \geqslant 0.05 + 1.5[(E_{3/4} - E_{1/4})_A + (E_{3/4} - E_{1/4})_B] \pm (E_{3/4} - E_{1/4})_B \log\left(\frac{i_B}{i_A}\right) \quad (110)$$

where $E_{1/4}$ and $E_{3/4}$ are the potentials at which one-fourth or three-fourths of
the limiting current i_A and i_B, respectively, are reached and the last term on the
right is added to the first two terms if both A and B are determined. For two
reversible two-electron waves of equal height a difference $\Delta E_{1/2} = 0.13$ V is
necessary for sufficiently accurate determination. In inorganic polarography,

an inconvenient situation is usually resolved by suitable complexation. With organic substances a pH change is often helpful if working with aqueous solutions; otherwise, another nonaqueous solvent and/or a different supporting electrolyte should be sought. The above relationships typify classic polarography; the $\Delta E_{1/2}$ requirements are much improved by DP polarography.

The current peak heights from this more advanced method are independent of each other if not overlapping, the measured currents are nonadditive, and the minimum distance between E_p—that is, ΔE_p—approaches an average of 40 mV.

5.6.3   Analytical Applications

Polarographic methods are applied in organic analysis most frequently in pharmaceutical chemistry and pharmacology, in polymer chemistry, in the foodstuff industry, in criminology, and, more recently, in environmental research (i.e., when solving ecological problems).

In pharmaceutical analysis [117, 143–147] polarography conveniently determines vitamins (vitamin B_1, B_2, B_6, B_{12}, C, E, and all forms of K), steroids (testosterone, progesterone, prednisone, hydrocortisone, and others directly, and seventeen ketosteroids indirectly), a long series of antibiotics (such as cephalosporine, streptomycine, rifamycine, tetracycline, oxytetracycline, chloramphenicol, griseofulvine, and their metabolites), chemotherapeutics (such as nitrofuran derivatives and sulfonamides), and several less important groups of pharmaceuticals. In psychopharmaceutics the numerous and very important groups of 1,4-benzodiazepine derivatives [144], 4-fluoro-substituted butyrophenones [106, 148] with a reduction at the DME in both groups, phenothiazines [149], amitriptyline [150], and related compounds, which are oxidized, are determined. Polarographic reduction of most 1,4-benzodiazepines is based on the two-electron reduction of the 4,5-C=N double bond in the seven-membered heterocycle. Many methods have been devised for assaying individual benzodiazepines in tablets, coated tablets, injections, and so on.

However, only DPP procedures enable assay of 1,4-benzodiazepines, such as chlorodiazepoxide, diazepam, oxazepam, flurazepam, or nitrazepam, in biological matrices such as blood, blood serum, blood plasma, urine, and saliva, where the substance concentration to be determined is too low (10–1000 ng/mL), often with their metabolites present and also to be determined, and, finally, with the shape of the dc polarographic wave deformed and decreased in height by surface-active substances in the sample.

Occasionally, DPP requires extraction followed by separation. Drug consumption by organisms, in different organs and in blood, and its excretion and metabolism in body tissue are followed and characterized by time profiles made possible by these techniques. Administration of too high or too low drug dosage is thus prevented.

Food analysis, closely related to the above problems, requires the assay of two different kinds of sample components. First, desirable compounds—for example, vitamins present in the food—have to be determined; this may be exemplified with ascorbic acid, which is assayed by means of its oxidation wave

[117, 118, 119], often without complicated prior separation. The other class of components are contaminants such as antibiotics like tetracycline in meat and meat products, or animal fodder growth factors like nitrovin [151] [1,5-di(5-nitro-20-furyl)-1,4-pentadien-5-one aminohydrodazone hydrochloride]. The sample is dissolved in triethanolamine (with 10% by volume methanol) and acetate buffer. DPP calculates concentration down to about 2×10^{-7} mol/L, with a detection limit of 7×10^{-8} mol/L. Certain plant-material foods may contain growth-factor-type contaminants, but more harmful is the presence of pesticide traces, such as herbicides, fungicides, insecticides, and perhaps rodenticides that may appear in plant-originated foodstuffs and also, of course, in meat.

Many of them are polarographically active, especially those containing an aromatic nitro group, for example, parathion, halogenated compounds of the DDT type, disulfides, bipyridylium salts, and heterocycles containing two or three nitrogen atoms. Polarographic procedures for their determination are reviewed in several articles [152–154] and in a book chapter [155]. Occasionally, indirect determination preceded by a chemical reaction such as nitrosation in carbaryl or reacting with iodine in warfarin must be performed [156].

Carcinogens are extremely dangerous foodstuff and environmental contaminants [152]. Polarographic methods exist for polycyclic hydrocarbons with four and more fused rings, for aromatic amines and related nitro compounds with two and more rings, azo compounds, nitroso compounds—particularly N-nitrosamines [156], and the group of aflatoxin [157]. Aflatoxin B_1 was determined in rice, milk, and corn by DPP after a Sephadex LH-20 separation. The concentrations were 16.0, 0.2, and 2.5 μg/mL, respectively. In the DPP determination of N-nitrosamines the detection limit was between 0.045 and 0.241×10^{-6} mol/L depending on pH.

5.6.4   Polarography and Chromatographic Separation Techniques

Despite the relatively high resolving power of the different polarographic methods described, it is often necessary to separate electroactive components if their half-wave potentials are adjacent. This was first attempted in the early 1950s as a modification of simple column-adsorption chromatography where the eluent was analyzed with a DME maintained at a constant potential [157, 158]. Nitro compounds and alkaloids were analyzed in this way.

In a later development the substances to be determined were separated by thin-layer chromatography [147, 159–162]. The position of the spots was determined in UV light, the spots were scratched off, and the substance to be determined was eluted, usually with N,N-dimethylformamide, which virtually yielded total extraction. The extract was mixed with a suitable supporting electrolyte, such as an aqueous buffer. This procedure proved to be successful with 1,4-benzodiazepine drugs and their metabolites, analgesics, and later, with some steroidal compounds.

Most recent and promising is the use of polarographic detectors in flow-through detection, by way of a constructed DME detector [163]. Later, polarographic continuous-flow detection was improved by synchronization, and by the use of pulse polarographic methods—normal and differential pulse—and phase-sensitive alternating current measurements. The detector was used in the HPLC (high-performance liquid chromatography) mode. The eluent background effects were eliminated with the so-called electrochemical eluent scrubber. By this technique the 1,4-benzodiazepine derivative, nitrazepam, was determined at the nanogram level, even after separation from blood serum. Selectivity was increased by phase-sensitive alternating current measurements, and four 1,4-benzodiazepines were determined in a mixture. The detection limit was 0.2 ng per injection of 25 μL. Occasionally the detection is made with multiple electrodes [164].

HPLC combined with polarographic detection can be applied in forensic science [165], and is useful for drug analysis of illicit drugs or in the detection of drugs in body fluids after a suspicious death, poisoning, or even when detecting a driver under the influence of drugs. Sensitivity is the decisive factor only in the latter cases and in controlled-potential electrochemical detection. Morphine, codeine, methadone, cannabis metabolites, tricyclic drugs, and cocaine with its metabolites can be detected at the 10 pg level with heterocyclic sulfur and at the 100 ng level for secondary amines in the oxidative mode of a "wall-jet" electrochemical detector.

A combination of liquid chromatography and reductive electrochemical detection led to the determination of nitro aromatic, nitramine, and nitrate esters explosives [166] in explosive mixtures and in smokeless gunpowders. However, from this a method has been developed for the detection of organic "gunshot residue" on the hands of persons who have discharged a weapon. The detection limit is in the vicinity of 1 pmol.

5.6.5 Conclusion

In the course of the last 15 years polarography has again become one of the most sensitive analytical methods, which, on the other hand, is only based on relatively inexpensive instrumentation. The number of polarographic analyzers which are produced by several firms particularly in the United States, France, Switzerland, and Czechoslovakia is still increasing. Whereas the sensitivity of classic dc polarography could be compared only with IR spectroscopy (about 1 μg/mL), differential pulse polarography, which is at present the most often used pulse polarographic method, possesses a sensitivity equal to that of atomic absorption or mass spectroscopy (1–10 ng/mL). This property still increases with stripping and pulse stripping methods where the lower sensitivity limit may be of the order of 10^{-10} mol/L (i.e., below 1 ng/mL). Adsorptive stripping polarography, a recently introduced electroanalytical method, not only increases the sensitivity, but also enables a polarographic determination of organic substances that are neither reducible nor oxidizable at the electrode. In such a case only the capacity current due to adsorption or desorption usually in the dc mode is measured.

The most important application fields of modern polarographic methods are pharmaceutical and pharmacological analyses, analyses in the food industry, in environmental research, and so on. Moreover, the knowledge of the relationships between the polarographic behavior and the chemical structure contributes not only, for example, to fundamental organic chemistry and elucidation of redox processes, but also to developing new electrosynthetic methods, as demonstrated in Baizer's monograph *Organic Electrochemistry* [118].

References

1. D. M. Mohilner, "The Electrical Double Layer. Part I: Elements of Double Layer Theory," in A. J. Bard, Ed., *Electroanalytical Chemistry. A Series of Advances*, Vol. 1, Dekker, New York, 1966, p. 241.

2. R. M. Reeves, "The Electrical Double Layer: The Current Status of Data and Models, With Particular Emphasis on the Solvent," in B. E. Conway and J. O'M. Bockris, Eds., *Modern Aspects of Electrochemistry*, Vol. 9, Plenum, New York, 1974, p. 239.

3. R. Payne, "The Electrical Double Layer in Nonaqueous Solutions," in P. Delahay, Ed., *Advances in Electrochemistry and Electrochemical Engineering*, Vol. 7, Interscience, New York, 1970.

4. R. Parsons, "The Structure of the Electrical Double Layer and Its Influence on the Rate of Electrode Reactions," in J. O'M. Bockris, Ed., *Modern Aspects of Electrochemistry*, Vol. 1, Butterworths, London, 1952, p. 1.

5. D. Ilkovič, *Collect. Czech. Chem. Commun.*, **6**, 498 (1934).

6. D. Ilkovič, *J. Chem. Phys.*, **35**, 129 (1938).

7. J. Koutecký, *Czech. J. Phys.*, **2**, 50 (1953).

8. T. Kambara and I. Tachi, *Bull. Chem. Soc. Japan*, **25**, 135 (1952).

9. H. Matsuda, *Bull. Chem. Soc. Japan*, **36**, 342 (1953).

10. T. E. Cummings and P. J. Elving, *Anal. Chem.*, **50**, 1980 (1978).

11. J. Heyrovský and D. Ilkovič, *Collect. Czech. Chem. Commun.*, **7**, 198 (1935).

12. J. Tomeš, *Collect. Czech. Chem. Commun.*, **9**, 12 (1937).

13. J. Koryta, *Electrochim. Acta*, **1**, 26 (1965).

14. M. v. Stackelberg and H. v. Freyhold, *Z. Elektrochem.*, **46**, 120 (1940).

15. J. Lingane, *Chem. Rev.*, **29**, 1 (1941).

16. A. A. Vlček, *Collect. Czech. Chem. Commun.*, **20**, 400 (1955).

17. J. Bjerrum, *Chem. Rev.*, **46**, 381 (1950).

18. D. D. De Ford and D. N. Hume, *J. Am. Chem. Soc.*, **73**, 5321 (1951).

19. J. Jacq, *Electrochim. Acta*, **12**, 1345 (1967).

20. J. Heyrovský and J. Kůta, *Principles of Polarography*, Czechoslovakian Academy of Science, Prague, 1965.

21. J. Tomeš, *Collect. Czech. Chem. Commun.*, **9**, 81 (1937).

22. M. Smutek, *Chem. Listy*, **45**, 241 (1951).

23. M. G. Evans and N. S. Hush, *J. Chim. Phys. Phys. Chim. Biol.*, **49**, C 159 (1952).

24. T. Kambara and I. Tachi, *Bull. Chem. Soc. Japan*, **25**, 135 (1952).

25. P. Delahay, *J. Am. Chem. Soc.*, **75**, 1430 (1953).

26. N. Mejman, *Zh. Fiz. Khim.*, **22**, 1454 (1948).

27. J. Koutecký, *Collect. Czech. Chem. Commun.*, **18**, 597 (1953).

28. J. Weber and J. Koutecký, *Collect. Czech. Chem. Commun.*, **20**, 980 (1955).

29. A. N. Frumkin, *Z. Phys. Chem.*, *Abt. A*, **164A**, 121 (1933).

30. L. Gierst, "The Double Layer and the Rate of Electrode Processes," in E. Yeager, Ed., *Transactions of the Symposium on Electrode Processes*, Wiley, New York, 1961, p. 109.

31. P. Delahay, *Double Layer and Electrode Kinetics*, Interscience, New York, 1965, p. 1.
32. J. Koutecký, *Chem. Listy*, **47**, 323 (1953).
33. J. Weber and J. Koutecký, *Chem. Listy*, **49**, 562 (1955).
34. H. Matsuda, *J. Phys. Chem.*, **64**, 336 (1960).
35. H. Matsuda, *Z. Elektrochem.*, **64**, 41 (1960).
36. H. Hurwitz, *Z. Elektrochem.*, **65**, 178 (1961).
37. L. Gierst and H. Hurwitz, *Z. Elektrochem.*, **64**, 36 (1960).
38. J. Koutecký, *Collect. Czech. Chem. Commun.*, **18**, 311 (1953).
39. J. Koutecký and J. Čížek, *Collect. Czech. Chem. Commun.*, **21**, 1063 (1956).
40. K. H. Hanke and W. Hans, *Z. Elektrochem.*, **59**, 676 (1955).
41. J. Koutecký, R. Brdička, and V. Hanuš, *Collect. Czech. Chem. Commun.*, **18**, 611 (1953).
42. J. Koutecký and J. Koryta, *Collect. Czech. Chem. Commun.*, **19**, 845 (1954).
43. J. Koutecký, *Collect. Czech. Chem. Commun.*, **20**, 116 (1955).
44. V. Hanuš, *Chem. Zvesti*, **8**, 702 (1954).
45. R. Brdička, *Z. Elektrochem.*, **48**, 278 (1942).
46. A. M. Bond and G. Hefter, *J. Electroanal. Chem.*, **35**, 343 (1972).
47. A. M. Bond and G. Hefter, *J. Electroanal. Chem.*, **42**, 1 (1973).
48. W. H. Reinmuth and K. Balasubramanian, *J. Electroanal. Chem.*, **38**, 79 (1972).
49. W. H. Reinmuth and K. Balasubramanian, *J. Electroanal. Chem.*, **38**, 2ýl (1972).
50. R. Guidelli, *J. Phys. Chem.*, **74**, 95 (1970).
51. E. Laviron, *J. Electroanal. Chem.*, **52**, 355 (1974).
52. M. Sluyters-Rehbach and J. H. Sluyters, *J. Electroanal. Chem.*, **65**, 831 (1975).
53. R. W. Schmid and C. N. Reilley, *J. Am. Chem. Soc.*, **80**, 2655 (1958).
54. J. Weber, J. Koutecký, and J. Koryta, *Z. Elektrochem.*, **63**, 583 (1959).
55. J. Kůta, J. Weber, and J. Koutecký, *Collect. Czech. Chem. Commun.*, **25**, 2376 (1960).
56. A. Ya. Gokhstein, *Dokl. Akad. Nauk SSSR*, **137**, 345 (1961).
57. J. Kůta and J. Weber, *Elektrochim. Acta*, **9**, 541 (1964).
58. J. Kipkowski and Z. Galus, *J. Electroanal. Chem.*, **61**, 11 (1975).
59. B. B. Damaskin and B. N. Afanas'ev, *Elektrokhimiya*, **13**, 1099 (1977).
60. R. Guidelli, M. L. Foresti, and M. R. Moncelli, *J. Electroanal. Chem.*, **113**, 171 (1980).
61. V. G. Levich, *Acta Physicochim. URSS*, **17**, 257 (1942); **19**, 133 (1944).
62. F. Opekar and P. Beran, *J. Electroanal. Chem.*, **69**, 1 (1976).
63. W. J. Albery and M. L. Hitchman, *Ring-Disk Electrodes*, Clarendon, Oxford, 1971.
64. G. C. Barker, "Faradaic Rectification," in E. Yeager, Ed., *Transactions of the Symposium on Electrode Processes*, Wiley, New York, 1961, p. 325.
65. G. C. Barker and A. W. Gardner, *Z. Anal. Chem.*, **173**, 70 (1960).
66. A. A. A. M. Brinkman and J. M. Los, *J. Electroanal. Chem.*, **7**, 171 (1964); **14**, 269, 285 (1967).
67. L. F. Loelevelt, B. J. C. Wetsema, and J. M. Los, *J. Electroanal. Chem.*, **75**, 831 (1977).
68. J. Galvez and A. Serna, *J. Electroanal. Chem.*, **69**, 133, 145, 157 (1976).
69. J. Galvez, A. Serna, and T. Fuente, *J. Electroanal. Chem.*, **96**, 1 (1979).
70. I. Ružić, *J. Electroanal. Chem.*, **75**, 25 (1977).
71. R. L. Birke, *Anal. Chem.*, **50**, 1489 (1978).
72. H. Matsuda, *Bull. Chem. Soc. Japan*, **53**, 3439 (1980).
73. E. P. Parry and R. A. Osteryoung, *Anal. Chem.*, **37**, 1634 (1965).
74. K. B. Oldham and E. P. Parry, *Anal. Chem.*, **40**, 65 (1968).
75. R. L. Birke, M-H. Kim, and M. Strassfeld, *Anal. Chem.*, **53**, 852 (1981).
76. G. C. Barker, *Anal. Chim. Acta*, **18**, 118 (1958).
77. B. Breyer, F. Gutmann, and S. Hacobian, *Aust. J. Sci. Res. Ser. A*, **A3**, 558 (1950).
78. B. Breyer and H. H. Bauer, *Rev. Polarogr.*, **8**, 157 (1960).

79. B. Breyer and H. H. Bauer, *Rev. Polarogr.*, **11**, 60 (1963).

80. D. E. Smith, "AC Polarography and Related Techniques: Theory and Practice," in A. J. Bard, Ed., *Electroanalytical Chemistry*, Vol. 1, Dekker, New York, 1966, p. 1.

81. M. Sluyters-Rehbach and J. H. Sluyters, "Sine Wave Methods in the Study of Electrode Processes," in A. J. Bard, Ed, Ed., *Electroanalytical Chemistry*, Vol. 4, Dekker, New York, 1970, p. 1.

82. D. D. MacDonald, *Transient Techniques in Electrochemistry*, Plenum, New York, 1977, p. 1.

83. R. De Levie and L. Pospíšil, *J. Electroanal. Chem.*, **22**, 277 (1969).

84. R. De Levie and A. A. Husovsky, *J. Electroanal. Chem.*, **22**, 29 (1969).

85. J. H. Sluyters, *Recl. Trav. Chim. Pays-Bas Belg.*, **79**, 1092 (1960).

86. R. Parsons, "Faradaic and Nonfaradaic Processes," in P. Delahay, Ed., *Advances in Electrochemistry and Electrochemical Engineering*, Vol. 7, Wiley, New York, 1970, p. 171.

87. L. Pospíšil, *J. Electroanal. Chem.*, **74**, 369 (1976).

88. M. Sluyters-Rehbach and J. H. Sluyters, *J. Electroanal. Chem.*, **136**, 39 (1982).

89. L. Novotný, *Proc. J. Heyrovský Memorial Congress on Polarography*, Vol. 2, Prague, 1980, p. 129.

90. A. Y. Gokhstein and Y. P. Gokhstein, *Zh. Fiz. Khim.*, **36** 651 (1962).

91. R. Kalvoda, *Operational Amplifiers in Chemical Instrumentation*, Halsted, London, 1975.

92. W. M. Schwarz and J. Shain, *Anal. Chem.*, **35**, 1770 (1963).

93. G. C. Barker and I. L. Jenkins, *Analyst (London)*, **77**, 685 (1952).

94. G. C. Barker and A. W. Gardner, *Z. Anal. Chem.*, **173**, 79 (1960).

95. S. A. Borman, *Anal. Chem.*, **54**, 698A (1982).

96. H. Blurstein and A. M. Bond, *Anal. Chem.*, **48**, 248 (1976).

97. R. Kalvoda, *Chem. Listy*, **71**, 530 (1977).

98. F. Vydra, K. Štulík, and E. Juláková, *Electrochemical Stripping Analysis*, Halsted, London, 1976.

99. J. Heyrovský and P. Zuman, *Practical Polarography*, Academic, London, 1968.

100. A. M. Bond, *Modern Polarographic Methods in Analytical Chemistry*, Dekker, New York, 1980.

101. A. A. Vlček, "Polarographic Behavior of Coordination Compounds," in F. A. Cotton, Ed., *Progress in Inorganic Chemistry*, Vol. 5, Interscience, New York, 1963, p. 211.

102. S. Mugnier, C. Noise, J. Tirouflet, and E. Laviron, *J. Organomet. Chem.*, **186**, C49 (1980).

103. M. K. Hanafey, Q. L. Scott, T. H. Ridgway, and C. N. Reilley, *Anal. Chem.*, **50**, 116 (1978).

104. J. Volke, *Ann. Chim. (Rome)*, 323 (1980).

105. J. Volke and M. Naarová, *Collect. Czech. Chem. Commun.*, **37**, 3361 (1972); M. Naarová and J. Volke, *Collect. Czech. Chem. Commun.*, **38**, 2670 (1973).

106. J. Volke, A. Kejharová-Ryvolová, O. Manoušek, and L. Wasilewska, *J. Electroanal. Chem.*, **32**, 445 (1971).

107. O. Manoušek, J. Volke, and J. Hlavatý, *Electrochim. Acta*, **25**, 515 (1980).

108. J. Volke and O. Manoušek, *Faraday Discuss. Chem. Soc.*, **56**, 300 (1973).

109. J. Holubek and J. Volke, *Collect. Czech. Chem. Commun.*, **27**, 680 (1962).

110. J. Volke and J. Holubek, *Collect. Czech. Chem. Commun.*, **28**, 1597 (1963).

111. A. M. Kardoš, P. Valenta, and J. Volke, *J. Electroanal. Chem.*, **12**, 84 (1966).

112. J. Volke and V. Skála, *J. Electroanal. Chem.*, **36**, 393 (1972)
113. J. Volke, O. Manoušek, and T. V. Troyepolskaya, *J. Electroanal. Chem.*, **85**, 163 (1977).
114. J. Holubek and J. Volke, *Collect. Czech. Chem. Commun.*, **25**, 3292 (1960).
115. L. Meites and P. Zuman, *CRC Handbook Series in Inorganic Electrochemistry*, Vols. 1 and 2, CRC Press, Boca Raton, FL, 1977 and 1982, pp. 503, 540.
116. J. Volke, "Electrochemical Properties in Solutions," in A. R. Katritzky, Ed., *Physical Methods in Heterocyclic Chemistry*, Vol. 1, Academic, New York, 1963, p. 217. J. Volke, *Talanta*, **12**, 1081 (1965).
117. M. Březina and J. Volke, "Polarography in Biochemistry, Pharmacology, and Toxicology," in G. P. Ellis and G. B. West, Eds., *Progress in Medicinal Chemistry*, Vol. 12, North-Holland, Amsterdam, 1975. M. Březina and P. Zuman, *Polarography in Medicine, Biochemistry, and Pharmacy*, Interscience, New York, 1958.
118. M. M. Baizer, *Organic Electrochemistry*, Dekker, New York, 1973.
119. J. Koutecký, *Proceedings of the First International Polarography Congress, Prague*, Vol. 1, Přírodověd. nakl., Prague, 1951, p. 826. J. Heyrovský, *Polarographie*, Springer, Wien, 1941, p. 109.
120. H. Oelschläger, "Advances in Electroanalytical Methods," in D. D. Breimer and P. Speiser, Eds., *Topics in Pharmaceutical Sciences*, Elsevier/North-Holland Biomedical, New York, 1981, p. 98.
121. M. A. Müller and H. Oelschläger, *Fresenius Z. Anal. Chem.*, **307**, 109 (1981).
122. P. Zuman, *Substituent Effects in Organic Polarography*, Plenum, New York, 1967, p. 384.
123. P. Zuman, *Fortschr. Chem. Forsch.*, **12**, 1 (1969).
124. A. Vincenz-Chodkowska and Z. R. Grabowski, *Electrochim. Acta*, **9**, 789 (1964).
125. A. Maccoll, *Nature*, **163**, 178 (1949).
126. C. Párkányi and R. Zahradník, *Bull. Soc. Chim. Belg.*, **73**, 57 (1964).
127. A. Streitwieser, *Molecular Orbital Theory for Chemists*, Wiley, New York, 1961, p. 195.
128. H. Berg and F. Kapulla, *Z. Elektrochem.*, **64**, 44 (1960).
129. S. G. Mairanovskii, *Dokl. Akad. Nauk SSSR*, **110**, 593 (1956).
130. S. G. Mairanovskii, *Izv. Akad. Nauk SSSR, Otd. Khim. Nauk*, 2140 (1961).
131. V. Hanuš, *Chem. Zvesti*, **8**, 702 (1954).
132. J. Kůta, *Chem. Listy*, **49**, 1467 (1955).
133. V. Hanuš, *Proceedings of the First International Polarography Congress, Prague*, Vol. 3, Přírodověd. vyd., Prague, 1952, p. 103.
134. J. Volke, *Collect. Czech. Chem. Commun.*, **22**, 1777 (1957).
135. B. Kastening, "Free Radicals in Organic Polarography," in P. Zuman, L. Meites, and I. M. Kolthoff, Eds., *Progress in Polarography*, Vol. 3, Wiley-Interscience, New York, 1972, p. 195.
136. D. T. Sawyer and J. L. Roberts, Jr., *Experimental Electrochemistry for Chemists*, Wiley, New York, 1974, p. 435.
137. A. A. Vlček, *Chem. Listy*, **50**, 400 (1956).
138. L. Meites and P. Zuman, *CRC Handbook Series in Organic Electrochemistry*, Vols. 1–6, CRC Press, Boca Raton, FL, 1977.
139. S. G. Mairanovskii, *J. Electroanal. Chem.*, **4**, 166 (1962).
140. J. Volke and P. Valenta, *Collect. Czech. Chem. Commun.*, **25**, 1580 (1960). J. Volke, *Chem. Listy*, **55**, 26 (1961).
141. K. Wiesner, *Collect. Czech. Chem. Commun.*, **12**, 64 (1947).
142. L. Meites, *Polarographic Techniques*, 2nd ed., Interscience, New York, 1965, p. 345.

143. J. Volke, *Bioelectrochemistry and Bioenergetics*, **10**, 7 (1983).
144. J. Volke, "Polarographic Determination of 1,4-Benzodiazepine Derivatives, Especially in Biological Fluids," in J. Koryta, Ed., *Medical and Biomedical Applications of Electrochemical Devices*, Wiley, London, 1980, p. 273.
145. G. J. Patriarche and J-C. Vire, "Electroanalytical Applications in Pharmacy and Pharmacology," in W. F. Smyth, Ed., *Electroanalysis in Hygiene, Environmental, Clinical and Pharmaceutical Chemistry*, Elsevier, Amsterdam, 1980, p. 209.
146. W. F. Smyth, Ed., *Electroanalysis in Hygiene, Environmental, Clinical and Pharmaceutical Chemistry*, Elsevier, Amsterdam, 1980.
147. J. V. Geil and F. Kintschel, *Die Polarographische und Voltammetrische Bestimmung von Wirkstoffen in Pharmazeutischen Zubereitungen*, Metrohm-Monographien, 9100 Herisau, 1982.
148. J. Volke, L. Wasilewska, and A. Ryvolová-Kejharová, *Pharmazie*, **26**, 399 (1971).
149. N. Šulcová, J. Němec, K. Waisser, and H. L. Kies, *Microchem. J.*, **25**, 551 (1980).
150. M. M. Ellaithy and J. Volke, *Collect. Czech. Chem. Commun.*, **43**, 813 (1978).
151. A. Rogsta and K. Høgberg, *Anal. Chim. Acta*, **94**, 461 (1977).
152. M. R. Smyth and J. Osteryoung, "Electroanalysis of Environmental Carcinogens," in W. F. Smyth, Ed., *Electroanalysis in Hygiene, Environmental, Clinical and Pharmaceutical Chemistry*, Elsevier, Amsterdam, 1980, p. 423.
153. J. Davídek, "Polarographic Analysis of Pesticides in Food Products," in W. F. Smyth, Ed., *Electroanalysis in Hygiene, Environmental, Clinical and Pharmaceutical Chemistry*, Elsevier, Amsterdam, 1980, p. 399.
154. J. Osteryoung, J. W. Whittaker, and M. R. Smyth, "Electroanalysis of Economic Poisons," in W. F. Smyth, Ed., *Electroanalysis in Hygiene, Environmental, Clinical and Pharmaceutical Chemistry*, Elsevier, Amsterdam, 1980, p. 413.
155. J. Volke and M. Slamník, "Polarography and Related Methods," in K. G. Das, Ed., *Pesticide Analysis*, Dekker, New York, 1980, p. 175.
156. R. Samuelson and T. Rydström, "Pulse Polarographic Studies of *N*-Nitrosamines," in W. F. Smyth, Ed., *Electroanalysis in Hygiene, Environmental, Clinical and Pharmaceutical Chemistry*, Elsevier, Amsterdam, 1980, p. 435.
157. M. R. Smyth, P. W. Lawellin, and J. D. Osteryoung, *Analyst*, **104**, 73 (1979).
158. W. Kemula, "Chromato-Polarography," in P. Zuman and I. M. Kolthoff, Eds., *Progress in Polarography*, Vol. 2, Interscience, New York, 1962, p. 397.
159. H. Oelschläger, J. Volke, and G. T. Lim, *Arch. Pharm. (Weinheim, Ger.)*, **298**, 213 (1965).
160. H. Oelschläger, J. Volke, and G. T. Lim, *Arzneim. Forsch.* **17**, 637 (1967).
161. J. Volke and H. Oelschläger, *Scientia Pharmaceutica*, Vol. 2, Butterworth, London, 1967, p. 105.
162. H. Oelschläger, S. Lumbantoruan, J. Volke, and G. Kraft, *Z. Anal. Chem.*, **279**, 257 (1976).
163. H. B. Hanecamp, "Polarographic Continuous-Flow Detection," unpublished doctoral dissertation, Vrije Universiteit, Amsterdam, 1981.
164. D. A. Rostow, R. E. Shoup, and P. T. Kissinger, *Anal. Chem.*, **54**, 1417A (1982).
165. I. Jane, "Application of HPLC with Electrochemical Detection in Forensic Science," Metropolitan Police Forensic Science Lab., London, 1981.
166. K. Bratin, P. T. Kissinger, R. C. Brines, and C. S. Bruntlett, *Anal. Chim. Acta*, **130**, 295 (1981).

INDEX